# ATMOSPHERIC CHEMISTRY
AND
GLOBAL CHANGE

TOPICS IN ENVIRONMENTAL CHEMISTRY
A SERIES OF ADVANCED TEXTBOOKS AND MONOGRAPHS


SERIES EDITOR
John W. Birks, University of Colorado

ASSOCIATE EDITORS
William Davison, Lancaster University
Michael R. Hoffmann, California Institute of Technology
William M. Lewis, University of Colorado
Richard D. Noble, University of Colorado
John H. Seinfeld, California Institute of Technology
Garrison Sposito, University of California, Berkeley


PUBLISHED VOLUMES
*Environmental Chemistry of Soils*, M. McBride
*NMR Spectroscopy in Environmental Chemistry*, M. Nanny,
    R. Minear, and J. Leenheer, editors
*Perspectives in Environmental Chemistry*, D. Macalady, editor
*Environmental Toxicology and Chemistry*, D. Crosby
*Atmospheric Chemistry and Global Change*, G. Brasseur,
    J. Orlando, and G. Tyndall, editors

# ATMOSPHERIC CHEMISTRY
AND
GLOBAL CHANGE

Edited by
Guy P. Brasseur
John J. Orlando
Geoffrey S. Tyndall

New York    Oxford
OXFORD UNIVERSITY PRESS
1999

Oxford University Press

Oxford  New York
Athens  Auckland  Bangkok  Bogotá  Buenos Aires
Calcutta  Cape Town  Chennai  Dar es Salaam
Delhi  Florence  Hong Kong  Istanbul  Karachi
Kuala Lumpur  Madrid  Melbourne  Mexico City
Mumbai  Nairobi  Paris  São Paolo  Singapore
Taipei  Tokyo  Toronto  Warsaw

*and associated companies in*
Berlin  Ibadan

Copyright ©1999 by Oxford University Press, Inc.

Published by Oxford University Press, Inc.
198 Madison Avenue, New York, New York 10016

Oxford is a registered trademark of Oxford University Press

All rights reserved. No part of this publication may be
reproduced, stored in a retrieval system, or transmitted,
in any form or by any means, electronic, mechanical,
photocopying, recording, or otherwise, without the prior
permission of Oxford University Press.

**Library of Congress Cataloging-in-Publication Data**

Atmospheric chemistry and global change / a textbook prepared by
  scientists at the National Center for Atmospheric Research, Boulder,
  Colorado, and other colleagues ; editors Guy P. Brasseur, John J.
  Orlando, and Geoffrey S. Tyndall.
    p.  cm. -- (Topics in environmental chemistry)
  Includes bibliographical references and index.
  ISBN 0-19-510521-4 (cloth)
    1. Atmospheric chemistry. I. Brasseur, Guy P. II. Orlando, John
J. (John Joseph) 1960-    . III. Tyndall, Geoffrey S. (Geoffrey
Stuart) 1955-    . IV. National Center for Atmospheric Research
(U.S.)    V. Series.
QC879.6.A85    1999
551.51'1 -- dc 21
                                                          98-33483
                                                               CIP

**Cover Image:** This image shows the distribution of ozone for December 10 at 300 hPa (9 km) as calculated by NCAR's global chemical transport model MOZART (Model of Ozone and Related species in the Troposphere). High concentration values (purple) over southern Africa are produced from biomass burning and lightning emissions of ozone precursors and are transported towards Australia across the Indian Ocean. Over the poles, the 300 hPa pressure level is located in the stratosphere and shows correspondingly high concentration values. The extremely low values (blue) over the northern Indian Ocean and western Pacific are associated with convection of low concentrations from the ocean surface.

9 8 7 6 5 4 3 2
Printed in the United States of America
on acid-free paper

# Contents

Preface     x

Contributing Authors     xii

List of Frequently Used Symbols     xv

1. ATMOSPHERIC CHEMISTRY AND THE EARTH SYSTEM     1

    1.1 Introduction . . . . . . . . . . . . . . . . . . . . . . . . . . . . 1
    1.2 The Earth System . . . . . . . . . . . . . . . . . . . . . . . . . 5
    Further Reading . . . . . . . . . . . . . . . . . . . . . . . . . . . 17
    Essay: Atmospheric Chemistry and the Earth System by Ralph J. Cicerone     19

## Part 1: Fundamentals

2. ATMOSPHERIC DYNAMICS AND TRANSPORT     23

    2.1 Introduction . . . . . . . . . . . . . . . . . . . . . . . . . . . 23
    2.2 The Governing Equations . . . . . . . . . . . . . . . . . . . . 25
    2.3 Constraints on Atmospheric Motion . . . . . . . . . . . . . . 32
    2.4 Zonal Means and Eddies . . . . . . . . . . . . . . . . . . . . 44
    2.5 Atmospheric Waves . . . . . . . . . . . . . . . . . . . . . . . 53
    2.6 Tropospheric Circulation and Transport . . . . . . . . . . . . 64
    2.7 Stratospheric Circulation and Transport . . . . . . . . . . . . 73
    2.8 Stratosphere-Troposphere Exchange . . . . . . . . . . . . . . 78
    Further Reading . . . . . . . . . . . . . . . . . . . . . . . . . . . 80
    Essay: Why Understand Dynamics — And What Is "Understanding"
        Anyway? by Michael E. McIntyre . . . . . . . . . . . . . . . 82

3. CHEMICAL AND PHOTOCHEMICAL PROCESSES     85

    3.1 Introduction . . . . . . . . . . . . . . . . . . . . . . . . . . . 85
    3.2 Radiation . . . . . . . . . . . . . . . . . . . . . . . . . . . . . 86
    3.3 Photophysical and Photochemical Processes . . . . . . . . . 94
    3.4 Chemical Reactions . . . . . . . . . . . . . . . . . . . . . . . 95
    3.5 Catalytic Cycles . . . . . . . . . . . . . . . . . . . . . . . . . 106
    3.6 Role of Excited States . . . . . . . . . . . . . . . . . . . . . . 108
    3.7 Measurement of Rate Coefficients . . . . . . . . . . . . . . . 109
    3.8 The Steady State Approximation . . . . . . . . . . . . . . . 112

3.9 Lifetimes in the Atmosphere . . . . . . . . . . . . . . . . . 112
Further Reading . . . . . . . . . . . . . . . . . . . . . . . . . 114
Essay: When Do We Know Enough about Atmospheric Chemistry?
    by Harold Schiff . . . . . . . . . . . . . . . . . . . . . . . 115

## 4. AEROSOLS AND CLOUDS    117

4.1 Introduction . . . . . . . . . . . . . . . . . . . . . . . . . . 117
4.2 Overview of the Atmospheric Aerosol . . . . . . . . . . . . 117
4.3 The Role of Clouds in Tropospheric Chemistry . . . . . . . 126
4.4 Single-Particle Physical Characteristics . . . . . . . . . . . 129
4.5 Gas-to-Particle Conversion . . . . . . . . . . . . . . . . . . 137
4.6 Acid-Base Reactions of Aerosol Particles . . . . . . . . . . 141
4.7 Removal Processes Associated with Aerosols . . . . . . . . 142
4.8 Solubility of Gases in Droplets . . . . . . . . . . . . . . . . 144
4.9 Mass Transfer Rates . . . . . . . . . . . . . . . . . . . . . . 148
4.10 Aqueous Reactions . . . . . . . . . . . . . . . . . . . . . . 151
Further Reading . . . . . . . . . . . . . . . . . . . . . . . . . 154
Essay: Aerosols and Clouds: A Postscript by Richard P. Turco . . . . . 155

## 5. TRACE GAS EXCHANGES AND BIOGEOCHEMICAL CYCLES    159

5.1 Introduction . . . . . . . . . . . . . . . . . . . . . . . . . . 159
5.2 Surface Exchanges . . . . . . . . . . . . . . . . . . . . . . . 161
5.3 The Global Water Cycle . . . . . . . . . . . . . . . . . . . . 165
5.4 The Global Carbon Cycle . . . . . . . . . . . . . . . . . . . 167
5.5 The Global Nitrogen Cycle . . . . . . . . . . . . . . . . . . 188
5.6 The Global Sulfur Cycle . . . . . . . . . . . . . . . . . . . . 195
5.7 Halogens . . . . . . . . . . . . . . . . . . . . . . . . . . . . 201
Further Reading . . . . . . . . . . . . . . . . . . . . . . . . . 203
Essay: The View from Outside by James Lovelock . . . . . . . . . 204

# Part 2: Chemical Families

## 6. HYDROGEN COMPOUNDS    209

6.1 Importance of Atmospheric Hydrogen Compounds . . . . . . . . 209
6.2 Scope and Definitions . . . . . . . . . . . . . . . . . . . . . 210
6.3 Sources of Hydrogen to the Atmosphere . . . . . . . . . . . 210
6.4 Chemistry of Hydrogen Species in the Middle Atmosphere . . . . 212
6.5 Chemistry of Hydrogen Compounds in the Troposphere . . . . . 216
6.6 Concentrations of Hydrogen Compounds in the Stratosphere . . . 220
6.7 Concentrations of Hydrogen Compounds in the Troposphere . . . 226
6.8 Summary . . . . . . . . . . . . . . . . . . . . . . . . . . . . 231
Further Reading . . . . . . . . . . . . . . . . . . . . . . . . . 232
Essay: Hydrogen Compounds by Dieter Ehhalt . . . . . . . . . . 233

## 7. NITROGEN COMPOUNDS    235

7.1 Importance of Atmospheric Odd Nitrogen . . . . . . . . . . 235
7.2 Scope and Definitions . . . . . . . . . . . . . . . . . . . . . 235
7.3 The Role of Odd Nitrogen in the Stratosphere . . . . . . . . 237

- 7.4 Odd Nitrogen in the "Contemporary" Stratosphere . . . . . . . . 248
- 7.5 Odd Nitrogen in the Troposphere . . . . . . . . . . . . . . . . 254
- 7.6 Experimental Summary of the Influence of Odd Nitrogen in the Continental Boundary Layer . . . . . . . . . . . . . . 276
- 7.7 $NO_3$ Chemistry . . . . . . . . . . . . . . . . . . . . . . . . . . 277
- 7.8 Gaseous Acid and Particulate Nitrate Formation . . . . . . . . 281
- 7.9 Chemistry of Organic Nitrates . . . . . . . . . . . . . . . . . 282
- Further Reading . . . . . . . . . . . . . . . . . . . . . . . . . . 287
- Essay: Time's Arrow by Ian Galbally . . . . . . . . . . . . . . 288

## 8. HALOGEN COMPOUNDS        291

- 8.1 Introduction . . . . . . . . . . . . . . . . . . . . . . . . . . . 291
- 8.2 Scope and Definitions . . . . . . . . . . . . . . . . . . . . . . 291
- 8.3 Sources of Halogens . . . . . . . . . . . . . . . . . . . . . . . 292
- 8.4 Loss Processes of Halogen Source Gases . . . . . . . . . . . . 298
- 8.5 Inorganic Chemistry of Halogen Species . . . . . . . . . . . . 301
- 8.6 Controlling the Detrimental Effects of Halogens on the Atmosphere: Future Outlook . . . . . . . . . . . . . . . . . . . . . . . . 316
- Further Reading . . . . . . . . . . . . . . . . . . . . . . . . . . 321
- Essay: CFCs and Stratospheric Ozone Depletion by Mario Molina . . . 322

## 9. CARBON-CONTAINING COMPOUNDS        325

- 9.1 Introduction . . . . . . . . . . . . . . . . . . . . . . . . . . . 325
- 9.2 Scope and Definitions . . . . . . . . . . . . . . . . . . . . . . 325
- 9.3 Atmospheric Photochemistry of Hydrocarbons . . . . . . . . . 326
- 9.4 Distribution of Hydrocarbons . . . . . . . . . . . . . . . . . . 338
- Further Reading . . . . . . . . . . . . . . . . . . . . . . . . . . 345
- Essay: Hydrocarbons by Hanwant Singh . . . . . . . . . . . . 346

## 10. SULFUR COMPOUNDS        349

- 10.1 Introduction . . . . . . . . . . . . . . . . . . . . . . . . . . 349
- 10.2 Scope and Definitions . . . . . . . . . . . . . . . . . . . . . 350
- 10.3 Sulfur Compounds . . . . . . . . . . . . . . . . . . . . . . . 351
- 10.4 Tropospheric Chemistry of Sulfur Compounds . . . . . . . . 351
- 10.5 Measurements of Sulfur Gas Abundances and Distributions . . . . 361
- 10.6 $SO_2$ and Acid Precipitation . . . . . . . . . . . . . . . . . . 367
- 10.7 Stratospheric Sulfur Chemistry . . . . . . . . . . . . . . . . 368
- 10.8 Gas-Phase Ionic Chemistry in the Stratosphere . . . . . . . 370
- Further Reading . . . . . . . . . . . . . . . . . . . . . . . . . . 370
- Essay: Sulfur, Aerosols, Clouds, and Rain by Robert J. Charlson . . . . 371

# Part 3: Tools

## 11. OBSERVATIONAL METHODS: INSTRUMENTS AND PLATFORMS        375

- 11.1 Introduction . . . . . . . . . . . . . . . . . . . . . . . . . . 375
- 11.2 Instrumentation for Constituent Measurements . . . . . . . . 375
- 11.3 Flux Measurements . . . . . . . . . . . . . . . . . . . . . . 399
- 11.4 Measurements of Atmospheric Radiation . . . . . . . . . . . 403

11.5 Instrumentation for Aerosol and Cloud Measurements . . . . . . 405
11.6 Observing Platforms . . . . . . . . . . . . . . . . . . . . . . 406
Further Reading . . . . . . . . . . . . . . . . . . . . . . . . . 419
Essay: From Individual Measurements to Scale Integration Strategies
   by Gérard Mégie . . . . . . . . . . . . . . . . . . . . . . . 420

## 12. MODELING 423

12.1 Introduction . . . . . . . . . . . . . . . . . . . . . . . . . 423
12.2 Model Equations . . . . . . . . . . . . . . . . . . . . . . . 424
12.3 Modeling Chemical Processes . . . . . . . . . . . . . . . . . 427
12.4 Modeling Atmospheric Transport . . . . . . . . . . . . . . . 433
12.5 Examples and Illustrations . . . . . . . . . . . . . . . . . . 440
12.6 Modeling Global Budgets and Biogeochemical Cycles . . . . . . 450
12.7 Data Assimilation . . . . . . . . . . . . . . . . . . . . . . 454
12.8 Inverse Modeling . . . . . . . . . . . . . . . . . . . . . . . 458
12.9 Chemical-Transport Models in the Future . . . . . . . . . . . 459
Further Reading . . . . . . . . . . . . . . . . . . . . . . . . . 460
Essay: How Complex Do Models Need to Be? by Henning Rodhe . . . 461

# Part 4: Ozone, Climate, and Global Change

## 13. TROPOSPHERIC OZONE 465

13.1 Introduction . . . . . . . . . . . . . . . . . . . . . . . . . 465
13.2 Distribution and Trends . . . . . . . . . . . . . . . . . . . 467
13.3 Production and Loss of Ozone . . . . . . . . . . . . . . . . 472
13.4 Major Uncertainties and Research Needs . . . . . . . . . . . 484
Further Reading . . . . . . . . . . . . . . . . . . . . . . . . . 485
Essay: Tropospheric Ozone by Paul Crutzen . . . . . . . . . . . . 486

## 14. MIDDLE ATMOSPHERIC OZONE 487

14.1 Introduction . . . . . . . . . . . . . . . . . . . . . . . . . 487
14.2 The Ozone Distribution . . . . . . . . . . . . . . . . . . . 491
14.3 Ozone Production . . . . . . . . . . . . . . . . . . . . . . 493
14.4 Ozone Destruction . . . . . . . . . . . . . . . . . . . . . . 494
14.5 Transport Effects . . . . . . . . . . . . . . . . . . . . . . 500
14.6 Polar Ozone . . . . . . . . . . . . . . . . . . . . . . . . . 501
14.7 Ozone Perturbations . . . . . . . . . . . . . . . . . . . . . 506
14.8 Impact of Ozone Depletion on UV Radiation . . . . . . . . . . 509
Further Reading . . . . . . . . . . . . . . . . . . . . . . . . . 511
Essay: Ozone Depletion: From Pole to Pole by Susan Solomon . . . . 513

## 15. ATMOSPHERIC CHEMISTRY AND CLIMATE 515

15.1 Introduction . . . . . . . . . . . . . . . . . . . . . . . . . 515
15.2 Radiation in the Atmosphere . . . . . . . . . . . . . . . . . 516
15.3 Natural Variations: Past Climates . . . . . . . . . . . . . . . 522
15.4 Impact of Anthropogenic Trace Gases on Climate . . . . . . . 523
15.5 Global Warming Potentials (GWPs) . . . . . . . . . . . . . . 528
15.6 Radiative Effects of Aerosols . . . . . . . . . . . . . . . . . 530

15.7 Response of the Climate System to Radiative Forcing . . . . . . . 534
Further Reading . . . . . . . . . . . . . . . . . . . . . . . . . . . 536
Essay: Can Climate Models Be Validated? by Stephen H. Schneider . . 537

## 16. ATMOSPHERIC EVOLUTION AND GLOBAL PERSPECTIVE 539

16.1 Introduction . . . . . . . . . . . . . . . . . . . . . . . . . . . 539
16.2 Atmospheric Evolution on Geological Timescales . . . . . . . . . 539
16.3 Human Influences on the Atmosphere . . . . . . . . . . . . . . 543
16.4 Future Trends . . . . . . . . . . . . . . . . . . . . . . . . . . 546
16.5 Global Perspective . . . . . . . . . . . . . . . . . . . . . . . 549
Further Reading . . . . . . . . . . . . . . . . . . . . . . . . . . . 550
Essay: The Atmospheric Humankind: Our Related Futures
 by Daniel L. Albritton . . . . . . . . . . . . . . . . . . . . . 551

## APPENDIXES 553

Appendix A: Physical Constants and Other Data . . . . . . . . . . . 555
Appendix B: Units, Conversion Factors, and Multiplying Prefixes . . . . 558
Appendix C: Atmospheric Parameters and Mixing Ratios of
 Chemical Constituents . . . . . . . . . . . . . . . . . . . . . 560
Appendix D: Chemical Species in the Atmosphere . . . . . . . . . . 567
Appendix E: Rate Coefficients for Second-Order Gas-Phase Reactions . . 570
Appendix F: Rate Coefficients for Association Gas-Phase Reactions . . . 574
Appendix G: Mass Accommodation Coefficients . . . . . . . . . . . 575
Appendix H: Surface Reaction Probability . . . . . . . . . . . . . . 576
Appendix I: Atmospheric Humidity . . . . . . . . . . . . . . . . . 578
Appendix J: Henry's Law Coefficients . . . . . . . . . . . . . . . . 580
Appendix K: Aqueous Equilibrium Constants . . . . . . . . . . . . 581
Appendix L: Rate Coefficients for Aqueous-Phase Reactions . . . . . . 582
Appendix M: Spectrum of Solar Extraterrestrial Actinic Flux (120-730 nm) 585
Appendix N: Photolysis Frequencies . . . . . . . . . . . . . . . . . 588

**Sample Problems**   597

**References**   615

**Index**   649

# Preface

> *Education is the only powerful weapon*
> *which you can use to change the world*
>
> NELSON MANDELA

Many of the documented changes that have affected the Earth's atmosphere on the global scale since the agricultural and industrial revolutions are of chemical origin. Anthropogenic perturbations including land use and industrial activities have profoundly modified the chemical composition of the lower and upper atmosphere, with potentially important consequences on future climate and on the evolution of living organisms. Examples of such changes include the formation of an ozone hole in Antarctica since the late 1970s, the observed trend in the atmospheric abundance of long-lived greenhouse gases, the change in the concentration of tropospheric ozone and of acidic deposition caused by the growing emissions of hydrocarbons, nitrogen oxides, and sulfur dioxide in industrialized regions, etc.

Although concerns about regional air quality remain important, especially in industrialized regions where the level of pollution has not been sufficiently reduced in spite of legislative measures, atmospheric chemists are now devoting a large amount of effort to the study of the global environment. They regard the global atmosphere as a complex chemical and dynamical system interacting both internally within the troposphere and stratosphere and externally with the oceans, land, and living organisms. Atmospheric chemistry has therefore become a central discipline of global change research. Because vital human activities such as energy and food production are directly involved, the subjects of atmospheric composition, air quality, biogeochemical cycles, and climate change constitute not only challenging scientific questions, but also policy problems of global significance.

The present book addresses several global environmental issues that are societally important and therefore require continued attention from the scientific community, as well as from economic and political leaders. It is the result of a collective effort undertaken by a group of scientists at the National Center for Atmospheric Research (NCAR) in Boulder, Colorado, and conducted jointly with colleagues in several universities and national laboratories. The purpose is to provide both a comprehensive textbook for students at the graduate level and a reference book for scientists. The focus of the book is on global-scale problems and their role in the currently observed and possible future evolution of the Earth system. The approach is largely interdisciplinary and emphasizes, for example, the interactions between the atmospheric chemical composition and physical, biological, and climatic processes, as well as anthropogenic perturbations associated with population growth and economic development.

The first part of this book (Chapters 2, 3, 4, and 5) describes the fundamental physical, chemical, and biological processes that affect the composition of the atmosphere. The following chapters of the book (Chapters 6, 7, 8, 9, and 10) present the chemical mechanisms that affect the production and the fate of hydrogen, nitrogen, halogen, carbon, and sulfur-containing species in the atmosphere and discuss their distribution in the troposphere and the stratosphere. Chapters 11 and 12 are devoted to the presentation of techniques used to investigate chemical processes in the atmosphere, including existing instrumentation and platforms and chemical-transport models. Finally, Chapters 13 and 14 present an overview of the problems related to tropospheric and stratospheric ozone, respectively. The climatic impact of changes in the abundance of radiatively active gases is discussed in Chapter 15. An overview of the long-term chemical evolution of the atmosphere is given in Chapter 16. A large number of tables are presented in several appendixes.

Finally, in order to place the different chapters in a broad scientific and societal perspective, we have invited several leading scientists to write brief essays that bring our attention to important environmental issues.

Approximately 40 colleagues have contributed their expertise to the 16 chapters of this book, and, although the editors have attempted to homogenize the various contributions as much as possible, some differences in the style and in the structure of the chapters may still be apparent.

Several other individuals extensively reviewed parts of the manuscript and efficiently helped the contributing authors. We owe thanks in particular to L. Avallone, T. Bates, S. Bekki, W. Brune, M.A. Carroll, P. Ciais, R. Cicerone, S. Cieslik, D. Fahey, J. Fishman, I. Fung, M. Hitchman, J. Holton, D. Jacob, J. Kaye, D. Möller, P. Novelli, U. Schmidt, J. Seinfeld, K. Shine, H. Singh, A. Thompson, M. Tolbert, D. Wuebbles, and R. Zander. Their remarks and constructive criticisms have certainly improved the quality of this work. We also appreciate the assistance of Elyse Dubin, Bob Rogers, and Karen Shapiro at Oxford University Press, and the support of Robert J. Serafin, Director of NCAR, Richard A. Anthes, President of the University Corporation for Atmospheric Research, and Anne-Marie Schmoltner and Jarvis Moyers at the National Science Foundation. Finally, we gratefully thank Ronna Terrell Bailey, Rachel Ginsburg, and Donna Sanerib for assistance in text preparation, and the NCAR Imaging and Design Center for their expert figure drafting. The National Center for Atmospheric Research is sponsored by the National Science Foundation.

*Guy P. Brasseur*
*John J. Orlando*
*Geoffrey S. Tyndall*

*National Center for Atmospheric Research*
*Boulder, Colorado*

# Contributing Authors

**Chapter 1: Atmospheric Chemistry and the Earth System**
*Authors:* Guy Brasseur (NCAR) and David Schimel (NCAR)
*Essay:* Ralph J. Cicerone (Univ. of California, Irvine)

**Chapter 2: Atmospheric Dynamics and Transport**
*Authors:* Rolando Garcia (NCAR), Peter Hess (NCAR), and Anne Smith (NCAR)
*Contributor:* Jean-François Lamarque (NCAR)
*Essay:* Michael E. McIntyre (Cambridge University)

**Chapter 3: Chemical and Photochemical Processes**
*Authors:* Geoffrey Tyndall (NCAR) and John Orlando (NCAR)
*Essay:* Harold Schiff (York University, Canada)

**Chapter 4: Aerosols and Clouds**
*Authors:* Sonia Kreidenweis (Colorado State Univ.), Geoffrey Tyndall (NCAR), Mary Barth (NCAR), Frank Dentener (Univ. of Wageningen, the Netherlands), Jos Lelieveld (Univ. of Wageningen, the Netherlands), and Michael Mozurkewich (York University, Canada)
*Contributor:* John Orlando (NCAR)
*Essay:* Richard Turco (Univ. of California, Los Angeles)

**Chapter 5: Trace Gas Exchanges and Biogeochemical Cycles**
*Authors:* Guy Brasseur (NCAR), Elliot Atlas (NCAR), David Erickson (NCAR), Alan Fried (NCAR), James Greenberg (NCAR), Alex Guenther (NCAR), Peter Harley (NCAR), Elisabeth Holland (NCAR), Lee Klinger (NCAR), Brian Ridley (NCAR), and Geoffrey Tyndall (NCAR)
*Contributors:* Julia Lee-Taylor (NCAR) and James Sulzman (NCAR)
*Essay:* James Lovelock (Visiting Fellow, Oxford University)

**Chapter 6: Hydrogen Compounds**
*Authors:* Christopher Cantrell (NCAR) and John Orlando (NCAR)
*Contributors:* Fred Eisele (NCAR), Didier Hauglustaine (NCAR), Richard Shetter (NCAR), and Geoffrey Tyndall (NCAR)
*Essay:* Dieter Ehhalt (Forschungszentrum Jülich, Germany)

**Chapter 7: Nitrogen Compounds**
*Authors:* Brian Ridley (NCAR) and Elliot Atlas (NCAR)
*Contributor:* Geoffrey Tyndall (NCAR)
*Essay:* Ian Galbally (CSIRO, Australia)

**Chapter 8: Halogen Compounds**
*Authors:* John Orlando (NCAR) and Sue Schauffler (NCAR)
*Essay:* Mario Molina (Massachusetts Institute of Technology)

**Chapter 9: Carbon-Containing Compounds**
*Authors:* Sasha Madronich (NCAR), James Greenberg (NCAR), and Suzanne Paulson (UCLA)
*Contributors:* John Orlando (NCAR) and Geoffrey Tyndall (NCAR)
*Essay:* Hanwant Singh (NASA/Ames Research Center)

**Chapter 10: Sulfur Compounds**
*Authors:* Alan Fried (NCAR) and Geoffrey Tyndall (NCAR)
*Contributor:* Mary Barth (NCAR)
*Essay:* Robert J. Charlson (Univ. of Washington)

**Chapter 11: Observational Methods: Instruments and Platforms**
*Authors:* William Mankin (NCAR), Elliot Atlas (NCAR), Christopher Cantrell (NCAR), Fred Eisele (NCAR), and Alan Fried (NCAR)
*Contributors:* David Edwards (NCAR), James Greenberg (NCAR), and Alex Guenther (NCAR)
*Essay:* Gérard Mégie (Service d'Aéronomie CNRS)

**Chapter 12: Modeling**
*Authors:* Guy Brasseur (NCAR), Boris Khattatov (NCAR), and Stacy Walters (NCAR)
*Contributors:* Paul Ginoux (NCAR) and Didier Hauglustaine (NCAR)
*Essay:* Henning Rodhe (Univ. of Stockholm, Sweden)

**Chapter 13: Tropospheric Ozone**
*Authors:* Shaw Liu (NOAA/Aeronomy Laboratory and Georgia Tech.) and Brian Ridley (NCAR)
*Contributor:* Geoffrey Tyndall (NCAR)
*Essay:* Paul Crutzen (Max-Planck-Institute for Chemistry)

**Chapter 14: Middle Atmospheric Ozone**
*Author:* Michael Coffey (NCAR) and Guy Brasseur (NCAR)
*Contributors:* Thomas Horst (NCAR)
*Essay:* Susan Solomon (NOAA/Aeronomy Laboratory)

**Chapter 15: Atmospheric Chemistry and Climate**
*Authors:* Claire Granier (NCAR), Guy Brasseur (NCAR), and David Erickson (NCAR)
*Essay:* Stephen H. Schneider (Stanford University)

**Chapter 16: Atmospheric Evolution and Global Perspective**
　　*Authors:* Guy Brasseur (NCAR) and Ed Martell (NCAR)
　　*Essay:* Daniel L. Albritton (NOAA/Aeronomy Laboratory)

**Appendixes and Sample Problems**
　　*Authors:* Guy Brasseur (NCAR), Daniel Jacob (Harvard Univ.), Gérard Mégie (Service d'Aéronomie CNRS), Volker Mohnen (SUNY, Albany), and John Orlando (NCAR)
　　*Contributors:* Mary Barth (NCAR), Simon Chabrillat (Inst. d'Aéronomie Spatiale, Brussels), Siri Flocke (NCAR), Theresa Huang (NCAR), Denise Mauzerall (NCAR), and Gary Rottman (Univ. of Colorado)

# List of Frequently Used Symbols

| | |
|---|---|
| $a$ | Earth's radius or aerosol radius |
| $a_D$ | Droplet's radius |
| $a_i$ | Instantaneous radiative forcing |
| $a_w$ | Activity of water |
| $\mathcal{A}$ | Surface area density (aerosol) |
| $A$ | Pre-exponential factor for reaction rate coefficient |
| $B_\lambda$ | Planck's function at wavelength $\lambda$ |
| $c$ | Phase velocity or speed of light |
| $\bar{c}$ | Mean molecular velocity |
| $C_p$ | Specific heat at constant pressure |
| $C_v$ | Specific heat at constant volume |
| $d$ | Displacement height |
| $D$ | Brownian diffusion coefficient |
| $D$ | UV-B daily dose |
| $D_g$ | Gas phase diffusion |
| $D_p$ | Diameter of aerosol particles |
| $E$ | Energy |
| $f$ | Coriolis parameter |
| $F$ | Radiant flux or mass flux |
| $\mathcal{F}$ | Eliassen-Palm (EP) flux |
| $F_R$ | Radiative forcing |
| $F_S$ | Solar incoming energy or solar irradiance |
| $F_T$ | Terrestrial radiative energy |
| $g$ | Gravitational acceleration |
| $G$ | Green's function |
| $\widehat{G}$ | Free energy |
| GWP | Global warming potential |
| $H$ | Scale height |
| $\mathcal{H}$ | Henry's law constant |
| $\mathcal{H}^*$ | Effective Henry's Law constant |
| $\mathcal{H}'$ | Dimensionless Henry's Law constant |
| $\widehat{H}$ | Enthalpy |
| $I$ | Radiance or intensity |

| | |
|---|---|
| $j$ | Photodissociation frequency |
| $J$ | Radiative source function |
| k | Reaction rate constant |
| $\tilde{k}$ | Von Karman's constant |
| $k$ | Zonal wavenumber |
| $K_a$ | Air-sea exchange coefficient *above* air-sea interface |
| $K_w$ | Air-sea exchange coefficient *below* air-sea interface |
| $K$ | Transfer coefficient |
| $K_p$ | Equilibrium constant |
| $K_z$ | Vertical eddy diffusion coefficient |
| $\ell$ | Mean free path |
| $l$ | Meridional wavenumber |
| $L$ | Chemical destruction rate |
| $\mathcal{L}$ | Dimensionless cloud liquid water fraction |
| $L_c$ | Latent heat of condensation |
| LWC | Liquid water content |
| $\overline{m}$ | Zonal-mean angular momentum |
| $m$ | Vertical wavenumber, refractive index |
| $M$ | Air molecule |
| $\mathcal{M}$ | Molecular mass or chemical eddy flux |
| $M_a$ | Mass associated with reservoir |
| $n_i$ | Number density of constituent $i$ |
| $n_a$ | Air number density |
| $n(D_p)$ | Density function of aerosol size distribution |
| $N$ | Aerosol number density |
| $N(O_3)$ | Ozone column abundance |
| $N_B$ | Buoyancy (Brunt-Väisälä) frequency |
| ODP | Ozone depletion potential |
| $p$ | Pressure |
| $p_s$ | Reference pressure, mean sea level pressure |
| $p_\lambda$ | Phase function at wavelength $\lambda$ |
| $P$ | Chemical production rate |
| $\widehat{P}$ | Ertel's potential vorticity |
| $q$ | Solar actinic flux density |
| $Q$ | Net diabatic heating rate (K s$^{-1}$) |
| $\widetilde{Q}$ | Net chemical production rate |
| $Q_e$ | Extinction efficiency |
| $R$ | Gas constant |
| $\widetilde{R}$ | Reaction rate |
| $\mathcal{R}$ | Revelle factor |

# List of Frequently Used Symbols

| | |
|---|---|
| $R_a$ | Aerodynamic resistance |
| $R_b$ | Quasilaminar resistance |
| $R_c$ | Canopy resistance |
| $R_p$ | Precipitation rate |
| RAF | Radiation amplification factor |
| RH | Relative humidity |
| $s_g$ | Standard deviation of particle size distribution |
| $S$ | Net chemical source |
| $\widehat{S}$ | Entropy |
| $\mathcal{S}$ | Supersaturation ratio |
| $S_i$ | Chemical source or sink for constituent $i$ |
| $t$ | Time |
| $T$ | Temperature |
| $T_r$ | Transmission |
| $u$ | Zonal wind velocity |
| $u^*$ | Friction velocity |
| $v$ | Meridional wind velocity |
| $\mathbf{v}$ | Wind velocity vector |
| $v_w$ | Partial molar volume of water in solution |
| $v_p$ | Particle volume |
| $w$ | Vertical wind velocity |
| $w_d$ | Deposition velocity |
| $z$ | Altitude |
| $\alpha$ | Albedo |
| $\alpha$ | Aerosol size distribution (Chapter 10) |
| $\widetilde{\alpha}$ | Dimensionless accommodation coefficient |
| $\alpha_C$ | Newtonian cooling coefficient |
| $\beta$ | First-order rate coefficient (s$^{-1}$) |
| $\beta_a$ | Absorption coefficient |
| $\beta_e$ | Extinction coefficient |
| $\beta_p$ | Meridional gradient in Coriolis parameter |
| $\beta_s$ | Scattering coefficient |
| $\beta_\lambda$ | Extinction coefficient at wavelength $\lambda$ |
| $\gamma$ | Reaction probability |
| $\delta$ | Solar declination |
| $\varepsilon$ | Atmospheric emissivity |
| $\zeta$ | Relative vorticity |
| $\widehat{\zeta}$ | Potential vorticity |
| $\theta$ | Potential temperature |
| $\lambda$ | Wavelength or longitude |
| $\lambda_c$ | Climate sensitivity factor |

| | |
|---|---|
| $\mu$ | Cosine of zenith angle |
| $\mu_i$ | Volume mixing ratio |
| $\widetilde{\mu}_i$ | Mass mixing ratio |
| $\mu_a$ | Air viscosity |
| $\nu$ | Light frequency |
| $\rho$ | Mass density |
| $\rho_0$ | Mass density in log-pressure coordinates |
| $\rho_a$ | Air mass density |
| $\rho_s$ | Reference mass density |
| $\sigma$ | Stefan-Boltzmann constant |
| $\sigma_a$ | Absorption cross section |
| $\sigma_e$ | Extinction |
| $\sigma_s$ | Scattering cross section |
| $\widehat{\sigma}$ | Pseudodensity in isentropic coordinates |
| $\widetilde{\sigma}$ | Surface tension of a solution |
| $\tau_i$ | Lifetime or time constant of constituent $i$ |
| $\tau_a$ | Optical depth for absorption |
| $\tau_s$ | Optical depth for scattering |
| $\varphi$ | Quantum yield for photolysis |
| $\phi$ | Latitude |
| $\Phi$ | Geopotential or quantum yield for photolysis |
| $\Psi$ | Montgomery streamfunction |
| $\chi$ | Mass-weighted stream function |
| $\omega$ | Wave frequency |
| $\Omega$ | Earth's angular velocity (rad s$^{-1}$) |
| $\Omega_\lambda$ | Albedo for single scattering at wavelength $\lambda$ |

# 1 Atmospheric Chemistry and the Earth System

*We the peoples of the world face a new responsibility for our global future. Through our economic and technology activity, we are now contributing to significant global changes on the Earth within the span of a few human generations. We have become part of the Earth System and one of the forces for Earth change.*

NASA: Earth System Science<br>
A Program for Global Change, 1986

## 1.1 Introduction

Many of the global environmental changes forced by human activities (see Box 1.1) are mediated through the chemistry of the atmosphere. Important changes include the global spread of air pollution, increases in the concentration of tropospheric oxidants (including ozone), stratospheric ozone depletion, and global warming (the so-called greenhouse effect). Since the agricultural and industrial revolutions, the delicate balance between physical, chemical, and biological processes in the Earth system has been perturbed as a result, for example, of the quasiexponential growth in the world population (Fig. 1.1a), the use of increasing amounts of fossil fuel and the related emissions of carbon to the atmosphere (Fig. 1.1b), and the intensification of

---

**Box 1.1 Examples of Global Environmental Problems**

- Degradation of air quality: Global pollution resulting from industrial combustion and biomass burning
- Increase in the abundance of tropospheric oxidants including ozone and related impacts on the biosphere and human health
- Changes in the self-cleaning capability of the atmosphere and in the residence time of anthropogenic trace gases
- Climatic and environmental impact of changes in land use including tropical deforestation, wetland destruction, etc.
- Perturbations of biogeochemical cycles of carbon, nitrogen, phosphorus, and sulfur
- Acidic precipitation
- Climatic changes (global warming) resulting from increasing emissions of $CO_2$ and other greenhouse gases
- Climatic impacts (regional cooling) of sulfate aerosols resulting from anthropogenic $SO_2$ emissions
- Depletion of stratospheric ozone, related increase in the level of UV-B solar radiation at the surface, and impacts on the biosphere and human health

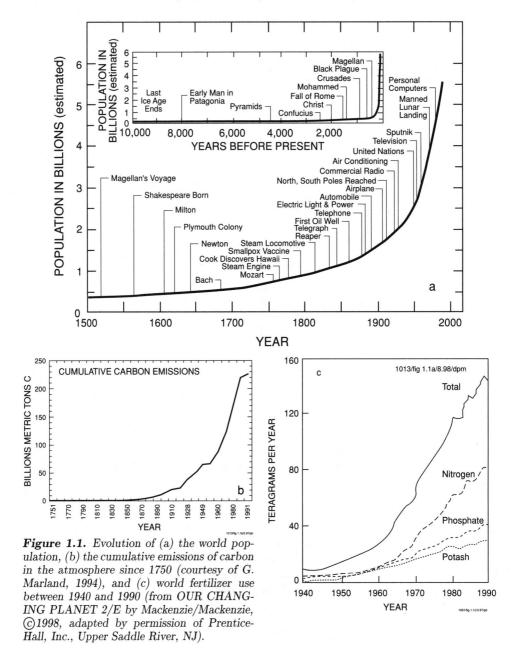

**Figure 1.1.** Evolution of (a) the world population, (b) the cumulative emissions of carbon in the atmosphere since 1750 (courtesy of G. Marland, 1994), and (c) world fertilizer use between 1940 and 1990 (from OUR CHANGING PLANET 2/E by Mackenzie/Mackenzie, ©1998, adapted by permission of Prentice-Hall, Inc., Upper Saddle River, NJ).

agricultural practices including the more frequent use of fertilizers (Fig. 1.1c). The observed increase in the atmospheric abundance of carbon dioxide ($CO_2$) (Fig. 1.2a) results mainly from fossil fuel burning, although biomass destruction is an important secondary source. Atmospheric concentrations are additionally influenced by exchanges of carbon with the ocean and the continental biosphere. The progressive modification and fertilization of the terrestrial biosphere are believed to have caused the observed increase in atmospheric nitrous oxide ($N_2O$), a tropospheric greenhouse gas and a source of reactive species in the stratosphere (Fig. 1.2b). Methane ($CH_4$), which also contributes to greenhouse forcing and, in addition, plays an important role in the photochemistry of the troposphere and the stratosphere, is produced by

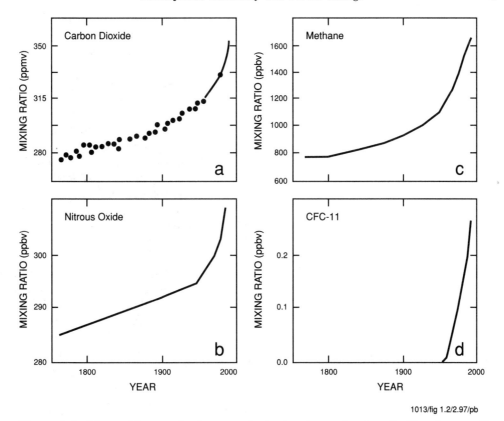

**Figure 1.2.** Observed increase in the atmospheric abundance of carbon dioxide, nitrous oxide, methane, and chlorofluorocarbon-11 in the surface level atmosphere (IPCC, 1990). See Box 1.4 for the definition and units of mixing ratio.

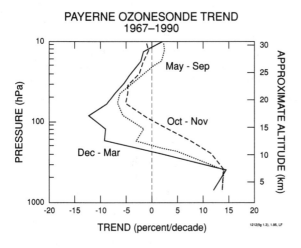

**Figure 1.3a.** Trends in tropospheric and stratospheric ozone (0-30 km) between 1967 and 1990 derived from ozonesonde records at Payerne, Switzerland. Values from three seasons are indicated. Substantial depletion (5 to 10 percent/decade) has been observed in the lower stratosphere (10-25 km), while an increase of approximately 10-15 percent per decade has been reported in the troposphere (4-8 km). Tropospheric trends are highly variable according to geographical location (WMO, 1992).

biospheric processes (wetlands, livestock, landfills, biomass burning); leakage from gas distribution systems is probably significant in the former Soviet Union and Eastern Europe. The global atmospheric concentration of methane has also grown in the past (Fig. 1.2c). Observed increases in the abundance of tropospheric ozone ($O_3$) (Fig. 1.3a), which contribute to deteriorating air quality, result from complex photochemical processes involving industrial and biological emissions of nitrogen

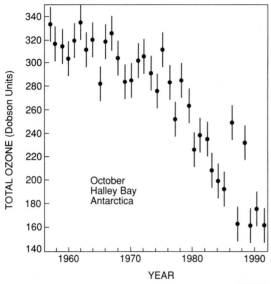

**Figure 1.3b.** Evolution between 1958 and 1992 of the October mean ozone column abundance (expressed in Dobson units; see Box 1.4) recorded at the British Antarctic Halley Bay Station. These measurements by Farman et al. (1985) led to the discovery of the springtime Antarctic ozone hole. It was later established that this dramatic destruction of polar ozone resulted from the release in the atmosphere of manmade chlorofluorocarbons. The higher values of 1986 and 1988 are associated with particular dynamic situations during these years (Andrews et al., 1996).

oxides, hydrocarbons, and certain other organic compounds. Ozone is a strong absorber of solar ultraviolet radiation; it also contributes to greenhouse forcing. Anthropogenic emissions of sulfur resulting from coal burning in highly populated and industrialized regions of the Northern Hemisphere, and the related increase in the aerosol load of the troposphere, have contributed to regional pollution and have probably produced a cooling of the surface in these regions by backscattering a fraction of the incoming solar energy. Finally, the rapid increase in the atmospheric abundance of industrially manufactured chlorofluorocarbons (Fig. 1.2d) has produced the observed depletion in stratospheric ozone (Fig. 1.3a) and the formation each spring (since the late 1970s) of an "ozone hole" over Antarctica (Fig. 1.3b).

The impact on the Earth system of these changes in the chemical state of the atmosphere is not yet fully elucidated, but could be significant in many cases. Changes in the atmospheric abundance of radiatively active gases could lead to substantial drift in the Earth's climate, including changes in temperature and precipitation, and in the frequency of occurrence of extreme events (*e.g.*, hurricanes). Reduction in the ozone column abundance leads to enhanced levels of UV-B radiation at the surface (see Chapter 14) with potentially harmful effects on living organisms, including phytoplankton in the ocean, and increased frequency of skin cancers affecting humans. Ecosystem damage and health problems also result from regional and global pollution. Acidic precipitation is believed to have suppressed life in several lakes of North America and Europe and, together with enhanced ozone levels, to have damaged forests in those same parts of the world.

A detailed understanding of the observed degradation in the "health of the planet" requires that atmospheric chemistry be studied in the broader context of "global change," and that the Earth system be viewed as a nonlinear interactive system consisting of the atmosphere, the ocean, and the continental biosphere. That the chemical composition of the atmosphere has been maintained far away from the thermodynamic equilibrium conditions encountered, for example, on Mars and Venus, and that the two major gases surrounding the planet are nitrogen (78%) and oxygen (21%), as opposed to $CO_2$, are a direct consequence of oxidation and reduction processes associated with the energy metabolism of various forms of life.

Changes encountered in the Earth system occur at different scales in time and space. For example, the formation of a tornado requires only a few minutes, while the response of the ocean to greenhouse warming is characterized by time scales of several decades. A variety of different time scales need also to be considered in the case of chemical processes in the atmosphere. For example, the chemical lifetime of a radical such as OH (which plays a key role in the chemistry of the atmosphere) is typically a few seconds, while that of most chlorofluorocarbon molecules lies in the range of several decades to a century. In the past, changes in the chemical composition of the atmosphere associated with glacial-interglacial transitions have occurred over hundreds to several thousands of years, although sometimes associated with rapid fluctuations indicative of the Earth system's nonlinear nature.

The focus of this book is on the basic chemical and photochemical processes occurring in the atmosphere (troposphere and middle atmosphere), with emphasis on changes that are expected to affect the Earth in the future over periods of time ranging from 10 to 100 years.

While atmospheric chemistry is at the center of global change, and while many global change effects of concern to society manifest themselves through changes in chemistry or its effects on climate, the field also maintains close connections with atmospheric physics, oceanography, and ecology.

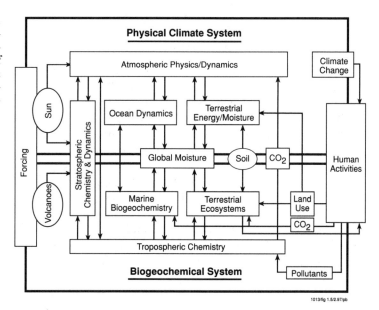

*Figure 1.4.* Diagram showing important interactions among different components of the Earth system with emphasis on the physical climate and biogeochemical cycles.

## 1.2 The Earth System

The role played by humans in global change requires that global Earth studies transcend the traditional disciplinary boundaries to investigate the interactions among the atmosphere, ocean, ice, solid Earth, and biological systems that shape the evolution of the Earth. What are the factors controlling the evolution of climate? How do chemical elements, such as carbon, nitrogen, sulfur, etc., which are crucial to the maintenance of life, circulate among the different components of the Earth system, and how are the chemical cycles perturbed by human activities? These key questions, which emphasize two major interacting components of the total system—the physical climate system and the biogeochemical cycles (Fig. 1.4)—are at the center of the rapidly developing Earth system science (see Box 1.2, and NASA, *A Program for Global Change*, 1986).

> **Box 1.2 Earth System Science**
> The objective of Earth system science is to obtain a scientific understanding of the Earth system as a whole by describing how its component parts (atmosphere, hydrosphere, biosphere, and lithosphere) and their interactions have evolved, how they function, and how they may be expected to continue to evolve on all time scales. An important challenge posed to Earth system scientists is to develop the capability to predict those changes that will occur in the next decade to century, both naturally and in response to human activity.

## 1.2.1 The Atmosphere

The atmosphere, the thin and fragile envelope of air surrounding the Earth, plays an important role because it greatly affects the environment in which we live. It is held around the Earth by gravitational attraction. The air pressure ($p$), of the order of 1000 mb (or hPa) near the surface ($z_0$), decreases quasiexponentially with height ($z$)

$$p(z) = p(z_o) \exp\left(-\frac{z - z_o}{H}\right) \quad (1.1)$$

where the atmospheric scale height

$$H = \frac{RT}{g}$$

of the order of 7 km, can be calculated from $R$ (287 J K$^{-1}$ kg$^{-1}$), the universal constant for ideal air, $g$ (9.80665 m s$^{-2}$) the gravitational acceleration, and $T$ the temperature (expressed in kelvin) (see Chapter 2 for more details). The total dry mass of the atmosphere (annual mean) is estimated to be $5.13 \times 10^{18}$ kg (Trenberth and Guillemot, 1994).

The atmosphere (see also Chapter 2) is generally described in terms of layers characterized by specific vertical temperature gradients (Fig. 1.5). The *troposphere*, which extends from the surface to the tropopause at the approximate altitude of 18 km in the tropics, 12 km at midlatitudes, and 6 to 8 km near the poles, is characterized by a decrease of the mean temperature with increasing altitude. This layer, which contains about 85-90% of the atmospheric mass, is often dynamically unstable with rapid vertical exchanges of energy and mass being associated with convective activity. Globally, the time constant for vertical exchanges is of the order of several weeks. Much of the variability observed in the atmosphere occurs within this layer, including the weather patterns associated, for example, with the passage of fronts or the formation of thunderstorms. The planetary boundary layer is the region of the troposphere where surface effects are important. Its depth is of the order of 1 km, but varies significantly with the time of day and with meteorological conditions. The exchange of chemical compounds between the surface and the free troposphere is directly dependent on the stability of the boundary layer. Above the troposphere, the atmosphere becomes very stable, as the vertical temperature gradient reverses in a second atmospheric region called the *stratosphere*. This layer, which extends up to 50 km (the stratopause), contains 90% of the atmospheric ozone. A typical residence time for material injected in the lower stratosphere is one to three years. The layer between 50 and 90 km altitude is called the *mesosphere* and its upper limit the mesopause. In this latter region, the temperature again decreases with height. Dynamical instability occurs frequently and is characterized by rapid vertical mixing (see Chapter 2). Above 85-90 km, in the

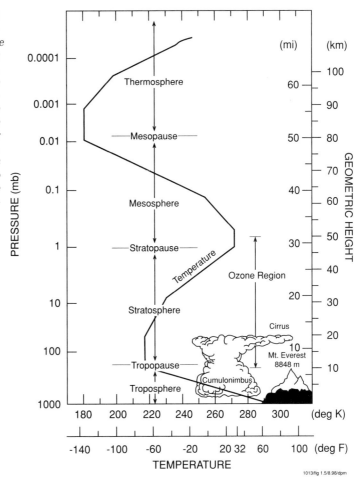

**Figure 1.5.** Vertical profile of the temperature between the surface and 100 km altitude as as defined in the U.S. Standard Atmosphere (1976) and related atmosphere layers. Note that the tropopause level is represented for midlatitude conditions. Cumulonimbus clouds in the tropics extend to the tropical tropopause located near 18 km altitude.

region called the *thermosphere*, the temperature increases to reach maximum values that are strongly dependent on the level of solar activity. Vertical exchanges associated with dynamical mixing become insignificant, but molecular diffusion becomes an important process that produces gravitational separation of species according to their molecular or atomic weight. The escape to space of hydrogen atoms (25 kilotons/year) from the *exosphere*, the atmospheric region in contact with the interplanetary medium, has most certainly contributed to substantial changes in the chemical composition of the atmosphere over geological history. This escape flux increases due to the increase in methane abundance (Ehhalt, 1986).

The uneven distribution of radiative heating in the Earth system produces a meridional circulation of air (circulation in the north-south direction, see Chapter 2), with rising motion at low latitudes and sinking motion at mid- and high latitudes. This meridional overturning of air masses is modified substantially by the Earth's rotation, especially outside the tropics, where the mean circulation is nearly circumpolar (along latitude circles). A small meridional component, however, transfers heat and transports trace constituents from equatorial to polar regions. The zonal flow (circulation in the west-east direction) is perturbed by orographic features at the Earth's surface (*e.g.*, mountains) and latent heat release associated with the formation of clouds, as well as synoptic weather systems in the troposphere. Chemical constituents are redistributed in the atmosphere by transport processes caused by

*Figure 1.6.* Schematic representation of the general circulation of the atmosphere during Northern Hemisphere summer. Because of maximum heating in the Northern Hemisphere tropics, the southern Hadley cell, from the summer to the winter hemisphere, is much stronger than its winter counterpart. (During southern summer the northern Hadley cell is much stronger.) Also shown in the figure as continuations of the tropical Hadley cell are the northern and southern mean extratropical circulations. These are the average result of highly structured circulations associated with extratropical synoptic waves (after Meehl, 1987; see Chapter 2 for details).

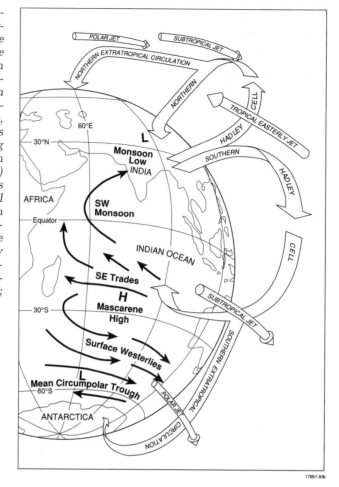

such dynamical disturbances. Figure 1.6 presents a schematic representation of the atmospheric circulation patterns (see Chapter 2 for more details) for Northern Hemisphere summer conditions. It shows the structure of the mean meridional circulation of Hadley cells (extratropical circulation), as well as surface winds over the Indian Ocean.

The Earth's atmosphere is a mixture of a multitude of chemical constituents (Box 1.3). The most abundant of them are nitrogen $N_2$ (78%) and oxygen $O_2$ (21%). These gases, as well as the noble gases (argon, neon, helium, krypton, xenon), possess very long lifetimes against chemical destruction and, hence, are relatively well mixed throughout the entire homosphere (below approximately 90 km altitude). Minor constituents, such as water vapor, carbon dioxide, ozone, and many others, also play an important role despite their lower concentration: They influence the transmission of solar and terrestrial radiation in the atmosphere and are therefore linked to the physical climate system; they are key components of biogeochemical cycles; in addition, they determine the "oxidizing capacity" of the atmosphere and, hence, the atmospheric lifetime of biogenic and anthropogenic trace gases. Table 1.1 provides information on important trace species of the atmosphere, while Figure 1.7 shows an approximate distribution with altitude of the mixing ratio of chemical constituents. Several useful definitions are provided in Box 1.4.

**Table 1.1**
*Chemical Composition of the Atmosphere*

| Constituent | Chemical Formula | Volume Mixing Ratio in Dry Air | Major Sources and Remarks |
|---|---|---|---|
| Nitrogen | $N_2$ | 78.084% | Biological |
| Oxygen | $O_2$ | 20.948% | Biological |
| Argon | Ar | 0.934% | Inert |
| Carbon dioxide | $CO_2$ | 360 ppmv | Combustion, ocean, biosphere |
| Neon | Ne | 18.18 ppmv | Inert |
| Helium | He | 5.24 ppmv | Inert |
| Methane | $CH_4$ | 1.7 ppmv | Biogenic and anthropogenic |
| Hydrogen | $H_2$ | 0.55 ppmv | Biogenic, anthropogenic, and photochemical |
| Nitrous oxide | $N_2O$ | 0.31 ppmv | Biogenic and anthropogenic |
| Carbon monoxide | CO | 50-200 ppbv | Photochemical and anthropogenic |
| Ozone (troposphere) | $O_3$ | 10-500 ppbv | Photochemical |
| Ozone (stratosphere) | $O_3$ | 0.5-10 ppm | Photochemical |
| Nonmethane hydrocarbons | | 5-20 ppbv | Biogenic and anthropogenic |
| Halocarbons (as chlorine) | | 3.8 ppbv | 85% anthropogenic |
| Nitrogen species | $NO_y$ | 10 ppt-1 ppm | Soils, lightning, anthropogenic |
| Ammonia | $NH_3$ | 10 ppt-1 ppb | Biogenic |
| Particulate nitrate | $NO_3^-$ | 1 ppt-10 ppb | Photochemical, anthropogenic |
| Particulate ammonium | $NH_4^+$ | 10 ppt-10 ppb | Photochemical, anthropogenic |
| Hydroxyl | OH | 0.1 ppt-10 ppt | Photochemical |
| Peroxyl | $HO_2$ | 0.1 ppt-10 ppt | Photochemical |
| Hydrogen peroxide | $H_2O_2$ | 0.1 ppb-10 ppb | Photochemical |
| Formaldehyde | $CH_2O$ | 0.1-1 ppb | Photochemical |
| Sulfur dioxide | $SO_2$ | 10 ppt-1 ppb | Photochemical, volcanic, anthropogenic |
| Dimethyl sulfide | $CH_3SCH_3$ | 10 ppt-100 ppt | Biogenic |
| Carbon disulfide | $CS_2$ | 1 ppt-300 ppt | Biogenic, anthropogenic |
| Carbonyl sulfide | OCS | 500 pptv | Biogenic, volcanic, anthropogenic |
| Hydrogen sulfide | $H_2S$ | 5 ppt-500 ppt | Biogenic, volcanic |
| Particulate sulfate | $SO_4^{2-}$ | 10 ppt-10 ppb | Photochemical, anthropogenic |

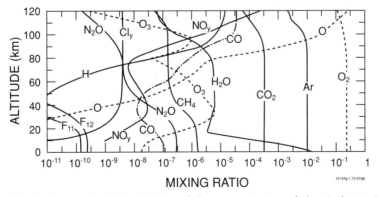

**Figure 1.7.** Typical vertical distribution of the concentration of chemical constituents in the atmosphere. Some lines are shown as dashes for clarity ($F_{11} = CFCl_3$ and $F_{12} = CF_2Cl_2$) (Goody, 1995).

## Box 1.3 The Discoveries of Air Composition

Our understanding of the chemistry of the atmosphere has greatly evolved with time. Greek philosophers like Empedocles (fifth century BC) believed that nature was composed of four universal elements: earth, water, fire, and air. Aristotle (384-322 BC), who wrote the first scientific treatise on meteorology, recognized water as a distinct component of air, and realized that this element was continously recycled between the atmosphere and the ocean. After the Greek period, very little progress was made until the period of the Renaissance, which corresponded to the revival of arts and sciences, and to the emergence of rational approaches. Meteorology became a scientific discipline and studies were conducted to better understand the atmosphere.

J. Priestley

A first significant advance was made when Leonardo da Vinci (1452-1519) in Italy and later John Mayow (1641-1679) in Great Britain suggested that air is composed of two distinct components: one ("fire-air") that supports combustion and life, and one ("foul-air") that does not. The first of these components, the "fire-air," was isolated in 1773 by the Swedish chemist Carl Wilhelm Scheele (1742-1786), one year later by the British scientist Joseph Priestley (1733-1804), and in 1789 by the French chemist Antoine-Laurent Lavoisier (1743-1794). Lavoisier, one of the fathers of modern chemistry (executed during the French revolution), named this gas oxygen [from the Greek *oxus* (acid) and *genan* (to beget)]. Jointly with the astronomer and mathematician Pierre-Simon de Laplace (1749-1827), he showed that animal respiration was a slow form of combustion with consumption of oxygen and release of carbon dioxide.

Carbon dioxide was discovered around 1750 by Joseph Black (1728-1799) and nitrogen was discovered several years later by Daniel Rutherford (1749-1819). Henry Cavendish (1731-1810) had suggested that air is a complex mixture of different gases, but it was not before the end of the nineteenth century that William Strutt (1842-1919, known as Lord Rayleigh) and Sir William Ramsay (1852-1916) identified argon, the most abundant of the noble gases in the atmosphere. Ramsay also discovered that air contained several other inert gases (neon, helium, krypton, and xenon), using a gas separation technique for which he received the Nobel Prize in Chemistry in 1904.

de Laplace        Lavoisier

Ozone research has been very active since the German chemist Christian Frederich Schönbein (1799-1868) discovered this gas and André Houzeau in France made the first atmospheric measurement in 1858 (see Box 14.1 for more details).

As new observational techniques became available, less abundant chemical constituents were discovered in the atmosphere. For example, in France J. B. Boussingault, aware that M. Theodore de Saussure had reported the presence of a natural "combustible gas" in the atmosphere in the late 1700s, established in 1862 that methane ($CH_4$) was a component of the atmosphere in urban areas and might also be present in rural areas. In 1898, Armand Gautier noted that this gas was released by swamps, decaying organic matter, and produced by wood and coal fires. Fifty years later, Marcel Migeotte from Belgium detected an absorption feature in the infrared solar spectrum that he attributed to the presence of methane in the atmosphere.

> Several other chemical compounds were discovered (and their abundance measured) from absorption features observed in the infrared solar spectrum. This is the case for nitrous oxide ($N_2O$), discovered in 1939 by Arthur Adel in Arizona, for carbon monoxide (CO), identified in 1949 by Migeotte, and for nitric acid ($HNO_3$), detected in 1968 by David Murcray and his collaborators at the University of Denver in Colorado.
>
> The presence of molecular hydrogen ($H_2$) in the atmosphere was suggested in the early 1900s by British chemist James Dewar and French physicist Georges Claude, but it was P. M. Schuftan who provided the first accurate estimate of the atmospheric abundance of this gas through experiments carried out in 1923 at the liquid air plant of Linde in Bavaria.
>
> Halocarbons were first detected in the atmosphere in the 1970s following the invention of an electron-capture detector by James Lovelock in the United Kingdom. Specifically, $CFCl_3$ and $SF_6$ were measured by Lovelock in 1971, $CF_2Cl_2$ by C. W. Su and E. D. Goldberg in 1973, $CCl_4$ and $CH_3I$ by Lovelock with collaborators in 1973, and $CH_3Cl$ by E. P. Grimsrund and R. A. Rasmussen in 1975. Methyl bromide ($CH_3Br$) was measured by H. B. Singh in 1977.

The spatial and temporal distribution of chemical species in the atmosphere is determined by several processes, including surface emissions and deposition, chemical and photochemical reactions, and transport. Surface emissions are associated with volcanic eruptions, biological activity on the continents as well as in the ocean, biomass burning, agricultural practices, industrial activity, etc. Chemical conversions are achieved by a multitude of reactions whose rate constants are measured in the laboratory (see Chapter 3). Transport is usually represented by large-scale advective motions (displacements of air masses in the quasihorizontal direction), and by smaller scale processes, including convective motions (vertical motions produced by thermal instability and often associated with the presence of large cloud systems), boundary layer exchanges, and mixing associated with turbulence. Wet deposition results from precipitation of soluble species, while the rate of dry deposition is affected by the nature of the surface (*e.g.*, type of soils, vegetation, ocean, etc.).

> **Box 1.4 Definitions**
>
> The abundance of the chemical constituent $i$ in the atmosphere is often expressed in terms of number density $n_i$ (particles m$^{-3}$ or particles cm$^{-3}$) or mass density $\rho_i$ (kg m$^{-3}$ or g cm$^{-3}$). Frequently, however, the concept of the dimensionless volume mixing ratio
>
> $$\mu_i = \frac{n_i}{n_a} \qquad (1)$$
>
> (where $n_a$ is the air number density) or mass mixing ratio
>
> $$\tilde{\mu}_i = \frac{\rho_i}{\rho_a} \qquad (2)$$
>
> (where $\rho_a$ is the air mass density) is used. Abundances expressed in parts per million (ppm), parts per billion (ppb), and parts per trillion (ppt) in volume (v) or in mass (m) correspond to mixing ratios of $10^{-6}$, $10^{-9}$, and $10^{-12}$, respectively. The vertically integrated number density (expressed, for example, in m$^{-2}$ or cm$^{-2}$) is called the column abundance. In the case of ozone, it is usually expressed in Dobson units (DU). One DU corresponds to the height (in $10^{-3}$ cm) the column ozone would have if all the gas were at standard pressure and temperature. One DU is equivalent to a column of $2.687 \times 10^{16}$ molecules cm$^{-2}$.

Chemical processes that affect minor constituents of the troposphere are reproduced schematically in Figure 1.8. Chemical compounds released at the surface by natural and anthropogenic processes are oxidized in the atmosphere before being removed by wet or dry deposition. Key chemical species of the troposphere include organic compounds (see Chapter 9) such as methane and nonmethane hydrocarbons as well as oxygenated organic species and carbon monoxide, nitrogen oxides (which are also produced by lightning discharges in thunderstorms; see Chapter 7) as well as nitric acid and peroxyacetyl nitrate (PAN), hydrogen compounds (and specifically the OH and $HO_2$ radicals as well as hydrogen peroxides; see Chapter 6), ozone, and sulfur compounds [dimethylsulfide (DMS), sulfur dioxide and sulfuric acid; see Chapter 10]. The hydroxyl (OH) radical deserves special consideration since it has the capability of efficiently destroying a large number of chemical compounds, and hence of contributing directly to the oxidation capacity of the atmosphere. Ozone also plays an important role in the troposphere: Together with water vapor it is the source of the OH radical; in addition, it contributes to climate forcing (see Chapter 15). The presence of this gas in the troposphere results not only from the intrusion of ozone-rich stratospheric air masses through the tropopause; it is also produced by photochemical reactions involving nitrogen oxides, hydrocarbons, and carbon monoxide (see Chapter 13). One major question is to what extent the oxidizing capacity of the atmosphere has changed as a result of human activities. Finally, the release at the Earth's surface of sulfur compounds and their oxidation in the atmosphere leads to the formation of small liquid or solid particles that remain in suspension in the atmosphere (see Chapter 4). These aerosol particles affect the radiative balance of the atmosphere directly, by reflecting and absorbing solar radiation, and indirectly, by influencing cloud microphysics. The release to the atmosphere of sulfur compounds has increased dramatically, particularly in regions of Asia, Europe, and North America as a result of human activities, specifically coal combustion.

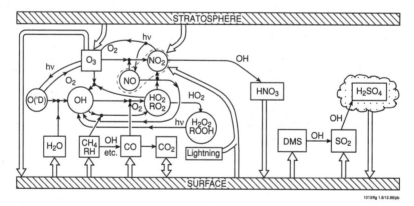

**Figure 1.8.** *Schematic diagram of the key chemical and photochemical processes affecting the composition of the global troposphere. The symbol R refers to an organic radical. hν refers to the absorption of light (photolytic reaction) (Prinn, 1994; reprinted with permission of Oxford University Press).*

Gases that are not rapidly destroyed (*e.g.*, by the OH radical) or removed by clouds and rain in the troposphere are transported upward into the stratosphere, where they are dissociated by short-wave ultraviolet radiation to produce fast-reacting radicals (see Fig. 1.9). Chlorofluorocarbons or nitrous oxide are examples of such long-lived gases that, when photolyzed in the stratosphere, provide a source of chlorine or

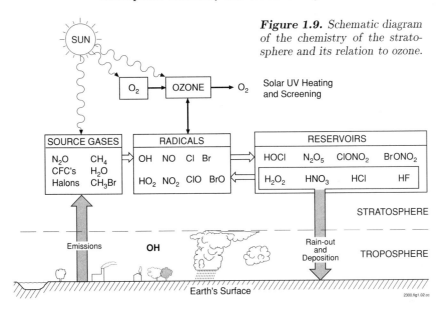

**Figure 1.9.** *Schematic diagram of the chemistry of the stratosphere and its relation to ozone.*

nitrogen oxides, respectively. Such fast-reacting radicals initiate catalytic cycles that lead to the destruction of ozone, before being converted into chemical reservoirs that are gradually removed from the stratosphere. The abundance of ozone results from a delicate balance between these destruction mechanisms and the natural production of $O_3$ through the photolysis of molecular oxygen (see Chapter 14). There is strong evidence that suggests that the depletion of stratospheric ozone observed over the past decade is the direct consequence of the release in the atmosphere of industrially manufactured chlorofluorocarbons.

The chemical system that affects the behavior of trace constituents in the atmosphere is complex and highly nonlinear. Most of the following chapters will describe this system in detail, and assess its role in the Earth's system. Since atmospheric ozone is a central theme of this book, several chapters will provide a detailed description of the chemistry affecting the various compounds that contribute to the formation or the destruction of this molecule in the troposphere and the stratosphere.

*1.2.2 The Ocean*

The ocean, which covers approximately 70% of the Earth's surface, is coupled to the atmosphere from both a physical and a biogeochemical perspective. The basic structure of the ocean is set by the geographic patterns of surface heating and freshwater input (precipitation minus evaporation), which influences the salinity distribution in the ocean. In general there is net warming of the ocean surface in the tropics and subtropics and net cooling at high latitudes. The ocean can be roughly divided into two regions: a warm, surface pool (on average 18°C) of about 1 km thickness and deep water (on average 3°C) that outcrops to the surface at high latitude and forms the bulk of the ocean volume. Unlike the atmosphere, heating of the ocean surface stabilizes the water column and prevents rapid exchange between the surface and deep water. Contact between the surface and deep waters is limited to localized polar regions (*e.g.*, North Atlantic, Antarctic), where losses of heat and freshwater lead to sinking and deep water formation. The resulting thermohaline circulation (see Fig. 1.10) is especially important over long timescales (*e.g.*, glacial cycles).

## OCEAN CIRCULATION

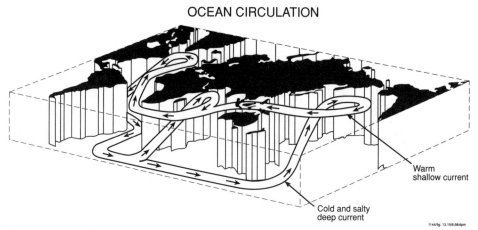

**Figure 1.10.** *Schematic representation of the global "conveyor belt" depicting the global thermohaline circulation of the ocean. Cold, salty water of the north Atlantic sinks to the deep ocean and moves southward to re-surface and be re-warmed in the Indian and north Pacific oceans. Surface currents return the water to the north Atlantic through the Pacific and south Atlantic. The time constant associated with a complete circuit is almost 1,000 years.*

The other component of the ocean circulation is produced by the drag of the surface winds on the ocean. The zonal wind patterns (Fig. 1.6) over the ocean result in rotating circulation patterns in the subtropical and subpolar regions and are also responsible for the circumpolar current in the southern ocean. On average, the wind-driven surface currents move heat and trace species (*i.e.*, chemical compounds, isotopes, etc.) from the tropics to the poles, and about an equal amount of solar energy received in the tropics is transported toward the pole by oceanic and atmospheric circulations. Wind forcing also causes divergence of the surface water and upwelling along both the equator and coastal regions on the eastern margin of ocean basins (*e.g.*, Peru, California). The upwelling of cooler, nutrient-rich waters in these areas greatly enhances ocean productivity (Fig. 1.11). The ocean circulation, which exhibits variations on different timescales (including perturbations such as the El Niño events that occur in the equatorial Pacific on an average of four years and produce massive warming of the coastal waters off Peru and Ecuador with torrential rainfall in the region), greatly affects biogeochemical cycles as well as the global climate.

From a chemical point of view, the ocean influences the atmosphere through the exchanges of trace gases across the air-sea interface. The transfer of $CO_2$ from the atmosphere to the ocean is controlled by the two competing factors of temperature: warming of surface waters, which releases $CO_2$ to the atmosphere, and biological productivity. Photosynthesis by marine phytoplankton converts dissolved $CO_2$ into organic carbon, leading to a reduction in surface $CO_2$ values and a $CO_2$ flux into the ocean. The amount of $CO_2$ dissolved in seawater is quite large due to its high solubility and its reactivity with water to form carbonic acid and its dissociation products. The ocean, therefore, serves as a major reservoir for $CO_2$, about 65 times larger than the atmosphere. It has played an important role in the evolution of atmospheric $CO_2$ over the geological history of the Earth and is a primary sink for anthropogenic $CO_2$ (see Chapter 5).

Other chemical species are released by the ocean, such as reduced sulfur, certain hydrocarbons, and carbon monoxide. The largest oceanic source of sulfur is provided by dimethyl sulfide, which is produced by specific types of phytoplankton

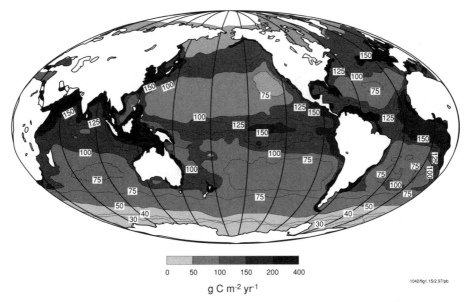

**Figure 1.11.** *Annual primary production within the ocean expressed in $gC\ m^{-2}\ yr^{-1}$ (Antoine et al., 1996).*

(coccolithophorids). These emissions appear to be most intense in regions where the net primary productivity of the ocean is highest, modified somewhat by poorly understood large-scale patterns in the distribution of phytoplankton species.

### 1.2.3 The Terrestrial Biosphere

The terrestrial biosphere is important to atmospheric chemistry as a source and sink for many compounds. A major activity within atmospheric chemistry has been the determination of such fluxes, which remain poorly known. The structure of the biosphere is controlled by the interaction of climate with the patterns of soils and topography resulting from geological processes on a range of time scales, and further modified by the biogeographic distribution of organisms. Figure 1.12 shows mean annual temperatures and precipitation for the globe. Climate patterns are reflected in productivity (annual carbon fixation through photosynthesis), with warmer and wetter regions having higher productivity (Fig. 1.13). Rates of nitrogen cycling follow similar trends, and as a result, trace gas emissions are usually higher—in some cases orders of magnitude higher—in the tropics than in mid- to high-latitude regions. This is clearly true for all soil trace gas fluxes, and may be true for plant-mediated fluxes. For example, large quantities of hydrocarbons such as isoprene ($C_5H_8$) are produced by the foliage of the abundant vegetation in productive ecosystems. Biomass burning fluxes are highest in tropical savanna ecosystems, which are warm and have sufficient rainfall during the wet seasons to accumulate significant biomass, which burns readily during the dry seasons. Large quantities of atmospheric $CO_2$, $CO$, hydrocarbons, and $NO_x$ are produced as a result of biomass burning. Soil NO fluxes are enhanced when soils are rapidly wet, dry, and wet again, and so may be higher in regions of sporadic rainfall, despite higher overall rates of nitrogen cycling, and $N_2O$ emissions in moist areas.

An important exception to this relationship between emissions and climatic conditions are wetlands, in which high rainfall, appropriate topography, or low evapotranspiration lead to permanent or seasonal flooding. Wetlands are disproportionately

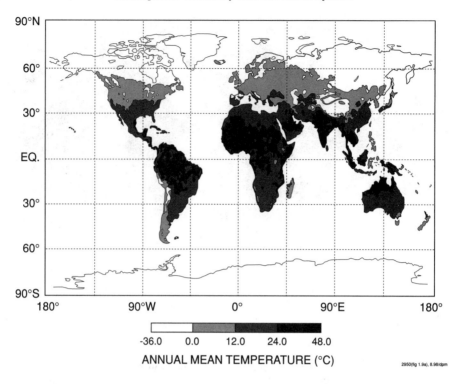

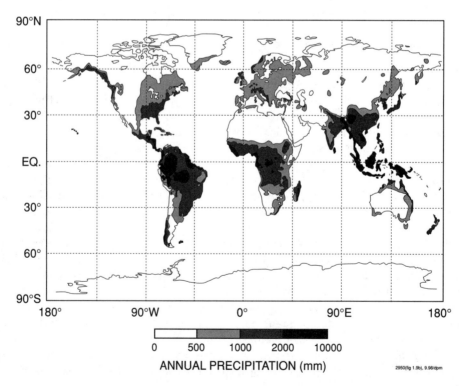

**Figure 1.12.** Mean annual surface temperature (degrees Celsius) and precipitation (mm of water) on the continents (created with data from Leemans and Cramer, 1992).

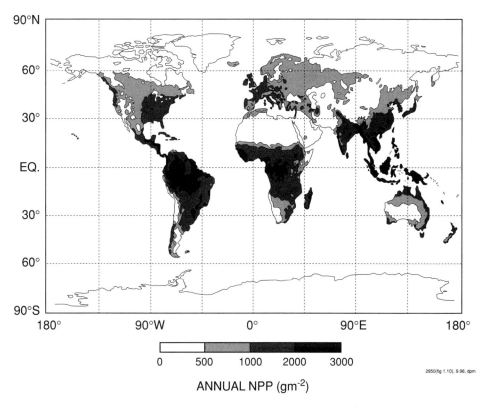

**Figure 1.13.** Net primary productivity on the continents in $gC\ m^{-2}\ yr^{-1}$ (calculated according to Leith, 1975, and created using climate data from Leemans and Cramer, 1992).

found in cooler latitudes, and northern wetlands are thought to share in importance with $CH_4$ emissions over tropical wetlands simply because of the greater area (albeit with lower fluxes) of northern wetlands. These areas become anoxic (deficient in oxygen) because microbial activity consumes all the oxygen. Anoxic conditions lead to reduced productivity, reduced organic matter decomposition, and greatly enhanced $CH_4$ production. Wetlands may also contribute significant amounts of reduced sulfur gases to the atmosphere.

The structure of the biosphere is important for the understanding of trace gas sources and sinks. However, the spatial structure of the biosphere cannot be described as the outcome of a series of physical calculations (unlike the atmosphere and oceans), but rather requires the use of large databases describing fine-scale structures within ecosystems. One of the great challenges of global change science has been to extrapolate trace gas and other biogeochemical processes from the scales of millimeters, at which the processes actually occur, to the global scale. This has required the development of new theories, remote sensing techniques, and data set assembly, but is resulting in a greatly improved capability to model global and regional trace gas exchange.

## Further Reading

Andrews, J. E., P. Brimblecombe, T. D. Jickells, and P. S. Liss (1996) *An Introduction to Environmental Chemistry*, Blackwell Science Publications, Oxford.

Calvert, J., ed. (1994) *The Chemistry of the Atmosphere: Its Impact on Global Change*, Blackwell Scientific Publications, Oxford.
Committee on Global Change (1988) *Toward an Understanding of Global Change*, National University Press, Washington, D.C.
Firor, J. (1990) *The Changing Atmosphere*, Yale University Press, New Haven.
Goody, R. (1995) *Principles of Atmospheric Physics and Chemistry*, Oxford University Press, New York.
Graedel, T. E. and P. J. Crutzen (1993) *Atmospheric Change: An Earth System Perspective*, Freeman and Co., New York.
Gurney, R. J., J. L. Foster, and C. L. Parkinson, eds. (1993) *Atlas of Satellite Observations Related to Global Change*, Cambridge University Press, Cambridge.
Junge, C. E. (1963) *Air Chemistry and Radioactivity*, Academic Press, New York.
Mackenzie, F. T. and J. A. Mackenzie (1995) *Our Changing Planet. An Introduction to Earth System Science and Global Environmental Change*, Prentice Hall, Upper Saddle River, NJ.
Rowland, F. S. and I. S. A. Isaksen (1988) *The Changing Atmosphere*, Report of the Dahlem Workshop on The Changing Atmosphere, Wiley-Interscience, Chichester, UK.
Seinfeld, J. H. and S. N. Pandis (1998) *Atmospheric Chemistry and Physics*, J. Wiley, New York.
Turco, R. P. (1997) *Earth Under Siege*, Oxford University Press, New York.

Ralph J. Cicerone

## Atmospheric Chemistry and the Earth System

In the opening chapter of this book, the Earth system and its atmosphere are described several times as "delicate" or "fragile." Why is it that among Earth scientists, atmospheric chemists are perhaps the group most likely to share this view? The answer to this question arises from the history of this field in the last three decades of the twentieth century, and it explains why atmospheric chemists feel that progress in research must accelerate as the century closes.

Consider first how our views of the atmosphere have changed since 1950. Then, atmospheric amounts of $N_2$, $O_2$, the noble gases, $CO_2$, $H_2O$ (below the tropopause), and $O_3$ (stratosphere only) were known quantitatively. $CH_4$ and $N_2O$ had already been identified in the atmosphere (in 1938 and 1948, respectively), but they were regarded as curiosities; amounts of $CH_4$, $N_2O$, and CO were known only to within about 50%. The photochemistry of urban air pollution was being fathomed, but stratospheric photochemistry had lain undeveloped since Chapman's work around 1930. Tropospheric ozone was thought to occur only when stratospheric air subsided. Airborne aerosol particles were known to exist, but even their bulk properties were poorly determined. Solar spectra indicated that stratospheric $O_3$ was important as a shield against ultraviolet light (UV), but other possible absorbers were unexplored. $CO_2$, $H_2O$, and $O_3$ were thought to play roles in climate and atmospheric dynamics, but were not well understood, and other greenhouse gases were not recognized. Similarly, the roles of sulfate, dust particles, and soot in the radiative energy balance of the Earth-atmosphere system were not yet identified.

In summary, the atmosphere in 1950 was viewed as inert chemically, and for good reason — most of the chemicals known to be present near the surface were essentially inert. Accordingly, the atmosphere near the surface was viewed as a fluid that moved moisture, heat, and momentum and that transported pollutants away from cities, factories, and fires.

Progress since then, actually since about 1970, has revealed a chemically dynamic atmosphere. First, new analytical instruments with impressive sensitivity and specificity have exposed the presence of many previously undetected chemicals, and time-series measurements have discovered temporal trends. New chemical reactions and relationships between atmospheric chemicals have been proposed and investigated; enormous advances in numerical modeling have allowed quantitative calculations to be performed on complex systems. Mechanistic studies of urban air-pollution photochemistry have demonstrated how organic gases, CO, and combustion-derived nitrogen oxides combine in sunlight to produce urban smog. The production of stratospheric ozone from solar UV dissociation of $O_2$ is now known to be counterbalanced by very efficient ozone-destroying catalytic chain reactions involving hydrogen oxides ($HO_x$), $NO_x$, chlorine oxides ($ClO_x$), and bromine oxides ($BrO_x$), and it is established that many of the reactive $HO_x$, $NO_x$, $ClO_x$, and $BrO_x$ radicals and molecules are formed in the stratosphere after stable source compounds from Earth's surface are photooxidized in the stratosphere — $CH_4$, $N_2O$, chlorofluorocarbons, and chlorocarbons, and halons and $CH_3Br$, respectively. Some of these source compounds did not exist in the 1970 atmosphere and all have grown in concentration.

We have also come to see that changing atmospheric chemical composition can exert strong leverage over global and regional climate through an enhanced greenhouse effect and separately through aerosol particles. Sorting out how climate will change is now a central research problem worldwide, as is the related question — what atmospheric chemical changes will follow climate change? Strong links between greenhouse gas amounts and climatic change over the past 150,000 years have emerged from beautiful experiments on dated ice cores, revealing much about the paleoatmosphere, paleoclimate, and global carbon and nitrogen cycles.

The atmosphere is also a window on the biosphere. Observing $CO_2$ and $O_2$ patterns and trends is illuminating the dynamics of the carbon cycle and the planet's photosynthetic primary productivity. Watching $N_2O$ amounts rise is telling us how far the nitrogen cycle is out of balance. Studying all of these gases plus dimethyl sulfide and methyl halides guides future research on oceanic processes. Quantifying atmospheric changes due to biomass burning is telling us much about consequences of land-use changes, as are observations of trace-gas exchange between the atmosphere, soils, and plants. Deposition of potential nutrients like nitrate and ammonium from the atmosphere to soils, plants, and water bodies is now recognized, as is deposition of toxins and other exotic chemical compounds. It is possible that these deposition fluxes are important ecologically, but temporal and regional patterns remain to be quantified. A prime example is the transport of continental dust to oceanic regions where iron may be a limiting nutrient; atmospheric inputs may provide essential amounts of iron.

Being a relatively small part (by mass) of the Earth system, the atmosphere responds relatively quickly to stimuli, both natural and anthropogenic. These short timescales provide at least two reasons why we must do our research quickly: Sampling theory alone places upper limits on the time interval between measurements to deduce trends or periodicities and separately, there is great need to determine unperturbed background states of the atmosphere before they are gone. Increasingly, quantitative research is needed to assess the impacts of specific human activities and generally in atmospheric chemistry. Many potentially important feedbacks between atmosphere, oceans, land surfaces, and the biota remain to be explored. Some of the feedback loops will turn out to be positive, some negative, and some negligible. We must find out which are which. How stable is the atmosphere-Earth system? What, if any, are the limits of stability? Let's get on with it!

*Ralph J. Cicerone is the Daniel G. Aldrich Professor of Earth System Science and of Chemistry and Chancellor at the University of California, Irvine. He is a member of the National Academy of Sciences and a former Director of the Atmospheric Chemistry Division at the National Center for Atmospheric Research.*

# Part 1

# Fundamentals

# 2 Atmospheric Dynamics and Transport

## 2.1 Introduction

The processes whereby air motions carry physical or chemical properties from one region of the atmosphere to another are collectively referred to as *transport*. Transport enables different chemical species, with different local sources, to interact. Through transport processes human activities have global consequences, changing the chemical balance in remote regions of the atmosphere. Atmospheric *dynamics* refers to the fluid physical phenomena that occur in the Earth's atmosphere and the physical laws that govern them. These laws place powerful constraints on transport processes by regulating the direction and time scales of atmospheric motions.

Without transport the local production and destruction of species would tend to balance and the chemical composition of the atmosphere would approach photochemical equilibrium. (Strictly speaking, such equilibrium can never be attained because the atmosphere is also subject to the effects of molecular diffusion. However, throughout the range of altitudes of concern to us—the troposphere and stratosphere—molecular diffusion is rather inefficient, and any significant departures from photochemical equilibrium are the result of bulk air motions.) Thus it is transport that determines the *net* source and sink regions of most atmospheric species by driving the atmosphere away from its local photochemical equilibrium state. For example, ozone is produced primarily in the tropical stratosphere and is then transported poleward and downward. The tropical stratosphere acts as a global source for ozone because of its transport out of this region. In other regions of the atmosphere (for example, in the polar lower stratosphere) transport increases the ozone concentration above its photochemical equilibrium value. In this way transport processes help determine both the local chemical time scale for a species (by driving the atmosphere away from its chemical equilibrium) and the global atmospheric lifetime of a species (by determining the time scale for the transport of a species from source regions to sink regions).

Atmospheric dynamics, through its influence on atmospheric temperature and pressure distributions, partially governs the time constants of chemical reactions. Through its influence on the distribution of various atmospheric species, with their different radiative absorption characteristics, it furthermore determines the transmissivity of the atmosphere to different wavelengths of radiation. This, in turn, helps determine the radiation spectrum available for breaking chemical bonds at different locations. The distribution of various atmospheric constituents, most important ozone and water vapor, also determines the atmospheric heating and cooling rates, thus influencing atmospheric dynamics.

It is useful to keep in mind some general features of atmospheric transport. The atmospheric layers described in Chapter 1 (Fig. 1.5) are based on the thermal structure, but they also are useful indicators of different transport regimes. The most important of the dividing lines for the purposes of this book is the tropopause, which separates the troposphere from the stratosphere. Two important subdivisions should be added to this, the planetary boundary layer (PBL) next to the Earth's surface and the free troposphere, between the top of the PBL and the tropopause.

Meteorologists have nomenclature for different scales of motion depending on the processes of interest. The largest is the global or planetary scale, which includes average descriptions and variations on the scale of continents. The synoptic scale refers to dynamical systems with a scale of about 1000 kilometers that make up most of the daily weather variations. Mesoscale refers to phenomena from tens to hundreds of kilometers, such as individual fronts or lines of thunderstorms. Small-scale processes range from the scale of an individual cloud or plume to the molecular level. Small-scale and mesoscale motions are most important in the PBL and the troposphere. Mesoscale systems are often associated with synoptic-scale systems (warm and cold fronts) in middle latitudes. Planetary-scale variations in the circulation are also present in the troposphere (*e.g.*, meanders in the jet stream, large-scale tropical circulations). Planetary-scale variations dominate the circulation in the stratosphere.

Many of the compounds in the atmosphere are emitted or produced at the Earth's surface. Transport begins in the PBL, the dynamically variable layer just above the surface (see Section 2.6.4). Air within the PBL is characterized by horizontal winds that are weak on average but have a high degree of turbulence. It often has relatively high levels of humidity and short-lived pollutants. The characteristic height of the PBL varies with time of day (shallower during night and early morning, deeper during afternoon and evening). Under certain meteorological conditions, a layer of haze or clouds at the top of the PBL can reflect or absorb incoming sunlight, leading to a low-level temperature inversion and a highly stable PBL (see Section 2.3.1). Temperature inversions can also be caused by a cold surface or by large-scale sinking motion. When an inversion exists, chemical compounds accumulate in the PBL and the atmosphere is polluted near the surface. In the contrasting situation, heating at the Earth's surface can lead to convection (Section 2.6.1), in which air moves vertically into the free troposphere and there is mixing of PBL and free tropospheric air. Convection is often associated with thunderstorms.

Convection is an important process for moving air vertically throughout the free troposphere. It is particularly important for the upward motion of air in the tropical regions. An individual convective cloud can move air from the surface to the upper troposphere in a matter of minutes to hours, which can lead to the intermittent presence of short-lived, surface-produced chemicals in the upper troposphere. Intrusion of convective clouds into the lower stratosphere is a mechanism of transfer of mass across the tropopause. Although localized intense downward motion (downbursts) can occur, overall the downward air movement occurs much more slowly and over wider horizontal areas than the upward convective motion. Because of this asymmetry, a representation of vertical transport as a simple transfer between the mean concentrations at adjacent levels will often be inadequate in the troposphere. Air that is moved upward (to lower pressure) will expand and cool (Section 2.2.3). The cooler air can hold less water vapor (Section 2.6.1). Once the temperature decreases below the point of saturation, water vapor will condense and form liquid or ice cloud particles. The particles grow by condensation or coagulation, and when they are large enough that their sedimentation rate exceeds the rate at which they evaporate or are lifted by the air motion, precipitation will occur. Precipitation removes water from the atmosphere and also removes a number of other substances, including soluble chemical compounds and airborne dust particles. Because of the heat needed to evaporate water, which is released when the water condenses, the precipitation process also redistributes thermal energy in the atmosphere.

On a large scale, variations of most meteorological parameters are larger in the vertical and latitudinal directions than in the longitudinal. Large-scale vertical winds are on the average quite small compared with horizontal winds. The tropospheric flow in midlatitudes is predominantly from the west in all seasons, although considerably stronger during winter. The air in the troposphere is continually stirred by weather events and convective activity. Chemical compounds with timescales of several months or more tend to be well mixed throughout the global troposphere. However, even for these there can be a difference between Northern Hemisphere and Southern Hemisphere concentrations if the sources or sinks have hemispheric differences or seasonal variability, since the cross-equatorial mixing time is on the order of a year.

The winds in the stratosphere are even more strongly horizontal. Vertical motion is very weak, and convection usually does not occur. Exchange of air vertically across the tropopause is therefore slow. The large-scale motion in the lower stratosphere is upward in the tropics and downward in middle and high latitudes, with a time scale for global turnover of a few months to more than a year. On the average, stratosphere-troposphere exchange introduces tropospheric air into the stratosphere in the tropics and stratospheric air into the troposphere in the middle latitudes. Because the stratospheric circulation is slow and mixing in latitude is weak, a gradient from equator to pole in the "age" of air can be seen in compounds that have temporal trends.

This chapter presents an introduction to some basic principles of atmospheric dynamics, with emphasis on how these principles can be used to gain insight into the nature of transport in the Earth's troposphere and stratosphere.

## 2.2 The Governing Equations

Much of the Earth's atmosphere can be considered a continuous, compressible fluid, whose motions are governed by fundamental principles of classical physics: Newton's laws of motion, conservation of energy, and conservation of mass. The discussion of atmospheric transport begins by considering the basic equations governing atmospheric flow.

The governing equations are most often expressed in the *Eulerian* frame of reference, that is, a frame of reference that is fixed with respect to the space coordinates. The Eulerian governing equations are introduced in Section 2.2.3. The Eulerian framework has the advantage that the variables describing the state of the fluid are related at every point in space by a set of partial differential equations. Solution of the Eulerian governing equations yields a description of the behavior of the fluid as a continuous field in space and time.

The Eulerian framework is to be contrasted with the *Lagrangian* framework, wherein the governing equations are expressed in a frame of reference moving with the fluid. The governing equations in the Lagrangian framework thus resemble the familiar equations of classical mechanics for point particles. The difference between the two frameworks can be appreciated by considering the statement of Newton's second law. In the Lagrangian frame, this is simply

$$\frac{d\mathbf{v}}{dt} = \mathbf{F}\left(\mathbf{x}\left(t\right);t\right) \qquad (2.1)$$

where $d\mathbf{v}/dt$ is the time derivative of the velocity of a fluid parcel, and $\mathbf{F}(\mathbf{x}(t);t)$ denotes the resultant of the forces per unit mass acting on that parcel as a function of its position, $\mathbf{x}$, and time, $t$, as the parcel follows its trajectory through space.

In the Eulerian frame, on the other hand, Newton's second law takes the form

$$\frac{\partial \mathbf{v}}{\partial t} + \mathbf{v} \cdot \nabla \mathbf{v} = \mathbf{F}(\mathbf{x}; t) \qquad (2.2)$$

which follows directly from differentiation of $\mathbf{v}(\mathbf{x}; t)$ when one considers the Lagrangian position vector to be a function of time, $\mathbf{x}(t)$, and notes that

$$\frac{d\mathbf{v}(\mathbf{x}; t)}{dt} = \left.\frac{\partial \mathbf{v}}{\partial t}\right|_{\mathbf{x}} + \frac{\partial \mathbf{v}}{\partial x}\frac{dx}{dt} + \frac{\partial \mathbf{v}}{\partial y}\frac{dy}{dt} + \frac{\partial \mathbf{v}}{\partial z}\frac{dz}{dt} \equiv \frac{\partial \mathbf{v}}{\partial t} + \mathbf{v} \cdot \nabla \mathbf{v} \qquad (2.3)$$

Equation (2.2) sacrifices the simplicity of the Lagrangian expression for the fluid acceleration in order to express the fluid velocity $\mathbf{v}$ and the force $\mathbf{F}$ as field variables, defined at every point $\mathbf{x}$ in space at time $t$. Note that in the Eulerian framework $\mathbf{x}$ is no longer considered to be a function of time. The Lagrangian framework is complicated by the fact that to calculate the force acting on a parcel it is necessary to know all the parcel trajectories, which are determined by the forces acting on the parcels in the first place. Furthermore, when it comes to evaluating $\mathbf{F}$, especially the contributions due to pressure gradients and turbulent stresses, the dependence on neighboring trajectories is complicated. Such contributions are expressed far more simply in Eulerian terms.

Although atmospheric motion is usually described in the Eulerian frame of reference, the Lagrangian approach also finds important applications in the derivation of some of the basic conservation laws, and in studies of chemical composition where one wishes to follow the chemical transformations occurring in an air parcel.

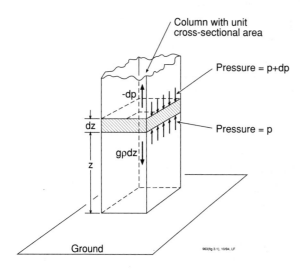

**Figure 2.1.** The vertical forces acting on a volume of air. Gravitation $(g\rho\,dz)$ exerts a force downward, while the pressure gradient (decreasing with altitude) exerts an upward force (from Wallace and Hobbs, 1977).

### 2.2.1 Hydrostatic Equilibrium

When Newton's second law is applied to the Earth's atmosphere, it is customary to make simplifying approximations based on neglecting certain small terms in the complete set of equations. The terms that are neglected depend on the space and time scales under consideration. The most important of these simplifying assumptions is the assumption of *hydrostatic equilibrium*. Under this assumption, the acceleration of the vertical velocity is neglected. Setting the left-hand side of (2.2) to zero in the vertical direction reduces the vertical component of this equation to a balance of vertical forces (per unit mass) acting on the fluid parcel. These forces (per unit mass) are the local vertical component of gravity, $g$, and the vertical gradient of pressure, $p$

(Fig. 2.1), so that

$$\frac{1}{\rho}\frac{\partial p}{\partial z_g} = -g \quad (2.4)$$

where $z_g$ is the geometric height and $\rho$ is the air density. Equation (2.4) is a very good approximation for atmospheric motions of horizontal scales greater than about 1000 km (see, e.g., Holton, 1992).

The prevalence of hydrostatic equilibrium has several important consequences for the mathematical treatment of atmospheric motions. In particular, the balance expressed in Eq. (2.4) implies that

$$p = p(x, y, z_g) = \int_\infty^{z_g} g\rho(x, y, z'_g) \, dz'_g \quad (2.5)$$

Equation (2.5) cannot be solved explicitly without knowing the distribution of density $\rho$. However, the assumption of a vertically isothermal atmosphere (i.e., an atmosphere at constant temperature) is a good first approximation. By doing so, it is possible to obtain an analytical expression for $p(x,y,z_g)$. Note first that the Earth's atmosphere also obeys the ideal gas law,

$$p = \rho RT \quad (2.6)$$

where $R = 287 \text{ m}^2 \text{ s}^{-2} \text{ K}^{-1}$ is the gas constant for air, that is, the gas constant evaluated for the mean molecular weight of air. Then, if $T$ is constant, Eqs. (2.4) and (2.6) can be combined to give

$$p(x, y, z_g) = p_s(x, y) \, e^{-z_g/H_o} \quad (2.7)$$

where $p_s$ is the pressure at $z_g = 0$ and

$$H_o = RT_o/g \quad (2.8)$$

is the e-folding height for pressure at atmospheric temperature $T_o$. This e-folding height varies between approximately 7 and 8 km for typical atmospheric temperatures. Thus, for example, $p$ at the stratopause ($\sim 50$ km) is approximately $e^{-7} \simeq 10^{-3}$ times the value of $p$ at sea level.

Combining Eqs. (2.6) and (2.7) yields

$$\rho(x, y, z_g) = \rho_s(x, y) \, e^{-z_g/H_o} \quad (2.9)$$

where $\rho_s = p_s/RT_o$. Thus in an isothermal atmosphere $H_o$ is also the e-folding height for density.

Equation (2.7) and, more generally, Eq. (2.5) imply that pressure can be used instead of geometric height as the vertical coordinate. The general relation (2.5) shows that pressure is a monotonic function of the height. Using pressure as the vertical coordinate has the important advantage that it simplifies the mathematical form of the Eulerian equations of motion, especially the equation for mass conservation.

*2.2.2 Coordinate System*

In the remainder of this chapter the spherical coordinate system $(x, y, z)$ will often be used (as shown in Fig. 2.2), where $dx = a\cos\phi \, d\lambda$, $dy = a \, d\phi$, $a$ is the radius of the Earth, and $\lambda$ and $\phi$ are longitude and latitude, respectively. The vertical coordinate can be taken as any variable that increases monotonically in height. Many coordinate systems used in atmospheric science are in fact not Cartesian. A particularly useful choice for the vertical coordinate is pressure. The use of pressure as the vertical coordinate introduces important simplifications in the equations governing atmospheric motions by removing explicit dependence upon the density (see Holton, 1992, for a discussion of the differences between geometric and constant-pressure, or

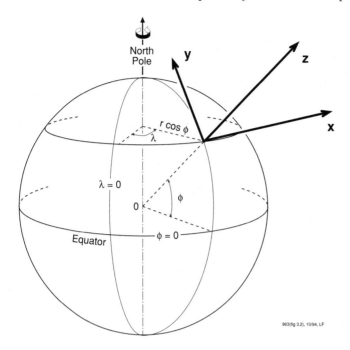

**Figure 2.2.** The spherical coordinate system, which remains fixed with respect to a point on the Earth's surface.

isobaric, vertical coordinates). Throughout much of this chapter, a type of isobaric coordinate known as *log-pressure* will be used, defined as:

$$z = H \ln(p_s/p) \tag{2.10}$$

where $p_s$ is any reference pressure, often taken equal to the mean sea-level pressure, and $H$ is a *scale height*. If the atmospheric temperature is a constant, $T = T_o$, then from Eq. (2.8) $H = H_o$, and the log-pressure altitude $z$ is equivalent to the geometric height. In general, of course, $T$ varies in both the horizontal and vertical directions, but if $H$ is evaluated at a representative global mean temperature, then $z$ does not differ greatly from $z_g$ throughout most of the atmosphere. Thus this vertical coordinate looks roughly like a geometric height coordinate, but retains the advantages of a pressure coordinate.

The components of the velocity vector, $\mathbf{v} = (u, v, w)$, are given by:

$$\begin{pmatrix} u \\ v \\ w \end{pmatrix} \equiv \frac{d}{dt} \begin{pmatrix} x \\ y \\ z \end{pmatrix} \tag{2.11}$$

where

$$\frac{d}{dt} \equiv \frac{\partial}{\partial t} + u \frac{\partial}{\partial x} + v \frac{\partial}{\partial y} + w \frac{\partial}{\partial z} \tag{2.12}$$

is the *material derivative*, or derivative following the motion [cf. Eq. (2.3)]. In spherical coordinates, $dx, dy$ stand for $dx = a \cos \phi \, d\lambda$, $dy = a \, d\phi$. As noted in Section 2.2, the material derivative arises when the time derivative is expressed in terms of fixed-space coordinates.

### 2.2.3 The Eulerian Primitive Equations in Log-Pressure Coordinates

The Eulerian set of equations derived under the assumption of hydrostatic equilibrium, and omitting certain other small terms, are known as the *primitive equations*. Details of their derivation are beyond the scope of this book, but can be found in several textbooks that provide excellent discussions of the subject (*e.g.*, Phillips, 1973; Lindzen, 1990; Holton, 1992).

In the log-pressure coordinate system introduced above, the hydrostatic equilibrium equation (2.4) takes the form

$$\frac{\partial \Phi}{\partial z} = \frac{RT}{H} \qquad (2.13)$$

where

$$\Phi = g \int_{z_s}^{z} dz' \qquad (2.14)$$

is the *geopotential*, and $z_s$ is a reference log-pressure altitude. It is easily verified that Eq. (2.13) follows directly from Eq. (2.4) plus the definitions (2.10) and (2.14).

Newton's second law yields the following equations for the rate of change of momentum per unit mass in the longitudinal and latitudinal directions:

$$\frac{du}{dt} - \left(f + \frac{u \tan \phi}{a}\right) v = -\frac{\partial \Phi}{\partial x} + X \qquad (2.15)$$

$$\frac{dv}{dt} + \left(f + \frac{u \tan \phi}{a}\right) u = -\frac{\partial \Phi}{\partial y} + Y \qquad (2.16)$$

where $d/dt$ is the material derivative (2.12).

The terms on the right-hand side of Eqs. (2.15) and (2.16) represent pressure gradient and nonconservative forces. The pressure gradient forces are actually given in terms of the geopotential (2.14), which, in the log-pressure system, is related to the pressure gradient force by

$$\nabla_p \Phi = \frac{1}{\rho} \nabla_z p \qquad (2.17)$$

where the suffixes $p, z$ on the gradient operators denote gradients taken on a constant-pressure surface and on a constant-altitude surface, respectively. Nonconservative forces (denoted by $X, Y$) include friction, which is important only near the Earth's surface, and molecular diffusion, which is negligible compared to bulk air motions, except at very high altitudes.

The terms involving the factors in parentheses on the left-hand side of Eqs. (2.15) and (2.16) denote "apparent" forces arising from the spherical coordinate system and the fact that the coordinate system is a rotating frame of reference. The terms due to the Earth's rotation $(-fu, fv)$ are also known as *Coriolis* forces. They are expressed in terms of the local vertical component of the Earth's angular velocity, the so-called *Coriolis parameter*:

$$f \equiv 2\Omega \sin \phi \qquad (2.18)$$

where $\Omega$ is the Earth's angular velocity. In a nonrotating reference frame the terms involving $f$ would be identically zero. The Coriolis terms exert apparent forces to the right of the velocity vector in the Northern Hemisphere and to the left of the velocity vector in the Southern Hemisphere.

The remaining primitive equations state the principles of conservation of mass and internal energy. Conservation of mass leads to the continuity equation:

$$\frac{\partial u}{\partial x} + \frac{1}{\cos \phi} \frac{\partial (v \cos \phi)}{\partial y} + \frac{1}{\rho_o} \frac{\partial (\rho_o w)}{\partial z} = 0 \qquad (2.19)$$

where

$$\rho_o = \rho_s e^{-z/H} \qquad (2.20)$$

and $\rho_s$ is a (constant) reference density.

Conservation of thermodynamic energy can be written as

$$\frac{d\theta}{dt} = \widehat{Q} \equiv \left(\frac{\theta}{T}\right) Q + Z \qquad (2.21)$$

where $\theta$ is the *potential temperature* (defined below), $T$ is the thermodynamic (absolute) temperature, $d/dt$ is the material derivative (2.12), $Q$ is the diabatic heating rate in kelvin per unit time, and $Z$ represents molecular diffusion. Molecular diffusion is usually negligible over the range of altitudes of interest to us (surface to lower thermosphere) and will therefore be omitted in all subsequent discussions. The primary diabatic heating sources are the atmospheric absorption of solar radiation, absorption and emission of infrared radiation, and the heating due to the condensation of water vapor (latent heating). The "frictional" heating due to viscous dissipation of the kinetic energy of moving air can contribute significantly to $Q$ at very high altitudes ($> 100$ km). A process that does not involve heating is called an *adiabatic process*.

The potential temperature is the temperature that a parcel of air in a compressible atmosphere would have if brought adiabatically to a constant standard pressure $p_s$. It is given by:

$$\theta = T(p_s/p)^\kappa \qquad (2.22)$$

where $\kappa = R/c_p$, $c_p$ is the specific heat of air at constant pressure, $R$ is the gas constant for air, and $p_s$ a reference pressure, often taken equal to the average sea level pressure (1000 hPa). In the log-pressure system, potential temperature is equivalent to

$$\theta = T \exp(\kappa z/H) \qquad (2.23)$$

as can be ascertained by substituting Eq. (2.10) into Eq. (2.22).

Substitution of Eq. (2.23) into Eq. (2.21) leads to the following form for the thermodynamic equation:

$$\frac{dT}{dt} + w\frac{\kappa T}{H} = Q \qquad (2.24)$$

Equations (2.21) and (2.24) are equivalent. Note that Eq. (2.21) states that potential temperature, $\theta$, is conserved following the motion of an air parcel in the absence of diabatic heat input or loss. On the other hand, the thermodynamic temperature, $T$, is not conserved even if $Q = 0$ because the temperature of a compressible fluid will change when it undergoes compression or expansion. This property of compressible fluids is embodied in the term $w\kappa T/H$ of Eq. (2.24), which vanishes for an incompressible atmosphere ($H \to \infty$).

In many applications, it is convenient to divide the temperature field into a global mean, $T_o(z)$, and deviations therefrom, $\widehat{T}(x, y, z; t)$. Thus $T = T_o + \widehat{T}$, and Eq. (2.24) can be written as:

$$\frac{d\widehat{T}}{dt} + w\left(\frac{\partial T_o}{\partial z} + \frac{\kappa T_o}{H}\right) = Q \qquad (2.25)$$

or

$$\frac{d\widehat{T}}{dt} + w\left(\frac{T_o}{\theta_o}\frac{\partial \theta_o}{\partial z}\right) = Q \qquad (2.26)$$

where $\theta_o = T_o \exp(\kappa z/H)$. In writing Eqs. (2.25)-(2.26) the term $w\kappa \widehat{T}/H$, usually small compared to $w\kappa T_o/H$, has been neglected.

Now, if $T_o \simeq gH/R$, per Eq. (2.8), Eq. (2.26) can be written as

$$\frac{d\widehat{T}}{dt} + w\left(\frac{HN_B^2}{R}\right) = Q \qquad (2.27)$$

where

$$N_B^2 = \frac{g}{\theta_o}\frac{\partial \theta_o}{\partial z} \qquad (2.28)$$

is the square of a frequency, known as the *Brunt-Väisälä frequency* or *buoyancy frequency*. The Brunt-Väisälä frequency is a measure of the restoring force exerted by buoyancy on fluid parcels in a stably stratified environment (see Section 2.3.1).

*2.2.4 Isentropic Coordinates*

Although the primitive equations in log-pressure coordinates will be used in much of the following discussion, it is important to mention a second coordinate system, wherein potential temperature is used as the vertical coordinate. As noted previously, any variable that increases monotonically with height may be used as a vertical coordinate. Since the Earth's atmosphere is usually stably stratified, potential temperature can serve this role while yielding certain important advantages in the formulation of the governing equations.

The coordinate system having potential temperature as the vertical coordinate is known as the *isentropic coordinate system*. The horizontal momentum equations in isentropic coordinates look much like Eqs. (2.15) and (2.16):

$$\frac{d_\theta u}{dt} - \left(f + \frac{u\tan\phi}{a}\right)v = -\frac{\partial \Psi}{\partial x} + X \qquad (2.29)$$

$$\frac{d_\theta v}{dt} + \left(f + \frac{u\tan\phi}{a}\right)u = -\frac{\partial \Psi}{\partial y} + Y \qquad (2.30)$$

where $\Psi$ is the *Montgomery streamfunction*, $\Psi = c_p T + \Phi$, and the material derivative is given by

$$\frac{d_\theta}{dt} = \frac{\partial}{\partial t} + u\frac{\partial}{\partial x} + v\frac{\partial}{\partial y} + \dot{\theta}\frac{\partial}{\partial \theta} \qquad (2.31)$$

The vertical velocity in the isentropic system is given by

$$\dot{\theta} \equiv \frac{d\theta}{dt} \qquad (2.32)$$

The hydrostatic relation is

$$\frac{\partial \Psi}{\partial \theta} = c_p \frac{T}{\theta} \qquad (2.33)$$

and the continuity equation takes the form

$$\frac{\partial \hat{\sigma}}{\partial t} + \frac{\partial}{\partial x}(\hat{\sigma}u) + \frac{1}{\cos\phi}\frac{\partial}{\partial y}(\cos\phi\,\hat{\sigma}v) + \frac{\partial}{\partial \theta}(\hat{\sigma}\dot{\theta}) = 0 \qquad (2.34)$$

where

$$\hat{\sigma} = -\frac{1}{g}\frac{\partial p}{\partial \theta} \qquad (2.35)$$

is a "pseudodensity" in the sense that, when multiplied by a "volume" element in isentropic coordinates, it yields an element of mass, that is, $\delta M = \hat{\sigma}\,\delta x\,\delta y\,\delta \theta$.

An important property of the isentropic system is that, by Eqs. (2.21) and (2.32), the velocity has no vertical component, $\dot{\theta}$, if the motion is adiabatic. Under these circumstances, the equations of motion and continuity are simplified considerably, although care must be taken in interpreting this result since the isentropic surfaces themselves move with respect to geometric height. The reader is referred to Dutton (1986) for a complete derivation and discussion of the primitive equations in isentropic coordinates.

*2.2.5 The Chemical Continuity Equation*

The continuity equation (2.19) is a statement of mass conservation for "air molecules." The Earth's atmosphere is a mixture of several gases, mostly chemically inert molecular oxygen and nitrogen, whose volume mixing ratios (21% and 78%, respectively) are essentially constant up to very great altitudes. The volume mixing ratio of an atmospheric constituent (see Box 1.4) is defined as:

$$\mu_i \equiv \frac{n_i}{n_a} \qquad (2.36)$$

where $n_i$ is the volume density (molecules per unit volume) of the $i$th constituent and $n_a$ is the volume density of air (*i.e.*, of the mixture of all atmospheric constituents).

In addition to the major constituents, hundreds of minor species are also present in the Earth's atmosphere. These minor species (whose combined volume mixing ratio is less than 1%) are chemically active to varying degrees, and their mixing ratios are highly variable in space and time. In the troposphere, water vapor is the most abundant minor constituent, followed by carbon dioxide, which is well mixed up to mesospheric altitudes. Both species are of great importance for the global thermal budget of the atmosphere. Species whose mixing ratios range from a few parts per million to a few parts per trillion by volume play an important role in the photochemistry and heat budget of the middle atmosphere, either directly (as in the case of ozone) or indirectly (as in the case of species that catalyze the destruction of ozone).

The role of dynamical and photochemical processes in determining the distribution of these minor chemical species is one of the major themes of this book. In mathematical form, this behavior can be described by a set of continuity equations for active chemical species, whose mixing ratio denoted by $\mu_i$. The continuity equation takes the form

$$\frac{d\mu_i}{dt} \equiv \frac{\partial \mu_i}{\partial t} + u\frac{\partial \mu_i}{\partial x} + v\frac{\partial \mu_i}{\partial y} + w\frac{\partial \mu_i}{\partial z} = S_i \qquad (2.37)$$

where $S_i$ denotes photochemical sources and sinks for the species, and molecular diffusion has been ignored, as noted in Section 2.2.3.

Equation (2.37) states that the mixing ratio of a chemical species is conserved following the motion of an air parcel provided that there are no photochemical sources or sinks. This type of equation is referred to as a *conservation equation*. Note that, while mixing ratio is a conserved quantity in the absence of photochemical processes, volume density is not. This can be verified by substituting Eq. (2.36) into Eq. (2.37). The resulting equation contains terms that arise from the compressibility of the atmosphere, as occurs for temperature when the thermodynamic equation is expressed in the form (2.24).

## 2.3 Constraints on Atmospheric Motion

Equations (2.13)–(2.24) describe three-dimensional, hydrostatic fluid motion in the Earth's atmosphere. If the equations are solved for the velocity field $(u, v, w)$, the three-dimensional evolution of the mixing ratio distribution of any chemical species, $\mu$, can be obtained from Eq. (2.37). General solutions of the governing equations are not available in closed analytical form; instead, such solutions require the use of numerical methods. However, important insights into the physical mechanisms that control the circulation can be gained by considering idealized or approximate problems for which analytical solutions are possible. In what follows this approach is used to illustrate how buoyancy and rotation impose very powerful constraints on the motions that can be realized in the Earth's atmosphere. These constraints in turn impose strong limitations on the timescales for transport between different locations in the atmosphere. In order to transport a fluid parcel either meridionally or vertically these constraints on its motion must be broken. A substantial portion of this chapter is devoted to showing how these constraints are broken. These ideas will also be mentioned when discussing various types of atmospheric wave motion.

## 2.3.1 Vertical Stratification and Static Stability

On average the Earth's atmosphere is stably stratified; that is, the variation of density with altitude is such that if a parcel of air is displaced vertically, buoyancy forces will act to return it to its original vertical position. This is just an example of the familiar Archimedean force as it applies to the atmosphere. A vertical displacement of a parcel in the atmosphere can result in the parcel oscillating about its neutral position. This situation can be analyzed mathematically by considering the vertical acceleration $dw/dt$ experienced by an air parcel when it is displaced adiabatically in a stably stratified environment.

The environment (denoted by the subscript $o$) is assumed to be in hydrostatic equilibrium (see Section 2.2.1):

$$\frac{1}{\rho_o}\frac{\partial p_o}{\partial z} = -g \qquad (2.38)$$

while the air parcel, initially at position $z_o$, is assumed to obey the full equation of motion in the vertical:

$$\frac{dw}{dt} = -g - \frac{1}{\rho}\frac{\partial p}{\partial z} \qquad (2.39)$$

When the parcel is at $z_o$, its pressure and density are equal to $p_o$ and $\rho_o$ and, from Eqs. (2.38) and (2.39), $dw/dt = 0$. Suppose now that the parcel is displaced vertically from $z_o$ to $z_o + \delta z$, and that its pressure adjusts instantaneously to the environment so that $p = p_o$ throughout the entire process. Then the vertical acceleration of the parcel is given by:

$$\left.\frac{dw}{dt}\right|_{z_o+\delta z} = -g\left[1 - \frac{\rho_o(z_o + \delta z)}{\rho(z_o + \delta z)}\right] \qquad (2.40)$$

which follows from Eqs. (2.38) and (2.39) if $p = p_o$.

Using the definition of potential temperature (2.22) and the ideal gas law (2.6), Eq. (2.40) can be rewritten as:

$$\frac{d^2(\delta z)}{dt^2} = -g\left[1 - \frac{\theta(z_o + \delta z)}{\theta_o(z_o + \delta z)}\right] \qquad (2.41)$$

where $w \equiv d(\delta z)/dt$. Note further that

$$\theta_o(z_o + \delta z) = \theta_o(z_o) + \delta z \frac{\partial \theta_o}{\partial z} \qquad (2.42)$$

and that the parcel temperature can be expressed as

$$\theta(z_o + \delta z) = \theta_o(z_o) \qquad (2.43)$$

since by Eq. (2.21) potential temperature is conserved following adiabatic motion. Using Eqs. (2.42) and (2.43) in Eq. (2.41) for small displacements yields:

$$\frac{d^2(\delta z)}{dt^2} \simeq -\frac{g}{\theta_o}\frac{\partial \theta_o}{\partial z}\delta z \qquad (2.44)$$

which has the simple solution

$$\delta z = C\exp(iN_B t) \qquad (2.45)$$

where $C$ is a constant and $N_B$ is the Brunt-Väisälä frequency (2.28). Under average tropospheric conditions, $N_B \simeq 10^{-2}$ s$^{-1}$, which gives a period of about 10 min for buoyancy oscillations. Clearly, buoyancy forces and more generally the conservation of potential temperature constitute a very strong constraint against vertical motion under normal conditions.

Vertical velocities are normally much weaker than horizontal motions, but important exceptions occur where the environmental stratification is weak or unstable. From Eq. (2.40), the restoring force acting on the air parcel vanishes as $N \to 0$. If $\partial \theta_o / \partial z$ becomes negative, the atmosphere is said to be *statically unstable*. Since parcel displacements would grow exponentially under such conditions according to Eqs. (2.44) and (2.45), unstable stratification must be rapidly destroyed by vertical motions. When solar heating of the Earth's surface is strong enough, a layer of varying thickness immediately above the surface can become statically unstable, with resulting rapid vertical motion, or *convection*. Convective motions generate turbulence and very effective vertical mixing of air masses. Without heating, on the other hand, an air parcel is constrained to stay on the same isentropic surface, and no vertical transport can occur. Heating relaxes the constraint on purely isentropic motion so that parcels can cross isentropic surfaces. This is crucial for vertical transport.

Convective overturning can also occur when certain types of large-amplitude wave motion create local regions of unstable stratification. This phenomenon, known as *gravity wave breaking* and discussed in more detail in Section 2.5.5, is locally important in the upper troposphere and lower stratosphere and apparently is very common throughout the mesosphere and lower thermosphere, where it has a major impact on the budgets of momentum and chemical constituents.

### 2.3.2 Inertial Motion

The discussion in the preceding section focused on the role of buoyancy in opposing vertical motion in the Earth's atmosphere. Atmospheric motion is also constrained in the latitudinal direction by the effect of the Earth's rotation. As in the vertical direction, the balance of forces in the horizontal direction introduces a restoring force to horizontal displacements. The action of this restoring force can be appreciated by considering a simple system, consisting initially of a steady-state background flow in the longitudinal, or zonal, direction, $u_o(y)$. Since the flow is in steady state, Eq. (2.16) implies that the following balance must hold approximately:

$$fu_o = -\frac{\partial \Phi_o}{\partial y} \quad (2.46)$$

where terms arising from the sphericity of the Earth, as well as nonconservative forces, have been neglected.

Suppose now that this flow is perturbed by the addition of an arbitrary horizontal velocity component, such that the new velocity field is given by $(u + u_o, v, 0)$. One can then write the horizontal equations of motion for an air parcel as:

$$\frac{du}{dt} = v\left(f - \frac{\partial u_o}{\partial y}\right) \quad (2.47)$$

$$\frac{dv}{dt} = -f(u_o + u) - \frac{\partial \Phi_o}{\partial y} = -fu \quad (2.48)$$

In writing Eq. (2.47) the fact that $du_o/dt \equiv v\, \partial u_o/\partial y$ has been used. In the second equality of Eq. (2.48), Eq. (2.46) has been used to eliminate the background state terms, ignoring terms associated with the sphericity of the Earth.

Integrating Eq. (2.47), following the motion gives the approximate result

$$u = \left(f - \frac{\partial u_o}{\partial y}\right) \delta y \tag{2.49}$$

where $f$ and $\partial u_o/\partial y$ have been assumed to remain approximately constant over the distance $\delta y$.

Substituting Eq. (2.49) into Eq. (2.48) then yields:

$$\frac{dv}{dt} \equiv \frac{d^2 y}{dt^2} = -f\left(f - \frac{\partial u_o}{\partial y}\right) \delta y \tag{2.50}$$

Assuming a solution of the form $y = \exp(i\omega t)$, Eq. (2.50) implies that

$$\omega = \sqrt{f\left(f - \frac{\partial u_o}{\partial y}\right)} \tag{2.51}$$

If the expression under the square root in Eq. (2.51) is positive, the air parcel undergoes stable oscillations about its initial position. Displacing the air parcel meridionally results in changes in its zonal velocity. An increase in the zonal velocity increases the Coriolis force on the parcel relative to the pressure gradient force, returning the parcel to its previous position; a decrease in the zonal velocity decreases the relative effect of the Coriolis force and again acts to return the parcel to its previous position. This type of motion is known as an *inertial oscillation* because it arises from the inertia of the air parcel in the Earth's frame of reference.

The meridional restoring force associated with inertial motions is much weaker than the buoyancy force; it has an oscillation frequency that is a function of latitude. In the absence of a meridional gradient in the zonal wind, $\omega = f$, which gives an inertial oscillation period ranging from less than one day at midlatitudes to several days in the deep tropics. It is clear from this example that the larger $f$, the larger the meridional stability of the atmosphere. Inertially stable motions are the rule in most of the Earth's atmosphere, with the notable exception of the tropics (where $f$ is small) and certain regions of the jet streams (where $u_{oy}$ can be quite large). Inertial instability occurs when $f(f - u_{oy}) < 0$, which implies that parcel motions grow exponentially at the rate $\sqrt{|f(f - u_{oy})|}$. Inertially unstable conditions will result in motions that will produce rapid horizontal mixing, reducing in the process the meridional shear, $u_{oy}$, until the flow is again stable.

*2.3.3 Geostrophic Balance*

The discussion of inertial oscillations assumed an initial state (2.46) such that the Coriolis and pressure gradient forces were in approximate balance. Indeed, scale analysis (*e.g.*, Holton, 1992) of the horizontal momentum equations (2.15)-(2.16) shows that, outside the tropics and for horizontal scales of order 1000 km and larger, the horizontal velocity field $(u, v)$ obeys the *geostrophic balance* equations:

$$u = -\frac{1}{f}\frac{\partial \Phi}{\partial y} \tag{2.52}$$

and

$$v = \frac{1}{f}\frac{\partial \Phi}{\partial x} \tag{2.53}$$

This approximate velocity field is called the *geostrophic wind*. It blows in a direction normal to the horizontal gradient of geopotential (see Fig. 2.3). This balance between the pressure gradient and Coriolis forces gives rise to the large-scale

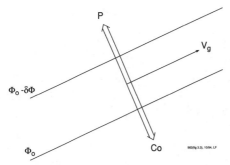

**Figure 2.3.** *A schematic representation of the geostrophic wind. The pressure gradient force, $P = (\partial\Phi/\partial x, \partial\Phi/\partial y)$, and the Coriolis force, $Co = (fv, -fu)$, are opposite in direction and comparable in magnitude (from Holton, 1972).*

quasihorizontal circulation systems common in the Earth's troposphere and stratosphere and explains the circulation characteristics of the cyclones and anticyclones familiar from weather maps. The balance expressed by these equations also helps explain why, on a zonal and time average, winds on Earth blow primarily in the east-west direction since the large-scale temperature gradients are in the meridional (north-south) direction (Fig. 2.4).

Equations (2.52) and (2.53) can be combined with the hydrostatic balance Eq. (2.13), to give the *thermal wind relations*

$$\frac{\partial u}{\partial z} = -\frac{R}{fH}\frac{\partial T}{\partial y} \quad (2.54)$$

$$\frac{\partial v}{\partial z} = \frac{R}{fH}\frac{\partial T}{\partial x} \quad (2.55)$$

which state that vertical shear of the zonal ($u$) component of the wind is associated with equatorward temperature gradients, and vertical shear of the meridional ($v$) component of the wind with eastward temperature gradients. The thermal wind relation can be appreciated from the global zonal wind and temperature distributions shown in Figure 2.4. Where the meridional temperature gradient is negative (*e.g.*, in the lower troposphere and throughout the winter stratosphere), the vertical shear of the zonal wind is positive; that is, $u$ increases with altitude. Similarly, where the meridional temperature gradient is positive (the upper troposphere, the summertime stratosphere), the vertical shear of $u$ is negative and $u$ decreases with altitude.

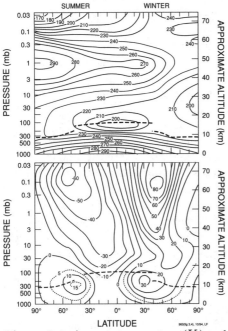

**Figure 2.4.** *Average temperature (K) and zonal wind (m s$^{-1}$) as functions of latitude and pressure (from Holton, 1975).*

### 2.3.4 Geostrophic Adjustment

The geostrophic balance equations are *diagnostic* relations; that is, they can be used to determine the horizontal wind field if the geopotential field is known, but they do not predict how an initially unbalanced flow field evolves towards geostrophic equilibrium. It turns out that this evolution involves the action of both the buoyancy and Coriolis forces discussed in Sections 2.3.1 and 2.3.2. A basic understanding of the adjustment process can be gained by considering a simplified version of the equations of motion, obtained by assuming that the advection terms in the material derivative (2.12), as well as terms arising from the sphericity of the Earth, can be neglected.

Under these assumptions, the horizontal momentum equations (2.15)-(2.16) become:

$$\frac{\partial u}{\partial t} - fv = -\frac{\partial \Phi}{\partial x} \tag{2.56}$$

$$\frac{\partial v}{\partial t} + fu = -\frac{\partial \Phi}{\partial y} \tag{2.57}$$

where, consistent with our neglect of the sphericity of the Earth, $f$ is a constant whose value depends on the latitude for which the analysis is carried out. At midlatitudes $f \simeq 10^{-4}$ s$^{-1}$.

The continuity equation (2.19) can be written as

$$\frac{\partial u}{\partial x} + \frac{\partial v}{\partial y} = -\left(\frac{\partial w}{\partial z} - \frac{w}{H}\right) \tag{2.58}$$

and the thermodynamic equation (2.27) becomes

$$\frac{\partial}{\partial t}\left(\frac{\partial \Phi}{\partial z}\right) = -wN_B^2 \tag{2.59}$$

where diabatic heating has been neglected and the hydrostatic relation (2.13) has been used to write the temperature in terms of the geopotential $\Phi$.

Taking $\partial(2.56)/\partial x + \partial(2.57)/\partial y$ gives the rate of change of the horizontal divergence:

$$\frac{\partial}{\partial t}\left(\frac{\partial u}{\partial x} + \frac{\partial v}{\partial y}\right) = f\zeta - \left(\frac{\partial^2 \Phi}{\partial x^2} + \frac{\partial^2 \Phi}{\partial y^2}\right) \tag{2.60}$$

where

$$\zeta = \left(\frac{\partial v}{\partial x} - \frac{\partial u}{\partial y}\right) \tag{2.61}$$

is the *vorticity*, a measure of the spin of a fluid parcel with respect to the coordinate system fixed to the Earth.

Using the continuity and thermodynamic equations, (2.58)-(2.59), to eliminate the divergence from Eq. (2.60) gives

$$\frac{\partial^2}{\partial t^2}\left(\Phi_{zz} - \frac{\Phi_z}{H}\right) = fN_B^2\zeta - N_B^2\left(\Phi_{xx} + \Phi_{yy}\right) \tag{2.62}$$

where the subscripts on $\Phi$ denote partial differentiation.

In the absence of rotation ($f = 0$), Eq. (2.62) reduces to

$$\frac{\partial^2}{\partial t^2}\left(\Phi_{zz} - \frac{\Phi_z}{H}\right) = -N_B^2\left(\Phi_{xx} + \Phi_{yy}\right) \tag{2.63}$$

which governs the behavior of the geopotential $\Phi$ under the influence of gravity alone; solutions of Eq. (2.63) describe *pure gravity waves* (see Section 2.5.1).

In a rotating fluid, on the other hand, Eq. (2.62) indicates that $\Phi$ responds to both buoyancy, through the term $N_B^2(\Phi_{xx} + \Phi_{yy})$, and to inertial forces, through $fN_B^2\zeta$. Note that Eq. (2.62) is not by itself sufficient to predict the evolution of $\Phi$. A second equation relating $\Phi$ and $\zeta$ is necessary, and it can be obtained by taking $\partial(2.57)/\partial x - \partial(2.56)/\partial y$. The result is a *vorticity equation*

$$\frac{\partial \zeta}{\partial t} + f\left(\frac{\partial u}{\partial x} + \frac{\partial v}{\partial y}\right) = 0 \tag{2.64}$$

which predicts the behavior of $\zeta$.

With the aid of Eqs. (2.58) and (2.59), Eq. (2.64) can be rewritten as

$$\frac{\partial}{\partial t}\left[\zeta + \frac{f}{N_B^2}\left(\Phi_{zz} - \frac{\Phi_z}{H}\right)\right] = 0 \qquad (2.65)$$

which implies that

$$\hat{\zeta} \equiv \zeta + \frac{f}{N_B^2}\left(\Phi_{zz} - \frac{\Phi_z}{H}\right) = \mathcal{C} \qquad (2.66)$$

where $\mathcal{C}$ is a constant. Equations (2.65)-(2.66) state that $\hat{\zeta}$ is conserved following the motion. The quantity $\hat{\zeta}$ is one of the simpler forms of *potential vorticity*, a measure of the vertical component of angular momentum of the parcel in the frame of reference fixed with respect to the Earth (see Section 2.3.5).

Equation (2.66) can be used to eliminate $\zeta$ in Eq. (2.62), yielding an equation that governs the evolution of $\Phi$ under both buoyancy and inertial forces:

$$\frac{\partial^2}{\partial t^2}\left(\Phi_{zz} - \frac{\Phi_z}{H}\right) + f^2\left(\Phi_{zz} - \frac{\Phi_z}{H}\right) + N_B^2\left(\Phi_{xx} + \Phi_{yy}\right) = fN_B^2\mathcal{C} \qquad (2.67)$$

The time-dependent solutions of Eq. (2.67) describe *inertia-gravity waves*, which are discussed in Section 2.5.2. The implications of Eq. (2.67) for the steady state, reached as the fluid adjusts to an initially unbalanced configuration of the geopotential field $\Phi$, are considered next.

Starting from the initial conditions

$$\begin{pmatrix} u \\ v \\ \Phi \end{pmatrix} = \begin{pmatrix} 0 \\ 0 \\ -\Phi_o(z)\,\text{sgn}(x) \end{pmatrix} \qquad (2.68)$$

which correspond to a fluid at rest with a step function in $\Phi$ at $x = 0$, as shown in the top panel of Figure 2.5. The potential vorticity of this initial configuration, which is conserved and therefore constant in time, is $-(f/N_B^2)(\Phi_{ozz} - \Phi_{oz}/H)\,\text{sgn}(x)$. Substituting this value for $\mathcal{C}$ in the steady-state version of Eq. (2.67) yields

$$f^2\left(\Phi_{zz} - \frac{\Phi_z}{H}\right) + N_B^2 \Phi_{xx} =$$
$$- f^2\left(\Phi_{ozz} - \frac{\Phi_{oz}}{H}\right)\,\text{sgn}(x) \qquad (2.69)$$

Note that derivatives with respect to $y$ are not present in Eq. (2.69). Since the initial state is independent of $y$ and vorticity must be conserved, the final state will also be independent of $y$.

Assuming that $\Phi_o(z)$ has the vertical dependence $\exp(imz + z/2H)$, where $m = 2\pi/\lambda_z$, the solution of Eq. (2.69) is:

$$\Phi = \begin{cases} \Phi_o\left[-1 + \exp(-\frac{x}{a_R})\right], & x > 0 \\ \Phi_o\left[\;\;1 - \exp(\frac{x}{a_R})\right], & x < 0 \end{cases} \qquad (2.70)$$

where

$$a_R \equiv \frac{N_B}{f\left(m^2 + \frac{1}{4H^2}\right)^{\frac{1}{2}}} \qquad (2.71)$$

The steady-state equilibrium described by Eq. (2.70) is shown in the bottom panel of Figure 2.5. Note that the initial step function in the geopotential field $\Phi$ has been smoothed out, but a horizontal gradient is still present. The tightness of this gradient depends on the parameter, $a_R$, which is known as the *Rossby radius of deformation*, after Carl-Gustaf Rossby, who first solved the geostrophic adjustment problem. The Rossby radius is a measure of the influence of rotation on the steady-state balance of the fluid. On a planet where inertial forces are much larger than buoyancy ($N_{BV}/f \ll 1$), $a_R \to 0$ and the equilibrium shape of $\Phi$ would differ little from the initial step-function configuration. On the other hand, in the limit of vanishing rotation, $a_R \to \infty$, and the final state no longer has a gradient in $\Phi$. Thus a nonrotating fluid adjusts to an unbalanced mass field by removing (to infinity) the gradient in surface elevation. In the Earth's atmosphere at midlatitudes, $N_{BV}/f \simeq 100$. For synoptic-scale motions $m^{-1} \simeq 3$ km, and Eq. (2.71) yields a value of about 1000 km.

The steady-state wind field corresponding to the geopotential configuration (2.70) can be found by substitution into the steady-state versions of Eq. (2.56)-(2.57), which gives:

$$\begin{pmatrix} u \\ v \end{pmatrix} = \begin{pmatrix} 0 \\ -\dfrac{\Phi_o}{fa_R} \exp\left(-\dfrac{|x|}{a_R}\right) \end{pmatrix} \quad (2.72)$$

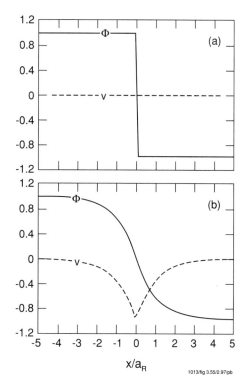

Figure 2.5. Geostrophic adjustment. The top panel shows the initial, unbalanced state, consisting of a step function in the geopotential $\Phi$ field in a fluid at rest. The bottom panel shows the final, steady-state solution. The initial step function in $\Phi$ has been spread out over a distance proportional to the Rossby radius of deformation; the geopotential field is now balanced by the Coriolis torque on the geostrophic wind. Note that the final configuration retains a gradient in the geopotential field and, therefore, some of the potential energy of the initial state (adapted from Holton, 1992).

The wind field (2.72) is in geostrophic balance with the geopotential distribution (2.70). Note that Eq. (2.72) cannot be obtained directly from the steady-state form of Eqs. (2.56)-(2.57), since such diagnostic relations between the velocity and the geopotential fields can be satisfied by any geopotential distribution. However, once it is required that potential vorticity be conserved, per Eq. (2.65), a unique solution (2.70) can be obtained.

A final point to be noted about the geostrophic equilibrium solution is that not all of the potential energy of the initial state can be converted into kinetic energy, since the final state retains a gradient in geopotential (Fig. 2.5). As seen above, in the limiting case of vanishing Rossby radius, the geopotential gradient of the final state approaches that of the initial state, and none of the potential energy is converted into kinetic energy.

The time evolution of an initially unbalanced flow toward geostrophic equilibrium involves the generation of inertia-gravity waves, which, as noted above, are governed by Eq. (2.67). The time-dependent solution of Eq. (2.67) will not be discussed here, although some of the properties of inertia-gravity waves are illustrated in Section 2.5.2. The interested reader is referred to Gill (1980) for a thorough discussion of all aspects of the geostrophic adjustment problem.

## 2.3.5 Potential Vorticity

It was shown above how inertial and buoyancy forces interact to produce a stable, geostrophically balanced state. The analysis assumed that the Coriolis parameter, $f$, was constant, which is a reasonable approximation as long as the scale of the motion is not much larger than about 1000 km. For motions on larger scales, the fact that $f$ actually varies with latitude must be taken into account. It turns out that this can be done in terms of a more general form of potential vorticity than the one introduced in Section 2.3.4.

Because potential vorticity is such a central concept in dynamic meteorology, it is useful to examine the derivation of the equation governing its conservation for the simple case of a homogeneous, incompressible fluid. Other forms of potential vorticity, applicable to a much wider class of fluids, have gained wide use in the analysis of transport processes. These can all be understood by analogy to this simpler case.

For a homogeneous, incompressible fluid, the continuity equation (2.19) takes the form

$$\left[\frac{\partial u}{\partial x} + \frac{1}{\cos\phi}\frac{\partial(v\cos\phi)}{\partial y}\right] = -\frac{\partial w}{\partial z} \tag{2.73}$$

and the horizontal momentum equations (2.74)-(2.75) are:

$$\frac{du}{dt} - \left(f + \frac{u\tan\phi}{a}\right)v = -\frac{\partial\Phi}{\partial x} + X \tag{2.74}$$

$$\frac{dv}{dt} + \left(f + \frac{u\tan\phi}{a}\right)u = -\frac{\partial\Phi}{\partial y} + Y \tag{2.75}$$

where

$$\frac{d}{dt} \equiv \frac{\partial}{\partial t} + u\frac{\partial}{\partial x} + v\frac{\partial}{\partial y} \tag{2.76}$$

denotes the (two-dimensional) material derivative. Note that, contrary to what was done in Section 2.3.3, the nonlinear terms in the material derivative and the terms due to the sphericity of the Earth are not neglected in these equations.

Equations (2.74)-(2.75) can be combined by cross-differentiation to yield a *vorticity equation*

$$\frac{d(\zeta+f)}{dt} = (\zeta+f)\left[\frac{\partial u}{\partial x} + \frac{1}{\cos\phi}\frac{\partial(v\cos\phi)}{\partial y}\right] + \frac{\partial Y}{\partial x} - \frac{1}{\cos\phi}\frac{\partial(X\cos\phi)}{\partial y} \tag{2.77}$$

where

$$\zeta = \frac{\partial v}{\partial x} - \frac{1}{\cos\phi}\frac{\partial(u\cos\phi)}{\partial y} \tag{2.78}$$

is the *relative vorticity*. The sum $(\zeta+f)$ is known as the *absolute vorticity*, and the Coriolis parameter, $f$, is referred to in this context as the *planetary vorticity*. The absolute vorticity of an air parcel is composed of its relative vorticity, or spin, with respect to the coordinate system fixed to the Earth, and the planetary vorticity, which arises because the coordinate system is rotating.

The absolute vorticity in Eq. (2.77) is not in general a conserved quantity, even in the absence of the nonconservative forces $X, Y$, since its tendency is proportional to the divergence of the motion field (the quantity in brackets on the right-hand side of the equation). However, it is possible to define a more general form of vorticity that is conserved for the fluid under consideration, as shown below.

The continuity equation (2.73) can be integrated vertically to give

$$\left[\frac{\partial u}{\partial x} + \frac{1}{\cos\phi}\frac{\partial(v\cos\phi)}{\partial y}\right]h = -w \quad (2.79)$$

where $h$ is the height of the free surface of the fluid, as shown in the top panel of Figure 2.6.

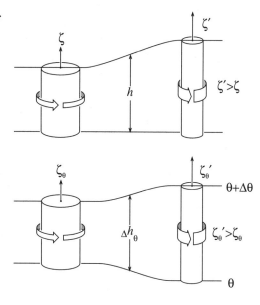

**Figure 2.6.** Schematic representation of the conservation of potential vorticity. The top panel shows the situation for the special case of a homogeneous, incompressible fluid. The depth of the fluid is denoted by the position of its free surface, $h$. Assuming $f$ is constant, the spin (vorticity) of the fluid parcel must change to compensate for the stretching or compression of the fluid element. According to Eq. (2.83), as $h$ increases, the relative vorticity must also increase to keep the potential vorticity constant, so that $\zeta' > \zeta$. The bottom panel shows the behavior of the Ertel potential vorticity in isentropic coordinates. In this case $\Delta h_\theta$ is the distance between isentropic surfaces, $\theta$ and $\theta + \Delta\theta$. As $\Delta h_\theta$ changes, a fluid element contained between the two isentropic surfaces will experience a corresponding change in vorticity. By Eq. (2.89), the isentropic relative vorticity must increase as $\Delta h_\theta$ increases, so $\zeta'_\theta > \zeta_\theta$ (adapted from Holton, 1992).

Noting that the vertical velocity is simply the motion of the free surface, $h$,

$$w \equiv \frac{dh}{dt} = \left(\frac{\partial}{\partial t} + \frac{u}{a\cos\phi}\frac{\partial}{\partial x} + \frac{v}{a}\frac{\partial}{\partial\phi}\right)h \quad (2.80)$$

Using Eq. (2.80), Eq. (2.79) can be rewritten as

$$\frac{dh}{dt} + \left[\frac{\partial u}{\partial x} + \frac{1}{\cos\phi}\frac{\partial(v\cos\phi)}{\partial y}\right]h = 0 \quad (2.81)$$

Finally, with the aid of Eq. (2.81), it is possible to eliminate the divergence from Eq. (2.77) to obtain:

$$\frac{d\hat{\zeta}}{dt} = \frac{1}{h}\left[\frac{\partial Y}{\partial x} - \frac{1}{\cos\phi}\frac{\partial(X\cos\phi)}{\partial y}\right] \quad (2.82)$$

where

$$\hat{\zeta} = \left(\frac{\zeta + f}{h}\right) \quad (2.83)$$

Equation (2.82) is the *potential vorticity equation* for an incompressible, homogeneous fluid, and Eq. (2.83) defines the *potential vorticity*. According to Eq. (2.82) the potential vorticity is conserved following the motion if the nonconservative forces $X, Y$ vanish, that is:

$$\frac{d}{dt}\left(\frac{\zeta + f}{h}\right) = 0 \quad (2.84)$$

The potential vorticity (2.83) is a measure of the component of angular momentum of fluid elements normal to the local horizontal plane. It includes contributions due to the motion of the fluid relative to the coordinate system (the relative vorticity, $\zeta$), the rotation of the Earth (the planetary vorticity, $f$), and the depth of the fluid, $h$. In order to conserve potential vorticity, the relative vorticity must change in response to the changing planetary vorticity as a parcel moves north or south. The relative vorticity will also change in response to changes in the depth of the fluid. This "stretching factor" takes into account the changes in angular momentum that take place as a fluid element narrows or thickens as it stretches or compresses (Fig. 2.6).

The potential vorticity equation derived above predicts the evolution of the potential vorticity field in a homogeneous, incompressible fluid, but the Earth's atmosphere is nonhomogeneous and compressible, with variable distributions of pressure and temperature. However, it turns out that it is possible to derive a conservation principle similar to Eq. (2.82) under rather general conditions, as shown by C.-G. Rossby (1936, 1940) and H. Ertel (1942). The form of the potential vorticity that enters into this equation is often referred to as the *Ertel* potential vorticity, although the name *Rossby-Ertel* potential vorticity would be more accurate. The Ertel potential vorticity equation is often written in isentropic coordinates (see Section 2.2.3). Manipulation of the primitive equations in isentropic coordinates (see, *e.g.*, Andrews *et al.*, 1987; Holton, 1992) leads to:

$$\frac{d_\theta \widehat{P}}{dt} = \frac{\widehat{P}}{\hat{\sigma}} \frac{\partial (\hat{\sigma} Q)}{\partial \theta} + F_p \qquad (2.85)$$

where $d_\theta/dt$ is the material derivative in isentropic coordinates, $F_p$ denotes nonconservative forces,

$$\widehat{P} = \left( \frac{\zeta_\theta + f}{\hat{\sigma}} \right) \qquad (2.86)$$

is the isentropic form of the Ertel potential vorticity, and $\hat{\sigma}$ is the "pseudodensity" defined in Eq. (2.35).

The Ertel potential vorticity is conserved if the diabatic heating rate, $Q$, and the nonconservative forces, $F_p$, vanish. That is, the only processes that can change the value of $\widehat{P}$ are heating and nonconservative forces along the parcel trajectory. In many situations of interest, the time scale of these processes is much longer than the advective time scale, so the Ertel potential vorticity is approximately conserved following the motion. For this reason, the Ertel potential vorticity is often an excellent tracer of fluid motion.

Isentropic coordinates offer many conceptual advantages in the interpretation of potential vorticity as parcels are also constrained to stay on the same isentropic surface ($\theta = $ constant) without the input of heat. Therefore, in the absence of heating (and nonconservative forces) potential vorticity is not only conserved, but it is advected only along isentropic surfaces; the vertical velocity in this coordinate system ($d\theta/dt$) is identically zero. Even when these assumptions are relaxed, the vertical velocity is often small in isentropic coordinates.

In the absence of heating the discussion can be restricted to the evolution of a fluid element between two isentropic surfaces, which can be regarded as elastic membranes. For simplicity it can also be assumed that the density is constant. Convergence of the fluid between these surfaces will increase the depth of the fluid ($\Delta h_\theta$) between the isentropes, but will not cause fluid to pass through the isentropic surfaces. Under these conditions, the pseudodensity (2.35) can be written simply as

$$\hat{\sigma} = \rho_o \frac{\Delta h_\theta}{\Delta \theta} \qquad (2.87)$$

where $dp = -\rho_o g \Delta h_\theta$ is used with $\rho_o$ a constant density. Then Eq. (2.86) can be written approximately as

$$\hat{P} \simeq \frac{\Delta\theta(\zeta_\theta + f)}{\rho_o \Delta h_\theta} \qquad (2.88)$$

Using the definition (2.88) for Ertel's potential vorticity, and in the absence of nonconservative processes, Eq. (2.85) becomes

$$\frac{d}{dt}\left(\frac{\zeta_\theta + f}{\Delta h_\theta}\right) = 0 \qquad (2.89)$$

where the constant factors $\Delta\theta$ and $\rho_o$ have been omitted.

Equation (2.89) is formally identical to (2.84). Therefore, the behavior predicted by Eq. (2.89) can be interpreted in the same way as for the potential vorticity in the homogeneous, incompressible case (top panel of Fig. 2.6), except that the depth $h$ is now the thickness of the fluid, $\Delta h_\theta$, between the two isentropic surfaces $\theta$ and $\theta + \Delta\theta$ (Fig. 2.6, bottom panel).

In the examples illustrated in Figure 2.6, the Coriolis parameter, $f$, was assumed to be constant. When this assumption is relaxed, conservation of potential vorticity can be used to show qualitatively that the north-south variation of $f$ acts to constrain the meridional movement of air parcels. If a group of parcels displaced from their equilibrium positions (as in Fig. 2.7) is considered, the resistance to latitudinal motion can be seen from a potential vorticity perspective. The air parcels are embedded in a background flow moving from the left to the right of the figure. To conserve absolute vorticity, $\zeta + f$, the relative vorticity $\zeta$ must decrease for parcels displaced north, and increase for parcels displaced south, since $f \sim \sin\phi$. A decrease in $\zeta$ implies an increase in the clockwise spin of the flow, while an increase in $\zeta$ means a greater counterclockwise spin. The net result is an oscillation of parcels about their reference latitude position, with the entire pattern moving in *retrograde* fashion, that is, against the direction of the flow.

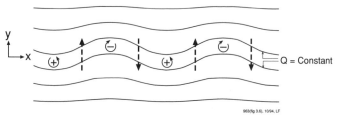

**Figure 2.7.** The variation in relative vorticity as air flows through a large-scale wave. The flow moves from left to right, along the streamlines denoted by the solid lines. Positive vorticity is induced when the air is displaced toward lower latitude, and the resulting motion tends to bring the flow back toward higher latitude; negative vorticity is induced when the air is displaced to higher latitude, and this tends to return the flow toward lower latitude. The entire oscillating pattern moves from right to left, opposite the direction of the flow (from Hoskins et al., 1985).

Conservation of potential vorticity implies that air parcels in adiabatic, frictionless motion are constrained to remain on constant-potential-vorticity surfaces. These potential vorticity surfaces resist being deformed in the meridional direction. Throughout much of the atmosphere, the potential vorticity is dominated by the planetary component $f$ so that air parcels are constrained to flow along surfaces of nearly constant latitude. Nonconservative processes are important in that they allow the migration of air parcels across surfaces of potential vorticity, thereby making possible meridional mass transport. Without such processes, meridional transport would not be possible except in regions where the meridional gradient of potential vorticity vanishes.

## 2.4 Zonal Means and Eddies

As discussed in Section 2.3.2, due to the Earth's rotation the predominant motion of the atmosphere is in the east-west, or zonal, direction. In a time-averaged sense, air moves predominantly in the east-west direction, and dynamical and chemical fields are often fairly uniform along constant-latitude circles. For this reason it is often useful to consider the equations governing atmospheric motion averaged in the zonal direction. Zonal averaging also clarifies how large-scale vertical and meridional circulation systems are generated.

Averaging entails a separation between a background state (in this case the zonal mean) and deviations from the mean, an approach that has a long history in fluid dynamics. The procedure is most useful when the deviations can be thought of as small-amplitude waves, or "eddies" superimposed on the background, zonal mean state. Then it is possible to develop a mathematical (and conceptual) framework that describes the evolution of atmospheric flows (and transports) in terms of the interaction between the zonal mean state and the eddies. However, it is essential to bear in mind that the zonal mean/eddy decomposition is essentially a mathematical procedure, and that there need not be in all cases a direct correspondence between mathematical constructs and physically meaningful entities.

The zonal mean, $\overline{X}$, is defined as the average value of $X$ along a latitude circle:

$$\overline{X}(\phi, z; t) = \frac{\int X(\lambda, \phi, z; t) \, d\lambda}{\int d\lambda} \tag{2.90}$$

The eddies, or zonally asymmetric fields, are then given by the deviation of the total field $X$ from the zonal mean:

$$X'(\lambda, \phi, z; t) = X(\lambda, \phi, z; t) - \overline{X}(\phi, z; t) \tag{2.91}$$

### 2.4.1 The Eulerian Zonal Mean Equations

One can obtain equations for zonal-mean quantities by writing all variables of interest as $X = \overline{X} + X'$, substituting into the primitive equations (2.15), (2.16), (2.19), and (2.21), and taking the zonal mean of the result (see, *e.g.*, Holton, 1975). The resulting zonal momentum equation is:

$$\frac{\partial \overline{u}}{\partial t} - \overline{v}\left(f - \frac{1}{\cos\phi}\frac{\partial \overline{u}\cos\phi}{\partial y}\right) + \overline{w}\frac{\partial \overline{u}}{\partial z} = \\ -\frac{1}{\cos^2\phi}\frac{\partial (\overline{u'v'}\cos^2\phi)}{\partial y} - \frac{1}{\rho_o}\frac{\partial (\rho_o \overline{u'w'})}{\partial z} \tag{2.92}$$

In Eq. (2.92) the continuity equation has been used to write the eddy correlation terms on the right-hand side in flux divergence form. In the meridional direction, the momentum equation can be approximated by the zonally averaged geostrophic balance:

$$\overline{u} = -\frac{1}{f}\frac{\partial \overline{\Phi}}{\partial y} \tag{2.93}$$

while the thermal wind equation becomes:

$$\frac{\partial \overline{u}}{\partial z} = -\frac{R}{fH}\frac{\partial \overline{T}}{\partial y} \quad (2.94)$$

The zonally averaged continuity and thermodynamic energy equations are:

$$\frac{1}{\cos\phi}\frac{\partial\left(\overline{v}\cos\phi\right)}{\partial y} + \frac{1}{\rho_o}\frac{\partial\left(\rho_o \overline{w}\right)}{\partial z} = 0 \quad (2.95)$$

and

$$\frac{\partial \overline{\theta}}{\partial t} + \overline{v}\frac{\partial \overline{\theta}}{\partial y} + \overline{w}\frac{\partial \overline{\theta}}{\partial z} = -\frac{1}{\cos\phi}\frac{\partial\left(\overline{v'\theta'}\cos\phi\right)}{\partial y} - \frac{1}{\rho_o}\frac{\partial\left(\rho_o\overline{w'\theta'}\right)}{\partial z} + \frac{\overline{\theta}}{\overline{T}}\overline{Q} \quad (2.96)$$

The form of the chemical continuity equation is the same as that of the thermodynamic energy equation:

$$\frac{\partial \overline{\mu}}{\partial t} + \overline{v}\frac{\partial \overline{\mu}}{\partial y} + \overline{w}\frac{\partial \overline{\mu}}{\partial z} = -\frac{1}{\cos\phi}\frac{\partial(\overline{v'\mu'}\cos\phi)}{\partial y} - \frac{1}{\rho_o}\frac{\partial\left(\rho_o\overline{w'\mu'}\right)}{\partial z} + \overline{S} \quad (2.97)$$

The eddy terms on the right-hand sides of Eqs. (2.92), (2.96), and (2.97) are known as *eddy flux divergences*. They can be thought to represent the effects of eddies, or waves, on the zonal mean quantities. Note that neither zonally averaged potential temperature (2.96) nor zonally averaged mixing ratio (2.96) is conserved unless the eddy flux divergences vanish. Of itself this is not surprising since the zonal average values $\overline{\theta}, \overline{\mu}$ are only part of the total potential temperature and mixing ratio fields. However, a peculiarity of the Eulerian mean equations is that the zonal mean tendencies introduced by the eddy flux terms do not vanish even when the motion is perfectly steady and conservative.

The problem is particularly acute for Rossby waves (Section 2.7.2). When Rossby waves dominate the eddy field, examination of the terms in the thermodynamic equation reveals a close balance between the eddy heat flux divergence and the vertical advection of mean potential temperature, such that:

$$\overline{w}\frac{\partial \overline{\theta}}{\partial z} \simeq -\frac{1}{\cos\phi}\frac{\partial\left(\cos\phi\,\overline{v'\theta'}\right)}{\partial y} \quad (2.98)$$

Similarly, the eddy momentum flux tends to balance the meridional advection of zonal mean momentum:

$$\overline{v}\left(\frac{1}{\cos\phi}\frac{\partial \overline{u}\cos\phi}{\partial y} - f\right) \simeq -\frac{1}{\cos^2\phi}\frac{\partial\left(\overline{u'v'}\cos^2\phi\right)}{\partial y} \quad (2.99)$$

The cancellation between these terms is often referred to as *eddy-mean cell cancellation*. In fact, it can be shown that the eddy flux divergences and the zonal mean advection terms will cancel exactly for waves that are steady and conservative (*i.e.*, waves whose amplitude does not change with time and that are not breaking or otherwise subject to dissipation). Such waves have no effect on the mean zonal wind or temperature structure even if they produce large eddy flux divergences. This result is known as the *nonacceleration theorem* (Boyd, 1976; Andrews and McIntyre, 1976; 1978b).

A similar cancellation occurs in the zonal mean chemical continuity equation, such that

$$\overline{w}\frac{\partial \overline{\mu}}{\partial z} \simeq -\frac{1}{\cos\phi}\frac{\partial\left(\cos\phi\,\overline{v'\mu'}\right)}{\partial y} \quad (2.100)$$

Not surprisingly, a *nontransport theorem*, similar to the nonacceleration theorem, applies to the transport equation so that $\partial \overline{\mu}/\partial t = 0$ when the species considered has no photochemical sources or sinks and the waves are steady and conservative.

Although in the real atmosphere waves are never completely steady and conservative, the limiting case addressed by the nonacceleration and nontransport theorems is conceptually useful because it highlights the fact that the eddy and mean transport arising in the conventional Eulerian system do not for the most part represent an actual physical processes. Even in situations where the waves are transient and undergoing dissipation, the *net* Eulerian mean transport is a small difference between the large mean advection and eddy flux divergence terms (see, *e.g.*, Mahlman *et al.*, 1980).

The cancellation between the tendency terms due to the eddies and to the mean meridional circulation is not a coincidence, but occurs because the eddy terms and the mean meridional circulation are closely related. A possible interpretation of this situation is that the momentum and heat flux divergences due to the waves "force" a mean meridional circulation $\bar{v}, \bar{w}$ such that the net tendencies $\partial \bar{u}/\partial t, \partial \bar{\theta}/\partial t, \partial \bar{\mu}/\partial t$ vanish. However, the mean meridional circulation obtained from the zonal mean equations when Rossby waves are present includes prominent "indirect," or *Ferrell*, cells in middle latitudes. Although such mean meridional circulations are portrayed in many older works on the general circulation, it would be quite misleading to think of these indirect cells as representing, even qualitatively, the mean motion of air parcels.

Closer examination of the situation reveals that the motion implied by the Ferrell cells is a mathematical artifact. Figure 2.8 represents schematically the three-dimensional structure of a Rossby wave. Streamlines of the flow in the latitude-longitude plane are denoted by the meandering lines, with arrows indicating the meridional and vertical velocities at selected points along the streamlines. The wave is assumed to be steady and conservative, so parcel trajectories are closed, their projection onto the latitude-height plane being denoted by the ellipses. (The actual three-dimensional trajectory is thus an elliptical corkscrew aligned in the longitudinal direction.) The size of the velocity vectors and the elliptical trajectories is intended to represent qualitatively their relative magnitude; thus the wave amplitude is largest in middle latitudes and decreases toward the pole and the equator. Temperature perturbations associated by the wave are denoted by the letters "W" (warm) and "C" (cold).

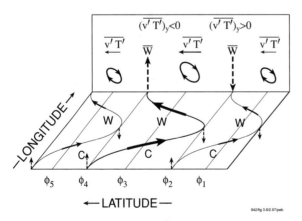

**Figure 2.8.** Schematic representation of the structure of a vertically propagating Rossby wave. Air flows along the streamlines, denoted by the solid lines. Horizontal velocity vectors (heavy arrows) and vertical velocity vectors (dashed lines) are shown at selected points along the streamlines. The wave amplitude is maximum in middle latitudes ($\phi_3$) and decays toward the pole and equator. The projections of parcel trajectories onto the latitude-height plane are shown as closed ellipses, which indicate that there is no net meridional or vertical parcel motion (reprinted with kind permission from Kluwer Academic Publishers.)

Nonetheless, the Eulerian zonal mean vertical velocity (computed at, say, latitudes $\phi_2$ and $\phi_4$) is <u>not</u> zero, and neither is the eddy heat flux divergence at the same latitudes. The transport effects of the mean vertical velocity and the eddy heat flux cancel, as indicated by Eq. (2.98). The result is a mathematically consistent but physically misleading picture of the effect of the planetary wave (adapted from Brasseur and Solomon, 1986).

The structure depicted in Figure 2.8 is typical of vertically propagating Rossby waves. Because the eddy velocity and temperature fields are in phase, the wave produces positive (poleward) eddy heat fluxes. Since these fluxes are largest at midlatitudes ($\phi_3$), where the wave amplitude is a maximum, there is eddy heat flux divergence at low latitudes ($\phi_2$) and convergence at high latitudes ($\phi_4$), as shown in the figure. At the same time, zonal averaging yields a mean vertical velocity ($\overline{w}$, denoted by the thick dashed arrows) directed downwards at low latitudes and upwards at high latitudes. This is the familiar Ferrell cell pattern. The effect of the Ferrell cell exactly cancels that of the eddy fluxes, with the result that the mean tendencies $\partial \overline{u}/\partial t, \partial \overline{T}/\partial t, \partial \overline{\mu}/\partial t$ all vanish. However, no physical meaning can be attached to the Ferrell cell since, as has been seen, air parcels do not in fact experience any net motion in the meridional plane. A detailed description of eddy mean flow cancellation is given by Matsuno (1980).

### 2.4.2 The Transformed Eulerian Mean Equations

It is now recognized that the complications introduced by the conventional Eulerian zonal mean formulation can be attributed to the zonal averaging procedure itself. Instead of Eq. (2.97), a zonally averaged conservation equation of the form:

$$\frac{\partial \overline{\mu}}{\partial t} + \overline{v}^L \frac{\partial \overline{\mu}}{\partial y} + \overline{w}^L \frac{\partial \overline{\mu}}{\partial z} = \overline{S}^L \qquad (2.101)$$

where $\overline{v}^L$ and $\overline{w}^L$ are suitably defined zonally averaged velocities, would offer a much simpler interpretation of zonally averaged chemical transport. An equation of this form can be formulated by taking averages along wave trajectories instead of along latitude circles. This averaging procedure is known as the *generalized lagrangian mean* (GLM; Andrews and McIntyre, 1978a). The GLM formulation eliminates the cancellation between mean and eddy transports, but in practice it is difficult to implement for large-amplitude wave motions. Discussion of the technical aspects of the GLM formulation is outside the scope of this chapter; the interested reader can consult a number of works on the subject (*e.g.*, Andrews *et al.*, 1987; McIntyre, 1980).

Although the GLM formulation is not commonly used in theoretical work on atmospheric transport, many of its desirable properties can be captured by a simple transformation of the Eulerian mean equations (Boyd, 1976; Andrews and McIntyre, 1976). The transformation is effected by defining the mean meridional and vertical velocities as:

$$\overline{v}^* = \overline{v} - \frac{1}{\rho_o} \frac{\partial}{\partial z}\left(\frac{\rho_o \overline{v'\theta'}}{\theta_z}\right) \qquad (2.102)$$

$$\overline{w}^* = \overline{w} + \frac{\partial}{\partial y}\left(\frac{\overline{v'\theta'} \cos\phi}{\theta_z}\right) \qquad (2.103)$$

These are the so-called *residual Eulerian mean*, or *transformed Eulerian mean* (TEM), velocities. With this transformation, the mean vertical residual velocity in the presence of conservative and steady waves tends to zero instead of being balanced by the eddy heat flux divergence, as was the case in the conventional Eulerian equation (2.97). The TEM vertical and horizontal velocities are related by the continuity equation:

$$\frac{1}{\cos\phi} \frac{\partial \cos\phi \overline{v}^*}{\partial y} + \frac{1}{\rho_o} \frac{\partial \rho_o \overline{w}^*}{\partial z} = 0 \qquad (2.104)$$

The zonal mean momentum in the TEM system is:

$$\frac{\partial \overline{u}}{\partial t} + \overline{v}^*\left(\frac{1}{\cos\phi}\frac{\partial \overline{u}\cos\phi}{\partial y} - f\right) + \overline{w}^*\frac{\partial \overline{u}}{\partial z} = \frac{1}{\rho_o \cos\phi}\boldsymbol{\nabla}\cdot\mathcal{F} \qquad (2.105)$$

where $\mathcal{F}$ is the *Eliassen-Palm flux*, a vector in the latitude/height plane defined as:

$$\mathcal{F} = (\mathcal{F}_y, \mathcal{F}_z) \tag{2.106}$$

with

$$\mathcal{F}_y = \rho_o \cos\phi \left( \overline{u}_z \frac{\overline{v'\theta'}}{\overline{\theta}_z} - \overline{u'v'} \right) \tag{2.107}$$

$$\mathcal{F}_z = \rho_o \cos\phi \left[ \left( f - \frac{1}{\cos\phi} \frac{\partial \overline{u} \cos\phi}{\partial y} \right) \frac{\overline{v'\theta'}}{\overline{\theta}_z} - \overline{u'w'} \right] \tag{2.108}$$

The components of the Eliassen-Palm flux vector (2.107)-(2.108) are proportional to momentum and heat fluxes, which in the conventional Eulerian mean system appeared to act independently on the momentum and thermodynamic equations, respectively. Therefore, in the TEM set of equations the effect of the temperature flux is explicitly included in the zonal mean momentum equation. The divergence of the Eliassen-Palm flux acts as a zonal mean force applied to the zonal mean flow. Equation (2.105) states that the eddies can produce an acceleration of the zonal mean flow, $\overline{u}_t$, only when the divergence of the Eliassen-Palm flux is nonzero. It can be shown that for steady, conservative waves the Eliassen-Palm flux divergence vanishes so the waves have no effect on the mean flow. This result is another formulation of the *nonacceleration theorem* discussed in Section 2.6.1. The Eliassen-Palm flux can be interpreted as a *wave stress* whose divergence acts to accelerate or decelerate the mean flow.

The TEM thermodynamic equation is given by:

$$\frac{\partial \overline{\theta}}{\partial t} + \overline{v}^* \frac{\partial \overline{\theta}}{\partial y} + \overline{w}^* \frac{\partial \overline{\theta}}{\partial z} = \frac{\overline{\theta}}{\overline{T}} \overline{Q} - \frac{1}{\rho_o} \frac{\partial}{\partial z} \left[ \rho_o \left( \overline{v'\theta'} \frac{\overline{\theta}_y}{\overline{\theta}_z} + \overline{w'\theta'} \right) \right] \tag{2.109}$$

The second term on the right-hand side of Eq. (2.109) is negligible for Rossby waves, although the contribution involving $\overline{w'\theta'}$ can be important for gravity waves (Section 2.5.1). Note that (aside from these small terms in the thermodynamic equation) wave forcing in the TEM system appears only in the zonal momentum equation. Further, it can be shown that the sum of these two small terms vanishes exactly for steady, conservative waves (Andrews and McIntyre, 1978b). These properties of the TEM system simplify considerably the physical interpretation of the interaction between the mean state and the waves.

The TEM circulation is similar to the generalized Lagrangian mean circulation and can be qualitatively interpreted as the relevant transport circulation for chemical constituents. The zonally averaged transport equation in the TEM system can be written as:

$$\frac{\partial \overline{\mu}}{\partial t} + \overline{v}^* \frac{\partial \overline{\mu}}{\partial y} + \overline{w}^* \frac{\partial \overline{\mu}}{\partial z} = S + \frac{1}{\rho_o} \boldsymbol{\nabla} \cdot \mathcal{M} \tag{2.110}$$

where $\mathcal{M}$ is a matrix depending on correlations of eddy amplitudes of the tracer ($\mu'$), temperature ($T'$), and wind fields ($u', v'$). Generally this term is small compared with the eddy flux terms on the right-hand side of the Eulerian mean transport equation (2.97) and therefore the eddy-mean cell cancellation discussed in connection with the Eulerian set of zonal mean equations is considerably reduced in the TEM equations. If the waves are steady and conservative, and the species $\mu$ has no sources or sinks, $\boldsymbol{\nabla} \cdot \mathcal{M}$ vanishes identically. This property of the TEM continuity equation is another formulation of the *nontransport theorem* (see, *e.g.*, Andrews *et al.*, 1987, for a detailed discussion).

For photochemically active chemical species there is a mechanism for transport that occurs independently of mass transport and is included in the $\mathcal{M}$ term. This is known as chemical eddy transport (Hartmann and Garcia, 1979; Garcia and Solomon,

1983) and occurs when the chemical sources or sinks of a species vary over the scale of a wave. It is most effective when the dynamical time scale with which the parcel traverses through the wave is approximately equal to its chemical time scale. A physical picture of chemical eddy transport can be gained by considering air flow in a Rossby wave where $S$ (the chemical source/sink term) changes with latitude, as illustrated schematically in Figure 2.9. Air flowing along the trajectory traverses regions with different photochemical characteristics. This will result in poleward-moving air having a higher mixing ratio than equatorward-moving air, and therefore a net zonal mean poleward transport of the species without any net mass transport.

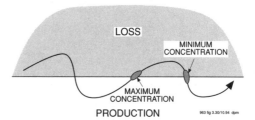

**Figure 2.9.** *The trajectory (wavy solid line) traverses regions with different photochemical production and loss rates. Concentrations increase in the production region and will be greater for air leaving this region than for that returning. The result is a net chemical flux.*

It should be noted that the TEM system is most useful for describing the mean flow in situations where Rossby waves are a major component of the circulation, since it is for such waves that the eddy-mean flow cancellation noted in connection with Eqs. (2.98)-(2.100) is greatest. Thus the advantages of the TEM system are less marked in the tropics, where gravity and Kelvin waves are a major component of the atmospheric circulation. For gravity waves in general, and particularly for Kelvin waves, eddy fluxes are dominated by the vertical flux divergence terms, $\overline{u'w'}$, $\overline{w'\theta'}$, and $\overline{w'\mu'}$, in Eqs. (2.92), (2.96), and (2.97). As a consequence, the problem of eddy-mean flow cancellation is much less acute in the tropics. Nonetheless, Rossby waves are the dominant type of wave motion in the extratropical troposphere and stratosphere, and the TEM system has proved extremely useful in providing a physically meaningful description of eddy-mean flow interactions in these regions of the atmosphere.

### 2.4.3 The Diabatic Circulation

For approximate steady-state conditions, the dominant balance in Eq. (2.109) is between the vertical advection of heat and the diabatic heating. Therefore, the residual vertical velocity resembles the vertical velocity solely due to heating or cooling of air parcels:

$$\overline{w}_{diab} = \frac{\overline{\theta}}{\overline{T}}\frac{\overline{Q}}{\overline{\theta}_z} \qquad (2.111)$$

This so-called *diabatic* vertical velocity is important because it indicates where and when vertical advection can transport chemical constituents through isentropic surfaces. The similarity between $\overline{w}_{diab}$ and $\overline{w}^*$, which, to a first approximation, is the relevant velocity for zonally averaged transport, illustrates the relation between vertical motion and the heating that must occur for movement across isentropic surfaces.

An example of the diabatic circulation in the middle atmosphere is shown in Figure 2.10. The circulation (bottom two panels) was computed from the net heating rates (top two panels) derived from satellite observations of ozone and temperature (Kiehl and Solomon, 1986; Solomon *et al.*, 1986b). This circulation is consistent with the observed distributions of water vapor and ozone in the stratosphere. Comparison of the streamlines and the heating distribution shows that parcels undergo net cooling as they descend near the north (winter) pole, and net heating as they ascend near the south (summer) pole. The diabatic velocity (and therefore the residual circulation, $\bar{v}^*, \bar{w}^*$) can explain to a large extent the gross features of the distribution of chemical constituents in the Earth's atmosphere, whereas it is clear that the mean Eulerian circulation does not.

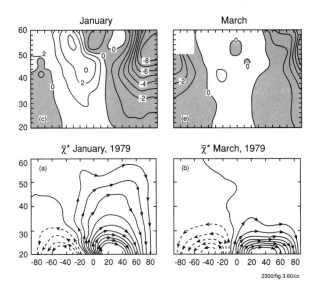

**Figure 2.10.** Meridional and vertical cross section of the net diabatic heating in the middle atmosphere (K/day) during winter solstice. The trajectories indicate the residual circulation (from Solomon et al., 1986b).

### 2.4.4 Wave Forcing and the Mean Meridional Circulation

Although large-scale heating gradients due to the absorption of solar radiation are the ultimate driver of atmospheric motions, the constraints on such motions imposed by the Earth's rotation (see Section 2.3) tend to suppress zonal-mean meridional circulations and produce instead a geotrophic balance (2.93) wherein the Coriolis force on the zonal-mean zonal wind is balanced by the meridional gradient of mass (geopotential). The constraints that produce such a geostrophic equilibrium are not equally effective in all situations. In the tropics, where the Coriolis force is weak, departures from geostrophic equilibrium are common; even in extratropical latitudes, the seasonal variation of solar heating will produce departures from geostrophic equilibrium because the steady-state balance (2.93) is never quite achieved. However, there are a number of important cases where zonal-mean meridional circulations can be said to arise principally from the effects of *wave forcing*, as seen below.

Atmospheric waves can force the circulation systems that are responsible for the global transport of chemical constituents and thus ultimately for the known chemistry of the atmosphere. In the TEM system, the waves affect the zonal-mean momentum budget through the divergence of the Eliassen-Palm flux. With the formalism introduced by the TEM system it is shown how atmospheric waves can break the constraints on vertical and meridional parcel movement imposed by the conservation of potential temperature and potential vorticity.

To understand the role of the Eliassen-Palm flux in forcing mean meridional circulations, and thus in the transport of trace constituents, it is instructive to write the zonal momentum equation in terms of the zonal-mean angular momentum:

$$\bar{m} = \cos\phi(\bar{u} + a\Omega\cos\phi) \qquad (2.112)$$

Equation (2.105) then becomes:

$$\frac{\partial \overline{m}}{\partial t} + \overline{v}^* \frac{\partial \overline{m}}{\partial y} + \overline{w}^* \frac{\partial \overline{m}}{\partial z} = \frac{1}{\rho_o} \boldsymbol{\nabla} \cdot \mathcal{F} \qquad (2.113)$$

The equation above is in conservation form; it states that the zonal-mean angular momentum is conserved following the TEM motion provided $\boldsymbol{\nabla} \cdot \mathcal{F}$ vanishes. Under *steady-state* conditions, and ignoring the generally small term $\overline{w}^* \overline{m}_z$, Eq. (2.113) reduces to

$$\overline{v}^* \frac{\partial \overline{m}}{\partial y} = \frac{1}{\rho_o} \boldsymbol{\nabla} \cdot \mathcal{F} \qquad (2.114)$$

According to Eq. (2.114), $\overline{v}^*$ must vanish when $\boldsymbol{\nabla} \cdot \mathcal{F} = 0$ unless the zonal-mean angular momentum gradient ($\overline{m}_y = 0$) also vanishes. In fact, the entire meridional circulation must vanish because $\overline{v}^*$ and $\overline{w}^*$ are related by the continuity equation (2.104). The conditions under which Eq. (2.114) is valid are approximated most closely at the solstices (when the time rate of change of the solar heating gradient approaches zero) and in extratropical latitudes (where $\overline{m}_y \neq 0$). Therefore, the extratropical, solstitial TEM circulation must be viewed as a *wave-driven* circulation.

An important corollary of these results is that if the TEM circulation vanishes, so must the *net* diabatic heating rate ($\overline{Q} = 0$, the sum of the solar heating rate and infrared emission from the atmosphere). The zonally averaged equilibrium approached by the extratropical atmosphere under steady-state conditions and in the absence of wave forcing is then

$$\overline{v}^* = \overline{w}^* = \overline{Q} = 0 \qquad (2.115)$$

Note that, in steady state, any net extratropical diabatic heating is then a *consequence* rather than a cause of the TEM circulation. This can be seen clearly by writing the net heating rate $\overline{Q}$ as the sum of the externally imposed solar heating $\overline{Q}_s$ and the longwave cooling response of the atmosphere:

$$\overline{Q} = \overline{Q}_s(\phi, z) - \alpha \overline{T}(\phi, z) \qquad (2.116)$$

The longwave cooling is approximately proportional to the local temperature ($-\alpha \overline{T}$), where $\alpha$ is a Newtonian cooling coefficient. It follows that, when $\overline{Q} = 0$, the radiative cooling and the solar warming exactly balance, or the temperature distribution approached by the atmosphere is in *radiative equilibrium*:

$$\overline{T}_e(\phi, z)_e = \frac{\overline{Q}_s(\phi, z)}{\alpha} \qquad (2.117)$$

This is in fact the steady-state balance that would be achieved in the extratropical atmosphere in the absence of wave forcing. In the radiatively balanced atmosphere the meridional gradient of $\overline{T}_e$ (and the associated meridional gradient in geopotential $\Phi$) are in balance with the zonal mean wind $\overline{u}$, per Eqs. (2.52), (2.54). Wave forcing can upset this radiative equilibrium so that air parcels are forced through isentropic surfaces and vertical transport can occur. A schematic representation of how wave forcing can drive a mean-meridional circulation is shown in Figure 2.11.

Figure 2.12 shows the mean surfaces of constant angular momentum and constant potential temperature in the atmosphere. Without wave driving, friction or heating, the movement of air would be constrained as these surfaces provide a barrier to vertical and meridional transport. However, a comparison of this figure with the mean residual circulation reveals that air does cross these mean surfaces and indeed must cross them to explain the observed distribution of chemical species in the atmosphere. It is clear from the above discussion that waves, through their associated Eliassen-Palm flux divergence, provide the necessary torque on the flow to decelerate a parcel sufficiently

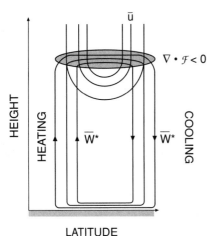

**Figure 2.11.** (left) Meridional and vertical cross section of the circulation induced by a momentum source ($\nabla \cdot \mathcal{F}$) in the atmosphere. The contours show the zonal wind $\overline{u}$, and the trajectories show the residual circulation. Heating and cooling occur in response to the meridional circulation (from McIntyre, 1992).

**Figure 2.12.** (below) Meridional and vertical cross section of the mean angular momentum (solid lines; in m s$^{-1}$) and potential temperature (dashed lines) in the stratosphere (from McIntyre, 1992).

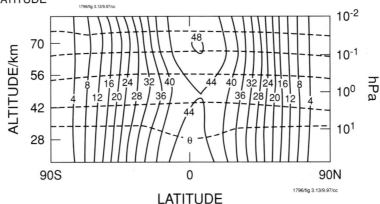

so that it can travel from the equator to the pole. The associated vertical motion then gives rise to the heating or cooling that drives air through isentropic surfaces. Therefore, a knowledge of atmospheric waves, their sources, and their propagation through the atmosphere can ultimately explain much of the observed characteristics of atmospheric transport.

It should be emphasized that the remarks above apply to the *extratropical* atmosphere near the solstices. In the tropics, where $\overline{m}_y$ is much weaker, it is possible to have a mean meridional circulation without substantial wave forcing. Such a circulation is akin to what would be expected from the familiar notion that "heated air rises and cooled air sinks." With mean heating near the equator it can be shown that a mean meridional circulation must exist in order to satisfy the steady-state thermal wind and angular momentum balances. Thus the tropical *Hadley cell* can be viewed as a circulation driven directly by tropical convective heating (see also Section 2.6.3). Similarly, strong, transient meridional circulations are driven by the rapid change of the solar heating gradient during the equinoxes.

It should also be noted that wave-driven circulations are by no means unimportant in the tropics, especially in the stratosphere and mesosphere. These tropical wave-driven circulations do not reach a steady state even approximately, but are instead characterized by alternating regimes of upwelling and downwelling associated with easterly and westerly zonal wind distributions. The zonal-mean circulation regimes of the stratosphere and mesosphere are dominated by such wave-driven circulations: the quasibiennial oscillation (QBO) in the lower stratosphere and the semiannual oscillation (SAO) in the upper stratosphere and mesosphere. These wave-driven, tropical circulation regimes play an important role in the transport of chemical constituents and volcanic aerosols in the middle atmosphere.

*2.4.5 The Downward Control Principle*

The fact that the extratropical mean meridional circulation, especially near the solstices, is wave driven has important consequences that must be understood in order to gain insight into the processes that control mean advective transport in the latitude-height plane. For steady-state conditions it is possible to relate the vertical residual velocity ($\overline{w}^*$) and the Eliassen-Palm flux divergence explicitly, using Eq. (2.114) and the continuity equation (2.104):

$$\overline{w}^*(y,z) = \frac{1}{\rho_o \cos\phi} \frac{\partial}{\partial y} \int_z^\infty \left[ \frac{\nabla \cdot \mathcal{F} \cos\phi}{\overline{m}_y} \right] dz' \qquad (2.118)$$

This is the *downward control principle* (Haynes et al., 1991) that relates the zonally averaged vertical motion at a given extratropical latitude to the vertically integrated wave forcing above that level. It follows from Eq. (2.118) that the mean vertical velocity at level $z$ is controlled by wave dissipation above that level. A particular, somewhat counterintuitive consequence is that extratropical zonal-mean mass exchange between the stratosphere and troposphere arises from wave forcing of the stratospheric circulation and not from the details of tropopause dynamics. This view of global, extratropical transport is emphasized in a number of recent works (Haynes et al., 1991; Holton et al., 1995).

## 2.5 Atmospheric Waves

The Eliassen-Palm flux can be interpreted as a wave stress, or wave drag, on the mean flow. As noted above, atmospheric waves through their associated Eliassen-Palm flux are the primary agents responsible for breaking the constraints on meridional and vertical transport. In addition, they are often directly responsible for creating atmospheric instabilities, thereby generating atmospheric turbulence. In this way waves can homogenize large regions of the atmosphere.

Since the angular momentum of the Earth-atmosphere system must be conserved, momentum stress due to the waves integrated over the atmosphere and the solid Earth is always zero. For example, waves can exert a stress on the Earth's surface, altering very slightly the Earth's rotation rate, while imposing an equal, but opposite, stress where they dissipate in the atmosphere, changing the zonal wind. In the propagation region separating the wave source and the wave sink, the atmosphere is essentially transparent to the passage of the wave; in this region, the wave will exert no net stress on the atmosphere and will not change the mean flow. Most waves are forced in the lower atmosphere and dissipate at higher altitudes. Such transfer of momentum by waves to remote regions of the atmosphere is one of the most important phenomena in atmospheric dynamics and has a profound effect on the circulation of the Earth's atmosphere.

Waves can be viewed as propagating through the atmospheric medium with a *phase speed*, $c$. Their characteristics at any point and time depend on the basic state of the atmosphere through which they are propagating. Properties such as the basic state velocity and wind shear, the static stability, and the potential vorticity gradient all affect the propagation of waves. Waves can often be refracted by the atmospheric basic state and in some cases the basic state is such that propagation is not allowed. Often it is possible to define a refractive index that determines the ray paths along which the wave propagates. These subjects are beyond the scope of this chapter.

However, it is important to recognize that the interaction between waves and the basic state is a nonlinear process. Waves can change the basic state through their Eliassen-Palm flux convergence, but at the same time changes in the basic state affect the propagation of waves.

The evolution of atmospheric waves can be described mathematically if they can be considered to be small disturbances on a basic state. The basic state is often taken to be the zonally averaged state of the atmosphere. The wave equations can then be obtained by writing all fields in the form (2.91), substituting into the primitive equations (2.13), (2.15), (2.16), (2.19), and (2.21), and subtracting from the result the zonally averaged equations (2.92)-(2.96). This procedure yields the horizontal wave momentum equations:

$$\frac{Du'}{Dt} + v'\left(\frac{1}{\cos\phi}\frac{\partial \bar{u}\cos\phi}{\partial y} - f\right) + w'\frac{\partial \bar{u}}{\partial z} = -\frac{\partial \Phi'}{\partial x} + X' \qquad (2.119)$$

and

$$\frac{Dv'}{Dt} + u'\left(\frac{2\bar{u}\tan\phi}{a} + f\right) = -\frac{\partial \Phi'}{\partial y} + Y' \qquad (2.120)$$

where

$$\frac{D}{Dt} \equiv \frac{\partial}{\partial t} + \bar{u}\frac{\partial}{\partial x} \qquad (2.121)$$

is the material derivative following the zonal-mean motion, and the Eulerian mean meridional and vertical velocities $(\bar{v}, \bar{w})$ are neglected. The hydrostatic equation is given by:

$$\frac{\partial \Phi'}{\partial z} = \frac{R}{H}T' \qquad (2.122)$$

The continuity equation is:

$$\frac{\partial u'}{\partial x} + \frac{1}{\cos\phi}\frac{\partial v'\cos\phi}{\partial y} + \frac{1}{\rho_o}\frac{\partial \rho_o w'}{\partial z} = 0 \qquad (2.123)$$

and the eddy thermodynamic equation can be written as:

$$\frac{D\theta'}{Dt} + w'\frac{\partial \bar{\theta}}{\partial z} = \frac{\theta}{T}Q' \qquad (2.124)$$

where the eddy advection of mean potential temperature, $v'\bar{\theta}_y$, has been neglected since it is small compared to the vertical advection. In terms of the temperature, Eq. (2.124) can be written as

$$\frac{DT'}{Dt} + w'\frac{HN_B^2}{R} = Q' \qquad (2.125)$$

### 2.5.1 Pure Gravity Waves

A complete description of the gravity waves that can exist in the Earth's atmosphere requires the use of the full set of wave equations (2.119)-(2.124). However, the basic mechanism of internal gravity waves, or *buoyancy waves*, can be appreciated from a simple subset of these equations.

Consider an adiabatic disturbance and assume that the effects of the Earth's rotation and sphericity can be neglected. Also assume that the basic state flow is independent of altitude. The horizontal momentum equations (2.119)-(2.120) then become

$$\left(\frac{\partial}{\partial t} + \bar{u}\frac{\partial}{\partial x}\right)u' = -\frac{\partial \Phi'}{\partial x} \qquad (2.126)$$

and
$$\left(\frac{\partial}{\partial t} + \bar{u}\frac{\partial}{\partial x}\right) v' = -\frac{\partial \Phi'}{\partial y} \qquad (2.127)$$

The continuity and thermodynamic equations are given by:
$$\frac{\partial u'}{\partial x} + \frac{\partial v'}{\partial y} + \frac{1}{\rho_o}\frac{\partial (\rho_o w')}{\partial z} = 0 \qquad (2.128)$$

$$\left(\frac{\partial}{\partial t} + \bar{u}\frac{\partial}{\partial x}\right) \theta' + w' \frac{HN_B^2}{R} = 0 \qquad (2.129)$$

Following the same procedure used in Section 2.3.4, Eqs. (2.126) and (2.127) can be combined and Eqs. (2.128)-(2.129) introduced to express the divergence in terms of the geopotential $\Phi'$:

$$\frac{D^2}{Dt^2}\left(\Phi_{zz} - \frac{\Phi_z}{H}\right) + N_B^2 \left(\Phi_{xx} + \Phi_{yy}\right) = 0 \qquad (2.130)$$

which is identical to Eq. (2.63), except that the partial time derivative has been replaced by the derivative following the mean motion, Eq. (2.121).

Assume a solution of the form
$$\Phi' = \Phi e^{i(kx+ly+mz-\omega t)} e^{z/2H} \qquad (2.131)$$

where $k, l$, and $m$ are the *zonal*, *meridional*, and *vertical* wavenumbers, respectively, and $\omega$ is the wave frequency. The wavenumbers are related to the horizontal and vertical wavelengths $(\lambda_x, \lambda_y, \lambda_z)$ by $\lambda_x = 2\pi/k$, $\lambda_y = 2\pi/l$, and $\lambda_z = 2\pi/m$. The frequency can also be written in terms of the *phase speed*, $c$, as $\omega = kc$. As shown in Figure 2.13, the wavelength measures the distance between successive wave crests (or troughs) and the phase speed is a measure of the rate at which these crests or troughs propagate with respect to the fixed (Eulerian) coordinate system.

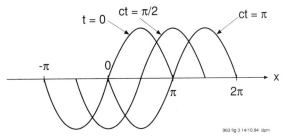

**Figure 2.13.** *The vertical displacements associated with an internal gravity wave propagating along the x direction according to* $\exp[i(kx-ct)]$, *as a function of x at three different times. The wave frequency $\omega$ is related to the phase speed c by $\omega = kc$ (Holton, 1972).*

Substituting Eq. (2.131) into Eq. (2.130) yields
$$-(k\bar{u} - \omega)^2 \left(m^2 + \frac{1}{4H^2}\right) + N_B^2(k^2 + l^2) = 0 \qquad (2.132)$$

which leads to the *dispersion relation*:
$$(k\bar{u} - \omega)^2 = \frac{(k^2 + l^2) N_B^2}{\left(m^2 + \frac{1}{4H^2}\right)} \qquad (2.133)$$

which must be satisfied by any solution of Eq. (2.130).

The complete solution to the set (2.126)-(2.129) can then be expressed as follows:

$$\begin{pmatrix} u' \\ v' \\ w' \end{pmatrix} = \begin{pmatrix} -\dfrac{k}{k\overline{u}-\omega} \\ -\dfrac{l}{k\overline{u}-\omega} \\ i\,\dfrac{k\overline{u}-\omega}{N_B^2}\left(im+\dfrac{1}{2H}\right) \end{pmatrix} \Phi e^{i(kx+ly+mz-\omega t)} e^{z/2H} \qquad (2.134)$$

where the amplitude $\Phi$ is determined by some appropriate boundary condition.

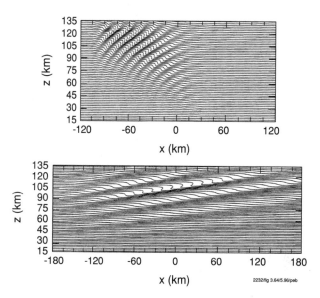

**Figure 2.14.** Potential temperature field disturbed by the presence of gravity waves. The potential temperature contours correspond to material surfaces. Wave troughs and ridges define lines of constant phase, their inclination being determined by the ratio of horizontal to vertical wavenumbers $k/m$. The wave in the top panel has $k \simeq m$, while that in the bottom panel has $k \ll m$ (adapted from Prusa et al., 1996).

Figure 2.14 shows two examples of gravity waves described by Eq. (2.134), with $l = 0$, that is, waves propagating only in the $x - z$ plane. The wave in the top panel has nearly equal horizontal and vertical wavenumbers ($k \simeq m$), so that its phase lines are inclined approximately 45° to the horizontal; the wave in the bottom panel has $k \ll m$ ($\lambda_x \gg \lambda_z$), and its phase lines are inclined at a shallow angle with respect to the horizontal.

Note that, according to Eq. (2.133), internal gravity waves depend on the buoyancy frequency, or vertical atmospheric stratification, to propagate. The wave frequency $\omega$ is proportional to (but much smaller than) $N_B^2$ because $k \ll m$ for the waves considered in this example. That is, air parcels move along slant trajectories, inclined with respect to the horizontal in the ratio $k/m$. The bottom panel of Figure 2.14 is an example of such waves.

The restriction $k \ll m$ can be shown to follow from the assumption of hydrostatic equilibrium made in the derivation. It is possible to derive a dispersion relation for nonhydrostatic internal gravity waves, in which case $\omega$ is given by (Gill, 1980):

$$(k\overline{u}-\omega)^2 = \frac{(k^2+l^2)\,N_B^2}{\left(k^2+l^2+m^2+\dfrac{1}{4H^2}\right)} \qquad (2.135)$$

which reduces to Eq. (2.133) when $k, l \ll m$. Note that for nonhydrostatic waves $\omega$ can be comparable to $N_B$ if $k, l \simeq m \gg 1/(4H^2)$. The top panel of Figure 2.14 shows a nonhydrostatic wave, with $k \simeq m$. The phase lines slant at a much steeper angle in this case. In the limit $k \gg l, m$ (and ignoring the background wind $\overline{u}$) the result $\omega = N_B$ can be recovered, derived in Section 2.3.1 for an air parcel. This is not surprising since this situation would correspond to phase lines aligned vertically, hence to vertical displacements only. Gravity waves that meet the assumptions implicit in the derivation of Eq. (2.135) have small horizontal scales (a few tens to a few hundred kilometers) and high frequencies (a few minutes to a few hours). These waves can also transport momentum efficiently from the lower to the upper atmosphere. The EP flux due to small-scale gravity waves is dominated by the $\overline{u'w'}$ term in Eq. (2.108).

## 2.5.2 Inertia-Gravity Waves

The restoring force that leads to the pure gravity waves discussed in the previous section is buoyancy, that is, gravity acting upon a stably stratified fluid. In inertia-gravity waves, both buoyancy and inertial forces play a role. The governing equation for these waves is obtained as in the previous section, but retaining the Coriolis acceleration terms in the horizontal momentum equations. The result is

$$\frac{d^2}{dt^2}\left(\Phi_{zz} - \frac{\Phi_z}{H}\right) + f^2\left(\Phi_{zz} - \frac{\Phi_z}{H}\right) + N_B^2\left(\Phi_{xx} + \Phi_{yy}\right) = 0 \qquad (2.136)$$

which is essentially the same as the homogeneous version of Eq. (2.67), except that here the time derivative is that following the mean motion.

Proceeding as in Section 2.5.1, substitute Eq. (2.131) into Eq. (2.136), which leads to the dispersion relation for inertia-gravity waves:

$$(k\overline{u} - \omega)^2 = f^2 + \frac{(k^2 + l^2)N_B^2}{\left(m^2 + \frac{1}{4H^2}\right)} \qquad (2.137)$$

which can be compared to its counterpart for pure gravity waves (2.133).

Note that Eq. (2.137) reduces to Eq. (2.133) when the second term on the right-hand side is much larger than the first. In general, this occurs when the horizontal scale of the waves is small so that the term $(k^2 + l^2)$ is large. This can be seen more precisely by rewriting Eq. (2.137) as

$$(k\overline{u} - \omega)^2 = \frac{(k^2 + l^2)N_B^2}{\left(m^2 + \frac{1}{4H^2}\right)}\left[1 + \left(\frac{L_H}{a_R}\right)^2\right] \qquad (2.138)$$

where $L_H^2 = (k^2 + l^2)^{-1}$ is a measure of the horizontal scale of the wave and $a_R$ is the Rossby radius of deformation (2.71). According to Eq. (2.138), rotational effects become unimportant when $(L_H/a_R)^2 \ll 1$. If $a_R \simeq 1000$ km, as in Section 2.3.4, $L_H \simeq 300$ km gives $(L_H/a_R)^2 \ll 1$. Thus waves of horizontal scale less than about 300 km will behave approximately like the pure gravity waves discussed in the previous section.

## 2.5.3 Rossby Waves

Rossby waves arise from the requirement that potential vorticity be conserved. The mechanism has been described qualitatively in Figure 2.7 in terms of the restoring force exerted on parcels due to the variation of the Coriolis parameter ($f$) with latitude. The basic properties of Rossby waves can be derived from the potential vorticity equation (2.84) if it is assumed that the fluid depth $h$ remains fixed. Assuming further that the motion is conservative and that the basic state is the zonal-mean wind field $\overline{u}$, the equation describing the evolution of deviations from this basic state follows from

Eq. (2.84):

$$\left(\frac{\partial}{\partial t} + \overline{u}\frac{\partial}{\partial x}\right)\zeta' + v'\beta = 0 \qquad (2.139)$$

where $\beta$ is the meridional gradient of the Coriolis parameter, $\partial f/\partial y$, the sphericity of the Earth has been neglected, and the material derivative has been written as $d/dt = \partial/\partial t + \overline{u}\partial/\partial x$.

Now, since the fluid depth $h$ is constant, it follows from Eq. (2.73) that the divergence vanishes, that is,

$$\frac{\partial u'}{\partial x} + \frac{\partial v'}{\partial y} = 0 \qquad (2.140)$$

which in turn implies that the perturbation velocity field $(u', v')$ can be expressed in terms of a streamfunction $\psi'$ such that the flow velocities:

$$u' = \frac{-\partial \psi'}{\partial y}$$
$$v' = \frac{\partial \psi'}{\partial x} \qquad (2.141)$$

satisfy Eq. (2.140).

Assuming that the wave fields are described by

$$\begin{pmatrix} u' \\ v' \end{pmatrix} = \begin{pmatrix} U \\ V \end{pmatrix} \exp i(kx + ly - \omega t) \qquad (2.142)$$

substitution of Eqs. (2.141) and (2.142) into Eq. (2.139) yields after some manipulation:

$$\omega = k\overline{u} - \frac{\beta k}{(k^2 + l^2)} \qquad (2.143)$$

The parcel displacements induced by the waves described by Eqs. (2.142)-(2.143) are predominantly in the horizontal $(x, y)$ plane, in contrast to the gravity waves described above, whose parcel displacements have a large vertical component. The difference can be understood in terms of the restoring forces responsible for the two types of wave. In the case of internal gravity waves, the restoring force is buoyancy, which acts in the vertical direction, whereas for Rossby waves the restoring force arises from the horizontal (latitudinal) variation of the Coriolis parameter.

If the assumptions leading to Eq. (2.143) are relaxed, a more general dispersion relation for Rossby waves can be obtained:

$$\omega = k\overline{u} - \frac{\beta k}{\left[k^2 + l^2 + \frac{f^2}{N_B^2}\left(m^2 + \frac{1}{4H^2}\right)\right]} \qquad (2.144)$$

where $m$ is now the vertical wavenumber (see, e.g., Holton, 1975; Andrews et al., 1987). The waves described by the dispersion relation (2.144) propagate both horizontally and vertically. Horizontal scales for Rossby waves range from a few thousand to tens of thousands of kilometers. The corresponding vertical wavelength can encompass many scale heights. Phase speeds relative to the zonal-mean wind are typically on the order of a few tens of meters per second. Rossby waves propagate against the zonal wind, as can be seen in Figure 2.7. In general, Rossby waves can transport momentum both horizontally and vertically.

It is rather easy to gain some insight into the transmission properties of Rossby waves in the atmosphere. The dependence of $m$ on other wave parameters can be derived from Eq. (2.144):

$$m^2 = \frac{N_B^2}{f^2}\left(\frac{\beta k}{k\bar{u}-\omega} - k^2 - l^2\right) - \frac{1}{4H^2} \quad (2.145)$$

In order for the Rossby wave to be vertically propagating, $m^2 \geq 0$; otherwise $m$ is imaginary and the wave decays with altitude. From Eq. (2.145), $m^2 \geq 0$ implies:

$$0 \leq (\bar{u}-c) \leq \frac{\beta}{\left(k^2 + l^2 + \frac{f^2}{N_B^2}\frac{1}{4H^2}\right)} \quad (2.146)$$

where $c = \omega/k$. From Eq. (2.146) it can be deduced that stationary Rossby waves ($c = 0$), or for that matter Rossby waves with small phase speed $c$, will not propagate into regions where the background zonal mean wind is easterly ($\bar{u} < 0$). Propagation into the regions where $\bar{u} > 0$ is possible if $k, l$ are sufficiently small, that is, if the spatial scales of the wave are large enough. Observations of Rossby waves in the stratosphere confirm these deductions: In the summer stratosphere the zonal mean winds are easterly, and Rossby waves are typically not observed; in the winter stratosphere the typical values of $\bar{u}$ are such that only the longest horizontal wavelengths can satisfy Eq. (2.146) for typical values of $\bar{u}$, and only the ultralong Rossby waves ($k = 1-3$) are typically observed. These constraints provide an effective filter for vertical propagation of Rossby waves into the stratosphere.

### 2.5.4 Equatorial Kelvin Waves

The Coriolis parameter changes sign at the equator and makes possible the existence of a special kind of wave motion, the equatorial Kelvin wave. The simplest description of the Kelvin wave can be obtained by using the equatorial beta-plane formulation, wherein the Coriolis parameter is approximated by

$$f = \beta y \quad (2.147)$$

where

$$\beta = \left(\frac{\partial f}{\partial y}\right)_{\phi=0} = \frac{2\Omega}{a} \quad (2.148)$$

is the latitudinal gradient of $f$ evaluated at the equator.

When the approximation (2.147) is used in Eqs. (2.119)-(2.120), the momentum equations become:

$$\frac{Du'}{Dt} - v'\beta y = -\frac{\partial \Phi'}{\partial x} \quad (2.149)$$

$$\frac{Dv'}{Dt} + u'\beta y = -\frac{\partial \Phi'}{\partial y} \quad (2.150)$$

where the sphericity of the Earth has also been neglected, consistent with the beta-plane approximation, and it has been assumed that the basic flow, $\bar{u}$, is uniform in latitude and height.

The thermodynamic and continuity equations are:

$$\frac{DT'}{Dt} + w'\frac{HN_B^2}{R} = Q' \quad (2.151)$$

$$\frac{\partial u'}{\partial x} + \frac{\partial v'}{\partial y} + \frac{\partial w'}{\partial z} = 0 \quad (2.152)$$

which are obtained from Eqs. (2.125) and (2.123), neglecting in the latter the variation of density with height compared with the vertical derivative of $w'$.

Equations (2.149)-(2.152) describe several classes of waves trapped in the vicinity of the equator (see Andrews *et al.*, 1987, for a thorough discussion). The Kelvin wave constitutes a special case, for which the meridional velocity perturbation, $v'$, vanishes. Accordingly, assume a solution of the form:

$$\begin{pmatrix} u' \\ \Phi' \end{pmatrix} = \begin{pmatrix} \hat{u}(y) \\ \hat{\Phi}(y) \end{pmatrix} \exp i(kx + mz - \omega t) \qquad (2.153)$$

and substitute into Eqs. (2.149)-(2.152) to obtain the following set of equations for $u_o$ and $\Phi_o$:

$$(k\bar{u} - \omega)\hat{u} + k\hat{\Phi} = 0 \qquad (2.154)$$

$$\beta y \hat{u} + \hat{\Phi}_y = 0 \qquad (2.155)$$

$$(k\bar{u} - \omega)\left(\frac{m}{N}\right)^2 \hat{\Phi} + k\hat{u} = 0 \qquad (2.156)$$

where the continuity equation (2.152) has been used to eliminate $w'$ in the thermodynamic equation (2.151), and the temperature $T'$ has been expressed in terms of the geopotential $\Phi'$, per Eq. (2.122).

The set (2.154)-(2.156) has the solution:

$$\begin{pmatrix} \hat{u}(y) \\ \hat{\Phi}(y) \end{pmatrix} = \begin{pmatrix} -\frac{k}{k\bar{u} - \omega} \\ 1 \end{pmatrix} A \exp\left[-\frac{\beta k y^2}{2(\omega - k\bar{u})}\right] \qquad (2.157)$$

where $A$ is a constant and

$$\omega = k\bar{u} - \frac{kN_{BV}}{m} \qquad (2.158)$$

The meridional structure of the Kelvin wave is Gaussian, centered on the equator, and its velocity field is purely along the zonal direction, as shown in Figure 2.15. The Kelvin wave exhibits an interesting separation in the balance of forces that maintain the oscillation: The force due to the geopotential gradient in the zonal direction is balanced by the advection of horizontal momentum, as in a pure gravity wave, while the force due to the geopotential gradient in the meridional direction is balanced by the Coriolis force on the velocity field.

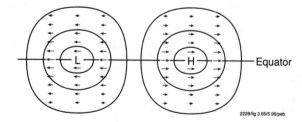

**Figure 2.15.** Geopotential and wind structure of an equatorial Kelvin wave. The zonal Coriolis torque on the velocity field balances the geopotential gradient (after Matsuno, 1966).

Kelvin waves are a very important component of the motion in the tropics. They play a role in the tropical tropospheric circulation and also in the circulation of the stratosphere, where they are believed to carry a significant fraction of the momentum that drives the quasibiennial and semiannual oscillations (see Section 2.7.1).

## 2.5.5 Wave Breaking

Dissipation of atmospheric waves can occur as a result of a number of mechanisms, including radiative relaxation of the perturbed temperature fields and friction. In recent years, the phenomenon known as *wave breaking* has gained increasing recognition as a major dissipation mechanism. Wave breaking refers to the breakdown that occurs when the amplitude of a quasilinear wave disturbance becomes large enough to render the disturbance unstable in some sense. The resulting instability acts to dissipate the wave as well as to mix the medium through which the wave is propagating. In many respects, breaking of atmospheric waves is analogous to the breaking of ocean waves on a beach; as in this familiar example, breaking atmospheric waves can induce mass motions in the background medium and lead to thorough mixing of the medium. Both these properties make wave breaking very important for atmospheric transport. Wave breaking is common in the atmosphere, and plays a central role in the stratosphere and especially in the mesosphere.

The two principal types of atmospheric waves discussed above are known to break in the atmosphere. It is easy to show formally that small-amplitude waves propagating in a steady mean flow in the absence of dissipative processes, such as thermal relaxation, will conserve their disturbance energy, which to a good approximation can be written as

$$E' = \frac{1}{2}\rho \left( \overline{u'u'} + \overline{v'v'} + \frac{\overline{\Phi'_z \Phi'_z}}{N_B^2} \right) \quad (2.159)$$

For $E'$ to remain constant requires that the wave fields must grow with altitude as $\exp(z/2H)$, since the density $\rho$ diminishes as $\exp(-z/H)$. In deriving the gravity wave solution (2.131) this exponential growth with height was adopted. The exponential growth with altitude of the *amplitude* of a conservative wave is thus a consequence of the density stratification of the Earth's atmosphere. (The decrease of atmospheric density with height plays a role analogous to the decrease of fluid depth as ocean waves approach a beach.)

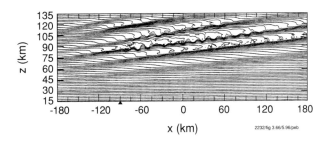

**Figure 2.16.** Temperature perturbations associated with a breaking internal gravity wave. The perturbations are small at the lowest levels, but grow with altitude due to the exponential drop in atmospheric density. In the mesosphere and lower thermosphere, the perturbations are so large that the vertical gradient of potential temperature is reversed over part of the wave, and the flow becomes convectively unstable (adapted from Prusa et al., 1996).

Amplitude growth cannot continue unabated as the wave propagates vertically. Eventually, the perturbation to the background state becomes large enough that the atmosphere becomes unstable. In the case of the gravity wave the vertical gradient of potential temperature becomes negative locally (Fig. 2.16). When this happens, the atmosphere is statically unstable, as discussed in Section 2.3.1, and the wave is rapidly dissipated. Gravity wave breaking can be reproduced successfully in numerical models provided the hydrostatic approximation is dropped. This must be done even when the wave appears to satisfy the condition for hydrostatic equilibrium ($k \ll m$, see Section 2.5.1) because the convectively unstable cells that develop during breaking are themselves nonhydrostatic. Observations and numerical modeling indicate that breaking occurs over a wide range of spatial scales, with some of the most vigorous breaking taking place in waves of horizontal wavelengths of a few tens to a few hundreds of kilometers. The convective cells associated with breaking are even smaller (on the order of a few kilometers; see Prusa et al., 1996). These scales are too small to include in comprehensive models of the atmosphere, whose grid spacing is much too coarse to resolve them. Therefore, gravity wave breaking is usually treated as a *subgrid scale* process, its effects being parameterized rather than explicitly computed (Lindzen, 1981; Garcia and Solomon, 1985).

Gravity wave breaking can have a major impact on the budgets of momentum and chemical constituents. Several recent studies show that General Circulation Models (GCMs) produce a more realistic wind distribution when parameterized gravity wave processes are included in the calculations (*e.g.*, Boville, 1993). Wave breaking is known to be of central importance for the momentum budget of the mesosphere. The basic features of mesospheric climatology, viz., the deceleration of the stratospheric wind systems and the associated reversed temperature gradient between summer and winter, can only be explained in terms of the zonal-mean accelerations and mean meridional circulation induced by breaking gravity waves (see Garcia and Solomon, 1985, and references therein).

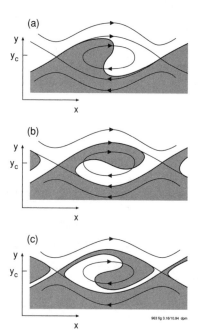

**Figure 2.17.** *A Rossby wave at the transition between high (white) and low (stippled) potential vorticity. As the wave amplitude grows (b and c), the local gradient of potential vorticity is reversed. The resulting circulation cell (c) tends to stretch out the displaced air in an irreversible manner (from Andrews et al., 1987).*

The work of McIntyre and Palmer (1983; 1984; 1985) and others has demonstrated that Rossby waves also dissipate through breaking, with important effects on constituent transport in the stratosphere. The nature of the Rossby wave restoring mechanism dictates that parcel displacements due to Rossby waves are quasihorizontal (Fig. 2.17). Breaking, and associated mixing, are also quasihorizontal; the instability involved is not the static instability of the gravity wave case but, it appears likely, *barotropic instability* (Haynes, 1989). The necessary condition for the occurrence of barotropic instability is that the local *potential vorticity gradient* be reversed by the wave (Charney and Stern, 1962). Rossby wave breaking takes place on relatively large scales; it can stir air parcels across spatial scales of thousands of kilometers. Although planetary wave breaking can be explicitly simulated in numerical models, it is still useful to parameterize the breaking process for use in two-dimensional models, or in very low-resolution three-dimensional models. It turns out that the same conceptual framework used for gravity wave breaking can be employed in the planetary wave breaking problem (Garcia, 1991; Garcia et al., 1992).

## 2.5.6 Mixing Processes

The foregoing discussion of atmospheric motions has focused mainly on large-scale global circulations and on the motions induced by waves that can be described as linearized departures from some background state. However, as noted in the preceding section, atmospheric flows can become turbulent when instabilities precipitate wave breaking. On small time and spatial scales, the air motions that transport chemical constituents are often turbulent in character: They vary in an apparent random and irregular fashion. The resulting turbulent motion appears to have a chaotic nature.

The ultimately turbulent nature of atmospheric flows mixes species of different origins. While transport by large-scale motions can be thought of as moving constituents in bulk from location to location in the atmosphere, mixing can be viewed as homogenizing the atmosphere. A familiar example of the importance of the wind field in mixing the atmosphere is the rapidity of mixing of a typical smoke plume: A smoke plume usually takes minutes to mix with its surroundings rather than hours as implied by the timescale for molecular diffusion. This mixing process is not primarily due to any organized circulation, but to a rather turbulent flow field tearing air parcels into finer and finer bits (Fig. 2.18).

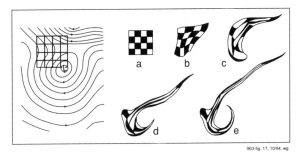

**Figure 2.18.** *A time sequence of the distortion of a "tracer" in a flow field (from Welander, 1955).*

Strong mixing also allows the barriers to vertical transport and horizontal transport to be relaxed. Instead of transporting through potential temperature or potential vorticity surfaces, one can simply mix across these surfaces. An example is provided by the Rossby wave breaking process in the stratosphere. Before the mixing process there is a continuous gradient of potential vorticity over a wide range of latitudes. After mixing the gradient has been relaxed, effectively eliminating the transport barrier imposed by the initial distribution of potential vorticity (Fig. 2.19).

In some flow regimes mixing is very rapid; in others, it occurs at a slow pace. Atmospheric instabilities are a very important means of initiating mixing processes over fairly large scales as the velocities grow exponentially within unstable regions until the flow has stabilized by mixing the background atmosphere across the unstable region. For example, if the vertical gradient of potential temperature becomes

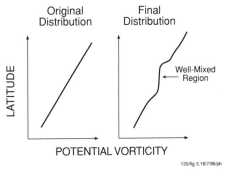

**Figure 2.19.** *A schematic representation of the variation of potential vorticity, $P$, with latitude. In the original distribution, $P$ primarily follows the planetary vorticity $f$. After mixing over a limited latitude band, $P$ has a weakened gradient within that band and enhanced gradients on either side of it. Note that the total $P$ is the same both before and after mixing (from McIntyre, 1992).*

negative, parcel displacements tend to grow exponentially until the unstable potential temperature gradient is eliminated. As discussed above, atmospheric waves are an important mechanism in the generation of atmospheric instabilities and therefore in the mixing of trace constituents in the atmosphere.

## 2.6 Tropospheric Circulation and Transport

There are fundamental differences between tropospheric and stratospheric transport processes. The influence of the Earth's surface and the high mixing ratios of water vapor are very important for determining transport in the troposphere. In the stratosphere (see Section 2.7), absorption of solar radiation by ozone creates a thermal structure quite different from that found below.

Whereas the temperature structure of the stratosphere is determined to a large extent by the competition between radiative forcing and wave driving (Section 2.4.4), in the troposphere the large-scale temperature structure is largely determined by the competition between radiative forcing and convection. The ability of the ground to absorb incoming radiation much more effectively than the atmosphere at large, and the net radiative cooling in the middle and upper troposphere, cause the troposphere to be heated from below and cooled from above. The release of latent heat by the condensation of water vapor further acts to destabilize the troposphere.

As a consequence of the vertical destabilization of the troposphere, the vertical temperature profile of the troposphere can be considered, on a global scale, to be determined by the radiative convective equilibrium temperature. That is, the equilibrium temperature distribution of the troposphere is approximately neutrally buoyant to moist parcels, as rather vigorous tropospheric convection continually acts to stabilize an atmosphere rendered unstable by the surface heating, radiative cooling of the upper troposphere, and the presence of water vapor (Fig. 2.20). *Convection* (see Section 2.6.1) is the process whereby the troposphere stabilizes itself, producing vertical stirring in the process.

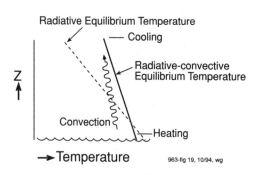

**Figure 2.20.** *Schematic representation of the maintenance of the radiative-convective temperature. Heating below and cooling above lead to the unstable radiative equilibrium temperature. Convection redistributes heat in the vertical until the air is no longer unstable.*

In addition to mixing parcels across isentropic surfaces, convection provides an internal diabatic heat source from the condensation of water vapor. Condensation and release of latent heat may also occur as moist parcels rise along vertically tilted isentropic surfaces. The latter process, typically associated with *baroclinic instability* (see Section 2.6.2), produces precipitation and clouds, and mixes the troposphere in both the horizontal and vertical directions.

Due to the presence of the Earth's surface and the prevalence of convection, the troposphere acts as an important wave source for the atmosphere as a whole. In addition, in the troposphere the lower boundary can act directly as a frictional force on air parcels, allowing meridional movement without explicit wave torque. The frictional effects of the surface are most strongly felt in an atmospheric layer of varying thickness in contact with the ground, the so-called *planetary boundary layer* (Sections 2.6.4 and 5.2). This layer is usually highly turbulent throughout its depth.

The influence of variations in the Earth's surface introduces zonal asymmetries into the tropospheric circulation. Both topographic variations and land/ocean contrasts are important. On global scales these features are responsible for longitudinal and latitudinal variations in the storm track (the predominant track of midlatitude synoptic-scale storms as they travel eastward), large-scale equatorial east-west circulations between the heated continents and the cooler oceans (the *Walker* cells), and the monsoon circulations. On more local scales variations in the Earth's surface drive local circulation regimes: The familiar land-sea breezes and mountain-valley breezes are driven by different rates of heating between various land surfaces.

Because of vigorous stirring by convection and baroclinic instability, and the associated heating through the condensation of water vapor, the troposphere is, on average, well mixed on a timescale of a few months. (But note that local transport times, *e.g.*, in deep convective clouds, can be much faster than this—on the order of a few hours.) As a result of the rather effective tropospheric mixing, both the meridional and vertical tropospheric gradients of chemical species are rather unremarkable for species with lifetimes of a few months or more.

## 2.6.1 Convection

Although the atmosphere is usually statically and inertially stable, regions of instability do exist. The atmosphere becomes unstable as the restoring force acting on an air parcel vanishes. As discussed in Section 2.3.1, when $\partial \theta_o / \partial z$ becomes negative, parcel motions tend to grow exponentially with time. Under these conditions, the atmosphere is said to be *statically unstable*.

Vertical instability can be induced by introducing a source of heat in an air parcel. A parcel heated in this way will become warmer than the surrounding environment and will accelerate upward. The heat released by the condensation of water vapor,

$$Q = \frac{L_c}{c_p} \frac{d\tilde{\mu}_w}{dt} \tag{2.160}$$

is the most important internal heat source for air parcels. In Eq. (2.160), $L_c$ represents the latent heat of condensation ($2.5 \times 10^6$ J kg$^{-1}$ at 0° C) and $\tilde{\mu}_w$ is the mixing ratio of water vapor. If a moist parcel is forced upward (by flow over a mountain, for example), adiabatic cooling of the ascending parcel may be sufficient to condense its water vapor, thereby releasing the internal heat of condensation. Once condensation begins, the parcel's temperature decreases more slowly than the adiabatic rate, since it is being heated by the latent heat of condensation. Depending on the temperature profile of the atmosphere, at some height the parcel may become warmer than the unperturbed atmosphere, after which it will be positively buoyant and will rise (Fig. 2.21).

Convection can occur on spatial scales of a few square kilometers (in small fair-weather cumulus clouds) to scales of thousands of square kilometers and can extend from the Earth's surface to the tropopause (in tropical convective complexes and in large organized mesoscale convection). Deep convective clouds have a tremendous potential for redistributing mass in the troposphere and for transporting trace constituents from the boundary layer to the overlying free atmosphere (Chatfield and Crutzen, 1984) as well as modifying the atmospheric chemical content due to aqueous phase and heterogeneous chemistry in the clouds.

On long time scales convection can be thought of as diffusively mixing the atmosphere between the ground and the tropopause. However, individual convective events provide a conduit whereby parcels in the boundary layer can rise rapidly in

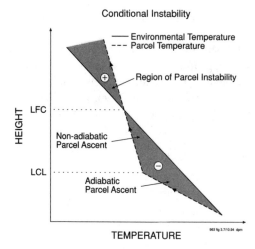

*Figure 2.21.* The solid line shows the background temperature structure and the dashed line shows that of an air parcel displaced vertically. The parcel temperature decreases adiabatically until the air is saturated (at the lifting condensation level - LCL). Above the LCL latent heat is released; so the parcel temperature decreases much more slowly than the adiabatic rate. If lifting continues as far as the level of free convection (LFC), where the parcel becomes warmer than the environment, the parcel will become buoyant and continue rising (after Wallace and Hobbs, 1977).

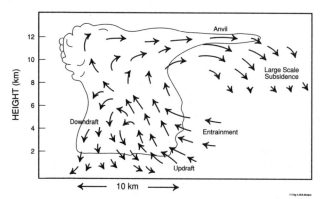

*Figure 2.22.* Schematic diagram of the mass fluxes in a convective cloud.

convective updrafts. This provides a mechanism whereby species with sources at the Earth's surface and rather short chemical lifetimes (too short to reach the tropopause diffusively) can interact chemically at the tropopause level. Downdrafts associated with convection also occur, as large downward buoyancy forces are created by the evaporative cooling of rain as it falls through the atmosphere. These downdrafts result in a rapid downward chemical transport. The mass fluxes in a convective cloud are shown in Figure 2.22.

In general, convection is characterized by small regions of rising motion (due to the primary convective instability) and large areas of compensatory sinking motion in the cloud's environment. In the extra-tropics, where atmospheric heating alone cannot drive parcels across angular momentum surfaces, the mixing due to convection is on a local scale: The upward motion driven by convection is balanced locally by downward motion. In the tropics the mixing is more global, as some of the air forced upward in the Inter-Tropical Convergence Zone (ITCZ) travels northwards to become part of the Hadley circulation. It is only in the tropical regions, where $\overline{m}_y \simeq 0$, that a meridional circulation can be forced by convective processes without explicit wave forcing of the mean flow.

### 2.6.2 Baroclinic Instability

Baroclinic instability is the other principal mechanism whereby the troposphere is stirred by unstable motions. Baroclinic instability allows Rossby waves to grow at the expense of the available potential energy associated with atmospheric flows in geostrophic balance. The mathematical analysis of even an idealized baroclinically unstable wave is rather involved and will not be presented here. The interested reader is referred to Holton (1992) for a simplified treatment, and to Charney and Stern (1962) or Pedlosky (1979) for a more complete analysis.

In Sections 2.3.3 and 2.3.4 it was shown that the equilibrium configuration of a flow in a rotating frame of reference involves *geostrophic balance* between the pressure gradient force and the Coriolis torque, Eqs. (2.52)-(2.53). As a consequence, the fluid retains potential energy associated with the geopotential gradient that balances the Coriolis force (Figs. 2.3, 2.5). Stable waves are not able to tap this potential energy, but under certain circumstances Rossby waves can extract potential energy and convert it into wave kinetic energy. For initially small perturbations on a background state, this can be shown to happen when the background gradient of potential vorticity reverses sign somewhere within the fluid. In practice, this occurs readily when there is a temperature gradient at the lower boundary, that is, the surface of the Earth. Strong temperature gradients are indeed present at the Earth's surface (Fig. 2.23), especially during the winter season.

Figure 2.24 shows schematically the parcel displacements in the latitude-height plane associated with stable and unstable Rossby waves superimposed on a typical background gradient of potential temperature $\theta$. The parcel trajectories of unstable waves have slopes smaller than those of the background $\theta$ surfaces. Such parcels are colder than the environment when moving downward, and warmer than the environment when moving

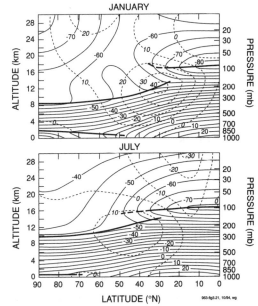

**Figure 2.23.** Temperature (solid lines; $°C$) and zonal wind (dashed lines; m s$^{-1}$) in the troposphere and lower stratosphere during January and July. The bold solid lines show the position of the tropopause (from Holton, 1972).

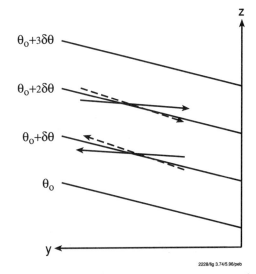

**Figure 2.24.** Schematic representation of parcel trajectories in the latitude-height plane with respect to the background potential temperature surfaces. Unstable waves have parcel trajectories (solid arrows) whose slopes are shallower than those of the isentropic surfaces, while stable waves have steeper slopes (dashed arrows) (from Holton, 1972).

upward. Thus buoyancy forces (Section 2.3.1) amplify the vertical motion of the parcels, contributing to the growth of the wave. In contrast, stable waves have parcel trajectories steeper than the background $\theta$ surfaces. The vertical motion of parcels along these trajectories is opposed by buoyancy, since the parcels are warmer than the environment when they are descending and colder when they are ascending.

Baroclinic instability contributes to the growth of atmospheric waves, primarily in the middle latitudes of both hemispheres. The force due to the divergence of the Eliassen-Palm flux associated with baroclinic waves is crucial in driving the tropospheric residual circulation outside the tropics (Section 2.6.3). There are two primary regions of Eliassen-Palm flux divergence due to baroclinic waves, one near the surface (a region of positive divergence) and one in the upper troposphere (a region of negative divergence). These act to decelerate the mean flow near 200 mb and to accelerate it at the surface. Large-scale horizontal mixing occurs near the surface, where the waves are generated, and in the upper troposphere, where they break (this breaking region resembles a smaller-scale version of the stratospheric surf zone).

A schematic of the circulation associated with baroclinic disturbances is shown in Figure 2.25. The horizontal component of the circulation is represented by streamlines at two levels, in the middle troposphere (600 mb) and at the surface (1000 mb). The midtroposphere horizontal flow is a wavelike disturbance in the background flow, but the surface circulation forms a closed low-pressure center with associated warm and cold fronts. Three-dimensional parcel trajectories are the result of advection by the horizontal and the vertical components of the flow. The latter is downward behind the surface cold front and upward at the surface warm front. As shown in the figure, this produces equatorward and downward flow of cold air and poleward and upward flow of warm air.

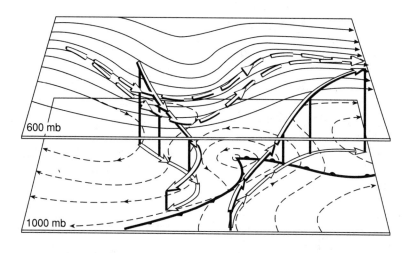

**Figure 2.25.** Schematic of air flow with respect to midlatitude baroclinic disturbance (from Palmén and Newton, 1969).

The circulation of a baroclinic wave releases available potential energy, as shown by the idealized initial and final configurations depicted in Figure 2.26. Because of the constraints imposed by geostrophic balance, this energy *cannot* be released by a zonally symmetric overturning of the fluid in the latitude-height plane, but instead involves the complex zonally asymmetric motion of air masses depicted in Figure 2.25.

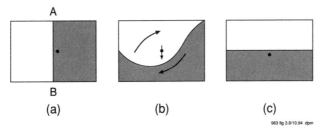

**Figure 2.26.** *Idealized representation of potential energy release due to meridional gradients of temperature. (a) Initial configuration with colder, dense fluid on the right (shading) separated by a removable partition from warmer lighter fluid on the left; (b) schematic of kinetic energy generation by removal of partition; (c) final configuration (from Wallace and Hobbs, 1977).*

## 2.6.3 The General Circulation of the Troposphere

A schematic view of the tropospheric circulation is shown in Figure 2.27. The upper boundary of the troposphere, the *tropopause*, occurs at 16 km in the tropics. The height of the tropopause decreases with latitude with a "break" located near 30°, just equatorward of the westerly midlatitude zonal wind maximum (the jet stream). In the extratropics, westerly winds generally occur throughout the depth of the troposphere so that the extratropical atmosphere is in super-rotation with respect to the Earth's surface (has higher angular momentum). This implies the continuous transport of high-angular-momentum air poleward from lower latitudes. In the tropics the zonal-mean wind is characterized by rather weak easterlies. The mean winds everywhere tend towards zero at the Earth's surface (Fig. 2.23), indicating the direct influence of friction at the ground.

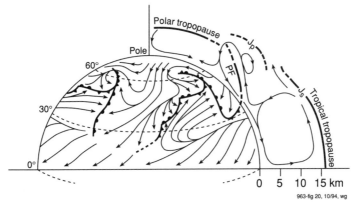

**Figure 2.27.** *Idealized schematic of the tropospheric circulation. Typical NH surface winds are denoted by arrows. In the middle latitudes these winds are drawn in relation to synoptic systems with their associated cold fronts (heavy lines with triangles), warm fronts (heavy lines with semicircles), and occluded fronts (portions of heavy lines with triangles and semicircles). On the right-hand side, the circulation in the latitude-height plane is depicted in relation to the polar tropopause, tropical tropopause, polar front (PF), polar jet stream ($J_p$), and subtropical jet stream ($J_s$) (from Defant and Defant, 1958).*

Between approximately 30° and 60° the circulation is dominated by baroclinic disturbances, the common weather producers familiar in the middle latitudes of both hemispheres. The surface winds in this regime are highly variable, with the different wind regimes separated by the warm fronts and cold fronts associated with *baroclinic disturbances*. The prevalence of these disturbances in the midlatitudes cause this region to be very well mixed both vertically and horizontally. Precipitation is usually associated with air rising above the warm fronts, as well as with the passage of the cold fronts.

The weather regime equatorward of approximately 30° is significantly different from that poleward. In most locations it is characterized by air slowly descending, although rapid rising motion occurs in the vicinity of the equator, along the Inter-Tropical Convergence Zone (ITCZ). Within the ITCZ deep convective clouds are prevalent and there is abundant rainfall. This rising motion near the equator in the ITCZ, the compensatory sinking motion in the subtropics (Fig. 2.27), poleward flow in the upper troposphere, and the equatorward meridional flow near the Earth's surface is referred to as the *Hadley cell*. The Hadley circulation cells, one in each hemisphere, cover about half of the Earth's surface area. As the flow returns toward the equator in the Hadley cell, it develops an easterly component due to the Coriolis torque on the equatorward flow, giving a characteristic easterly and equatorward component to the near surface air flow in the subtropics (trade winds).

Strong seasonal differences occur in the tropospheric circulation regimes. The increased meridional temperature gradients in the winter hemisphere (due primarily to the strong seasonal cycle in temperatures poleward of the subtropics) imply stronger mean zonal winds during the winter months, according to the thermal wind equations (Section 2.3.3), and more vigorous baroclinic disturbances.

The Hadley circulation is also more vigorous in the winter hemisphere. In the summer hemisphere the flow is more quiescent and mixing on the hemispheric scale is consequently slower. Large temporal variations in the tropospheric circulation on timescales longer than the seasonal scale are also observed, driven by the internal dynamics of the ocean and the atmosphere. Perhaps the best known of these interannual variations are the circulation anomalies associated with changes in the Pacific equatorial sea surface temperatures known as El Niño.

The mean motion of air in the latitude-height plane is known as the *mean meridional circulation*. It can be described mathematically in terms of the transformed Eulerian *mean meridional streamfunction equation*, which is obtained by taking $f\partial(2.105)/\partial z + (R/H)\partial(2.109)/\partial y$ and assuming that the zonal-mean wind and geopotential fields are in geostrophic balance, per Eq. (2.52). The resulting equation can be used to calculate the mean meridional mass streamfunction $\chi^*$ in terms of the zonal-mean diabatic heating $\overline{Q}$ and the Eliassen-Palm flux divergence:

$$\frac{\partial^2 \chi^*}{\partial y^2} + \frac{\rho_o}{N_B^2} f \left( f - \frac{\partial \overline{u}}{\partial y} \right) \frac{\partial}{\partial z} \left( \frac{1}{\rho_o} \frac{\partial \chi^*}{\partial z} \right) = \frac{\rho_o}{N_B^2} \frac{R}{H} \frac{\partial \overline{Q}}{\partial y} + \frac{\rho_o}{N_B^2} f \frac{\partial}{\partial z} \left( \frac{\nabla \cdot \mathcal{F}}{\rho_o} \right) \quad (2.161)$$

where (generally small) terms related to the sphericity of the Earth have been neglected, the vertical advection of zonal-mean momentum, and the horizontal advection of zonal-mean temperature. Details of the derivation can be found, for example, in Holton (1992).

The streamfunction $\chi^*$ is related to the transformed Eulerian mean velocities by:

$$\overline{v}^* = -\frac{1}{\rho_o} \frac{\partial \chi^*}{\partial z}, \quad \overline{w}^* = \frac{1}{\rho_o} \frac{\partial \chi^*}{\partial y} \quad (2.162)$$

An example of the mass streamfunction computed from climatological values of $\overline{Q}$ and $\nabla \cdot \mathcal{F}$ is shown in Figure 2.28. In each hemisphere the circulation extends from the equator to the pole, with mean rising motion in the tropics ($\partial \chi^*/\partial y > 0$) and sinking motion over the pole. The motion is primarily poleward in the upper troposphere ($\partial \chi^*/\partial z < 0$). The Hadley portion of the residual circulation is clearly visible between the equator and approximately 30° in each hemisphere. In the

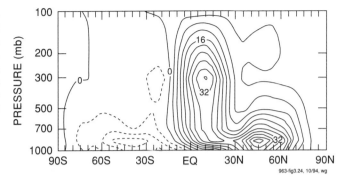

**Figure 2.28.** The streamfunction for the residual circulation in the troposphere for northern hemisphere winter. Air flows between the contours, in the direction such that higher contours are to the right (from Holton, 1992).

extratropics the mean residual circulation is wave driven. The total troposphere EP flux divergence includes the effects of gravity waves, stationary orographically and thermally forced waves, and baroclinically unstable waves.

Air parcels rise through the mean isentropes in the equatorial zones (mean heating) associated with the rising branch of the Hadley circulation, and slowly sink through the isentropes in the descending region of the residual circulation (mean cooling). A large percentage of the net rising motion occurs rapidly in the convective towers in the ITCZ, while the sinking motion occurs more slowly due to the radiative cooling of air parcels (of about 1° per day). This slow cooling gives a transit time from the tropopause to the surface in the Hadley cell of approximately two months if the parcel remains unaffected by further convective processes.

The tropical Hadley cells extend to about 30° in both hemispheres, with the winter hemisphere cell being much stronger than the summer hemisphere cell (Fig. 2.28). It was noted in Section 2.4.4 that the maintenance of the tropical Hadley circulation does not require wave driving, since the mean angular momentum gradient in the tropics approaches zero, and Eq. (2.114) can be approximately satisfied with nonzero $\bar{v}^*$ even when $\nabla \cdot \mathcal{F} \to 0$. Further, it can be shown that, in the presence of even small amounts of viscosity, the radiative equilibrium state (2.117) with geostrophically balanced zonal wind (2.93) is *not* a solution of the zonal-mean equations (Held and Hou, 1980). The actual solution has a nonzero Hadley circulation and, remarkably, does not approach the radiative equilibrium solution as the viscosity is allowed to vanish. The solution approached in the limit of vanishing viscosity is in fact an angular-momentum-conserving Hadley cell. Held and Hou have calculated the meridional structure of such a circulation and shown it to be consistent with the observed width of the tropical Hadley cell.

Poleward of the Hadley cells, the mean meridional circulation of the troposphere is driven by the baroclinically unstable disturbances discussed in Section 2.6.2. The *net* meridional circulation is still direct, with poleward motion aloft and equatorward motion near the surface, but the actual three-dimensional circulation is much more complex, as discussed in connection with Figure 2.23. Zonally asymmetric motions are thus an integral part of the extratropical tropospheric general circulation, as shown schematically in Figure 2.29. A clear and insightful discussion can be found in Palmén and Newton (1969).

Note that in the troposphere and lower stratosphere each hemisphere has its own circulation cell, as opposed to the upper stratosphere, where the circulation extends from one hemisphere to the other, and switches direction with season. Without horizontal mixing at the equator a parcel will not easily cross from the NH circulation cell to the SH cell, or vice versa. Consequently the interhemispheric transport time of approximately a year is significantly longer than the intrahemispheric transport

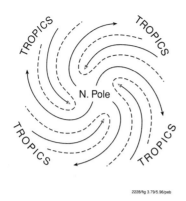

**Figure 2.29.** *A simplified representation of the extratropical general circulation in the troposphere. Warm air moves upward and poleward while cold air moves downward and equatorward. The pattern is zonally asymmetric, but the resultant mean meridional circulation is direct, as shown in Fig. 2.27 (from Bjerknes and Solberg, 1922).*

timescale. The slow transport time across the equator is responsible for the time delay between the hemispheres in the concentrations of long-lived anthropogenic pollutants emitted primarily in the NH; the SH concentrations often lag the NH values by over a year (Prather *et al.*, 1987).

The tropospheric mean circulation cells are punctuated by frequent convective events that mix the troposphere vertically. This provides a conduit whereby parcels in the upper and lower troposphere are continually exchanged, regardless of the direction of the mean residual circulation. Due to the frequency of convection, a tropospheric parcel may sometimes find itself in the lower branch of the mean circulation cell traveling equatorward, sometimes in the upper branch of the circulation cell traveling poleward. It can be shown that the resulting net meridional transport is diffusive in nature (Taylor, 1953). The exception to this might be in the Hadley regime, where for the most part convection is confined to the vicinity of the equator.

### 2.6.4 The Planetary Boundary Layer

The Earth's surface is important in terms of its frictional drag on the atmosphere, as a source or sink of water vapor, as a source or sink of heat, and as an emitter and/or absorber of atmospheric trace substances. Due to the presence of the Earth's surface, a planetary boundary layer (PBL) develops through the depth of the atmosphere most directly affected by the Earth's surface. The depth of the PBL is related to the atmospheric structure and the surface fluxes of water and heat. For typical conditions the boundary layer can range from a few hundred meters to 1-2 kilometers. For average middle-latitude conditions, the PBL contains about 10% of the mass of the atmosphere.

Turbulent motion is almost always excited at the surface by the heating of the land surface and by frictional processes. To maintain the turbulent motion the boundary layer extracts energy from the vertical wind shear of the atmosphere, or from atmospheric thermal gradients. Convectively driven turbulence can be readily generated as the heating of the surface by solar radiation tends to destabilize the lowest atmospheric layers. Even without convection, turbulence is generated through the large vertical wind shears that develop near the Earth's surface due to frictional constraints at the ground. Turbulent mixing by eddies in the PBL is very effective in transferring heat and chemical constituents away from the surface many orders of magnitude faster than could be done by molecular processes. The result is a rather rapid homogenization of constituents within the boundary layer, often modeled as a diffusive process.

The top of the PBL is often marked by a layer of high static stability (when the temperature increases with height, this is called an *inversion*). Above this altitude vertical mixing is usually significantly smaller than within the boundary layer as the turbulent eddies are suppressed by the stability of the atmosphere. The inversion at the top of the boundary layer is resistant to vertical motions and acts to cap the turbulent motion generated within the PBL. It therefore isolates the boundary layer from the troposphere at large and slows the transport of chemical species from the boundary layer into the free troposphere above. The rapid mixing time within the boundary layer, coupled with the slow exchange time between the boundary layer and the free atmosphere, implies that chemical species emitted within the PBL are often trapped there for a time period much longer than the time scale associated with the vertical redistribution of the species within the boundary layer.

Typical time scales range from several hours to a few days for transport out of (or into) the boundary layer, assuming entrainment velocities into the boundary layer in the range 0.01-0.20 m s$^{-1}$ (Stull, 1988) and typical layer depths of 100 m to 3 km. However, active convection can completely vent the boundary layer much more rapidly (less than one hour). Very stable boundary layers occur over the continents during winter when the land surface is cold, in maritime regions of cold ocean temperatures (*i.e.*, off the coast of California), and in the summer season under a large contintental stable air mass. In the maritime subtropics a very stable marine boundary layer is created by the descending and warming air in the downward branch of the Hadley circulation. Under special conditions chemical species can remain trapped in the boundary layer for 3-5 days, resulting in large pollution events over the eastern United States (for example).

## 2.7 Stratospheric Circulation and Transport

Many of the approximations used elsewhere in this chapter to help understand dynamical concepts and transport have important applications to the stratosphere. Because the stratosphere lacks several of the processes that make the troposphere so complex (phase changes of water, planetary boundary layer, convection), the processes that maintain the general circulation are more readily apparent. The concept of wave disturbances superimposed on a zonal-mean background (Section 2.4) is more closely approximated, in part because the waves achieve a more regular structure once they are away from the source region in the troposphere. The transformed Eulerian circulation (Section 2.4.2) is quite successful in explaining the distributions of many trace species.

In one way the stratosphere is more complex than the troposphere. The chemistry of the stratosphere, much of it involving ozone, is very active and is driven by intense solar radiation and a continuous influx of compounds from the troposphere. The ozone concentration has a strong effect on the diabatic heating, which can then affect both the temperatures and the winds. The winds in turn affect the ozone distribution through transport, while the temperatures influence many of the reaction rates important for ozone chemistry (see Chapter 14).

### 2.7.1 The General Circulation of the Stratosphere

The zonal average state of the stratosphere is characterized by a strong seasonal cycle at middle and high latitudes with significant interhemispheric asymmetries. The temperature increases with altitude (see Fig. 2.4), which makes the stratosphere highly stable against vertical motion. In the lower stratosphere the coldest temperatures occur in the tropics, while above about 30 mb there is a temperature gradient from the summer to the winter pole. The thermal gradient leads to strong zonal winds, as expected from the thermal wind relation (Section 2.5.4). These winds reach peak speeds near the solstices and reverse after the equinoxes. The change from winter westerlies to summer easterlies is variable, but on average occurs during March-April in the north and October-November in the south. Stronger wintertime westerlies and lower polar temperatures are found in the Southern Hemisphere.

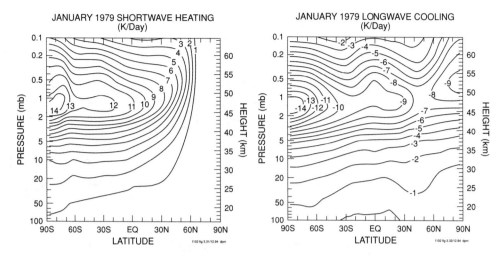

**Figure 2.30.** Diabatic heating in the stratosphere, as derived from LIMS data. (a) The solar heating (K/day) and (b) the longwave cooling due to $CO_2$, $H_2O$, and $O_3$ (K/day) (courtesy of L. Lyjak. Values calculated using the approach described in Gille and Lyjak, 1986).

Diabatic heating in the stratosphere results primarily from absorption of solar radiation by ozone in the Hartley and Huggins bands, which maximizes at ~10 K/day near the summer stratopause. In the upper stratosphere absorption by molecular oxygen in the Schumann-Runge bands and continuum becomes important (see Section 3.2.3). The total shortwave heating for the month of January is shown in the left panel of Figure 2.30; cooling occurs through infrared radiative transfer by $CO_2$, $H_2O$, and $O_3$, shown in the bottom panel. Cooling is much larger than heating at the winter pole, where the temperatures are maintained above the radiative equilibrium values by the sinking branch of the meridional circulation cell (Section 2.6.2).

Despite the large diabatic heating, the mean circulation in the stratosphere is predominantly wave driven. Rossby waves in the winter hemisphere (see Section 2.7.2) overcome the constraints on meridional motion and transport mass poleward. They also are quite effective in transporting trace constituents that have meridional gradients, such as ozone. Conservation of mass requires that the air transported poleward sinks; the adiabatic heating associated with compression then warms the air in the vicinity of the winter pole. The strong infrared cooling (Fig. 2.30) is a response to the fact that temperatures are well above their radiative equilibrium values.

Conservation of mass also leads to rising motion away from the winter hemisphere, in the summer hemisphere and tropics. Throughout the middle and upper stratosphere, a global circulation cell has slow rising of the air in the summer hemisphere and tropics and more rapid sinking, but over a smaller area, in the winter (Fig. 2.10).

The effect of this circulation on constituent distributions has several distinctive features. The rising in the tropics brings source gases from the troposphere (Section 2.8). These are transported poleward in the lower stratosphere, strongly into the winter hemisphere and less so into the summer. The effect can be seen in the distribution of $N_2O$ (Fig. 2.31), which has sources in the troposphere and is destroyed in the stratosphere.

Rossby waves not only drive the mean circulation, but are also responsible for quasihorizontal mixing of chemical species. Air motions due to Rossby waves are approximately adiabatic, and therefore tend to occur along potential temperature surfaces, which are approximately horizontal. Mixing of chemical constituents by

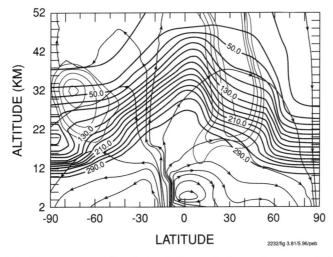

**Figure 2.31.** *The mean meridional circulation and the surf zone, and their relation to the distribution of $N_2O$ in a dynamical/chemical model of the middle atmosphere (Garcia et al., 1992). The contours with arrows represent the TEM mass streamfunction, which is related to the TEM velocities by Eq. (2.162). The thick contours show the $N_2O$ mixing ratio in ppbv. The thin contours superimposed on the $N_2O$ distribution denote the region of strong wave breaking (the surf zone). The mean meridional circulation produces a $N_2O$ peak in the tropics, where there is ascent, and depressed values at high latitudes, where there is strong descent, especially in the winter hemisphere. The intervening region of relatively flat $N_2O$ gradient in midlatitudes is a result of mixing by planetary waves breaking in the surf zone.*

breaking Rossby waves is an efficient transport mechanism in the stratospheric surf zone.

Consider $N_2O$, which is produced in the troposphere and destroyed in the upper stratosphere. Mean upwelling motions will tend to increase its concentration in the tropics, while mean downwelling will tend to decrease the concentration at high latitudes. Thus the mean meridional circulation will produce a meridional gradient, with higher values in the tropics. Isolines of mixing ratio will slope downward from equator to pole, as can be seen in Figure 2.31. Mixing due to Rossby waves, on the other hand, occurs primarily in the meridional direction. The net effect of such mixing will tend to smooth latitudinal gradients of mixing ratio. Figure 2.31 shows that, in regions where Rossby wave mixing is dominant, the meridional gradient is much reduced. A sharp gradient in $N_2O$ occurs in high latitudes where the Rossby wave activity diminishes. The sharp gradient at high latitudes is also indicative of the edge of the polar vortex (Section 2.7.2). Such a gradient also occurs in potential vorticity because it, too, is mixed by Rossby wave transport at lower latitudes.

### 2.7.2 Extratropical Dynamics

During winter, planetary Rossby waves cause deviations from zonal symmetry that are often large and transient. Waves of longest zonal scales, primarily wavenumbers 1-3, are able to propagate into the stratosphere when the mean zonal winds are westerly (Section 2.5.3). Planetary waves are always present in the winter stratosphere, but can be highly variable in amplitude and phase on weekly to interannual time scales. In the Northern Hemisphere quasistationary wavenumber 1 has the largest amplitude on average, while in the Southern Hemisphere the overall amplitudes are smaller and traveling waves account for a larger portion of the total.

Planetary waves undergo dissipation by diabatic heating at a rate that is maximum near the stratopause. In addition, strong dissipation occurs through wave breaking (Section 2.5.5) in low and middle latitudes, where the background potential vorticity gradient weakens. The subtropical region of the winter hemisphere where wave breaking is prevalent has been given the name the *surf zone* to emphasize the rapid horizontal mixing that occurs there (see Fig. 2.31). Poleward of the surf zone, there is a band of stronger potential vorticity gradient that coincides with stronger wind speeds and represents the edge of the polar vortex. Although the vortex edge often encloses the polar region, it is normally somewhat distorted and is sometimes sufficiently displaced, especially in the Northern Hemisphere, that the vortex does not enclose the polar region. The large potential vorticity gradient at the vortex edge constrains air parcels located inside the vortex from moving outward across the edge, and vice versa. The link between potential vorticity and horizontal winds, which provides the restoring force preventing movement across the vortex edge, becomes less effective at smaller scales. When air parcels are reduced to smaller scales, such as the narrow features associated with high wind shear shown in Figure 2.18, high-potential-vorticity air can be separated from the vortex. Once an air parcel is separated, it can more easily mix with the surrounding air.

### 2.7.3 Tropical Dynamics

Stratospheric tropical dynamics is characterized by long-period variations in the mean zonal wind, as seen in the time series of the zonal wind in Figure 2.32. In the lower stratosphere the wind shifts direction between easterlies and westerlies with an irregular time scale that averages about 27 months in what is called the quasi-biennial oscillation (QBO). The QBO is characterized by descending regions of shear that dominate the wind variability from the tropopause to about 30 km, and extend poleward to 20-25 degrees in both hemispheres. Above the QBO, the zonal wind undergoes a semiannual oscillation (SAO), with predominant easterlies during the solstices and westerlies during the equinoxes. The SAO, like the QBO, has descending regions of shear that alternate with time between easterly and westerly. The SAO dominates wind variability in the upper stratosphere.

At the equator the Coriolis force vanishes, which affects the impact that wave dissipation has on the mean circulation. From the transformed zonal momentum balance (Eq. 2.113), wave dissipation gives an EP flux divergence that, because of the vanishing Coriolis torque, tends to lead directly to a change in the zonal-mean wind. The waves that drive the QBO and SAO are equatorially trapped waves, Kelvin waves, Rossby-gravity waves, and inertia-gravity waves. Eastward-moving waves carry westerly momentum and can propagate with little damping through easterly winds; when they encounter a westerly shear zone, they will dissipate and cause the shear zone to move downward. The reverse holds for westward-moving waves. If both eastward and westward waves are generated, the resulting cycle of wave-mean-flow interactions leads to an oscillation in the mean wind.

The SAO, unlike the QBO, is locked to the annual cycle. The easterly phase of the oscillation, near the solstices, is caused at least in part by the advection of easterly momentum across the equator from the summer hemisphere. This occurs near solstice as planetary waves in the winter hemisphere drive a global circulation with net summer to winter flow. The westerly acceleration of the SAO is similar to the QBO in that it appears to be driven by Kelvin waves, in this case fast Kelvin waves that can propagate through the alternating winds below, and by eastward-traveling internal gravity waves.

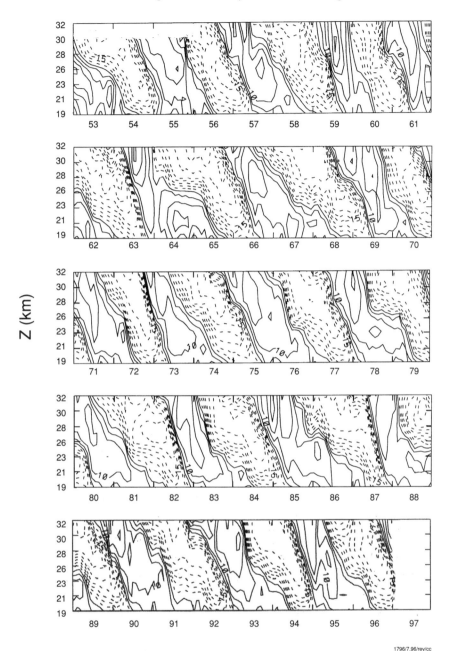

**Figure 2.32.** Time-height cross section of the monthly average zonal wind in the lower stratosphere at Singapore, with the seasonal cycle removed (courtesy of S. Pawson, Stratospheric Research Group, Free University of Berlin).

Global analysis reveals that the quasibiennial oscillation is linked to variability in high latitude winter. Winters tend to be colder and have less planetary wave activity during years when the QBO is westerly. The mechanism for this appears to be the dependence of planetary wave propagation and breaking on the structure and speed of the mean zonal wind (Holton and Austin, 1991; O'Sullivan and Young, 1992; O'Sullivan and Dunkerton, 1994). When the tropical wind is easterly, waves cannot propagate as far toward the equator, and the surf zone is shifted to higher latitudes.

The QBO has a period of ∼27 months and is not tied to a harmonic of the annual cycle. The mechanism for coupling to midlatitudes operates only during winter, since it is only in this season that planetary waves can propagate into the stratosphere (see Section 2.5.3). For any given winter, a variety of QBO situations are possible, ranging from predominantly easterly or westerly to easterly overlaying westerly or the reverse. The relation between the QBO and high latitude winter conditions can be quite variable depending on the relative phases of the QBO and annual cycle. In addition to extratropical signals in zonal wind and planetary wave activity, there is also a substantial QBO in mean ozone amount (Hasebe, 1983; Dunkerton and Baldwin, 1991; Tung and Yang, 1994).

An SAO related to that in zonal wind also appears in the zonal mean distributions of temperature and several trace species. In order to maintain mass balance, a temperature oscillation accompanies the wind oscillation, and is characterized by cooler temperatures during the easterly phase and descending warmer temperatures with the descending westerly phase. Thermal balance is maintained by a secondary mean circulation cell. This SAO circulation has downward motion (warming) over the equator during the westerly phase of the SAO and upward motion during the easterly phase. The vertical motion produces a local minimum in the concentrations of long-lived tracers such as $CH_4$ and $N_2O$ during the equinox seasons (Jones and Pyle, 1984). An ozone variation also occurs that, because of the dependence of chemical reaction rates on temperature, is out of phase with the temperature oscillation (Ray et al., 1994).

## 2.8 Stratosphere-Troposphere Exchange

The exchange of mass between the stratosphere and troposphere is important to the chemistry of both regions, as it brings chemical species with sources in the troposphere (such as CFCs) into the stratosphere, while species with stratospheric sources can be brought into the troposphere. However, just as the boundary layer is isolated from the atmosphere at large, typically by an inversion, the troposphere is isolated from the stratosphere by the high static stability of the stratosphere. Similarly, just as the boundary layer is turbulent and well mixed compared to the troposphere at large, the troposphere is well mixed vertically and horizontally compared to the stratosphere. The mixing time in the troposphere (on the order of months within each hemisphere; on the order of a year between the hemispheres) is much shorter than the time required to exchange the mass of the entire troposphere with the stratosphere (on the order of 18 years, although, due to the difference in mass, the entire stratosphere mixes with the troposphere on the order of every two years).

The simplest qualitatively plausible model for stratosphere-troposphere exchange consists of bulk advection by a single mean cell in each hemisphere with uniform rising motion across the tropical tropopause, poleward drift in the stratosphere, and by continuity of mass, a return flow into the troposphere in the extratropics (Fig. 2.31). Such a circulation was first proposed by Brewer (1949) to explain the observed low water vapor mixing ratios in the stratosphere. The only place near the tropopause where the temperature is low enough to cause such low values of humidity is in the tropics, where the tropopause is high and cold. Dobson (1956) pointed out that poleward and downward advection by this type of mean circulation was qualitatively consistent with the observed high concentration of ozone in the lower polar stratosphere, far from the region of photochemical production. Although this Brewer-Dobson model does not provide a complete description of the exchange processes, it is believed to be substantially correct.

The Brewer-Dobson circulation cell is now known to be primarily wave driven. The principle of downward control states that the time-averaged transport through a pressure surface (or an isentropic surface) depends only on the wave forcing above that surface (Section 2.4.5). The distribution of stratospheric wave forcing implies that upward movement of air into the stratosphere occurs in the tropics and downward movement of air into the troposphere occurs preferentially in winter in the middle to high latitudes.

Holton et al. (1995) have presented a unified picture of the processes that are believed to be important for stratosphere-troposphere exchange. For conceptual purposes, they divide the troposphere and stratosphere into three regions: the troposphere proper, delimited by the 380 K isentrope in the tropics, and the 2 PVU potential vorticity surface in the extratropics (1 PVU = $10^{-6}$ $m^2$ $s^{-1}$ K $kg^{-1}$); the "overworld," the region above the 380 K isentrope, wherein isentropes lie entirely in the stratosphere; and the extratropical "lowermost stratosphere," lying below the 380 K isentrope and the 2 PVU potential vorticity surface (Fig. 2.33).

Wave dissipation in the winter stratosphere drives the mean meridional circulation that controls mass flux through the lower boundary of the overworld. Even near the equator remote wave forcing is responsible for lofting tropospheric air into the stratosphere (Holton et al., 1995). Indeed, Yulaeva et al. (1994) have argued that the annual cycle of tropical lower stratospheric temperatures is the result of interhemispheric differences in wave driving of the stratospheric circulation. Thus the coldest temperatures in the tropical lower stratosphere occur in late January/early February due to more vigorous upwelling during northern winter, when planetary wave driving of the stratospheric circulation is strongest.

Further evidence for wave driving of the stratospheric mean meridional circulation comes from the work of Mote et al. (1996), who have used satellite observations to show that a minimum in water vapor mixing ratio appears at the tropical tropopause (380 K) in NH winter (when temperatures are lowest and freeze-drying of air parcels most effective). This minimum is then advected by the meridional circulation so that six months later it is found near the 450 K isentrope and overlies air with higher water vapor mixing ratio, corresponding to warmer temperatures at the tropopause at that time (see Fig. 8 of Holton et al., 1995). The observed seasonal evolution of water vapor mixing ratio at and above the tropical tropopause is consistent with mean vertical advection estimated from calculations of the mean heating rate (Rosenlof, 1995).

The lowermost stratosphere differs from the overworld in that it is the only part of the stratosphere that can exchange air with the troposphere by transport along isentropic surfaces (cf. the wavy arrows denoting quasihorizontal transport in Fig. 2.33). Exchange between the lowermost stratosphere and the troposphere tends to occur in association with dynamical events known as tropopause folds, in which the tropopause on the poleward side of the jet stream is distorted during the development of synoptic-scale weather systems such as that shown in Figure 2.25. Large extrusions of stratospheric air extend into the troposphere, and much of the air becomes trapped in the troposphere, and eventually mixes with tropospheric air. Because the air in tropopause folds is dry, stratospheric tracers such as ozone are not immediately destroyed, and can have an impact on tropospheric chemistry. Air from the troposphere can also become trapped within the stratosphere during tropopause folds, but because of the large-scale net downward motion, it can have only local impact.

The details of exchange between the lowermost stratosphere and the troposphere are complex, since they depend on the spatial and temporal distribution of relatively

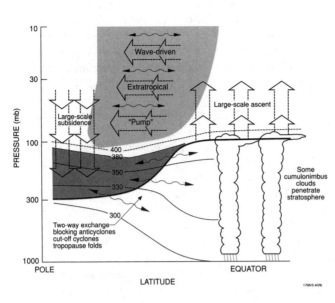

*Figure 2.33.* Schematic view of the processes that control stratosphere-troposphere exchange. The tropopause is denoted by the thick solid line; it is defined by the 380 K surface in the tropics and the 2 PVU potential vorticity surface in the extratropics. The wave-driven, mean meridional circulation (large dashed arrows) controls mass flux through the 380 K surface, the lower boundary of the "overworld." Transport in the "lowermost stratosphere," the region between the overworld and the troposphere, is influenced both by the mean meridional circulation and by quasihorizontal mixing due to wave motions (denoted by the two-headed wavy arrows) (from Holton et al., 1995).

short-lived, synoptic-scale dynamical structures. Nonetheless, Holton *et al.* (1995) note that, for chemical species whose sources or sinks lie entirely in the overworld (or in the troposphere), stratosphere-troposphere exchange can be evaluated most simply and effectively in terms of the mass flux through the lower boundary of the underworld, which, as noted above, is controlled by the wave-driven meridional circulation. On the other hand, detailed evaluation of the effect of smaller-scale processes may be necessary for species that have sources or sinks in the lowermost stratosphere itself, such as chemical compounds emitted by high-flying aircraft, or ozone in regions where heterogeneous chemical processing is important. The exchange of such species represents one of the major challenges to our conceptual understanding of transport between the stratosphere and the troposphere.

## Further Reading

Andrews, D. G., J. R. Holton, and C. B. Leovy (1987) *Middle Atmosphere Dynamics*, Academic Press, Orlando.
Brasseur, G. P. and S. Solomon (1986) *Aeronomy of the Middle Atmosphere*, D. Reidel Publishing Co., Dordrecht.
Brewer, A. W. (1949) Evidence for a world circulation provided by the measurements of helium and water vapor distribution in the stratosphere. *Quart. J. Roy. Meteor. Soc.*, 75, 351.
Gill, A. (1980) *Atmosphere-Ocean Dynamics*, Academic Press, New York.
Holton, J. R. (1992) *An Introduction to Dynamic Meteorology*, 3rd ed., Academic Press.
Holton, J. R., P. H. Haynes, M. E. McIntyre, A. R. Douglass, R. B. Rood, and L. Pfister (1995) Stratosphere-troposphere exchange, *Rev. Geophys.*, 33, 403.
Houghton, J. T. (1986) *The Physics of Atmospheres*, 2nd ed., Cambridge University Press, Cambridge, UK.
James, I. N. (1994) *Introduction to Circulating Atmospheres*, Cambridge University Press, Cambridge, UK.
McIntyre, M. E., and T. N. Palmer (1983) Breaking planetary waves in the stratosphere, *Nature*, 305, 593.

McIntyre, M. E. (1992) Atmospheric dynamics: some fundamentals, with observational implications, in *Proc. Internat. School Phys.* "Enrico Fermi," CXV Course, ed. J. C. Gille, G. Visconti; North-Holland, Amsterdam, 313.

Riegel, C. A. and A. F. C. Bridger (1992) *Fundamentals of Atmospheric Dynamics and Thermodynamics*, World Scientific Publishing, Singapore.

Visconti, G., and R. Garcia (eds.) (1987) *Transport Processes in the Middle Atmosphere*, NATO ASI Series, D. Reidel Publishing Co., Dordrecht, The Netherlands.

Wallace, J., and P. Hobbs (1977) *Atmospheric Science: An Introductory Survey*, Academic Press, San Diego.

Michael E. McIntyre

## Why Understand Dynamics — and What Is "Understanding" Anyway?

What is the most important, exciting, and intellectually challenging task that a scientist can undertake? Many would now say that it is to help understand what is happening to our planetary life-support system. The international action now being taken to limit chlorofluorocarbon (CFC) emissions, for example, has come about through an increasingly clear scientific understanding of stratospheric ozone depletion, including the reasons why CFC emissions in the Northern Hemisphere have their greatest effect in the Southern Hemisphere. Like most other atmospheric-chemical phenomena, this depends on the interplay of chemistry, radiation, and, last but not least, atmospheric fluid dynamics.

It hardly needs pointing out that much of our planetary environment is fluid, from the upper atmosphere to the depths of the oceans, and that our survival depends on that fact. Fluid motions, both air and water, are essential to all life as we know it. Fluid motions prevent climatic extremes, by transporting heat energy at petawatt rates, dwarfing the combined outputs of the world's electric power systems. It is to fluid motions that we owe a tolerable local chemical environment, despite the production of waste substances by ourselves and by other species. Fluid motions powerfully influence the rate at which CFCs, for example, are removed from the environment. It is fluid motions that make it possible for Northern Hemispheric pollution to cause Southern Hemispheric ozone holes. They do so by carrying air from near the Earth's surface in the Northern Hemisphere up into regions bombarded by hard solar ultraviolet in the tropical stratosphere and thence, over times of the order of several years, via complicated, chaotic pathways, into the Antarctic lower stratosphere and elsewhere. It is fluid motions, involving an extremely subtle interplay between the shearing effects illustrated in Figure 2.18 and the Rossby-wave mechanism illustrated in Figure 2.7—an interplay first explicitly recognized, I believe, by my former student Martin Juckes—it is fluid motions that are responsible for that remarkable and much-studied phenomenon, the chemical near-isolation of the wintertime stratospheric polar vortices. This is a close cousin of the near-isolation of the smoke in smoke rings, and of the moist updraft in hurricanes and the cores of Atlantic Ocean Mediterranean eddies.

Fluid dynamics is intellectually challenging as well as phenomenologically interesting. Richard Feynman thought so, as did Werner Heisenberg. Knowing the equations is not enough; computing solutions is not enough; there is the challenge of understanding what the equations and solutions mean, of seeing to what extent they are related to physical reality. As with any other physico-chemical system, you have to do thought experiments and, where possible, numerical or laboratory experiments, and check for mutual consistency. Let me mention just two examples that are relevant to this book. One is the fact that the rising branch of the lower stratospheric circulation is observed to be over the tropics, even at solstice. It is not (in the case of the stratosphere) observed to be over the summer pole, where solar heating is maximal; see Figure 2.30. I shall come back to this point. The other example, at first sight quite unrelated, is the famous experiment done in the mid-1970s by Alan Plumb and Angus McEwan, a laboratory counterpart of the quasibiennial oscillation observed in the real atmosphere (QBO; see Section 2.7.3). It is an important variation on a theme pioneered earlier in the seminal papers of James Holton, Richard Lindzen, and John M. Wallace; and I think it proves at a stroke that fluid behavior can be truly fascinating.

Imagine a large annular container, about a meter across and half a meter tall, the gap width being a few tens of centimeters, filled with a low-viscosity fluid such as brine having a strong vertical density gradient, with the densest fluid at the bottom. Gently but persistently oscillate the bottom boundary in a standing wave, equivalent to the sum of two progressive waves of equal amplitude travelling clockwise and counterclockwise around the annulus. Make the oscillation frequency equal to some modest fraction of the buoyancy frequency $N_B$ of the stable stratification, Eq. (2.28), and keep the oscillation going. Coriolis effects are negligible, and the system is mirror symmetric. There is no distinction between the clockwise and counterclockwise directions. What will happen? Textbook physics will tell you, nothing much. Textbook physics says that there will just be some small oscillations of the fluid.

In reality—when the amplitude and frequency of the oscillations are in certain ranges—the symmetry is spontaneously broken. You do have small oscillations, in the form of internal gravity waves. But you also have a mean flow circulating around the annulus, growing from imperceptible beginnings and evolving on a much longer timescale. The fluid circulates first clockwise, then counterclockwise, with different timings at different levels, showing qualitatively the same kind of pattern as we see in the real atmosphere (Fig. 2.32). This well illustrates the ability of fluid motions to surprise us. I doubt if even a scientific genius would have guessed *a priori* that this whole complicated, yet remarkably regular, response would arise from doing just one thing to the system, gently oscillating its bottom boundary. To understand how it works, you need to think about the behavior of internal gravity waves, and their ability to set up a radiation stress, equivalently a vertical flux of horizontal momentum. The word flux is crucial. Contrary to a popular myth, based on a very restricted set of examples, the waves do not, in any unique or natural sense, have momentum; what matters is that they set up a momentum *flux*, an entirely different thing. The importance of this distinction was pointed out long ago, in the 1920s, by Léon Brillouin.

It turns out that exactly such a wave-induced momentum flux is crucial to understanding not only the QBO, but also the entire global-scale circulation of the middle atmosphere, with far-reaching consequences for ozone replenishment rates, CFC lifetimes, and stratosphere–troposphere mass exchange rates. In particular, it is crucial to explaining the abovementioned fact that the rising branch of the lower stratospheric circulation is always observed to be over the tropics, regardless of whether solar heating is maximal there. This is an effect of extratropical wave-induced momentum fluxes together with the Coriolis effect of the Earth's rotation. It can be thought of as a kind of global-scale "gyroscopic pumping." The extratropical stratosphere and mesosphere, up to altitudes of about 80 km, act on the tropical stratosphere like a gigantic, seasonally and interannually variable suction pump. That is why the global-scale circulation is often referred to as a "wave-driven" circulation. One part of this story—involving considerable fluid-dynamical subtlety and, at the time of writing, still at the cutting edge of research—is that the Earth's rotation makes the tropics fluid dynamically very different from the extratropics, through the relative weakness of Coriolis effects. Another is that, in the extratropics, one can apply the "downward control" principle summarized in Figure 2.11.

The downward control principle is a statement about a particular thought experiment, one in which attention is confined to the extratropics and the wave-induced momentum fluxes are regarded as given. As with other good thought experiments, this helps us understand an aspect of the real atmosphere, even though the real atmosphere has all kinds of other aspects, one might say feedback loops, among which one of the most important is the *upward* influence resulting from upward wave propagation. To see more clearly the implications of the downward control principle itself, you might like to consider the following thought experiment after reviewing Figure 2.11 and the associated text.

Imagine a perpetual-winter stratosphere in which the wave-induced momentum fluxes are artificially kept constant, and apply some extra cooling to the polar stratosphere.

More precisely, do something—anything—that increases the infrared cooling-to-space throughout the polar stratosphere. One could imagine filling the polar vortex with some chemical constituent that radiates efficiently in some of the available infrared spectral windows. Does the extra cooling cause more descent in the polar vortex? The answer is no! After the system adjusts to a new steady state, there is a slightly weaker descent than before. Most people find this counterintuitive at first. It turns out that the extra cooling makes the gyroscopic pumping action slightly weaker. It is this that controls the descent rate. Some publications that discuss these issues are McIntyre (1992) and Holton et al. (1995).

An important question is why thought experiments are so important in science, indeed what is meant by scientific understanding and scientific uncertainty, why scientific understanding is humanly as well as practically important, why it is cross-cultural, and why it need not conflict with the artistic and the spiritual—the so-called philosophical issues, or whatever you want to call them, that are so widely and dangerously misunderstood today. I dare to mention those wider issues, despite their controversial nature, and despite not having the space here for an adequate discussion, because it is clear that tomorrow's scientists are going to need a broader, more interdisciplinary, and generally better-informed outlook than has been traditional in science. That outlook will have to include a very clear appreciation of what scientific understanding is—its strengths and equally its limitations—and a far better training in how to achieve lucidity, and humility, in explaining these things to others, both professional and lay people, as well as to oneself. The fate of our species will depend on how tomorrow's scientists meet this most formidable challenge of all.

*Michael E. McIntyre is Professor of Atmospheric Dynamics at the University of Cambridge, UK, and co-director of the Centre for Atmospheric Science there. He describes himself as a fluid dynamicist interested in the workings of the atmosphere.*

# 3 Chemical and Photochemical Processes

## 3.1 Introduction

The Earth's atmosphere can be thought of as a low-temperature combustion system in which energy from the sun is employed to drive a wide variety of oxidative processes. For example, the photolysis of $O_2$ in the stratosphere is responsible for initiating the chemistry involved in the production of ozone,

$$O_2 + h\nu \rightarrow O + O \tag{3.1}$$
$$O + O_2 + M \rightarrow O_3 + M \tag{3.2}$$

Photolysis of ozone throughout the atmosphere leads to the formation of OH,

$$O_3 + h\nu \rightarrow O\,(^1D) + O_2 \tag{3.3}$$
$$O\,(^1D) + H_2O \rightarrow OH + OH \tag{3.4}$$

which in turn is responsible for initiating the oxidation of a wide variety of atmospheric trace constituents, for example, methane:

$$OH + CH_4 \rightarrow CH_3 + H_2O \tag{3.5}$$

The exact composition of the atmosphere is determined by a complex chemical mechanism that consists of many thousands of elementary reactions such as the ones listed above. The rates at which these reactions occur, along with the appropriate product distributions, can be used to quantify each of the chemical constituents in the atmosphere and to predict future changes in atmospheric composition through chemical modeling (see Chapter 12). This chapter summarizes the principles required to understand how photochemical and chemical reactions occur, the factors controlling the rates at which they occur, and the methods used to measure and parameterize these rates.

### 3.1.1 Scope and Definitions

Reaction Kinetics: The branch of physical chemistry that deals with the determination of the rates and mechanisms of chemical reactions.

Rate Coefficient: A constant of proportionality relating the rate of a chemical reaction to the concentrations of the chemical species involved in the reaction.

Arrhenius Equation: A relation used to describe the temperature dependence of the rate coefficient for a chemical reaction. The form of the equation is $k = A\exp(-E_a/RT)$, where $k$ is the rate coefficient, $A$ is the pre-exponential factor, $E_a$ is the activation energy for the reaction, $R$ is the gas constant, and $T$ is the absolute temperature.

Reaction Order: In an elementary chemical reaction, the reaction order is the number of species that actually participate in the reaction. Reactions involving one, two, or three species are thus known as uni-, bi-, or termolecular. In a complex reaction scheme, the overall *apparent* reaction order may be fractional, or may even change with time.

Free Radical: A molecule containing an unpaired electron in its outer shell, as a consequence of which it is usually very reactive.

Ion: Charged species (atomic or molecular) occurring in solution or in the gas phase (usually the upper atmosphere).

Mole: The standard amount of a substance, containing Avogadro's number ($6.02 \times 10^{23}$) of individual particles.

Mass Accommodation Coefficient, $\alpha$: The probability that reversible uptake of a gas-phase species will occur upon collision of that species with a given surface (liquid or solid). This is often equated with a "sticking" probability.

Gas-Surface Reaction Probability, $\gamma$: The probability that irreversible uptake of a gas molecule will occur upon collision with a surface.

Quantum Yield: The probability that a particular photochemical event will occur following the absorption of a photon by a molecule, usually given the symbol $\Phi$. For example, a molecule may photodissociate, fluoresce, phosphoresce, or be quenched by collisions. The quantum yield for each process then gives the fractional occurrence of that process. Quantum yields are in general dependent on the wavelength of excitation.

$j$ Value: A first-order rate coefficient (in units of inverse time) for the occurrence of a photochemical reaction; also known as photolysis rate. The $j$ value is calculated from the product of the absorption cross section of the molecule being photolyzed, the quantum yield for the process, and the actinic flux, all integrated over the wavelength region of interest.

## 3.2 Radiation
### 3.2.1 Regions of the Electromagnetic Spectrum

Radiation comes in various "colors," extending from $\gamma$- and X-rays at the high-frequency end of the spectrum to radio-waves at the lowest frequencies. The entire electromagnetic spectrum is shown in Figure 3.1. The energy ($E$) of a photon is directly proportional to the frequency ($\nu$),

$$E = h\nu \tag{3.6}$$

with the constant of proportionality being Planck's constant, $h$. Before discussing the effects of various types of radiation on the chemistry of the atmosphere, it is important to understand the principles involved in the interaction of light with molecules.

### 3.2.2 Absorption and Emission of Radiation

The processes of absorption and emission of a photon by a molecule result in a change in the energy level of the molecule. In absorption, the molecule is transferred from a lower to a higher energy level (*i.e.*, from $E_l$ to $E_u$), with the energy gap between the levels equivalent to the energy of the absorbed photon. The same principle applies for emission, with the energy of the emitted photon determined by the energy difference between the upper and lower states. Thus the frequency of the transition between the upper and lower levels can be expressed as

$$\nu = \frac{E_u - E_l}{h} \qquad (3.7)$$

Transitions can occur between rotational levels in a molecule, which generally involve photons in the microwave or millimeter-wave region of the spectrum; between vibrational levels (typically in the infrared region of the spectrum); or between electronic states (which occur in the near IR, visible, and ultraviolet regions). Additionally, vibrational transitions usually involve a change in the rotational energy, while electronic transitions may also be accompanied by changes in the rotational and vibrational energy. This leads to the characteristic "structure" observed in these spectra.

Molecular energy levels are quantized, and transitions thus occur at discrete frequencies. Classically, a molecule will interact with an external electromagnetic field of frequency $\nu$ (and hence absorb or emit a photon) when it possesses a transient dipole oscillating at that frequency. Quantum mechanically, the

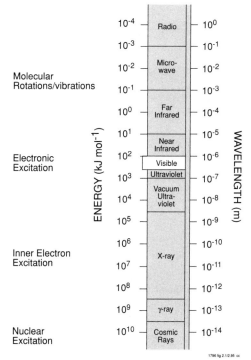

**Figure 3.1.** Diagram of the electromagnetic spectrum, showing the molecular and atomic transitions associated with each wavelength range.

intensity of an absorption or emission transition is proportional to the square of the transition dipole ($\mu_{1,2}$), which is defined as follows:

$$\mu_{1,2} = \int \psi_1^* \, \mu \, \psi_2 \, d\tau \qquad (3.8)$$

where $\psi_1$ and $\psi_2$ are the wavefunctions for the two energy levels and $\mu$ is the dipole moment operator. For a transition to occur this integral must be nonzero. The circumstances under which this condition will hold for the various types of transitions will now be discussed. More detailed treatments of the energy levels are to be found in textbooks such as those by Atkins (1986) or Herzberg (1945).

For rotational transitions, this condition will hold only when the molecule has a permanent electric dipole. Thus molecules such as $H_2O$ and $CO$ possess allowed rotational spectra, while $O_2$, $N_2$, and $CO_2$ do not. Absorption and emission of microwave and millimeter-wave radiation have no direct effect on the chemistry of the atmosphere, but have been employed for the determination of the concentration of various trace atmospheric constituents, and for the determination of the temperature of the atmosphere (see, for example, Fig. 11.11).

A molecule will possess a vibrational spectrum only when the dipole moment of the molecule changes with the occurrence of the vibration. Thus, as in the case of rotation, $O_2$ and $N_2$ possess no allowed vibrational transitions. However, $CO_2$ (which has no permanent dipole) does possess some allowed vibrational transitions, since the asymmetric stretch and bending motions lead to a transient dipole (see Fig. 3.2 for details). The absorption and emission of infrared absorption by molecular trace gases, as in the case of molecular rotation, can be used for the atmospheric

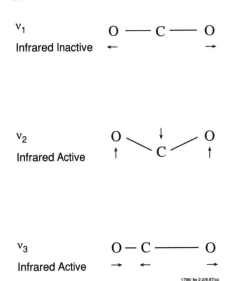

**Figure 3.2.** *Illustration of the infrared-active and -inactive vibrational modes of $CO_2$. Modes $\nu_2$ and $\nu_3$ lead to a change in the dipole moment, and are thus infrared active.*

monitoring of various infrared-active trace species, using remote sensing techniques (see Chapter 11 for more details). Absorption of IR radiation has another major consequence for atmospheric chemistry, that of global warming. Heat, in the form of infrared radiation, that would normally be dissipated to outer space, is absorbed by atmospheric constituents (such as $H_2O$, $CO_2$, $CH_4$, and CFCs), leading to heating of the Earth's lower atmosphere. When the atmospheric concentration of some of these species increases as the result of anthropogenic activity (for example, $CO_2$), there exists the potential for altering the climate of the Earth. This concept will be explored in Chapter 15.

In the case of electronic transitions, the transition dipole moment of Eq. (3.8) is a measure of the dipole moment associated with the shift of charge occurring as a result of electronic redistribution. The energy associated with electronic transitions (*i.e.*, with visible and UV photons) is comparable to (or greater than) that of a molecular bond, and hence it is this type of process that leads to photochemical change in the Earth's atmosphere. For example, both $O_3$ (105 kJ mol$^{-1}$) and $NO_3$ (210 kJ mol$^{-1}$) contain relatively weak molecular bonds, and the photodissociation processes

$$O_3 + h\nu \rightarrow O\left(^3P\right) + O_2 \tag{3.9}$$

$$NO_3 + h\nu \rightarrow NO_2 + O\left(^3P\right) \tag{3.10}$$

can occur readily upon the absorption of visible radiation. Stronger chemical bonds, such as in the $O_2$ molecule (500 kJ mol$^{-1}$), can only be broken by more energetic ultraviolet photons. More details of photochemical processes are presented in Section 3.3. Remote sensing of atmospheric constituents in the UV/visible region of the spectrum is again a viable option, but only for species possessing structured spectra (like OClO and $NO_2$).

The usual method of parameterizing the intensity of a particular transition is the Beer-Lambert Law, which states that the attenuation of radiation of frequency $\nu$ on passing through a sample of thickness $d\ell$, containing an absorber of concentration $n$, will be proportional to the thickness and the concentration:

$$\frac{dI}{I} = -\sigma\left(\nu\right) n\, d\ell \tag{3.11}$$

The proportionality constant, $\sigma$, is referred to as the absorption cross section and is related to the intensity of the transition (it is actually proportional to $\mu_{1,2}^2$). If the pathlength, $\ell$, is expressed in units of cm, and the concentration, $n$, in molecule cm$^{-3}$, it follows that $\sigma$ will have units of cm$^2$ molecule$^{-1}$. The Beer-Lambert Law is more often expressed in its integrated form,

$$\frac{I\left(\nu\right)}{I_o\left(\nu\right)} = \exp\left[-\sigma\left(\nu\right) n\ell\right] \tag{3.12}$$

where $I_0(\nu)$ and $I(\nu)$ are the intensity of incoming and outgoing radiation at frequency $\nu$. Other forms of the equation are also employed. For example, a useful parameter is the absorbance, $A$, which is defined as follows:

$$A(\nu) = \ln\left(\frac{I_0(\nu)}{I(\nu)}\right) = \sigma(\nu)\, n\ell \qquad (3.13)$$

This quantity is often referred to as the optical depth in remote sensing applications, and is then given the symbol $\tau$. An absorption spectrum is simply a plot of $\sigma(\nu)$ against frequency $\nu$. Sometimes the cross section is plotted against the wavelength or the wavenumber, which is the inverse wavelength or the frequency divided by the speed of light.

Absorption spectra can be determined quantitatively in laboratory experiments. The basic experiment consists of a lamp, which emits photons over a range of frequencies under investigation; a cell, in which a known amount of the gas under investigation can be placed; and a detector that monitors the light intensity. The detector usually is placed at the exit of a monochromator or similar device for dispersing the radiation from the lamp into its component frequencies. In such an experiment, from a measure of the light intensity in the absence of sample ($I_0$) and in the presence of a sample ($I$) through a vessel of known pathlength ($\ell$) containing a known concentration ($c$) of absorber, one can obtain the absorption cross section.

Some examples of absorption spectra are now presented. The infrared spectrum of CFC-11 is given in Figure 3.3. Like most chlorofluorocarbons, CFC-11 absorbs strongly

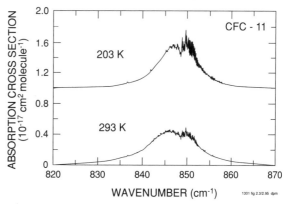

**Figure 3.3.** Infrared spectrum of chlorofluorocarbon-11 ($CFCl_3$) at temperatures of 203 K and 293 K (from McDaniel et al., 1991).

in the so-called "atmospheric infrared window" between 800 and 1200 $cm^{-1}$, where other major infrared-active atmospheric constituents (such as $H_2O$ and $CO_2$) do not absorb strongly. This absorption spectrum can be useful for two purposes: It can be used to calculate the contribution of CFC-11 to global warming, and it can also be used to quantify the atmospheric abundance of CFC-11 from IR spectra of the atmosphere obtained via remote sensing techniques. The transition corresponds to the absorption of one quantum of vibrational energy. The fine structure, which is due to changes in the rotational energy, gets sharper as the temperature is lowered. An example of an electronic spectrum, that of $NO_2$, is given in Figure 3.4. Since $NO_2$ absorbs strongly in the near-UV/visible region of the spectrum and dissociates with unit quantum yield at wavelengths shorter than about 400 nm, it is subject to rapid photodissociation in the atmosphere. In addition, the structured nature of the spectrum can be used to identify $NO_2$ quantitatively in UV/visible spectra of the atmosphere.

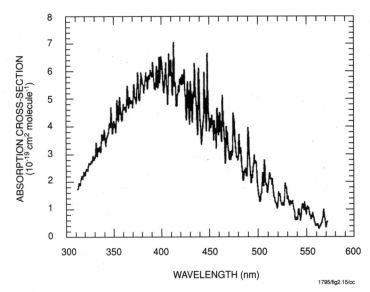

**Figure 3.4.** The UV/visible absorption spectrum of $NO_2$ at room temperature (Harwood and Jones, 1994). The absorption is caused by an electronic transition in the $NO_2$. The fine structure is due to rotational and vibrational transitions excited simultaneously. The absorption features have been used to detect $NO_2$ in the atmosphere. Photodissociation occurs for wavelengths less than about 400 nm.

### Box 3.1 Ozone Photolysis

Among the many important roles played by ozone in the atmosphere is the role its photolysis plays in the generation of OH radicals. This OH production occurs from the formation of the excited $O(^1D)$ species in the UV photolysis of ozone, followed by the reaction of $O(^1D)$ with $H_2O$, $H_2$, and $CH_4$ (see Chapter 5 for more details). As discussed in detail throughout this book, these OH radicals play a critical role in initiating the oxidation of a variety of trace gases (*e.g.*, organic compounds, reduced sulfur species, etc.) and thus in removing them from the atmosphere.

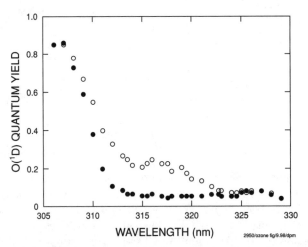

**Figure 3.5.** $O(^1D)$ quantum yields from the photolysis of ozone as a function of wavelength and temperature. Filled circles = 203 K; open circles = 298 K (created from data in Talukdar et al., 1998).

The photochemistry of ozone is very complex, as the relatively weak chemical bonds in ozone allow excited states of the O and $O_2$ photoproducts to be accessed. In fact, only now is a consensus being reached on the quantitative aspects of this photochemistry, particularly near the energy threshold for O($^1D$) production near 310 nm. It is now believed (Ravishankara et al., 1998, and references therein) that four different photolysis channels are occurring in the 290-350 nm spectral region, the region of importance for the generation of OH in the lower atmosphere, with the quantum yields for these four channels varying with wavelength and temperature:

$$O_3 + h\nu \rightarrow O(^3P) + O_2(^3\Sigma) \qquad (A)$$
$$\rightarrow O(^3P) + O_2(^1\Delta) \qquad (B)$$
$$\rightarrow O(^1D) + O_2(^1\Delta) \qquad (C)$$
$$\rightarrow O(^1D) + O_2(^3\Sigma) \qquad (D)$$

For atmospheric considerations, the key channels are those which produce O($^1D$), that is, those which lead to OH production. Interestingly, the energy threshold for channel $(C)$ is at about 308 nm, yet this process still occurs to longer wavelengths as a result of internal energy (vibrational and rotational) contained in the ozone molecule itself, while channel $(D)$ is a quantum-mechanically "spin-forbidden" process whose occurrence was thought to be unlikely. The figure below shows the O($^1D$) quantum yields near 200 K and 300 K. Channel $(C)$ dominates below about 320 nm at 300 K, but is strongly temperature dependent, while the near-constant contribution of the spin-forbidden channel (quantum yield $\approx 0.06$) is evident beyond about 320 nm.

## 3.2.3 Solar Radiation

In order to calculate the rate at which atmospheric molecules are excited, it is necessary to know what radiation is available at various altitudes in the atmosphere. For this, a knowledge of the solar spectrum is required (see Appendix M), as well as an understanding of how this radiation is attenuated as it penetrates the Earth's atmosphere. An account is given by Brasseur and Solomon (1986).

The radiation emitted by the sun corresponds, to a first approximation, to the emission of a black body at a temperature of about 6000 K. This emission, which emanates from the Sun's photosphere, has a maximum intensity in the visible region of the spectrum. The infrared emission is actually somewhat higher than would be predicted for a 6000 K black body, and more closely approximates a 7000 K black body. The intensity of ultraviolet radiation reaching Earth is less than that from a 6000 K body, due to absorption by atoms and molecules in the Sun's photosphere. The Sun's output more closely approximates a color temperature of 5100 K between 200 and 250 nm, and 4600 K between 130 and 170 nm.

The solar spectrum at the top of the Earth's atmosphere is shown in Figure 3.6. Wavelengths shorter than about 100 nm, which are absorbed by $O_2$, O, and $N_2$, do not penetrate below about 100 km. $O_2$ absorbs strongly in the 100-175 nm region, the Schumann-Runge continuum, and attenuates this radiation in the thermosphere (i.e., at altitudes above 80 km). Absorption by $O_2$ in the Schumann-Runge bands, 175-205 nm (see Fig. 3.7), occurs predominantly in the mesosphere and upper stratosphere. Because of the structured nature of these absorption bands, wavelengths near band minima penetrate deeper into the stratosphere than wavelengths near band maxima. $O_2$ also possesses a number of overlapping absorption features in the 200-245 nm, collectively known as the Herzberg continuum. However, these bands are quite weak, and ozone becomes the dominant attenuator of incoming solar radiation in the 210-300 nm region. The near-UV absorption spectrum of $O_3$ (known as the Hartley band) is shown in Figure 3.8; its effect on the solar flux can clearly be seen in Figure 3.6. Also apparent in Figure 3.6 is a narrow "window" of radiation located

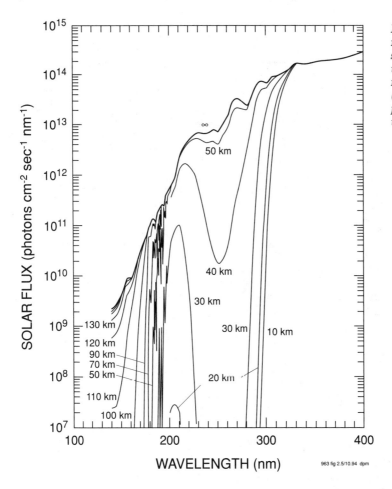

**Figure 3.6.** Solar fluxes in the atmosphere as a function of wavelength for different altitudes (K. Minschwaner, private communication).

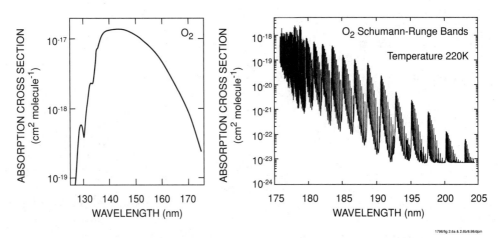

**Figure 3.7.** Left: Absorption spectrum of $O_2$ in the Schumann-Runge continuum (from Watanabe et al., 1953). Right: Molecular oxygen absorption cross sections for a temperature of 220 K in the wavelength region of the Schumann-Runge bands (from Kockarts, 1994, using the technique described by Minschwaner et al., 1993). The band structure is clearly visible and the effect of the underlying Herzberg continuum is apparent at wavelengths greater than 190 nm. Kockarts' approximation can be safely used down to a solar attenuation of the order of $10^{-10}$.

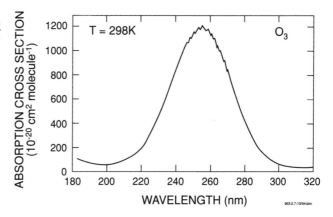

*Figure 3.8.* UV absorption spectrum of $O_3$ at 298 K (Molina and Molina, 1986).

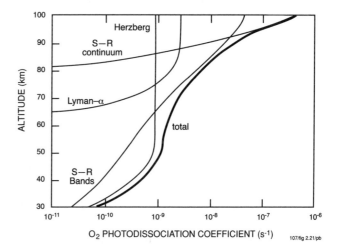

*Figure 3.9.* Contributions to the photolysis rate of $O_2$ at various altitudes due to Schumann-Runge (bands and continuum), Herzberg continuum and Lyman-$\alpha$ atomic radiation for overhead sun (from Goody, 1995).

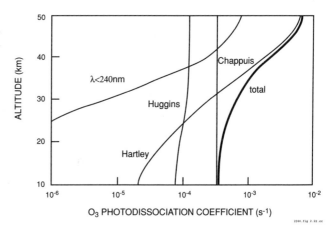

*Figure 3.10.* Contributions of different absorption bands to ozone photodissociation for overhead sun (from Goody, 1995).

at 200-225 nm between the Schumann-Runge bands of $O_2$ and the Hartley band of $O_3$, which is allowed to penetrate to the lower stratosphere. Photolysis of many key stratospheric species, including the chlorofluorocarbons and nitrous oxide, occurs to a large extent in this wavelength region. Wavelengths longer than about 290 nm are allowed to penetrate to the Earth's surface, and are only partially attenuated by ozone absorption in the Huggins bands between 310 and 400 nm, and the Chappuis bands between 400 and 850 nm. This attenuation by ozone limits photochemical activity in the troposphere to wavelengths longer than about 290 nm. The photolysis of ozone, which drives tropospheric chemistry, is particularly sensitive to the overhead column of stratospheric ozone. Figures 3.9 and 3.10 show contributions to the photolysis rates of $O_2$ and $O_3$ in the various absorption bands as a function of altitude. The contribution marked Lyman-$\alpha$ is due to the emission by atomic hydrogen in the Sun's spectrum. Due to a coincidence with a minimum in the oxygen absorption spectrum, Lyman-$\alpha$ radiation penetrates relatively far into the atmosphere.

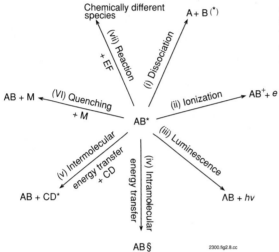

**Figure 3.11.** Illustration of photochemical and photophysical processes that can occur upon electronic excitation of a molecule (from Wayne, 1991; reprinted by permission of Oxford University Press).

## 3.3 Photophysical and Photochemical Processes
### 3.3.1 General Considerations

Upon absorption of a UV or visible photon, the absorbing species is, in general, promoted to an excited electronic state. Once in this excited state, any of a number of processes can occur (as shown in Fig. 3.11). The excited species can re-emit the photon (via fluorescence or phosphorescence) and return to a lower electronic state; it can collide with another gas, and transfer energy to the collision partner (quenching); it can undergo chemical reaction with a collision partner; or it can undergo unimolecular chemical change (by photodissociation or photoisomerization, for example). The probability that any of these processes will occur upon absorption of a photon is referred to as the quantum yield (usually designated as $\varphi$) for that process. By definition, the sum of the quantum yields for all possible occurrences following photon absorption is unity. The most common photochemical event of relevance to atmospheric chemistry is photodissociation. For example, photolysis of $NO_2$ in the near UV occurs with a quantum yield near unity,

$$NO_2 + h\nu \rightarrow NO + O\,(^3P) \tag{3.14}$$

and is the dominant source of tropospheric ozone

$$O\,(^3P) + O_2 + M \rightarrow O_3 + M \tag{3.2}$$

However, at wavelengths above about 400 nm, the photon energy is insufficient to break the NO-O bond of $NO_2$, and the observed quantum yield for photodissociation decreases rapidly to zero. Thus $NO_2$ excited by visible radiation in the atmosphere is susceptible to quenching via collisional processes.

### 3.3.2 Atmospheric Photolysis Rates

In order to assess the importance of an atmospheric photochemical reaction, it is important that its rate be known. For example, in the photochemical conversion of $NO_2$ to NO, referred to above,

$$NO_2 + h\nu \rightarrow NO + O\,(^3P) \tag{3.14}$$

the rate of disappearance of $NO_2$ (or, equivalently, the rate of appearance of products NO and O) is parameterized as follows:

$$-\frac{d\,[NO_2]}{dt} = j\,[NO_2] \tag{3.15}$$

where square brackets indicate concentration and $j$ is the unimolecular rate constant for photolytic loss of $NO_2$. In the atmosphere, $j$ will be determined by the number of photons available (the solar flux $q$; see Box 3.2), the ability of the molecule to absorb these photons (the absorption cross section $\sigma_a$), and the probability that the molecule will be photochemically destroyed following photon absorption (*i.e.*, the quantum yield for photodissociation $\varphi$). This product is integrated over all wavelengths for which all three terms are nonzero:

$$j = \int_\lambda \sigma_a\,(\lambda)\,\varphi\,(\lambda)\,q\,(\lambda)\,d\lambda \tag{3.16}$$

As defined above, the absorption cross sections and quantum yields are fundamental properties of the molecule under investigation, and can be determined via laboratory experiment. In general, the absorption cross sections will vary with temperature. The quantum yields can also vary with temperature and may also vary with pressure if a competing collisional process, such as quenching, can occur. The solar flux, and hence the rate of photochemical processes, will of course vary as a function of altitude (see Fig. 3.6) and with solar zenith angle. More general considerations regarding the calculation of photolysis rates are given in Box 3.2. An important effect is the shielding of ultraviolet radiation by overhead ozone around 310 nm. A list of some atmospherically important photolysis processes is given in Table 3.1, along with the altitude range, wavelength, and the approximate rate at which they occur. Rates of a number of photolysis reactions are also given in Appendix N.

## 3.4 Chemical Reactions
### 3.4.1 Generalities

Reaction schemes can in general be broken down into the categories elementary and complex. Complex schemes, such as those describing the decomposition of organic molecules, involve many consecutive elementary reactions, and simple laws may not be available to describe the temporal behavior of all the reactants and products. The entire atmosphere can be regarded as such a system from a chemical standpoint. Fundamentally, though, all complex schemes can be broken down into a series of individual or elementary steps, and if one knows the rates at which all of these occur, it is possible to predict what will happen. A reaction for which a simple mechanism can be written, and in which no unrealistic bond breaking or "molecular gymnastics" takes place, can be termed elementary. These processes usually take place under the single collision regime, in which the interaction time is of the order of the duration of a molecular collision ($\sim$1 ps). In the atmosphere, where temperatures and total gas

### Box 3.2 Calculation of the Photodissociation Frequency

The photodissociation frequency $j$ [s$^{-1}$] of molecule $X$ at altitude $z$, for a solar zenith angle $\chi$ is given by the integral over wavelength $\lambda$

$$j(X; z, \chi) = \int_\lambda \sigma_a(X; \lambda) \, \varphi(X; \lambda) \, q(\lambda, z, \chi) \, d\lambda \qquad (1)$$

where $\sigma_a(X; \lambda)$ is the absorption cross section (cm$^2$), $\varphi(X; \lambda)$ the quantum yield for photodissociation and $q(\lambda, z, \chi)$ the solar actinic flux (photons cm$^{-2}$ s$^{-1}$ nm$^{-1}$, if the wavelength interval $d\lambda$ is expressed in nm). This integral is calculated over the spectral domain for which solar radiation is available and radiative energy is sufficiently high to photolyze the molecule. The solar actinic flux at altitude $z$ and for solar zenith angle $\chi$ can be expressed as the product of the extraterrestrial flux $q_\infty(\lambda)$ and an effective transmission function $T_r$

$$q(\lambda, z, \chi) = q_\infty(\lambda) \, T_r(\lambda, z, \chi) \qquad (2)$$

Assuming that the attenuation of solar radiation is due only to the absorption by atmospheric gases (*e.g.*, ozone and molecular oxygen) and by aerosols, the transmission can be expressed by the simple Beer-Lambert law

$$T_r(\lambda, z, \chi) = \exp\left\{-\left[\tau(O_3; \lambda, z, \chi) + \tau(O_2; \lambda, z, \chi) + \tau_{\text{aerosols}}(\lambda, z, \chi)\right]\right\} \qquad (3)$$

where $\tau$ represents the (dimensionless) optical depth. In the case of the ozone absorption, for example, and assuming a plane-parallel atmosphere

$$\tau(O_3; \lambda, z, \chi) = \sec\chi \int_z^\infty \sigma_a(O_3; \lambda, z') \, n(O_3; z') \, dz' \qquad (4)$$

where $n(O_3; z)$ is the ozone number density at altitude $z$. Note that cross-section $\sigma_a$ is assumed to be altitude dependent to account for its sensitivity to temperature.

If radiative transfer in the atmosphere is affected by scattering processes and reflections, the calculation of the radiation field (or of the transmission function) becomes more complex. In this case, the angular dependence of the diffuse radiation must be derived by solving the integro-differential radiative transfer equation (see Box 15.2) in which the scattering phase function (probability that a photon incoming from a given direction is scattered in another given direction) is specified. For gases (Rayleigh scattering), the scattering probability is small, and is important only in the lower atmosphere where the air density is greatest. For aerosols and cloud particles (Mie scattering), the phase functions are complex and depend on the size distribution of these particles. Note that photons may be scattered several times as they propagate through the atmosphere (multiple scattering).

Integral (1) is often expressed by the finite sum over $N$ wavelength intervals

$$j(X; z, \chi) \simeq \sum_i \sigma_i(X) \, \varphi_i(X) \, Q_{i,\infty} \, T_{r,i}(z, \chi) \qquad (5)$$

where $\sigma_i \varphi_i$ represents an average of product $\sigma_a \varphi$ over wavelength interval $\Delta \lambda_i$, and

$$Q_{i\infty} = \int_{\Delta\lambda_i} q_\infty(\lambda) \, d\lambda \qquad (6)$$

Appendix M provides values of $Q_{i,\infty}$ for wavelength intervals of 5 nm. Values of the absorption cross sections and quantum yields for various molecules can be found in DeMore *et al.* (1997).

The illumination of the Earth and hence the local zenith angle $\chi$ (angle between the local vertical and the direction toward the center of the Sun) varies with the time of the day, season, and geographic location. At latitude

$\phi$, and for a solar declination $\delta$, the solar zenith angle varies with the local hour angle $t_h$ according to

$$\cos \chi = \sin \phi \sin \delta + \cos \delta \cos \phi \cos t_h \qquad (7)$$

Note that, at local noon, $t_h = 0$, while at sunrise or sunset

$$t_h = \pm \cos^{-1}(-\tan \phi \tan \delta)$$

The solar declination (angle between the Sun's direction and the Earth's equatorial plane) varies between +23.45° on 21 June and −23.45° on 21 December, and is equal to zero at the spring and fall equinoxes. To a first-order approximation, it varies with the daynumber $d_n$ of the year ($d_n = 1$ for 1 January) according to

$$\delta(\text{radians}) = -0.4 \cos\left[\frac{2\pi (d_n + 10)}{365}\right] \qquad (8)$$

Neglecting the so-called equation of time (which accounts for nonuniformities in the apparent angular speed of the Sun in the sky), the local hour angle (expressed in radians) is given by

$$t_h = \pi \left[\frac{t(\text{GMT})}{12} - 1 + \frac{\text{lon}}{180°}\right] \qquad (9)$$

where $t$ (GMT) is the Greenwich meridian time (expressed in hours) and lon is the longitude (expressed in degrees).

Finally, it should be noted that the approximation referring to a plane parallel atmosphere (see Eq. A.4) is appropriate only for solar zenith angles smaller than 75°. For larger zenith angles, the effect of the Earth's sphericity should be taken into account, and the calculation becomes more complex. For an absorber that is uniformly distributed in the atmosphere (such as molecular oxygen) the "sec $\chi$" factor can be replaced by the Chapman function $Ch(\chi, x)$. According to Smith and Smith (1972), this function can be approximated as follows if

$$x = \frac{a + z}{H}$$

$z$ being the altitude, $a$ the Earth's radius, and $H$ the atmospheric scale height:

For $\chi \leq 90°$:

$$Ch(\chi, x) = \sqrt{\frac{\pi x}{2}} \operatorname{erfc}\left(\sqrt{\frac{x}{2}} \cos \chi\right) \exp\left(\frac{x}{2} \cos^2 \chi\right) \qquad (10)$$

For $\chi > 90°$:

$$Ch(\chi, x) = \sqrt{2\pi x} \left\{\sin \chi \exp[x(1 - \sin \chi)] - \frac{1}{2} \exp\left(\frac{x}{2} \cos^2 \chi\right) \operatorname{erfc}\left(\sqrt{\frac{x}{2}} \cos \chi\right)\right\} \qquad (11)$$

where $\operatorname{erfc}(y) = 1 - \operatorname{erf}(y)$ is the complementary error function.

densities are fairly low, this is usually the case, and most atmospheric interactions are described in terms of elementary, bimolecular (*i.e.*, involving only two species) collisions. Occasionally, three-body (termolecular) reactions occur, and these invariably involve the participation of $N_2$ or $O_2$, the bath gases in the atmosphere. Such reactions are very important because they can lead to the formation of stable reservoir species.

**Table 3.1**
*Rates and mechanisms of
some atmospherically important photolysis processes*

| Process | Altitude | Wavelength (nm) | Approx. Rate |
|---|---|---|---|
| $O_2 + h\nu \to O(^1D) + O(^3P)$ | >50 km | < 174 | $10^{-12}$ s$^{-1}$ |
| $O_2 + h\nu \to O(^3P) + O(^3P)$ | stratosphere and above | < 246 | $10^{-10}$ s$^{-1}$ |
| $O_3 + h\nu \to O(^1D) + O_2(^1\Delta)$ | all | < 310 | $10^{-5}$ s$^{-1}$ at 10 km $10^{-3}$ s$^{-1}$ at 40 km |
| $NO_2 + h\nu \to NO + O(^3P)$ | all | 250-400 | $8 \times 10^{-3}$ s$^{-1}$ at surface $10^{-2}$ s$^{-1}$ in stratosphere |
| $NO_3 + h\nu \to NO_2 + O(^3P)$ | all | 400-625 | 0.2 s$^{-1}$ at surface 0.25 s$^{-1}$ in stratosphere |
| $NO_3 + h\nu \to NO + O_2$ | all | 585-625 | ~0.02 s$^{-1}$ independent of altitude |
| $CF_2Cl_2 + h\nu \to CF_2Cl + Cl$ | lower stratosphere | <220 | $10^{-8}$ s$^{-1}$ at 30 km $10^{-6}$ s$^{-1}$ at 50 km |
| $N_2O_5 + h\nu \to NO_3 + NO_2$ | stratosphere | <320 | $2 \times 10^{-5}$ s$^{-1}$ at 30 km |
| $ClONO_2 + h\nu \to$ products | stratosphere | <320 | $6 \times 10^{-5}$ s$^{-1}$ |
| $CH_2O + h\nu \to HCO + H$ $\to H_2 + CO$ | all | <350 | $3 \times 10^{-5}$ s$^{-1}$ $4 \times 10^{-5}$ s$^{-1}$ |

### 3.4.2 Bimolecular Reactions

The important bimolecular reactions between atmospheric species are typically metathesis reactions of the type:

$$AB + C \to A + BC \tag{3.17}$$

where $C$ can be an atom or molecule. Furthermore, reactions are favored in which there is a smooth transition from the reactants to the products. In the current JPL evaluation of rate coefficients for use in stratospheric modeling (DeMore et al., 1997) approximately 300 bimolecular reactions are listed. A fully explicit tropospheric model would contain around 10,000 individual reactions. The rates at which chemical reactions occur are parameterized in terms of rate coefficients. For a simple bimolecular reaction (such as the generic one above), the rate of disappearance of the reactants, which is also equal to the rate of appearance of products, is given by:

$$-\frac{d[AB]}{dt} = \frac{-d[C]}{dt} = \frac{d[A]}{dt} = \frac{d[BC]}{dt} = k[AB][C] \tag{3.18}$$

where the square brackets denote the concentrations of the reactants, in appropriate units, and $k$ is the reaction rate coefficient, with units (concentration)$^{-1}$(time)$^{-1}$. In general, a bimolecular rate coefficient is a function of temperature, but it is usually independent of pressure.

An upper limit to the rate of a bimolecular reaction can be obtained from the gas-kinetic collision rate. This can be estimated from the kinetic theory of gases, and is around $2 \times 10^{-10}$ cm$^3$ molecule$^{-1}$ s$^{-1}$ at room temperature. In practice, most reactions are slower than this for two reasons. First, there are entropy considerations, which can be of the form of restrictions on the geometry of the reaction path, or steric limitations caused by the size of groups participating in the reaction. Second, many reactions have an activation energy ($E_{act}$) that must be overcome before the reaction can proceed. The rate coefficient is usually represented in terms of the Arrhenius expression $k = A \exp(-E_{act}/RT)$. Rate coefficients for a number of reactions occurring in the background troposphere and stratosphere are given in Appendix E (from DeMore et al., 1997).

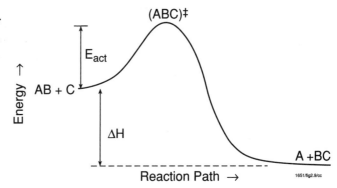

**Figure 3.12.** Depiction of the transition state for a bimolecular reaction.

Reactions are often described in terms of an activated complex or transition state, which can be envisaged as some particular molecular configuration defining a separation between reactants and products. If the atoms pass through this configuration, then the reaction is assumed to occur irreversibly. This is depicted schematically in Figure 3.12. The activation energy is then identified as the energy required to reach the transition state. Alternatively, it can be defined as the mean energy of a reacting molecule minus the mean thermal energy of a molecule.

*3.4.3 Equilibrium Considerations*

The driving force behind any chemical reaction is the free energy change. Quite simply, for a change to occur spontaneously, the free energy change must be negative. The free energy is given by:

$$\Delta \widehat{G} = \Delta \widehat{H} - T \Delta \widehat{S} \qquad (3.19)$$

where $\Delta \widehat{H}$ is the enthalpy (heat) change, and $\Delta \widehat{S}$ the entropy change. If $\Delta \widehat{H} > 0$, the reaction is endothermic, while if $\Delta \widehat{H} < 0$, the reaction is exothermic. For atmospheric reactions the enthalpy change usually dominates and endothermic reactions are not very common. For any reaction system the equilibrium constant is given by $\Delta \widehat{G} = -RT \ln(K_p)$, where $K_p$ is given in terms of standard states (1 atmosphere pressure for gases). Thus, for the generic bimolecular gas phase reaction, $K_p = P_A P_{BC}/P_{AB} P_C$. Since the partial pressure and the concentration are linearly related, an equivalent expression in terms of concentration can also be written, $K_c = [A][BC]/[AB][C]$. [If there are the same number of species on both sides of the chemical equation, then $K_p = K_c$; if the change in the number of species is $\Delta n$, then $K_c = K_p (RT)^{\Delta n}$.] However, at equilibrium the rates of forward and reverse reactions must be equal, and so:

$$k_f[AB][C] = k_r[A][BC] \qquad (3.20)$$

Therefore, $K_c = k_f/k_r$, and the equilibrium constant can be expressed in terms of the individual rate coefficients for the forward and reverse reactions. This can be visualized in terms of the barrier to reaction (see Fig. 3.12), in which the activation energy is identified with the enthalpy associated with achieving the transition state, while the pre-exponential factor $A$ is closely related to the entropy of activation, $\Delta \widehat{S}^{\ddagger}$:

$$A = \frac{kT}{h} \exp\left(\Delta \widehat{S}^{\ddagger}/R\right) \tag{3.21}$$

Formally, one can describe the activation energy as the difference between the mean enthalpy of reacting molecules and the mean enthalpy of the bulk molecules. If a very restricted geometry is required to form products (tight transition state), the lower $\Delta S^{\ddagger}$ will be, and this slows the reaction down (low $A$ factor). Rapid reactions between free radicals are usually not hindered by geometric considerations (loose transition states) and have high pre-exponential factors. The calculation of rate coefficients based on the knowledge of thermochemical and molecular parameters was developed by Benson and co-workers, and accounts of the methods are detailed in Benson's books (1960; 1976).

### 3.4.4 Magnitudes of Activation Energies

Many reactions of importance in the atmosphere have small activation energies. This is understandable, because at thermal energies even a small barrier of 12-16 kJ mol$^{-1}$ can slow a reaction down dramatically [$\exp(-2000/T) = \exp(-6.7) = 0.0013$]. For example, the reaction of OH with $CH_4$, Reaction (3.5), has an activation energy of 15 kJ mol$^{-1}$, leading to a rate coefficient of about $6.5 \times 10^{-15}$ cm$^3$ molecule$^{-1}$ s$^{-1}$ at 298 K. A rate coefficient this small would normally limit the importance of the reaction, except for the fact that it is the major removal process for $CH_4$ in the troposphere, and the mixing ratio of $CH_4$ thus builds up to about 1.7 ppm. Generally, the activation energy increases with the strength of the bond being broken.

Many of the reactions between free radicals in the atmosphere occur with little or no activation barrier, and many have been determined to have *effective* activation energies that are less than zero. This can be easily interpreted in the terms described above; it simply means that the average energy of reacting molecules is less than the energy of the bulk gas. This normally means that one of the modes of internal energy (translation, rotation, or vibration) inhibits the probability of reaction, usually by affecting the entropy of reaction. For instance, if the molecules have to achieve a particular orientation in space, rotational energy will inhibit the reaction. If the molecules have to "stick together" and form a complex, then translational energy will be detrimental (see next section). On a molecular level, the existence of a negative activation energy is usually associated with an attractive potential energy surface, upon which there is a gradual decrease in energy as the molecules approach. Often, a molecular complex is formed, but this is not necessary, and the complex need not be able to be stabilized or have an independent existence past the duration of the collision (a few psec). For example, in the case of the reaction $HO_2 + NO \rightarrow OH + NO_2$, it can be imagined that a transient HOONO molecule is formed, which exists for maybe a few rotational periods before dissociating to OH + $NO_2$. The acid HOONO (peroxynitrous acid) is known from liquid-phase chemistry, but has never been isolated in the gas phase. However, not all reactions for which $E_{\text{act}} < 0$ can be interpreted in terms of a known intermediate. More detailed discussions on the theory of reaction rates can be found in general physical chemistry textbooks (*e.g.*, Atkins, 1986) or in more specialized monographs such as those by Smith (1980) or Pilling and Seakins (1995).

## 3.4.5 Unimolecular and Termolecular Reactions

It is convenient to treat these reactions together, since they mostly involve association and dissociation mechanisms. Not all reactions are simple bimolecular collisions. Some of the most important reactions in atmospheric chemistry are actually termolecular in nature

$$A + B + C \rightarrow \text{prod} \tag{3.22}$$

This means that the rate expression is of the form:

$$d\,(\text{prod})/dt = k\,[A][B][C] \tag{3.23}$$

Due to restrictions of density in the atmosphere, and the overall paucity of three-way collisions, this invariably means that one of the reactants has to be either $O_2$ or $N_2$. Nitrogen is very unreactive, and the only reaction that really ever has to be considered in which all three molecules undergo chemical change is the termolecular reaction between NO and $O_2$

$$NO + NO + O_2 \rightarrow NO_2 + NO_2 \tag{3.24}$$

However, there is a whole class of reactions in which $N_2$ and $O_2$ do not participate chemically, but serve to remove excess energy in an association reaction. These reactions are vitally important in the atmosphere since they lead to the formation of reservoir species that remove free radicals that have the potential to destroy ozone. Some examples are given in Appendix F. The general form of the interaction is:

$$A + B + M \rightarrow AB + M \tag{3.25}$$

where $M$ is a third body (usually $O_2$ or $N_2$) that carries away excess energy. Mechanistically, it can be imagined that an excited $AB$ molecule is formed with large amounts of vibrational energy that must be removed by collisions before the $AB$ molecule dissociates back to $A$ and $B$.

In such cases the effective rate expression actually varies with pressure, from being termolecular at low pressures to being bimolecular at high pressures. This is a result of the fact that, as the frequency of collisions increases, the stabilization of $AB$ occurs almost instantaneously, and so further increase in pressure leads to no increase in rate. At the low pressure limit the frequency of collisions is not enough to stabilize the newly formed molecules before they redissociate, and the overall rate increases linearly with pressure. In between, there exists a falloff region where the order of the reaction changes from 3 to 2. Typical behavior of a pressure-dependent reaction is illustrated in Figure 3.13.

The earliest treatment of this was by Hinshelwood and Lindemann, who considered the excited molecules, denoted $(AB)^*$, to be in steady state

$$A + B \rightarrow (AB)^* \tag{3.26a}$$

$$(AB)^* \rightarrow A + B \tag{3.26b}$$

$$(AB)^* + M \rightarrow AB + M \tag{3.26c}$$

Then the effective rate coefficient can be expressed in terms of the elementary steps as:

$$k = k_a\,k_c\,[M]/(k_b + k_c\,[M]) \tag{3.27}$$

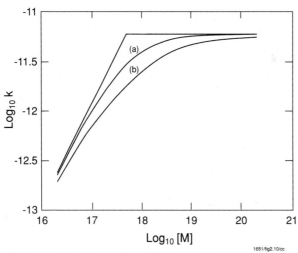

**Figure 3.13.** Schematic depiction of the falloff curve for a unimolecular reaction with $k_0 = 1.2 \times 10^{-29}$ $cm^6$ $molecule^{-2}$ $s^{-1}$ and $k_\infty = 6.0 \times 10^{-12}$ $cm^3$ $molecule^{-1}$ $s^{-1}$. (a) Lindemann representation; (b) Troe representation with $F_c = 0.6$. The straight lines are the asymptotes for low- and high-pressure behavior (from Tyndall et al., 1998; reprinted with permission from J. Phys. Chem., copyright 1998, American Chemical Society).

The termolecular rate coefficient at low pressure is thus given by $k_0 = k_a k_c / k_b$, while the bimolecular high-pressure limiting rate coefficient is $k_\infty = k_a$. One can therefore recast the Hinshelwood-Lindemann equation as

$$k = k_0 [M] k_\infty / (k_0 [M] + k_\infty) \qquad (3.28)$$

It should be noted that, due to the relationship between forward and reverse rate coefficients and the equilibrium constant, molecular dissociation also follows the same kind of falloff behavior as addition reactions. These "unimolecular" reactions follow bimolecular kinetics at low pressure (rate is proportional to bath gas pressure and concentration of $AB$) where collisions are infrequent, and apparent unimolecular kinetics at high pressure, where collisional activation is rapid enough to populate excited levels faster than the molecules can dissociate.

The Hinshelwood-Lindemann mechanism gives a qualitative dependence on pressure as observed in experiments, but in reality several other factors have to be considered. First, the collisions of $(AB)^*$ with bath gas $M$ do not necessarily remove all the excess energy at one time. In fact, the average energy removed is typically of the order of $kT$, while the bond strengths of molecules are many times this value (otherwise they would keep falling apart!). Also, the internal energy of an excited molecule is constantly flowing between its various quantized modes, and this means that the simple collisional picture of deactivation is complicated further since the energy must become localized in order for a bond to break. The mathematical model for this behavior was formulated by Kassel and was developed for real systems by Marcus and co-workers, Troe and co-workers, and others. Many parameterized versions have been presented in order to facilitate the description of such reactions for practical purposes, but the one that is most widely used is the factorization method of Troe (1979). This takes a semiempirical fit to the curves of Kassel and allows experimental data to be fitted in terms of 3 or 4 adjustable parameters. The most-used form of the equation is:

$$k = \frac{k_\infty k_0 [M]}{k_\infty + k_0 [M]} F_c^{\left(1+\left\{\log_{10}(k_0[M]/k_\infty)^2\right\}\right)^{-1}} \qquad (3.29)$$

where $F_c$ is known as the broadening factor. A good representation of the pressure dependence of many atmospheric reactions can be obtained with $F_c$ set at 0.6. A representation of the pressure dependence predicted by the Hinshelwood-Lindemann and Troe formulae is included in Figure 3.13.

In its simplest form, the Troe equation is used purely as an empirical representation of measured data in a form that can be easily calculated and used.

With a little extra work, it allows complete falloff curves to be generated from readily available molecular parameters such as bond enthalpies and rotational and vibrational frequencies.

Several important reactions display mixed bimolecular and termolecular behavior; that is, the rate coefficient is nonzero at low pressure but also increases with pressure. The reactions OH + CO, HO$_2$ + HO$_2$, and OH + HNO$_3$ are the most important ones (see Appendix F). The interpretation of this phenomenon is that an intermediate exists that can be stabilized by collisions, but that there is also a low-energy direct pathway to the observed products. The pressure dependence arises from competition in the activated complex between collisional stabilization and fragmentation to products.

## 3.4.6 Rates of Reactions in Solution

Homogeneous liquid-phase reactions play a large part in the oxidation of trace gases in droplets. Many of the concepts encountered are similar to those in the homogeneous gas phase (*e.g.*, activation energies). Two major differences are encountered, however, that affect the energetics. The first difference is that diffusion effects are much more severe in the liquid phase. First, motion through the droplet is restricted since the density of the liquid phase is much higher than the gas phase. On the other hand, the slow rate of diffusion prevents reaction partners from separating, which facilitates the reactions. This so-called cage effect is also responsible for reducing the quantum yields of many solution-phase photolytic processes compared to the gas phase, since the photolysis products can recombine if they have sufficient internal energy to overcome any activation energy. The maximum rate coefficient with which two neutral species can react in solution is termed the diffusion rate.

Second, liquid-phase mechanisms also have to take account of the presence of ions in solution, particularly since water has a high dielectric constant and is thus a good ionizing medium. Many atmospheric ions are formed from the reversible, dissociative loss of a proton from stable molecules that are often transferred from the gas phase.

$$H_2O \rightleftharpoons H^+ + OH^- \tag{3.30}$$

$$H_2O_2 \rightleftharpoons H^+ + HO_2^- \tag{3.31}$$

$$H_2SO_4 \rightleftharpoons H^+ + HSO_4^- \tag{3.32}$$

For an ionizing species H$A$, the equilibrium between the ionic and neutral forms is given by:

$$K_a = [H^+][A^-]/[HA] \tag{3.33}$$

The p$K_a$ value, given by $-\log_{10}K_a$, is the pH at which half the species exists in its ionized form. The degree of ionization is dependent on the pH of the solution, that is, the concentration of H$^+$, since pH = $-\log_{10}$[H$^+$]. Atmospheric droplets normally have a pH less than 5.6, which is the value obtained from the dissolution of CO$_2$, an acidic gas (see Box 4.2)

$$CO_{2(g)} + H_2O_{(l)} \rightleftharpoons H_2CO_{3(l)} \tag{3.34}$$

$$H_2CO_3 \rightleftharpoons H^+ + HCO_3^- \tag{3.35}$$

The dissolution of more acidic gases, such as H$_2$SO$_4$ and HNO$_3$, normally lowers the pH further than this.

Other examples of rapid liquid-phase reactions include electron transfer reactions that interconvert species,

$$NO_3 + Cl^- \rightarrow NO_3^- + Cl \tag{3.36}$$

$$NO_2^- + OH \rightarrow NO_2 + OH^- \tag{3.37}$$

and reactions of ions with dissolved neutral species

$$HSO_3^- + O_3 \rightarrow HSO_4^- + O_2 \tag{3.38}$$

In extremely concentrated solutions, kinetic salt effects exist that cause the rate to depend on the concentrations.

### 3.4.7 Multiphase Chemistry

A further class of reactions that has received much attention of late is that in which a gas-phase molecule reacts with an aerosol particle or a liquid droplet. Although heterogeneous chemistry strictly refers to a reaction occurring at the interface between two phases, in atmospheric chemistry the term is often used to describe reactions occurring inside, or at the surface of, a droplet, or uptake into a droplet. The rate of reaction is derived from the rate of collisions with the particle, which is a function of the surface area, and the reaction probability $\gamma$. Two parameters are commonly used to describe uptake on particles. The accommodation coefficient, $\alpha$, is the probability of physical uptake (which may be reversible) at a surface. As the droplet (or, strictly, the surface layer) becomes saturated, $\alpha$ tends toward zero. The reaction probability, $\gamma$, combines physical uptake and chemical reaction, to give the overall loss rate. The formalism used to describe uptake is again based on the kinetic theory of gases. At low pressures (the diffusion limit) the rate of collision of a molecule $A$ with a defined unit area is $\bar{c}\,[A]/4$, where $\bar{c}$ is the mean speed, or $(8kT/\pi\mathcal{M})^{0.5}$, $\mathcal{M}$ is the mass, $T$ the temperature, and $k$ the Boltzmann constant. Thus, from a knowledge of $\alpha$ and either the surface density (area per unit volume) or the number of particles per unit volume and their average radius, one can form a rate coefficient analogous to those used for gas-phase reactions. For reaction of a gas $A$ with an aerosol of surface area density $\mathcal{A}$, that is, number density $N$ and mean radius $a$:

$$-\frac{d[A]}{dt} = \frac{\alpha\bar{c}\mathcal{A}[A]}{4} = \alpha\bar{c}\pi a^2 N[A] \tag{3.39}$$

since the surface area of a sphere is $4\pi a^2$. This equation adequately describes the uptake when the mean free path of the gas-phase molecules is larger than the radius $a$ of the particle, for example, at the low pressures of the stratosphere. At higher pressures and/or larger particles one also has to consider the bulk diffusion of the molecules. Schwartz has solved the three-dimensional diffusion equation and shown (Schwartz and Freiberg, 1981) that the rate of diffusion of a gas-phase molecule to a particle of radius $a$ is given by:

$$-\frac{d[A]}{dt} = 4\pi a D_g N[A] \tag{3.40}$$

where $D_g$ is the gas-phase diffusion constant. Values of $D_g$ at 1 atm pressure range from around 1 cm$^2$ s$^{-1}$ for atoms and small free radicals to less than 0.1 for larger molecules. $D_g$ varies inversely with the total pressure.

If we ascribe a characteristic time $\tau_I$ to interfacial transport, and a time $\tau_{GD}$ to gas-phase diffusion, then the overall time for uptake, $\tau_G$, is given by the sum of $\tau_I$ and $\tau_{GD}$:

$$\tau_G = \tau_I + \tau_{GD} \tag{3.41}$$

Since the times are inversely related to the rate coefficients, Eqs. (3.39) and (3.40) may be combined to give the overall rate for diffusion and uptake:

$$-\frac{d[A]}{dt} = \frac{4\pi a^2 \alpha \bar{c} D_g N[A]}{a\alpha\bar{c} + 4D_g} \tag{3.42}$$

Fuchs and Sutugin (1970) derived an equation analogous to Eq. (3.42) that takes into account the fact that uptake depletes the more energetic molecules:

$$-\frac{d[A]}{dt} = \frac{\alpha \pi a^2 \bar{c} N [A]}{1 + \frac{3\alpha (1 + 0.47 K_n)}{4 K_n (1 + K_n)}} \qquad (3.43)$$

where $K_n$ is the Knudsen number $= 3D_g/a\bar{c}$. Equation (3.43) is appropriate when considering tropospheric particles. In Chapter 4 the concepts of reaction in a droplet and diffusion in the gas and liquid phases are discussed further.

### 3.4.8 Liquid Phase Diffusion and Reaction

Once inside a liquid droplet, a molecule will diffuse toward the center until it encounters a reactive partner. The concentration profiles inside the droplet and the time rates at which liquid- and gas-phase concentrations change are a complex function of the diffusion rate and the reaction rate. The diffusion equation can again be solved to give the rate of reaction inside the droplet. If the pseudo-first-order rate of the liquid-phase reaction is given by $k_R$ (s$^{-1}$) and the liquid-phase diffusion rate by $D_L$ (cm$^2$ s$^{-1}$), then a useful concept is the dimensionless parameter $\tilde{q} = a(k_R/D_L)^{0.5}$, which describes the relative rates of diffusion and chemical reaction. Using $\tilde{q}$ as the variable, the overall rate of liquid-phase reaction is given by

$$-\frac{d[A]}{dt} = 4\pi a^3 \mathcal{H} k_R RT \left( \frac{\tilde{q} \coth \tilde{q} - 1}{\tilde{q}^2} \right) N[A] \qquad (3.44)$$

where $\mathcal{H}$ is the Henry's Law solubility constant (Section 4.8.1). This equation has two limiting cases. If liquid-phase diffusion is rapid compared to chemical reaction, then $\tilde{q} \to 0$, and the rate of reaction is given by:

$$-\frac{d[A]}{dt} = \frac{4}{3} \pi a^3 \mathcal{H} RT k_R N[A] \qquad (3.45)$$

If the chemical reaction is very rapid, then

$$-\frac{d[A]}{dt} = 4\pi a^2 \mathcal{H} RT (D_L k_R)^{\frac{1}{2}} N[A] \qquad (3.46)$$

In this case the reaction occurs before diffusion has time to occur, and appears as a surface reaction [rate is proportional to $a^2$ in Eq. (3.46)]. The effects of uptake and liquid-phase reaction can be combined into one constant, $\gamma$, given at low pressure by

$$\frac{1}{\gamma} = \frac{1}{\alpha} + \frac{\bar{c}}{4\mathcal{H} RT (D_L k_R)^{\frac{1}{2}} (\coth \tilde{q} - \tilde{q}^{-1})} \qquad (3.47)$$

Equation (3.47) describes the size dependence of the reaction probability in terms of the droplet radius $a$, the solubility $\mathcal{H}$, the liquid-phase rate coefficient $k_R$, and the liquid-phase diffusion coefficient, $D_L$. The combined effects of gas-phase and liquid-phase reactions are discussed by Schwartz and Freiberg (1981) for tropospheric aerosols and clouds, and for stratospheric reactions on sulfuric acid aerosols by Hanson et al. (1994). Values of accommodation coefficients and reaction probabilities for atmospheric gases on various surfaces are given in Appendices G and H.

### 3.4.9 Kinetic Isotope Effects

Measurements of isotopes (different forms of the same element caused by different numbers of neutrons in the nucleus) have been exploited in atmospheric chemistry, because they can give information as to the source of the atmospheric emission (Box 5.1). The reaction rates of isotopically different forms can vary significantly, both in the atmosphere and within the source (thus varying the emission rates).

Kinetic isotope effects arise because otherwise identical species with different mass have different energy levels (vibrational and rotational), leading to differences in spectra and bond strengths. Kinetic isotope effects are found in both primary and secondary forms. In the primary isotope effect the isotopically labeled atom is involved directly in the reaction

$$OH + CH_4 \rightarrow H_2O + CH_3 \tag{3.5}$$

$$OH + CH_3D \rightarrow H_2O + CH_2D \tag{3.48}$$

Here, the reaction of the deuterated form is about 15% slower than the normal form. Most of this reduction is due to reaction at the C-D bond, since the C-D bond is somewhat stronger than the C-H bond and consequently less reactive. Note that the rate coefficient *per C-H bond* is found to remain almost constant, while the reduction is due to the presence of deuterium at one site.

In the case of secondary isotope effects, the isotopically labeled atoms are remote from the reaction site and exert an effect only through the dependence of the overall energy levels on mass. In this case, the isotope effects are much smaller, for example, for $^{12}CH_4$ and $^{13}CH_4$ reacting with OH the effect is approximately 0.5%

$$OH + {}^{12}CH_4 \rightarrow H_2O + {}^{12}CH_3 \tag{3.49}$$

$$OH + {}^{13}CH_4 \rightarrow H_2O + {}^{13}CH_3 \tag{3.50}$$

The range of observed isotope effects is thus very wide, and current computational methods are not able to reproduce the observed magnitudes of kinetic isotope effects, due to difficulties in defining the exact energetics of the transition state and the effects of quantum mechanical tunneling. Furthermore, some reactions that apparently proceed without a barrier exhibit primary isotope effects (*e.g.*, OH + $H_2O_2$, $D_2O_2$). Subtle energetic effects at the transition state thus play a role in determining the magnitudes of rate coefficients, and make exact calculation difficult, even at the 10% level.

Large isotope effects can also be found in solution-phase reactions. In fact, enzyme-catalyzed reactions can be very isotope specific. Such reactions are responsible for the large isotopic fractionation found in some biogeochemical systems, which make source identification through isotope signatures possible.

## 3.5 Catalytic Cycles

The concentration of ozone in the stratosphere is given by a balance between its production and loss terms. In a pure $O_2$-$N_2$ atmosphere four reactions could describe the steady state of ozone

$$O_2 + h\nu \rightarrow O + O \tag{3.1}$$

$$O + O_2 \rightarrow O_3 \tag{3.2}$$

$$O_3 + h\nu \rightarrow O_2 + O \tag{3.51}$$

$$O + O_3 \rightarrow O_2 + O_2 \tag{3.52}$$

This reaction scheme was formulated by Chapman in the late 1920s to explain the observed distribution of ozone in the stratosphere. However, the observed levels of ozone are lower than would be predicted by this simple mechanism. The stratosphere is a somewhat isolated region of the atmosphere, and it is difficult for enough material to be transported there to remove ozone continuously and irreversibly. However, certain trace species are present that can destroy ozone catalytically. In this case the species are constantly regenerated through multiple reactions, and so do not have to be present at levels comparable to ozone. In fact, the overall rate of ozone destruction only has to be comparable with the rate of the (O + $O_3$) reaction to be important. The general form of a catalytic cycle is

$$X + O_3 \rightarrow XO + O_2 \tag{3.53}$$
$$O + XO \rightarrow O_2 + X \tag{3.54}$$
$$\text{overall}: \quad O + O_3 \rightarrow O_2 + O_2 \tag{3.52}$$

Various pairs $X/XO$ that fulfill this requirement are H/OH, OH/$HO_2$, NO/$NO_2$, Cl/ClO, Br/BrO. In addition, catalytic cycles occur involving members of different families

$$Cl + O_3 \rightarrow ClO + O_2 \tag{3.55}$$
$$Br + O_3 \rightarrow BrO + O_2 \tag{3.56}$$
$$ClO + BrO \rightarrow Cl + Br + O_2 \tag{3.57}$$

Thermodynamically, in order for both parts of the catalytic cycle to be exothermic, the bond strength of $XO$ must lie in the range $170 < D(X\text{–}O) < 500$ kJ mol$^{-1}$. This is not a particularly stringent requirement, though, since most X-O bonds fall comfortably within this range. More detailed discussion of the catalytic cycles are to be found in Chapter 14 and in the chapters dealing with the relevant chemical families (Chapters 6, 7, and 8).

An important concept in the analysis of catalytic cycles is that of a rate-limiting (or rate-determining) step, which is the reaction in a cycle with the slowest overall reaction rate (or reactive flux) expressed in molecule cm$^{-3}$ s$^{-1}$. If a process is purely cyclic, then all the steps occur with the same rate, and there is no rate-limiting step. If, however, competition occurs for one of the reactants, then quantification of the cycle depends on identification of the critical reaction. Consider the photolysis of $NO_2$ in the lower stratosphere, which constitutes part of an ozone destruction cycle (with $X$=NO).

$$NO_2 + h\nu \rightarrow NO + O \tag{3.14}$$
$$O + O_2 + M \rightarrow O_3 + M \tag{3.2}$$
$$NO + O_3 \rightarrow NO_2 + O_2 \tag{3.58}$$
$$O + NO_2 \rightarrow O_2 + NO \tag{3.59}$$

The first three reactions, which all occur rapidly, lead to no net change in odd oxygen, and are said to constitute a "null cycle." However, the occurrence of Reaction (3.59) leads to catalytic destruction of odd oxygen. Since the rate of reaction is usually much slower than the others, it is the rate-limiting step in the catalytic cycle. In the lower stratosphere, the photolysis rate of $NO_2$ is about $1.5 \times 10^{-2}$ s$^{-1}$ × [$NO_2$]. Since the mixing ratio of O atoms at 25 km altitude is around $1 \times 10^{-12}$, corresponding to a number density of $1 \times 10^6$ molecule cm$^{-3}$, and the rate coefficient for Reaction (3.59) is $9 \times 10^{-12}$ cm$^3$ molecule$^{-1}$ s$^{-1}$, the rate of Reaction (3.59) is given by $(9 \times 10^{-12})$ × $(1 \times 10^6)$ × [$NO_2$]. Thus the rate of the catalytic cycle is approximately $6 \times 10^{-4}$ times that of the null cycle. The destruction of ozone by nitrogen species is discussed further in Chapters 7 and 14.

It is important to be able to distinguish rate-limiting steps in various cycles. In many of the stratospheric cycles, reactions involving O atoms are involved, since they are present in very low abundance. For example, the rate-limiting steps in ozone destruction by $NO_x$, $ClO_x$, and $HO_x$ (above 40 km) are the reactions O + $NO_2$, O + ClO, and O + $HO_2$. The efficiency of an ozone-destroying cycle is given by the number of times the cycle repeats before the active molecule (or atom) is lost, usually to a stable reservoir species such as $HNO_3$, HCl, or $H_2O$.

## 3.6 Role of Excited States

Both electronically and vibrationally excited states play a role in atmospheric processes. The extra energy can lead to an enormous enhancement of reactivity compared to the ground state. Also, the excess energy can be radiated and contribute to the observed emission from the atmosphere. This can be used as a diagnostic for measuring atmospheric concentrations or also as a marker for certain reactions in the atmosphere.

### 3.6.1 Electronic Excitation

Oxygen, in both its molecular and atomic forms, possesses several low-lying, metastable electronic states, and due to its high abundance in the atmosphere, these excited states can themselves play a role in the chemical and radiative balance of the atmosphere. The most important electronically excited species in the atmosphere is $O(^1D)$. This excited state, which lies approximately 270 kJ mol$^{-1}$ above ground-state oxygen atoms, is extremely reactive, particularly toward molecules that contain hydrogen atoms. Many reactions of ground-state $O(^3P)$ atoms have large activation barriers, but this is not the case for $O(^1D)$. The reaction of $O(^1D)$ with water vapor is responsible for forming OH radicals throughout the atmosphere, while the reaction of $O(^1D)$ with nitrous oxide is the major source of active nitrogen in the stratosphere

$$O(^1D) + H_2O \rightarrow OH + OH \tag{3.4}$$

$$O(^1D) + N_2O \rightarrow NO + NO \tag{3.60}$$

$$O(^1D) + N_2O \rightarrow N_2 + O_2 \tag{3.61}$$

Electronically excited states of $O_2$ in the upper atmosphere contribute to the atmospheric emission, for example, $O_2(^1\Delta)$. The excited form of molecular oxygen $O_2(^1\Delta)$ has a very long radiative lifetime ($> 10$ sec), but is quenched slowly, and so can build up to appreciable levels, approaching $10^{10}$ molecule cm$^{-3}$ at 50 km (Mlynczak et al., 1993). It is formed in one of the channels for the photolysis of ozone

$$O_3 + h\nu \rightarrow O_2(^1\Delta) + O(^1D) \tag{3.3}$$

The second excited state of molecular oxygen, $O_2(^1\Sigma)$, can be populated directly by absorption of solar radiation. This state has been implicated in the formation of ozone at high altitudes.

Sodium is present in the upper atmosphere from the ablation of meteors, and can also be involved in a catalytic destruction of $O_3$

$$Na + O_3 \rightarrow NaO + O_2 \tag{3.62}$$

$$NaO + O \rightarrow O_2 + Na^* \tag{3.63}$$

The sodium atoms are formed in an excited electronic state, and emit the characteristic $D$ line at 589 nm. This emission actually comes from a relatively narrow altitude band and is a very sensitive probe of upper atmospheric dynamics, for example, as a marker for the passage of gravity waves (Qian and Gardner, 1995).

### 3.6.2 Vibrational Excitation

Reactions of vibrationally excited species do not in general play a great role in the lower atmosphere, since these species are quenched via collisions or emit radiation and return to the ground state before reaction can occur. A recent field of interest is the question of whether it is possible to produce ozone from the photolysis of vibrationally excited oxygen molecules (Miller et al., 1994). Since these molecules possess excess energy, it should be easier to photodissociate them than is the case with ground-state

$O_2$. This extra source of ozone could go part of the way towards explaining why modeled ozone is consistently less than measured ozone around 40 km.

Vibrationally excited species play a greater role in emission processes. One of the most important of these is the formation of vibrationally excited OH from the reaction of H with ozone

$$H + O_3 \rightarrow O_2 + OH \ (v \leq 9) \tag{3.64}$$

The vibrational emission from OH, the Meinel bands, is a very prominent feature of the atmospheric emission spectrum (Box 14.2). The emission involves both strong transitions ($\Delta v = 1$), and also many "forbidden" transitions where more than one quantum of energy is lost.

## 3.7 Measurement of Rate Coefficients

This topic is much too detailed to be covered in any depth in a chapter of this length. Many ingenious methods have been developed over the past few years to measure rate coefficients of atmospheric reactions, and the increasingly common use of lasers to initiate reactions and to detect transient species has led to a revolution in the science. Rate coefficient measurements fall into the categories of direct measurements, where the concentration of a reactant is measured in real time, or indirect methods where the rate coefficient is inferred from bulk properties of the gas mixture (loss of reactants or buildup of products). The time-resolved experiments fall into two broad categories: flow experiments and pulsed experiments. In flow experiments, the reactor is usually a long tube with a well-defined diameter, along which the gases are flowed at a known velocity. Concentration measurement is made at one end of the tube, and variation of the reaction time is accomplished by varying the addition point of one of the gases, usually through a movable inlet port. Since the flow velocity is known, conversion of distance into time is trivial. The method essentially "freezes" the profile of the reacting molecules in space along the tube, which has the advantage that detection methods with relatively long time constants can be used. Due to mass transport considerations, flow tubes are usually restricted to pressures of a few Torr, although recently developed systems have overcome this restriction (Seeley et al., 1996). A comparison of the advantages and drawbacks of the pulsed and flow methods is given by Howard (1979).

In pulsed methods, a flash of light from a laser or flashlamp is used to dissociate some precursor and thus initiate the reaction. These are true time-resolved experiments, and require fast detection electronics, which are often required to respond on a time scale of microseconds. The pressure range for photolysis experiments is not limited, and reactions have been carried out between a few Torr and hundreds of atmospheres (Ravishankara et al., 1983; Forster et al., 1995). Both flow and pulsed systems have had great success in measuring atmospheric rate parameters, and the appropriateness of each depends on the reaction being studied.

Much effort has been expended recently in the development of methods to measure the rates of heterogeneous reactions. The problem is that the "excess additive" has to be defined in terms of its surface area rather than its concentration, and in the case of liquid droplets a distribution of sizes is usually encountered. Until now, the methods used have all involved flow techniques of one form or another. Either flat surfaces or droplets can be used. In the former case, experiments have involved either a static surface (Williams and Golden, 1993) or the wetted wall of a flow reactor (Utter et al., 1992). Methods using droplets have used anything from submicron aerosols suspended in a gas (Mozurkewich and Calvert, 1988) to a train of droplets falling through a gas stream (van Doren et al., 1990).

Indirect methods include product studies and relative rate studies. In product studies the concentrations of as many reactants and products as possible are measured, and rate coefficients and mechanistic details are derived from the concentration-time profiles of these species. These reactions take place over time frames of minutes to hours. In the relative rate method, the compound of interest is allowed to react in the presence of a compound whose rate coefficient is known, and the relative rates of decay give the ratio of the rate coefficients (Wallington et al., 1988). This method is very useful, since it circumvents many of the systematic uncertainties inherent in direct measurements. In the case of product studies and relative rate methods Fourier transform infrared spectrometry is often used for analysis, although gas chromatography is also useful.

As well as providing a representation of the kinetics of a reaction, measurement of the temperature and pressure dependence of a reaction rate coefficient can also provide useful information about the mechanism of a reaction. For example, the identification of a negative activation energy points to the intermediacy of an activated complex. When coupled with modern, sophisticated potential energy surface calculations, this enables the extrapolation of the rate coefficient outside the range of the measurements, and also allows the prediction of as yet unknown rate coefficients. This is a field that is still in its infancy, but that promises to be very useful.

### 3.7.1 Pseudo-First-Order Kinetics

In all the systems described above, the analysis is simplified if the reaction is carried out in what is known as the pseudo-first-order kinetic regime. This is when one of the reactants is effectively held constant (in practice, in large excess), and the time dependence of the other is studied. In the general example earlier (Eq. 3.17), assume that $AB$ is in large excess, and thus essentially does not change. Equation (3.18) can be integrated to yield $[C] = [C]_0 \exp(-k[AB]t)$, or $[C] = [C]_0 \exp(-k't)$, where $k' = k[AB]$ is the apparent pseudo-first-order rate coefficient. Most experimental determinations are organized so that pseudo-first-order conditions are satisfied. This is often easy to achieve if, for example, $C$ is a free radical, and $AB$ a stable molecule. Then the rate of decay of $C$ is monitored as a function of the concentration of $AB$, and a plot of $k'$ against $[AB]$ yields the desired rate coefficient $k$. It should be noted that the lifetime of $C$ (the time for it to decay by a factor of $e$) is simply given by $1/k'$. Note also that it is not necessary to know the absolute concentration of $C$, since only ratios $[C]/[C]_0$ are involved. This is particularly advantageous in the case of free radicals, for which absolute calibration can be difficult.

### 3.7.2 Reaction Mechanisms

Measurement of the rate coefficient for a reaction only provides a part of the overall picture of a reaction. Many reactions have more than one possible product channel, and in order to describe a reaction fully, the distribution of products must also be known quantitatively. This is most often achieved through the use of chamber experiments that use final product analysis, coupled with time-resolved measurements of a free radical reactant to give the overall rate coefficient. However, the determination of overall reaction mechanisms from product observations is a painstaking task, and an unambiguous description of the system is not always possible.

More modern techniques are taking advantage of the fact that quantitative time-resolved measurements of products contain both kinetic and product information. Mass spectrometry has been used for many years to furnish quantitative product information in both flow tubes and chambers. Optical spectroscopic techniques using tunable diode lasers and color center lasers are beginning to enjoy widespread use in kinetics measurements. Quantification of stable products is a relatively easy task, since known amounts can be introduced into the reaction vessel for calibration.

An example of how direct and indirect information can be melded into a complete mechanism is illustrated by the laboratory oxidation of $CH_4$ in the absence of $NO_x$. Chamber studies using infrared analysis show that the major products are $CH_3OH$, $CH_2O$, and $CH_3OOH$. Following abstraction of a hydrogen atom from methane, the following reactions are thought to occur

$$CH_3 + O_2 + M \to CH_3O_2 + M \qquad (3.65)$$
$$CH_3O_2 + CH_3O_2 \to CH_3OH + CH_2O + O_2 \qquad (3.66)$$
$$CH_3O_2 + CH_3O_2 \to CH_3O + CH_3O + O_2 \qquad (3.67)$$
$$CH_3O + O_2 \to CH_2O + HO_2 \qquad (3.68)$$
$$CH_3O_2 + HO_2 \to CH_3OOH + O_2 \qquad (3.69)$$

From the amounts of $CH_3OH$ and $CH_2O$ formed, the relative branches into channels (3.66) and (3.67) can be determined (Tyndall et al., 1998). Figure 3.14 illustrates the formation of products in the photooxidation of $CH_4$ in air.

**Figure 3.14.** Time profiles of the products formed in the oxidation of $CH_3$ radicals in air (Tyndall et al., 1998).

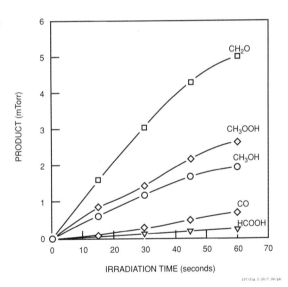

Time-resolved measurements of $CH_3O_2$ and $HO_2$ using UV spectrometry have been used to determine rate coefficients for the overall $CH_3O_2$ self-reaction (3.66)+(3.67), and the secondary reaction (3.69) (Lightfoot et al., 1992). In practice, the situation is complicated since the measurements of the $CH_3O_2$ self-reaction have to be corrected for the occurrence of (3.69), because the $HO_2$ produced in Reaction (3.68) leads to an increased $CH_3O_2$ loss. Once these rate coefficients are sufficiently well established, it is possible to extract more information from the product measurements by computer simulation of the time evolution of both the major and minor products (which include CO, HCOOH, and $CO_2$). This interplay between various sorts of experiments is vital, as neither can give the full picture, and as new analytical techniques are developed mechanisms will constantly be refined.

## 3.8 The Stationary-State Approximation

Atmospheric concentrations do not in general build up indefinitely, but tend to reach a steady value. This occurs when the source and sink terms for an atmospheric component are equal. This can be seen from the simple continuity equation:

$$\frac{\partial n}{\partial t} = P - \beta n + \text{transport terms} \tag{3.70}$$

where $n$ is the density (concentration), $P$ the production rate, and $\beta$ the loss coefficient (the loss rate is $L = \beta n$). For species undergoing rapid chemical loss, such as free radicals, the concept of a stationary state is an important aid to calculating concentrations. In general, if the lifetime for the loss of a species is shorter than the timescale on which the sources are varying, then a steady-state concentration for that species may be calculated by:

$$n = \frac{P}{\beta} \tag{3.71}$$

In the case of free radicals, a photostationary state is usually achieved on a very rapid timescale. This corresponds to a situation where the (photo-)chemical terms are balanced at any moment, since the chemical lifetime is shorter than any transport processes. For the OH radical, the atmospheric lifetime is on the order of 1 sec, while for $O(^1D)$ it is around 1 nsec. Thus the assumption of stationary state is justified, and it is possible to calculate their concentrations by balancing their production and loss terms. Since the production terms are usually dependent on the solar photolysis rate, the actual steady-state concentration will vary over the course of the day, but will track the solar intensity very closely.

In atmospheric models (see Chapter 12 for more details), compounds of the same chemical family that are in a photostationary state are often grouped together and transported as one entity, since this eases the constraints imposed by having to calculate the concentration of every species at every timestep.

## 3.9 Lifetimes in the Atmosphere

The concept of atmospheric lifetimes is one that is frequently encountered, and also one that can lead to confusion. Consider a first-order reaction, that is, one in which the apparent reaction rate varies linearly with the concentration of the species of interest. Such processes can be unimolecular decompositions at their high-pressure limit, photolytic reactions, or bimolecular reactions in which one of the reactants is held constant (referred to as pseudo-first-order conditions, Section 3.7.1). Then, in the absence of any production, the species of interest would react away, its kinetics being described by

$$n = n_0 \exp(-kt) \tag{3.72}$$

where $n_0$ and $n$ are the concentrations initially and at time $t$. The lifetime, $\tau$, is then given simply by the reciprocal of $k$ (see also Section 12.6). Of course, in the atmosphere it is not possible to simply switch off the sources of a molecule, but the concept applies just the same. If several processes are responsible for removing a molecule, then the overall lifetime is given by

$$\tau^{-1} = \tau_a^{-1} + \tau_b^{-1} + \ldots \tag{3.73}$$

since the overall rate of reaction is given by the sum of the individual rates. The lifetime of any species can be defined at any point in space or time in this way. Figure 3.15 shows that the lifetime of chemical constituents in the atmosphere varies typically from seconds (in the case of chemical radicals) to centuries (in the case of stable molecules), and that the spatial and temporal variabilities of species are inversely related to their chemical lifetime. Long-lived gases such as the CFCs tend to be relatively well mixed throughout the troposphere, while the concentration of a radical like OH exhibits much more temporal and spatial variations. Long-lived compounds are usually referred to as tracers, since their behavior is directly and strongly influenced by atmospheric dynamics. The concentration of fast-reacting species is determined by the local photochemistry (*e.g.*, intensity of the solar flux) and is only indirectly influenced by dynamics (through the transport of source and sink compounds).

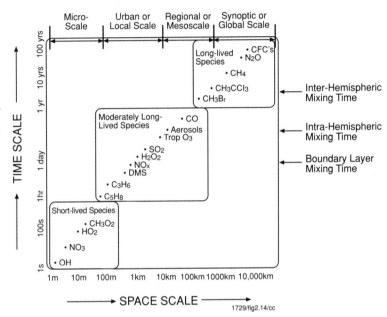

**Figure 3.15.** The spatial and temporal scales of variability for atmospheric constituents in the atmosphere (courtesy of W. L. Chameides).

For certain types of compounds such as the source molecules, we are often interested in the *global* or *integrated* atmospheric lifetime. This concept, which applies, for example, to methane, nitrous oxide, or CFCs, accounts for the fact that these molecules, after having been injected into the atmosphere, are transported through regions with spatially varying loss processes and loss rates. The global lifetime of these source gases is defined as their atmospheric burden (total density) divided by their integrated loss rate

$$T = \frac{\int_V n\, dV}{\int_V \beta n\, dV} \qquad (3.74)$$

where $n$ is the density, $\beta$ the loss coefficient, and $V$ the atmospheric volume over which the integrals are performed. If steady-state conditions are assumed, the integrated loss rate can be replaced by the global emission. In the case of methane, for example, the total atmospheric burden is of the order of 5000 Tg and the global emission is estimated to be close to 500 Tg/yr, so that the global steady-state lifetime is approximately 10 years. Note that global lifetimes can be broken down into their tropospheric and stratospheric parts. This distinction is often made when the major loss processes vary between these regions. If a molecule with a long tropospheric lifetime is destroyed rapidly in the stratosphere, then transport across the tropopause is essentially irreversible. In this case, an upper limit to the tropospheric lifetime is given by the time constant associated with transport across the tropopause (usually thought to be about 1.5 years).

The tropospheric lifetimes of many species are governed by reaction with the hydroxyl radical, OH. Since the global concentration of OH is not known, an empirical approach is sometimes taken. That is, a compound whose inventory is well known, such as $CH_3CCl_3$, and which is long-enough lived to be well mixed, is used to infer an OH concentration, and other molecules are scaled according to their rate coefficients with OH.

The response timescale of a chemical system to a perturbation in the concentration of a given chemical compound is often mistakenly thought to be the chemical lifetime of this compound. For example, if a perturbation is applied to the atmospheric abundance of methane (*e.g.*, through an injection at the Earth's surface), the concentration of other species including OH (which provides the major destruction of methane) will also change, and the timescale to return to equilibrium will be different from the lifetime of methane calculated for a given (fixed) concentration of OH. For a coupled nonlinear chemical system, the response to a perturbation must be diagnosed in terms of the natural modes (eigenvectors and eigenvalues of the Jacobian matrix) of this system. Prather (1994) has applied this approach to the nonlinear [$CH_4$, CO, OH] system and has shown that the *e*-folding time for the response to a methane perturbation is about 15.5 years, about 1.6 times the calculated lifetime of 9.6 years.

## Further Reading

Atkins, P. W. (1986) *Physical Chemistry*, 3rd ed., W. H. Freeman and Company, New York.
Benson, S. W. (1960) *The Foundations of Chemical Kinetics*, McGraw-Hill, New York.
Benson, S. W. (1976) *Thermochemical Kinetics*, Wiley, New York.
Brasseur, G. and S. Solomon (1986) *Aeronomy of the Middle Atmosphere*, 2nd ed., D. Reidel, Dordrecht.
Finlayson-Pitts, B. J. and J. N. Pitts, Jr. (1986) *Atmospheric Chemistry: Fundamentals and Experimental Techniques*, Wiley, New York.
Goody, R. (1995) *Principles of Atmospheric Physics and Chemistry*, Oxford University Press, UK.
Herzberg, G. (1945) *Infrared and Raman Spectra*, Van Nostrand Reinhold Company, New York.
Warneck, P. (1988) *Chemistry of the Natural Atmosphere*, Academic Press, New York.

Harold Schiff

## When Do We Know Enough about Atmospheric Chemistry?

The simple Chapman mechanism given in Section 3.5 of this chapter was too successful in explaining the main features of the ozone layer. Not only did it account for the presence of a small steady-state concentration of ozone in the Earth's atmosphere, but also its location and distribution in the stratosphere. Since the solar intensity increases with altitude while the concentration of molecular oxygen decreases, there will be an altitude at which the ozone concentration reaches its maximum. Nature, knowing the rate constants for each of the reactions, knew that the optimal place to locate the ozone factory was in the stratosphere. In fact, the mechanism can account for the very existence of the stratosphere. The energy abstracted from the solar radiation to dissociate molecular oxygen and ozone in the first and third reactions is deposited locally by the second and fourth reactions, resulting in the temperature inversion that defines the stratosphere.

So successful was this simple four-step mechanism that the stratosphere remained relatively unpopulated by atmospheric scientists from 1930, when Chapman first proposed his mechanism, until the 1970s. Everything was known about the stratosphere!! The higher territories of the mesosphere and ionosphere were relatively highly populated, mostly by physicists, who were attracted to the emissions from the electronically and vibrationally excited molecules and ions that occurred there. The lower territory, the troposphere, was populated by chemists, who, under the pioneering leadership of A. J. Haagen-Smit, were attracted by the photochemistry involved in urban pollution areas such as Los Angeles.

A handful of chemists remained in their laboratories measuring the rate constants of the Chapman reactions. As these rate constants became well established, it became evident that the simple four-step mechanism was quantitatively inadequate. It predicted ozone concentrations roughly twice what was observed to be present. Obviously other reactions must be destroying ozone. Since the reactions of ozone with the major and minor gases in the atmosphere were too slow to account for the discrepancy, the missing reactions had to be with species present in minute, trace quantities, which, therefore, required them to initiate catalytic chain reactions. The realization that the concentration of stratospheric ozone is controlled by gases that are present in amounts even smaller than ozone itself is the basis of the concern that human activity could alter its steady-state concentration.

The first of these activities that could affect the steady-state ozone layer was the operation of a fleet of supersonic aircraft, which was designed to fly in the stratosphere. The exhausts from these aircraft would inject water vapor and oxides of nitrogen, which could form HO, $HO_2$, and NO and $NO_2$ catalytic chain carriers. The next was the emission of chlorofluorocarbons from aerosol spray cans and refrigerants, which could form Cl and ClO chain carriers. This was considered to be an even more insidious threat to the ozone layer because of the long lifetime of the halocarbons in the atmosphere.

These concerns resulted in a mass immigration of scientists into the stratosphere, including chemists, physicists, engineers, meteorologists, and computer modelers. Laboratory scientists developed new and powerful techniques for studying the mechanisms and rate constants of the reactions, which were then used by the modelers to describe the existing atmosphere as well as to predict the future according to various anthropogenic emission scenarios. Another group of scientists developed sophisticated techniques and instruments to measure the composition and distribution of key species in the atmosphere. These instruments were either used at ground level or carried aboard high-flying aircraft or stratospheric balloons. These measurements were then compared with the computer models. A highly productive community flourished where results of each group were used to suggest improvements in the others. By the end of the 1980s the four Chapman reactions involving three chemical species had grown to more than 1000 reactions involving hundreds of reactive species. By the mid-1980s, aside from a few minor discrepancies, models and measurements agreed pretty well. Virtually everything was known about the stratosphere!!! Many of us then migrated out of the stratosphere to more rewarding employment in the troposphere, where acid deposition and other tropospheric pollution problems were gaining interest.

And then in 1985 the Antarctic ozone hole was discovered which caught the entire stratospheric community by surprise. Nothing in the existing models could explain these observations. For a short period of time there were almost as many explanations as there were scientists. But to the great credit of the community, it took a remarkably short time for the laboratory chemists to find the correct explanation and for observations to confirm it. The explanation was that of surface reactions, which are referred to briefly in this chapter and in greater detail in a later chapter.

Those of us who have moved back into the troposphere are quite aware of the importance of heterogeneous reactions. The oxidation of $SO_2$ occurs largely in cloud droplets and involves free radicals and ions. The importance of reactions on the surfaces of aerosol particles in urban pollution is only beginning to be understood. Many of the major production and loss processes of atmosphere species occur at the surface of soils and vegetation and involve biological processes. Biological scientists are joining the atmospheric science community in increasing numbers. There appears to be no danger of overpopulation in this community. We can never be complacent that we know everything about any region of the atmosphere. The only surprise would be if there were no other surprises. This is what makes the Earth's atmosphere such a fascinating place to live and work in.

*Harold Schiff was the founding Dean of the Faculty of Science at York University in Toronto and the founding Director of the Canadian Institute for Research in Atmospheric Chemistry. His research interests include laboratory studies of reactions of atmospheric interest and the measurement of key species in the atmosphere. He is currently the President of Unisearch Associates, Inc., an R&D company involved in developing new gas measuring instruments and in using these instruments to make atmospheric measurements.*

# 4 Aerosols and Clouds

## 4.1 Introduction

The atmosphere is not simply a mixture of gases. Every cubic centimeter of air contains up to thousands of suspended particles; most of these are only a fraction of a micrometer in diameter. Particulate matter in the atmosphere is commonly referred to as "aerosol", although the term properly applies to the multiphase *mixture* of solid or liquid particles dispersed in a gas. Atmospheric particulate matter may consist of a large variety of species, from naturally and anthropogenically derived material such as suspended smoke, soot, minerals, soil dust, and fly-ash to biogenically derived particles such as pollens and spores. The composition of atmospheric particles varies widely not only from place to place but also from particle to particle within a given parcel of air, and is a major factor in determining how particles interact with the gas phase. An overview of the atmospheric aerosol is given by Prospero *et al.* (1983).

The atmosphere is also filled with clouds that are composed of cloud drops of tens to hundreds of micrometers in diameter, as can readily be seen from satellite pictures. Clouds and aerosols are not separate entities, since clouds depend on the existence of aerosols for their formation, and clouds can re-evaporate, leaving behind an aerosol nucleus.

Atmospheric particles are important for several reasons. First, they serve to transport nonvolatile material from one place to another. An important example of this is the transport of crustal minerals, such as iron, to the world's oceans, where they may serve as sources of nutrients to marine organisms. Second, they affect the optical properties of the Earth's atmosphere. This not only makes spectacular sunsets possible, but also affects the Earth's climate by altering the amount of sunlight that penetrates the atmosphere. Third, aerosol particles act as the nuclei on which cloud or fog drops form. Fourth, some trace species in the atmosphere may be present in either the gas phase or the particle phase. The partitioning between the phases affects the atmospheric transport of these species since gases and particles are deposited to the Earth's surface at different rates. Finally, aerosol particles have a significant effect on chemical reactions occurring in the atmosphere.

## 4.2 Overview of the Atmospheric Aerosol

The sources, sinks, and characteristics of aerosols in the troposphere and stratosphere are markedly different. The tropospheric aerosol is more variable in spatial and temporal distribution and in chemical nature; this variability arises in part from the link between aerosols and the hydrological cycle. Physical and chemical sources of stratospheric aerosol are fewer in number, and stratospheric particles have a relatively long lifetime, a consequence of the lack of wet removal processes and the stable thermal structure of the stratosphere, which inhibits dynamic mixing processes.

> **Box 4.1 Definitions and Units**
>
> An important role of the water contained in the Earth system (and in the atmosphere) is its ability to dissolve many substances. A homogeneous mixture of substances is called a *solution*. The component of this mixture that is present in the largest quantity is called the *solvent*, while the other components are called the *solutes*. In an aqueous solution, the solvent is water. When the solute dissolves and precipitates at equal rates, the solution is said to be *saturated*. If the amount of solute is higher (lower) than in the saturated solution, the solution is said to be *supersaturated* (*unsaturated*).
>
> The amount of solute present in a given quantity of solvent is often expressed by the *molar concentration* (or *molarity*) defined as the number of moles of solute per liter of solution. The corresponding units are mol L$^{-1}$ (often written as M for short). Note that, for a given amount of solute in a solution, the molarity changes with temperature, due to the expansion of the liquid; for this reason, the *molality* (moles of solute per kg of solution) is sometimes used. A measure of the acidity of a solution is the pH, defined by pH = $-\log_{10}$ [H$^+$], where [H$^+$] is the concentration of hydrogen ions. For pure water [H$^+$] = [OH$^-$] = $10^{-7}$ M, and thus the pH = 7.

The distribution of atmospheric particle number concentration with respect to size exhibits one or more "modes"; that is, particles are grouped within several diameter subranges. Whitby and Cantrell (1976) suggested, based on observations, the idealized distribution of particle surface area shown in Figure 4.1, and proposed the commonly used descriptors for the subpopulations: nucleation or Aitken (diameter $\lesssim 0.1$ $\mu$m), accumulation (diameter between $\sim$0.1 and 2.0 $\mu$m), and coarse (diameter > 2.0 $\mu$m) modes. Additionally, particles with diameters less than about 2 $\mu$m constitute the fine particle fraction.

A major loss process (for number) of nucleation mode particles is coagulation with larger particles (see Section 4.7.1). (Note that this process does not remove particulate mass from the atmosphere.) Coarse particles have high enough sedimentation velocities for deposition to serve as an important removal mechanism. The accumulation mode occurs at diameters for which the diffusional and sedimentation loss mechanisms are least effective, and consequently particles tend to aggregate in this size range until removed by wet deposition processes. The atmospheric lifetime of submicron particulate matter is thus very strongly linked to the hydrological cycle. Particles for which sedimentation is negligible (diameter $\leq$ 5 $\mu$m) could also be removed by dry deposition, analogous to gases. However, little is known about how the deposition velocity varies with size and composition.

A wide variety of chemical species may be found in the atmospheric aerosol. When individual particles differ in composition from each other, the aerosol is said to be externally mixed; when individual particles have all the same composition, the aerosol is said to be internally mixed.

A fundamental distinction between the fine and coarse particle fractions lies in the source mechanisms that produce them. Mechanical processes (anthropogenic and natural) produce the relatively large particles in the coarse mode; examples are sea spray, windblown dust, fly-ash, volcanic ash, and particles from tire and brake wear. The chemical composition of the coarse fraction reflects these sources: crustal elements (Fe, Ca, Si, Al, etc.) and seawater species (Na, Cl, etc.) are commonly detected.

The mechanically produced particles are examples of primary emissions, that is, particles that have been directly emitted to the atmosphere. Aerosol that has formed in the atmosphere by physical or chemical transformations is termed secondary aerosol. An example is sulfate, which forms downwind of an industrial source, as emitted sulfur gases are chemically converted to condensable species, which are then incorporated into particles. This process is known as gas-to-particle conversion and is the mechanism by which aerosol particles serve as sinks in the chemical cycles of several important tropospheric trace gases. Condensational growth, compound formation, homogeneous nucleation from the gas phase, and aqueous-phase conversion of gases to nonvolatile species are gas-to-particle conversion processes that will be discussed in more detail in following sections.

Soot, semivolatile hydrocarbons, and metals are often found in the fine-particle mode. Soot is a composite of primary emissions of black carbon, formed in combustion processes, and organic species that partition to the particle phase as emitted gases cool. As suggested in Figure 4.1, these processes tend to produce small spherical primary particles (on the order of 0.01 $\mu$m in diameter), which agglomerate into the aggregate soot structure. Other semivolatile organic species, both natural and anthropogenic in origin, may be emitted directly or produced by reactions in the atmosphere; for example, polycyclic aromatic hydrocarbons (PAHs), such as the carcinogen benzo(a)pyrene, are emitted from combustion sources. Metals such as lead, arsenic, and vanadium are present in fuels and are volatilized during combustion, subsequently condensing onto atmospheric particles as the gases cool.

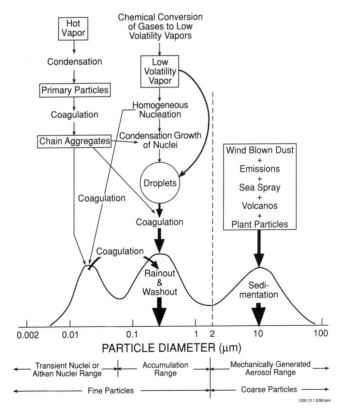

*Figure 4.1.* Idealized schematic of the distribution of particle surface area in the atmosphere. Modes, sources, and removal mechanisms are indicated (Whitby and Cantrell, 1976).

Ionic species are important components of the fine-particle fraction, and commonly include sulfate ($SO_4^{2-}$), nitrate ($NO_3^-$), and ammonium ($NH_4^+$). Sulfate and nitrate, derived from sulfuric acid ($H_2SO_4$) and nitric acid ($HNO_3$), are fully oxidized end products of the atmospheric cycles of sulfur and nitrogen compounds, and they partition to the particulate phase because of their low volatilities and/or their high solubility in water. Ammonia dissolves in wetted particles and serves as a neutralizing species, so that, in the lower troposphere, sulfate and nitrate anions are usually found as fully or partially neutralized species such as ammonium sulfate [$(NH_4)_2SO_4$] or ammonium bisulfate ($NH_4HSO_4$), or as salts such as sodium nitrate ($NaNO_3$) if cations such as sodium or calcium are present. The ionic character of a large fraction of the mass of accumulation-mode aerosols is a key feature in understanding the role of such aerosols in the hydrologic cycle, climate effects, and visibility: The soluble salts are hygroscopic, and dry particles swell as they are exposed to humid conditions and take up water to form solution droplets. The humidity at which deliquescence occurs (Section 4.5.3), and the mass of water added per unit mass of particle, depends strongly upon the aerosol chemical composition. Theoretical calculations of water uptake for atmospheric aerosols are complicated by the fact that most accumulation mode aerosol is probably an internal mixture of some nonhygroscopic material and a variety of hygroscopic salts. Also, the thermodynamics of a concentrated mixture of ions is difficult to treat, and insufficient laboratory data exist to support the development of more than a limited number of empirical relationships.

Accumulation-mode aerosols (0.1-2 $\mu$m diameter) are those that penetrate most deeply into the lung. The observations that acidic ionic species and toxic and mutagenic compounds such as metals and PAHs are found primarily in this size fraction raise issues with respect to the health effects of fine particulate matter.

**Table 4.1**
*Global Sources of Atmospheric Aerosols (in megatonnes per year)*
(Adapted by L. Barrie)

| Aerosol Sources | Total Emissions $D < 25\ \mu m$ | Emissions $D < 1\ \mu m$ |
|---|---|---|
| **Manmade** | | |
| *Primary:* | | |
| Industrial dust | 40-130 | 20-65 |
| Soot | 10-30 | 10-30 |
| Biomass burning | 50-190 | 50-190 |
| Windblown dust | 820 | 140 |
| *Gas-Particle Conversion of:* | | |
| $SO_2$: Smelters/power plants | 120-180 | 120-180 |
| $NO_x$: Autos/power plants | 20-50 | 5-10 |
| Anthropogenic VOCs | 5-25 | 5-25 |
| **Total Manmade Sources** | 1065-1325 | 565-640 |
| **Natural** | | |
| *Primary:* | | |
| Windblown dust | 1000-3000 | 265 |
| Forest fires | 3-150 | 2-75 |
| Sea salt | 1000-10000 | 20-100 |
| Volcanoes | 4-10000 | 0.4-100 |
| Organics | 26-50 | — |
| *Gas-Particle Conversion of:* | | |
| DMS, $H_2S$ | 60-110 | 60-110 |
| Volcanic $SO_2$ | 10-30 | 10-30 |
| Biogenic $NO_x$ | 10-40 | 10-40 |
| Biogenic VOCs | 40-200 | 40-200 |
| $NH_3$ to $NH_4^+$ salts | 80-270 | 80-270 |
| **Total Natural Sources** | 1363-3550 | 397-1390 |

### 4.2.1 Composition of Tropospheric Aerosols

Tropospheric aerosol within the planetary boundary layer can be broadly classified as marine, continental, or polar; free tropospheric aerosol also has unique characteristics. The continental aerosol has components derived from soils, minerals, biogenic sources, combustion, and gas-to-particle conversion. The continental aerosol may be classified into rural (remote), urban, and desert dust storm aerosol. Characteristics of each category are described in the reviews by Jaenicke (1988) and d'Almeida et al. (1991), while Table 4.1 shows a recent estimate of the various contributions to the aerosol source. In general, higher number concentrations than those observed in clean marine regions are observed in regions influenced by anthropogenic continental sources; concentrations may be even higher in urban locations due to the proximity of industrial and transportation sources. Aerosol of anthropogenic origin has components arising from combustion of fossil fuels, such as soot, trace metals, and partially oxidized organic matter. Figure 4.2 presents some representative measurements of tropospheric fine particles, demonstrating the contrast between chemical composition of urban and remote particulate matter.

Fitzgerald (1991) has compiled a review of the characteristics of particles within the marine boundary layer (MBL) that are of marine origin. This aerosol is considered to define the "background"; that is, it is representative of air masses that have resided over the ocean for a sufficient time period (on the order of 10 days) to minimize influences from anthropogenic aerosol. Near the east coast of the United States, measurements indicate that total particle concentrations are typically near 6000 cm$^{-3}$, reflecting continental influences. The background concentrations range from less than 100 cm$^{-3}$ to a mean of about 200 cm$^{-3}$ throughout the tropical trade wind regions, and similar marine background concentrations have been measured over the Pacific Ocean. Sea salt is the most important component of larger particles (diameter > 1 $\mu$m), with a variable fraction of continental mineral dust. About 90-95% of the number concentration is accounted for by particles with diameters less than 0.6 $\mu$m, although they represent only about 5% of the total aerosol mass.

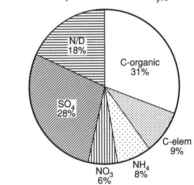

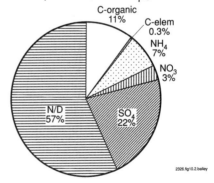

**Figure 4.2.** Average chemical composition of fine particles in urban and remote regions (Heintzenberg, 1989). N/D = not determined.

These small particles consist primarily of non-sea-salt sulfate, presumably generated from oxidation of reduced sulfur compounds, principally dimethylsulfide (DMS), emitted from the ocean.

Desert dust results from weathering processes, with potential source regions covering about a third of the global land surface. Transport of particles with radius less than 5 μm has been observed to occur over marine regions for distances up to 5000 km. For example, Saharan dust is observed at island sampling sites in the western Atlantic Ocean (see Color Plates 6 and 12, and Prospero et al., 1987). This long-range transport is initiated by Saharan dust storms that lift material out of the boundary layer; as vertical temperature profiles evolve over Africa and the ocean, the dust "layer" becomes trapped between two inversions as it is transported westward, preventing significant mixing and removal of smaller particles to the surface. Atmospheric dust loadings can be perturbed by anthropogenic activity: changes in land use, such as deforestation and desertification, increase dust emissions.

Because of the spatial and temporal variability of aerosol sources and sinks, it has been difficult to establish a basis for a global aerosol model. Recent work has summarized large numbers of observations to provide a useful first attempt at a global aerosol climatology (d'Almeida et al., 1991). Jaenicke (1988) has also proposed size distributions for several classifications of aerosol and gives parameters for representing these as lognormal functions (see Fig. 4.3).

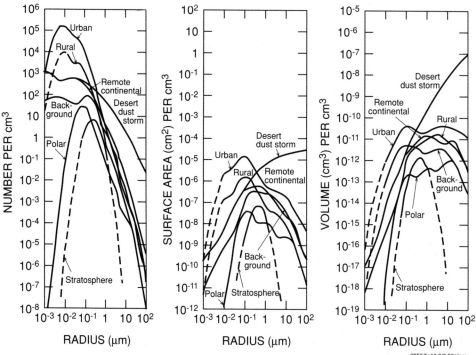

**Figure 4.3.** Number, surface area, and volume distribution function for model aerosols (Jaenicke, 1988).

There is evidence that particles in the free troposphere have significantly different characteristics than particles found at lower levels. The calculations of Delmas (1992) suggest that these particles have a high sulfur content and an acidic character, a conclusion supported by the measurements of Clarke (1993). Observations of regions of elevated concentrations of ultrafine aerosol (diameter less than 0.01 μm) (Clarke, 1993) also suggest that significant new particle formation from the gas phase may occur in the free troposphere.

### 4.2.2 Composition of Stratospheric Aerosols

The lower stratosphere contains a significant concentration of small sulfuric particles (see Color Plate 11). Junge et al. (1961) were the first to make measurements of particles at altitudes above 20 km and showed the presence of an apparently stable and globally distributed layer above the tropopause (the so-called Junge layer). A unique feature of the stratospheric aerosol is its long lifetime, due to the absence of downward transport and precipitation removal processes. The stability of the stratosphere allows transport processes to mix particulate matter longitudinally, while generally keeping it in the "layer" in which it was formed or injected.

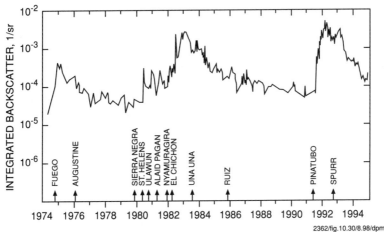

**Figure 4.4.** Long-term record of integrated aerosol backscatter using lidar scattering at 694 nm. The integrated backscatter is a measure of the total stratospheric column of aerosol. The major volcanic eruptions that increased northern midlatitude aerosol loading are noted on the time axis (Osborn et al., 1995).

The nonvolcanic source of the stratospheric aerosol material is largely carbonyl sulfide (OCS) emitted at the Earth's surface (Crutzen, 1976; Khalil and Rasmussen, 1984). OCS is not readily attacked by OH and, as a result, has a very long lifetime in the troposphere. It is eventually transported to the stratosphere, where it is oxidized by O atoms and dissociated by short-wavelength UV radiation; the products are oxidized to form $CO_2$ and $H_2SO_4$ (see Chapter 10). In addition, large volcanic eruptions inject $SO_2$ directly into the stratosphere; the oxidation of this $SO_2$ results in a rapid buildup of $H_2SO_4$ aerosol in the months following the eruption. Figure 4.4 shows the effect of volcanic eruptions on the loading of stratospheric particles (see also Color Plate 10). The spectacular sunsets and climatic effects that may follow major eruptions, such as those of El Chichón and Mt. Pinatubo, are due to this $H_2SO_4$ aerosol and not the direct injection of volcanic ash into the stratosphere (Hofmann, 1987). It has recently been demonstrated that both background and perturbed levels of particulate matter can affect stratospheric chemistry through heterogeneous reactions. Proposed high-altitude aircraft could also inject both soot and sulfur into the stratosphere.

The exact composition of the stratospheric aerosol at a given altitude is a complex function of the temperature and water vapor concentration. At the coldest temperatures (in the lower stratosphere) water condenses into the aerosol and the $H_2SO_4$ weight percentage of the aerosol can be as low as 50%. Higher up, as the temperature increases, the water evaporates and the sulfuric acid approaches 80% by weight. The change in liquid water content affects the chemistry occurring in the aerosols, as described later. Nitric acid is not very soluble in pure $H_2SO_4$, but at the very low temperatures characteristic of the polar vortex sufficient water is present in the aerosols that nitric acid uptake can occur. It has been suggested that these ternary mixtures of $H_2SO_4/H_2O/HNO_3$ are the precursors to polar stratospheric cloud (PSC) formation or are PSCs themselves (see Section 4.2.3).

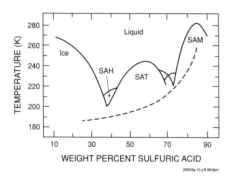

**Figure 4.5.** *Phase diagram for the $H_2SO_4$-$H_2O$ system. The solid line gives the equilibrium composition as a function of temperature and sulfuric acid content (SAM, sulfuric acid monohydrate; SAT, sulfuric acid tetrahydrate; SAH, sulfuric acid hexahydrate). The dashed line shows the expected composition of liquid sulfuric acid at typical stratospheric temperatures and water vapor mixing ratios (Anthony et al., 1995).*

Figure 4.5 shows a phase diagram for the freezing of sulfuric acid-water mixtures, along with a temperature-composition plot for the liquid system under stratospheric conditions. The solid line shows the temperature at which the indicated solid phase is in equilibrium with the liquid. The dashed line was calculated using typical stratospheric water vapor mixing ratios to show the approximate composition of liquid aerosol at a given temperature. Note that the liquid $H_2SO_4/H_2O$ aerosol is often substantially supercooled below the equilibrium curves for the solid hydrates.

### 4.2.3 Polar Stratospheric Clouds

During the winter the stratosphere over the poles becomes very cold; in the Antarctic, temperatures can reach 185 K and very sparse "clouds" form. These are called polar stratospheric clouds, or PSCs (McCormick *et al.*, 1982). The stratosphere contains small amounts of $H_2O$ vapor, either transported from the troposphere or produced by methane oxidation. A typical $H_2O$ partial pressure at an altitude of 15 to 20 km is about $4 \times 10^{-7}$ bar; this is equal to the vapor pressure of ice at 191 K. Thus, if the temperature falls below 191 K, as occurs in the Antarctic winter, the formation of dense clouds of ice particles is observed. Rapid cooling occurs when air flows over mountains, creating a standing wave. At the crest of the wave, a nacreous cloud, composed of water ice, forms when temperatures drop below 190 K. Nacreous clouds have a large number of ice crystals that are small, typically about 2 $\mu$m, and that are distributed over a range of sizes, creating an iridescent glow. Slow cooling can also occur when the stratosphere radiates heat to space during polar winter or when the air is lifted by air mass movement below (*e.g.*, from weather systems).

However, PSCs are often observed when the temperature is too high to allow the formation of pure water ice. These clouds appear to consist of $HNO_3(H_2O)_3$, nitric acid trihydrate, or NAT (Kawa *et al.*, 1992). This compound is in equilibrium with gas-phase $HNO_3$ and $H_2O$:

$$HNO_3(H_2O)_{3(s)} \rightleftharpoons HNO_{3(g)} + 3\,H_2O_{(g)} \qquad (4.1)$$

When the vapor pressure product, $p_{HNO_3}(p_{H_2O})^3$, exceeds the equilibrium constant for this reaction, it becomes possible to form the solid compound. With a typical stratospheric $HNO_3$ partial pressure of $5 \times 10^{-10}$ bar, this occurs if the temperature falls below 195 K. Thus this compound may form at temperatures that are too high to allow the formation of water ice. PSCs consisting of $HNO_3$-$H_2O$ are referred to as "Type I." Type I PSCs tend to be around 0.5-1 μm in diameter. Evidence that nitric acid dihydrate, $HNO_3(H_2O)_2$, may form has also been obtained (Worsnop et al., 1993). Figure 4.6 illustrates the regions of stability associated with the trihydrate. The solid, straight lines to the left side of the figure indicate the variation with temperature of the vapor pressures over liquid $HNO_3$-$H_2O$ mixtures containing the mole fraction of $HNO_3$ indicated. The solid curve shows the envelope of solid-liquid equilibrium temperatures for such liquids. The dashed lines indicate the temperatures where two solid compounds such as nitric acid tri- and monohydrate can co-exist, and also show the vapor pressures above these solids.

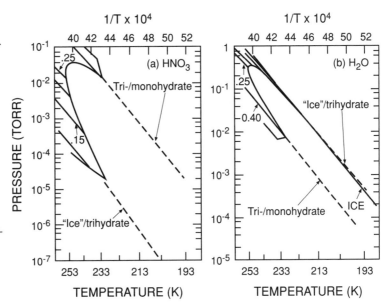

**Figure 4.6.** Vapor pressures of (a) $HNO_3$ and (b) $H_2O$ for solid and partially frozen mixtures. The co-existence curves are labeled. The ice line is shown for reference (Hanson and Mauersberger, 1988; reprinted with permission from J. Phys. Chem., copyright 1988, American Chemical Society).

The details of the PSC formation process are still unclear (Tolbert, 1994), but it is suspected that sulfuric acid aerosols may provide the required nuclei. It is not yet clear whether the growth of PSCs involves the deposition of $HNO_3$ onto solid nuclei, the freezing of ternary solutions, or whether the supercooled droplets remain liquid even below the frost point. When the air surrounding NAT particles continues to cool to temperatures below 190 K, water vapor will rapidly deposit onto these particles, creating "Type II" PSCs. Particles in Type II PSCs are primarily composed of water. These PSCs have fewer particles than nacreous clouds, but grow to much larger sizes of the order 10-100 μm. Because of the cold temperatures required to form Type II PSCs, they are observed much less frequently than Type I PSCs.

The formation of polar stratospheric clouds is important because chemical reactions occurring on the surfaces of the cloud particles convert "reservoir" chlorine compounds, such as HCl and $ClONO_2$, into "active" compounds, such as $Cl_2$. Granier and Brasseur (1992) summarize these findings and discuss the role of heterogeneous chemistry in ozone depletion. Also, the settling of the cloud particles removes $HNO_3$ and $H_2O$ from the stratosphere. The result is to set the stage for the dramatic ozone depletions observed during the Antarctic spring (Hamill and Toon, 1991).

## 4.3 The Role of Clouds in Tropospheric Chemistry
### 4.3.1 Introduction

The global mean cloud cover is close to 60%; the global average volume fraction of clouds in the troposphere is 7-8% (Section 4.3.2). Since most clouds occur in the lower atmosphere, which involves most of its mass, about 10% of the total mass of the troposphere is contained in cloud systems; the cloud volume fraction in the lower troposphere is about 15%. On the average, only one-tenth of these clouds is in a precipitating state. Moreover, the global average precipitation efficiency is about 50%, which means that half of the amount of water entering a precipitating system reaches the Earth's surface (Cotton and Anthes, 1989), while half of the water that has condensed in the atmosphere evaporates again. The mean time that air spends inside clouds per cloud event is several hours (averaged over all cloud types), whereas the subsequent time outside clouds is a few days. Water vapor condenses on aerosol particles that act as cloud condensation nuclei (CCN) (Section 4.5). The mean CCN lifetime in the atmosphere is about one week (Section 4.7), so that these particles go through about 5-10 cloud condensation-evaporation cycles before being removed from the troposphere. Similarly, trace gases may be repeatedly exposed to a cloud environment during their presence in the troposphere.

The presence of clouds influences photochemical processes by modifying the incoming ultraviolet radiation, and thus photodissociation rates of trace gases (Section 4.4.4). Further, a large fraction (about 80%) of the solar radiation absorbed at the Earth's surface is returned to the atmosphere via infrared radiation, while most of the remainder is utilized to evaporate water and is transferred to latent heat (see Chapter 15). The heating of the atmosphere from below and rising warm, moist air leads to the formation of cumulus clouds. Tropospheric chemistry is influenced by this process, especially by deep convection in cumulonimbus clouds, because trace gases emitted at the surface are carried from the planetary boundary layer into the overlying free troposphere (Chatfield and Crutzen, 1984). This is particularly important for surface-emitted trace gases with a relatively short lifetime, which otherwise would not participate in the chemistry of the upper troposphere.

Clouds themselves have a substantial influence on tropospheric chemistry, since they remove soluble species from the gas phase, essentially separating them from insoluble species. The aqueous-phase reactions occurring inside cloud droplets have a large effect on the oxidation of certain species. Another feature of clouds that affects tropospheric chemistry is that they cause precipitation, which returns particulate matter and water-soluble gases to the Earth's surface (Sections 4.7-4.8).

### 4.3.2 Occurrence of Clouds
#### 4.3.2.1 Cloud Formation

When an air parcel ascends, it cools by adiabatic expansion. Moist air may then reach the saturation point, and, in the presence of CCN, water vapor starts condensing into a cloud. Uplift is more effective in cooling air than other physical mechanisms, such as radiation and conduction, so that this is the common cause of cloud formation. The CCN are necessary for the vapor-liquid transition, a process called heterogeneous nucleation. It appears that sulfate, which originates mostly from $SO_2$ oxidation in the atmosphere, is a major constituent of CCN, both in continental and in maritime environments (Twomey, 1977).

Cloud condensation is accompanied by the release of latent heat, which inhibits the cooling and enhances the uplift. The condensation process is counteracted by entrainment of relatively dry air into the cloud from its environment. While ascending, the cloud may pass the 0°C level. Ice formation, however, is dependent on the presence of ice nuclei (IN). In the absence of IN, supercooled cloud droplets freeze at temperatures as low as −35 to −40°C. In contrast to CCN, IN can be insoluble particles, although their exact nature is not well known. IN are much less abundant in the troposphere than CCN, so that lifting to between −5 and −25°C may occur before significant ice formation takes place. Consequently, the presence of supercooled liquid water clouds is quite common in the lower and middle troposphere.

Cloud droplets are quite stable with respect to their number and size (they grow only slowly), until the air parcel descends again and the droplets evaporate. Sometimes the droplet size distribution changes rapidly during the formation of rain droplets. Often, the precipitation process is initiated when sufficient ice particles are present. The equilibrium water vapor pressure over ice particles is lower than over liquid droplets, so that the ice particles grow by deposition at the expense of the droplets. Most clouds evaporate rather than precipitate, making the CCN available as condensation sites when the air parcel is lifted again.

There are differences in CCN number between maritime and continental environments, which reflect on the cloud microphysics. In clean marine air less than 100 CCN cm$^{-3}$ are usually observed, while continental air generally contains 100-1000 CCN cm$^{-3}$ (Twomey, 1977). Hence the droplet number in maritime clouds is up to an order of magnitude smaller than in comparable cloud types over the continents. The range of droplet sizes (the droplet spectrum) of maritime clouds is usually wider than over the continents. The relatively large number of CCN over land compete for the moisture, so that in continental clouds many small droplets are formed. As clouds mature, their droplet spectra broaden, because larger droplets arise from the coalescence of small droplets. Stratiform clouds generally have somewhat smaller droplets than cumuliform clouds, since water vapor supersaturations are lower and because turbulence, and thus coalescence, are less important in stratus.

The liquid water content (LWC) is quite variable among different cloud types, but also within clouds. This is related to fluctuations in the vertical air velocity. Strong updrafts, which produce rapid cooling and thus high supersaturation, cause high liquid water contents. In most cases the maximum liquid water content is observed in the upper half of the cloud (Pruppacher and Klett, 1980). Despite the high LWC variability, averaging over many observations, it appears that LWCs over land are roughly equal to those over the oceans (comparing the same cloud types). Nevertheless, as indicated above, there are pronounced differences in droplet spectra, radii, and number concentrations. For example, the average droplet radius in marine clouds is approximately twice as large as in continental clouds. Thus marine clouds are characterized by relatively large droplet radii and small droplet number concentrations.

### 4.3.2.2 Average Cloud Characteristics

In the NCAR cloud atlases (Warren *et al.*, 1986; 1988) routine weather observations from meteorological stations and ships have been compiled into cloud maps, distinguishing six cloud types. Although relatively few observations are reported from remote areas, such as the polar regions and the Southern Ocean, a global cloud coverage data set with a resolution of 5° × 5° latitude and longitude is available. Some averages of these cloud coverages are given in Table 4.2. Combining these data with observed vertical extents of the different cloud types, the volume fractions of the troposphere that are covered by clouds were obtained. In many cases the residence time of air parcels in clouds is considerably shorter than the lifetime of clouds. This is a consequence of the rapid vertical transport of air through the clouds. This is important for atmospheric chemistry because air parcels are "processed" repeatedly by clouds. As an example, the typical lifetime of a Cu cloud is about an hour; thus updraft velocities of several m s$^{-1}$ indicate an in-cloud residence time of 20-30 min. Hence the in-cloud air mass is replenished several times during the lifetime of the cloud.

**Table 4.2**
*Zonally Averaged Cloud Properties*

| Latitude | 90-60S | 60-30S | 30-0S | 0-30N | 30-60N | 60-90N |
|---|---|---|---|---|---|---|
| Cb coverage (%) | 0.1 | 0.2 | 0.3 | 0.3 | 0.2 | 0.1 |
| Cu coverage (%) | 1 | 4 | 7 | 7 | 3 | 1 |
| As+Ac coverage (%) | 26 | 25 | 23 | 23 | 22 | 23 |
| Ns coverage (%) | 14 | 9 | 3 | 3 | 7 | 13 |
| St+Sc coverage (%) | 35 | 35 | 22 | 19 | 27 | 31 |
| Ci coverage (%) | 22 | 26 | 15 | 18 | 20 | 23 |
| Mean cloud volume (%) | 8.2 | 6.8 | 4.6 | 4.7 | 5.6 | 7.4 |
| Mean time in clouds: $t_c$ (h) | 3.4 | 3.2 | 2.8 | 2.8 | 3.2 | 3.5 |
| Mean time outside clouds: $t_{nc}$ (h) | 38 | 44 | 58 | 59 | 56 | 44 |

$t_c$ and $t_{nc}$ refer to the mean time spent by air inside and outside clouds in the lower part of the atmosphere, where clouds are present. Total cloud coverage may add up to more than 100% due to the occurrence of multiple cloud layers.

Cb = cumulonimbus, Cu = cumulus, As = altostratus, Ac = altocumulus, Ns = nimbostratus, St = stratus, Sc = stratocumulus, Ci = cirrus

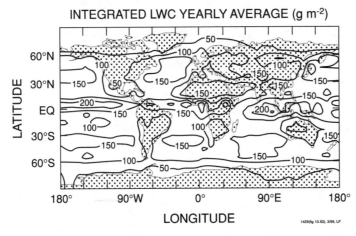

**Figure 4.7.** Model calculated annual mean cloud liquid water column (LWC) (g m$^{-2}$) (courtesy of J. Lelieveld).

To study the combined effect of chemical processes in alternating cloudy and cloud-free periods, assessments of updraft velocities through convective clouds and lifetime estimates of stratiform clouds were used, yielding the average time that air spends in clouds and the time between successive cloud encounters (Table 4.2). Also, the LWC for the average clouds was computed by weighting typical LWCs for different cloud types and locations according to the cloud type coverage. Cb clouds typically have an LWC of about 1-2 g m$^{-3}$, Cu about 0.5 g m$^{-3}$, while stratiform clouds usually have an LWC of 0.1-0.3 g m$^{-3}$. The calculated global average cloud LWC appears to be about 0.3 g m$^{-3}$. Integrating these calculated LWC values over all altitudes, we obtain liquid water columns that appear to vary strongly latitudinally, from less than 50 g m$^{-2}$ at high latitudes up to 200 g m$^{-2}$ in the tropics (Fig. 4.7); this approximately agrees with liquid water column observations from satellites.

## 4.4 Single-Particle Physical Characteristics

Many important particle properties are size dependent (Fig. 4.8). Although atmospheric particles can have complicated morphologies (for example, soot is often present as a chain agglomerate), particles are generally described in terms of equivalent spherical properties, so that particle size is expressed as the diameter of a sphere possessing the same properties, $D_p$. The terminal settling velocity of such a spherical particle is

$$v_t = \frac{1}{18} \frac{D_p^2 \rho_p g C_c}{\mu_a \psi} \tag{4.2}$$

where $\rho_p$ is the density of the particle, $\mu_a$ is the viscosity of the air, $g$ is the gravitational acceleration, $\psi$ is the shape factor (described below), and $C_c$ is the Cunningham correction factor, which accounts for noncontinuum effects as particle diameter approaches the mean free path in the gas. The settling velocity increases approximately with the square of the particle diameter.

The shape factor applies to nonspherical particles, and is defined as the ratio of the resistance force for that particle to the force acting on a sphere of the same volume and velocity. For example, the shape factor for a cube is 1.08. Discussions of the shape factor, the Cunningham correction factor, and details of other aerosol physics concepts briefly presented here are found in the text by Fuchs (1964).

The following formula is used to compute the Cunningham correction factor for spherical particles:

$$C_c = 1 + \frac{2\ell}{D_p} \left[ 1.246 + 0.42 \exp\left(-0.87 \frac{D_p}{2\ell}\right) \right] \tag{4.3}$$

where $\ell$ is the mean free path of the medium,

$$\ell = \frac{\mu_a}{0.499 \, N \mathcal{M} \bar{c}} \tag{4.4}$$

In the above formulas, $\ell$ refers to the mean free path, $N$ is the number concentration of molecules, $\mathcal{M}$ is the molecular mass, and $\bar{c}$ is the mean molecular velocity. For air, $\ell = 0.065$ $\mu$m at 25°C and 1 atm.

The Brownian diffusion coefficient for the particle is computed from

$$D = \frac{kTC_c}{3\pi \mu_a D_p \psi} \tag{4.5}$$

The diffusivity of the particle increases inversely with particle diameter (and varies as $D_p^{-2}$ for very small particles because of the functional form of $C_c$). Thus diffusion (and adherence) to surfaces, including other particles, is an efficient removal mechanism for small particles ($< 0.1$ $\mu$m), whereas gravitational settling and turbulent impaction efficiently remove larger particles ($> 2$ $\mu$m) from lower levels of the atmosphere. For particles with diameters between about 0.1 and 1 $\mu$m, neither mechanism operates very efficiently; particles in this size range thus "accumulate" in the atmosphere.

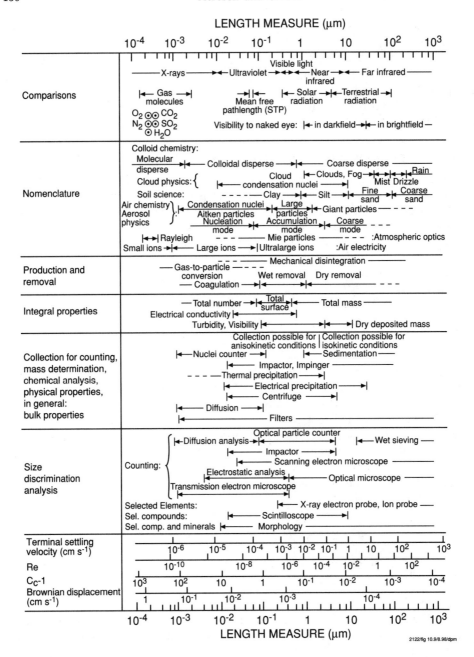

**Figure 4.8.** Characteristics of particles and aerosols as a function of their size (Jaenicke, 1988). $C_c$ is the Cunningham factor, and Re is the Reynold's number.

### 4.4.1 Aerosol Size Distributions

The size range of interest in the characterization of atmospheric aerosol (0.001 to 10 $\mu$m) spans four decades in diameter, and the mass (or volume) range thus spans twelve decades, complicating descriptions of the physical nature of the aerosol. The size distribution density function, $n(D_p)$, is an efficient means for describing some aerosol physical aspects; $n(D_p)\,dD_p$ represents the number of particles in the population having diameters in the range $D_p$ to $D_p + dD_p$ per unit volume. The integral over all diameters yields the total number concentration of particles, $N$:

$$N = \int_0^\infty n(D_p)\, dD_p \qquad (4.6)$$

and thus one may also express $n(D_p) = dN/dD_p$.

The distributions of particle surface area $[n_s(D_p)]$ and volume $[n_v(D_p)]$ with respect to particle size may be derived from the size distribution function: $n_s(D_p) = \pi D_p^2 n(D_p)$ and $n_v(D_p) = (\pi/6) D_p^3 n(D_p)$. If all the particles have the same density, the volume distribution function may be multiplied by the density to give the mass distribution function. The total aerosol surface area, volume, and mass are the integrals (over all diameters) of the respective distribution functions. Although written here with diameter $D_p$ as the independent variable, other choices are possible. The particle volume and particle mass are often used; the form of the surface area, volume, and mass distribution functions are modified as appropriate. Figure 4.9 shows the number, surface area, and volume distributions for a representative aerosol population. The smallest particles dominate the number distribution, while the largest particles contribute most heavily to the volume (or mass) distribution (cf. Fig. 4.3). This is an important consideration in aerosol sampling strategies, since a few large particles may constitute a large percentage of the sampled particulate mass, and subsequent chemical analyses thus describe the chemical nature of only a small fraction of the total number of particles. Processes that are related to the total aerosol surface area (for example, heterogeneous chemical reactions occurring on particle surfaces) will be most heavily influenced by intermediate-sized particles in the population.

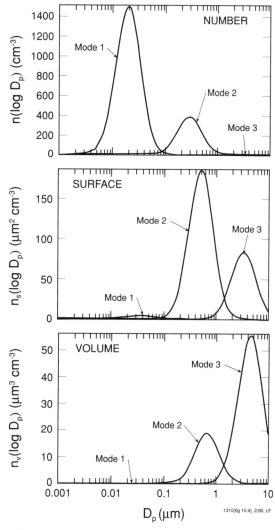

**Figure 4.9.** Number, surface area, and volume distributions for a trimodal aerosol; each mode is lognormal with a geometric standard deviation of 1.7. Geometric mean diameters used: Mode 1: $\overline{D_{pg}} = 0.02$ μm; Mode 2: $\overline{D_{pg}} = 0.3$ μm; Mode 3: $\overline{D_{pg}} = 2$ μm.

Mean diameters are computed as follows:

$$\overline{D}_{p,n} = \frac{1}{N} \int_0^\infty D_p n(D_p)\, dD_p \qquad (4.7)$$

$$\overline{D}_{p,s} = \left[\frac{1}{N}\int_0^\infty D_p^2 n(D_p)\, dD_p\right]^{\frac{1}{2}} \tag{4.8}$$

$$\overline{D}_{p,v} = \left[\frac{1}{N}\int_0^\infty D_p^3 n(D_p)\, dD_p\right]^{\frac{1}{3}} \tag{4.9}$$

where $\overline{D}_{p,n}$ is the number mean diameter, $\overline{D}_{p,s}$ is the surface area mean diameter, and $\overline{D}_{p,v}$ is the volume mean diameter.

The median diameters are those diameters above and below which one-half of the total integrated property (*e.g.*, total surface area of the aerosol) lies. They are denoted here by subscripts without overbars. For a given aerosol population, $D_{p,v} > D_{p,s} > D_{p,n}$, as seen in Figure 4.9 for Mode 2.

Since particle size spans several orders of magnitude, the logarithm of the diameter is often used as the independent variable in the size distribution function. The size distribution function based on $\log D_p$ is related to $n(D_p)$ as follows:

$$n(D_p) dD_p = \tilde{n}(\log D_p)\, d\log D_p \tag{4.10}$$

### 4.4.2 Representation of Particle Size Distributions

Particle size distributions may be represented as histograms (most closely related to measurements, which usually obtain integrated properties over some size range) or as continuous functions (Fig. 4.10), which may be derived from fits to measurements or assumed from a theoretical basis. One of the most commonly used representations is the lognormal distribution function:

$$n(\ln D_p) = \frac{N}{(2\pi)^{\frac{1}{2}} \ln \sigma_g} \exp\left[-\frac{(\ln D_p - \ln \overline{D}_{pg})^2}{2 \ln^2 \sigma_g}\right] \tag{4.11}$$

The equation describes a Gaussian distribution, with the logarithm of particle diameter as the normally distributed variable. The two parameters of the distribution are the geometric standard deviation, $\sigma_g$, which determines the breadth of the distribution ($\sigma_g = 1$ for a monodisperse distribution, *i.e.*, all particles are the same size), and the geometric mean (or median) diameter, $\overline{D}_{pg}$. The popularity of the lognormal distribution function for representing atmospheric aerosols stems from both observations, which generally support such a functionality for many types of aerosol, and from its mathematical tractability, which makes the function attractive for use in modeling studies. The two parameters of the distribution are sufficient to describe

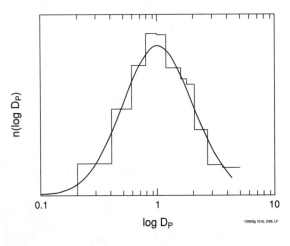

**Figure 4.10.** Histogram of aerosol number concentration data with best-fit lognormal distribution superimposed.

## Table 4.3
### Properties of Distribution Functions and Parameters for Lognormal Distributions

| Property | Defining Relation | $\nu$ for Lognormal Aerosol $\overline{D}_{p\nu} = \overline{D}_{pg} \exp\left(\nu \ln^2 \sigma_g\right)$ | Description |
|---|---|---|---|
| Number mean diameter | $\overline{D}_p = \dfrac{1}{N} \displaystyle\int_0^\infty D_p n(D_p)\, dD_p$ | 0.5 | average diameter of the population |
| Number median diameter | $\dfrac{N}{2} = \displaystyle\int_0^{D_{p,\text{median}}} n(D_p)\, dD_p$ | 0 | diameter below which one-half of the number of particles lie and above which one-half of the number of particles lie |
| Number mode | $\left[\dfrac{dn(D_p)}{dD_p}\right]_{D_p,\text{mode}} = 0$ | $-1$ | diameter at which the largest number of particles lie, i.e., the most frequently occurring number |
| Surface area mean diameter | $D_{p,\text{avg surface area}}^2 = \dfrac{1}{N} \displaystyle\int_0^\infty D_p^2 n(D_p)\, dD_p$ | 1 | diameter of that particle whose surface area equals the mean surface area of the population |
| Volume mean diameter | $D_{p,\text{avg volume}}^3 = \dfrac{1}{N} \displaystyle\int_0^\infty D_p^3 n(D_p)\, dD_p$ | 1.5 | diameter of that particle whose volume equals the mean volume of the population |
| Surface area median diameter | $\displaystyle\int_0^{D_{p,\text{surf median}}} D_p^2 n(D_p)\, dD_p = \dfrac{1}{2} \displaystyle\int_0^\infty D_p^2 n(D_p)\, dD_p$ | 2 | diameter below which one-half of the particle surface area lies and above which one-half of the particle surface area lies |
| Volume median diameter | $\displaystyle\int_0^{D_{p,\text{vol median}}} D_p^3 n(D_p)\, dD_p = \dfrac{1}{2} \displaystyle\int_0^\infty D_p^3 n(D_p)\, dD_p$ | 3 | diameter below which one-half of the particle volume lies and above which one-half of the particle volume lies |

its shape, allowing easy comparison between aerosol samples. Table 4.3 summarizes some definitions of distribution parameters and useful relationships between such parameters for lognormal distributions.

### 4.4.3 Interactions with Radiation

Electromagnetic radiation is scattered and absorbed by molecules and particles in the atmosphere. A comprehensive discussion of these interactions is given in the text by Bohren and Huffman (1981); some basic relationships are reviewed here.

The size parameter, $\alpha$, is generally used to express the relationship between particle diameter, $D_p$, and wavelength, $\lambda$, of the radiation with which the particle is interacting

$$\alpha = \frac{\pi D_p}{\lambda} \qquad (4.12)$$

For visible light ($0.4 \leq \lambda \leq 0.7$ $\mu$m), the average wavelength is about 0.53 $\mu$m. The scattering of light by particles less than about 0.05 $\mu$m in diameter ($\alpha \ll 1$) is described by the theory of Rayleigh scattering; the scattering due to large particles (larger than about 100 $\mu$m) ($\alpha \gg 1$) is described by geometric optics. In the intermediate region, for which particle diameter and wavelength are of roughly the same order of magnitude, scattering phenomena are complicated. The computation of aerosol radiative properties in this regime is accomplished using Lorenz-Mie theory, again under the assumption of sphericity. Atmospheric particles may both absorb and scatter solar radiation; optical properties are dependent upon size and upon the real and imaginary parts of the refractive index. The refractive index depends upon wavelength, especially near absorption bands.

Table 4.4 lists the refractive indices at $\lambda$ near 0.53 $\mu$m of several species of atmospheric interest. Only those species with nonzero imaginary parts of their refractive index can contribute to absorption. All other species contribute to light extinction by scattering, with the magnitude of the contribution dependent upon the size of the particle, the value of the refractive index, and the amount of suspended material present.

Extinction is the attenuation of light along a beam by scattering and absorption due to particles present in the beam. In the absence of multiple scattering, the ratio of the light intensity observed at a receptor ($I$) to that incident upon the sample ($I_0$) is given by the Beer-Lambert Law,

$$\frac{I}{I_0} = \exp\left(-\beta_e L\right) \qquad (4.13)$$

where $L$ is the path length through the sample and $\beta_e$ is called the extinction coefficient for the aerosol. The extinction efficiency of a single particle, $Q_e$, is the ratio of the radiant power scattered and absorbed by a particle to the radiant power incident on the geometric cross section of the particle. For a single particle, the extinction

**Table 4.4**
Index of Refraction ($\lambda = 0.53$ $\mu$m) and Densities (g cm$^{-3}$) at 25°C

| Species | Refractive Index | Density |
|---|---|---|
| Vacuum | 1.0 | |
| Water (vapor) | 1.00025 | $2.3 \times 10^{-5}$ |
| Water (liquid) | 1.33 | 1.0 |
| Ammonium sulfate | 1.53 | 1.76 |
| Sodium chloride | 1.544 | 2.165 |
| Soot | 1.96-0.66i | 2.0 |
| Organic carbon | 1.55 | 1.4 |
| Benzene (liquid) | 1.5012 | 0.877 |
| Ice | 1.309 | 0.98[a] |
| Polystyrene latex | 1.59 | |

[a] At $-30°$C.

efficiency, $Q_e$, and extinction coefficient, $\beta_e$, are related by

$$\beta_e = \frac{\pi D_p^2 Q_e}{4} \qquad (4.14)$$

and $\beta_e$ is the sum of two terms, the scattering ($\beta_s$) and the absorption ($\beta_a$) coefficients. Since the atmospheric aerosol is polydisperse, the contributions over all sizes of particles must be included, so that (for one wavelength)

$$\beta_e = \int_0^\infty \frac{\pi}{4} D_p^2 Q_e (m, \alpha) \, n \, (D_p) \, dD_p \qquad (4.15)$$

where $n(D_p)$ is the aerosol size distribution function, discussed in Section 4.4.1, and $m$ is the refractive index. If $Q_e$ can be computed, then light attenuation caused by a population of aerosol particles may be calculated from Eqs. (4.15) and (4.13). Values of $Q_e$ generally range from 0 to 5 (see Fig. 4.11). However, there is no simple approximation for the computation of $Q_e$ for all particle sizes, although approximations exist for the limiting cases.

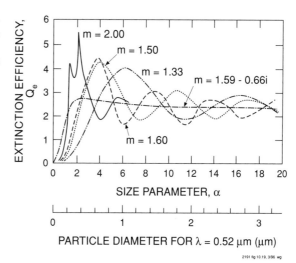

**Figure 4.11.** Extinction efficiency versus particle size for spheres, m = index of refraction (after Hodkinson, 1966; reprinted by permission of the publisher Academic Press).

The aerosol optical depth is the integral of the aerosol extinction coefficient over a path length through the atmosphere:

$$\tau_A (\lambda) = \int_{z_1}^{z_2} \beta_e (\lambda, z) \, dz \qquad (4.16)$$

Since the extinction coefficient $\beta_e$ depends upon the aerosol chemical composition, size distribution, and mass concentration, as seen from Eq. (4.15), all these parameters in turn affect the optical depth. The aerosol optical depth is the quantity generally derived from satellite information, and clearly shows variations that can be attributed to phenomena such as volcanic eruptions.

"Rayleigh scattering" is the special case for which the particle size is much smaller than the wavelength of the radiation, that is, $\alpha \ll 1$. The scattered intensity is symmetric in the forward and backward directions, proportional to the square of the particle volume, and inversely proportional to the fourth power of the wavelength (Twomey, 1977); however, absorption is proportional to volume, so that the absorption per unit volume (or mass) becomes independent of size (and is nonzero) for very small absorbing particles. Such small particles ($D_p \ll 0.05$ μm) are generally assumed to contribute negligibly to light scattering. At sea level, the Rayleigh extinction coefficient due to background atmospheric gases is approximately $13 \times 10^{-6}$ m$^{-1}$ at $\lambda = 0.53$ μm. For particles with $\alpha > 20$, $Q_e$ approaches the limiting value of 2; that is, the particle removes twice as much light as expected from simply considering its projected area; the additional extinction arises from Fresnel diffraction.

A discussion of the use of Lorenz-Mie theory to compute extinction efficiencies for $\alpha \sim 1$ is found in Twomey (1977). A notable feature is the shift from angular symmetry in scattered intensity observed for $\alpha \ll 1$ to a peak in forward scattering for $\alpha \sim 1$. For this reason, illuminated haze appears different when viewed from different "sides." That is, if the haze is between the observer and the sun, the forward scattering of light causes visibility degradation, and the haze may appear bright against the background sky. If the sun is behind the observer viewing the haze, the haze backscatters less light into the observer's sight path, visibility is less degraded, and the haze may appear dark against skylight. The potentially large values of $Q_e$ in the Lorenz-Mie regime, due primarily to contributions from scattering, are also the reason that particles with diameters in the range from 0.1 to 2 μm have the greatest impact with respect to visibility reduction. Figure 4.11 gives examples of the extinction efficiency as a function of particle size and refractive index.

Computed values of $Q_e$ allow estimates of the visual range $(x)$ to be made using the Koschmieder equation, $x = 3.912/\beta_e$, which assumes that a threshold contrast of 2% is detectable by the "average" observer. The visual range in air containing a given mass of water present in different forms is given in Table 4.5. Although several of the computed values are unrealistically low, the large variations in estimated visual range demonstrate the sensitivity of light scattering and absorption to particle size; that is, the visual range depends strongly on the size distribution into which the total suspended aerosol mass concentration is apportioned.

**Table 4.5**
*Extinction and Visual Range in Air Containing 18 g m$^{-3}$ of Water[a] in Different Forms* (from Hinds, 1982)

| Particle Size (μm) | Extinction[b] $1 - I/I_0$ | Visual Range[c] (km) |
|---|---|---|
| Vapor | $1.8 \times 10^{-7}$ | 220 |
| 0.01 | $3.8 \times 10^{-5}$ | 1.0 |
| 0.1 | 0.29 | $1.1 \times 10^{-4}$ |
| 1.0 | 0.64 | $3.8 \times 10^{-5}$ |
| 10 | 0.052 | $7.4 \times 10^{-4}$ |
| 1 mm (rain) | $5.3 \times 10^{-4}$ | 0.074 |

[a] Equivalent to saturation at 20°C
[b] L = 1.0 cm and $\lambda = 0.5$ μm.
[c] From Koschmieder equation.

### 4.4.4 Photodissociation Frequencies

Solar radiation in the ultraviolet (UV) range, especially at wavelengths $\lambda < 315$ nm, constitutes the driving force of tropospheric photochemistry. The presence of clouds has several consequences for the calculation of actinic fluxes and photolysis rates. Incoming radiation can be partly backscattered owing to the difference in refractive index between cloud droplets and air. Clouds may backscatter up to about 85% (on average 50%) of the incident short-wave radiation, which has important consequences for the photochemistry in the subcloud layers, but also for the regions above clouds, which are subject to enhanced doses. Further, multiple scattering of radiation within clouds causes diffusion, so that the radiance $I(\theta, \phi)$ becomes independent of direction. (When one cannot locate the Sun through the cloud deck, light has been completely diffused. Some thin clouds allow location of the Sun as a bright spot; they only cause partial diffusion.) Diffusion causes an increase of the effective path length of radiation through the clouds, and thus of the photon density, which enhances the chance of collisions between photons and molecules that can photodissociate. Note that the total radiant energy is decreased by the amount of radiation that is reflected by clouds. Despite the radiant energy decrease, actinic fluxes in the higher levels in clouds may be enhanced by a factor 2 to 5 compared to cloud-free conditions as a consequence of the longer path length of diffuse radiation. On the other hand, in the lower cloud levels and below clouds decreases of up to 80-90% can occur compared to clear-sky values, which can be included in models by assuming actinic flux profiles as a function of cloud thickness and liquid water content (Madronich, 1987).

## 4.5 Gas-to-Particle Conversion

The exchange of material between the gas phase and the particle phase can take several different forms; these include the nucleation and growth of fine particles, the adsorption of gases onto the surfaces of solid particles, the uptake of water by particles to form aqueous solutions and cloud drops, and the dissolution of soluble trace gases in these drops. These processes have several effects. First, they alter the number and size of atmospheric particles and therefore the optical properties of the atmosphere and the properties of clouds. Also, the transport and removal of trace species depend on these processes. Finally, the chemical reactions that occur on surfaces or in solution are often very different from those that occur in the gas phase; as a result, the particles may act as catalysts for chemical reactions that otherwise would not occur in the atmosphere.

### 4.5.1 Homogeneous Nucleation

Some of the products of gas-phase chemical reactions, including sulfuric acid and highly oxygenated organic compounds, have very low vapor pressures, and thus partition preferentially into the liquid or solid state. One consequence of the formation of these species in the gas phase is the nucleation of new particles; however, the details of this process are poorly understood.

Classical homogeneous nucleation theory describes nucleation in terms of the formation of small clusters that must reach a critical size before they are stable. The free energy of formation of a small cluster is positive, since the decrease in entropy more than offsets the decrease in enthalpy. At this point, the growth is reversible. However, after a certain critical number of molecules cluster together, the overall free energy change becomes negative due to the decreased surface tension, and the droplets grow spontaneously. More details can be found in Seinfeld and Pandis (1998).

Very small particles of a pure substance are not stable unless they are in a supersaturated environment. That is, a criterion for stability is that the partial pressure of the substance over the drop surface is equal to the partial pressure of the substance in the gas phase. However, the vapor pressure over a curved surface is greater than that over a flat surface, with the enhancement given by the Kelvin equation:

$$p_A = p_A^\circ \exp\left[\frac{4\widetilde{\sigma}\bar{v}_w}{RTD_p}\right] \qquad (4.17)$$

where $\widetilde{\sigma}$ is the surface tension, $v_w$ is the partial molar volume, and $p_A^\circ$ and $p_A$ are the vapor pressures over the flat and curved surfaces, respectively. For a pure water drop of diameter 0.01 $\mu$m to be stable and not evaporate, the ambient water vapor

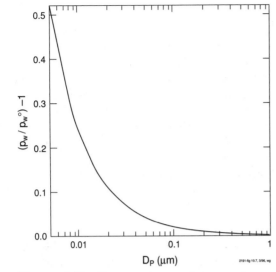

**Figure 4.12.** Vapor pressure enhancement over a pure water droplet at 298 K as a function of droplet diameter. The curve is given by the Kelvin equation (4.17).

concentration must be about 120% of its saturation value (Fig. 4.12). Here, the supersaturation $S$ is defined as the ratio of the vapor pressure of the nucleating species $A$ in the gas phase, $p_A$, to the saturation vapor pressure of that species over a flat surface and at the ambient temperature:

$$S = \frac{p_A}{p_A^\circ} \qquad (4.18)$$

For equilibrium over a drop surface, $S$ must be equal to the exponential term in the Kelvin equation:

$$S = \frac{p_A}{p_A^\circ} = \exp\left[\frac{4\tilde{\sigma}\bar{v}_w}{RTD_p}\right] \qquad (4.19)$$

If supersaturated conditions are generated in the gas phase (for example, by expansion processes which cool a particle-free gas), it may be possible to form a stable droplet from the vapor. Such spontaneous particle formation is termed homogeneous nucleation; the case of a pure substance nucleating (such as a water droplet) is called homomolecular homogeneous nucleation. An important example of heteromolecular homogeneous nucleation is the formation of sulfuric acid/water particles. Classical nucleation theory combines thermodynamic and kinetic approaches to predict the conditions for onset of spontaneous particle formation and the rate of new particle formation. It should be mentioned that many approximations are involved (*e.g.*, the application of bulk physiochemical properties such as surface tension to clusters of molecules), and values of nucleation rates determined from experimental investigations do not necessarily agree with those computed from theory. In clean regions of the atmosphere, rates of nucleation of particles are observed that are much more rapid than can be explained by simple bimolecular nucleation of $H_2SO_4/H_2O$ (Weber *et al.*, 1997). Coffmann and Hegg (1995) have suggested that a third molecule, ammonia, is involved in the nucleation process by lowering the $H_2SO_4$ vapor pressure.

Since the formation of new particles requires that a considerable energy barrier be overcome, low-vapor-pressure species may instead condense on existing particles. This results in the growth of the extremely small particles (a few nanometers in diameter) that are formed by nucleation. Such low-vapor-pressure species may also be collected by particles that are directly injected into the atmosphere, such as sea salt or soot. Modern aerosol models include parameterizations of nucleation and growth of particles, with explicit treatment of both the aerosol size distribution and chemical composition (Lurmann *et al.*, 1997; Meng *et al.*, 1998; Fitzgerald *et al.*, 1998). In the continental troposphere there is usually sufficient aerosol present to scavenge low volatility gases such as those formed from the oxidation of organic compounds. The partitioning of oxygenated organics between the gas and particle phases has been examined in some detail by Seinfeld and coworkers (Bowman *et al.*, 1997; Odum *et al.*, 1997; Hoffmann *et al.*, 1997). Once adsorbed, the organics can modify the physical properties of the aerosol, including their ability to take up water and hence become cloud condensation nuclei (Saxena *et al.*, 1995).

## 4.5.2 Gas-Solid Equilibria

In some cases, gas-phase species that are individually quite volatile may together form low-vapor-pressure solids; the classic example of this is the formation of $NH_4NO_{3(s)}$ from $NH_{3(g)}$ and $HNO_{3(g)}$. When large amounts of both $NH_3$ and $HNO_3$ are present in the atmosphere, the formation of $NH_4NO_3$ limits the amounts of these species that are present in the gas phase. It is expected that when $NH_4NO_3$ is present, the partial pressures of $NH_3$ and $HNO_3$ should be determined by the equilibrium constant for the reaction

$$NH_4NO_{3(s)} \rightleftharpoons NH_{3(g)} + HNO_{3(g)} \quad (4.20)$$

that is, that

$$p_{HNO_3} p_{NH_3} = K_p(NH_4NO_3) \quad (4.21)$$

where the equilibrium constant, $K_p$, is a function of temperature and, if an aqueous solution forms, relative humidity. Thermodynamic data may be used to compute this equilibrium constant accurately (Mozurkewich, 1993). Figure 4.13 shows a comparison of measured vapor pressure products (points) to $K_p$ based on data from various sources (lines). Some of the data are in reasonable agreement with the equilibrium constant, while others are in substantial disagreement. Wexler and Seinfeld (1992) have shown that the establishment of equilibrium should not be regarded as instantaneous, and that both the atmospheric transport and the kinetics of aerosol formation should be considered. Such effects may explain the lack of equilibrium in the data of Figure 4.13.

When $NH_4NO_3$ is present, the dry deposition of $NH_3$ and $HNO_3$ (both very rapid) has less of an effect on the gas-phase concentrations of these species than might otherwise be expected. This is because as these species are deposited, they are replaced by evaporation of solid $NH_4NO_3$. This substantially alters the apparent deposition rates for both the gases and the particles (Huebert et al., 1988; Pandis and Seinfeld, 1990).

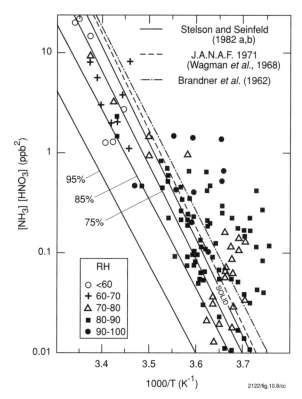

**Figure 4.13.** Measured concentration products of $[NH_3][HNO_3]$ as a function of temperature and relative humidity, compared with calculated values (lines).

## 4.5.3 Particle Hygroscopicity

Certain ionic salts are able to take up water vapor from the atmosphere, a property known as hygroscopicity. If the amount of water taken up is sufficient for a solution

**Table 4.6**
*Relative Humidity of Deliquescence (at 25°C) of Various Salts*

| Salt | Deliquescence RH |
|---|---|
| $Pb(NO_3)_2$ | 98 |
| $K_2SO_4$ | 97 |
| $Na_3PO_4 \cdot 12H_2O$ | 95 |
| $Na_2SO_4 \cdot 10H_2O$ | 93 |
| $NaBrO_3$ | 92 |
| $KH_2PO_4$ | 92 |
| $KNO_3$ | 92 |
| $Na_2CO_3 \cdot 10H_2O$ | 90 |
| $ZnSO_4 \cdot 7H_2O$ | 90 |
| $K_2CrO_4$ | 88 |
| $KHSO_4$ | 86 |
| $KBr$ | 84 |
| $KCl$ | 84 |
| $NH_4HSO_4$ | 40 |
| $(NH_4)_2SO_4$ | 80 |
| $NH_4Cl$ | 80 |
| $NaCl$ | 76 |
| $NaNO_3$ | 74 |
| $CH_3COONa \cdot 7H_2O$ | 78 |
| $(NH_4)_3H(SO_4)_2$ | 69 |
| $NH_4NO_3$ | 62 |
| $NaBr \cdot 2H_2O$ | 58 |
| $Mg(NO_3)_2$ | 53 |
| $K_2CO_3 \cdot 2H_2O$ | 43 |
| $MgCl_2 \cdot 6H_2O$ | 33 |
| $CaCl_2 \cdot 6H_2O$ | 32 |

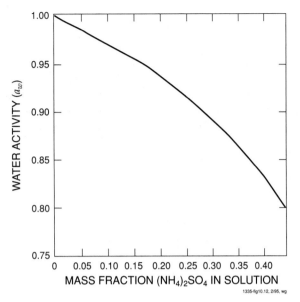

**Figure 4.14.** Water activity of an ammonium sulfate solution as a function of the mass fraction of ammonium sulfate in solution. Figure is based on relationship given by Tang (1976).

to form spontaneously, the salt is said to deliquesce. The transition from solid to solution occurs at a well-defined humidity. Table 4.6 lists some common salts and the relative humidity (at 25°C) at which they deliquesce. The ambient relative humidity, RH, varies with temperature for a given partial pressure of water vapor in the ambient air, $p_w$, and is defined by

$$\text{RH}(T) = \frac{p_w}{p_w^\circ(T)} \tag{4.22}$$

where $p_w^\circ(T)$ is the saturation vapor pressure of water at the ambient temperature $T$. As the RH is increased beyond the deliquescence point, the salt continues to add water to satisfy the equilibrium relationship

$$\text{RH} = a_w = \frac{p_{w,s}}{p_w^\circ(T)} \tag{4.23}$$

where $a_w$ is the activity of water in the saline solution, and the subscript $s$ denotes that the water vapor pressure is that above the solution. The quantity $a_w$ has been measured and tabulated for a variety of single-salt and single-acid or -base water solutions as a function of the amounts of solute and water in the solution; by finding the appropriate $a_w$ that satisfies Eq. (4.23), one may then look up the solution concentration that produces this $a_w$, and thus determine the amount of water associated with the equilibrium condition. Figure 4.14 gives this relationship for ammonium sulfate.

Equation (4.23) must be modified to account for the fact that the vapor pressure over a curved surface (*e.g.*, that of a spherical particle) always exceeds that of the same substance over a flat surface. Use of the Kelvin equation (4.17) leads to the exponential term in the modified equation,

$$\text{RH} = a_w \exp\left(\frac{4\widetilde{\sigma}\overline{v}_w}{RTD_p}\right) \tag{4.24}$$

where $\widetilde{\sigma}$, the surface tension of the solution, and $\overline{v}_w$, the partial molar volume of water in the solution, are both a function of composition.

Soluble liquids, such as sulfuric acid, $H_2SO_4$, do not exhibit a sharp deliquescence point near room temperature, but rather take up water continuously at all (even very low) relative humidities. Figure 4.15 compares particle growth behavior at 25°C for sulfate [S(VI)] species of various degrees of neutralization; it is seen that, the more acidic the particle, the more water will be incorporated at a given RH.

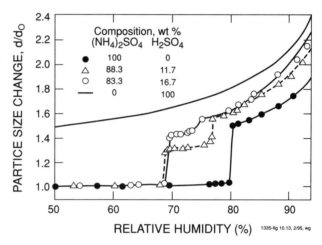

**Figure 4.15.** Ratio of wetted diameter (d) to dry diameter ($d_0$) ("growth curves") for sulfate aerosols at 25°C (Tang et al., 1978). The deliquescence points for the ammonium salts are clearly visible at 68% and 80% relative humidity, where large increases in volume occur.

## 4.6 Acid-Base Reactions of Aerosol Particles

The acidity in aerosol particles is an important component of acid deposition. Virtually all the $H_2SO_4$ formed in the atmosphere ends up in the particle phase. This is eventually deposited to the surface, either as particles (dry deposition) or in rainfall (wet deposition). However, in the lower troposphere a great deal of this acidity is neutralized by $NH_3$; atmospheric particles typically contain 1.5 to 2 $NH_4^+$ ions for each $SO_4^{2-}$ (*e.g.*, Quinn et al., 1992).

The uptake of acidic or basic gases alters the pH of aerosol particles and cloud drops and therefore the chemical environment in which condensed-phase reactions take place. This affects the solubilities of gases that are themselves acidic ($SO_2$, $CO_2$, and $HCOOH$) or basic ($NH_3$). Even if the solution concentrations are not affected by pH, the rates of chemical reactions may be affected since many reactions are acid or base catalyzed. The properties of particles can be significantly impacted by uptake of acidic species. For example, above 75% RH, sea salt particles form solution droplets; the $Cl^-$ ions in these droplets should be in equilibrium with gas-phase HCl according to Eq. (4.25).

$$\text{HCl}_{(g)} \rightleftharpoons \text{H}^+{}_{(aq)} + \text{Cl}^-{}_{(aq)} \tag{4.25}$$

The equilibrium constant at 15°C for this reaction is $5.5 \times 10^6$ mol$^2$ L$^{-2}$ atm$^{-1}$ (Brimblecombe and Clegg, 1988). Because of the presence of dissolved $CO_3^{2-}$ salts, sea water is basic with a pH near 8. As a result, the hydrogen ion concentration is low ($\sim 10^{-8}$ mol L$^{-1}$) and the partial pressure of HCl, given by

$$p_{\text{HCl}} = \frac{[\text{H}^+][\text{Cl}^-]}{K_{eq}} \tag{4.26}$$

is only about $5 \times 10^{-15}$ atm for a reasonable particle Cl$^-$ concentration of 3 mol L$^{-1}$. The particles will first pick up ambient acidic gases until the ambient HCl is in equilibrium with the H$^+$ in the particle, at about pH 3.7. As further HNO$_3$, SO$_2$, or H$_2$SO$_4$ dissolves, the pH drops, and HCl is lost from the particles. Eventually, a substantial amount of HCl is driven off the particles. When sea salt particles are collected and analyzed, the ratio of Cl/Na is generally found to be substantially less than in sea water (*e.g.*, Sievering *et al.*, 1990). Much of the missing Cl$^-$ can be accounted for by increased amounts of NO$_3^-$ and SO$_4^{2-}$. The additional SO$_4^{2-}$, in excess of what is normally present in sea water, is usually referred to as non-sea-salt sulfate, nss-SO$_4$.

## 4.7 Removal Processes Associated with Aerosols

The mean residence time of an aerosol in the troposphere may generally be computed from

$$\tau = \frac{1}{\beta_d + \beta_w + \beta_t} \tag{4.27}$$

where $\beta_d$ is the first-order loss rate constant for dry deposition, $\beta_w$ is that for wet deposition, and $\beta_t$ is that for transport to the stratosphere. Transport to the stratosphere is a slow process and is not important for tropospheric aerosols. Figure 4.16, from Jaenicke (1988), presents estimates of the residence time of aerosol particles in the atmosphere as a function of particle radius. Three regimes are apparent—thermal diffusion, wet removal, and dry removal. The residence time can vary over several orders of magnitude, from a few minutes to a month, depending upon particle size and other properties such as hygroscopicity, that affect wet removal.

### 4.7.1 Dry Removal of Aerosols

Holsen and Noll (1992) have reviewed models for size-dependent dry particle deposition. The dry deposition is dominated by large particles, since models predict that deposition velocities increase rapidly with particle diameter (Figs. 4.16 and 4.17). The variation of the deposition velocity is reflected in the lifetime estimates shown in Figure 4.16 (Curve F) for three levels in the atmosphere. The solid lines are "overall" mean residence time estimates. The shaded regions represent the range of estimates of aerosol lifetime due to different processes, denoted by the letter symbols, and are used to bound the solid-line estimates; in other words, a single process may indicate a long residence time, but for the particle size in question there may be other faster processes that actually limit the residence time. The shaded regions show this conveniently, and also help show the range of values associated with the solid line estimates.

### 4.7.2 Wet Removal of Aerosols and Gases

Precipitation is an important process for cleaning the atmosphere of trace constituents. For aerosol particles with diameters between 0.1 and 1 $\mu$m, precipitation scavenging is the primary removal mechanism from the atmosphere. Clouds and precipitation

*Figure 4.16.* The residence time of aerosols ($\tau$) in seconds (left axis) and days (right axis), as a function of particle radius, $r$ ($\mu m$). The shaded areas represent estimates of the lifetimes made as follows: I, molecular or ionic clusters; C, for coagulation of particles of the various sizes, this gives an idea of whether coagulation is a fast or slow removal mechanism, and relates to particle diffusivity, which is one of the quantities shown in Fig. 4.7, "Brownian displacement." The Brownian displacement decreases rapidly with particle diameter, so lifetime increases rapidly; P, range of estimates for removal by precipitation; F, gravitational settling; A, derived from the spatial distribution of Aitken particles; R, derived from the distribution of small radioactive particles.

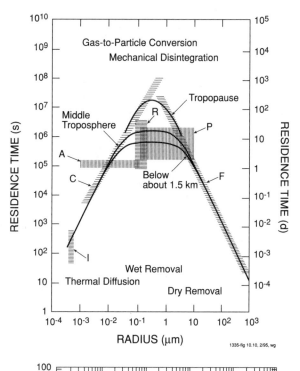

*Figure 4.17.* Comparison of deposition velocities computed from several different models, as indicated. Assumed parameters: particle density, 1.5 g cm$^{-3}$; wind speed, 6.7 m s$^{-1}$; friction velocity, 34.1 (Holsen and Noll, 1992; reprinted with permission from *Environ. Sci. Technol.*, copyright 1992, American Chemical Society).

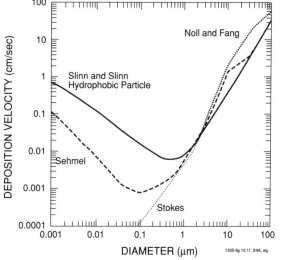

may scavenge aerosols through nucleation scavenging, Brownian diffusion, or impaction scavenging. Nucleation scavenging is the process where particles (or cloud condensation nuclei, CCN) serve as nuclei for cloud droplet formation when water vapor supersaturates. In the absence of CCN, the condensation process would require a water vapor supersaturation of a few hundred percent, whereas in reality supersaturation rarely exceeds 2% (Pruppacher and Klett, 1980) because the vapor condenses on the CCN, and this process requires only about 0.1% supersaturation. For aerosols having diameters in the range 0.1-10 $\mu m$, nucleation scavenging is a very efficient removal process (Fig. 4.16). Interstitial aerosol may be collected by cloud drops or ice through diffusional processes, but this mechanism is insignificant compared to other scavenging processes. Interstitial and below-cloud aerosol also may be captured through impaction by falling precipitation. Impaction scavenging, while not insignificant, is relatively unimportant for CCN removal compared with nucleation scavenging. Trace gases are absorbed into the cloud and rain drops and may be removed by the falling precipitation.

Cloud-scale and mesoscale models that parameterize the cloud microphysics are able to describe the scavenging processes of aerosols and trace gases (Hegg et al., 1986; Chaumerliac et al., 1987). The calculation of wet deposition of trace gases and aerosols to the ground may be explicitly made in these models since the mass of the species (or aerosol) in precipitation is explicitly known. However, for regional and global-scale models, the calculation of wet removal of aerosols and gases must be simply parameterized since complex microphysics schemes are not incorporated in these models, and little detail is known of the characteristics of subgrid-scale clouds.

A simplified approach to describing the removal of aerosol and soluble gases by rain is to approximate the pseudo-first-order wet removal coefficient as a function of precipitation rate at the surface, the efficiency of uptake, the liquid water content of the cloud, and the fraction of precipitation released at a certain height interval (Junge and Gustafson, 1957). Giorgi and Chameides (1985) modified the first-order rainout parameterization by taking into account the species' solubility and the intermittent nature of precipitation. For regional and global models that take into account the hydrological cycle, wet removal of species may be estimated through evaluation of the flux of precipitation in that layer and in-cloud scavenging (Roelofs and Lelieveld, 1995). For these large-scale models, the fractional cloudiness in a grid box and the mass fluxes in subgrid (convective) clouds also must be accounted for.

Calculations from the first-order loss method show that typical timescales connected with the removal by rain of soluble trace species range from 2 days in the tropics to more than a week in high latitudes. As a consequence of the continuous cycling of air between clear and cloudy conditions in the troposphere on a time scale of 2-3 days, the chemical composition of the troposphere is significantly affected by cloud chemical processes. Characteristic time scales of the cloud processes involved are given in Table 4.7.

**Table 4.7**
*Characteristic Timescales in Atmospheric Chemistry Associated with Clouds*

| | |
|---|---|
| Cloud cycling | few days |
| Wet removal of soluble species | ~1 week |
| Gas-aqueous-phase exchange | 0.1 s |
| Aqueous-phase diffusion | 0.01 s |
| Aqueous chemical reactions: | |
| OH radicals | $10^{-4}$ s to minutes |
| S[IV] | minutes to hours |

## 4.8 Solubility of Gases in Droplets

The oxidation of trace species in the aqueous phase is important to the formation of species associated with "acid rain," to gas-to-particle conversion processes that modify atmospheric aerosols, and to the chemical cycles of various species; for example, Langner and Rodhe (1991) estimate that more than 70% of the global oxidation of $SO_2$ to $SO_4^{2-}$ occurs in the aqueous phase. The reactions that occur in atmospheric droplets are basically the same as those that occur in bulk aqueous solutions in the laboratory.

There are, however, some additional considerations that must be taken into account. First, except for cloud, rain, and fog droplets, liquid atmospheric particles are highly concentrated solutions. The very high ionic strengths of sea-salt aerosol and haze particles (sometimes in excess of 10 molal) can cause equilibrium and rate constants to be substantially different from dilute solutions. Since most laboratory data are for dilute solutions, this creates considerable uncertainty in applying these data to certain atmospheric particles.

A second difference is that the concentration of a bulk solution in the laboratory is determined by the amounts of the various substances that one places in it. Atmospheric particles, on the other hand, can readily exchange material with a large reservoir of reactants present in the gas phase. Therefore, in order to understand the reactions that occur in these particles, both the equilibria of material between the phases and the rate at which material is transferred between the phases need to be considered. The overall mechanism can be broken down into several steps: diffusion of gases to the surface of the particle or droplet; dissolution of the species (that is, transfer across the gas-liquid interface); diffusion in the aqueous phase; and chemical reaction. Some species also ionize in solution, a process that is generally regarded as being instantaneous and governed by equilibrium expressions. A good general discussion of the effect of these processes on cloud water reactions is given by Schwartz (1986). Applications to $SO_2$ and $NO_2$ oxidation are given by Schwartz (1984a; 1988) and $HO_2$ reactions in clouds are discussed by Schwartz (1984b). An informative discussion of some of the effects of cloud water reactions on tropospheric ozone is given by Lelieveld and Crutzen (1991).

### 4.8.1 Henry's Law

The solubility of a gas in a liquid is governed by an equilibrium of the form:

$$A_{(g)} \rightleftharpoons A_{(aq)} \tag{4.28}$$

$$\mathcal{H}_A = \frac{[A_{(aq)}]}{p_A} \tag{4.29}$$

where $\mathcal{H}_A$ is the equilibrium constant, the activity of the dilute species $A$ in the aqueous phase has been approximated by its concentration, and the activity of $A$ in the gas phase has been approximated by its partial pressure. The equilibrium constant $\mathcal{H}_A$ is also known as the Henry's Law constant. The usual units for $\mathcal{H}_A$ are M atm$^{-1}$, where M is the molarity of the solution (moles $A$ L$^{-1}$). Appendix J presents values of $\mathcal{H}_A$ for several gases of atmospheric interest. Note that the solubility of gases increases with decreasing temperature, as indicated by the positive values of $-\Delta \widehat{H}/R$ in Appendix J.

Note that Henry's Law is usually applicable only to rain or cloud drops, which consist of dilute solutions. For more concentrated solutions, such as those expected to be found in haze particles, this expression is not appropriate. In such applications, the activity of the species must be used. The use of activities and activity coefficients is discussed in detail in texts on chemical thermodynamics. In general, these can only be determined from experimental data. The determination of activity coefficients in the multicomponent, high-ionic-strength solutions characteristic of atmospheric aerosols requires methods for computing mixture properties from available thermodynamic data for simpler systems. An important application is the determination of the amount of water contained in a particle at equilibrium with a particular relative humidity.

Acidic gases dissolve in water and dissociate to produce the $H^+$ ion, while basic species attach $H^+$, thereby increasing the concentration of the basic $OH^-$ ion

$$HA_{(l)} \rightleftharpoons H^+ + A^- \tag{4.30}$$

$$B + H_2O \rightleftharpoons BH^+ + OH^- \tag{4.31}$$

The extent of dissociation can be expressed as an equilibrium constant, $K_a$ (see also Eq. 3.33):

$$K_a = \frac{[H^+][A^-]}{[HA]} \tag{4.32}$$

Some common aqueous equilibrium constants are given in Appendix K. The occurrence of such dissociation reactions increases the total solubility of the gas above what would be obtained by physical solubility alone. This increased solubility is usually expressed as an effective Henry's Law constant $\mathcal{H}_A^*$, which relates the total amount of dissolved substance to the pressure of the gas-phase precursor. For an acid HA,

$$[HA]_{tot} = [HA] + [A^-] = [HA]\left(1 + K_a/[H^+]\right) \tag{4.33}$$

and the effective Henry's Law constant is given by

$$\mathcal{H}_A^* = \mathcal{H}_A \left(1 + \frac{K_a}{[H^+]}\right) \tag{4.34}$$

The effective solubility is thus a function of the pH of the particle. The implications of the increased solubility of $SO_2$ are discussed in Chapter 10.

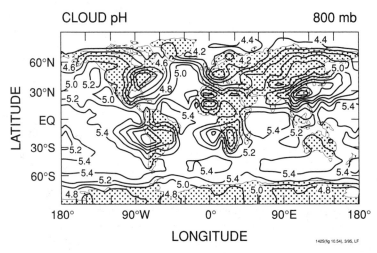

**Figure 4.18.** Model calculated global and annual average pH of boundary layer clouds (courtesy of J. Lelieveld).

Cloud water in the atmosphere does not have a pH of 7; the pH is usually reduced due to dissolved acidic gases. In the remote atmosphere the pH is seldom greater than about 5.6 due to the presence of $CO_2$ (see Box 4.2). In the lower troposphere of the Northern Hemisphere $H_2SO_4$ and $HNO_3$ from anthropogenic combustion sources can cause the pH to be as low as 3.0. Even in the remote troposphere pH values are often encountered that are lower than that due to dissolved $CO_2$, and it has been suggested that this is caused by the uptake of $H_2SO_4$ or CCN containing $SO_4^{2-}$ produced in the oxidation of reduced sulfur gases (see Chapter 10). Figure 4.18 shows modeled values of pH in boundary layer clouds. Note the relatively low values above industrial areas of the Northern Hemisphere.

For strong acids or bases, there is essentially no undissociated acid in the aqueous phase, so that the Henry's Law equilibrium is expressed in terms of the gas and ionic

> **Box 4.2 The Natural Acidity of Cloudwater**
>
> Carbon dioxide dissolves in water to give H$^+$ and the bicarbonate ion HCO$_3^-$, which can further dissociate to the carbonate ion (see Section 5.4.1.2).
>
> $$CO_{2(g)} \rightleftharpoons CO_{2(l)}$$
> $$CO_{2(l)} + H_2O \rightleftharpoons HCO_3^- + H^+$$
> $$HCO_3^- \rightleftharpoons H^+ + CO_3^{2-}$$
>
> The equilibria involved are
>
> $$\frac{CO_{2(l)}}{CO_{2(g)}} = \mathcal{H}_{CO_2}$$
>
> $$\frac{[HCO_3^-][H^+]}{CO_2(l)} = K_{a_1}$$
>
> $$\frac{[CO_3^{2-}][H^+]}{[HCO_3^-]} = K_{a_2}$$
>
> Current CO$_2$ mixing ratios are around 360 ppm. Thus, using the equilibrium constants in Appendices J and K,
>
> $$[CO_{2(l)}] = \mathcal{H} P_{CO_2} = 3.4 \times 10^{-2} \times 360 \times 10^{-6} = 1.2 \times 10^{-5} \text{ M}$$
> $$[H^+][HCO_3^-] = K_{a1}[CO_{2(l)}] = 4.5 \times 10^{-7} \times 1.2 \times 10^{-5}$$
> $$= 5.5 \times 10^{-12} \text{ M}^2$$
>
> But, if the acidity is entirely due to the dissolved CO$_2$, the concentrations of H$^+$ and HCO$_3^-$ must be equal; hence,
>
> $$[H^+] = [HCO_3^-] = \left(5.5 \times 10^{-12}\right)^{\frac{1}{2}} = 2.3 \times 10^{-6} \text{ M}$$
>
> Since pH is defined as -log$_{10}$[H$^+$],
>
> $$\text{pH} = \log_{10}\left(4.5 \times 10^5\right) = 5.6$$
>
> At this pH the contribution of CO$_3^{2-}$ can be shown to be negligible (see Fig. 5.14).

species, for example, for HNO$_3$:

$$HNO_{3(g)} \rightleftharpoons H^+_{(aq)} + NO_3^-{}_{(aq)} \qquad (4.35)$$

To get an idea of the importance of different atmospheric gases in droplet-phase chemistry, we can calculate the phase ratio, $P_x$, for a typical cloud. $P_x$ is the amount of gas in a cloud volume that resides in the aqueous phase relative to the gas phase. Thus, if $P_x = 1$, half of the gas is dissolved in the droplets and half remains in the cloud interstitial gas phase.

The phase ratio is defined by

$$P_x = \mathcal{L}\mathcal{H}^* RT \qquad (4.36)$$

where $\mathcal{L}$ is the cloud liquid water fraction (cm$^3$ H$_2$O/cm$^3$ air), $R$ the universal gas constant (0.0821 atm L/mol/K), and $T$ the temperature (K). Some values for $P_x$, in a typical cloud with a liquid water fraction of $3 \times 10^{-7}$, are given in Table 4.8, in which the rather strong temperature dependence of solubility is also expressed. From Table 4.8 we see that soluble gases, in particular HNO$_3$, go almost completely into the aqueous phase, while nonsoluble gases such as CO stay in the interstitial gas

te that this can strongly change the gas-phase chemistry; insoluble gases
ively separated from the soluble ones.

**Table 4.8**
Temperature-Dependent Phase Ratios of Gases in Clouds ($P_x$),
Indicating the Relative Amounts Dissolved in the Aqueous Phase

| Gas | 268 K | 293 K |
| --- | --- | --- |
| $HNO_3$ | $1.6 \times 10^9$ | $6.4 \times 10^6$ |
| $NH_3$ | $3.7 \times 10^2$ | 17.0 |
| $H_2O_2$ | 6.4 | 0.8 |
| HCOOH | 4.4 | 0.5 |
| $SO_2$ | $7.2 \times 10^{-2}$ | $1.4 \times 10^{-2}$ |
| $CO_2$ | $6.3 \times 10^{-7}$ | $2.9 \times 10^{-7}$ |
| $O_3$ | $1.9 \times 10^{-7}$ | $9.1 \times 10^{-8}$ |
| $NO_2$ | $1.2 \times 10^{-7}$ | $5.4 \times 10^{-8}$ |
| NO | $2.4 \times 10^{-8}$ | $1.5 \times 10^{-8}$ |
| CO | $1.4 \times 10^{-8}$ | $6.8 \times 10^{-9}$ |

## 4.9 Mass Transfer Rates

The equations in the preceding section were based on the assumption that the drops are in equilibrium with the gas phase. This amounts to assuming that material is transported between the phases at a rate that is fast compared to the rate of chemical reaction. Because the particles are so small, this is often true; however, the assumption sometimes breaks down. Determining whether this assumption is valid requires determining which step is rate limiting. There are several possibilities for the rate-limiting step: (1) the rate at which molecules diffuse through the gas phase up to the surface of the drop, (2) the rate at which molecules cross the gas-liquid interface, (3) the rate of diffusion within the liquid phase, and (4) the rate of the liquid-phase chemical reaction.

For reactions between two dissolved gases, the rate at which either gas is consumed by a liquid-phase chemical reaction is proportional to the cube of the drop radius. The rate at which molecules cross the gas-liquid interface is proportional to the square of the drop radius (i.e., the surface area). As a result, as drops get larger, this rate increases more slowly than the rate of reaction, and it becomes more and more difficult to provide the reactants from the gas phase as fast as they are consumed in the liquid phase. The rate at which gas-phase molecules diffuse toward the surface of the drop is proportional to the drop radius; this result is obtained from solution of the diffusion equations in spherical coordinates (Schwartz, 1986). Thus gas-phase diffusion tends to limit the rates of gas-particle reactions for very large particles, such as cloud drops.

The determination of whether mass transfer effects reduce the rate of a gas-particle reaction can be made by comparing the characteristic times for the various processes. The largest characteristic time identifies the slowest, or rate-limiting, step in the process. If this is not the characteristic time associated with the chemical reaction, then mass transfer limitations significantly slow down the rate of reaction.

### 4.9.1 Transfer between the Gas and Liquid Phases

Transfer of gases to the droplets through the gas-liquid interface and gas volatilization from the droplets can be described with a single coefficient, $\beta_t$, which can be understood as a first-order rate constant (s$^{-1}$), describing the loss of reactants from one phase and the gain in the other phase (Schwartz, 1986). The inverse of $\beta_t$ is the characteristic exchange time of gases between the interstitial air and the droplets, given by

$$\frac{1}{\beta_t} = \frac{1}{\beta_{dg}} + \frac{1}{\beta_i} \qquad (4.37)$$

in which $\beta_{dg}$ is the gas-phase transport coefficient and $\beta_i$ the interfacial transfer coefficient. Note that the gas transfer to the aqueous phase is confined to the volume fraction of air that actually contains liquid water; for aqueous- to gas-phase transfer this distinction is not relevant.

In Section 3.4.8 it was shown that the rate of diffusion to a particle of radius $a_D$ is given by $\beta_{dg} = 4\pi a_D D_g N$, where $D_g$ (m$^2$ s$^{-1}$) is the gas-phase diffusion constant and $N$ the number of particles per unit volume. Since the volume per particle is $\frac{4}{3}\pi a_D^3$, this can be rewritten in terms of the liquid water fraction, $\mathcal{L}$ (volume of water per unit volume of air), as

$$\beta_{dg} = \frac{3 D_g \mathcal{L}}{a_D^2} \qquad (4.38)$$

since $\mathcal{L} = (4\pi a_D^3 N)/3$. The diffusion constant, $D_g$, is a function of the mean free path of molecules and the mean molecular speed ($\bar{c} \sim 300$ m s$^{-1}$). A typical value for $D_g$ at STP is $10^{-5}$ m$^2$ s$^{-1}$. Although droplet radii can vary widely in clouds, an average value for $a_D = 10^{-5}$ m may be adopted for illustrative purposes. Using a representative value of the dimensionless liquid water fraction, $\mathcal{L}$, of $5 \times 10^{-7}$ ($\sim 0.5$ g m$^{-3}$), a characteristic time for diffusion from the gas to the aqueous phase of 0.15 s is obtained.

Equation (3.39) for interfacial transport can also be rewritten in terms of the liquid water fraction. The dimensionless accommodation coefficient, $\tilde{\alpha}$, is an expression for the fraction of molecules that is successfully scavenged by the droplets after molecule-droplet collisions. Gases that do not dissolve well usually have $\alpha$ values below 0.01, while soluble gases appear to have $\alpha$ values roughly between 0.01 and 1. Interfacial transport of gases, represented by the coefficient $\beta_i$, is a function of the accommodation coefficient, the droplet radius and the mean molecular speed

$$\beta_i = \bar{c} \tilde{\alpha} \pi a_D^2 N = \frac{3 \bar{c} \tilde{\alpha} \mathcal{L}}{4 a_D} \qquad (4.39)$$

As can be seen from the definition of $\beta_i$, gas transfer to small droplets is more efficient than to large droplets, which is due to the larger surface-to-volume ratio of smaller droplets. Using $\alpha = 0.01$, and applying the dimensionless liquid water fraction of $5 \times 10^{-7}$, a characteristic time for gas transfer from the gas to the liquid phase is about 0.1 s. Combining the expressions for $\beta_{dg}$ and $\beta_i$ yields

$$\beta_t = \left[ \frac{a_D^2}{3 D_g \mathcal{L}} + \frac{4 a_D}{3 \bar{c} \tilde{\alpha} \mathcal{L}} \right]^{-1} \qquad (4.40)$$

For $\alpha$ values smaller than 0.01 interfacial gas transfer of gases is less efficient than gas transfer to the droplets, while for gases that have accommodation coefficients that exceed 0.01 gas-phase diffusion may become the rate-limiting process in the troposphere. Generally, the calculations are not very sensitive to the droplet size, because gas transfer is generally much faster than the chemical processes involved. Only some very fast chemical processes in clouds, for example, OH radical reactions, take place on similar or shorter time scales. Hence, from the point of view of gas exchange processes it is often reasonable to assume a monodisperse cloud (*i.e.*, a cloud with one "effective" droplet radius), and Henry's Law equilibrium. Nevertheless, different droplet sizes may reflect different chemical compositions of the CCN upon

### 4.9.2 Aqueous-Phase Diffusion

Once a species is adsorbed at the surface of a droplet it can diffuse to its interior. The time scale involved with this process is given by Schwartz (1986)

$$\tau_{ad} = \frac{a_D^2}{\pi^2 D_a} \tag{4.41}$$

in which $D_a$ is the aqueous-phase diffusion coefficient. A representative value of $D_a$ is $2 \times 10^{-9}$ m$^2$ s$^{-1}$, associated with a timescale $\tau_{ad}$ of 0.005 s. Aqueous-phase diffusion is in general much faster than chemical reactions in the aqueous phase, and fast enough to maintain homogeneous reaction mixtures within the droplets. However, for some very fast reactions in the aqueous phase (*e.g.*, those of the OH radical), a correction for aqueous-phase diffusion limitation needs to be applied (Schwartz, 1986).

### 4.9.3 Surface Chemistry in the Troposphere

In comparison with soluble gases, insoluble gases must have either a reduced rate of mass transfer into the drops and/or an increased rate of transfer out of the drops; this is because the gas-liquid equilibrium is a balance between the two rates. In either case, the efficiency with which the gas is taken up by the drops is reduced. As a result, the rates of chemical reactions in droplets can be limited by the reduced rates of uptake of relatively insoluble gases. Thus for the oxidation of $SO_2$ in droplets we expect ozone to be the species most subject to mass transfer limitations since it is much less soluble than either $SO_2$ or $H_2O_2$. Detailed calculations show that in cloud drops that are not too acidic, the liquid-phase ozone concentration is substantially less than predicted by Henry's Law. It turns out that this is due to the effect of the rate of liquid-phase diffusion. Several milliseconds are required for diffusion to mix reactants throughout a cloud drop. If the dissolved gas reacts in a time that is less than this, it is unlikely to reach the center of the drop. As a result, the average liquid-phase concentration is less than expected on the basis of Henry's Law, and the overall rate of reaction is reduced. Since the molecule does not need to diffuse to the center of the drop before it reacts, the rate is never entirely controlled by liquid-phase diffusion. Instead it depends on a combination of diffusion and reaction (Schwartz and Freiberg, 1981).

An extreme example of this occurs in the reaction of $N_2O_5$ with liquid water

$$N_2O_5 + H_2O \rightarrow 2\ HNO_3 \tag{4.42}$$

Although this reaction does not occur at a significant rate in the gas phase, it occurs readily in aqueous drops (Mozurkewich and Calvert, 1988). In fact, the rate is so fast that the reaction appears to occur as soon as the $N_2O_5$ molecule enters solution. Thus the rate is proportional to the surface area of the drop, as if it were occurring on the surface.

### 4.9.4 Surface Reactions in the Stratosphere
#### 4.9.4.1 Reactions on Sulfate Aerosols

In the lower stratosphere sulfate aerosols provide enough surface area to allow heterogeneous reactions to occur. These reactions can liberate chlorine from long-lived reservoir species. The pressures are low and the aerosols sufficiently small that gas-phase diffusion is not usually limiting and the rate of uptake is thus proportional to the surface area of the aerosol. The major reactions thought to occur on sulfate aerosols are:

$$N_2O_5 + H_2O \rightarrow 2\ HNO_3 \tag{4.42}$$

$$ClONO_2 + H_2O \rightarrow HNO_3 + HOCl \tag{4.43}$$

$$N_2O_5 + HCl \rightarrow HNO_3 + ClNO_2 \tag{4.44}$$

$$ClONO_2 + HCl \rightarrow HNO_3 + Cl_2 \tag{4.45}$$

$$HOCl + HCl \rightarrow Cl_2 + H_2O \tag{4.46}$$

$$BrONO_2 + H_2O \rightarrow HNO_3 + HOBr \tag{4.47}$$

The overall effect of these reactions is to convert $N_2O_5$, a thermally unstable $NO_x$ reservoir, to the longer-lived $HNO_3$, and to convert $HCl$ and $ClONO_2$ to the more photolabile species $Cl_2$, $HOCl$, and $ClNO_2$. The chemical effects of these reactions are discussed in the chapters on nitrogen, halogens, and stratospheric ozone. Here, a few of the physical aspects are highlighted. In terms of the global stratosphere the first reaction is by far the most important. The reaction is very rapid and is thought to occur essentially at the liquid surface; thus the rate of reaction is proportional to the surface area present. Under conditions of high aerosol loading, such as in the period following a volcanic eruption, the reaction is so rapid that the rate of production of $HNO_3$ is controlled by the rate of formation of $N_2O_5$

$$NO_2 + NO_3 + M \rightarrow N_2O_5 + M \tag{4.48}$$

Under these conditions the reaction is often said to be saturated.

The reaction of chlorine nitrate with liquid water (4.43) is thought to be very slow, and for small droplets the rate of reaction becomes dependent on the liquid water content and the total volume of the aerosol. Since the liquid water is a strong function of the temperature, the reaction only becomes important at temperatures around 200 K (Hanson *et al.*, 1994).

The reaction of $ClONO_2$ with $HCl$ is strongly dependent on temperature due to the low solubility of $HCl$ in concentrated sulfuric acid aerosol. As the temperature approaches 200 K, however, the liquid water content of the aerosol increases sufficiently that $HCl$ can dissolve and Eq. (4.45) occurs. Hanson *et al.* (1994) have shown that Reaction (4.45) can have a significant role in chlorine activation after a volcanic eruption.

### 4.9.4.2 Reactions on Polar Stratospheric Clouds

PSCs also provide very efficient surfaces upon which heterogeneous reactions may occur. The reactions are analogous to those described for the sulfuric acid aerosols. The most important ones are the reactions of chlorine nitrate with water and HCl. While the reaction of $N_2O_5$ with water on water ice (PSC I) is quite fast, the corresponding reaction on NAT particles is rather slow. The reaction of HOCl with HCl has also been shown to be rapid on NAT particles. Since HOCl is a product of the reaction of $ClONO_2$ with water, the conversion of HCl to more labile species can be accelerated dramatically by this reaction. Since PSCs only form at very low temperatures, the reactions on liquid droplets may in some circumstances precede those on PSCs in precipitating the destruction of ozone in the polar stratosphere.

## 4.10 Aqueous Reactions

Some of the most important reactions that occur in cloud and raindrops involve the oxidation of aqueous $SO_2$ to form sulfate. These reactions and their effect on the chemical composition of the atmosphere are discussed in detail in Chapter 10. Here,

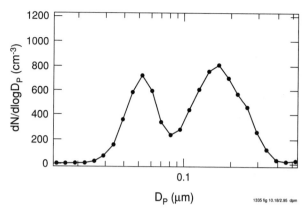

**Figure 4.19.** Bimodal aerosol size distribution measured in June 1992 in the boundary layer over the northeastern Atlantic Ocean (courtesy of T. L. Jensen and S. M. Kreidenweis).

an illustration is given of how the same reactions can affect the aerosol distribution. In clouds that do not precipitate, the in-cloud conversion of $SO_2$ to sulfate is responsible for creating bimodal size distributions after the clouds evaporate. Figure 4.19 shows a measured aerosol size distribution in the marine troposphere. It is assumed that the size distribution was originally single-mode, and that CCN in the size range 0.07-0.09 $\mu$m were scavenged upon condensation of the cloud drops (Jensen et al., 1996). Inside the drops, the sulfate mass grew due to in-cloud oxidation of $SO_2$. After re-evaporation of the cloud, the same aerosols increased to the size-range 0.1-0.4 $\mu$m. If these aerosols served as CCN again, they would form cloud drops at a lower supersaturation than before, due to their larger size, and hence lower Kelvin effect (Eq. 4.19).

Reactions in droplets can influence gas-phase concentrations. Both the uptake of soluble gases and the subsequent supply of liquid-phase products to the gas phase need to be considered. Some dissolved gases interact with liquid water. For example, when formaldehyde dissolves in water, it forms a gem-diol:

$$CH_2O_{(aq)} + H_2O_{(l)} \rightleftharpoons CH_2(OH)_{2_{(aq)}} \tag{4.49}$$

The aqueous concentration of $CH_2(OH)_2$ is proportional to the concentration of $CH_2O$ and therefore to the partial pressure of $CH_2O$. Thus this equilibrium may be incorporated into the gas solubility without altering the basic form of Henry's Law, as discussed in Section 4.8.1. It should also be pointed out that formation of the gem-diol removes the aldehyde functionality, and protects the $CH_2O$ from photolysis.

Free radicals such as OH and $HO_2$ play an important role in aqueous-phase chemistry in the atmosphere as well as in the gas phase. The major source of free radicals in cloud drops is the transport of $HO_2$ radicals from the gas phase. OH transport from the gas phase is much less important due to its much lower solubility and gas-phase concentration. Once in solution, $HO_2$ is a weak acid with a $pK_a$ of 4.7,

$$HO_2 \rightleftharpoons H^+ + O_2^- \tag{4.50}$$

The conjugate base of $HO_2$ is the superoxide radical anion. These radicals form $H_2O_2$ in solution via the reactions

$$HO_2 + O_2^- \rightarrow HO_2^- + O_2 \tag{4.51}$$

$$HO_2^- + H^+ \rightleftharpoons H_2O_2 \tag{4.52}$$

Superoxide is also converted to OH by its reaction with dissolved ozone,

$$O_3 + O_2^- \rightarrow O_3^- + O_2 \tag{4.53}$$

$$O_3^- + H^+ \rightleftharpoons OH + O_2 \tag{4.54}$$

Below pH 9.3, this equilibrium favors OH. In spite of the low solubility of $O_3$, model calculations show that this is the major source of OH in cloud water. These reactions have the potential to cause tropospheric ozone to be significantly less than what would result from gas-phase chemistry alone (Lelieveld and Crutzen, 1990). However, the magnitude of the perturbation is uncertain (Liang and Jacob, 1997).

Aqueous OH is converted back to $HO_2$ by its reactions with hydrocarbons. Formaldehyde is especially important because it has such a high solubility due to its reaction with water to form $CH_2(OH)_2$. The reaction of this species with OH produces formic acid and $HO_2$,

$$OH + CH_2(OH)_2 \rightarrow H_2O + HC(OH)_2 \quad (4.55)$$

$$HC(OH)_2 + O_2 \rightarrow HO_2 + HC(O)OH \quad (4.56)$$

The formic acid produced is in rapid equilibrium with the formate anion,

$$HC(O)OH \rightleftharpoons H^+ + HC(O)O^- \quad (4.57)$$

Both formic acid and formate react further with OH,

$$OH + HC(O)OH \rightarrow H_2O + HCO_2 \quad (4.58)$$

$$OH + HC(O)O^- \rightarrow OH^- + HCO_2 \quad (4.59)$$

$$HCO_2 + O_2 \rightarrow HO_2 + CO_2 \quad (4.60)$$

As in the gas phase, a free radical chain reaction oxidizes hydrocarbons to $CO_2$ (see Chapter 9). However, whereas the gas-phase reaction sequence results in the oxidation of NO to $NO_2$ with subsequent formation of ozone, the aqueous-phase sequence consumes ozone.

Not all reactants for aqueous-phase chemistry are provided from the gas phase. For example, $HO_2$ radicals may also be converted to hydrogen peroxide in particles by a catalyzed reaction sequence involving dissolved Cu (Mozurkewich et al., 1987; Ross and Noone, 1991)

$$HO_2 + Cu^{2+} \rightarrow H^+ + O_2 + Cu^+ \quad (4.61)$$

$$HO_2 + Cu^+ \rightarrow HO_2^- + Cu^{2+} \quad (4.62)$$

$$HO_2^- + H^+ \rightarrow H_2O_2 \quad (4.63)$$

The rate of production of $H_2O_2$ by this mechanism is proportional to the aqueous phase concentrations of both Cu and $HO_2$. $[HO_{2(aq)}]$ is proportional to the partial pressure of $HO_2$ in the gas phase. However, $[Cu_{(aq)}]$ is inversely proportional to the volume of the liquid phase since all the Cu is in the particles. As a result, the rate of consumption of gas-phase $HO_2$ is proportional to the total mass of Cu present in the particles, not the volume of the liquid phase. This reaction should therefore not be greatly enhanced in clouds as compared with aerosols in clear air.

Another example where dissolved ions play an important role is sea-salt aerosol, which is formed in the marine boundary layer from the bursting of bubbles at the ocean surface. As mentioned in Section 4.6, the aerosol is a highly concentrated ionic solution. A series of reactions analogous to those in the stratosphere (Section 4.9.4) can occur with the potential to release photolabile halogen compounds ($Cl_2$, BrCl, HOBr) into the atmosphere (Section 8.5.3). The halogen atoms that are formed from photolysis of the above compounds are thought to play a role in affecting ozone on a regional scale in the Arctic boundary layer, but it is not known whether they can affect ozone globally (Fan and Jacob, 1992; Vogt et al., 1996). Sea-salt aerosol is also known to be an efficient medium for the oxidation of $SO_2$ to $SO_4^{2-}$ (Chameides and Stelson, 1992), leading to the displacement of HCl from the aerosol particles (Section 4.6).

The total volume of liquid aerosol particles in tropospheric air is probably too small to allow significant aqueous-phase chemistry to occur (Warneck, 1988). This can easily be appreciated from consideration of the values in Table 4.8. With typical liquid water fractions of $10^{-11}$-$10^{-10}$ associated with tropospheric aerosols, even highly soluble species like $H_2O_2$ will be partitioned heavily in favor of the gas phase. However, some reactions are so rapid that uptake of reactive molecules on aerosol surfaces is almost immediately followed by reaction. This applies to $N_2O_5$ (Dentener and Crutzen, 1993), where the reaction taking place at the particle or droplet surface, as mentioned earlier, is

$$N_2O_5 + H_2O_{(cloud,\ aerosol)} \rightarrow 2\ HNO_3 \tag{4.42}$$

This reaction is significant on the abundantly present continental tropospheric aerosols, though of somewhat less importance in clean marine air. Nitrogen oxides and aerosols have relatively short lifetimes, 1-2 days to about a week, respectively, so that their transport from pollutant source areas, mostly the industrialized continents of the Northern Hemisphere, is limited to several thousand kilometers (see Chapter 7). Thus Reaction (4.42) on aerosols is most important in the industrially affected Northern Hemisphere. Since $NO_x$ is an important ozone precursor in the troposphere, the heterogeneous removal of $NO_x$ through Reaction (4.42) on aerosols reduces the global lower atmospheric $O_3$ concentration by about 10-15% (Dentener and Crutzen, 1993).

The last consideration of aqueous chemistry is the effect of liquid particles on photolysis rates of dissolved species. A simple physical effect is the increased pathlength associated with the refraction of light inside a particle, which can increase photolysis rates by 10-25%. A much more important effect is the change in quantum yield of dissolved species compared to gas-phase species. A solvent cage effect analogous to that described in Section 3.4.6 means that the photolysis products cannot escape each other after photolysis, and tend to recombine readily. Consequently, the quantum yields of some compounds, most notably HONO, are reduced by an order of magnitude compared to the gas phase. Finally, it should be noted that the occurrence of ionized forms can lead to different cross sections and quantum yields from those of the dissolved neutral species

$$HNO_3 + h\nu \rightarrow NO_2 + OH \tag{4.64}$$

$$HNO_3 \rightleftharpoons H^+ + NO_3^- \tag{4.65}$$

$$NO_3^- + h\nu \rightarrow NO_2 + O^- \tag{4.66}$$

$$NO_3^- + h\nu \rightarrow ONOO^- \tag{4.67}$$

Thus the overall photolysis rates and product distributions may both vary as a function of the pH of the droplet.

## Further Reading

Brimblecombe, P. (1986) *Air Composition and Chemistry*, Cambridge University Press, Cambridge.

Finlayson-Pitts, B. J., and J. N. Pitts (1986) *Atmospheric Chemistry*, Wiley & Sons, New York.

Hobbs, P. V. (1995) *Basic Physical Chemistry for the Atmospheric Sciences*, Cambridge University Press, Cambridge.

Junge, C. E. (1963) *Air Chemistry and Radioactivity*, Academic Press, New York.

Seinfeld, J. H. and S. N. Pandis (1998) *Atmospheric Chemistry and Physics*, J. Wiley, New York.

Warneck, P. (1988) *Chemistry of the Natural Atmosphere*, Academic Press, San Diego.

Richard P. Turco

## Aerosols and Clouds: A Postscript

Aerosols create beautiful sunsets, while at the same time fouling the air. For the most part, aerosols are considered a nuisance. Dust and smoke irritate eyes, sulfate haze reduces visibility, and acidic fog causes respiratory distress. Indeed, until recently, aerosols were widely neglected by the atmospheric sciences community, which focused on meteorology, climate, and photochemistry. However, owing to new research and startling discoveries concerning recent climate change and the ozone hole, we are beginning to realize just how important these microscopic airborne particles can be in controlling the state of the environment.

Without aerosols, the Earth's climate and weather would be very different from what is considered normal today. The water droplets that comprise clouds always condense on pre-existing aerosols, or "cloud condensation nuclei." Without aerosols, clouds as we know them could not form. The atmosphere—lacking the stratus, cumulus, and cirrus clouds familiar to us, and completely devoid of haze and fog—would be crystal clear. Moist clean air would churn upward in turbulent towers; but rain would come only in local drenching downbursts. The surface would likely be a tropical hothouse baking under an unattenuated sun. The other extreme might be as bad. If humans continue to pollute the atmosphere with increasing quantities of fine particles, clouds could take on a quite different character. The number of water droplets forming in these contaminated clouds would increase dramatically. Then rainfall could decrease. The clouds might become hydrologically "constipated," as the water in them evaporated rather than fell as precipitation. Fortunately, the atmosphere always contains enough small particles to form the every-day types of clouds we depend on to provide gentle rains and a stable climate. Just as fortunately, the air rarely holds enough particles to cause rains to stop or merely drizzle.

Imagine the many sources of particles for the atmosphere: emissions from the surface, an influx from space, injection by aerospace vehicles, and formation within the air itself. At the highest altitudes, small meteors continuously bombard the Earth's atmosphere creating microscopic dust particles that eventually settle to the ground. Winds blowing over dry soils raise mineral dust and carry it long distances. Natural fires in forests and grasslands generate countless smoke particles that waft in plumes. The ever-active biogeochemical cycles of sulfur, nitrogen and carbon all have aerosol phases. Sulfur is the most ubiquitous precursor of airborne particles, forming the common sulfate-based aerosols. Carbon forms organic hazes, however; and nitrogen, in its oxidized form, produces nitrate that readily condenses with ammonia, its reduced form. In addition to these natural contributions, there are numerous parallel sources of aerosols related to human activities, ranging from cooking to driving. In populated regions of the world, anthropogenic emissions of particulates and their precursors dominate the total aerosol mass.

Water vapor does not easily condense into droplets spontaneously, or "homogeneously," even when the vapor is supersaturated—or exists at concentrations exceeding that which would be in equilibrium with the bulk material. To form a particle, the vapor must undergo a phase transition to a liquid (or, less likely, directly to a solid). To do so, however, small clusters of molecules, so-called "embryos," must be present for the vapor to collect on. It turns out that, to collect vapor on such an embryo, considerable extra energy must be expended to create additional surface area (compared, say, to condensation onto a fixed surface of water in a beaker). This condensation tax levied by small objects is referred to as the Kelvin effect; as a result, the vapor pressure of a small water droplet is greater than that of a flat water surface. Accordingly, it is highly improbable that embryos will form and grow spontaneously into drops unless the water vapor supersaturation is quite large—up to a factor of ten in some circumstances.

By stark contrast, in today's atmosphere, clouds form at water vapor supersaturations of only a few percent at most. The easiest way for water to condense is to be absorbed into an existing aerosol particle that is composed of "hygroscopic," or easily wettable or soluble, compounds—such aerosols are referred to as cloud condensation nuclei, or CCN. Common soluble materials include sulfates and nitrates. Their attraction for moisture reduces the energy barrier for water condensation and lowers the supersaturation required for stable cloud or fog droplets to form in a very slightly supersaturated environment. The thermodynamics of the atmosphere is thus controlled by CCN.

Aerosols can affect our environment in many other ways. Hundreds of thousands of fine particles are drawn into our lungs with each breath we take. Their cargo of chemicals are deposited in the mouth, nose, throat, bronchia, and lungs. Identification of aerosol-borne toxins has recently identified this source as a critical health issue. Those same particles are responsible for reducing visibility in cities and national parks by scattering sunlight and creating background radiation that reduces contrast and blocks our view. Similarly, scattering and absorption of sunlight by aerosols can alter the Earth's energy balance and climate; this is referred to as the "direct" effect of aerosols on climate.

Perhaps most startling of all is the influence that aerosols have on certain chemical reactions that control the composition of the atmosphere. In particular, the discovery that the "ozone hole" is caused by the formation of reactive chlorine compounds on the surfaces of unusual ice particles in the winter polar regions of the stratosphere has focused a major research effort to determine the exact nature of these "heterogeneous" chemical processes.

The direct radiative effects of aerosols are augmented by their "indirect" radiative effects. As mentioned earlier, the microphysical properties of clouds are determined to a large degree by the character of the particles that act as CCN. In turn, the reflective power of a cloud depends on the number and size of its water droplets (and ice crystals). With more CCN present, clouds tend to be more reflective at visible wavelengths. That is, aerosols act to enhance the Earth's albedo both through their direct and indirect processes. Aerosols, however, also affect the longwave, or thermal infrared, radiation emitted by Earth to space. Depending on their sizes, particles can have direct effects ranging from negligible to significant. Further, clouds modified by aerosols will have altered infrared properties. Generally speaking, after accounting for all of the likely complexities, it is found that an increase in the aerosol burden most likely to reduce the net radiative forcing at the Earth's surface. Further complications associated with the environmental impacts of aerosols can be surmised. Thus reductions in stratospheric ozone caused by aerosol-catalyzed heterogeneous chemical reactions would allow additional solar ultraviolet radiation to penetrate into the lower atmosphere, where it can impact the biosphere. Ozone is also a "greenhouse" gas that modulates the escape of terrestrial longwave radiation through the atmospheric spectral "window." Aerosol-induced variations in ozone therefore affect surface temperatures. There are many interesting particle-driven processes such as these that have important environmental implications.

Aerosols are both a necessity and a nuisance. In the troposphere, for example, aerosols orchestrate the formation of clouds that, in the short term, bring essential rains and, in the long term, provide a thermostat for Earth's climate. In the stratosphere, on the other hand, they act to destroy the layer of ozone that protects all life from destruction by solar ultraviolet rays. In urban smog, aerosols transport lethal compounds into the deepest recesses of our bodies, while at the same time creating spectacular sunsets. Many aspects of these ubiquitous objects remain uncertain or unknown. The obvious importance of aerosols to our well-being, however, points to more intense research aimed at understanding their sources, properties, and effects.

*Richard P. Turco is Professor and Chair of Atmospheric Sciences at the University of California, Los Angeles, a member of the UCLA Institute of Geophysics and Planetary Physics, and Director of UCLA's new interdisciplinary Institute of the Environment. Turco investigates the chemistry and microphysics of gases and aerosols and their relationship to atmospheric chemistry and global climate change. His research activities include studies of volcanic particulates and their effects on radiation, polar stratospheric clouds and ozone depletion, urban air pollution, and global biogeochemical cycles. Turco is co-discoverer of the "nuclear winter" effect of smoke and has received a Macarthur Foundation Fellowship.*

# 5 Trace Gas Exchanges and Biogeochemical Cycles

## 5.1 Introduction

Throughout the entire history of our planet, the chemical composition of the atmosphere has evolved in response to biological and geological processes occurring at the surface of the Earth. The chemical composition of the Earth's atmosphere is fundamentally different from that of other planets (*e.g.*, Mars and Venus) because it is intimately linked to the presence of living organisms. For example, oxygen ($O_2$) and nitrogen ($N_2$) are the major gases of the Earth's atmosphere only because these gases are constantly regenerated by the biota (*e.g.*, photosynthesis, microbial activity). The atmosphere of a lifeless Earth would be characterized by large abundances of carbon dioxide, and only traces of oxygen (Lovelock, 1979). Life is sustained on Earth by the energy received from the Sun, and by the interactions and feedbacks among the physical and biological systems within our planet. Proponents of the so-called Gaia hypothesis (Lovelock, 1979) maintain that the Earth itself, consisting of the biota, the atmosphere, the oceans, and the lithosphere, is an evolving and self-regulating system.

The chemical composition has changed dramatically since the formation of the atmosphere more than four billion years ago (see Chapter 16). Ice core records of the atmospheric composition show that the abundance of such gases as carbon dioxide, methane, and nitrous oxide has fluctuated significantly in the past in conjunction with changes in the Earth's climate, especially temperature. Since the agricultural and industrial revolutions the abundance of several gases has increased as a result of human activities.

Trace gases are produced by biological and geochemical processes on the continents and in the ocean. Volcanic eruptions also release chemical compounds into the atmosphere. An accurate estimate of the magnitude and geographical distribution of these natural emissions requires that the biological and geological processes involved be well understood. The natural cycles of chemical compounds in the Earth's system are, however, perturbed by several socio-economic factors such as resource use, industrial activity, and land conversion (*e.g.*, agricultural activity, deforestation, biomass burning). Large quantities of pollutants are also released in regions of high energy production and/or high population density, such as North America, Europe, and Eastern Asia. Biomass burning, mostly in the tropics (but also in northern forests), is another significant source of atmospheric pollution.

The deposition of chemical compounds to the surface (*e.g.*, acid rain, ozone deposition on vegetation) is an important process that directly affects the biosphere, and, in several regions of the world, has contributed to major environmental problems.

The circulation of chemical elements through the global environment, which simultaneously affects the chemistry and the biology of the Earth system, produces *biogeochemical cycles*, which are described in terms of reservoirs (*e.g.*, the atmosphere, the ocean, terrestrial ecosystems) and exchange fluxes between these reservoirs. To a

### Table 5.1a
*Reservoir Masses in the Global Biogeochemical Cycles of Carbon, Nitrogen, Phosphorus, Sulfur, and Oxygen*

| Reservoir | \multicolumn{5}{c}{Element ($10^{15}$ g of element)} |
|---|---|---|---|---|---|
| | C | N | P | S | O |
| Atmosphere | 760 | 3,950,000 | 0.00003 | 0.003 | 1,216,000 |
| Ocean | 38,400[a] | 570[b] | 80[c] | 1,248,000[d] | 4100[e] |
| Land biota | 600 | 10 | 3 | 2.5 | 800 |
| Marine biota | 3 | 0.5 | 0.07 | 0.1 | 4.2 |
| Soil organic matter | 1600 | 190 | 5 | 95 | 850 |
| Sedimentary rocks | 78,000,000 | 999,600 | 4,030,000 | 12,160,000 | 1,250,000,000 |

[a] Dissolved inorganic carbon. [b] $NO_3^-$. [c] $PO_4^{3-}$. [d] $SO_4^{2-}$. [e] Dissolved $O_2$.
From Mackenzie, 1997.

### Table 5.1b
*Fluxes of Carbon, Nitrogen, Phosphorus, Sulfur, and Oxygen in Their Global Biogeochemical Cycles*

| Flux | \multicolumn{5}{c}{Element ($10^{12}$ g yr$^{-1}$ of element)} |
|---|---|---|---|---|---|
| | C | N | P | S | O |
| River dissolved[a] | 400 | 40 | 3 | 115 | |
| Net primary production | | | | | |
|   Land | 63,000 | 580 | 320 | 265 | 168,000 |
|   Ocean | 45,000 | 7925 | 1097 | 1925 | 120,000 |
| Respiration and decay | | | | | |
|   Land | 61,400 | 560 | 310 | 260 | 163,700 |
|   Ocean | 45,200 | 7960 | 1100 | 1930 | 120,500 |
| Nitrogen fixation | | | | | |
|   Land | | 270 | | | |
|     Natural | | 130 | | | |
|     Anthropogenic | | 140 | | | |
|   Ocean | | 40 | | | |
| Denitrification | | | | | |
|   Land | | 115 | | | |
|   Ocean | | 70 | | | |
| Combustion fossil fuel | 6000 | 30 | | 80 | |
| Land-use activities[b] | 1600 | 15-46 | | 1-4 | |
| Burial and uplift[c] | 400 | 15 | 3 | 40 | |
| Metamorphism and volcanism | 120 | | | 10 | |
| Weathering | 220 | | | | 380 |

[a] Inorganic flux to ocean. [b] Deforestation and biomass burning. [c] Steady-state flux.
From Mackenzie, 1997.

very close approximation, the Earth system can be regarded as a closed system with no flow of material out of or into the system (accretion on the Earth from cosmic dust and meteors or escape of light molecules to space represent a very small fraction of the total mass of the Earth). Table 5.1a provides an estimate of the reservoir masses of key chemical elements, and Table 5.1b provides the fluxes of these elements in their global biogeochemical cycles (Mackenzie, 1997). A major challenge for scientists involved in "global change" research is to assess the potential variations in the large, globally connected components of the Earth system caused by natural fluctuations as well as by the activities of the expanding human population.

This chapter considers several biogeochemical cycles (water, carbon, nitrogen, sulfur, halogens, etc.) that directly affect the chemical composition of the global atmosphere and ultimately the climate system (Fig. 5.1). Within these cycles, we identify sources, atmospheric sinks, surface deposition, global lifetimes, etc. A more detailed description of chemical transformations affecting these gases in the atmosphere is presented in subsequent chapters. We will first provide a brief description of the physical mechanisms involved in trace gas exchanges at the surface, specifically surface emissions, as well as dry and wet deposition.

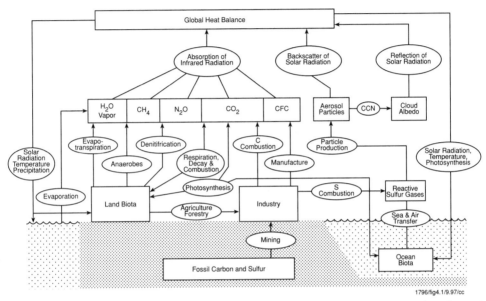

*Figure 5.1.* Schematic representation of the processes that relate biogeochemical cycles and the climate system (Charlson et al., 1992a; reprinted by permission of the publisher Academic Press).

## 5.2 Surface Exchanges

For the atmosphere, the Earth's surface acts as both a source and sink for trace gases and particles. Exchanges of chemical constituents between the ocean and the atmosphere are calculated from empirical relationships, while an estimation of gas release by soils and vegetation has to rely in most cases on field measurements. The extrapolation of local measurements (*e.g.*, emissions from a tree or a square meter of soil) to broader ecosystems under different meteorological conditions remains a major challenge (Schimel and Potter, 1995).

Key to our understanding of the coupling between the surface and the atmosphere is the planetary boundary layer. This layer (Fig. 5.2), whose thickness is variable in time and space (typically hundreds of meters to a few kilometers), is defined as that region of the troposphere that responds to surface forcings (*e.g.*, frictional drag, heat transfer, evaporation and transpiration, terrain-induced flow modification, pollutant emission) with a timescale of about an hour or less (Stull, 1993). The meteorological conditions in the boundary layer substantially affect the exchanges of chemical compounds between the surface and the free atmosphere. For example, if the temperature near the ground increases with height (*e.g.*, temperature inversion when the surface cools), vertical mixing is impeded, and gases released at the surface

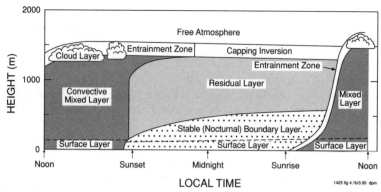

**Figure 5.2.** *The evolution of the boundary layer over 24 hours, with the surface layer, the daytime mixed layer, and the nighttime stable layer (from Stull, 1993).*

accumulate in the boundary layer. Pollution events are often observed under such stable conditions. If the temperature decreases rapidly with height, shallow or deep convection occurs, producing rapid mixing of air; under these conditions, exchange of trace gases between the surface and the free troposphere is rapid. Under strongly convective conditions, chemical compounds released at the surface can reach the upper troposphere in only a few hours.

The vertical flux $F$ of a trace gas in the boundary layer is often represented by a diffusion equation

$$F = -K(z)\, n_a\, \frac{\partial \mu}{\partial z} \qquad (5.1)$$

where $K$ is a transfer coefficient, $n_a$ is the air density, and $\partial \mu/\partial z$ is the vertical gradient of the mixing ratio $\mu$ ($z$ is the height). In the case of neutral stability of the boundary layer (*i.e.*, no thermal exchanges are taking place along the vertical and the turbulence is produced solely by the horizontal wind), the transfer coefficients for trace gas concentration can be written as

$$K(z) = k u_* (z - d) \qquad (5.2)$$

where $k$ (=0.4) is the von Karman constant, $d$ is the so-called displacement height (which results from the canopy acting as a displaced lower boundary and whose value is typically 70-80% of the canopy height), and $u_*$ is the so-called friction velocity, which is a measure of the drag exerted by the wind on the surface. The horizontal wind velocity $u(z)$ near the surface generally increases logarithmically with height in the neutral case, so that

$$u(z) = \frac{u_*}{k} \ln\left[\frac{z-d}{z_0}\right] \qquad (5.3)$$

where $z_0$ is an empirical parameter called the surface roughness length. Note that these relations provide a method for determining experimentally the surface fluxes of trace gases, by measuring meteorological parameters near the surface as well as the vertical gradient in the concentration of these gases (Lenschow, 1995). This method is often referred to as the aerodynamic method. Another technique, called the eddy-correlation method, relies on the measurement of the fluctuations in the vertical component of the wind ($w'$) and of the associated fluctuations in the trace gas concentration ($n'$). The vertical flux averaged over a given period of time $T$ (typically 10-30 min) is given by

$$F = \frac{1}{T} \int_0^T w' n'\, dt \qquad (5.4)$$

Irreversible deposition of chemical constituents to the Earth's surface (continents and oceans) is often described in terms of deposition velocities. The deposition occurs either continuously as direct uptake by the surface (dry deposition, $F_d$) or intermittently in precipitation (wet deposition, $F_w$). The deposition flux can be expressed by the product of total deposition velocity (the sum of dry and wet deposition velocity, $v_d$ and $v_w$), and the concentration of molecules or particles $n_A$ in the atmosphere at a reference height $z_T$ (typically 1 m above the surface or canopy)

$$F = F_d + F_w = (v_d + v_w)n_A(z_T) \tag{5.5}$$

When considering wet deposition associated with precipitation, the simple formulation presented here is generally replaced by a more complex parameterization that accounts for the spatial distribution of clouds, the frequency of occurrence and intensity of precipitation (see Section 4.3.2), the potential evaporation of water droplets in the atmosphere, and the solubility of molecules in water (Section 4.8.1). The concept of deposition velocity is most commonly used to represent dry deposition (removal of chemical compounds from the atmosphere by turbulent transfer and uptake by the surface). In this case, the deposition velocity is often regarded as the inverse of a resistance to transfer between the atmosphere and the surface. By analogy with Ohm's law in electricity, the total resistance to transport can be represented by a number of subresistances connected either in series or in parallel (Fig. 5.3). The simplest representation is provided by serial resistances including an aerodynamic resistance ($R_a$) associated with atmospheric turbulence (common to all compounds), a quasilaminar sublayer resistance ($R_b$) accounting

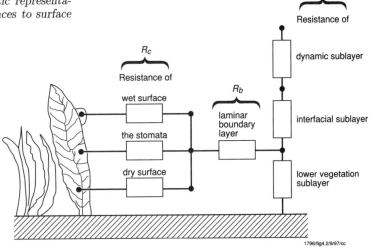

**Figure 5.3.** *Schematic representation of various resistances to surface deposition.*

for molecular diffusion in the layer adjacent to the surface, and bulk surface resistance ($R_c$) associated with the uptake of gases and particles by the canopy. Thus

$$v_d = \frac{1}{R_a + R_b + R_c} \tag{5.6}$$

The gas-phase resistance ($R_g = R_a + R_b$) depends essentially on wind friction, and can be modeled empirically as a function of the type of surface; the canopy or bulk surface resistance ($R_c$) needs to be determined from field experiments. Wesely (1989) has provided a detailed parameterization of surface resistances to gaseous dry deposition for various land types and different seasons. Other factors considered include surface temperature, stomatal response to environmental parameters, the wetting of surfaces by dew and rain, and the covering of surfaces by snow. Wesely's approach is particularly suitable for regional and global chemical transport models.

Chemical compounds can also be released or taken up by the ocean. The net flux through the air-sea interface (expressed, for example, in molecules cm$^{-2}$ s$^{-1}$) is generally assumed to be proportional to the difference in concentration $\Delta n$ (molecule cm$^{-3}$) between the gas and liquid phases (see, *e.g.*, Liss and Merlivat, 1986)

$$F = w_P \Delta n \quad (5.7)$$

where $w_P$ (cm s$^{-1}$) is a transfer coefficient, which has the dimension of a velocity, and is often referred to as the transfer or piston velocity.

The concentration difference driving the flux can be expressed as

$$\Delta n = n_w - \frac{n_a}{\mathcal{H}'} \quad (5.8)$$

if $n_w$ and $n_a$ are the gas concentrations (expressed in the same units) in water and air, respectively, and $\mathcal{H}'$ the dimensionless Henry's Law constant (expressed as the ratio of the concentration of gas in air to its concentration in unionized form dissolved in water, *at equilibrium*). By expressing $\Delta n$ as in Eq. (5.8), the net flux is positive when directed from the ocean to the atmosphere. Note that the dimensionless Henry's Law constant used here is related to that used in Chapter 4 by

$$\mathcal{H}' = (\mathcal{H}RT)^{-1}$$

where R = 0.0821 L atm mol$^{-1}$ K$^{-1}$.

The inverse of the piston velocity $R$ (s cm$^{-1}$) corresponds to a resistance to transfer across the air-sea interface, and can be broken into its two component parts

$$R = R_w + R_a$$

where $R_w = 1/\alpha k_w$ and $R_a = 1/-\mathcal{H}' k_a$ are the resistances in the water and air phases, respectively, $k_w$ and $k_a$ are the individual transfer velocities for chemically unreactive gases in the two phases, and $\alpha$ quantifies any enhancement in the value of $k_w$ due to chemical reactivity of the gas in water ($\alpha = 1$ for unreactive gases). For gases such as $H_2O$, $SO_2$, $NH_3$, and $HCl$, which partition dominantly into the water (low value of $\mathcal{H}'$), the value of $R_a$ is much larger than that of $R_w$. In contrast, for the majority of the gases that are important regarding biogeochemical cycles [*e.g.*, $O_2$, $N_2$, $N_2O$, $CO_2$, $CO$, $CH_4$, $CH_3I$, $(CH_3)_2S$], the solubility is low (high values of $\mathcal{H}'$), and $\alpha \simeq 1.0$, so that the dominant resistance is $R_w = 1/k_w$, and the piston velocity is $w_P = k_w$.

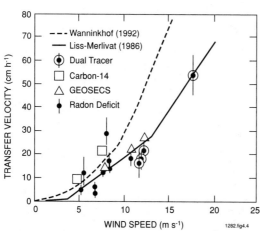

**Figure 5.4.** *Transfer velocity (cm h$^{-1}$) at the surface of the ocean as a function of the wind velocity (m s$^{-1}$).*

Empirical values of the transfer velocities as a function of wind speed have been derived from wind tunnel experiments, field measurements, and boundary layer modeling. Figure 5.4 provides different estimates of $k_w$ (expressed in cm h$^{-1}$). Wind tunnel experiments show that for wind speeds ($u$) smaller than about 3-5 m s$^{-1}$

(smooth surface regime), the value of $k_w$ increases only very gradually with $u$. Between about 3-5 and 12-14 m s$^{-1}$ (rough sea surface regime), the presence of waves produces a significant increase in $k_w$. Finally, above 12-14 m s$^{-1}$ (breaking wave regime), the large number of bubbles near the sea surface enhances gas transfer with high values of $k_w$. The transfer velocity $k_w$ (here in cm h$^{-1}$) can be estimated as a function of wind speed $u$ (m s$^{-1}$), as measured 10 m above the sea surface, by the following three relationships (see Liss and Merlivat, 1986; Monfray, 1987):

$$k_w = 0.17[A(\theta)]^{-\frac{2}{3}} u \quad (5.9a)$$

for wind speed $u \leq 3.6$ m,

$$k_w = 0.17[A(\theta)]^{-\frac{2}{3}} u + 2.68[A(\theta)]^{-\frac{1}{2}} (u - 3.6) \quad (5.9b)$$

for wind speed $3.6 < u < 13$ m s$^{-1}$, and

$$k_w = 0.17[A(\theta)]^{-\frac{2}{3}} u + 2.68[A(\theta)]^{-\frac{1}{2}} (u - 3.6) + 3.05[A(\theta)]^{-\frac{1}{2}} (u - 13) \quad (5.9c)$$

for wind speed $u > 13$ m s$^{-1}$. In these equations $A(\theta)$ represents the ratio between the Schmidt number of the gas under consideration at $\theta$°C and the Schmidt number of $CO_2$ at 20°C (595 according to Liss and Merlivat, 1986). The Schmidt number is defined as the ratio between the kinematic viscosity and the molecular diffusivity of the gas. Both factors, and hence the Schmidt number, are a function of temperature. Relationships between wind speed and gas exchange over the ocean have been further discussed by Wanninkhof (1992), while Erickson (1993) has proposed a parameterization for $k_w$ based on the thermal stability of the air-seawater interface. In addition, Frew (1997) has reviewed laboratory results that suggest that a unique relationship between wind speed and transfer velocity is unlikely to exist, due to the ubiquitous presence of organic surfactant films at the air-sea interface, but that correlation between $k_w$ and the variance of wave slope is linear and surprisingly robust for a variety of surfactant types over a wide range of wind speeds. If this correlation applies to in situ wave fields, it may be possible to estimate $k_w$ directly from remotely sensed multiwavelength radar backscatter measurements.

## 5.3 The Global Water Cycle

Water is a key element of the Earth system. It allows life to exist and plays a fundamental role with regard to climate. As it circulates between the ocean, the atmosphere, and the biosphere, it is constantly switching between liquid, solid, and vapor phases. The hydrological cycle is dominated by the oceans, which contain 97 percent of the total water supply (see Table 5.2). The global circulation of ocean water in large-scale currents is critical to climate because it transports heat around the Earth, as well as salt and other chemicals. Surface currents, such as the Gulf Stream, are driven by the winds above them. Currents deep in the ocean are driven by gradients in density, temperature, and salinity. An immense deep current formed by dense, salty water in the North Atlantic propagates southwards along Africa, through the Indian Ocean, and finally northwards in the deep

Table 5.2
Volume of Water in the Earth System ($10^6$ km$^3$)

| | |
|---|---|
| Oceans | 1350 |
| Glaciers and ice caps | 29 |
| Ground water | 8 |
| Freshwater lakes | 0.1 |
| Salt lakes and inland seas | 0.1 |
| Soil moisture and seepage | 0.07 |
| Atmosphere | 0.013 |
| Rivers and streams | 0.001 |

Pacific Ocean. This water wells up in the North Pacific and in the northern Indian Ocean, and is eventually transported back to the North Atlantic in the surface layers. Upwelling of deep waters often brings nutrient-rich waters to the surface, giving rise to regions of high biological productivity and elevated emissions of trace gas species.

On land, most of the water (2.1 percent of the total water supply) is contained in glaciers and ice caps. Aquifers represent less than 1 percent of the water available on Earth. The atmosphere contains only a tiny fraction of the water present in the Earth system. Although the most visible manifestation of this element is provided by clouds, fog, and precipitation, the largest fraction of water in the atmosphere is found in the vapor phase. Clouds, however, play major roles in the radiation, water, and chemical budgets of the troposphere. When associated with convective activity and precipitation, they are accompanied by intense vertical exchanges of mass and chemical constituents.

Water is provided to the atmosphere by evaporation, mostly over the oceans, but also by evapotranspiration of plants over the continents. Globally, this flux is balanced by the return of water to the surface through various types of precipitation. However, in the vapor phase, water can be transported several thousands of kilometers before condensing; as a consequence, precipitation on the continents exceeds evaporation, but the budget is balanced by a surface flow of water (rivers, etc.) between land and sea. Figure 5.5 shows a representation of the global water budget with an estimate of related pools and flows. The lifetime of a water molecule in the atmosphere is estimated to be about 10 days.

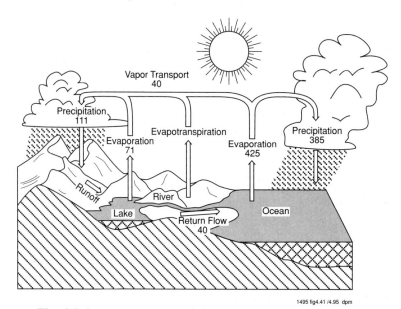

**Figure 5.5.** *The global water cycle. Water fluxes are expressed in thousands of cubic kilometers per year (adapted from ATMOSPHERE, CLIMATE, AND CHANGE by Thomas E. Graedel and Paul J. Crutzen. ©1995 by Scientific American Library. Used with permission of W. H. Freeman and Company).*

The hydrologic cycle is closely linked to patterns of atmospheric circulation and temperature. The distribution of water vapor in the troposphere therefore varies considerably with meteorological conditions. On the average, however, the mixing ratio of $H_2O$ is typically 1 percent at the surface, but is only a few ppmv in the stratosphere. The dryness of the stratosphere arises from the fact that, as air is transported upward by the Hadley cell in the tropics, condensation and precipitation produce an efficient dehydration mechanism.

Finally, water plays an important role for the chemistry of the atmosphere. For example, the oxidation of water vapor molecules provides the major source of the hydroxyl radical (OH). Water vapor affects the budget of ozone in the troposphere, lower stratosphere, and mesosphere. In the troposphere, chemistry and photochemistry are directly affected by water droplets within clouds (see Chapter 4), while in the stratosphere, polar stratospheric clouds initiate heterogeneous chemical reactions that lead, for example, to the formation of the Antarctic ozone hole.

## 5.4 The Global Carbon Cycle
### 5.4.1 Carbon Dioxide

Measurements of the $CO_2$ concentration of air trapped in Antarctic ice cores indicate that over the past 200,000 years, atmospheric concentrations of $CO_2$ have fluctuated between approximately 200 and 280 ppm, until the past century. Data for the period 1000 to 1800 AD indicate that the concentration was quite stable, averaging 280 ppm and varying over that period by only about 10 ppm (Barnola *et al.*, 1994), indicating that the $CO_2$ cycle was nearly in equilibrium in the centuries prior to the industrial revolution. Over the past two hundred years, however, the concentration has increased from about 280 to the 1994 value of 358 ppm, a 30% increase attributed primarily to burning of fossil fuels, with a smaller contribution from deforestation. Changes in the isotopic abundance of $CO_2$ (see Box 5.1) have also been reported.

---

**Box 5.1 Isotopes in the Environment**

Isotopes are often used to constrain the global budget of chemical elements in the Earth system. For example, $CO_2$ has two stable isotopes of carbon, $^{12}C$ and $^{13}C$. The natural abundance of $^{13}C$ is roughly 1% of the lighter $^{12}C$. Analytical instruments are able to measure slight differences in the relative abundances of these two isotopes in a variety of materials. The abundance of $^{12}C$ and $^{13}C$ relative to each other is defined on a "per mil" scale using differential notation and can be written as:

$$\delta(\text{\textperthousand}) = \frac{R_{\text{sample}} - R_{\text{std}}}{R_{\text{std}}} \times 1000 \quad (5.10)$$

where $R_{\text{sample}}$ is the absolute isotope ratio of the sample and $R_{\text{std}}$ is the corresponding ratio of a standard traceable to a marine limestone, Pee Dee

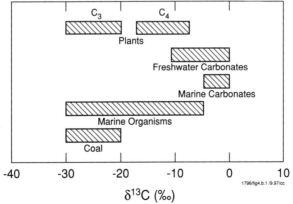

**Figure 5.6.** *Observed ranges of carbon isotope ratios from various substances.*

Belemnite (PDB). Figure 5.6 shows the characteristic isotopic ratios of various components of the carbon cycle. Note that coal ranges between −20 and −30 per mil, meaning that coal has a "lighter" isotopic composition than marine carbonates and consequently less $^{13}$C and more $^{12}$C. Figure 5.7 shows the evolution of the atmospheric $\delta^{13}$C from 1740 until 1985. The data prior to the 1970s are obtained from ice core bubbles that retain the signature of the atmospheric $^{13}$CO$_2$ concentration when they were formed. Clearly, the trend toward "lighter" atmospheric CO$_2$ over the past two hundred years, from −6.4 to −7.8, reflects the addition of isotopically "lighter" fossil fuel CO$_2$ to the atmosphere.

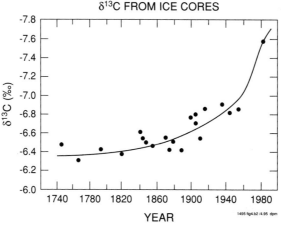

**Figure 5.7.** Evolution of the $\delta^{13}$C isotopic ratio (‰) in the atmosphere from 1740 to present (adapted from Leuenberger et al., 1992).

The radioactive $^{14}$C isotope (which decays into nitrogen over a half-time period of 5730 years) is present in the atmosphere only in very small quantities. Its production (about 10 g of $^{14}$C per year) is the result of the action (mostly in the stratosphere) of particle radiation from the Sun. $^{14}$C isotopes were also produced in large quantities by nuclear explosions in the atmosphere during the 1960s. Because they have been stored in the Earth's sedimentary layer for very long periods of time, fossil fuels do not contain any $^{14}$C.

Measurements of the $^{13}$C/$^{12}$C isotopic ratio have also proved useful in differentiating between various biological and fossil fuel related sources of atmospheric methane (Tyler, 1989). Most sources of methane (wetlands, ruminants, rice paddies, etc.—see Section 5.4.2) are depleted ($\delta^{13}$C less than −50 per mil) relative to methane present in the atmosphere ($\delta^{13}$C = −47.7 per mil). This discrepancy may be explained by the contribution of biomass burning emissions ($\delta^{13}$C values in terrestrial vegetation are clustering around −14 and −29 per mil for C$_4$ and C$_3$ plants, respectively) and, more important, by the fact that methane oxidation tends to remove the lighter isotopes preferentially (Cantrell et al., 1990), causing the remaining methane to be isotopically heavy. Measurements of $^{14}$C in atmospheric methane have demonstrated that 20-30% of it is radioactively dead (and thus has been provided by fossil fuel—see, e.g., Lowe et al., 1988; Wahlen et al., 1989).

Isotopic signatures are also recorded to assess the relative importance of the various sources affecting atmospheric sulfur. As for carbon, the ratio of the $^{34}$S to $^{32}$S stable isotopes is expressed by the $\delta$ notation (as parts per mil), but, in this case, the international standard is an iron sulfide mineral known as the Canyon Diablo Triolite (CDT). As suggested by Figure 5.8, the value of $\delta^{34}$S for sulfur released by power stations is much lower than that for sulfate in seawater (from which phytoplankton produce dimethylsulfide—

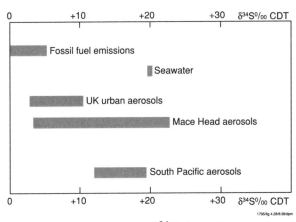

**Figure 5.8.** *Sulfur isotope ratios ($\delta^{34}S$) for various sources of sulfur and in atmospheric aerosols for several localities (from Andrews et al., 1996).*

see Section 5.6). The figure also shows the isotopic ratio corresponding to atmospheric aerosols collected in different geographical regions of the world. The isotopic signatures of human activity (fossil fuel burning) is visible in samples collected in urban areas.

Isotopic ratios such as $^{18}O/^{16}O$ or D/H (deuterium/hydrogen) are generally expressed relative to standard mean ocean water (SMOW). Their measurement in ice cores provides crucial information on the Earth's temperatures over time, and is fundamental to understanding the evolution of the climate system. For more details on isotopes, see Schimel (1993). Table 5.3 provides the relative abundance (in percent) of stable isotopes commonly employed in biogeochemical studies.

**Table 5.3**
*Stable Isotopes Commonly Employed in Biogeochemical Studies*

| | | | | | | | | |
|---|---|---|---|---|---|---|---|---|
| Carbon | $^{12}C$ | 98.89 | $^{13}C$ | 1.11 | | | | |
| Oxygen | $^{16}O$ | 99.763 | $^{17}O$ | 0.0375 | $^{18}O$ | 0.1995 | | |
| Hydrogen | H | 99.9844 | D | 0.0156 | | | | |
| Nitrogen | $^{14}N$ | 99.64 | $^{15}N$ | 0.36 | | | | |
| Sulfur | $^{32}S$ | 95.02 | $^{33}S$ | 0.75 | $^{34}S$ | 4.21 | $^{36}S$ | 0.02 |
| Calcium | $^{40}Ca$ | 96.94 | $^{42}Ca$ | 0.65 | $^{43}Ca$ | 0.14 | $^{44}Ca$ | 2.08 |
| | $^{46}Ca$ | 0.003 | $^{48}Ca$ | 0.19 | | | | |
| Potassium | $^{39}K$ | 93.08 | $^{40}K$ | 0.0119 | $^{41}K$ | 6.91 | | |
| Magnesium | $^{24}Mg$ | 78.8 | $^{25}Mg$ | 10.15 | $^{26}Mg$ | 11.06 | | |

From Hoefs, 1980, and Schimel, 1993.

In 1993, approximately 6 GtC were released to the atmosphere as a result of fossil fuel burning. Figure 5.9 shows that between 1850 and the early 1970s, the release of $CO_2$ increased exponentially at a remarkably constant rate (4.3% $yr^{-1}$), with the exception of limited periods (the economic crisis of 1929; the two World Wars). After the oil crisis of 1973, development of nuclear reactors and efforts to reduce energy consumption have slowed the trend in $CO_2$ release. It is interesting to note that the type of fossil fuel consumed has evolved over the past century (Fig. 5.10). Coal was the largest contributor until the mid-1960s, and oil and natural gas have played an

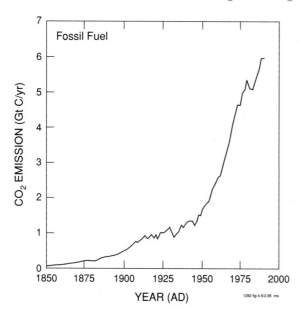

**Figure 5.9.** Estimated emission of carbon to the atmosphere resulting from fossil fuel burning between years 1850 and 1993.

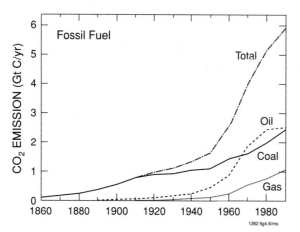

**Figure 5.10.** Relative contribution of coal, oil, and natural gas to the total amount of carbon released to the atmosphere since the preindustrial era for different types of fossil fuel consumed.

increasingly important role since World War II, with oil consumption dropping during and after the energy crisis of the 1970s.

Analyses of air bubbles trapped inside ice cores collected in Antarctica and Greenland suggest that the atmospheric concentration of $CO_2$ started to increase dramatically at the beginning of the industrial era (Fig. 5.11a). Accurate measurements of $CO_2$ concentrations made on Mauna Loa, Hawaii (Fig. 5.11b) since 1958 and elsewhere in the world indicate that the average rate of $CO_2$ increase (1980-1989) is approximately 1.5 ppmv yr$^{-1}$. This corresponds to an increased burden of 3.3 GtC yr$^{-1}$, that is, considerably less than the amount currently released as a result of anthropogenic activities (7.1 GtC yr$^{-1}$). An important research objective is to determine in which component or components of the Earth system the remaining carbon is stored.

### 5.4.1.1 Global Budget

In order to improve predictions about the future evolution of atmospheric $CO_2$ concentration, atmospheric scientists, oceanographers, and ecologists have attempted for several decades to develop an accurate global budget of carbon in the Earth

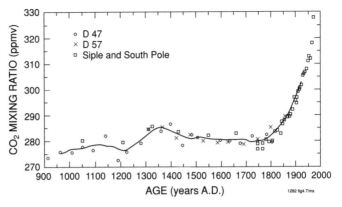

**Figure 5.11a.** Concentration of atmospheric $CO_2$ over the past 1,000 years deduced from the analysis of air trapped inside ice cores (Barnola et al., 1994).

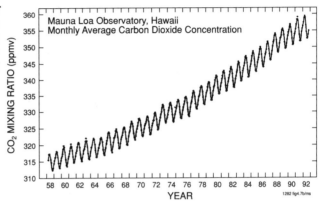

**Figure 5.11b.** Observation of atmospheric $CO_2$ mixing ratio at Mauna Loa, showing trends and seasonal cycles (courtesy of NOAA/Climate Monitoring and Diagnostics Laboratory).

system. Although great strides have been made, balancing the carbon budget remains problematic, largely because of the difficulty of measuring carbon fluxes accurately over large scales, and of modeling oceanic and atmospheric transport of carbon species.

Atmospheric, oceanic, and terrestrial pools of carbon are extremely large and fluxes between pools are bidirectional and nearly (but not exactly) in balance; thus the net exchange of atmospheric $CO_2$ with the oceans and the biosphere is extremely small relative to the pool sizes and hence difficult to determine. Flux estimates for both sources and sinks of $CO_2$ are based less on measurement *per se* than on global surveys or complex models of physical and biological processes. Flux estimates tend to be quite uncertain. Despite these difficulties, significant progress has been made. A global carbon budget is illustrated in Figure 5.12, which presents current best estimates of the amount of C held in several important reservoirs and the magnitude of annual fluxes between them, as indicated by the arrows. Figure 5.12 shows that the total anthropogenic input of $CO_2$ to the atmosphere (7.1 GtC yr$^{-1}$) represents a significant perturbation of the natural carbon cycle.

Instead of attempting to deal with the complete carbon budget, it is perhaps easier to limit our discussion to the *perturbation budget* for $CO_2$ (Table 5.4), that is, what happens to the $CO_2$ injected into the atmosphere by the combination of fossil fuel combustion and tropical land-use change. The single value in which there is the greatest confidence is the change in atmospheric $CO_2$ concentration, which is measured continuously at a large number of sites. Although there are large seasonal changes in $CO_2$ concentration (Fig. 5.11b) and significant site-to-site differences, the average increase in $CO_2$ in the atmosphere corresponds to $3.3 \pm 0.2$ GtC yr$^{-1}$. Fossil fuel combustion, which can also be estimated with considerable accuracy,

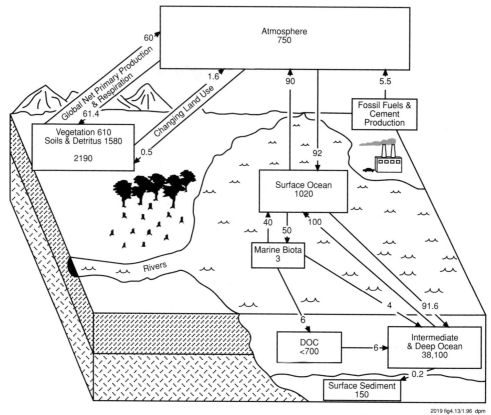

**Figure 5.12.** Global carbon cycle showing carbon pools (GtC) and average fluxes (GtC yr$^{-1}$) for the period 1980-1989 (adapted from IPCC, 1996, and Schimel, 1995).

**Table 5.4**
Global Budget for the Anthropogenic Perturbation of $CO_2$, Based on IPCC (1996). Estimates Are Based on Ten-Year Averages (1980-1989)

|  | $CO_2$ (GtC yr$^{-1}$) |
|---|---|
| *Sources* | |
| Fossil fuel combustion and cement production | $5.5 \pm 0.5$ |
| Tropical deforestation and land use changes | $1.6 \pm 1.0$ |
| Total known sources | $7.1 \pm 1.1$ |
| *Sinks* | |
| Retained in atmosphere | $3.3 \pm 0.2$ |
| Oceanic uptake | $2.0 \pm 0.8$ |
| Uptake by Northern Hemisphere forest regrowth | $0.5 \pm 0.5$ |
| Total known sinks | $5.8 \pm 1.0$ |
| *Net imbalance* | |
| (Inferred additional terrestrial sink) | $[1.3 \pm 1.5]$ |
| (including $CO_2$ fertilization, nitrogen fertilization, and climatic effects) | |

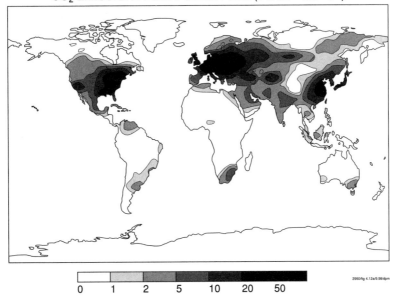

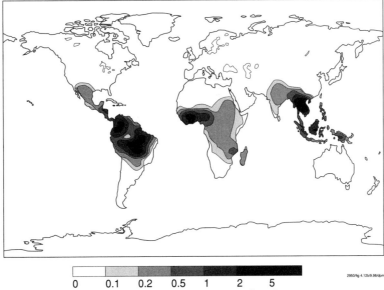

**Figure 5.13.** Exchange of carbon ($CO_2$ molecules $cm^{-2}$ $s^{-1}$) between the surface and the atmosphere: emission from (a) fossil fuel combustion and (b) deforestation.

releases approximately 5.5±0.5 GtC to the atmosphere each year, with emissions in the Northern Hemisphere dominating (Fig. 5.13a). Estimates of $CO_2$ contributions resulting from deforestation, primarily in the tropics (Fig. 5.13b), are quite uncertain, but using a value of 1.6±1.0, IPCC (1996) estimates the average annual sources of anthropogenic $CO_2$ to the atmosphere as 7.1±1.1 GtC. Since 3.3±0.2 GtC $yr^{-1}$ accumulate in the atmosphere, the remaining 3.8 GtC $yr^{-1}$ must be reabsorbed either by the oceans or by the terrestrial biosphere. Current models calculate an oceanic uptake of 2.0±0.8 GtC $yr^{-1}$ and an uptake by Northern Hemisphere regrowth of

0.5±0.5 GtC yr$^{-1}$, leaving an imbalance (or "missing sink") of 1.3±1.5 GtC yr$^{-1}$ (see Table 5.4). A large amount of effort has been directed towards partitioning this so-called "missing sink" into ocean and land components, and discovering an additional mechanism for the uptake of such large amounts of $CO_2$. The observed changes in the isotopic abundance of $CO_2$ (see Box 5.1) provides information on the evolution with time of carbon sources and sinks.

### 5.4.1.2 Oceanic Surface Fluxes

Although, on the global scale, the exchanges of $CO_2$ between the ocean and the atmosphere are nearly equilibrated, large imbalances can occur locally. The net exchange flux at a given location depends on the degree of saturation of $CO_2$ in the surface waters, and, by anaology with Eqs. (5.7) and (5.8), can be expressed by

$$F_{net} = k\left(p_{CO_2}^{ocean} - p_{CO_2}^{atm}\right) \tag{5.11}$$

where $p_{CO_2}^{atm}$ represents the partial pressures of $CO_2$ in the air and $p_{CO_2}^{ocean}$ represents an equilibrium $CO_2$ partial pressure in the surface ocean. $k$ is an exchange coefficient (see Fig. 5.4) that increases with wind speed (turbulence) at the air-sea interface. (In the above expression, the net flux is assumed to be positive when it is directed from the ocean into the atmosphere.) $p_{CO_2}^{ocean}$ varies with a variety of factors including temperature and oceanic biological activity.

The dissolution of $CO_2$ in sea water is achieved through the following reactions

$$CO_{2(g)} \to CO_{2(aq)} \tag{5.12a}$$
$$CO_{2(aq)} + H_2O \rightleftharpoons H_2CO_3 \tag{5.12b}$$
$$H_2CO_3 \rightleftharpoons HCO_3^- + H^+ \tag{5.12c}$$
$$HCO_3^- \rightleftharpoons CO_3^{2-} + H^+ \tag{5.12d}$$

where (g) and (aq) stand for gas and aqueous phase, respectively. The relative abundance of $CO_{2(aq)}$, $CO_3^{2-}$, and $HCO_3^-$ is determined by the concentration of hydrogen ions (pH) in water and the equilibrium constants of the above reactions, which vary with temperature, water salinity, and to a lesser extent, pressure. Figure 5.14 shows the partitioning of inorganic carbon species as a function of water pH.

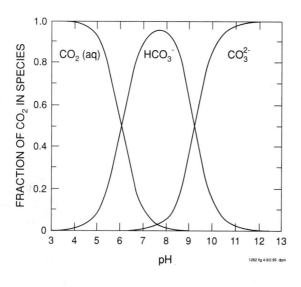

**Figure 5.14.** Relative concentration of dissolved carbon species in sea water calculated as a function of pH for a temperature of 15°C and a salinity of 35%. The average oceanic pH is approximately 8.2 (based on data from Mehrbach et al., 1973).

Since the pH of the ocean is usually close to 8.0, the predominant compound is the bicarbonate ion ($HCO_3^-$). Note, however, that in some fresh water ecosystems the pH of rivers and streams may vary between 3 and 8 with a shifting in the inorganic carbonate ion chemistry. The release and deposition of anthropogenic sulfur and nitrogen acids ("acid rain") has been shown to have a large impact upon chemistry of some natural water systems.

An important property of the ocean is its buffering capacity with respect to $CO_2$ uptake. A measure of this effect is provided by the Revelle factor $\mathcal{R}$ (Revelle and Suess, 1957), defined as

$$\mathcal{R} = \frac{\Delta p_{CO_2}/p_{CO_2}}{\Delta \Sigma C/\Sigma C} \qquad (5.13)$$

which represents the ratio between an increase in the partial pressure $p_{CO_2}$ of atmospheric $CO_2$ and the corresponding increase in the concentration of inorganic carbon in the ocean. $\Sigma C$ is defined as the sum of the $CO_2(aq)$, $CO_3^{2-}$, and $HCO_3^-$ concentrations in sea water. It can be shown that, on the average, this ratio is approximately equal to 10. Therefore, at equilibrium, a 10% increase in the atmospheric abundance of $CO_2$ leads to an increase of only 1% in the concentration of oceanic inorganic carbon.

The transfer of carbon to the depths of the ocean is achieved by sedimentation of calcium carbonate ($CaCO_3$). There is little evidence for inorganic precipitation of $CaCO_3$, except perhaps in warm tropical waters. The predominant formation of $CaCO_3$ is provided by shell-forming organisms (algae, protozoa, mollusks, corals, etc.) residing in sunlit surface waters. A large fraction of the carbonate sediments is therefore produced in nutrient-rich regions, especially in shelf zones where detritus of continental origin is advected by rivers. As these carbonates, as well as the calcareous remains of organisms, settle toward the bottom of the ocean, they often dissolve in deep waters, and only a small fraction reaches the ocean floor. That fraction of material that does reach the ocean floor carries various isotopic signatures, which can be used to assess climates of the past.

Organic forms of carbon are also present in the ocean. Their dominant source is the photosynthetic fixation of $CO_2$ by algae in the sunlit surface waters. The resulting growth of phytoplankton is, however, kept in balance by grazing species. Dissolved organic carbon compounds (hydrocarbons, aromatic compounds, carbohydrates, organic acids, etc.) are produced by the decay of dead cells and by direct release of these molecules by phytoplankton. The ocean probably acts as a source of several hydrocarbons and other organic compounds, which affect the chemical composition of the atmosphere. Figure 5.15 presents a schematic representation of the carbon cycle in the ocean.

In summary, the net transfer of carbon between the atmosphere and the ocean results from a disequilibrium of $CO_2$ gas concentration between these two large carbon pools. Figure 5.16 shows an estimate of the global distribution of this disequilibrium. The high positive values ($CO_2$ release to the atmosphere) in equatorial regions are related to sea water upwelling from the deep ocean, while the negative values ($CO_2$ uptake by the ocean) at mid- and high latitudes are the result of high $CO_2$ solubility in cold waters. Some of the negative values may also be due to local deficits of $CO_2$ in surface waters, due to the fixation of carbon by marine biota. Global models suggest that the net global uptake of $CO_2$ by the ocean represents $2.0 \pm 0.8$ GtC yr$^{-1}$. A major difficulty in reducing the uncertainty in the estimate of the oceanic $CO_2$ sink remains the paucity of data from high-latitude oceans.

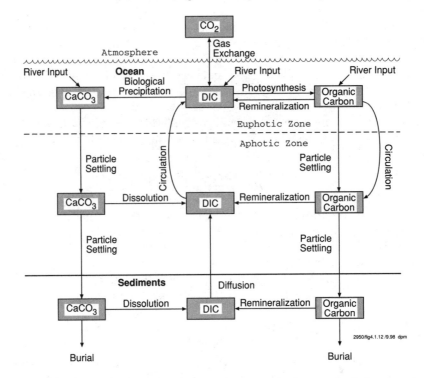

**Figure 5.15.** The carbon cycle in the ocean: In the euphotic zone (well-lit upper layers of the ocean), dissolved inorganic carbon (DIC) is converted to calcium carbonate ($CaCO_3$) and organic carbon. This material is transported to the aphotic zone (dark layers of the ocean) and to the sediments by particle settling and by ocean circulation. In these deep layers, carbonate calcium is dissolved by chemical and biological processes, while organic carbon is remineralized to DIC. DIC is transported to the upper ocean layers in regions where upwelling is intense (from Najjar, 1992; reprinted with the permission of Cambridge University Press).

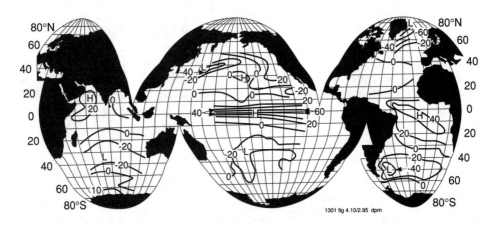

**Figure 5.16.** Difference in $CO_2$ partial pressure (ppmv) between the ocean and the atmosphere. A positive value refers to a flux from the ocean to the atmosphere.

### 5.4.1.3 Terrestrial Surface Fluxes

Exchanges of carbon and oxygen between the terrestrial biosphere and the atmosphere result from photosynthesis and respiration by plants. While the mechanisms involved are composed of a multitude of elementary reactions, the sum of these processes is generally expressed by single stoichiometric reactions. For example, the assimilation of $CO_2$ by plants through photosynthesis (conversion of $CO_2$ into organic matter in the presence of light) is represented stoichiometrically by

$$CO_2 + H_2O \xrightarrow{h\nu} CH_2O + O_2 \qquad (5.14a)$$

where $CH_2O$ is used as a shorthand notation to represent organic matter. The corresponding gross influx of $CO_2$ from the atmosphere to the biosphere is referred to as gross primary production (GPP). A fraction of the carbon is re-emitted by the plants to the atmosphere through (autotrophic) respiration processes, which, in a stoichiometric form, can be expressed by the inverse reaction

$$CH_2O + O_2 \rightarrow CO_2 + H_2O \qquad (5.14b)$$

The net carbon flux, that is, the accumulation of carbon in the phytomass, is called net primary production (NPP). It represents all the carbon stored in leaves, branches, and roots of plants. When vegetation is eaten, much of the organic matter is oxidized back to $CO_2$ in respiration (Reaction 5.14b). Likewise, when plants or animals die, the material is decomposed (oxidized) by microorganisms, and $CO_2$ is remineralized and released to the atmosphere. Over a sufficiently long time period, respiration of this organic matter (heterotrophic respiration) approximately balances net primary production. Table 5.5 provides an estimate of the surface covered by several types of ecosystems in the world, the corresponding NPP, and the amount of carbon stored in the phytomass and in the soils. Figure 5.17 provides a schematic overview of the global carbon cycle in relation to the continental biosphere.

**Table 5.5**
*Surface Area, NPP, and Carbon Contained in the Phytomass and Soils for Different Ecosystems*

| Ecosystem Type | Surface ($10^{12}$ m$^2$) | NPP (GtC yr$^{-1}$) | Phytomass (GtC) | Soils (GtC) |
|---|---|---|---|---|
| Seasonal tropical forest | 10.3 | 10.5 | 193 | 82 |
| Evergreen tropical forest | 4.5 | 3.2 | 51 | 41 |
| Temperate forest | 7.0 | 4.6 | 88 | 72 |
| Boreal forest | 9.5 | 3.6 | 96 | 135 |
| Savanna | 27.0 | 19.9 | 90 | 236 |
| Grassland | 12.5 | 4.4 | 9 | 295 |
| Tundra | 9.5 | 0.9 | 6 | 121 |
| Semi-desert | 21.0 | 1.3 | 7 | 168 |
| Desert | 24.5 | 0.1 | 1 | 23 |
| Cultivated | 16.0 | 6.8 | 3 | 128 |
| Wetlands | 3.5 | 4.0 | 15 | 225 |
| Other | 4.0 | 0.6 | 1 | 10 |
| Total | 149.3 | 59.9 | 560 | 1536 |

After Ajtay et al., 1979.

Prior to the onset of large-scale human perturbations, the terrestrial biosphere is assumed to have been in a steady-state condition with respect to carbon. On a global scale, and averaged over several years, release of C in one locality, by natural fires, for example (see Box 5.2 for recent estimates of the amount of carbon and other

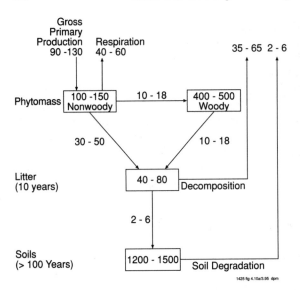

**Figure 5.17.** Carbon cycle in the continental biosphere. Reservoirs and fluxes are expressed in GtC and $GtC\ yr^{-1}$, respectively (after Fung, 1993).

**Box 5.2 Biomass Burning**

Approximately a half-billion hectares of land are burned globally every year. Most of the burning occurs in the tropics for shifting cultivation, deforestation, grazing in the savannas, firewood, and clearing of agricultural residues (Fig. 5.18). Fires are also used as a tool in forest management, removing undergrowth, and helping to prevent insect infestation and disease. Carbon dioxide, many other trace gases, and particulate matter are released during the fires. These gases and aerosol particles impact the health of local residents and fire fighters. They also degrade air quality and affect the chemistry of the troposphere and stratosphere, and global climate. The amount and identity of emissions from biomass fires is dependent on ambient temperature, humidity, wind velocity, topography, the type of vegetation, and the moisture content and elemental composition of biomass. The combustion processes of forest fires are often less efficient than those of savanna fires,

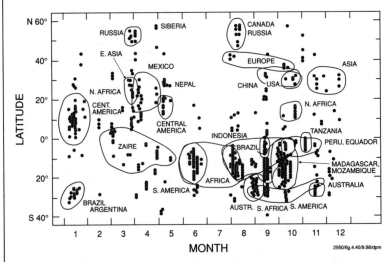

**Figure 5.18.** Fires detected as a function of season and latitude (from Andreae, 1993).

and thus result in production of larger quantities of CO, $CH_4$ and other hydrocarbons. Estimates of the amount of biomass burned per year and the resulting release of carbon to the atmosphere (Andreae, 1993; Hao and Liu, 1994) are summarized in the Table 5.6.

**Table 5.6**
Amount of Biomass Burned ($Pg\ yr^{-1}$)

|  | Trop. America | Trop. Africa | Trop. Asia | Australia | Temperate Regions | Boreal Regions | Total |
|---|---|---|---|---|---|---|---|
| Deforestation | 0.2 | 0.2 | 0.1 |  |  |  | 0.5 |
| Shifting cultivation | 0.6 | 0.5 | 0.2 |  |  |  | 1.3 |
| Savanna fires | 0.7 | 1.6 | <0.1 | 0.3 |  |  | 2.7 |
| Firewood and charcoal | 0.1 | 0.2 | 0.3 | <0.1 | 0.2 | 0.1 | 0.9 |
| Agricultural residues | 0.07 | 0.03 | 0.15 | 0.03 | 0.21 |  | 0.49 |
| Wild fires |  |  |  | 0.02 | 0.11 | 0.14 | 0.27 |
| Prescribed fires |  |  |  | <0.01 | 0.02 |  | 0.03 |
| Total | 1.7 | 2.5 | 0.8 | 0.4 | 0.6 | 0.2 | 6.2 |
| Carbon released ($PgC\ yr^{-1}$) | 0.8 | 1.2 | 0.4 | 0.2 | 0.3 | 0.1 | 3.0 |

Emissions of other compounds are estimated on the basis of emission ratios to $CO_2$ measured in the laboratory or in the field. Extensive research has been conducted to characterize emissions from biomass burning, both in tropical and in temperate regions, based mostly on measurements made during large integrated expeditions. Table 5.7 provides an estimate (Andreae, 1993) of the emissions associated with biomass burning as well as their fractional contribution to global emissions by all sources.

**Table 5.7**
Global Biomass Burning Emissions

| Compound | Biomass Burning ($Tg\ yr^{-1}$)[a] | Percentage Relative to Total Surface Emission |
|---|---|---|
| CO | 290 | 26 |
| $NO_x$ | 9.6 | 24 |
| $CH_4$ | 29 | 8 |
| NMHC[b] | 36 | 36 |
| $CH_3Cl$ | 1.28 | 55 |
| $H_2$ | 16 | 45 |

[a] Emissions are given in $TgC\ yr^{-1}$ for CO, $CH_4$, and nonmethane hydrocarbons (NMHC), in $TgN\ yr^{-1}$ for $NO_x$ and in $TgCl\ yr^{-1}$ for $CH_3Cl$.
[b] Nonmethane hydrocarbons excluding isoprene and terpenes.

Biomass burning in the tropics is regarded as a major "forcing mechanism" for the photochemistry of regional and even global troposphere. For example, the photochemical oxidation of carbon monoxide and hydrocarbons emitted by fires, in the presence of nitrogen oxides, leads to the production of large amounts of ozone in the tropics.

compounds released to the atmosphere as a result of biomass burning), was largely balanced by ecosystem growth and regeneration elsewhere. As humans began to alter natural ecosystems on a large scale, however, this situation changed, as destruction of plant communities, with concomitant losses of C to the atmosphere, was no longer balanced by regrowth. Only if ecosystems containing large amounts of carbon are converted to systems containing relatively little carbon (or vice versa) will land conversion have a significant impact on the global carbon budget. Approximately 80% of the carbon held in above-ground biomass and 40% held below ground (in roots, litter, and soil) is contained in forests (see Dixon et al., 1994). Thus conversion of forests to systems containing relatively little carbon, such as grazing and agriculture, can significantly influence carbon cycling. Likewise, regrowth of forests on previously cleared land can provide a significant sink for atmospheric $CO_2$.

Most of the emphasis on land-use change has focused on low-latitude forests, which contain approximately 59% and 27% of global forest and soil C, respectively (Dixon et al., 1994), and which are being cleared for grazing or agriculture at an alarming rate. Estimates of carbon losses in the conversion of tropical forests depend on estimates of total land area currently under forest, the carbon density (gC ha$^{-1}$) of forests in different regions, and the rates of conversion to pasture or cultivation. The carbon flux associated with deforestation (Fig. 5.19) is believed to be about 1.6 GtC yr$^{-1}$ (40% in Latin America, 40% in southern and southeastern Asia, 20% in Africa), but this number is subject to an uncertainty of at least a factor of 2. For the 1850-1990 period, the cumulative flux could be as high as 120 GtC.

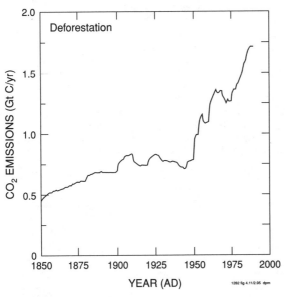

**Figure 5.19.** Estimated emission of carbon associated with deforestation between 1850 and 1993.

Tropical deforestation and the resulting release of $CO_2$ have received a great deal of attention. What has been less commonly appreciated until recently is the fact that large amounts of land in mid- and high latitudes, primarily in the Northern Hemisphere, which had been previously cleared for grazing or agriculture, are now reverting to forest, providing a possible sink for some of the excess $CO_2$ arising from anthropogenic sources. Pioneering studies of Keeling et al. (1989) and Tans et al. (1990) used atmospheric transport models coupled with data indicating a significant north-south gradient in $CO_2$ concentrations to infer a large previously unidentified Northern Hemisphere sink for $CO_2$. Because photosynthesis discriminates against $^{13}CO_2$ (Box 5.1) while isotopic fractionation when $CO_2$ enters the ocean is small, $^{13}C/^{12}C$ ratios provide a good fingerprint of terrestrial biosphere fluxes of $CO_2$. Ciais et al. (1995) used $^{13}C/^{12}C$ data from 43 sites worldwide to infer the existence of a strong terrestrial $CO_2$ sink in temperate northern latitudes. This implies a significant amount of carbon entering Northern Hemisphere forests, and several ideas have been put forth to explain this phenomenon.

Table 5.5 includes a sink for $CO_2$ from northern forest regrowth of 0.5±0.5 GtC yr$^{-1}$. Two recent studies (Kauppi et al., 1992; Sedjo, 1992) report carbon accumulation in Northern Hemisphere forests on the order of 1 GtC yr$^{-1}$. Dixon et al. (1994) estimate that high-latitude forests comprise an annual sink for 0.48±0.10 GtC and midlatitude forests another 0.26±0.09 GtC. These numbers will doubtless be revised as increasingly accurate forest inventory data and data on rates of land-use change are collected, but several lines of evidence suggest that C may be accumulating in different ecosystem types, compensating for a significant fraction of the carbon lost through tropical forest destruction (Wisniewski and Lugo, 1992).

Increasing levels of atmospheric $CO_2$ exert a direct influence on photosynthesis and other plant processes (Amthor, 1995). Many experimental studies, performed in controlled environments with increased $CO_2$ abundances, have reported enhanced photosynthesis, and some of these have also reported an increase in plant growth (Strain and Cure, 1985; Drake, 1992). Quantifying these possible "fertilization" effects on plant growth under natural conditions has proven difficult, however. In addition to possible $CO_2$ fertilization, there is growing evidence that the productivity of forests has increased due to the fertilizing effect of nitrogen deposition, which has increased several fold from preindustrial levels. Schindler and Bayley (1993) suggested that large amounts of deposited N were retained by several ecosystems. Assuming a constant C:N ratio in plant tissues, increased N input implies an increase in the total amount of C sequestered in the same systems.

Finally, it is interesting to note that climate change may influence the C cycle through a variety of feedbacks. Changes in wind patterns may alter air-sea $CO_2$ exchange rates, and changes in surface temperature affect the solubility of $CO_2$ in the ocean. Changes in temperature would also potentially alter the distribution and productivity of terrestrial ecosystems, with implications for carbon storage, and projected temperature increases at high latitudes might lead to a net release of large amounts of carbon now sequestered in peatlands. Should the reduction of stratospheric $O_3$ result in increased UV-B fluxes, resulting stresses in both terrestrial and marine ecosystems would also affect $CO_2$ uptake rates.

### 5.4.2 Methane

Methane ($CH_4$) is the most abundant organic trace gas in the atmosphere, with an average tropospheric mixing ratio (1984-1993) of approximately 1740 ppb in the Northern Hemisphere and 1620 ppb in the Southern Hemisphere (Dlugokencky et al., 1994). After water vapor and $CO_2$, methane is the most abundant greenhouse gas in the troposphere, and, on a per molecule basis, methane has a much greater climate warming potential than $CO_2$. Methane is also quite reactive and plays an important role in both tropospheric and stratospheric chemistry. Oxidation of methane by OH in the troposphere leads to the formation of formaldehyde, of carbon monoxide, and, in environments with sufficient $NO_x$, of ozone. In the stratosphere, methane acts as a sink for chlorine atoms and is therefore important in stratospheric ozone chemistry. Methane oxidation by OH is also a major source of water vapor in the stratosphere. The lifetime of methane in the atmosphere is approximately 8-10 years (Watson et al., 1992).

The atmospheric $CH_4$ concentration had remained relatively constant at less than half its present value for thousands of years before beginning to increase about two hundred years ago from 700 to 1700 ppb (Fig. 5.20). Until recently, global atmospheric methane concentrations were increasing at a rate of 1.3% per year (Blake and Rowland, 1988). However, the rate of methane increase has appeared to slow in recent years, to about 0.6% per year (Steele et al., 1992) or less. The cause of this decline in the $CH_4$ growth rate is not clear. It may be due to an increase in the chemical sink for $CH_4$

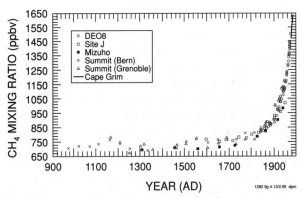

**Figure 5.20.** Mixing ratio of atmospheric methane over the past 1,000 years, as observed in air trapped inside ice cores from Antarctica and Greenland. Data from Cape Grim, Tasmania, from the recent past are also represented.

due to an increase in the concentration of tropospheric OH, although decreasing $CH_4$ source strengths cannot be ruled out.

Methane is produced in oxygen-deficient wetland habitats such as swamps, lakes, rice paddies, tundra, boreal marshes, etc. Methane production in soils and oceans is the end product of a variety of reductive pathways during the decomposition of organic matter. In inland terrestrial ecosystems with low sulfate and nitrate concentrations, the dominant pathway for methane production is the reduction of acetate, but other pathways include the reduction of $CO_2$ and other organic compounds including formate and amino acids. Oxidation processes at the interface between oxygen-poor and oxygen-rich environments limit the emission of methane to the atmosphere. The rate of methane production is highest in tropical wetlands, reflecting a positive relationship between temperature, net ecosystem production, and the rate of methanogenesis in many wetlands. Management of wetland areas, including the increase in the area of rice cultivation, may have contributed to the observed increase in the abundance of atmospheric methane over the past century.

Methane is also released by anaerobic microbial activity in the stomachs of cattle, termites, and perhaps other insects, whereas coal mining, natural gas losses, and solid waste burning are important anthropogenic sources. Large amounts of $CH_4$ are also released by biomass burning. Releases of natural gas and biomass burning appear to have increased the $^{13}C$ isotopic ratio of atmospheric methane from a preindustrial value of about $-50‰$ to the present value of $-47‰$.

Table 5.8 presents a recent estimate of the global methane budget (IPCC, 1994; 1996) that demonstrates that the anthropogenic contribution to methane sources (375 Tg $CH_4$ yr$^{-1}$) is more than twice the contribution of natural sources (160 Tg yr$^{-1}$). Large emissions associated with human activities (typically 100 Tg yr$^{-1}$) are provided by inadvertant releases of fossil fuel $CH_4$ during mining operations or in conjunction with the use of coal and natural gas. Waste management including the decomposition of animal waste and of landfill material leads to a production of about 90 Tg $CH_4$ yr$^{-1}$. Finally, in addition to rice cultivation (producing 60 Tg $CH_4$ yr$^{-1}$), digestive fermentation by grazing animals and insects produces perhaps 85 Tg $CH_4$ yr$^{-1}$. Table 5.8 also shows that the largest methane sink results from OH oxidation, and that stratospheric removal accounts for less than 10 percent of the total loss. A small amount of methane (30 Tg yr$^{-1}$) diffuses into soils, where it is consumed by methanotrophic bacteria.

### 5.4.3 Nonmethane Hydrocarbons

Hundreds of nonmethane hydrocarbons (NMHC) are released into the atmosphere from a variety of anthropogenic and biogenic sources, with emissions exceeding

**Table 5.8**
Estimated Sources and Sinks of Methane
in the Atmosphere (Tg $CH_4$ $yr^{-1}$)

| Sources or Sinks | Range | Likely |
|---|---|---|
| *Natural* | | |
| Wetlands | | |
| Tropics | 30-80 | 65 |
| Northern latitude | 20-60 | 40 |
| Others | 5-15 | 10 |
| Termites | 10-50 | 20 |
| Ocean | 5-50 | 10 |
| Freshwater | 1-25 | 5 |
| Geological | 5-15 | 10 |
| Total | | 160 |
| *Anthropogenic* | | |
| Fossil fuel related | | |
| Coal mines | 15-45 | 30 |
| Natural gas | 25-50 | 40 |
| Petroleum industry | 5-30 | 15 |
| Coal combustion | 5-30 | 15 |
| Waste management system | | |
| Landfills | 20-70 | 40 |
| Animal waste | 20-30 | 25 |
| Domestic waste treatment | 15-80 | 25 |
| Enteric fermentation | 65-100 | 85 |
| Biomass burning | 20-80 | 40 |
| Rice paddies | 20-100 | 60 |
| Total | | 375 |
| Total sources | | 535 |
| *Sinks* | | |
| Reaction with OH | 405-575 | 490 |
| Removal in stratosphere | 32-48 | 40 |
| Removal by soils | 15-45 | 30 |
| Total sinks | | 560 |
| Atmospheric increase | 35-40 | 37 |

From IPCC, 1994; 1996.

1000 TgC $yr^{-1}$. Many of these compounds are highly reactive and play an important role in tropospheric chemistry, affecting the regional and global oxidant balance, and contributing to organic acid production and deposition (Fehsenfeld et al., 1992). Of particular interest is the role of NMHC in tropospheric ozone formation in both urban and rural environments. Although anthropogenic emissions dominate in urban and industrial areas, natural emissions account for over half the total NMHC emissions in the United States and over three-fourths globally (Guenther et al., 1995). Sources of these compounds are variable in time and space, and emission estimates remain quite uncertain, but recent estimates of the magnitude of NMHC sources are listed in Table 5.9. Emission estimates and sources for specific NMHC of importance are given in Table 5.10.

### Table 5.9
### Estimates of Global NMHC Emissions

| | Emissions (TgC yr$^{-1}$) |
|---|---|
| *Sources* | |
| *Anthropogenic* | |
|    Transportation | 22 |
|    Stationary source fuel combustion | 4 |
|    Industrial processes including | |
|       natural gas production | 17 |
|    Biomass burning, forest fires, incineration | 45 |
|    Organic solvents | 15 |
| Anthropogenic subtotal | 103 |
| *Natural* | |
|   *Oceanic* | |
|     Light hydrocarbons | 5-10 |
|     $C_9$-$C_{28}$ *n*-alkane | 1-26 |
|   *Terrestrial* | |
|     Microbial production | 6 |
|     Emissions from vegetation | |
|       isoprene | 500 |
|       monoterpenes | 125 |
|       other | 520[a] |
| Natural subtotal | ~1170 |
| Total emissions | ~1273 |

[a] Extremely uncertain.
Adapted from Singh and Zimmerman, 1992, and Guenther *et al.*, 1995.

### Table 5.10
### Global Sources of Individual NMHC Species

| Hydrocarbon | Emission (TgC yr$^{-1}$) | Major Source |
|---|---|---|
| ethane | 10-15 | natural gas emissions, biomass burning, oceans, vegetation |
| ethene | 20-45 | fuel combustion, biomass, terrestrial ecosystems |
| acetylene | 3-6 | fuel combustion, biomass burning |
| propane | 15-20 | natural gas, biomass burning, oceans, vegetation emissions |
| propene | 7-12 | fuel combustion, biomass burning, oceans |
| *n*-butane | 1-2 | fuel combustion, natural gas, biomass burning, oceans |
| *i*-butane | 1-2 | fuel combustion, natural gas, biomass burning, oceans |
| butenes | 2-3 | fuel combustion, biomass burning, oceans |
| *n*-pentane | 1-2 | fuel combustion, natural gas, biomass burning |
| *i*-pentane | 2-3 | fuel combustion, natural gas, biomass burning |
| benzene | 4-5 | fuel combustion, biomass burning |
| toluene | 4-5 | fuel combustion, biomass burning, solvents |
| xylenes (*m, p, o*) | 2-3 | fuel combustion, biomass burning, solvents |
| isoprene | 500 | forest/plant emissions |
| monoterpenes | 125 | forest/plant emissions |

Adapted from Singh and Zimmerman, 1992, and Guenther *et al.*, 1995.

Virtually all anthropogenic activities related to energy use or transfer result in the release of NMHC into the atmosphere. In industrialized countries, fossil fuel combustion (including automobiles), natural gas emissions, and industrial processing of chemicals and waste are major sources of anthropogenic NMHC, while in less developed parts of the world, biomass burning is an important component.

A small percentage of natural emissions arises from oceanic sources, but these are dwarfed by the large emissions from terrestrial vegetation, particularly trees. By far the single most important NMHC, in terms of both magnitude of emissions and reactivity, is the five-carbon compound, isoprene ($C_5H_8$), which is emitted in large amounts by many deciduous tree species. Another major source of biogenic NMHC is the large class of ten-carbon compounds known as monoterpenes, which are emitted by many trees and shrubs, especially conifers, and which give rise to the characteristic fragrance of pines, mints, and eucalypts. Attention has focused primarily on these two classes of compounds, but recent evidence indicates that a variety of substituted hydrocarbons, including alcohols, ketones, and aldehydes, are also emitted in large quantities by vegetation, though these fluxes are only beginning to be quantified. For completeness, these emissions are included in Table 5.9 as "other" but should be regarded as highly uncertain. The impact of these hydrocarbons on regional photochemistry is discussed in Chapter 13.

Estimation of the role of vegetation in overall NMHC emissions is complicated by a number of factors. Emission rates of individual NMHC vary by several orders of magnitude, and whether a tree emits isoprene, monoterpenes, or other compounds is very species specific. Thus, to make an accurate estimate of the flux of various compounds from a given region requires detailed knowledge of the vegetation species composition, and the emission characteristics and amount of leaf biomass of each species. Furthermore, all emissions from vegetation are very sensitive to temperature. Isoprene emissions exhibit a distinct temperature optimum, while monoterpene emissions, generally much lower than isoprene emissions, increase exponentially without an optimum. Isoprene emissions are also dependent on light intensity. These light and temperature dependencies have been determined experimentally (Guenther et al., 1993) and are shown in Figure 5.21. This information has been incorporated into models of biogenic NMHC emissions for use on regional and global scales (Guenther et al., 1995). The distribution of isoprene emission rates during two seasons, generated by one such model, is shown in Figure 5.22.

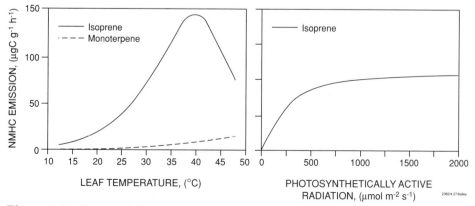

**Figure 5.21.** *Response of isoprene and monoterpene emissions to changes in temperature, and response of isoprene emission to changes in photosynthetically active radiation (based on the work of Guenther et al., 1993).*

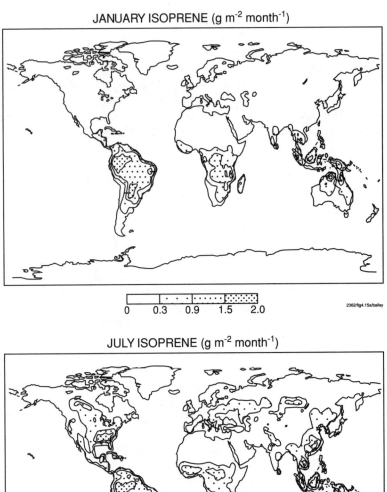

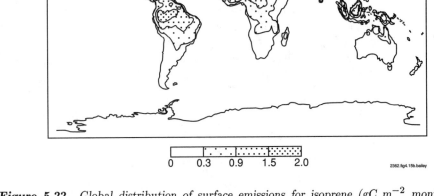

**Figure 5.22.** *Global distribution of surface emissions for isoprene ($gC\ m^{-2}\ month^{-1}$) in January and July, estimated by Guenther et al. (1995).*

Although light and temperature are the primary controls of NMHC emission from vegetation, recent studies have shown that leaf age and growth environment, for example, temperature, affect NMHC emission and emission of specific NMHC have been correlated with flowering and bud break. Additional factors that may significantly influence natural emissions include leaf hydrocarbon concentrations, nitrogen levels, water status, insects, disease, and other stresses, but these controls are not yet well understood.

### 5.4.4 Carbon Monoxide

Although not a significant greenhouse gas, carbon monoxide plays a central role in tropospheric chemistry via its reaction with the OH radical. In many areas, concentrations of OH are controlled by CO levels (Logan et al., 1981). In the atmosphere, CO is produced by the oxidation of methane and other hydrocarbons such as isoprene. It is also released at the surface by incomplete combustion associated with biomass burning and fossil fuels. Sources of secondary importance include emission by vegetation and microorganisms on the continents and by photochemical oxidation of dissolved organic matter in the oceans. Table 5.11, which provides an estimate of the global budget for CO, indicates that biomass burning comprises perhaps a third of the total source (see Box 5.2), and that the oxidation of nonmethane hydrocarbons contributes approximately the same amount as the oxidation of methane. About 50-60 percent of the CO emission results from human activities. Figure 5.23 shows an estimate of the global distribution of CO emissions in December. Once released in the atmosphere, CO is oxidized by the hydroxyl (OH) radical. Soil uptake of CO is a sink of secondary importance.

The observed mixing ratio of CO in the remote regions of the troposphere is typically 60-70 ppbv in the Southern Hemisphere and 120-180 ppbv in the Northern Hemisphere. Much higher concentrations can be observed in localized urban or industrial areas where pollution is intense. If a global atmospheric burden of 400 Tg and a global loss of 2400 Tg yr$^{-1}$ are adopted as likely values, the global lifetime of CO in the atmosphere should be approximately two months. This value is too short to allow complete mixing of this gas in the global atmosphere. The distribution of CO is therefore far from uniform in space and time; the highest concentrations are found in industrialized regions (Europe, eastern United States, eastern Asia) and in the tropics, where the surface sources are the most intense.

Because of the high variability associated with the distribution of CO, global temporal trends in atmospheric abundance are difficult to estimate. Since anthropogenic sources were small until this century, the concentration of CO may have doubled in the Northern Hemisphere over the past 50 to 80 years. A comparison of solar infrared spectra recorded in the early 1950s and in the late 1970s at Jungfraujoch (Switzerland) suggests, for example, that CO has increased on the average by 2% per year during this period. Changes in the Southern Hemisphere are not well documented. A recent analysis of data collected at a large number of observing stations suggests that the abundance of CO has decreased significantly over the past decade (Novelli et al., 1994) in both hemispheres. Such a negative trend could be explained by a reduction in surface emissions (e.g., pollution control, reduction in biomass burning) or by an increase in the oxidation capacity of the atmosphere (e.g., increase in the OH density).

**Table 5.11**
*Global Budget for Carbon Monoxide (Tg yr$^{-1}$)*

|  | Magnitude |
|---|---|
| **Sources** |  |
| Biomass burning | 300-900 |
| Fossil fuel burning | 300-600 |
| Vegetation | 50-200 |
| Oceans | 6-30 |
| Methane oxidation | 400-1000 |
| NMHC oxidation | 300-1000 |
| Total | 1400-3700 |
| **Sinks** |  |
| Chemical loss (OH) | 1400-2600 |
| Uptake by sinks | 150-500 |
| Total | 1550-3100 |

Based on Khalil and Rasmussen, 1990, and Bates et al., 1995.

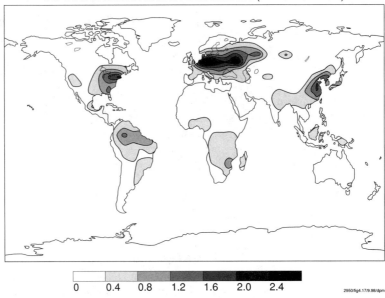

**Figure 5.23.** Global distribution of total CO emissions ($10^{12}$ molecules cm$^{-2}$ s$^{-1}$) in December, including the contribution of fossil fuel combustion, biomass burning, soils, and oceans, at a spatial resolution of 5° × 5° (based on data from Müller, 1992b, and Hao and Granier, personal communication, 1995).

## 5.5 The Global Nitrogen Cycle

*Nitrogen* ($N_2$) is the most abundant element in the Earth's atmosphere, comprising approximately 80% of the total. The $N_2$ molecule is extremely stable, and plays almost no chemical role, except in the thermosphere (above 90 km altitude) where it can be photolyzed or ionized. Minor constituents such as nitrous oxide ($N_2O$), nitric oxide (NO), nitrogen dioxide ($NO_2$), nitric acid ($HNO_3$), and ammonia ($NH_3$) are chemically reactive, and play important roles in contemporary environmental problems, including the formation of acid precipitation, photochemical smog and tropospheric aerosols, and the destruction of stratospheric ozone. The nitrogen oxides, NO and $NO_2$, are rapidly interconvertible and exist in dynamic equilibrium. For convenience, the sum of the two species is referred to as $NO_x$ ($NO_x = NO + NO_2$). Large amounts of $NO_x$ are released by combustion processes, particularly as a result of industrial activity and the intensive use of automobiles. Biomass burning and microbial production in soils are other important sources of $NO_x$. In relatively "clean" environments, the emission of some nitrogen compounds (*e.g.*, $N_2O$, NO, $NH_3$) results primarily from a wide range of complex biological processes in soils (Fig. 5.24).

*Nitrogen fixation* is any process in which atmospheric $N_2$ reacts to form any other nitrogen compound. An example of biological nitrogen fixation is provided by the enzyme-catalyzed reduction of $N_2$ to $NH_3$ or $NH_4^+$. Nitrogen fixation occurs also as a result of anthropogenic activities (combustion, industrial production of ammonia or nitric acid, agricultural management). In the absence of fertilizers, biological N fixation is the ultimate source of nitrogen for living organisms and is carried out exclusively by bacteria. Once fixed, nitrogen is assimilated into biomass, both plant and microbial. It can also be oxidized to nitrites ($NO_2^-$) and nitrates ($NO_3^-$), a process called *nitrification* that is facilitated by the presence of certain bacteria, and

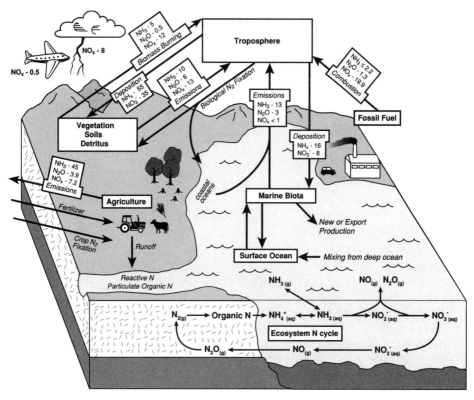

**Figure 5.24.** The global N cycle including ammonia and ammonium from Table 5.10, nitrous oxide from Table 5.11, and $NO_x$ and nitrate from Table 5.14. All units are provided in $Tg(10^{12})$ $yr^{-1}$. Stratospheric sources and sinks of the gases are neglected, but appear in the individual tables.

is summarized by the two consecutive reactions

$$2\,NH_4^+ + 3\,O_2 \rightarrow 2\,NO_2^- + 2\,H_2O + 4\,H^+ \tag{5.15a}$$

$$2\,NO_2^- + O_2 \rightarrow 2\,NO_3^- \tag{5.15b}$$

Nitrous and nitric oxides are byproducts of these reactions, so that significant amounts of $N_2O$ and NO can be released to the atmosphere as a result of nitrification. The reduction of $NO_3^-$ to any gaseous nitrogen species (mostly $N_2$, $N_2O$, NO) is called *denitrification* and occurs both chemically and biologically. In the absence of denitrification, all biologically available nitrogen that has been released from the igneous rocks of the Earth's original crust and mantle would have been converted long ago to its more thermodynamically stable form of $NO_3^-$ in the oceans. Instead, as a result of denitrification, $N_2$ in the atmosphere remains by far the Earth's largest reservoir of nitrogen. Denitrification is carried out by a diverse array of bacteria and is widespread throughout the Earth's surface. It is summarized by the following series of reductive processes

$$NO_3^-(+5) \rightarrow NO_2^-(+3) \rightarrow NO(+2) \rightarrow N_2O(+1) \rightarrow N_2(0)$$

where the respective oxidation states are indicated inside the parentheses.

Denitrification occurs in environments with restricted oxygen availability, sufficient quantities of a suitable reductant (usually organic carbon), and sufficient concentrations of the required nitrogen oxides. Under most conditions when oxidant is limited, nitrogen is reduced all the way to $N_2$. However, when the availability of oxidant outweighs the supply of reductant, nitrogen reduction is incomplete, and the $N_2O/N_2$ ratio increases. The use of nitrogen fertilizers, required for intensive agriculture, is believed to enhance the flux of $N_2O$ to the atmosphere, and to account in part for the observed increase (0.2-0.3% per year) in the atmospheric abundance of $N_2O$ (see Fig. 5.25). Denitrification in the ocean remains poorly understood. Water is usually supersaturated with $N_2O$; so oceans should be a source of nitrous oxide to the atmosphere.

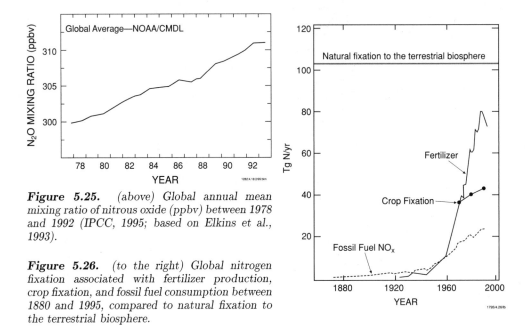

**Figure 5.25.** (above) Global annual mean mixing ratio of nitrous oxide (ppbv) between 1978 and 1992 (IPCC, 1995; based on Elkins et al., 1993).

**Figure 5.26.** (to the right) Global nitrogen fixation associated with fertilizer production, crop fixation, and fossil fuel consumption between 1880 and 1995, compared to natural fixation to the terrestrial biosphere.

With the intensification of agricultural practice and the development of industrial activities in many regions of the world, the global nitrogen cycle has been modified substantially (see, e.g., Vitousek, 1994). Under natural conditions, biological nitrogen fixation is estimated to represent approximately 44-200 TgN yr$^{-1}$ (the most likely value is approximately 100 TgN yr$^{-1}$), while fixation by marine ecosystems is probably on the order of 5-30 TgN yr$^{-1}$. Fixation by lightning, another component of the natural nitrogen cycle, represents 2-20 TgN yr$^{-1}$. Today, several additional factors related to human activities must be taken into account: Industrial nitrogen fixation in the production of fertilizers probably exceeds 80 TgN yr$^{-1}$ (see Fig. 1.1c), while fixation by legume crops is on the order of 30-50 TgN yr$^{-1}$. An additional 20 TgN yr$^{-1}$ are fixed in internal combustion engines and released into the atmosphere as nitrogen oxides. Human activities have also affected long-term storage pools of nitrogen through biomass burning, forest clearing, and wetland drainage. The extent of human alteration of the global nitrogen cycle is shown in Figure 5.26. Note that the total amount of nitrogen fixed as a result of human activities has most likely become larger than natural fixation, and that fertilizer production accounts for more than half the nitrogen fixed anthropogenically. The consumption of nitrogenous fertilizers alters the nitrogen cycle through enhanced emissions of ammonia, nitrous oxide, and nitric oxide to the atmosphere (Matthews, 1994), nitrate pollution of ground water, and increased deposition of ammonium to terrestrial ecosystems (leading to soil acidification, forest decline, loss of biodiversity as well as stimulation of carbon uptake). As noted by Chameides *et al.* (1994), the three regions of the world that currently dominate global agricultural and industrial activity, and account for most of the world fertilizer use, food-crop production, and commercial energy consumption, are responsible for more than half of the world's $NO_x$ emissions, and, hence, are prone to ground-level ozone pollution during the summer. Since repetitive episodes of ozone pollution during the growing season are known to reduce crop yields, increasing $NO_x$ emissions will probably result in significant agricultural losses and reduce food production capacity.

### 5.5.1 Ammonia

Ammonia is the third most abundant nitrogen gas in the atmosphere (after $N_2$ and $N_2O$) despite its relatively short residence time of approximately 10 days. The mixing ratio is highly variable in space and time, but typical values at the surface over continents are 0.1-10 ppbv. Table 5.12 provides an estimate of the global budget of $NH_3$. Dominant sources include enzymatic decomposition of urea in animal urine and emanations from decomposing excrement, emissions by soils, biomass burning, and losses during the production and application of fertilizers. A variety of industrial activities release ammonia, including paper manufacture, wastewater treatment, manufacturing processes, petroleum refining, cooking, and coal and oil combustion.

Ammonia is thus primarily a product of biological activity as well as a byproduct of agriculture and waste production and processing (both human and animal waste). Atmospheric concentrations of ammonia are greater over the continents than over the oceans and there is little transport of ammonia ($NH_3$) or ammonium ($NH_4^+$) from the ocean to land masses although the transport from land masses to oceans can be substantial (Quinn *et al.*, 1987). Oceanic emissions of $NH_3$ may be as high as 13 TgN yr$^{-1}$ (Schlesinger and Hartley, 1992; based on Quinn *et al.*, 1990) but are nearly balanced by the deposition of $NH_3/NH_4^+$ on the sea surface (Duce *et al.*, 1991).

Ammonia is released via the mineralization of organic material in animals, soils, and the ocean, and intracellular reactions within plant tissues. Most of the ammonia mineralized is utilized. However, a small pool remains in solution and is subject to volatilization at sufficiently high pH. Inorganic fertilizers and animal excreta accelerate soil emissions by increasing the $NH_3$ or $NH_4^+$ content of the soil, particularly in areas where livestock are concentrated, like feedlots. Ammonia can

**Table 5.12**
*Global Budget of Ammonia*

| Ammonia | TgN yr$^{-1}$ |
|---|---|
| *Sources*[1] | |
| Domestic animals | 21.3 (20-40) |
| Human excrement | 2.6 (2.6-4) |
| Soil emissions | 6 (6-45) |
| Biomass burning | 5.7 (1-9) |
| Wild animals | 0.1 (0.1-6) |
| Industry | 0.2 |
| Fertilizer losses | 9 (5-10) |
| Fossil fuel combustion | 0.1 (0.1-2.2) |
| Ocean | 8.2 (5-15) |
| *Sinks*[2] | |
| Wet precipitation (land) | 11 (11-80) |
| Wet precipitation (ocean) | 10 (6-26) |
| Dry deposition (land) | 11 (10-150) |
| Dry deposition (ocean) | 5 |
| Reaction with OH | 3 (1-9) |

[1] The best estimates are from Bouwman *et al.* 1997, and the ranges adapted from Warneck, 1988; Quinn *et al.*, 1990; Schlesinger and Hartley, 1992; and Dentener and Crutzen, 1994.
[2] The best estimates are from MOGUNTIA simulations (Dentener and Crutzen, 1994) and the ranges adapted from Warneck, 1988, and Duce *et al.*, 1991.

also be released by plants grown in areas with abundant nitrogen or during senescence when plants are transporting a great deal of N within their tissues.

Ammonia is the only natural alkaline gas in the atmosphere and thus is an important neutralizer of anthropogenic acidity by reactions such as

$$2\,NH_3 + H_2SO_4 \rightarrow (NH_4)_2\,SO_4 \tag{5.16}$$

The major sink for atmospheric $NH_3$ is conversion to ammonium-containing aerosols, which are deposited via dry deposition or in precipitation. Note that ammonium sulfate [$(NH_4)_2SO_4$] aerosols have different radiative properties from sulfuric acid ($H_2SO_4$) aerosols.

Ammonia release to the atmosphere has increased in recent years (Asman and Janssen, 1987), and deposition is commonly restricted to local areas downwind of the source (Draajers et al., 1989). Many of these areas have exhibited forest decline (Nihlgard, 1985; Aber et al., 1989) and declining soil pH resulting from the nitrification of $NH_4^+$ (4 $H^+$ ions are released to the soil solution for each mole of $NH_4^+$ nitrified).

## 5.5.2 Nitrous Oxide

Nitrous oxide ($N_2O$) is released to the atmosphere predominantly from soils and water. As shown in Table 5.13, tropical soils are probably the most important natural source. In the regions of the tropics where land-use changes have led to the intensification of agriculture, $N_2O$ sources are probably increasing (Matson and Vitousek, 1990). Soils in the temperate regions are believed to release approximately half as much nitrous oxide as tropical soils on a global scale.

The net ocean source of 3 TgN $yr^{-1}$ quoted in Table 5.13 (see also Weiss, 1994) is uncertain because it is calculated as the difference between the relatively large upward and downward gross fluxes that are not well quantified.

Emissions of nitrous oxide from fertilized agricultural systems are probably the single largest anthropogenic contribution to the global budget of $N_2O$. Several other anthropogenic sources (e.g., industrial activity, biomass burning, degassing of ground water used for irrigation, etc.) are believed to be smaller, but are not well quantified.

The removal of $N_2O$ from the atmosphere results primarily from its photolysis [and to a lesser extent from its oxidation by the electronically excited oxygen atom $O(^1D)$] in the

**Table 5.13**
Estimated Sources and Sinks of $N_2O$, TgN $yr^{-1}$

| Sources | |
|---|---|
| *Natural* | |
| Oceans | 3.0 (1-5) |
| Tropical soils | |
|     Wet forests | 3.0 (2.2-3.7) |
|     Dry savannas | 1.0 (0.5-2.0) |
| Temperate soils | |
|     Forests | 1.0 (0.5-2.3) |
|     Grasslands | 1.0 |
| Total natural sources | 9.0 |
| *Anthropogenic* | |
| Cultivated soils | 3.5 (1.8-5.3) |
| Biomass burning | 0.5 (0.2-1.0) |
| Industrial sources | 1.3 (0.7-1.8) |
| Cattle and feed lots | 0.4 (0.2-0.5) |
| Total anthropogenic sources | 5.7 |
| *Total sources* | 14.7 (10-17) |
| Sinks | |
| Stratospheric photochemistry | 12.3 (9-16) |
| Removal by soils | ? |
| *Total sinks* | 12.3 (9-16) |
| *Atmospheric increase* | 3.9 (3.1-4.7) |

Adapted from IPCC, 1994.

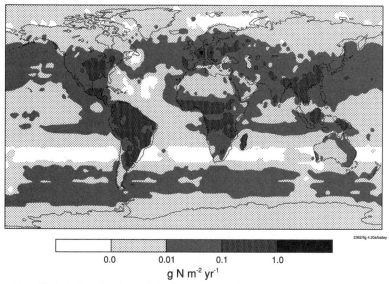

**Figure 5.27.** Global distribution of $N_2O$ emissions for June; based on data from Nevison (1994).

stratosphere. The consumption of nitrous oxide by some soils is another potential sink whose magnitude is unknown.

As demonstrated above, the global $N_2O$ budget remains uncertain as new sources are identified and already identified sources are revised. The dominant sources, however, seem to be undisturbed and cultivated soils, and the oceans. An estimation of the global distribution of $N_2O$ emissions is given in Figure 5.27. Nitrous oxide provides a source of reactive nitrogen oxides in the stratosphere (see Chapters 7 and 14 and Color Plate 17) and contributes to the radiative forcing of the climate system (see Chapter 15). Its global lifetime in the atmosphere is 130-150 years.

### 5.5.3 Active Oxides of Nitrogen

The largest production of nitric oxide in the present atmosphere most likely results from fossil fuel combustion (approximately 20 TgN yr$^{-1}$), biomass burning (12 TgN yr$^{-1}$), and emission from soils (20 TgN yr$^{-1}$) (see Table 5.14). Nitric oxide is also formed by lightning (see Color Plate 9). The magnitude of this source is uncertain, but estimates range from 2 to 20 TgN yr$^{-1}$. The release of NO by aircraft engines in the troposphere and lower stratosphere leads to an additional source of approximately 0.5 TgN yr$^{-1}$ (in 1990). As the size of the fleet of subsonic aircraft is increasing and as development of high-altitude (15-20 km) supersonic transport is planned, this source is likely to increase in the future. Emissions of NO by soils are poorly quantified, but could represent approximately 20 TgN yr$^{-1}$. Finally, the production of NO in the stratosphere from the oxidation of nitrous oxide represents less than 1 TgN yr$^{-1}$. Figure 5.28 shows an estimation of the total $NO_x$ emission at the surface, including the contributions of fossil fuel and biomass combustion, and biological processes in soils. Fossil fuel combustion dominates the $NO_x$ budget in industrialized areas, while the contribution of soil-biogenic emissions is important in rural areas and may dominate in tropical regions (*e.g.*, the savanna) or near heavily fertilized soils.

While soil sources of NO are subject to large variability, the single source of $NO_x$ with the highest uncertainty is lightning. This source is significant not only in its

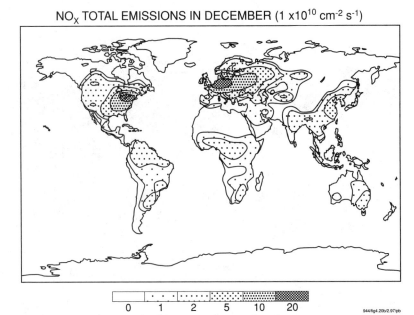

**Figure 5.28.** Surface emission ($10^{10}$ molecules $cm^{-2}$ $s^{-1}$) of $NO_x$ on the global scale in December (based on data from Müller, 1992b). Contributions of fossil fuel combustion, biomass burning, and soil emissions are included at a spatial resolution of $5° \times 5°$.

**Table 5.14**
*Global Budget of $NO_x$ in the Troposphere*

|  | TgN $yr^{-1}$ |
|---|---|
| *Sources* | |
| Fossil fuel combustion | 19.9 (14-28) |
| Biomass burning | 12.0 (4-24) |
| Release from soils | 20.2 (4-40) |
| Lightning discharges | 8.0 (2-20) |
| $NH_3$ oxidation | 3 (0-10) |
| Ocean surface | <1 |
| Aircraft | 0.5 |
| Injection from the stratosphere | 0.1 (0.6 total $NO_y$) |
| Total sources | 64 (25-112) |
| *Sinks* | |
| Wet deposition of $NO_3^-$ (land) | 19 (8-30) |
| Wet deposition of $NO_3^-$ (ocean) | 8 (4-12) |
| Dry deposition of $NO_x$ | 16 (12-22) |
| Total sinks | 43 (24-64) |

Based on Logan, 1983; IPCC, 1994; and Davidson, 1991.

magnitude, but also because lightning produces $NO_x$ in the upper troposphere where its lifetime is relatively long. Although recent assessments (IPCC, 1994; WMO, 1995) indicate that the lightning source is probably between 2 and 20 TgN $yr^{-1}$, a modeling study by Levy et al. (1996) suggests that a global source strength ranging from 2 to 6 TgN $yr^{-1}$ is sufficient to explain most levels of $NO_x$ measured in the middle and upper troposphere of the tropics and subtropics. Calculations of global $NO_x$ production from lightning by different investigators have been based on different assumptions of lightning flash frequency, energy, and channel length. Liaw et al. (1990), for example, calculate mean values of 72.2±96.2 TgN $yr^{-1}$ from theoretical estimates, 19.1±10.0 TgN $yr^{-1}$ from laboratory studies, and 152±59.9 TgN $yr^{-1}$ from field observations. In all cases, these estimates are larger than those assumed in previous budget estimates of $NO_x$ production from lightning, and in current global chemical-transport models of the troposphere.

## 5.5.4 Wet and Dry Deposition

Nitrous oxide is a relatively inert molecule in the troposphere and is not readily deposited to the surface. It is, however, transported to the stratosphere, where it is destroyed by photolysis and oxidation. In contrast, $NO_x$ is rapidly oxidized in the troposphere to nitrates that are soluble and removed from the atmosphere by wet deposition. Nitric oxide (NO) is taken up by plants over the continent, but more slowly than nitrogen dioxide ($NO_2$) and nitric acid ($HNO_3$). Transport, followed by wet and dry deposition, can thus perturb the nitrogen cycle found in undisturbed "natural ecosystems."

Constraints on the $NO_x$ budget, and particularly the lightning source, are based on measurements of the total nitrate deposition at surface sites. These measurements are from rainfall, snowfall, and ice core studies (Logan, 1983; Clausen and Langway, 1989). Ice core studies from the Northern Hemisphere (Greenland) show an increase in $NO_3^-$ over the past century. The increase is related to anthropogenic $NO_x$ emission and correlates with increases in non-sea-salt sulfate in the same core samples. Roughly speaking, the levels of nitrate in surface ice are 2-3 times preindustrial concentrations. This trend suggests that Northern Hemisphere anthropogenic emissions of $\approx 20$ TgN $yr^{-1}$ are 2/3 of the total input to this site, which puts the "natural" background at $\approx 10$ TgN $yr^{-1}$ in the Northern Hemisphere ($\approx 20$ TgN $yr^{-1}$ globally). This value is similar to estimates shown in Table 5.14, and suggests an upper limit for soil emission + lightning of 20 TgN $yr^{-1}$. Still, a major uncertainty in this analysis is the distribution of sources and the paucity of deposition measurements. Both lightning and soil NO emissions tend to be located in the tropics, and reliable measurement of $NO_3^-$ deposition in these areas is very limited. Furthermore, the relationship between $NO_x$ production, transformation, transport, and eventual surface deposition is highly complex and not well understood at present. This uncertainty in the "natural" sources of $NO_x$, particularly in the upper troposphere, is a problem for assessing the impact of future increases in $NO_x$ emissions from anthropogenic activities. For example, increased air traffic in the upper troposphere/lower stratosphere may have a significant impact on $NO_x$-related chemical processes, but the importance of this anthropogenic perturbation cannot be reasonably assessed without better constraints on the sources of $NO_x$ to the middle and upper troposphere from lightning and redistribution of $NO_x$ by convection.

## 5.6 The Global Sulfur Cycle

Sulfur is a chemical element essential to life on Earth. Living organisms, including plants, assimilate sulfur, while at the same time sulfur in various forms is released by living organisms as an end product of metabolism. The major sulfur gases include dimethyl sulfide ($CH_3SCH_3$ or DMS), carbonyl sulfide (OCS), hydrogen sulfide ($H_2S$), dimethyl disulfide (DMDS), carbon disulfide ($CS_2$), and sulfur dioxide ($SO_2$). Over the past centuries, the sulfur cycle has been increasingly perturbed by human activities. Today, globally, anthropogenic emissions constitute almost 75% of the total sulfur emission budget with 90% occurring in the Northern Hemisphere. Excluding biomass burning, natural emission sources (marine + terrestrial + volcanic) constitute 24% of the total emission budget with 13% in the Northern Hemisphere and 11% in the Southern Hemisphere. Estimates of global emissions from each source are listed in Table 5.15. Although anthropogenic emissions (primarily $SO_2$) clearly dominate in certain regions, particularly in the northeastern United States and sections of Europe and Asia, both natural and anthropogenic sources affect the global distribution of sulfur in the atmosphere.

**Table 5.15**
*Summary of Global Sulfur Fluxes in TgS yr$^{-1}$*

|  | Compound | Flux |
|---|---|---|
| *Sources* | | |
| Volcanoes | SO$_2$, H$_2$S, OCS | 7-10 |
| Vegetation and soils | H$_2$S, DMS, OCS CS$_2$, DMDS | 0.4-1.2 |
| Biomass burning | SO$_2$, OCS, H$_2$S | 2-4 |
| Ocean | DMS, OCS, CS$_2$, H$_2$S | 10-40 |
| Anthropogenic | SO$_2$, sulfates | 88-92 |
| *Sinks* | | |
| Dry deposition | SO$_2$, sulfates | 50-75 |
| Wet deposition | SO$_2$, sulfates | 50-75 |

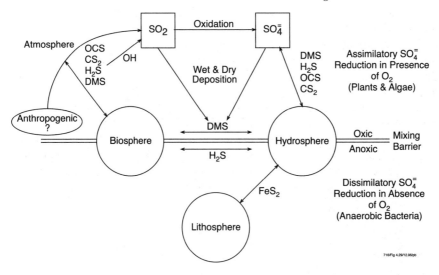

**Figure 5.29.** Exchanges of sulfur in the global environment.

The sulfur cycle is conceptually shown in Figure 5.29. As can be seen, there are two independent environments separated by a mixing barrier: an anoxic region where oxygen is absent, and an oxic region where atmospheric oxygen is present. In the reducing environment, sulfide, primarily H$_2$S, is formed by anaerobic bacteria from sulfate (SO$_4^{2-}$). This reduction is necessary for anaerobic respiration because oxygen is absent. The sulfide is then partially taken up by the lithosphere in the form of rocks and sediments by reaction of the sulfide with metal ions like iron. Some of the sulfide in the lithosphere may return to other "spheres" in the form of sulfate by the action of weathering. However, because of the mixing barrier, which is both physical and microbial in nature, very little of the sulfide produced in the anoxic regime escapes directly into the atmosphere by other means. In the oxic environment, bacteria reduce sulfate (assimilatory sulfate reduction) to sulfide in order to form amino acids and proteins. In the biosphere and the hydrosphere, the reduced sulfur compounds thus produced, DMS, $H_2S$, $CS_2$, and OCS, are volatile and are easily exchanged with the atmosphere.

Once in the atmosphere, DMS, $H_2S$, OCS, and $CS_2$ are oxidized by reactions primarily with OH to form $SO_2$ and eventually $H_2SO_4$ or $CH_3SO_3H$ (methane sulfonic acid or MSA), as further discussed in Chapter 10. Sulfate is returned to the surface by both wet and dry deposition. The reaction of OCS with OH is very slow, and thus transport to the stratosphere occurs. Once in the stratosphere, OCS is photolyzed; $SO_2$ and ultimately $H_2SO_4$ aerosols are formed. This process is thought to be the primary mechanism by which the Junge layer, a layer of stratospheric sulfate particles that may influence climate, is sustained during nonvolcanic periods.

### 5.6.1 Dimethyl Sulfide

Dimethyl sulfide is the dominant biogenic sulfur compound (Bates et al., 1992). Under natural conditions, this gas is thus the primary contributor to the atmospheric burden of non-sea-salt (NSS) sulfate. It is formed in the ocean from the breakdown of dimethyl sulfoniopropionate (DMSP), which is thought to be used by some marine plankton to control their internal osmotic pressure. However, under conditions of more or less constant salinity, there should be very little net DMSP transfer. Release of DMSP can occur through excretion from an organism, senescence of an organism, or following grazing of one species by another. DMSP is hydrolyzed chemically or enzymatically to DMS, creating a reservoir in the ocean. The most dramatic production of DMS occurs in late spring and summer at high latitudes as a result of intense phytoplankton productivity coupled with zooplankton grazing. In addition, coastal seaweed beds provide very intense, localized DMS sources.

DMS is only marginally soluble in water, and most surface ocean waters are found to be supersaturated. Thus the emission of DMS is limited by its rate of transfer from the ocean to the atmosphere. Probably only about 3-10% of the DMS in seawater is released to the atmosphere, the rest being consumed chemically or biologically while in the ocean. Current estimates put the oceanic DMS flux at about 16 TgS $yr^{-1}$.

The oceans account for about 99% of the global DMS flux (Fig. 5.30b and c). The remainder is associated with terrestrial emissions from both vegetation and soils. Terrestrial plants contain an average sulfur content of 0.25% on a dry weight basis, and thus various mechanisms including organic decomposition may act to release DMS and other sulfur compounds into the atmosphere. The major loss of gas-phase DMS is its reaction with the hydroxyl radical, or, in some coastal regions, with the nitrate radical during nighttime.

### 5.6.2 Carbonyl Sulfide

OCS fluxes of about 0.3 TgS $yr^{-1}$ have been estimated for both oceans and terrestrial sources. However, as discussed by Weiss et al. (1995), the global oceanic source may be much smaller (see Section 10.5.1). This represents a very small portion of the total oceanic sulfur source, but a very large part of the total terrestrial source. Vegetation is reportedly both a source and sink of OCS. Measurements by Berresheim and Vulcan (1992) and by Rennenberg et al. (1990) on spruce and pine trees suggest OCS emissions rather than uptake. Trees, like other plant species, may thus serve as either an OCS (and perhaps $CS_2$) source or sink depending upon the ambient concentrations of these gases as well as other gases and the presence of stress from heat, light, and soil moisture. Addition of nitrogen compounds to forest soils, as occurs by acid deposition, could also significantly alter the fluxes of sulfur gases emitted by

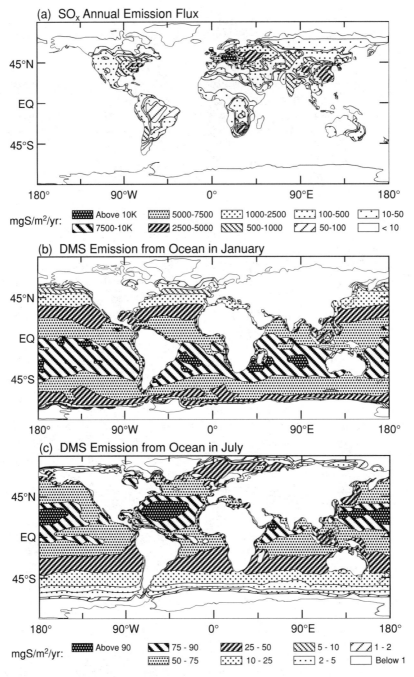

**Figure 5.30.** Geographical distributions of sulfur emissions (based on data from Pham et al., 1995); (a) annual emission of $SO_2$; (b) oceanic emission of DMS in January; (c) same as (b) but in July.

soils and vegetation into the atmosphere. Enhanced UV radiation at the ocean surface resulting from stratospheric ozone depletion may also act to change future OCS fluxes, since this gas is thought to be produced in the oceans by UV-initiated breakdown of dissolved organic matter.

Biomass burning currently represents the major direct anthropogenic OCS source (12% of the total). All other known direct anthropogenic sources comprise only 4% of the total. Present estimates indicate that the sum of known OCS sources is at least 50% greater than known sinks.

Due to its low chemical reactivity, OCS is the most abundant sulfur gas in the atmosphere. Its major tropospheric sink appears to be uptake by soils and hydrolysis by natural waters. As the global lifetime of OCS is approximately 1.5 years, it can be transported into the stratosphere where it is photolyzed by solar UV radiation and acts as a source of $SO_2$ and sulfate particles (Junge layer).

### 5.6.3 Hydrogen Sulfide

Sources of $H_2S$ include volcanoes, terrestrial plants and soils, biomass burning, oceans, and industrial activities. $H_2S$ represents about 1% (0.1 TgS yr$^{-1}$) of the total volcanic source of sulfur. $H_2S$ is one of the primary sulfur species emitted by soils and vegetation. Guenther *et al.* (1989) demonstrate that emissions of $H_2S$ and other sulfur gases from soils and vegetation can be modeled using landscape and climatological databases. Global natural emissions from volcanoes, terrestrial plants and soils, and oceans are all of a similar magnitude (Bates *et al.*, 1992). In the atmosphere, $H_2S$ is rapidly oxidized by the OH radical.

### 5.6.4 Carbon Disulfide

In the case of $CS_2$ the major source is from chemical processing, especially in the cellulose industry. However, very little information exists regarding this source as well as other potential anthropogenic emissions. By contrast, the reaction of $CS_2$ with OH to form OCS has been firmly established as the major $CS_2$ sink.

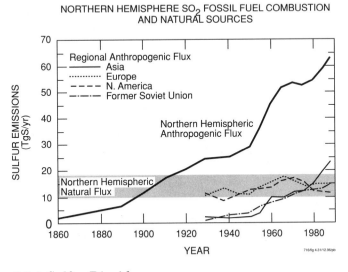

*Figure 5.31.* Natural and fossil fuel emissions of $SO_2$ in the Northern Hemisphere from 1860 to 1990 (IPCC, 1996).

### 5.6.5 Sulfur Dioxide

Anthropogenic emissions of $SO_2$, primarily from fossil fuel combustion, represent the largest contribution to the global sulfur budget. Coal and oil combustion constitute approximately 53% and 28%, respectively, of this $SO_2$ source. Present estimates of the total $SO_2$ production from anthropogenic activities range from 80 (Bates *et al.*, 1992) to 90 TgS yr$^{-1}$ (Möller, 1984) for the year 1985. Möller (1984) estimates that this source will increase to 99 TgS yr$^{-1}$ by the year 2000, primarily a result of increased coal usage. Figure 5.31 shows the increasing anthropogenic $SO_2$ emissions from 1860 to 1990. Figure 5.30a shows the global annual emission of $SO_2$, as estimated by Pham *et al.* (1995).

Sulfur gas emissions from volcanoes and biomass burning are primarily in the form of $SO_2$. These sources account for about 10% and 2%, respectively, of the global $SO_2$ flux (Bates *et al.*, 1992). However, Berresheim and Jaeschke (1983) report about a factor of 10 lower yearly average emission rate from volcanoes. Although sulfur dioxide has been assumed to be the dominant product from biomass burning, OCS, $H_2S$, and sulfate aerosol have also been measured. In addition to emission rates, uncertainties still remain as to the identity of all the sulfur compounds emitted during biomass burning. Bates *et al.* (1992) estimate a total sulfur emission rate of 1.4 to 2.9 TgS yr$^{-1}$ from this source, while Andreae (1985) estimates a flux of 7 TgS yr$^{-1}$.

## *5.6.6 Nonvolatile Sulfur Compounds*

In addition to the above compounds, nonvolatile sulfur compounds are also released into the atmosphere from the oceans by surface bubble bursting. Although relatively high production rates are calculated for the first 20 meters above the ocean from this sea-salt aerosol source, the sulfur has a short atmospheric lifetime and is not uniformly mixed throughout the boundary layer. At present, therefore, large uncertainties exist in assessing the importance of this source. Airborne sulfur can also originate from wind-blown erosion of soils. Sulfur compounds can be transported over large distances on particles larger than 1 micron in diameter, which are termed "aeolian dust." The contribution of this source to the global sulfur budget is highly uncertain, with estimates ranging from 0.19 to 19 TgS yr$^{-1}$ (Aneja, 1990).

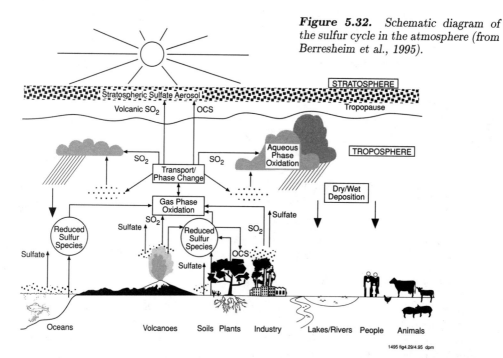

**Figure 5.32.** Schematic diagram of the sulfur cycle in the atmosphere (from Berresheim et al., 1995).

## 5.6.7 Sulfur Conversion and Deposition

The most important processes affecting sulfur in the atmosphere are represented in Figure 5.32. As the atmosphere acts as an oxidizing medium, reduced sulfur is converted into $SO_2$ and further into sulfuric acid ($H_2SO_4$) (see Chapter 10 for details). Sulfate particles, produced by homogeneous and heterogeneous nucleation of $H_2SO_4$, are believed to be the major source of cloud condensation nuclei. The removal of sulfur from the atmosphere results from dry and wet deposition at the surface with potentially severe environmental consequences. The acidification of precipitation and related impact on living organisms, observed in several regions of the Northern Hemisphere, has been caused by increasing anthropogenic emissions of sulfur and nitrogen oxides. As discussed in Chapter 15, sulfur chemistry is also believed to have implications for the global and regional climates of the planet.

## 5.7 Halogens

The origin of chlorine, bromine, fluorine, and iodine in the atmosphere is related to several natural and anthropogenic processes at the Earth's surface. Large quantities of chlorine in the form of salt particles (NaCl) are released from the ocean. Since most sea-salt particles are larger than 1.0 $\mu$m, they are rapidly returned to the sea and little $Cl^-$ is released to the atmosphere. The release of hydrogen chloride (HCl) by passively degassing volcanoes and during volcanic eruptions is another important natural source of atmospheric chlorine. HCl is not rapidly destroyed by chemical reactions, but it is highly soluble in water, and therefore rapidly rained out from the troposphere. Hence, it is unlikely that large quantities of HCl are injected into the stratosphere even during major volcanic eruptions. Enhanced stratospheric abundance of hydrogen chloride was, however, reported after the eruption of El Chichón (Mexico, 1982; Mankin and Coffey, 1984), but not after the eruption of Mt. Pinatubo (Philippines, 1991; Mankin et al., 1992).

Methyl chloride ($CH_3Cl$), with a total emission of approximately 2.6 Tg yr$^{-1}$, plays a significant role in the budget of chlorine. A major source of this compound is provided by the ocean; $CH_3Cl$ is produced by photochemical reactions in seawater containing dissolved organic compounds related to the presence of algae and phytoplankton, before being released to the atmosphere. Another major source is associated with biomass burning including forest and savanna fires. Fungi also release methyl chloride. $CH_3Cl$ destruction results mainly from its reaction with the hydroxyl radical and from photolysis. As anthropogenic sources of $CH_3Cl$ are probably small, the atmospheric abundance of this compound has remained nearly constant during the twentieth century, with a mean mixing ratio of about 650 pptv. The global lifetime of $CH_3Cl$ is close to 1.5 years, so that the distribution of this gas in the troposphere is relatively uniform. Methyl chloride is believed to be the largest natural source of chlorine in the stratosphere.

Over the past decades, the budget of chlorine (and fluorine) has been dramatically perturbed by the release into the atmosphere of industrially manufactured chlorofluorocarbons (CFCs). These very stable compounds are used as refrigerants, as inflating agents in the manufacturing of foam materials, as solvents and, less and less frequently, as aerosol spray propellants. The chemical lifetime of the CFCs varies typically from 10 to 100 years, and is therefore long enough to allow these gases to penetrate into the stratosphere. Near 20 km altitude, CFCs are photolyzed by short-wave ultraviolet radiation and release ozone-destroying chlorine atoms. It is estimated that, in 1990, approximately 85% of stratospheric chlorine was of anthropogenic origin. At that time, the emission of CFCs represented approximately 1 Tg yr$^{-1}$. International agreements

(Montreal Protocol, 1987, and subsequent amendments) have limited and eventually phased out the production of CFCs. Alternative chlorine- and fluorine-containing products, such as the hydrogenated chlorofluorocarbons (HCFCs), are expected to be produced in increasing quantitites and released into the atmosphere. Their potential to destroy stratospheric ozone is, however, substantially smaller than that of the CFCs, because they are destroyed relatively efficiently by the OH radical in the troposphere.

Methyl bromide ($CH_3Br$), which is believed to be the major reservoir of bromine in the atmosphere (typical Southern Hemispheric concentration of 9 ppt, 11 ppt in the Northern Hemisphere) is produced in the ocean (approximately 50 Gg yr$^{-1}$ oceanic release), and as a result of biomass burning (10 to 50 Gg yr$^{-1}$ emissions). $CH_3Br$ is also used as a pesticide (preplanting and postharvesting fumigant); the corresponding anthropogenic source ranges from 20 to 60 Gg yr$^{-1}$. The uptake of $CH_3Br$ by the ocean represents approximately 75 Gg yr$^{-1}$, so that the ocean is believed to be a net sink for this gas. Methyl bromide is destroyed by the hydroxyl radical and by photolysis in the atmosphere; if OH represents the only sink of $CH_3Br$, its lifetime should be close to 2 years. Taking into account the ocean uptake and uptake by soils (estimated as approximately 40 Gg yr$^{-1}$, Shorter et al., 1995) in addition to this, the global lifetime of methyl bromide is of the order of 0.6-0.9 years (Yvon-Lewis and Butler, 1997). Note, however, that with the currently known sources and sinks there is a significant imbalance in the global budget of $CH_3Br$, with sinks exceeding sources. Because methyl bromide penetrates into the stratosphere where it releases bromine atoms (which, on a per atom basis, have an ozone-depleting efficiency 40 to 100 times larger than chlorine), regulations will most probably limit the future production of this compound for agricultural purposes. Bromine is also delivered to the stratosphere by other organic compounds of anthropogenic origin. Halons, whose lifetime can reach several decades, are used in fire extinguishers, but their production ceased in 1994 as a result of the Montreal Protocol and its amendments.

Table 5.16
Estimated Budgets of $CH_3Cl$, $CH_3Br$, and $CH_3I$

|  | $CH_3Cl$ (Tg yr$^{-1}$) | $CH_3Br$ (Tg yr$^{-1}$) | $CH_3I$ (Tg yr$^{-1}$) |
| --- | --- | --- | --- |
| Typical surface mixing ratio (pptv) | 650 | 10 | 2 |
| *Sources* | | | |
| Biomass burning source | 0.7 | 0.01-0.05 | – |
| Ocean source | 0.4 | ~0.03 | 1-2 |
| Terrestrial ecosystems | 0.1 | – | – |
| Anthropogenic source | 0.05 | 0.02-0.06 | – |
| *Total sources* | ~1.2 | ~0.05-0.15 | ~1.5 |
| Loss to soils | 0.1 | 0.01-0.07 | – |
| Atmospheric loss | $\simeq 3$ | 0.06-0.11 | $\simeq 5$ |
| Atmospheric lifetime | 1.5 years | 0.7 years | 5 days |

Finally, methyl iodide ($CH_3I$) is also produced by marine algae and phytoplankton, and probably by biomass burning. The global source is estimated to be on the order of 1-2 Tg yr$^{-1}$. Its atmospheric abundance is relatively small because the lifetime of $CH_3I$ (which is rapidly photodecomposed by solar light) is only on the order of five days; typical mixing ratios are 1-3 pptv in the boundary layer, and decrease rapidly with altitude. Over ocean regions of high biological productivity, values of 10-20 pptv are found. It has been recently speculated (Solomon *et al.*, 1994b) that if iodine were transported to the stratosphere during strong convective events, it could potentially destroy significant amounts of ozone in the lower stratosphere. Inorganic forms of iodine, however, have not been detected above the tropopause. Table 5.16 summarizes the estimated sources of $CH_3Cl$, $CH_3Br$, and $CH_3I$.

## Further Reading

Andrews, J. E., P. Brimblecombe, T. D. Jickells, and P. S. Liss (1996) *An Introduction to Environmental Chemistry*, Blackwell Science, Oxford, UK.

Buat-Ménard, P. (1986) *The Role of Air-Sea Exchange in Geochemical Cycling*, D. Reidel, Dordrecht.

Butcher, S. S., R. J. Charlson, G. H. Orians, and G. V. Wolfe, eds. (1992) *Global Biogeochemical Cycles*, Academic Press, San Diego.

Calvert, J., ed. (1994) *The Chemistry of the Atmosphere: Its Impact on Global Change*, Blackwell Scientific Publications, Oxford, UK.

Chameides, W. L. and E. M. Perdue (1997) *Biogeochemical Cycles, A Computer-Interactive Study of Earth System Science and Global Change*, Oxford University Press, New York.

Clark, W. C. and R. E. Munn, eds. (1986) *Sustainable Development of the Biosphere*, Cambridge University Press, Cambridge.

Graedel, T. E. and P. J. Crutzen (1995) *Atmosphere, Climate, and Change*, Scientific American Library, New York.

Liss, P. S. and R. A. Duce, eds. (1997) *The Sea Surface and Global Change*, Cambridge University Press, Cambridge, UK.

Lovelock, J. E. (1979) *Gaia: A New Look at Life on Earth*, Oxford University Press, Oxford, UK.

Lovelock, J. (1988) *The Ages of Gaia*, W. W. Norton and Company, New York.

Lovelock, J. E. (1991) *Healing Gaia*, Harmony Books, New York.

Mackenzie, F. T. (1997) *Our Changing Planet: An Introduction to Earth System Science and Global Environmental Change*, 2nd ed., Prentice Hall, Upper Saddle River, New Jersey.

Matson, P. A. and R. C. Harriss, eds. (1995) *Biogenic Trace Gases: Measuring Emissions from Soil and Water*, Blackwell Science Ltd. (Methods in Ecology Series).

Oremland, R. S., ed. (1993) *Biogeochemistry of Global Change*, Chapman and Hall, New York.

Pruppacher, H. R., R. G. Semonin, and W. G. N. Slinn (1983) *Precipitation Scavenging, Dry Deposition and Resuspension*, Elsevier Press, New York.

Schlesinger, W. H. (1997) *Biogeochemistry: An Analysis of Global Change*, 2nd ed., Academic Press, San Diego.

Schneider, S. H. and P. J. Boston (1991) *Scientists on Gaia*, The MIT Press, Cambridge, Massachusetts.

Sharkey, T. D., E. A. Holland, and H. A. Mooney, eds. (1991) *Trace Gas Emissions by Plants*, Academic Press, San Diego.

Singh, H. B. (1995) *Composition, Chemistry, and Climate of the Atmosphere*, Van Nostrand Reinhold, New York.

Stull, R. B. (1993) *An Introduction to Boundary Layer Meteorology*, Kluwer Academic Publishers, Dordrecht.

Warneck, P. (1988) *Chemistry of the Natural Atmosphere*, Academic Press, San Diego.

Wollast, R., F. T. McKenzie, and L. Chou, eds. (1993) *Interactions of C, N, P and S Biogeochemical Cycles and Global Change*, NATO ASI Series, Springer-Verlag.

James Lovelock

## The View from Outside

The devices and mechanisms of the sci-fi starship *Enterprise* are all but indistinguishable from magic. Except for one instrument in the control room, the telebioscope. This, the exobiological officer uses to find out if the planet in view bears life. NASA reduced it to practice at the Jet Propulsion Laboratory over thirty years ago. Their purpose was to discover if Mars or Venus harbored life. They found that they probably did not. The device was an IR telescope combined with a multiplex interferometer. A spacecraft equipped with it could detect and measure most of the gases in a planet's atmosphere. The analysis could be made from interplanetary distances. It works because lifeless planets have atmospheres close to the state of chemical equilibrium. In contrast, large departures from chemical equilibrium characterize planets with life. A practical version would imply a simple algorithm to calculate the degree of disequilibrium. This could then be used to assign a probability for the presence of life and display the answer in whatever way was fashionable.

The view of the Earth from space must be one of the outstanding achievements of the century. That immaculate blue-white sphere has become an icon as important as the crescent or the cross. For the first time we saw the Earth as it is and were dazzled. It was a sight so splendid that we hardly noticed the crucial new scientific evidence that also came from this top-down view. We saw a whole planet, not only visually, but also through the more objective and discerning eyes of scientific instruments.

When NASA and its Soviet rivals were making these first steps to space research in the 1960s, mainstream science was still geocentric. We had discovered that our planet was not the center of the universe, but we still thought of it as the center of things. E. O. Wilson has reminded us that we are by nature tribal carnivores. Our tribal tendency is so strong that we have divided science itself. Each separate discipline has its own territory and department, coexisting more or less peacefully on the campus. The biologists were sure that life adapted to an Earth that was adequately described by their colleagues in the geology department. The geologists were equally certain that life was a well-behaved passenger and did not interfere with the air conditioning equipment. Everyone knew that plants photosynthesized oxygen and organic matter from carbon dioxide and water and that animals recycled these products. Biologists and geologists were happy with the anecdote of a balance of nature. They rarely asked awkward questions, such as: If nature balances, why is oxygen 600 times more abundant than $CO_2$? Looking back, the papers published in the 1960s about the Earth are astonishing to read. Some claimed that oxygen in the air came mainly from the photolysis of water vapor in the upper atmosphere; others, that methane was a product of outgassing from primeval reservoirs. The biochemist G. E. Hutchinson was almost alone when he wrote that methane, nitrous oxide, and other gases probably came from bacterial sources.

The atmosphere is an almost negligible part of the Earth's mass, yet it profoundly affects conditions at the surface. We now know that the biota metabolizes all but the rare gases of the air. Oxygen, methane, ammonia, and DMS are almost entirely direct biological products. Carbon dioxide and nitrogen are products of outgassing from the interior, but organisms probably regulate the abundance of these gases. The large questions, still unanswered, are: To what extent is the atmosphere a chance product of separate biological and geological processes? How much has the atmosphere evolved as part of a system made from the organisms and their material environment? How much is Darwinian natural selection among individual organisms involved in the fine tuning of the Earth's climate and composition to a state comfortable for the contemporary biota?

It may be a long time before we have satisfactory answers to these questions, but I think it worth continuing to explore the top-down view of the Earth. The preceding chapter clearly shows how much has been gained from the interdisciplinary science of biogeochemistry. Going further and taking the geophysiological approach has been justified by the discoveries it has stimulated, such as that of the several roles of DMS in the natural environment, as sulfur carrier, pheromone, and source of CCNs. Thinking of the Earth as a whole has also led to new understanding of community ecology and to models that explain the phenomenon of biodiversity. Something that previously was obscure.

The opening of our minds, the astronaut's harvest, was at first no more than a radical idea. Now we accept the need for a whole Earth science and begin to practice it. Climatologists may soon be routinely including the ecosystems at the Earth's surface in their models. Not long ago they would have instinctively rejected such an idea as absurd. The chapter just read is mainly about the biochemistry of the atmosphere. It shows how the atmosphere, the complection of the Earth, reveals as much about life as it does about the inorganic materials of the Earth. We need the wisdom that came with the view from space to see how distorted is the fly's eye view of fragmented science. The narrow focus of each separate discipline sees only a small feature of the whole planet. The journeys to space, whether real or in the mind, have forced us to recognize that science is not naturally divided. Separate disciplines are not fundamental to science, merely the consequence of human nature. The view from space has done more to make us aware of our shortcomings; it offers a top-down physiological approach to Earth science. A view that complements, and reintegrates, mainstream science.

*James Lovelock is a Visiting Fellow of Green College, Oxford, United Kingdom, and a Fellow of the Royal Society. He is a practicing geophysiologist. His Gaia hypothesis, the idea that the Earth functions as a single organism, has dramatically changed our view of the evolution of the Earth system.*

# Part 2

# Chemical Families

# 6 Hydrogen Compounds

## 6.1 Importance of Atmospheric Hydrogen Compounds

Molecules containing hydrogen perform key roles in most chemical cycles throughout the atmosphere. These compounds are involved in the determination of the oxidation capacity of the troposphere, contribute to the Earth's radiative balance, and participate in homogeneous and heterogeneous reactions that determine levels of stratospheric ozone.

In the troposphere, peroxy radicals oxidize nitric oxide, a process that leads to photochemical ozone production. Hydroxyl and hydroperoxy radicals also react with ozone; thus, depending on the concentration of active nitrogen compounds, regions of the troposphere can be net ozone sources or sinks. Hydroxyl radicals react with hydrogen-containing compounds such as hydrocarbons, as well as with other common pollutants, including carbon monoxide and sulfur dioxide. Hydrogen peroxide formed in the gas or condensed phase can lead to aqueous-phase oxidation of sulfur dioxide and other compounds. Water vapor is important radiatively, but also chemically, since it reacts with excited oxygen atoms formed in the photolysis of ozone to produce hydroxyl radicals. In the stratosphere, hydroxyl and hydroperoxy radicals react with ozone in catalytic ozone destruction cycles and also contribute to other destruction cycles involving nitrogen and halogen compounds. Stratospheric water vapor, formed primarily from the oxidation of methane, is much less abundant than in the troposphere. Condensed-phase processes are less well understood, but reactions on sulfuric acid aerosol or ice particles clearly play a role in the stratosphere.

The different roles of the various hydrogen compounds may be made clearer by considering the very different chemical lifetimes involved. The very short-lived hydroxyl radical, with a lifetime of 1 second or less, reacts readily with nearly every atmospheric trace gas, whereas relatively unreactive compounds such as molecular hydrogen or water vapor have very long chemical lifetimes. Hydroperoxy radicals have lifetimes of minutes, and the lifetime for hydrogen peroxide is typically a few hours. It is important to note that the production rates for these species are similar (typically about $10^5$ or $10^6$ molecules $cm^{-3}$ $s^{-1}$), but the typical atmospheric mixing ratios range from about 0.1 pptv for hydroxyl radicals to 0.5 ppmv for molecular hydrogen. This span of 5,000,000 in abundance of these species reflects the marked difference in lifetime.

In the first part of this chapter, the budgets of various hydrogen-containing chemical compounds in the atmosphere are examined, and the important gas-phase and aqueous-phase chemical reactions are indicated. At the end of the chapter, the results of measurement and modeling efforts related to these species are presented in order to provide an overview of their stratospheric and tropospheric abundances.

## 6.2 Scope and Definitions

For the purposes of this discussion, the hydrogen compounds are those compounds of hydrogen and oxygen that exist in the atmosphere. This definition includes atomic and molecular hydrogen (H and $H_2$, respectively), water vapor ($H_2O$), hydrogen peroxide ($H_2O_2$), hydroperoxy and hydroxyl radicals ($HO_2$ and $OH$). Because of their chemical similarity, it is useful to discuss the chemistry of the organic peroxy radicals ($RO_2$) here as well. Compounds of hydrogen and carbon are discussed in Chapter 9. The compounds discussed in this chapter are:

H: Hydrogen atoms are extremely short-lived in the lower atmosphere because they quickly combine with molecular oxygen to produce $HO_2$.

OH: The hydroxyl radical reacts with H-containing compounds and other trace gases to serve as the primary atmospheric oxidant.

$HO_2$: The hydroperoxy radical is formed from reaction of OH with CO and other molecules and is an important oxidant of NO (see Chapter 7). The main source of atmospheric $H_2O_2$ is reaction between two $HO_2$ molecules.

$H_2$: Molecular hydrogen is a relatively unreactive hydrogen compound formed through surface biological and chemical processes.

$H_2O_2$: Hydrogen peroxide is an important aqueous-phase oxidant for $SO_2$ and other species. It is formed from the reaction of $HO_2$ radicals.

$H_2O$: Water vapor is a compound critical to the Earth's radiative balance that is also involved in trace gas chemistry through its reaction with excited oxygen atoms to form OH radicals, and through the effects of condensed-phase processes.

## 6.3 Sources of Hydrogen to the Atmosphere

Hydrogen enters the atmosphere from the biosphere in three main forms: water, molecular hydrogen, and hydrocarbons.

### 6.3.1 Water

The hydrological cycle was discussed in some detail in Chapter 5, where it was shown that only a small fraction of the Earth's water supply is present in the atmosphere. The presence of water vapor in the atmosphere is of importance for many reasons— its reaction with $O(^1D)$ atoms provides the major source of OH radicals to the atmosphere, it is a major greenhouse gas and thus plays a role in the atmospheric radiation budget, and, finally, condensed forms of water (clouds, aerosols, etc.) mediate the occurrence of multiphase reactions.

### 6.3.2 Molecular Hydrogen, $H_2$

Molecular hydrogen ($H_2$) is the second most abundant reactive trace gas in the troposphere, after methane. A budget of tropospheric $H_2$ is presented in Table 6.1 (Schmidt et al., 1980). $H_2$ enters the atmosphere in part through biological processes, including anaerobic bacterial activity and emission from ocean waters. Other possible natural sources include geothermal steam, volcanoes, and biogenic activity in waterlogged soils. The total flux from all identified natural sources was estimated at $5 \times 10^{12}$ g yr$^{-1}$.

A major source of H$_2$ is associated with anthropogenic pollution, probably due to incomplete combustion or as intermediates in industrial processes, in ammonia or methanol synthesis, for example. The connection with fossil fuel burning is primarily due to the shift in the water-gas equilibrium at combustion temperatures

$$CO + H_2O \rightleftharpoons CO_2 + H_2 \tag{6.1}$$

There are also photochemical processes that produce H$_2$, of which the molecular channel in the photolysis of formaldehyde is the most important:

$$CH_2O + h\nu \rightarrow H_2 + CO \tag{6.2}$$

**Table 6.1**
*Tropospheric Budget of H$_2$*

| Sources ($10^{12}$ g yr$^{-1}$) | Global | Northern Hemisphere | Southern Hemisphere |
|---|---|---|---|
| Natural | 2.0-5.0 | 1.2-3.0 | 0.8-2.0 |
| Photochemical | | | |
|   CH$_4$[a] | 10.8-16.0 | 5.4-8.0 | 5.4-8.0 |
|   Isoprene/terpenes | 10.0-35.0 | 6.0-20.0 | 4.0-15.0 |
| Anthropogenic | | | |
|   Automobiles | 11.5-57.1 | 9.8-48.5 | 1.7-8.6 |
|   Industry | — | — | — |
|   Biomass burning | 9.0-21.0 | 4.5-10.5 | 4.5-10.5 |
| Total sources | 43.3-134.1 | 26.9-90.0 | 16.4-44.1 |
| Sinks $10^{12}$ g yr$^{-1}$ | | | |
| Photochemical | | | |
|   Stratosphere | 0.6-1.6 | 0.3-0.8 | 0.3-0.8 |
|   Troposphere | 10.0-23.7 | 5.0-11.8 | 5.0-11.9 |
| Soil surface | 19.9-107.4 | 12.9-69.6 | 7.0-37.8 |
| Total sinks | 30.5-132.7 | 18.2-82.2 | 12.3-50.5 |
| Average mixing ratio (ppmv) | 0.560 | 0.575 | 0.550 |
| Abundance ($10^{12}$ g) | 170.7 | 87.2 | 83.5 |

[a]Based on an average tropospheric [OH] of $(6.5\pm2.5) \times 10^5$ molecule cm$^{-3}$ (Volz et al., 1981). Note that current estimates (Prinn et al., 1995) are about 50% higher.
After Schmidt et al., 1980.

Formaldehyde is produced from the oxidation of methane and many other biogenic and anthropogenic organic compounds. Schmidt *et al.* estimated that $16 \times 10^{12}$ g yr$^{-1}$ could be produced in the troposphere from methane oxidation.

Loss of H$_2$ from the atmosphere is likely dominated by deposition to soil surfaces, at rates that depend on the temperature and type of vegetation growing in the soil. It is estimated that $(20\text{-}100) \times 10^{12}$ g yr$^{-1}$ could be removed globally due to surface uptake. Chemical destruction, primarily by reaction with hydroxyl radicals, or to a lesser extent with O($^1$D), also contributes to the removal of H$_2$ from the atmosphere

$$OH + H_2 \rightarrow H_2O + H \tag{6.3}$$
$$O(^1D) + H_2 \rightarrow OH + H \tag{6.4}$$

*6.3.3 Hydrocarbons*

Hydrocarbons enter the atmosphere from a variety of sources, some anthropogenic and some biogenic. While these sources were presented in Chapter 5 and their chemistry is described in detail in Chapter 9, they are mentioned here to emphasize their role as a source of reactive hydrogen to the atmosphere. In some cases [*e.g.*, reaction of $O(^1D)$ with methane], the oxidation of hydrocarbon species leads to the direct production of hydrogen oxide radicals. In other cases (for example, the reaction of OH with methane), hydrocarbon oxidation results in the cycling of radicals species within the $HO_x$ family.

## 6.4 Chemistry of Hydrogen Species in the Middle Atmosphere

Interest in the chemistry of hydrogen-containing compounds in the stratosphere and mesosphere revolves around the reactive species OH, $HO_2$, and H (collectively known as $HO_x$), particularly with regard to the role these species play in catalytic ozone destruction cycles. Reactions that interconvert OH, H, and $HO_2$ occur at rates that are rapid compared to reactions that either produce or destroy these species, and hence it is convenient to treat them as a family. Reactions can then be separated into three categories: those that produce $HO_x$, those which destroy $HO_x$, and those that result in the interconversion of members of the $HO_x$ family.

*6.4.1 Production of $HO_x$ Species*

Production of $HO_x$ radicals in the stratosphere results from reaction of $O(^1D)$ atoms with the hydrogen source gases $H_2O$, $H_2$, and $CH_4$. The $O(^1D)$ is generated predominantly from $O_3$ photolysis:

$$O_3 + h\nu \to O(^1D) + O_2 \qquad (6.5a)$$
$$\to O(^3P) + O_2 \qquad (6.5b)$$

While most of the $O(^1D)$ is quenched by collisions with $N_2$ or $O_2$,

$$O(^1D) + O_2 \to O(^3P) + O_2 \qquad (6.6)$$
$$O(^1D) + N_2 \to O(^3P) + N_2 \qquad (6.7)$$

a small fraction (about 1 out of every 15,000 at 25 km) reacts with $H_2O$, $H_2$, or $CH_4$ to generate $HO_x$ radicals:

$$O(^1D) + H_2O \to 2\ OH \qquad (6.8)$$
$$O(^1D) + H_2 \to OH + H \qquad (6.4)$$
$$O(^1D) + CH_4 \to OH + CH_3 \qquad (6.9a)$$
$$\to CH_3O + H \qquad (6.9b)$$
$$\to CH_2O + H_2 \qquad (6.9c)$$

Throughout the lower and middle stratosphere, H atoms from Reactions (6.4) and (6.9b) quickly react with $O_2$ to generate $HO_2$

$$H + O_2 + M \to HO_2 + M \qquad (6.10)$$

Thus the H-atom concentration is kept very low and only OH and $HO_2$ need be considered. Further reactions of the $CH_3$ and $CH_3O$ radicals produced in Reaction (6.9) also lead to $HO_2$. The three reactions (6.4, 6.8, and 6.9) all have very similar rate coefficients, of order $(1\text{-}2) \times 10^{-10}$ cm$^3$ molecule$^{-1}$ s$^{-1}$. Because $H_2O$ is the most abundant of the three source gases in the stratosphere (3-6 ppmv, compared to about 0.5 ppmv for $H_2$, and 1-2 ppmv for $CH_4$), Reaction (6.8) provides the largest source of $HO_x$ radicals. As will be shown in more detail later, $H_2O$ mixing ratios increase and $CH_4$ mixing ratios decrease with altitude (the oxidation of $CH_4$ leads to $H_2O$ production), while $H_2$ mixing ratios remain roughly constant. Thus the relative importance of the $O(^1D) + H_2O$ source in the stratosphere increases with increasing altitude.

### 6.4.2 Destruction of $HO_x$ Species

Loss processes for stratospheric $HO_x$ involve the recombination of OH and $HO_2$ radicals, either directly or through interactions with nitrogen oxide species (Wennberg et al., 1990). In the middle to upper stratosphere, the destruction of $HO_x$ is dominated by the direct reaction of OH with $HO_2$

$$OH + HO_2 \rightarrow H_2O + O_2 \tag{6.11}$$

At lower altitudes, recombination of OH with $HO_2$ occurs through more complex mechanisms involving oxides of nitrogen. For example, reaction of OH with $NO_2$ leads to production of nitric acid, $HNO_3$. Subsequent reaction of OH with $HNO_3$ leads to a net loss of $HO_x$

$$OH + NO_2 + M \rightarrow HNO_3 + M \tag{6.12}$$
$$OH + HNO_3 \rightarrow NO_3 + H_2O \tag{6.13}$$

A similar reaction sequence, involving reaction of $HO_2$ with $NO_2$ (shown below), may also play an important role in $HO_x$ destruction in the lower stratosphere. However, the products of Reaction (6.15) have yet to be conclusively determined

$$HO_2 + NO_2 + M \rightarrow HO_2NO_2 + M \tag{6.14}$$
$$OH + HO_2NO_2 \rightarrow H_2O + O_2 + NO_2 \tag{6.15}$$

Reactions of $HNO_3$ and $HO_2NO_2$ with OH occur in competition with the photolysis of these reservoir species, which regenerates the $HO_x$ and $NO_x$ radicals lost in their formation

$$HNO_3 + h\nu \rightarrow OH + NO_2 \tag{6.16}$$
$$HO_2NO_2 + h\nu \rightarrow HO_2 + NO_2 \tag{6.17a}$$
$$\rightarrow OH + NO_3 \tag{6.17b}$$

From the loss processes above, the lifetime of stratospheric $HO_x$ can be determined (Wennberg et al., 1990). Despite the changing mechanisms for $HO_x$ loss and formation and the changing atmospheric abundance with altitude, the lifetime remains relatively constant (of order $10^3$ s) throughout the 20-40 km range.

### 6.4.3 Interconversion of $HO_x$ Radicals — Ozone Destruction Cycles

As mentioned earlier, interconversion of OH and $HO_2$ occurs on a very rapid time scale compared to the rate of their formation and loss. In many cases, the reactions that interconvert OH and $HO_2$ result in net ozone destruction. As will be shown in Chapter 14, ozone destruction is dominated by $HO_x$-based chemistry in both the lower stratosphere (below 20 km) and the upper stratosphere and mesosphere (above 45 km).

Conversion of OH into $HO_2$ in the lower stratosphere is dominated by its reaction with $O_3$, which occurs on a time scale of about 5-10 sec at 20 km:

$$OH + O_3 \rightarrow HO_2 + O_2 \tag{6.18}$$

Reaction with CO, followed by addition of the H-atom product to $O_2$, is also of some importance in converting OH to $HO_2$:

$$OH + CO \rightarrow H + CO_2 \qquad (6.19)$$

$$H + O_2 + M \rightarrow HO_2 + M \qquad (6.10)$$

Conversion of $HO_2$ back to OH occurs largely via reaction with NO or with $O_3$

$$HO_2 + NO \rightarrow OH + NO_2 \qquad (6.20)$$

$$HO_2 + O_3 \rightarrow OH + 2O_2 \qquad (6.21)$$

The time scale for this conversion of $HO_2$ back to OH is on the order of 1 min in the lower stratosphere. Reactions (6.18) and (6.20), combined with $NO_2$ photolysis and combination of O with $O_2$,

$$NO_2 + h\nu \rightarrow NO + O \qquad (6.22)$$

$$O + O_2 + M \rightarrow O_3 + M \qquad (6.23)$$

result in a null cycle with respect to ozone. However, Reaction (6.18) coupled with Reaction (6.21) results in a net conversion of $O_3$ to $O_2$ without any loss of $HO_x$ — that is, catalytic ozone destruction occurs (see Section 3.5). The rate-limiting step in this cycle is reaction of $HO_2$ with $O_3$, which must compete with Reaction (6.20).

Two other reactions are of some importance in conversion of $HO_2$ to OH. These involve reaction of $HO_2$ with ClO and BrO. The HOCl and HOBr products of these reactions are rapidly photolyzed, and lead to catalytic ozone destruction:

$$HO_2 + ClO \rightarrow HOCl + O_2 \qquad (6.24)$$

$$HO_2 + BrO \rightarrow HOBr + O_2 \qquad (6.25)$$

$$HOBr + h\nu \rightarrow OH + Br \qquad (6.26)$$

$$HOCl + h\nu \rightarrow OH + Cl \qquad (6.27)$$

From a consideration of the Reactions (6.18) through (6.21), (6.24), and (6.25), and assuming steady state for OH and $HO_2$, the ratio of the OH to $HO_2$ concentrations can be expressed as follows (Cohen et al., 1994):

$$\frac{[OH]}{[HO_2]} = \frac{k_{20}[NO] + k_{21}[O_3] + k_{24}[ClO] + k_{25}[BrO]}{k_{18}[O_3] + k_{19}[CO]} \qquad (6.28)$$

Values in the lower stratosphere for the ratio of the OH to $HO_2$ mixing ratio vary with latitude and altitude, but typically range between 0.1 and 0.4 (Cohen et al., 1994). Furthermore, as will be detailed later in this chapter, values of $[OH]/[HO_2]$ calculated from this expression are in excellent agreement with measurements made aboard the NASA ER-2 aircraft, suggesting that the $HO_x$ chemistry of the lower stratosphere is well characterized by the reaction sequence given above. An even simpler expression, in which only the dominant reactions (6.18), (6.20), and (6.21) are considered,

$$\frac{[OH]}{[HO_2]} = \frac{k_{20}[NO] + k_{21}[O_3]}{k_{18}[O_3]} \qquad (6.29)$$

provides a useful, semiquantitative approximation to the full expression (6.28).

At higher altitudes in the stratosphere and in the mesosphere (see Color Plate 23), $HO_x$-based chemistry again plays a dominant role in ozone destruction, although the chemistry involved is somewhat different from that presented above. At the higher altitudes, the abundance of $O_3$ decreases while the O atom abundance increases as a result of the reduced efficiency of Reaction (6.23) at the lower pressures encountered at higher altitudes, and the result of a more rapid conversion of $O_3$ to O atoms via photolysis (Reaction 6.5). Thus an additional reaction for conversion of OH to $HO_2$

is of importance, that being reaction with O atoms:

$$O + OH \rightarrow H + O_2 \tag{6.30}$$

$$H + O_2 + M \rightarrow HO_2 + M \tag{6.10}$$

The rate of Reaction (6.30) becomes equivalent to that of Reaction (6.18) at about 35-40 km, and becomes more rapid at higher altitudes. Similarly, conversion of $HO_2$ to OH via reaction with O atoms becomes important, and occurs in competition with Reaction (6.21)

$$O + HO_2 \rightarrow OH + O_2 \tag{6.31}$$

As was the case for the competition between Reactions (6.20) and (6.21) at low altitudes, Reaction (6.31) leads to net ozone depletion, while Reaction (6.20) results in a null cycle with respect to ozone loss. A simple consideration of Reactions (6.18), (6.20), (6.30), and (6.31) leads to the following steady-state approximation for the OH to $HO_2$ ratio:

$$\frac{[OH]}{[HO_2]} = \frac{k_{20}[NO] + k_{31}[O]}{k_{30}[O] + k_{18}[O_3]} \tag{6.32}$$

For the midlatitude, sunlit stratosphere near 40 km, the ratio of OH to $HO_2$ is of order 1.5 (reactions converting OH to $HO_2$ are now slower than those converting $HO_2$ back to OH). Interconversion between OH and $HO_2$ occurs on a timescale of a few tens of seconds.

In addition, some further chemistry involving hydrogen atoms must be considered in the upper portion of the stratosphere and in the mesosphere (see also Box 14.2). Because of the reduced pressure and $O_2$ concentrations, Reaction (6.10), which converts H to $HO_2$, occurs at a slower rate, and reaction of H with $O_3$ becomes non-negligible (H reacts with $O_3$ about 3-5% of the time at altitudes between 35 and 50 km)

$$H + O_3 \rightarrow OH + O_2 \tag{6.33}$$

Reaction (6.33) coupled with (6.30) constitutes an additional $HO_x$-based $O_3$ depletion cycle.

### 6.4.4 Interaction of $HO_x$ with Other Families in the Stratosphere

The above discussion has served to illustrate the processes involved in the production, interconversion, and destruction of OH and $HO_2$ radicals. It is important to stress, however, that these $HO_x$ reactions do not occur in isolation, and that interaction does occur between the $HO_x$ species and the odd nitrogen and halogen families, which are also involved in $O_3$ destruction in the stratosphere. These interactions will become more apparent in later chapters, but are briefly noted here. For example, reaction of $HO_2$ with NO serves to alter the ratio of NO to $NO_2$, while reactions of OH and $HO_2$ with $NO_2$ convert the $NO_x$ species into less reactive reservoir compounds ($HONO_2$ and $HO_2NO_2$). Similarly, the partitioning of Cl and ClO is altered by reaction of ClO with NO. Also, reaction of OH with HCl

$$OH + HCl \rightarrow Cl + H_2O \tag{6.34}$$

is the major reaction involved in the conversion of an inactive form of inorganic chlorine (HCl) to more reactive forms (Cl and ClO). Finally, reaction of OH with CO and $CH_4$ initiates the oxidation of these species. The role of OH as an important atmospheric oxidant will become even more apparent in the next section dealing with tropospheric $HO_x$ chemistry.

## 6.5 Chemistry of Hydrogen Species in the Troposphere

The chemistry of hydrogen compounds is central to tropospheric chemistry. Discussion of hydrogen compounds in the troposphere often revolves around the hydroxyl radical, OH, which is the major reactant for many of the compounds (*e.g.*, hydrocarbons, sulfur species) emitted at the surface. Thus the level of OH in both polluted and remote regions essentially determines the burden of those compounds.

However, it is not possible to talk about the level of OH as if it were a fixed quantity. Reactions of OH with hydrocarbons, for example, generate organic peroxy radicals (commonly referred to as $RO_2$). The reactions of these radicals with nitrogen oxides lead to recycling of the OH radical and to the production of ozone, the photolysis of which leads in turn to the generation of more OH. Thus the troposphere is a closely coupled system involving $HO_x$, $NO_x$, and ozone. The interactions of these families will be introduced in this chapter, and developed in more detail in subsequent chapters (7, 9, and 13).

Due to the rapid interconversion of OH, $HO_2$, and organic peroxy radicals ($RO_2$), and their cumulative effects on ozone production, it is often useful to consider organic peroxy radicals as members of the odd hydrogen family. In this sense, reactions of hydroxyl radicals with organic species should not strictly be regarded as sinks of hydrogen radicals, since the total number of radicals is not changed. As was the case in the discussion of the stratosphere, the chemistry of $HO_x$ in the troposphere will be discussed in terms of the reactions that create and destroy the $HO_x$ species, and those that interconvert the members of the family.

### 6.5.1 Sources of $HO_x$ in the Troposphere

The primary source of odd hydrogen compounds in the troposphere (as was the case in the stratosphere) is the reaction of excited oxygen atoms, $O(^1D)$, with water vapor (Levy, 1971; Talukdar *et al.*, 1998; and references therein).

$$O(^1D) + H_2O \rightarrow 2\,OH \tag{6.8}$$

This process is most effective at low altitudes, where the water vapor mixing ratio is high, but its efficiency falls off with altitude as the progressively lower temperature limits the amount of water vapor available. The source of $O(^1D)$ is again the photolysis of ozone, Reaction (6.5a), which occurs at wavelengths shorter than about 320 nm. The large majority (>90%) of $O(^1D)$ atoms are quenched to ground-state $O(^3P)$ by collisions with nitrogen and oxygen and re-form ozone, Reactions (6.6), (6.7), and (6.23). Thus the rate of formation of OH can be expressed approximately as:

$$\frac{d[OH]}{dt} = \frac{2k_8[H_2O]j_{O_3}[O_3]}{(k_6[O_2] + k_7[N_2])} \tag{6.35}$$

where $j_{O_3}$ is the first-order rate constant for photolysis of ozone into $O(^1D)$. Typical OH production rates near the surface under sunlit conditions are of order $10^6$ molecules cm$^{-3}$ s$^{-1}$.

A second major source of odd hydrogen radicals is the photolysis of carbonyl compounds, particularly formaldehyde

$$CH_2O + h\nu \rightarrow H + HCO \tag{6.2a}$$

$$CH_2O + h\nu \rightarrow H_2 + CO \tag{6.2b}$$

$$HCO + O_2 \rightarrow HO_2 + CO \tag{6.36}$$

$$H + O_2 + M \rightarrow HO_2 + M \tag{6.10}$$

Carbonyl compounds are produced in the oxidation of organic molecules. Formaldehyde, the simplest carbonyl, is formed from methane oxidation, and is thus expected to be present even in remote areas. Acetone, produced in the oxidation of a number of nonmethane hydrocarbons (particularly propane), has a sufficiently long lifetime that its photolysis may also be a key contributor to the production of odd-hydrogen radicals in the free troposphere (Singh *et al.*, 1995)

$$CH_3C(O)CH_3 + h\nu \rightarrow CH_3 + CH_3CO \quad (6.37)$$

Other shorter-lived carbonyls mainly contribute to the photochemistry of continental areas, close to the sources of their parent hydrocarbons.

A mechanism for radical generation that is receiving more attention is the reaction of ozone with alkenes (Atkinson, 1997b). The exact mechanism of these reactions differs for the various alkenes, but it is thought that a large fraction of the products are radicals. The reaction of $O_3$ with the simplest alkene (ethene, $C_2H_4$) is believed to proceed via the formation of a biradical species, $\cdot CH_2OO \cdot$, the subsequent decomposition of which produces $HO_x$ radicals at least part of the time

$$O_3 + C_2H_4 \rightarrow \cdot CH_2OO \cdot + CH_2O \quad (6.38)$$

For reaction of larger, substituted alkenes with $O_3$, direct formation of OH is often a major pathway. The reaction of ozone with alkenes has been shown to be appreciable in continental areas, even during daytime, when ozone production is most active. However, it can also occur during the hours of darkness, and consequently can lead to continuing oxidation at night (Hu and Stedman, 1995; Paulson and Orlando, 1996).

A second mechanism for generating peroxy radicals at night is from the $NO_3$ radical (Wayne *et al.*, 1991). The nitrate radical undergoes rapid photolysis in the daytime, and consequently does not build up to very high levels. Furthermore, its reactions tend to be slower than those of the hydroxyl radical with the same species. However, at night $NO_3$ may reach several parts per trillion in continental areas, and its reactions with aldehydes and alkenes can become important sources of peroxy radicals

$$NO_3 + CH_3CHO \rightarrow HNO_3 + CH_3CO \quad (6.39)$$
$$NO_3 + R_2C=CR_2 \rightarrow O_2NOCR_2CR_2 \rightarrow O_2NOCR_2CR_2O_2 \quad (6.40)$$

### 6.5.2 Cycling of Hydrogen Radicals — Effect on Tropospheric Ozone

The rate of interconversion of $HO_x$ radicals is generally very rapid in the troposphere. Depending on the level of pollution (specifically, the NO concentration), there is a change in the mechanism of cycling and also in the overall effect on the ambient ozone level. In the clean, background troposphere the oxidation of methane or carbon monoxide is the major reaction of OH

$$OH + CH_4 \rightarrow H_2O + CH_3 \quad (6.41)$$
$$OH + CO \rightarrow H + CO_2 \quad (6.19)$$

Both reactions lead to the production of peroxy radicals

$$CH_3 + O_2 + M \rightarrow CH_3O_2 + M \quad (6.42)$$
$$H + O_2 + M \rightarrow HO_2 + M \quad (6.10)$$

Under conditions of low $NO_x$ the predominant mechanism for the return of peroxy radicals to OH is through the reaction of $HO_2$ with $O_3$

$$HO_2 + O_3 \rightarrow OH + 2O_2 \quad (6.21)$$

Such conditions pertain in the marine boundary layer, far from sources of $NO_x$, or in the free troposphere, where the air has not been subject to anthropogenic influence for several days. The reaction of OH with $O_3$, Reaction (6.18), is also an efficient mechanism for the cycling of OH to $HO_2$ in the upper troposphere. Consequently, the presence of $HO_x$ radicals tends to lead to a loss of $O_3$ under background conditions. This is essentially the same chemistry discussed earlier with regard to catalytic ozone destruction in the lower stratosphere.

In more polluted regions, the presence of NO provides a different mechanism for the interconversion of peroxy radicals:

$$HO_2 + NO \to OH + NO_2 \qquad (6.20)$$
$$RO_2 + NO \to RO + NO_2 \qquad (6.43)$$

These reactions lead to the formation of $O_3$ following the photolysis of $NO_2$

$$NO_2 + h\nu \to NO + O \qquad (6.22)$$
$$O + O_2 + M \to O_3 + M \qquad (6.23)$$

Throughout the troposphere, OH can react with hydrocarbons and other organic species, either by abstraction of a hydrogen atom or addition to an olefinic double bond (see Chapter 9). These reactions in fact account for most of the loss of organic species, sulfur species, and entirely manmade compounds such as the hydrochlorofluorocarbons and hydrofluorocarbons in the atmosphere. However, there is no overall loss of radicals from the atmosphere, since peroxy radicals are inevitably produced, and these are rapidly converted through $HO_2$ and back to OH.

The tropospheric lifetime of OH is usually a few seconds or less. This upper limit to its lifetime is determined by its reaction with $CH_4$, which is present throughout the troposphere at a level of about 1.7 ppm, and with CO, which is also a ubiquitous component of the troposphere. With a production rate of about $10^6$ molecule $cm^{-3}$ $s^{-1}$, and a lifetime of about 1 s, it can be seen that the concentration of OH in the sunlit troposphere is typically near $10^6$ molecule $cm^{-3}$. The cycling of $HO_2$ back to OH, on the other hand, takes much longer, on the order of 100 sec; thus the ratio $HO_2$:OH is typically around 100:1 — that is, $HO_2$ concentrations of order $10^8$ molecule $cm^{-3}$ are often observed.

The concentration of OH is clearly a highly varying quantity, depending very strongly on the solar flux, the $O_3$ concentration, levels of hydrocarbon species, etc. In addition, measurements of OH are extremely difficult to make (see later in this chapter), and one can never expect to make sufficient measurements of OH mixing ratios at enough locations under enough conditions to obtain global OH concentration fields. However, the determination of an average global OH concentration is important in arriving at tropospheric lifetimes for long-lived (greater than a few months), fairly well-mixed compounds, such as methane. For these purposes, global average OH concentrations have been established from global measurements of methylchloroform (Prinn et al., 1995). Sources of methylchloroform to the atmosphere are wholly anthropogenic and well quantified, while its destruction is almost exclusively through reaction with OH. This information, coupled with atmospheric methylchloroform concentration measurements at a variety of locations, has been used to establish a global average OH concentration of $9.7 \times 10^5$ molecule $cm^{-3}$, which does not appear to have changed significantly over the past decade or more (Prinn et al., 1995).

### 6.5.3 Sinks of Hydrogen Radicals

Sinks of $HO_x$ not only have a direct effect on the levels of the oxidants OH and $HO_2$, but also influence the efficiency of the ozone and $NO_x$ cycles described above. The simplest loss mechanism for hydrogen compounds is the direct termination reactions

$$HO_2 + HO_2 \rightarrow H_2O_2 + O_2 \tag{6.44}$$

$$OH + HO_2 \rightarrow H_2O + O_2 \tag{6.11}$$

The first of these reactions is effective throughout the troposphere. The reaction is not just a simple bimolecular reaction, as it might first appear, and is in fact rather complicated. The reaction does have a bimolecular component (6.44a), but also has a termolecular component involving atmospheric $O_2$ or $N_2$ (6.44b), and an enhancement due to water vapor (6.44c):

$$HO_2 + HO_2 \rightarrow H_2O_2 + O_2 \tag{6.44a}$$

$$HO_2 + HO_2 + M \rightarrow H_2O_2 + O_2 + M \tag{6.44b}$$

$$HO_2 + HO_2 + H_2O \rightarrow H_2O_2 + O_2 + H_2O \tag{6.44c}$$

The water enhancement is about a factor of 2.6 at 100% relative humidity and 298 K. The rate of Reaction (6.44b) is about 70% that of (6.44a) at 298 K and one atmosphere pressure. Thus the rate of formation of hydrogen peroxide is a complex function of pressure, temperature, and water vapor concentration, as shown in Figure 6.1.

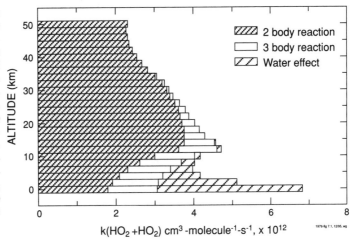

**Figure 6.1.** Change in rate coefficients for the disproportionation reaction of $HO_2$ with altitude using the U.S. Standard Atmosphere (1976) temperature, water vapor, and pressure conditions. Reaction (6.44a) is the bimolecular part, Reaction (6.44b) is the termolecular part; the enhancement due to the "chaperon" effect of water vapor, Reaction (6.44c), is also shown. Kinetic data from DeMore et al. (1997).

The reaction of $HO_2$ with organic peroxy radicals also leads to the formation of stable peroxides, for example,

$$CH_3O_2 + HO_2 \rightarrow CH_3OOH + O_2 \tag{6.45}$$

If the peroxides are lost, for example, by deposition, then the reaction is effectively a sink for both the peroxy radicals involved. Reaction of OH with peroxides also results in a net loss of radical species,

$$OH + H_2O_2 \rightarrow H_2O + HO_2 \tag{6.46}$$

$$OH + CH_3OOH \rightarrow H_2O + CH_3O_2 \tag{6.47}$$

$$OH + CH_3OOH \rightarrow H_2O + CH_2O + OH \tag{6.48}$$

while photolysis of the peroxides returns the $HO_x$ radicals.

$$H_2O_2 + h\nu \rightarrow 2OH \tag{6.49}$$

$$CH_3OOH + h\nu \rightarrow CH_3O + OH \tag{6.50}$$

$$CH_3O + O_2 \rightarrow CH_2O + HO_2 \tag{6.51}$$

In polluted areas, the reactions of OH with $NO_2$ and $HNO_3$ become major losses for $HO_x$ species:

$$OH + NO_2 + M \to HNO_3 + M \qquad (6.12)$$

$$OH + HNO_3 \to H_2O + NO_3 \qquad (6.13)$$

Removal of $HNO_3$ by wet deposition processes also results in the loss of $HO_x$ species from the gas phase.

Hydrogen peroxide is a relatively soluble species and thus is readily incorporated into the liquid phase. This is important not only as a removal mechanism for gas-phase $HO_x$, but also because hydrogen peroxide is important in its own right as a liquid-phase oxidant. For example, dissolved $SO_2$ (sulfur IV) can be oxidized to sulfuric acid (sulfur VI) by hydrogen peroxide

$$SO_2 + H_2O \rightleftharpoons HSO_3^- + H^+ \qquad (6.52)$$

$$HSO_3^- + H_2O_2 \to HSO_4^- + H_2O \qquad (6.53)$$

The rate coefficient for Reaction (6.53) is very pH dependent, maximizing near pH 1. However, due to the pH dependence of Reaction (6.52), the overall reaction rate does not depend strongly on the acidity, as is shown in Chapter 10. Reaction (6.53) is effective because of the large solubility of $H_2O_2$; its Henry's Law constant is about $10^5$ M atm$^{-1}$ at $20°C$.

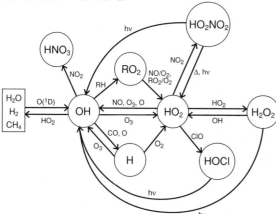

**Figure 6.2.** Schematic diagram of the important reactions involving hydrogen compounds in the troposphere and stratosphere. Interconversion of the reactive radicals results in the oxidation of reduced species of carbon, nitrogen, and sulfur. Loss of these reactive radicals occurs through physical and chemical removal of such species as $HNO_3$, $HO_2NO_2$, and $H_2O_2$ (not shown).

The above discussion has focused on the sources of hydrogen compounds to the atmosphere, the chemistry they undergo, and the effects this chemistry has, particularly with regard to levels of stratospheric and tropospheric ozone. The set of chemical reactions just discussed is summarized in Figure 6.2. The remainder of the chapter will provide an overview of various measurements and modeling studies that will help to provide a general picture of the distribution of hydrogen compounds in the stratosphere and troposphere.

## 6.6 Measured and Modeled Concentrations of Hydrogen Compounds in the Stratosphere

There have not been sufficient measurements made of all the hydrogen compounds to establish a good picture of their diurnal, seasonal, latitudinal, and altitudinal profiles. Several of the compounds are difficult to measure, namely the radical species OH and $HO_2$ and water vapor. What data are available, however, have been synthesized into a few summary graphs that are presented below. These are compared with model results, when available, and serve to illustrate the general behavior of these species.

Atomic hydrogen increases from less than 0.1 atoms cm$^{-3}$ in the lower stratosphere to a peak of about $10^9$ atoms cm$^{-3}$ in the upper mesosphere (80 km). The absolute concentration gradually decreases above this altitude, but the fraction of hydrogen in the atomic form (compared to $H_2$ and $H_2O$) increases until in the upper thermosphere it is all atomic (see discussion in McEwan and Phillips, 1976).

Molecular hydrogen chemistry and measurements are reviewed in Ehhalt et al. (1977) and Schmidt et al. (1980). A summary of measurements made by gas chromatographic techniques during balloon, aircraft, and rocket flights from ground level to 50 km at midlatitudes (30-40° N) is shown in Figure 6.3. The measurements can be understood in terms of a surface source of $H_2$ along with molecular hydrogen production from formaldehyde photolysis in the troposphere. The mixing ratio is nearly constant at 0.50 (±0.01) ppmv in the troposphere in regions removed from local sources. An increase in the mixing ratio is seen in the midstratosphere due to contribution from the photolysis of formaldehyde produced in the oxidation of methane, which offsets the loss of $H_2$ to reaction with OH. The mixing ratio of $H_2$ increases to about 0.6 ppmv at 27 km, and decreases above that altitude as its methane source decreases and its destruction rate by OH increases. There are no measurements available above 50 km.

Typical values of the $H_2O$ volume mixing ratio in the stratosphere are in the range 2-7 ppmv, although observations as high as 12 ppmv have been reported in the past. It is not known with certainty whether the observed variability is entirely due to atmospheric effects or whether instrumental artifacts played a role.

Representative water vapor mixing ratios versus latitude and altitude measured from the HALOE satellite are presented in Figure 6.4. The lowest mixing ratios are observed at the tropical tropopause, where the temperature is low and air entering the stratosphere is "freeze-dried"; mixing ratios of about 3 ppmv are typical. Because of increased

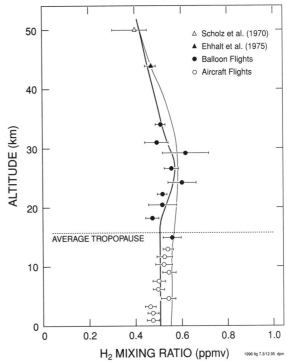

**Figure 6.3.** Average vertical profiles of the $H_2$ mixing ratio in the troposphere and stratosphere (heavy line). Triangles represent results of two rocket flights, and solid and open circles represent results of samples collected over Texas during balloon and aircraft flights, respectively. Error bars indicate the standard error of measurement in the case of the rocket samples and the mean standard deviation in the case of the balloon and aircraft samples. The light line gives the result of theoretical calculations with a one-dimensional model (after Ehhalt et al., 1977).

vertical motion in the Northern Hemisphere winter, the tropical tropopause is coldest during these months. These colder temperatures at the tropical tropopause lead to lower water mixing ratios, while the higher summer temperatures allow for higher mixing ratios. Because the air maintains this seasonal signal for months as it is

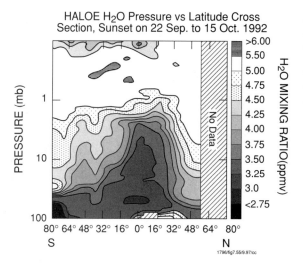

**Figure 6.4.** Pressure altitude versus latitude cross-section HALOE satellite observations of water vapor volume mixing ratio (ppmv) for 22 September to 15 October 1992 (after Mote et al., 1993, and Tuck et al., 1993).

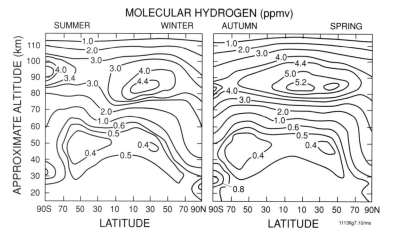

**Figure 6.5.** Model-calculated molecular hydrogen mixing ratio distribution in January and March (ppmv) (after Le Texier et al., 1988).

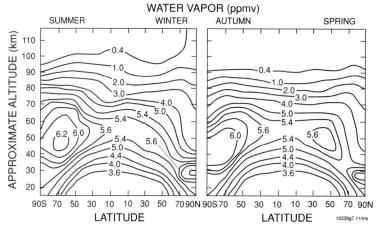

**Figure 6.6.** Model-calculated water vapor mixing ratio distribution in January and March (ppmv) (after Le Texier et al., 1988).

transported upward by the large-scale circulation, alternating layers of high and low water mixing ratio are observed (Mote *et al.*, 1995). Because the imprinting of this seasonal signal on an air parcel passing through the tropical tropopause is much like a strip of magnetic tape passing over a tape recorder head, the term "tape recorder effect" has been coined (Holton et al., 1995). The analysis of these "tape-recorder" signals allows for the determination of transport rates in the tropical lower stratosphere (the so-called tropical pipe), and for the determination of the degree of mixing with extratropical air.

Away from the tropical tropopause, water vapor mixing ratios generally increase with altitude throughout the lower and middle stratosphere (see Fig. 6.4), largely the result of the oxidation of $CH_4$ to $H_2O$. Satellite observation of water vapor and methane have allowed a "hydrogen balance" to be calculated throughout the stratosphere (Mote et al., 1993; Tuck et al., 1993). This analysis has indicated that the "total hydrogen," defined as $2 \times [CH_4] + [H_2O]$, is relatively constant at 6.2 ppmv throughout the stratosphere. These observations are consistent with near-quantitative conversion of $CH_4$ to $H_2O$ in the stratosphere, with the mixing ratio of $H_2$ remaining nearly constant.

Stratospheric molecular hydrogen and water vapor concentrations have been calculated by Le Texier et al. (1988), and the results are presented as functions of latitude and altitude for January and March in Figures 6.5 (for $H_2$) and 6.6 (for $H_2O$). These model results are in reasonable accord with the satellite observations discussed above, showing the importance of methane oxidation in the stratospheric water vapor distribution, as well as the photochemical conversion of $CH_2O$ and water vapor to $H_2$ in the upper stratosphere and mesosphere.

The next species to be considered are the radicals, OH and $HO_2$. Hydrogen peroxide, $H_2O_2$, will be considered simultaneously as it is formed from the self-reaction of $HO_2$ radicals, and thus measurements of this species present a good test for our understanding of $HO_x$ chemistry. A number of measured and modeled stratospheric profiles for OH, $HO_2$, and $H_2O_2$ have been reported, and representative examples are given in Figures 6.7-6.12. Measurements of hydrogen peroxide mixing ratios have been

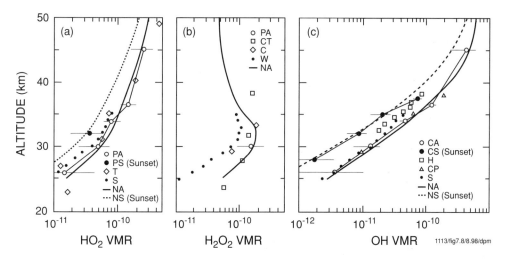

**Figure 6.7.** *Comparison of measurements of stratospheric $HO_2$, $H_2O_2$, and OH for similar conditions of latitude and solar zenith angle. Left panel, $HO_2$: Park and Carli, 1991, afternoon, open circles; Park and Carli, 1991, sunset, filled circle; Traub et al., 1990, open diamonds; Stimpfle et al., 1990, small filled circles. Middle panel, $H_2O_2$: Park and Carli, 1991, open circle; Chance et al., 1991, open squares; Chance and Traub, 1987, open diamonds; Wennberg et al., 1990, small filled circle, inferred using $HO_2$ and OH measurements. Right panel, OH: Carli et al., 1989, afternoon, open circles; Carli et al., 1989, sunset, filled circles; Heaps and McGee, 1985, open squares; Carli and Park, 1988, open triangles; Stimpfle et al., 1990, small filled circles. The solid and dashed lines are theoretical profiles using DeMore et al. (1990) chemistry for the afternoon and sunset case, respectively (after Park and Carli, 1991).*

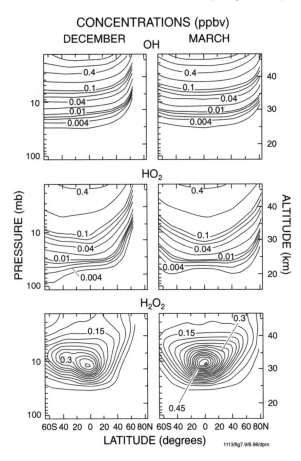

**Figure 6.8.** Contour plots showing the latitude-altitude calculated distributions of daytime OH (top), $HO_2$ (middle), and $H_2O_2$ (lower) mixing ratios (ppbv) for December and March. Contours for OH and $HO_2$ are drawn at values $A \times 10^E$ ppbv where $A = 2, 4, 6, 8, 10$ and $E = -3, -2, -1$. Contours for $H_2O_2$ are drawn every 0.03 ppbv (after Kaye and Jackman, 1986).

made using millimeter and microwave remote sensing techniques. The available data have been summarized by Park and Carli (1991) and are presented in Figure 6.7. The $H_2O_2$ mixing ratio apparently peaks near 32 km at about 0.2 ppbv, although there is not a great deal of data to verify this. Clearly, further measurements are needed. Mixing ratios of $H_2O_2$ obtained from a photochemical model constrained by satellite observations (Kaye and Jackman, 1986) show a similar pattern, peaking near 32 km, with a maximum value of about 0.5 ppbv near the equator (see Fig. 6.8).

Mixing ratios of OH and $HO_2$ from balloon-borne instruments (Park and Carli, 1991) and from a photochemical model constrained by satellite observations (Kaye and Jackman, 1986) have also been reported (see again Figs. 6.7 and 6.8). For $HO_2$, the balloon-borne measurements indicate mixing ratios of about 0.01 ppbv near 25 km, gradually increasing to 0.2 ppbv near 45 km. The satellite measurements are similar. For OH, mixing ratios obtained from the balloon-borne instruments range from about 3 pptv near 25 km to 400 pptv at 45 km. Satellite measurements range from 3 pptv near 25 km to about 600 pptv at 45 km, in reasonable agreement with the balloon instruments (see also Color Plate 23).

A number of modeled distributions of the species OH, $HO_2$, and $H_2O_2$ are also available; representative examples are given in Figure 6.9 (Connell et al., 1985) and Figures 6.10-6.12 (Fabian et al., 1982). The modeled distributions are in general accord with the measured ones. OH and $HO_2$ mixing ratios generally increase with altitude above 20 km, while the peak in the $H_2O_2$ distribution near 30 km is again evident. The increase in $HO_x$ mixing ratios with altitude can largely be attributed to

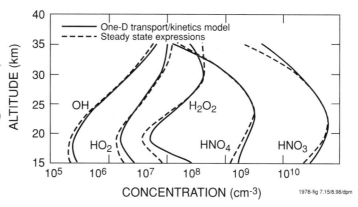

**Figure 6.9.** Stratospheric vertical concentration profiles of $HO_x$ and some nitrogen species from a 1D transport-kinetics model (solid curves) compared to the photochemical approximations (dashed curves) (after Connell et al., 1985).

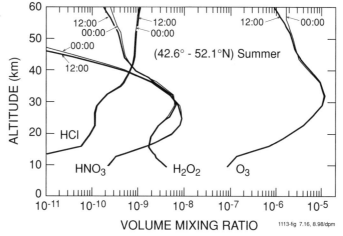

**Figure 6.10.** Modeled diurnal variation of upper tropospheric and stratospheric mixing ratios of $H_2O_2$ and some other species for $47°N$ in summer at local times labeled (after Fabian et al., 1982).

an increased formation rate due to the large increase in the abundance of $O(^1D)$ at higher altitudes (see Reactions 6.4 to 6.9).

The measurements and model results presented above provide a general indication of the mixing ratios of OH and $HO_2$ as a function of altitude in the stratosphere. Mixing ratios of OH and $HO_2$ are also available from in situ measurements made in the lower stratosphere aboard the ER-2 aircraft (Cohen et al., 1994). There have been a number of programs conducted with this aircraft. One of the strengths of these studies is that simultaneous measurements of a whole suite of reactive species are possible, providing a very stringent test of our understanding of lower stratospheric chemistry. Representative $HO_x$ data from the SPADE (Stratospheric Photochemistry, Aerosols, and Dynamics Expedition) mission are given in Figure 6.13, where the ratio $[OH]/[HO_2]$ is plotted as a function of the measured [NO]. A general increase of this ratio with increasing NO is observed, as would be expected from a consideration of Eq. (6.28) developed earlier. The agreement between the measured ratio and that obtained from Eq. (6.28) indicates that our understanding of the $OH/HO_2$ chemistry of the lower stratosphere is well understood. In fact, Cohen et al. were able to conclude that the relative and absolute rate coefficients for the important reactions involved in the interconversion of OH and $HO_2$, Reactions (6.20) and (6.18), have uncertainties that are lower than described in the review of DeMore et al. (1992).

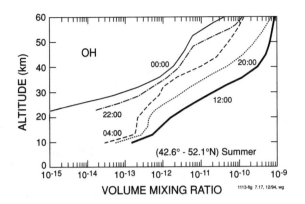

*Figure 6.11.* Same as Figure 6.10 for OH (after Fabian et al., 1982).

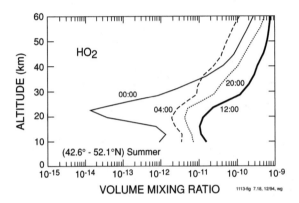

*Figure 6.12.* Same as Figure 6.10 for HO$_2$ (after Fabian et al., 1982).

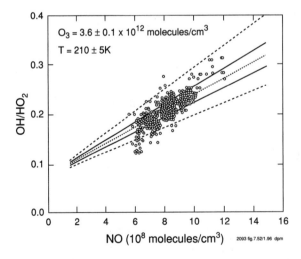

*Figure 6.13.* Measured (o) and modeled OH/HO$_2$ versus simultaneously measured NO during SPADE. The dotted line is a calculation of the ratio using DeMore et al. (1992) kinetic data and measurements of O$_3$, NO, ClO, and BrO. Dashed lines (- - -) are calculations at the DeMore et al. (1992) error limits for the rate of HO$_2$ + NO (Reaction 6.20). Solid lines correspond to error limits for measured NO (after Cohen et al., 1994).

## 6.7 Measured and Modeled Concentrations of Hydrogen Compounds in the Troposphere

The distributions of the source gases, H$_2$ and H$_2$O, in the troposphere have been presented earlier in this chapter and are only briefly summarized here. Molecular hydrogen, with its lifetime of about 10 years, is well mixed throughout the troposphere, with a measured mixing ratio of about 0.55 ppmv (see Fig. 6.3), although one may expect some variability if measurements are taken near sources. The modeling results shown in Figure 6.3 indicate that, although the agreement is not perfect, the tropospheric hydrogen budget seems to be reasonably well understood (see Table 6.1). Tropospheric water vapor is highly variable, but is present in general at levels considerably higher than the few ppmv levels found in the stratosphere. The mixing ratios can range from a few ppmv near the tropopause up to 4% by volume in a warm, humid environment near the surface. Typical levels are in the range of 0.1 to 2% by volume.

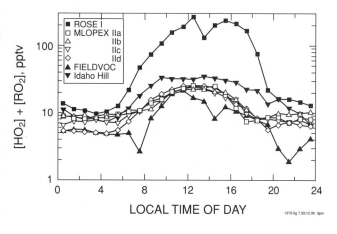

*Figure 6.14.* Concentrations of the sum of the peroxy radicals ($HO_2 + RO_2$) versus time of day for seven field campaigns conducted at various Northern Hemisphere sites. ROSE I measurements taken in a pine forest in west-central Alabama in 1991. MLOPEX II was four campaigns at the Mauna Loa Observatory on the northern slope of the Mauna Loa volcano in Hawaii. FIELD-VOC measurements were taken near sea level on the Brittany coast of France, and Idaho Hill measurements were taken in the Colorado Rocky Mountains (after Cantrell et al., 1995).

Measurements of the $HO_x$ radical species (OH, $HO_2$, $RO_2$) are very difficult to make, owing to the low abundance of these species. Also, because of their short chemical lifetimes, concentrations of these species display high temporal and spatial variability. Tropospheric $HO_2$ and $RO_2$ have been reported, for example, by Mihelcic and co-workers using a matrix-isolation technique (Mihelcic et al., 1990) and Cantrell and co-workers using a "chemical amplifier" (Cantrell et al., 1993a,b). These are primarily ground-based measurements, although some data collected using aircraft-based instrumentation have also been reported. A summary of peroxy radical data collected from several ground-based field campaigns is shown in Figure 6.14. The concentrations of total peroxy radicals show a strong diurnal variability, expected due to their photochemical sources. Midday peak mixing ratios reported range from 25 pptv for clean free tropospheric conditions to 260 pptv for rural continental conditions near sea level. Peroxy radical mixing ratios in urban areas are also typically in the many tens of pptv at midday (Hu and Stedman, 1995). Of course, these midday peak concentrations are expected to be strong functions of latitude and altitude due to variations in insolation, and ozone and water vapor concentrations, which control $HO_x$ production. Nighttime mixing ratios are usually nonzero at levels of about 5 to 20 pptv. These nighttime radicals are thought to arise from the $O_3$- or $NO_3$-initiated chemistry mentioned earlier.

There have been attempts to measure tropospheric hydroxyl radicals for nearly two decades. Early measurements were plagued by problems with intereferences and artifact signals. More recent remote sensing techniques (Mount, 1992) and in situ

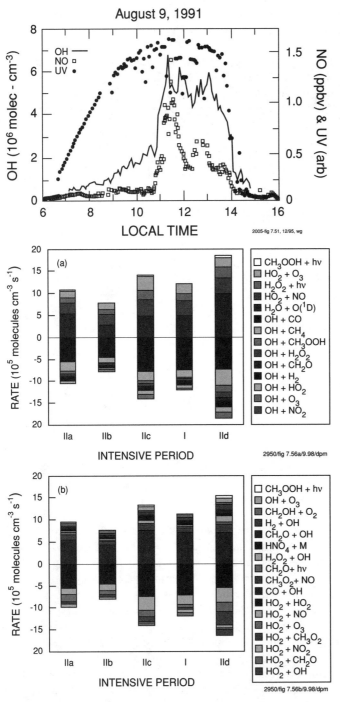

**Figure 6.15.** Diurnal profile of ground-level OH concentrations along with ultraviolet intensity and concentration of NO (after Eisele, 1995). The pollution event at 11 a.m., evidenced by a large increase in NO, results in enhanced OH levels in the afternoon.

**Figure 6.16.** Calculated formation and destruction rates for (a) OH and (b) $HO_2$ for free tropospheric conditions in the central Pacific Ocean on Hawaii. The chemical box models were constrained by the average measured quantities during a given intensive period. The intensive periods correspond to Autumn (IIa), Winter (IIb), Spring (I and IIc), and Summer (IId) that were part of the Mauna Loa Observatory Photochemistry Experiments I and II in 1988 and 1991-1992, respectively (from Hauglustaine et al., 1998).

measurements (Eisele and Tanner, 1991) appear to have overcome these problems and side-by-side measurements using the two techniques have been made (Mount and Eisele, 1992). Ground-level concentrations of hydroxyl for clear-sky, midlatitude, summer conditions are typically $(1-5) \times 10^6$ molecules cm$^{-3}$ (0.04 to 0.2 pptv). Nighttime levels are usually quite low ($< 1 \times 10^4$ molecules cm$^{-3}$) (see Fig. 6.15), again reflecting the photochemical source of $HO_x$.

The relationships between the various formation and destruction processes for hydrogen radicals have been calculated by Hauglustaine et al. (1998) for conditions representative of the free troposphere during the Mauna Loa Observatory Photochemistry Experiment 1 and 2 in Hawaii in 1988 and 1991-1992. Shown in Figure 6.16a are the processes that remove and produce OH as determined by a chemical box model constrained by the actual measurements of many of the controlling species. Figure 6.16b is the same for $HO_2$. These figures show that while there are dominant processes in the formation and destruction of these radicals, other chemical reactions when grouped together can be nearly as important.

Recently, aircraft-based measurements of odd hydrogen radicals (OH and $HO_2$) have been made in the upper troposphere using the technique of laser-induced fluorescence (for OH) with chemical conversion for $HO_2$. These data (Fig. 6.17) reveal a minimum in $HO_2$ mixing ratio of about 2 pptv near the tropopause with increasing concentrations down into the upper troposphere. The mixing ratio of OH decreases monotonically from the lower stratosphere into the upper troposphere. These data are in reasonable accord with stratospheric measurements by other techniques, and provide new data to test photochemical theories in the chemically different environment of the troposphere.

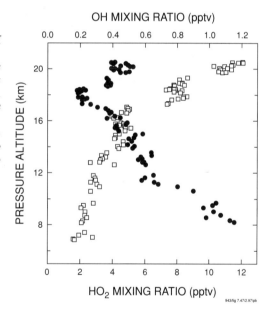

**Figure 6.17.** The mixing ratio of OH (open squares) and $HO_2$ (closed circles) plotted vs. altitude. The measurements were obtained from the $HO_x$ instrument (Wennberg et al., 1994) aboard the NASA ER-2 aircraft during descent into Barber's Point NAS, Oahu, Hawaii on Feb. 12, 1996. The mixing ratio of $HO_2$ reaches a minimum near the tropopause (~17 km) before rapidly increasing in the upper troposphere (P. O. Wennberg, T. F. Hanisco, and J. G. Anderson, personal communication).

Levels of hydrogen peroxide in the troposphere are quite variable; mixing ratios are typically in the 1-3 ppbv range near the surface. However, scavenging of $H_2O_2$ by clouds can result in very low gas-phase mixing ratios (<0.1 ppbv). A number of measurements of $H_2O_2$ have been made in the free troposphere (Heikes, 1992; Staffelbach et al., 1996), and mixing ratios of about 1 ppbv are generally found (see Figs. 6.18 and 6.19). The measurements show very little diurnal variation, reflecting the fairly long lifetime of $H_2O_2$ (a day or two). Measurements have also been made in the marine boundary layer as a function of latitude above the Atlantic Ocean (Weller and Schrems, 1993); see Figure 6.20. The maximum near the equator may reflect increasing photochemical activity there, leading to larger $HO_x$ concentrations and hence larger $H_2O_2$ values.

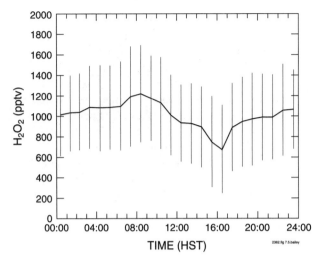

**Figure 6.18.** Composite diurnal variation of $H_2O_2$ measured at Mauna Loa Observatory (3.4 km) in May 1988 by Heikes (1992) binned into one-hour average values. The solid curve is the average using the entire data set.

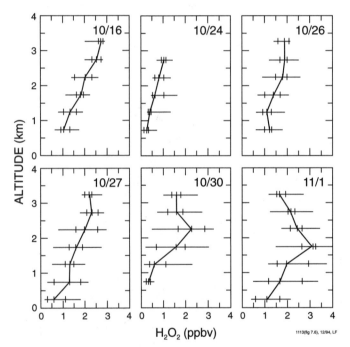

**Figure 6.19.** Aircraft-based $H_2O_2$ measurements over the continental US made by Heikes et al. (1987) summarized for six days and grouped into 500 m altitude increments. The ends of the horizontal lines represent the lowest and highest concentrations measured at that level; the tick marks represent the 25th percentile, the median, and the 75th percentile, respectively, from left to right. Curves connect the median values at each altitude (after Heikes et al., 1987).

A set of modeled peroxide and peroxy radical concentrations was presented by Logan et al. (1981). While some of the details of the chemical scheme may be somewhat outdated, these results give a good qualitative picture of the diurnal, latitudinal, and altitudinal dependence of these species. The upper panels of Figure 6.21 show low-latitude ground-level and 6 km calculated concentrations. The lower panels of Figure 6.21 show high-latitude results for the same altitudes. In comparing these model results with free tropospheric, clear-sky measured levels at Mauna Loa Observatory (3.4 km altitude, 19° N), one sees that the ground-level, 15° N, midday peroxy radical levels (63 pptv) are much higher than the average observations (25 pptv). The 15° N, 6 km and the 45° N, 0 km calculations are closer to the observations (29 and 15 pptv, respectively). Clearly, more tropospheric peroxy radical observations at several latitudes and altitudes are required to test the modeled levels.

*Figure 6.20.* Latitude-dependent gas-phase $H_2O_2$ mixing ratio in the marine troposphere of the Atlantic Ocean. Values measured during wet deposition are indicated by open squares (after Weller and Schrems, 1993).

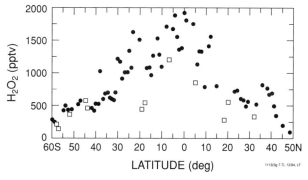

*Figure 6.21.* Diurnal variations of $HO_x$ in the tropics and midlatitudes for cloud-free equinoctial conditions at the surface and at 6 km. The lines are derived from the standard model of Logan et al., 1981, which does not include the chemistry of peroxynitric acid. The diurnally averaged concentration of OH is indicated by the arrows (after Logan et al., 1981).

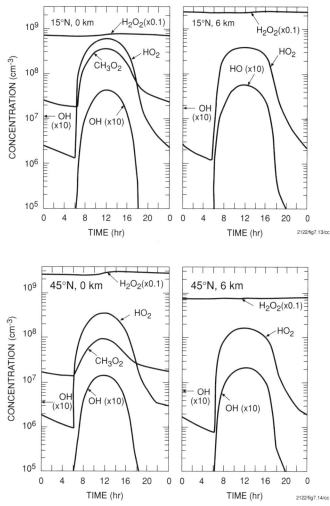

## 6.8 Summary

This chapter has shown that the hydrogen compounds have important atmospheric roles. They are among the most important oxidants for many atmospheric compounds, and they are coupled into nearly every other chemical family (as will become even more apparent in subsequent chapters). It appears that the concentrations of peroxy radicals, hydroxyl radicals, and hydrogen peroxide can be modeled with reasonable accuracy, although there are not enough measurements to allow a full evaluation of the model results. Clearly, more measurements of these important atmospheric species need to be made at a number of representative latitudes, altitudes, and seasons in order to judge whether their chemistry is fully understood.

## Further Reading

Finlayson-Pitts, B. J., and J. N. Pitts, *Atmospheric Chemistry*, 1098 pp., J. Wiley & sons, New York, 1986.

Thompson, A. (1995) Measuring and modeling the tropospheric hydroxyl radical (OH), *J. Atmos. Sci.* 52, 3315.

Warneck, P., *Chemistry of the Natural Atmosphere*, 757 pp., Academic Press, San Diego, CA, 1988.

Wayne, R. P., *Chemistry of Atmospheres*, 2nd edition, 447 pp., Clarendon Press, Oxford, 1991.

Dieter Ehhalt

# Hydrogen Compounds

Hydrogen is the most abundant element in the universe. Thus it is no wonder that some of the compounds treated in this chapter, which includes the oxidized hydrogen species as well, have also been found in stellar atmospheres, others in the interstellar space.

More important here, hydrogen compounds are found everywhere in the Earth's atmosphere, from the planetary boundary layer to the exosphere, in polluted and unpolluted air. Moreover, they are known to play important roles, sometimes the key parts in the chemistry of the various layers of our atmosphere.

In the troposphere and stratosphere, the atmospheric layers nearest and of most concern to biosphere and mankind, virtually all the compounds discussed in this chapter are of importance. There, the oxidized species, water vapor ($H_2O$), and hydroxyl radical (OH) play the key parts and are more important than the others, whose roles, however, are by no means negligible.

It is easy to recognize the role of water vapor in the atmosphere. It is the most abundant of the trace gases, and its turnover in the atmosphere of about $3 \times 20$ g/yr is the largest of all gases. Evaporation, condensation, cloud formation, and precipitation, which form the atmospheric path of the life-supporting hydrological cycle, are experienced in everyday life making its manifold impacts on meteorology and climate obvious to everyone.

Water and water vapor also have several roles in atmospheric chemistry. These vary somewhat with altitude. In the troposphere the following two are the most important: Upon condensation, water vapor forms liquid water, which acts as a solvent, or ice, which acts as an adsorbent of polar molecules. Thus cloud formation scavenges these molecules, an important class of products of the oxidizing gas-phase chemistry, from the atmosphere and deposits them with the precipitation at the ground level. The best known manifestation of that scavenging is the formation of acid rain, which deposits nitric and sulfuric acid, the final reaction products of atmospheric chemistry acting on nitrogen monoxide and sulfur dioxide released into the atmosphere mostly by human activities. But it is an important removal pathway for many other oxidized molecules as well.

Interestingly, former precipitation stored in large ice sheets of Greenland and Antarctica preserves some of its chemical composition and provides an archive — albeit a fragmentary one — of the chemical history of the atmosphere.

The other important chemical role of tropospheric water vapor is that of a primary precursor for the hydroxyl radical, OH. The most important chemical scavenger of the atmosphere, OH is the other key hydrogen species in the troposphere. In contrast to water vapor, it is present in the atmosphere only in minute concentrations, and it has taken a long time before its role in tropospheric chemistry was recognized, and even longer before its presence was demonstrated by measurements.

Despite OH's low concentration, its chemical turnover in the troposphere is one of the largest, about $5 \times 10^{15}$ g/yr, as OH reacts with most of the important natural and anthropogenic pollutants. In virtually all cases, reaction with OH initiates the chain of oxidizing reactions that degrade the primarily emitted molecules, and in most cases that reaction is the rate-determining step. Thus the global OH concentration and its distribution determine the lifetime of most trace gases.

All the other oxidized hydrogen molecules mentioned in this chapter are linked to OH reactions. They, like many of the longer-lived degradation products of OH chemistry, which are partly or fully oxidized, are water soluble and thus subject to scavenging by precipitation. The two hydrogen species OH and $H_2O$ can be viewed as forming two interlocking cycles, namely, a chemical gas-phase mechanism governed by OH, which breaks down the emitted gases and converts them to water-soluble products and a liquid-phase mechanism that eventually removes the soluble species physically from the atmosphere by precipitation. In that sense one could argue that tropospheric chemistry is dominated by the chemistry of hydrogen compounds. It certainly experiences the largest turnover of hydrogen species anywhere in the Earth's atmosphere.

*Dieter H. Ehhalt is the Director of the Institute of Atmospheric Chemistry at the Research Center Jülich, Germany, and Professor of Geophysics at the University of Cologne. His research interests are atmospheric trace gas cycles and the chemical degradation of trace gases in the troposphere by radical reaction.*

# 7 Nitrogen Compounds

## 7.1 Importance of Atmospheric Odd Nitrogen

Chemical interactions of odd nitrogen constituents with other trace species are important in oxidation processes throughout the Earth's atmosphere. Active nitrogen constituents, primarily NO and $NO_2$, are necessary for the photochemical production of $O_3$ in the troposphere. Indeed, in the absence of these odd nitrogen species tropospheric oxidation processes involving methane, other hydrocarbons, or carbon monoxide would not lead to significant $O_3$ production, and photochemical smog episodes as we know them today would be nonexistent. Since the sources of $O_3$ in the troposphere are from photochemical production and transport from the stratosphere, possible trends in tropospheric $O_3$ are closely coupled to the distribution of odd nitrogen and its transport across atmospheric boundaries. The chemistry of odd nitrogen constituents also directly and indirectly affects the concentration of the hydroxyl radical, which, together with $O_3$, constitute the primary tropospheric oxidants responsible for converting many primary emissions to secondary products that are more readily removed from the atmosphere by wet and dry deposition.

In contrast, in the stratosphere, the emphasis of odd nitrogen chemistry has been on its ability to destroy $O_3$ photochemically in so-called catalytic cycles. The combination of these destruction cycles with those involving odd chlorine and bromine, odd oxygen, and odd hydrogen determines the photochemical balance of stratospheric $O_3$. Since observed and very large depletions of $O_3$ in austral spring over Antarctica have been of headline importance, and the current downward trend of stratospheric $O_3$ globally is on record, it is clearly important to understand in detail the chemical interactions of odd nitrogen with other trace species.

This chapter focuses mainly on the atmospheric chemistry of odd nitrogen constituents and especially the dramatic changes that have occurred in our understanding over the past decade or so.

## 7.2 Scope and Definitions

A large number and variety of nitrogen-containing molecular species are involved in the chemistry of the atmosphere. However, three main groups of nitrogen-containing compounds are the most important in the atmospheric environment. These groups are: nitrous oxide ($N_2O$), ammonia and organic amines, and NO and related compounds. There is a link between these different species because they interconvert and interact in chemical and radiative processes that can affect the chemical state of the atmosphere. For example, nitrous oxide is relatively inert in the troposphere, but it is partially linked to the soil emission of the more reactive nitric oxide (see Chapter 5), and $N_2O$ also is the major source of reactive nitrogen in the stratosphere. Ammonia is also a major component of surface nitrogen emissions, but its chemistry in the atmosphere is largely related to acid-base equilibria in aerosols rather than gas-phase photochemistry. There are often complex links between the major nitrogen species, for example, through the effect of $N_2O$ on radiative properties of the atmosphere or through the effect of ammonia on heterogeneous processes and aerosol chemistry. Chapter 5 presents a discussion of the sources and biogeochemistry of the major nitrogen compounds. The scope of this chapter will be limited to the stratosphere and troposphere, with an emphasis on reactive nitrogen and its relation to atmospheric oxidants. Sources of odd nitrogen from the mesosphere and thermosphere will be only briefly considered. Further detail on upper atmospheric chemistry can be found in Brasseur and Solomon (1986). The major reactive nitrogen species in the atmosphere are:

- N: Nitrogen atoms are a major source of NO in the thermosphere through the photolysis of molecular $N_2$. N atoms are insignificant in the lower stratosphere and troposphere.

- $NO + NO_2$: Nitric oxide and nitrogen dioxide are closely tied through photolytic and chemical equilibria in the atmosphere. The primary surface emission is NO, but because of rapid interconversion the two species are often grouped together as $NO_x$ ($= NO + NO_2$). $NO_x$ is also referred to as "active nitrogen."

- $NO_3$: The nitrate radical, formed from the $NO_2 + O_3$ reaction, is readily photolyzed, but may play an important role in the nighttime oxidation of organic compounds.

- $N_2O_5$: Dinitrogen pentoxide is related to the $NO_3$ radical through an equilibrium with $NO_2$. $N_2O_5$ hydrolyzes on aerosol surfaces to form nitric acid.

- $HNO_3$: Nitric acid is one of the major oxidation products of $NO_x$ through reaction with the OH radical, or heterogeneous conversion of $N_2O_5$. The subsequent washout and dry deposition of nitric acid to the surface are the major mechanisms for removal of $NO_x$ from the atmosphere. Nitric acid uptake on particles produces an aerosol reservoir of $NO_3^-$ ion in the form of ammonium or other nitrate salt.

- HONO: Nitrous acid is easily photolyzed and is a potentially important source of OH radicals in atmospheres polluted with high levels of $NO_x$. Organic nitrites ($R$ONO, where $R$ designates an organic moiety) are also formed but are rapidly photolyzed.

- $HO_2NO_2$: Peroxynitric acid is formed from reaction of $HO_2$ with $NO_2$ and is thermally stable only at temperatures characteristic of the upper troposphere and lower stratosphere. Alkyl peroxynitrates, $RO_2NO_2$, are formed from $RO_2 + NO_2$ reaction and have similar thermal properties to peroxynitric acid.

- $RC(O)O_2NO_2$: Peroxyacyl nitrates are formed by the reaction of organic acyl radicals ($R$CO) with oxygen and $NO_2$. The most abundant peroxyacyl nitrate is PAN [peroxyacetyl nitrate: $CH_3C(O)O_2NO_2$], though there are other potentially important peroxyacyl nitrates formed from the oxidation of biogenic hydrocarbons. This class of organic nitrogen compound is important in the transport and recycling of $NO_x$.

- $RONO_2$: Organic nitrates are formed during the oxidation of most hydrocarbons by OH or $NO_3$ radical. The organic fragment $R$ can be a simple alkyl chain or a multifunctional organic group, depending on the parent hydrocarbon.

$X$ONO$_2$: Halogen nitrates, where $X$=Cl or Br, are significant reservoir species of reactive nitrogen in the stratosphere.

NO$_y$: Taken together, all the species from the above list can be considered as "total reactive nitrogen," "total odd nitrogen," or simply as "NO$_y$." N$_2$O and NH$_3$ are not considered to be components of NO$_y$. The concept of total reactive nitrogen is useful in considering the budget of odd nitrogen and in evaluating the partitioning among the individual species.

The main objectives of the chapter are to describe the relationships between odd nitrogen species and to summarize the interactions of odd nitrogen with other trace gases and with radical species. The catalytic role of NO$_x$ in the formation and loss of ozone, and the influence of NO$_x$ in the partitioning of other reactive species is highlighted. The details of the chemistry of odd nitrogen in the stratosphere and in the troposphere are presented in separate sections. As a guide to the text, schematic representations of the main odd-nitrogen interactions in the stratosphere and in the troposphere are presented in Figures 7.1 and 7.2. For the stratosphere, a description of the chemical processes is developed, starting from the relatively simple case of gas-phase photochemistry and progressing through more complex interactions involving heterogeneous processes. Similarly, the discussion of tropospheric odd nitrogen proceeds from the relatively simple case of a low-NO$_x$ atmosphere (representative of much of the troposphere) to the more complex relationships involving a variety of peroxy radicals in the typical high-NO$_x$ atmosphere. Throughout the discussion the effects of both chemistry and transport processes on shaping the measured distributions of odd-nitrogen species throughout the atmosphere will be highlighted.

## 7.3 The Role of Odd Nitrogen in the Stratosphere

The gas-phase chemistry of odd nitrogen species in the stratosphere, when treated in isolation, is relatively simple. However, the overall chemistry that determines the budget of O$_3$ (or any trace constituent) in the stratosphere is a highly interactive system involving the oxygen, nitrogen, hydrogen, halogen, sulfur, and carbon families in a solar insolation field that varies with season, altitude, and wavelength. This marvelous complexity is compounded by the range of characteristic chemical reaction timescales, times that span values where transport is either negligible or very important in determining the spatial distribution of trace constituents.

The coupling of the solid earth-atmosphere system is well illustrated by the relation between N$_2$O and odd nitrogen constituents. As discussed previously, the primary source of active nitrogen to the stratosphere on a global basis is from N$_2$O (Color Plate 17), which is released at the Earth's surface through biological processes and to a smaller extent through anthropogenic activity. Although most of the N$_2$O transported through the planetary boundary layer and the free troposphere to the stratosphere is photolyzed at ultraviolet (UV) wavelengths to N$_2$ and O($^1D$) atoms (7.1), a small fraction reacts with excited oxygen atoms O($^1D$), produced mainly by UV photolysis of O$_3$, to yield NO (7.2)

$$N_2O + h\nu \rightarrow N_2 + O(^1D) \qquad (7.1)$$
$$N_2O + O(^1D) \rightarrow NO + NO \quad (58\%) \qquad (7.2a)$$
$$\rightarrow N_2 + O_2 \quad (42\%) \qquad (7.2b)$$

This production of active nitrogen occurs predominantly in the middle and upper stratosphere. Because NO$_y$ constituents are formed at the expense of N$_2$O, a negative

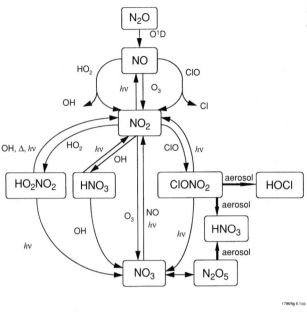

**Figure 7.1.** Schematic diagram of odd-nitrogen interactions in the stratosphere.

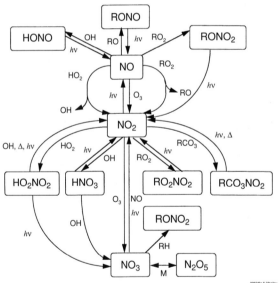

**Figure 7.2.** Schematic diagram of odd-nitrogen interactions in the troposphere.

relation might be expected between these two species. Figure 7.3a is an example of this relationship in the lower stratosphere and also serves as an example of the advances made in measurement capability using the NASA ER-2 aircraft over the past few years. Figure 7.3b shows the average anticorrelation from a number of flights in the unperturbed lower stratosphere. The slope of this regression shows that ~3% of the $N_2O$ molecules arriving in the stratosphere are converted to $NO_y$. Furthermore, the relationship (and the absolute mixing ratios) can be used to ascertain aspects of vertical transport since lower values of $N_2O$ (and higher $NO_y$) observed at a particular altitude would indicate that an air mass descended from higher altitudes. The $N_2O/NO_y$ correlation is one example of compact relationships that have been observed to exist between long-lived stratospheric species that are ultimately lost

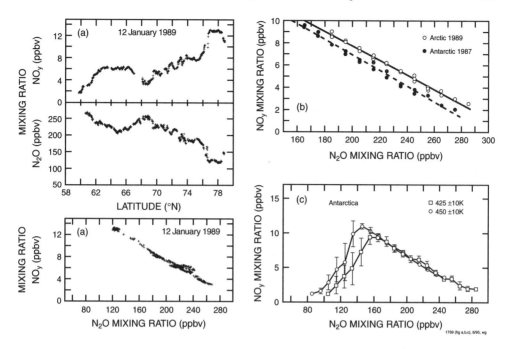

**Figure 7.3.** (a) An example of the strong anticorrelation between total reactive nitrogen ($NO_y$) and $N_2O$ in the lower stratosphere. The data were averaged to 10-sec intervals from a flight of the NASA ER-2 aircraft in the lower Arctic stratosphere. $NO_y$ was measured using a catalytic converter and an $NO$-$O_3$ chemiluminescence detector. $N_2O$ was measured using a tunable diode laser absorption spectrometer. (b) Averaged data that show that in the unperturbed stratosphere the anticorrelation is linear in the Arctic and Antarctic stratosphere. (c) An example of strong deviation from linearity that is attributed to heterogeneous processes leading to disproportionate loss of $NO_y$, i.e., denitrification of the lower Antarctic stratosphere in August and September (from Fahey et al., 1990c; reprinted with permission from Nature, copyright 1990, Macmillan Magazines Limited).

through photochemical processes. Plumb and Ko (1992) have provided the theoretical understanding (see Chapter 2). Basically, these relations can develop when transport and mixing times between the upper and lower stratosphere are short compared to photochemical lifetimes. These relationships, not necessarily linear ones, allow good estimates to be made of the concentration of one constituent from measurements of the other. Deviations from the usual relationships have also been observed and were recognized as extremely useful indicators of the occurrence of abnormal chemical processing. An important example is shown in Figure 7.3c, where for $N_2O$ mixing ratios less than about 160 pptv much lower than expected mixing ratios of $NO_y$ were observed over Antarctica in winter during an "ozone hole" episode. In this case, the deviation marks denitrification of the Antarctic lower stratosphere, a process that is discussed further in Section 7.4.

Another example of these compact relationships is the positive correlation between $NO_y$ and $O_3$ in the lower stratosphere. Figure 7.4 is an example. This close coupling is qualitatively understood by both constituents having their major photochemical source in the 30-40 km region and having a common sink through transport to the troposphere. Thus both $O_3$ and $NO_y$ usually have a strong positive gradient with increasing altitude in the lower and middle stratosphere as shown in Figure 7.5. Combined with each constituent having similar and long photochemical lifetimes in this altitude region, longer than that required for horizontal or vertical

240  Nitrogen Compounds

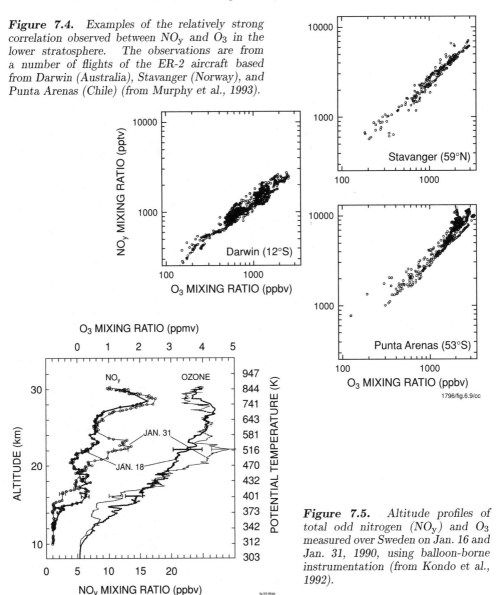

**Figure 7.4.** Examples of the relatively strong correlation observed between $NO_y$ and $O_3$ in the lower stratosphere. The observations are from a number of flights of the ER-2 aircraft based from Darwin (Australia), Stavanger (Norway), and Punta Arenas (Chile) (from Murphy et al., 1993).

**Figure 7.5.** Altitude profiles of total odd nitrogen ($NO_y$) and $O_3$ measured over Sweden on Jan. 16 and Jan. 31, 1990, using balloon-borne instrumentation (from Kondo et al., 1992).

exchange, a positive association is expected at least for what might be considered the normal stratosphere. Again departures from a good correlation are indicative of unusual and fast chemical or dynamical processing.

The preceding figures have given limited information on the magnitude of mixing ratios of $O_3$, $NO_y$, and $N_2O$ in the lower stratosphere below about 20 km altitude. Before discussing some of the detailed chemistry, it is worthwhile to have an idea of the magnitude and vertical distribution of other odd-nitrogen species in the stratosphere. Figure 7.6 shows the partitioning of $NO_y$ into components as derived from the Atmospheric Trace Molecule Spectroscopy (ATMOS) interferometer on board the Space Shuttle. Figure 7.7 shows model calculations of profiles of constituents in units of both mixing ratio and concentration. Photodissociation time constants are also given for some of these species. The partitioning of $NO_y$ depends upon latitude, season, and the time of day. In summer $HNO_3$ is the dominant odd-nitrogen constituent in

Atmospheric Chemistry and Global Change 241

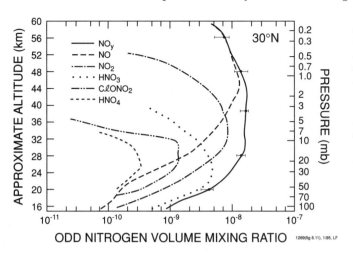

**Figure 7.6.** Vertical distribution of $NO_y$ and its constituents in the stratosphere and lower mesosphere determined using an interferometer (ATMOS) flown on the Space Shuttle. The observations are for May 1, 1985, at sunset (from Russell et al., 1988).

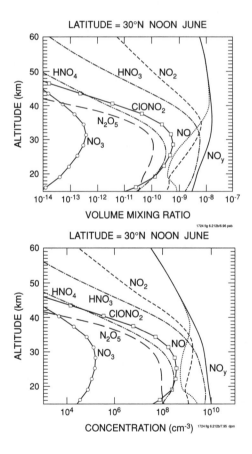

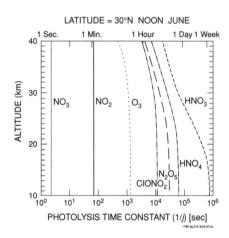

**Figure 7.7.** Top left panel, the vertical distribution between 15 and 60 km of the mixing ratio of odd-nitrogen species calculated at noon for June 30 at $30°N$. Lower left panel, same as top left panel but expressed in terms of concentration ($cm^{-3}$). Top right panel, the time constant associated with the photolysis of several compounds between 10 and 40 km altitude, for noon on June 30 at $30°N$. Background aerosol concentrations were assumed in the calculations (courtesy of X. X. Tie and G. Brasseur).

the lower stratosphere while $NO_x$ is dominant at higher altitudes in part because of the increase in the rate of photolytic conversion of $HNO_3$ to $NO_x$ with increasing altitude. (See Color Plate 16 for a recent satellite-based measurement of stratospheric $HNO_3$ mixing ratios and Plate 18 for a satellite measurement of the $NO_2$ vertical column.) Although $N_2O_5$ and $ClONO_2$ have relatively small mixing ratios in the lower stratosphere, they are key species in our understanding of chemical interactions.

Examples of dramatically different wintertime distributions in polar regions as well as examples of the large spatial-scale distributions achievable by satellite instruments will be given in subsequent sections.

### 7.3.1 "Catalytic" Destruction of Odd Oxygen

NO is readily oxidized to $NO_2$ by atmospheric ozone. Photolysis of $NO_2$ at wavelengths shorter than 420 nm returns NO and a ground-state O atom according to:

$$NO + O_3 \to NO_2 + O_2 \qquad (7.3)$$

$$\underline{NO_2 + h\nu \to NO + O} \qquad (7.4)$$

$$\text{Overall}: \quad O_3 \to O + O_2$$

Reactions (7.3) and (7.4) constitute a null cycle with respect to changing odd oxygen ($O_x = O + O_3$) or with respect to $O_3$ destruction since the O atom formed in (7.4) can recombine with molecular $O_2$ to return $O_3$. (Throughout the stratosphere the concentration of O atoms is small relative to $O_3$; so destruction of $O_x$ is numerically nearly equivalent to destruction of $O_3$.) Note also that NO and $NO_2$ are interconverted in this cycle but that $NO_x$ ($= NO + NO_2$) is conserved.

However, at the lower atmospheric density of the stratosphere, and especially in the middle and upper altitude regions, the concentration of O atoms, produced mainly from photolysis of $O_2$ and $O_3$, becomes high enough that their reaction with species other than $O_2$ becomes important.

This is the basis of the catalytic cycle:

$$NO + O_3 \to NO_2 + O_2 \qquad (7.3)$$

$$\underline{NO_2 + O \to NO + O_2} \qquad (7.5)$$

$$\text{Overall}: \quad O + O_3 \to 2\,O_2$$

Here odd oxygen is "destroyed," but again $NO_x$ is conserved. Many cycles of (7.3) + (7.5) can occur before other processes transform active $NO_x$ into less active components of odd nitrogen, $NO_y$.

Reactions (7.3) through (7.5) also illustrate one of the multitude of reaction competitions that occur in the atmosphere. The relative importance of the null cycle (7.3) + (7.4) compared to the destruction cycle (7.3) + (7.5) clearly depends upon the competition between the rate of $NO_2$ photolysis and the rate of reaction with atomic oxygen, the abundance of the latter being determined in part by the rate of photolysis of precursors.

Reactions (7.3) + (7.5) are but one example of the "classic" cycles of odd oxygen destruction found in the stratosphere. Replacing NO by HO, Cl, or Br yields the catalytic processes involving $HO_x$, $ClO_x$, and $BrO_x$ (see Chapters 6, 8, and 14). These cycles and others determine the overall chemical destruction of odd oxygen in the stratosphere.

There are a number of less important catalytic destruction cycles involving odd nitrogen constituents. One cycle that demonstrates a link with the chlorine family is illustrated by the following set of reactions:

$$ClONO_2 + h\nu \to Cl + NO_3 \qquad (7.6)$$

$$NO_3 + h\nu \to NO + O_2, \quad \lambda < 640 \text{ nm} \qquad (7.7a)$$

$$Cl + O_3 \to ClO + O_2 \qquad (7.8)$$

$$NO + O_3 \rightarrow NO_2 + O_2 \qquad (7.3)$$
$$\underline{ClO + NO_2 + M \rightarrow ClONO_2 + M} \qquad (7.9)$$

Overall: $\quad 2\,O_3 \rightarrow 3\,O_2$

It is usually a minor cycle in comparison to Reactions (7.3)-(7.5) because ~90% of $NO_3$ molecules photolyze to give the products $NO_2 + O$ rather than the products $NO + O_2$ (7.7a), resulting in no loss of odd oxygen

$$NO_3 + h\nu \rightarrow NO_2 + O, \quad \lambda < 580 \text{ nm} \qquad (7.7b)$$

The above cycles emphasize the role of odd nitrogen in the destruction of odd oxygen or ozone in the stratosphere. However, there are also cycles involving $NO_x$ and CO, $CH_4$, and other hydrocarbons that lead to the catalytic production of $O_3$. These cycles are critically important to understanding possible trends in tropospheric $O_3$ (see Section 7.5) and can also be important in the lower stratosphere.

### 7.3.2 Partitioning of Active Nitrogen

Because the catalytic cycles involving $NO_x$ depend on solar radiation, it is clear that NO and $NO_2$ have a strong diurnal variation, with NO being depleted at night. Upon sunrise, $NO_2$ will be photolyzed (with a time constant $1/j_4$ of a few minutes at high solar zenith angle) to NO. Figure 7.8 is an example of one of the earliest determinations of the variation through sunrise at an altitude near 26 km during a high-altitude balloon flight over New Mexico.

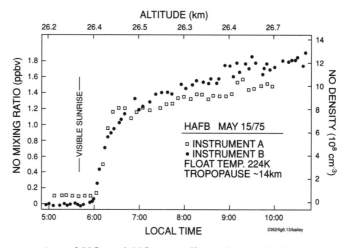

**Figure 7.8.** Nitric oxide mixing ratio during sunrise determined by two balloon-borne instruments near 26 km over New Mexico on May 15, 1975 (from Ridley et al., 1977).

In daytime, the interconversion of NO and $NO_2$ is sufficiently rapid throughout the stratosphere (and troposphere) that photochemical steady-state assumptions give an accurate expression for the $NO_2/NO$ ratio. If odd nitrogen and oxygen reactions in (7.3)-(7.5) are treated in isolation, then equating $d[NO_2 \text{ or } NO]/dt$ to zero gives

$$[NO_2]/[NO] = k_3[O_3]/(j_4 + k_5[O]) \approx k_3[O_3]/j_4 \qquad (7.10)$$

where the simplest expression on the right-hand side is quite accurate below about 30 km altitude. At higher altitudes the O-atom concentration strongly increases and the $k_5[O]$ term cannot be neglected. The initial rapid growth in NO shown in Figure 7.8 results primarily from the increase in the photolysis of $NO_2$ ($j_4$) since it depends sensitively on the solar zenith angle near sunrise. Similar but reversed behavior for NO occurs through sunset.

The importance of the relative partitioning of $NO_x$ in daytime to chemical odd-oxygen destruction is found by examining (7.3)-(7.5). Substituting the full expression (7.10) into $d[O_x]/dt$ yields

$$\begin{aligned}-d[O_x]/dt &= -d[O_3 + O]/dt \\ &= k_3[NO][O_3] + k_5[NO_2][O] - j_4[NO_2] \\ &= 2k_5[NO_2][O]\end{aligned} \quad (7.11)$$

The destruction of odd oxygen by odd nitrogen is dominated by twice the rate of (7.5), or (7.5) is said to be the rate-limiting process. Thus, for a given O atom concentration, other reactions that tend to increase the daytime $NO_2/NO$ ratio also indirectly enhance the rate of destruction of odd oxygen by active nitrogen.

Two other important examples that demonstrate the strong chemical interaction between different families in the atmosphere are:

$$NO + HO_2 \to NO_2 + OH \quad (7.12)$$
$$NO + ClO \to NO_2 + Cl \quad (7.13)$$

Including these reactions in the steady-state expression yields

$$[NO_2]/[NO] = (k_3[O_3] + k_{12}[HO_2] + k_{13}[ClO]) / (j_4 + k_5[O]) \quad (7.14)$$

Because ClO and $HO_2$ are the analogs of $NO_2$ in their "classic" odd-oxygen destruction cycles, the occurrence of (7.12) or (7.13) enhances the importance of the $NO_x$ destruction cycle but diminishes that of the $ClO_x$ and $HO_x$ catalytic cycles (O-atom attack on ClO and $HO_2$ is also the rate-limiting reaction in their destruction cycles). In other words, just as ClO and $HO_2$ affect the partitioning of $NO_x$, $NO_x$ affects the partitioning of $ClO_x$ between Cl and ClO and the partitioning of $HO_x$ between HO and $HO_2$. For example, processes that increase the concentration of NO increase the abundance of OH and Cl or decrease the $HO_2/OH$ and $ClO/Cl$ ratios.

### 7.3.3 Nighttime Chemistry

$NO_2$ can be oxidized by $O_3$ to form $NO_3$, a strong atmospheric oxidant and a precursor to the formation of dinitrogen pentoxide, $N_2O_5$. Because $NO_3$ is rapidly photolyzed at visible wavelengths, both its daytime concentration and chemistry are of relatively minor importance in the lower and middle stratosphere. In contrast, $N_2O_5$ can have morning concentrations in the lower stratosphere comparable to $NO_x$, even though its production requires the formation of $NO_3$

$$NO_2 + O_3 \to NO_3 + O_2 \quad (7.15)$$
$$NO_3 + h\nu \to NO + O_2 \quad (7.7a)$$
$$\to NO_2 + O \quad (7.7b)$$
$$NO_3 + NO_2 + M \rightleftharpoons N_2O_5 + M \quad (7.16)$$

The explanation results primarily from the difference in photolysis rates of $N_2O_5$ and $NO_3$. During the night $N_2O_5$ builds at the expense of $NO_x$, or more precisely at the expense of $NO_2$. For example, if $\tau_N$ is the length of the night, then the ratio of $NO_2$ present just before sunrise to that present just after sunset is $\exp(-2k_{15}[O_3]\tau_N)$, since (7.15) is the rate-limiting reaction.

After sunrise, $N_2O_5$ can be photolyzed to essentially two $NO_x$ molecules

$$N_2O_5 + h\nu \to NO_3 + NO_2 \quad (7.17)$$

but the photolysis time constant is dependent on both the usual factors (zenith angle, altitude, albedo, etc.) and the ambient temperature. In summer typical photolysis times range from 7-24 h in the 20-30 km region. Thus at sunrise, the remaining

NO$_2$ and the N$_2$O$_5$ formed are photolyzed, the former rapidly to give the steady-state NO$_2$/NO ratio of Eq. (7.14) and the latter more slowly. This slower release of NO$_x$ from N$_2$O$_5$ formed at night is partially responsible for the continued growth of NO through the morning period shown in Figure 7.8. Figure 7.9 shows the expected diurnal cycle for these and other odd-nitrogen constituents for two different altitudes in the stratosphere.

Thus the steady-state concentration of N$_2$O$_5$ will depend on the balance between the extent of nighttime formation and daytime destruction. But at high latitudes in winter the photolysis time constant can become much longer than typical transport times and steady state may never be achieved. This NO$_3$, NO$_x$, N$_2$O$_5$ chemistry describes the primary gas-phase cycle of N$_2$O$_5$ in the stratosphere. Additional, more significant, reactions of N$_2$O$_5$ will be discussed in Section 7.4.

This nighttime chemistry can be extended to introduce the first example of the peculiarities of northern latitude winter stratospheric chemistry, which is controlled to a large extent by dynamics. The column abundance of NO$_2$ in the stratosphere can be measured by spectroscopic techniques from the ground or from aircraft flying in the troposphere. Figure 7.10 shows observations of the column abundance of NO$_2$ as a function of latitude over the continental United States and Canada during the winter of 1977, well before the discovery of the "ozone hole" in Antarctica. They were made by J. Noxon, who pioneered the use of a ground-based technique for determining the latitudinal and seasonal behavior of the column abundance of NO$_2$. The striking feature of these observations is the

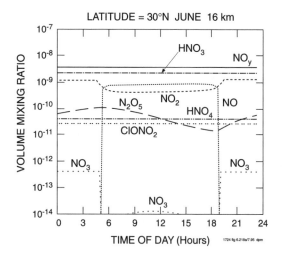

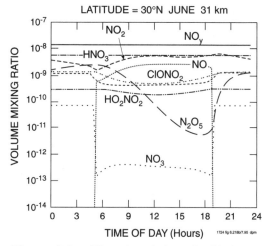

**Figure 7.9.** Diurnal variation of odd-nitrogen species in the stratosphere at (a) 16 km and (b) 31 km altitude calculated at 30°N for June conditions assuming background aerosol conditions (courtesy of X. X. Tie and G. Brasseur).

sharp decline in the column amount of NO$_2$ north of ∼50° latitude, a feature that became known as the Noxon "cliff," but a feature that was not always present. In situ measurements of NO by balloon-borne instrumentation at 54°N in conjunction with the Noxon 1977 winter survey revealed that daytime NO [and NO$_2$ by inference from Eq. (7.14)] was strongly reduced relative to summer mixing ratio profiles in the lower and middle stratosphere. Figure 7.11 gives an example from a balloon flight made in 1982 at 51°N, where both NO and NO$_2$ were directly measured. At these latitudes in

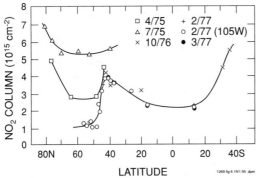

**Figure 7.10.** *The latitude dependence of the overhead column abundance of $NO_2$. In the winter of 1977 especially a very sharp drop in the abundance was observed north of $\sim 50°$, a feature that is now called the "Noxon cliff" (from Noxon, 1979).*

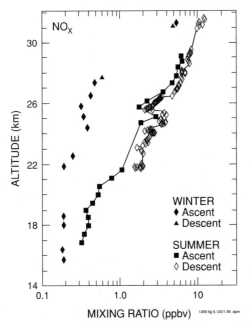

**Figure 7.11.** *The contrast in the vertical distribution of $NO_x$ in winter and summer. The balloon-borne instrument was launched from Gimli, Manitoba ($51°N$), in August, and again in December 1982. A strong reduction of $NO_x$ occurred in the 20-28 km altitude region (from Ridley et al., 1987).*

winter $NO_x$ is reduced by up to a factor of $\sim 10$ relative to summer observations in the 20-27 km altitude region.

In winter, the polar jet at latitudes of $50°$-$60°N$ is often intense and can effectively isolate air masses to the north from those to the south in what is called the polar vortex. Air masses to the north have no or very small insolation, and $NO_x$ can be nearly completely converted to $NO_3$ and $N_2O_5$ if this isolation lasts for several days or more and if the flow is nearly zonal. However, planetary wave structure in the vortex often exists, and air masses may be carried to lower latitudes and increased insolation in the asymmetric flow. Thus the $NO_x$ or $N_2O_5$ abundance should depend sensitively on the time spent in near dark versus low light conditions, that is, on the details of the transport. At that time some groups believed these ideas were sufficient to explain the formation of the "Noxon cliff." In contrast, Evans et al. (1985) proposed that the reduction in $NO_x$ was caused by conversion to $N_2O_5$ and subsequent transformation to $HNO_3$ in gas/aerosol interactions. In the former case, $NO_x$ could be returned relatively quickly if the polar air mass moved to regions of higher insolation, whereas in the latter case the return would be much slower due to the substantially longer photolysis time constant for $HNO_3$ compared to $N_2O_5$. We shall see that chemical reactions occurring on stratospheric aerosols are extremely important. Indeed, such processing has revolutionized our understanding of stratospheric chemistry.

### 7.3.4 Transformations of $NO_x$ to Other Reactive and Reservoir Constituents of $NO_y$

The formation of $N_2O_5$ at night or in polar winter is one example of production of a longer-lived odd-nitrogen reservoir. Other important reservoir constituents are nitric acid ($HNO_3$), chlorine nitrate ($ClONO_2$) and the bromine analog ($BrONO_2$), peroxynitric acid ($HO_2NO_2$), and aerosol nitrate ($NO_3^-$). The homogeneous formation of the reservoir species requires termolecular chain termination reactions (7.16) and (7.18)-(7.20).

$$NO_2 + OH + M \rightarrow HNO_3 + M \qquad (7.18)$$
$$NO_2 + HO_2 + M \rightarrow HO_2NO_2 + M \qquad (7.19)$$
$$NO_2 + ClO + M \rightarrow ClONO_2 + M \qquad (7.20)$$

The occurrence of these reactions not only curtails $O_3$ destruction through temporary removal of active $NO_x$, but they also sequester $HO_x$ and $ClO_x$ radicals and thus decrease the contribution of their catalytic cycles to $O_3$ destruction. Alternatively, any other stratospheric process that diminishes $NO_x$ directly or indirectly will enhance the abundance of active $HO_x$ and $ClO_x$ by decreasing the rates of (7.18)-(7.20) and therefore enhance their ozone destruction efficiency. Section 7.4 will discuss newly discovered stratospheric chemical processes that indeed lead to a reduction in the $NO_x/NO_y$ ratio or in some regions nearly complete removal of $NO_x$ from the lower stratosphere.

However, all these reservoirs can return active nitrogen and radicals through homogeneous processes including photolysis, reaction with radical or atomic species, or by unimolecular decomposition

$$HNO_3 + h\nu \rightarrow OH + NO_2, \quad < 320 \text{ nm} \qquad (7.21)$$
$$HNO_3 + OH \rightarrow NO_3 + H_2O \qquad (7.22)$$

$$ClONO_2 + h\nu \rightarrow Cl + NO_3, \quad < 400 \text{ nm} \qquad (7.6)$$
$$\rightarrow ClO + NO_2 \qquad (7.23)$$

$$HO_2NO_2 + h\nu \rightarrow HO_2 + NO_2, \quad < 330 \text{ nm} \qquad (7.24)$$
$$\rightarrow OH + NO_3 \qquad (7.25)$$
$$HO_2NO_2 + OH \rightarrow H_2O + O_2 + NO_2 \qquad (7.26)$$
$$HO_2NO_2 + M \rightarrow HO_2 + NO_2 + M \qquad (7.27)$$

Photolysis and/or reaction with OH are often the major processes, but numerous factors determine the overall lifetime of these reservoirs. Because UV wavelengths are required for photolysis, the lifetime against photodissociation alone is a strong function of altitude, latitude, and season (cf. Fig. 7.7c). Depending upon location in the stratosphere, some of the reservoirs exhibit a significant diurnal variation, as shown in Figure 7.9. The lifetime is also sensitive to the distribution of OH, which can be strongly variable in space and time. Local temperature also affects the lifetime of some reservoirs either through the temperature dependence of Reactions (7.16) and (7.27) or through the temperature dependence of photolysis rates. Finally, and as will be discussed in the next section, heterogeneous processes involving chemical reactions of reservoir species on stratospheric aerosols contribute to determining the overall lifetime of these species. Indeed, the major changes in our understanding of the stratosphere that have occurred over the past 10 years have centered on pathways for either the production and transformations of reservoir species or their heterogeneous (gas-aerosol) chemistry. Although many factors are involved in determining the overall lifetime of these species, in the lower stratosphere they are often considerably longer than characteristic transport times. Thus the distribution of many of the reservoir species is controlled largely by dynamics and transport, including the sedimentation of aerosols from the stratosphere, rather than by local photochemical processes.

## 7.4 Odd Nitrogen in the "Contemporary" Stratosphere:
The "Ozone Hole," Polar Stratospheric Clouds, Sulfuric Acid Aerosols, and Heterogeneous Chemistry

The formation of the Antarctic "ozone hole" is an excellent example of the inseparability of atmospheric chemistry, dynamics, and radiation. Chapter 14 gives a detailed description of the evolution of the winter and spring polar atmospheres and the overall chemistry and dynamics involved. Since most of the ozone depletion is observed in the 12-23 km altitude region, this section focuses on the observations and influence of odd-nitrogen species in this region of the polar stratosphere. The interaction of odd-nitrogen species with liquid and frozen sulfuric acid aerosol, also called sulfate aerosol, is then discussed. These aerosols are pervasive throughout the lower and middle stratosphere and comprise what is known as the Junge layer (see Chapter 4). Sulfuric acid aerosol concentrations can be greatly enhanced in the stratosphere by some volcanic eruptions.

### 7.4.1 Odd Nitrogen and the "Ozone Hole"

Observations combined with modeling and laboratory studies have shown convincingly that the ozone depletion over Antarctica in spring is primarily due to chlorine chemistry, with a lesser but significant role for bromine chemistry (see Chapters 8 and 14). Anthropogenic activity has been responsible for the increase in the chlorine and bromine loading of the stratosphere over the past decades. The depletion is triggered by heterogeneous chemical reactions that occur on the surface of stratospheric aerosols and transform reservoir halogen species into active species which can catalytically destroy ozone with the return of solar insolation in spring.

Of the five heterogeneous reactions identified to be important on liquid or solid PSCs, four involve the odd nitrogen reservoirs $N_2O_5$ and $ClONO_2$ (7.28)-(7.31). Variants of (7.30) were considered for some time to be important in the polluted troposphere and it is also similar to the mechanism proposed in the mid 1980s to be involved in the formation of the "Noxon cliff"

$$ClONO_{2(g)} + HCl_{(s)} \rightarrow Cl_{2(g)} + HNO_{3(s)} \tag{7.28}$$

$$ClONO_{2(g)} + H_2O_{(s)} \rightarrow HOCl_{(g)} + HNO_{3(s)} \tag{7.29}$$

$$N_2O_{5(g)} + H_2O_{(s)} \rightarrow 2\ HNO_{3(s)} \tag{7.30}$$

$$N_2O_{5(g)} + HCl_{(s)} \rightarrow ClNO_{2(g)} + HNO_{3(s)} \tag{7.31}$$

Subscripts $(g)$ and $(s)$ are used to differentiate the gas-phase and solid PSC particle, respectively. Three of these reactions (7.28, 7.29, and 7.31) activate chlorine species directly since $Cl_2$, $HOCl$, and $ClNO_2$ are released to the gas phase and on the reappearance of solar insolation in spring are readily photolyzed to form active $ClO_x$. All four reactions lead to removal or reduction of $NO_x$ in polar winter air masses through conversion to $ClONO_2$ or $N_2O_5$ and ultimately to $HNO_3$, since on PSCs the $HNO_3$ product is not released to the gas phase, but is sequestered within the particle. It is this heterogeneous processing that explains the formation of the "Noxon cliff" shown previously by Figures 7.10 and 7.11.

The reduction or removal of $NO_x$ is a critical requirement of polar ozone depletion and represents indirect chlorine activation. For example, if significant concentrations of $NO_2$ remained in the vortex, the chain termination reaction

$$ClO + NO_2 + M \rightarrow ClONO_2 + M \qquad (7.20)$$

would be sufficiently rapid to keep active $ClO_x$ that is ultimately formed in Reactions (7.28)-(7.31) in spring tied up as reservoir $ClONO_2$. This formation of $ClONO_2$ would severely limit chlorine-catalyzed $O_3$ destruction. The low to zero solar insolation of winter and early spring at high latitudes would also reinforce the stability of this reservoir by limiting photolysis of $ClONO_2$ (Reactions 7.6 and 7.23).

The rates of Reactions (7.28)-(7.31) depend upon the specific surface area ($\mu m^2$ $cm^{-3}$) of the liquid or solid PSC particle, which is related to their size and concentration (see Chapter 4). However, the detailed kinetic parameters for the different types of stratospheric aerosols are very complex relative to those for homogeneous or gas-phase processes. For details the reader is referred to DeMore et al. (1997), WMO (1995), Solomon (1990), and Hanson et al. (1994). Briefly, the rates can depend upon the composition, phase, temperature, and, as well, the relative humidity of the air mass. The PSC composition may also depend upon the temperature history of the air mass, especially in the northern polar region. Reaction (7.28) occurs efficiently on both types of PSCs. Reactions (7.29) and (7.30) occur readily on Type-II PSCs, but less efficiently on Type-I PSCs. Kinetic parameters for Reaction (7.31) are not as well known.

Observations of odd nitrogen and other species made inside and outside the polar vortices both guided and confirmed the unique role of aerosol chemistry in determining ozone abundances. Figures 7.12 through 7.14 show observations made from NASA ER-2 aircraft flights in the lower stratosphere of the Antarctic or Arctic. Figure 7.12 shows that NO declined sharply as the aircraft crossed from outside to inside the Arctic vortex. By inference from Eq. (7.14), $NO_2$ and consequently $NO_x$ was also strongly reduced in the interior of the vortex. The low mixing ratios of NO and $NO_2$ are reminiscent of those of Figures 7.10 and 7.11 found in winter at high latitude in the Northern Hemisphere during investigations of the "Noxon cliff." However, the striking feature shown in Figure 7.13 is that the reduction in $NO_x$ is accompanied by a sharp reduction in total odd nitrogen, $NO_y$, in the

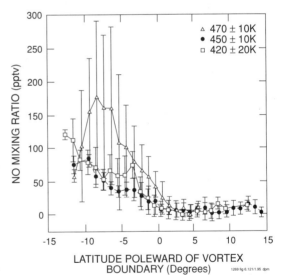

**Figure 7.12.** The decline in NO across the Arctic polar vortex as determined with instrumentation on board the NASA ER-2 aircraft. The data from five flights have been averaged over $1°$ latitude intervals referenced to the boundary of the polar vortex. Potential temperatures of 470, 450, and 420 K correspond roughly to geometric altitudes of 21, 19, and 17 km, respectively (from Fahey et al., 1990a).

interior of the vortex. Recalling Section 7.3.2, a strong anticorrelation between $N_2O$ and $NO_y$ is expected in the "normal" stratosphere. The solid line in the figure is the abundance of $NO_y$ calculated by means of the regression fit of Figure 7.3b from the simultaneous measurements of $N_2O$. The shaded region thus represents the difference between expected and measured gaseous $NO_y$. As shown by the figure,

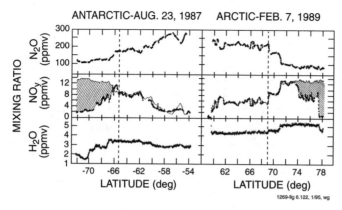

**Figure 7.13.** The changes in $N_2O$, $NO_y$, and $H_2O$ observed on flights of the ER-2 aircraft across the polar vortex boundary (indicated by the vertical dashed line) in the Antarctic and Arctic. The shaded region marks the difference between the observed $NO_y$ and that expected on the basis of the $N_2O$ mixing ratios and represents denitrification in air masses poleward of the vortex boundary (from Fahey et al., 1990b; reprinted with permission from Nature, copyright 1990, Macmillan Magazines Limited).

gaseous reservoir and active nitrogen constituents can be severely depleted in air masses interior to the polar vortex due to irreversible sequestering of $HNO_3$ in the stratospheric aerosols as expected from Reactions (7.28)-(7.31). Denitrification of the vortex region occurs when the aerosols become large enough to sediment appreciably, indeed to sediment from the lower stratosphere to the troposphere. As shown very nicely by the springtime observations of Figure 7.14, the odd-nitrogen removal or reduction allows ClO, the active constituent of ozone destruction, to build within the Antarctic vortex to ~1.3 ppbv or 50-100 times larger than observed at midlatitudes or in the stratosphere unperturbed by PSC formation. Satellite observations of ClO have confirmed these large increases in ClO in spring (Color Plate 21), but over much larger spatial scales than those shown by the in situ observations of Figure 7.14 (Waters et al., 1993).

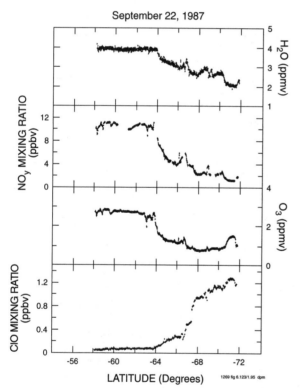

**Figure 7.14.** The variation of $O_3$, $H_2O$, ClO, and $NO_y$ as the ER-2 aircraft flew across the Antarctic vortex boundary in September of 1987. Within the vortex, at latitudes south of ~-64°S, the region is perturbed chemically, and a sharp increase in ClO, the active constituent responsible for the observed $O_3$ reduction, is observed (based on data from Fahey et al., 1989).

Instruments on board satellites have extended the in situ aircraft observations of odd nitrogen to much larger spatial scales and have provided remarkable temporal information for both the northern and southern polar regions. Color Plate 19 shows results for nitric acid (the major $NO_y$ constituent in the lower stratosphere) from the

Microwave Limb Sounder (MLS) instrument onboard the Upper Atmosphere Research Satellite (Santee et al., 1995). Equally impressive measurements of $ClONO_2$ have been reported by Roche et al. (1994). The presentation of Plate 19 illustrates that the Antarctic vortex is much more zonal, obtains (and maintains) the low temperatures required for PSC formation in late fall, and as a consequence nitric acid is sequestered or removed (denitrification) over much larger spatial scales than in the Arctic. Indeed, only a small area of the northern polar region (Feb. 22 panel) shows significant reduction of nitric acid in the area enclosed by air temperatures low enough for PSC formation. The Nov. 1 panel for the Southern Hemisphere shows that the denitrification, and therefore the "ozone hole," extends to late in the austral spring. With denitrification, the $NO_x$ necessary for sequestering ClO as reservoir $ClONO_2$ cannot be regained by springtime photolysis of $HNO_3$. Instead, recovery requires breakup of the Antarctic vortex (mixing with midlatitude air masses) to replenish odd nitrogen over the polar region.

### 7.4.2 Sulfate Aerosol and Volcanic Emissions

Liquid sulfuric acid aerosol (or sulfate aerosol) is always present throughout the lower stratosphere due to oxidation of biogenic and anthropogenic sulfur species emitted from the Earth's surface (see Chapter 4). With the eruption of the Mount Pinatubo volcano in the Philippines in June 1991, 15-30 MT of $SO_2$ were injected into the stratosphere. Rapid oxidation to liquid sulfuric acid aerosol increased the loading in much of the stratosphere by factors of 50-100 compared to typical background values characteristic of the Junge layer (see Color Plates 10 and 12). Large enhancements reached polar latitudes in the Southern Hemisphere in the winter of 1992 and in the Northern Hemisphere in the winter of 1992/1993. Fortuitously, this natural event occurred during a period of intense observational research in the stratosphere, and so led to new and important discoveries. With the focus on chemical processing by PSCs in winter polar regions it was natural to return to earlier proposals (Hofmann and Solomon, 1989; Solomon et al., 1986a) that increases in sulfate aerosol from some volcanoes could also perturb ozone concentrations. Indeed, satellite and ground-based observations had revealed a small downward trend in stratospheric ozone at midlatitudes in addition to the ozone depletion in spring polar regions. Studies have now shown that heterogeneous chemical reactions involving odd nitrogen and other reservoirs on sulfuric acid aerosol can also lead to enhanced rates of ozone loss in the lower stratosphere. However, a fundamental difference relative to polar ozone loss is that this destruction can occur near globally and over the lifetime of the increased sulfate loading in the stratosphere (typically 1-3 years), not just at high latitudes in spring. Furthermore, when the volcanic sulfate enhancements reached high latitudes in 1992/1993 they not only provided new reactive surfaces, but also promoted PSC formation to augment ozone destruction in the polar vortices.

The heterogeneous reactions on sulfuric acid aerosol involving odd-nitrogen reservoirs are analogous to those occurring on PSCs. Those requiring HCl are not rewritten, but the two hydrolysis reactions (7.32) and (7.33) will illustrate the basic differences from PSC reactions

$$N_2O_{5(g)} + H_2O_{(sa)} \to 2\ HNO_{3(g)} \qquad (7.32)$$
$$ClONO_{2(g)} + H_2O_{(sa)} \to HOCl_{(g)} + HNO_{3(g)} \qquad (7.33)$$

The subscript (sa) denotes the liquid or solid sulfuric acid aerosol. At temperatures below about 205 K the aerosol takes up $HNO_3$ to form a liquid ternary solution that, with further cooling, can become a supercooled solution. Below about 195 K PSC formation may result. In the absence of significant $HNO_3$ further cooling can also form solid sulfuric acid tetrahydrate (SAT). Again the reaction rates depend on the specific surface area and can be complicated functions of composition, phase, water content, and temperature. The reader is referred to earlier references given for the discussion of PSC reactions and in Tolbert (1994). Reaction (7.32) is relatively insensitive to temperature and is efficient on the liquid phase, but much less so on the solid phase. The rate of Reaction (7.33) depends strongly on the water fraction in the aerosol or, for stratospheric conditions, the ambient temperature. An important difference between Reactions (7.32), (7.33) and the analogs of the PSC reactions involving HCl compared to the PSC reactions (7.28)-(7.31) is that product $HNO_3$ is not retained within the aerosol, but is released to the gas phase, at least for liquid sulfuric acid aerosol. It is not currently known whether this is true for frozen sulfate aerosol.

The gross effect of the above reactions is similar to that from the reactions that occur on PSCs. Chlorine is activated directly by conversion of $ClONO_2$ to more readily photolyzed constituents and indirectly by the reduction of $NO_x$ levels by the formation of $HNO_3$. In this case, however, the distinguishing difference is that $HNO_3$ released in gaseous form is subject to photolysis. Thus, particularly at low and midlatitudes, and in polar regions depending upon season, $NO_x$ and $HO_x$ can be returned to the air mass. The net effect of (7.32), (7.33), and the sulfate analogs of (7.28) and (7.31) is to decrease the $NO_x/NO_y$ ratio, which, as discussed earlier, decreases the destruction of ozone by the $NO_x$ catalytic cycle, but enhances the destruction by the $ClO_x$ and $HO_x$ cycles.

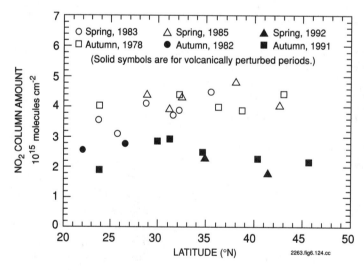

**Figure 7.15.** *The decline in the column abundance of $NO_2$ attributed to volcanic aerosol depletion of $NO_x$. The solid symbols represent observations made after the eruption of El Chichón in March and April of 1982 and after the eruption of Mt. Pinatubo in June, 1991 (from Coffey and Mankin, 1993).*

Observations confirm the importance of sulfate aerosols at all latitudes. Figure 7.15 shows that the column abundance of $NO_2$ was substantially reduced after the Pinatubo eruption compared to years when the aerosol loading was more representative of background conditions. This is indirect support for the importance of (7.32) and (7.33) and the HCl analogs of (7.28).

Since heterogeneous reactions affect the partitioning of $NO_y$ between reservoir and active constituents, their occurrence can also be inferred from observations of the $NO_x/NO_y$ ratio. Figure 7.16 shows the $NO_x/NO_y$ ratio determined from a number of flights of the ER-2 aircraft as a function of the specific surface area of aerosols. The observations are compared to model calculations that include or exclude heterogeneous processes on sulfate aerosols. Although the agreement is not perfect,

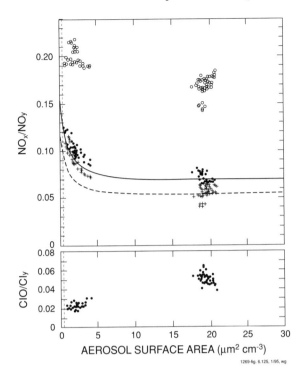

**Figure 7.16.** The $NO_x/NO_y$ and $ClO/Cl_y$ ratios versus the aerosol surface area observed in the lower stratosphere. The observations are the solid circles. The open circles and crosses are the expected ratio from model calculations appropriate to individual aircraft flights without and with the heterogeneous reactions (7.32) and (7.33). The vertical dashed line indicates background aerosol surface area. The solid and dashed curves denote model calculations for averaged flight conditions for September and March, respectively. The anticorrelation between the $NO_x/NO_y$ and $ClO/Cl_y$ ratios is direct evidence for the importance of the heterogeneous processes on sulfate aerosol. Because the formation of $N_2O_5$ is the limiting rate process, the $NO_x/NO_y$ ratio "saturates" at a constant ratio for surface area larger than $\sim 5$ $\mu m^2$ $cm^{-3}$ (from Fahey et al., 1993; reprinted with permission from Nature, copyright 1993, Macmillan Magazines Limited).

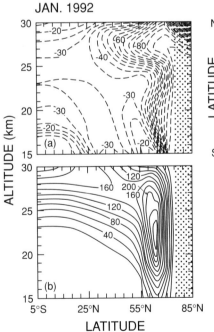

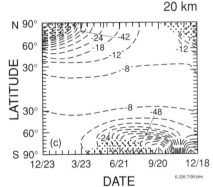

**Figure 7.17.** Perturbations in the stratosphere due to heterogeneous reactions on the surface of sulfate particles after the eruption of Mt. Pinatubo determined using a two-dimensional chemical/transport model. (a) The change in $NO_2$ in percent relative to background aerosol conditions versus altitude and latitude for January 1992, (b) the same as (a) but for ClO in percent, (c) the change in $O_3$ at 20 km versus latitude and season (reprinted with permission from Brasseur, G. and C. Granier (1992) Mount Pinatubo aerosols, chlorofluorocarbons, and ozone depletion, Science 257, 1239, ©1992, American Association for the Advancement of Science).

the observations are in much better agreement with the calculations that include aerosol reactions. For the stratospheric conditions of the flights of Figure 7.16, Reaction (7.32) was most important. Most important, there is not a runaway depletion of $NO_x$ at very large specific surface area that would allow extensive $O_3$ destruction during volcanic enhancements of sulfate because the $ClONO_2$ reservoir would not be formed. Instead the ratio approaches a constant value dictated by the rate of formation of $N_2O_5$ (mostly at night), which as discussed earlier is limited by the rate of (7.15). Thus formation of $ClONO_2$ by remaining $NO_2$ limits the destruction of $O_3$ by $ClO_x$. Since the stratospheric loading of chlorine (and bromine) species is expected to increase until the limitations on halocarbon release agreed to in the Montreal Protocol result in a significant decrease, future effects of the enhanced role of $ClO_x$ in $O_3$ depletion at midlatitudes remain uncertain, since volcanic eruptions are sporadic. Figure 7.17 shows the predicted change in $NO_2$, $ClO$, and $O_3$ due to Mount Pinatubo aerosols from a two-dimensional chemistry transport model. The reference scenario included typical background aerosol surface area. The relative change in $NO_2$ and $ClO$ is large due to the heterogeneous processing. The change in $O_3$ near 20 km is less but significant at midlatitudes. Larger reductions are seen in the polar regions in winter, mostly due to the increased rate of (7.33) at lower temperature.

## 7.5 Odd Nitrogen in the Troposphere

Based on global budget estimates, anthropogenic activity is responsible for a major fraction of the release of odd nitrogen, mostly in the form of NO, to the troposphere (see Chapter 5). Often these emissions are concentrated in urban/industrial centers, and we are all at least casually familiar with the link between these emissions and others (*e.g.*, nonmethane hydrocarbons) that can lead to photochemical smog (see Chapter 13). Indeed, some of the basic chemical interactions responsible were first proposed by Haagen-Smit in 1952 to account for Los Angeles' increasingly poor air quality. It is then perhaps surprising, after more than 40 years have passed, that programs focused on these issues still consume large fractions of research budgets and generate considerable scientific and political debate over what is the best control strategy weighed against economic and social impacts. In part, this reflects the complexity of the processes.

Odd-nitrogen chemistry in the troposphere is not limited to this "pollutant" role. On regional and global scales, interactions with the odd-hydrogen family, which are similar or identical to those discussed in the stratosphere, can have a strong influence on the concentration of OH. OH is the primary oxidant in the troposphere and is responsible for much of the transformation of primary emissions to secondary species that are more susceptible to atmospheric removal by wet and dry deposition processes.

Active nitrogen constituents play fundamentally different roles in the stratosphere and troposphere. Because the O atom concentration is negligibly small in the troposphere, the reaction of O atoms with $NO_2$ (7.5) cannot compete with $NO_2$ photolysis (7.4), and catalytic destruction of $O_3$ by $NO_x$ is negligible. Thus catalytic cycles of $O_3$ or odd-oxygen destruction are most important in the stratosphere, while catalytic production is most important in the troposphere. Without $NO_x$ there would be no, or at least much less, photochemical production of $O_3$ in the troposphere. Although $O_3$ is usually considered a pollutant due to its toxicity at elevated concentrations, without $O_3$ the concentration of OH, the major atmospheric cleanser, would be substantially reduced, and the tropospheric composition as we know it today would be very different due to the growth of emissions commensurate with population over the past 100 years. On the regional scales of heavily populated urban

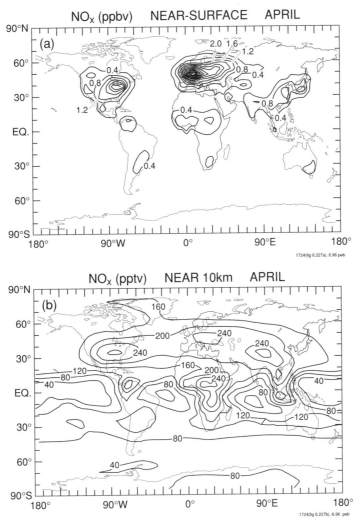

**Figure 7.18.** Distribution of the $NO_x$ ($NO + NO_2$) mixing ratio calculated by the NCAR IMAGES model (a) near the surface (in ppbv) and (b) in the upper troposphere near 10 km (in pptv) during the month of April (courtesy of J.-F. Müller and G. Brasseur).

and industrial centers, curbing $O_3$ episodes and photochemical smog in general is necessary. Whether a global increase in tropospheric $O_3$ of the order of 10 ppbv is bad or good, at a time when stratospheric $O_3$ is decreasing and some primary emissions are increasing, is controversial.

### 7.5.1 Observations of $NO_y$ Constituents in the Troposphere

Considering the variety of sources of active nitrogen to the atmosphere, it is clear that both near-surface and higher-altitude sources of active nitrogen are geographically nonuniform. Factors affecting the lifetime of $NO_x$ will be discussed in detail later, but in summer it is of the order of a few hours to a day in the lower troposphere (surface to 1-2 km) and perhaps 4-7 days in the upper troposphere. Thus in the planetary boundary layer the distribution of $NO_x$ strongly reflects the distribution of near-surface sources while the distribution in the middle and upper troposphere is more strongly affected by transport processes. For example, Figures 7.18a,b give the global

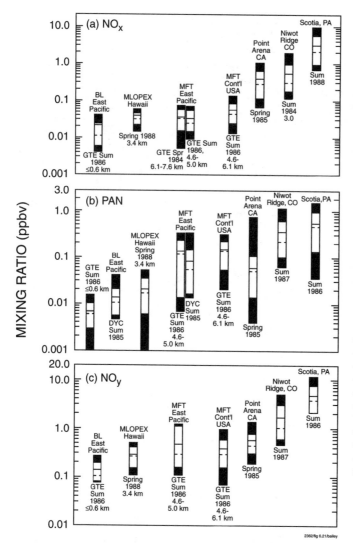

**Figure 7.19.** $NO_x$, PAN, and $NO_y$ mixing ratios obtained at three continental surface sites (Scotia, Penn., Niwot Ridge, Colo., Point Arena, Calif.) compared to aircraft observations made in the middle free troposphere and marine boundary layer and measurements made at 3.4 km on the Island of Hawaii. The outer rectangles and unshaded rectangles represent the central 90% and 67% of the observations. The bar is the average, and the dashed bar is the median. The continental surface site measurements were made by the NOAA Aeronomy Laboratory group, the aircraft measurements were part of the NASA Global Tropospheric Experiment (GTE), and the observations in Hawaii were part of the Mauna Loa Observatory Photochemistry Experiment. In panel (b), DYC was an NCAR Electra aircraft experiment conducted over the northeastern Pacific Ocean (adapted from Ridley, 1991).

distribution of $NO_x$ in the planetary boundary layer and near 10 km estimated using a 3D chemistry/transport model, which uses current surface emission inventories and estimated contributions from aircraft, lightning, and the stratosphere. Independent of the validity of the model, the large-scale vertical and horizontal distribution of active nitrogen is expected to be highly variable depending on both chemical and meteorological-transport events. Herein lies part of the complexity of understanding trends in global tropospheric $O_3$ or oxidizing capacity.

Figure 7.19 summarizes some observations of $NO_y$ and several of its constituents. Figure 7.19a shows that the range of $NO_x$ mixing ratios spans almost four orders of magnitude as the measurement site is removed from the rural continental United States (Scotia PA surface site) to the more remote Pacific Ocean region. Broadly speaking, this strong reduction reflects the short photochemical lifetime versus distance from the major source regions and air mass mixing. Because the figure does not include data from urban centers where mixing ratios can be another order of magnitude larger than observed at Scotia, one might well wonder whether values of 30-40 pptv found in the remote atmosphere are important at all.

Figure 7.19b shows the corresponding measurements of PAN, an important temporary $NO_y$ reservoir that is found to be in abundance in urban and rural locations. Its lifetime can vary from a few hours in the warm planetary boundary layer to a few weeks to a month in the colder middle and upper troposphere. Consequently, if PAN reaches higher altitudes, it can be transported over large distances. This behavior is reflected in the figure, where mixing ratios in the middle continental and marine troposphere are only about a factor of 5 lower than typically observed at Scotia. Most important is the observation revealed by a comparison of the $NO_x$ and PAN mixing ratios of Figure 7.19a and 7.19b, which show that PAN can exceed active nitrogen by a factor of roughly 5. Thus processes that can convert PAN to active nitrogen (Section 7.9.3) can be an important source of $NO_x$ to the more remote troposphere.

Figure 7.19c gives the corresponding measurements of total odd nitrogen, $NO_y$. The trend with distance from source regions is similar to that for $NO_x$, but is not nearly as dramatic. $NO_y$ is not conserved in the atmosphere, since some of its constituents (e.g., $HNO_3$) can be removed efficiently by wet and dry deposition. Mixing with "cleaner" air masses can also lead to dilution. However, $NO_y$ is more conservative than $NO_x$, as is illustrated by the slower "decay" in Figure 7.19c compared to 7.19a. It is then common to attribute a longer "lifetime" to $NO_y$ as if it were a single species. Clearly, its lifetime is controlled by its longest-lived component. A comparison of Figures 7.19a and 7.19c reveals that the partitioning of $NO_y$ into $NO_x$ changes with location. For example, the $NO_x/NO_y$ ratio can approach unity in locations in proximity to source regions (Scotia site), whereas it is much lower, $\sim 0.1$, at the MLOPEX (Mauna Loa Observatory Photochemistry Experiment) site on Hawaii. A decrease in the ratio is often attributed to "aging" of the air mass, since

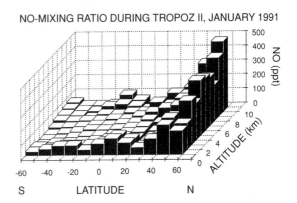

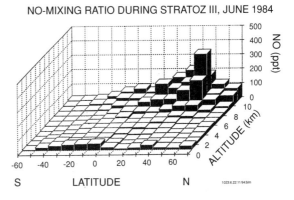

**Figure 7.20.** Meridional representation of the average NO mixing ratio measured during the Tropospheric Ozone Programme (TROPOZ) on flights of the French Caravelle aircraft in January 1991. The flights were concentrated over the coasts of Eastern North America, South America, North Africa, and Europe (Wahner et al., 1994; reprinted by permission of Oxford University Press).

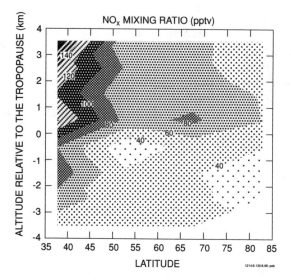

**Figure 7.21.** Distribution of $NO_x$ in the meridional plane derived from measurements during 10 flights of the NASA DC-8 in February and March of 1992 during the Airborne Arctic Stratospheric Expedition II (Weinheimer et al., 1994).

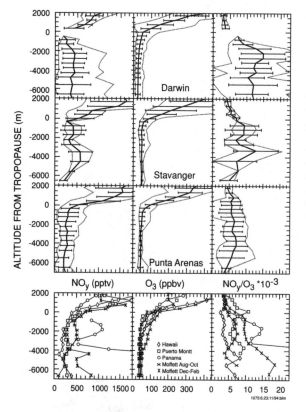

**Figure 7.22.** Vertical profiles of $NO_y$, $O_3$, and the $NO_y/O_3$ ratio obtained using the NASA ER-2 aircraft. The data are plotted relative to the altitude of the tropopause. Mean tropospause altitudes were 16.5 km at Darwin, 10.8 km at Punta Arenas, and 11.8 km at Stavangar. The flights were made from Darwin (Australia), Stavanger (Norway), Punta Arenas (Chile), and other locations. Note the strong increase in $NO_y$ and $O_3$ on entering the stratosphere, but the decrease in the magnitude and variation of the $NO_y/O_3$ ratio (Murphy et al., 1993).

photochemistry transforms active $NO_x$ into $NO_y$ reservoirs faster than $NO_y$ is removed. However, there are several other factors that cause the $NO_x/NO_y$ ratio to change; the ratio is at best a poor "clock." For example, the $NO_x/NO_y$ ratio is expected to increase with altitude in the free troposphere so that vertical mixing can confuse this simple interpretation. Or soluble $HNO_3$ can be removed in clouds and precipitation while $NO_x$ is not removed. This differential scavenging results in

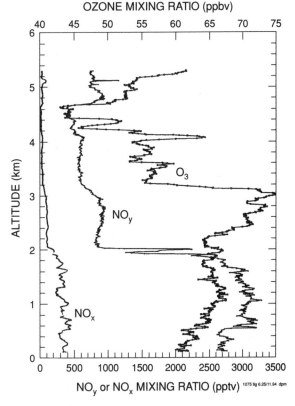

**Figure 7.23.** Profiles of $NO_y$, $NO_x$, and $O_3$ over Alabama acquired using the NCAR Saberliner aircraft during the 1990 Southern Oxidant Study. Note the strong decrease in odd-nitrogen constituents above the planetary boundary layer, the top of which is located near 2 km altitude.

an increase in the $NO_x/NO_y$ ratio ("younger" air), although the air mass would be considered to have continued to age.

Figure 7.20 gives a large-scale view of the NO mixing ratio measured in aircraft flights mostly over the North and South Atlantic in summer and winter. Mixing ratios are generally smaller in the Southern Hemisphere in both seasons, and increase with latitude and altitude in the Northern Hemisphere. This suggests the influence of larger sources commensurate with larger populations in the Northern Hemisphere. Figure 7.21 gives another large-scale view of the $NO_x$ distribution in winter in the Northern Hemisphere from a large number of flights mostly over North America and the North Atlantic. These flights reveal considerably lower mixing ratios of $NO_x$ at high latitudes than just the NO measurements of Figure 7.20, even though they were made in the same season.

Figures 7.22 and 7.23 are presented mainly to give a flavor for the vertical distributions of odd nitrogen and ozone at various locations. They emphasize the variability that typifies the atmospheric distribution, and they also illustrate the dynamic range that is required to obtain the measurements.

Since some species like peroxynitric acid have yet to be measured in the troposphere, this section concludes with Figure 7.24, taken from a current model for a remote maritime region. The profiles are plotted in both concentration and mixing ratio units for 25°N over the Pacific ocean for solar noon in April.

### 7.5.2 Ozone Production and Loss in the Troposphere: Role of Odd Nitrogen

Considering the entire troposphere, there are two major sources of $O_3$: transport from the stratosphere and in situ photochemical production. Because at present there are

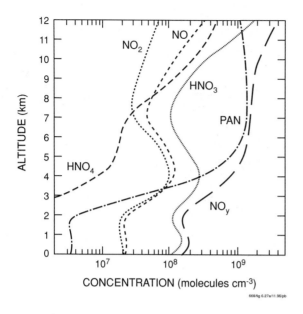

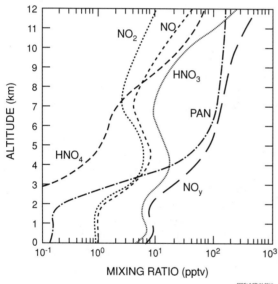

**Figure 7.24.** (a) Vertical distribution (0-12 km) of odd-nitrogen species calculated using the NCAR IMAGES 3D model for July at 25° N and 180° longitude, top panel (molecule cm$^{-3}$), bottom panel (mixing ratio, pptv) (courtesy of J.-F. Müller and G. Brasseur).

not dramatic trends in global tropospheric $O_3$, these processes are roughly balanced by two major loss processes: destruction at the Earth's surface and in situ photochemical loss. Trends in tropospheric ozone globally are governed by the overall competition between these processes. If the troposphere is partitioned into planetary boundary layer and free troposphere components, then transport of primary and secondary chemical species across these boundaries becomes a particularly important process in determining the balance of $O_3$ within these regions. The same is true for export across lateral boundaries, such as that separating a photochemically productive continental boundary layer coastal region from the marine boundary layer.

At the high concentrations of molecular $O_2$ and $N_2$ in the troposphere, the lifetime of atomic oxygen is very short due to the recombination reaction

$$O + O_2 + M \to O_3 + M \tag{7.34}$$

But (7.34) is also the only major reaction that leads to $O_3$ formation. Because wavelengths shorter than ~290 nm are absorbed by $O_2$ and $O_3$ in the stratosphere, production of O atoms from photolysis of $O_2$ is negligible in the troposphere. Another source of O atoms must be available to allow $O_3$ formation in the troposphere. The major source is from photolysis of $NO_2$ at wavelengths less than ~420 nm, which is followed immediately by (7.34)

$$NO_2 + h\nu \rightarrow NO + O \quad (7.4)$$

Since NO is readily oxidized by $O_3$,

$$NO + O_3 \rightarrow NO_2 + O_2 \quad (7.3)$$

Reactions (7.3) + (7.4) + (7.34) constitute a null cycle with respect to odd oxygen or ozone just as it was in the stratosphere.

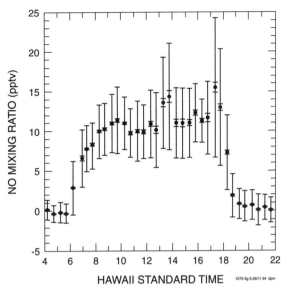

**Figure 7.25.** The diurnal variation of NO observed during the Mauna Loa Observatory Photochemistry Experiment (MLOPEX) conducted at an elevation of 3.4 km on the Island of Hawaii. The data are averaged over almost six weeks beginning in May of 1988. Only clear sky values are considered during daytime. The central dot is the mean, the inner bars the standard error of the mean, and the outer bars the standard deviation (from Ridley et al., 1992).

All this cycle does is interconvert NO and $NO_2$, or define the partitioning of $NO_x$ according to Eq. (7.35), which is analogous to Eq. (7.10) for the stratosphere when the O atom concentration is small

$$[NO_2]/[NO] = k_3[O_3]/j_4 \quad (7.35)$$

With this limited reaction scheme, the NO and $NO_2$ interconversion is controlled by the local concentration of $O_3$, temperature (through $k_3$), and the diurnal variation of $j_4$. Figure 7.25 shows the average diurnal variation of NO at a 3.4 km altitude site in the Pacific. NO is oxidized to $NO_2$ and other species at night and returned during the day by photolysis. In order to have a net production of $O_3$ photochemically, NO has to be oxidized to $NO_2$ during the day by reactions other than (7.3). The species responsible are a variety of peroxy radicals.

There are three basic solar-driven chemical cycles involving $NO_x$ that lead to a net production of $O_3$ in the troposphere. The chain cycles are all initiated by the primary oxidant OH. In each case, the cycles are "catalytic" with respect to $NO_x$; there is only interconversion of NO and $NO_2$ with no loss of $NO_x$. However, CO, $CH_4$, or nonmethane hydrocarbons (NMHC, also called volatile organic carbon species, VOCs) are consumed in the cycles and as such are regarded as the chemical fuels. The dominance of one or the other cycle in various regions of the atmosphere will depend upon the abundance and reactivity of the fuels, and in the case of NMHC, on the relative proportions of alkanes, alkenes, aromatics, etc. Whether the oxidation processes lead to net photochemical $O_3$ production inefficiently (high-$NO_x$ case), efficiently (low-$NO_x$ case), or at all (very low-$NO_x$ case) depends sensitively on the concentration of $NO_x$. For example, in the absence of $NO_x$ photochemical oxidation processes lead to a destruction of $O_3$. As usual the net outcome is dictated by competition. The first focus is on the simplest case, the role of odd nitrogen in the chemistry of the troposphere remote from continental sources. As an example, results from a field study conducted at 3.4 km elevation on the island of Hawaii in the spring of 1988, the Mauna Loa Observatory Photochemistry Experiment (MLOPEX) will be used (Ridley and Robinson, 1992, and subsequent papers).

### 7.5.3 Odd Nitrogen in the Remote Troposphere
#### 7.5.3.1 Low-$NO_x$ Case

Even in the remote atmosphere there is an abundance of oxidizable fuel. Methane is typically 1.7 ppmv and carbon monoxide is usually in the range of 50-150 ppbv. In the remote atmosphere the cycle involving CO is dominant in spite of the larger abundance of $CH_4$ because the rate coefficient for (7.36) is much larger than for (7.38). For example, at midlatitudes at 3 km, $k_{36}/k_{38} \sim 30$. When $NO_x$ is sufficiently large (the meaning of sufficiently large will be made clear later), then the catalytic cycle involving carbon monoxide is:

$$OH + CO \to CO_2 + H \tag{7.36}$$
$$H + O_2 + M \to HO_2 + M \tag{7.37}$$
$$HO_2 + NO \to NO_2 + OH \tag{7.12}$$
$$NO_2 + h\nu \to NO + O \tag{7.4}$$
$$\underline{O + O_2 + M \to O_3 + M} \tag{7.34}$$

Overall :   $CO + 2\, O_2 + h\nu \to CO_2 + O_3$

In this case CO is consumed to form $CO_2$ and $O_3$. Note that there is no loss of $NO_x$ ($= NO + NO_2$) or $HO_x$ ($= OH + HO_2$) in this cycle. Other processes, which will be discussed later, lead to a loss of $NO_x$ over longer timescales.

The cycle involving methane is:

$$OH + CH_4 \to CH_3 + H_2O \tag{7.38}$$
$$CH_3 + O_2 + M \to CH_3O_2 + M \tag{7.39}$$
$$CH_3O_2 + NO \to CH_3O + NO_2 \tag{7.40}$$
$$CH_3O + O_2 \to HO_2 + CH_2O \tag{7.41}$$
$$HO_2 + NO \to OH + NO_2 \tag{7.12}$$
$$2\,(NO_2 + h\nu \to NO + O) \tag{7.4}$$
$$\underline{2(O + O_2 + M \to O_3 + M)} \tag{7.34}$$

Overall :   $CH_4 + 4\, O_2 + 2\, h\nu \to CH_2O + H_2O + 2\, O_3$

Methane is oxidized to formaldehyde ($CH_2O$), water, and $O_3$. Because two peroxy radicals are generated for each initial OH reaction with $CH_4$ (7.38), two molecules of $O_3$ can be formed. Oxidation of formaldehyde by OH or photolysis can also lead to additional production of $O_3$ in the presence of $NO_x$ (see Section 7.5.4).

In contrast to the reaction of NO with $O_3$ (7.3), the $HO_2$ and $CH_3O_2$ peroxy radicals can oxidize NO to $NO_2$ without consuming $O_3$. Thus, in the low-$NO_x$ case, net photochemical $O_3$ production occurs only above levels of NO or $NO_x$ that make (7.40) and (7.12) dominate over reactions between peroxy radicals. Indeed, (7.40) and (7.12) are prerequisites for $O_3$ formation in the troposphere from CO or methane oxidation. In such circumstances, the local or in situ gross rate of formation of $O_3$ can be written as

$$P(O_3) = (k_{12}[HO_2] + k_{40}[CH_3O_2])\,[NO] \tag{7.42}$$

(see Chapter 13 for further details concerning this expression). The NO and peroxy radical concentrations can be dependent on one another, especially in the high-$NO_x$ case (Section 7.5.4), and expression (7.42) can then be much more complicated than it appears. However, one definition of the low-$NO_x$ troposphere is when the NO and peroxy radical concentrations are not strongly dependent on one another. The net result is that, in the low-$NO_x$ case, the gross rate of production of $O_3$ increases as a nearly linear function of NO or $NO_x$.

The presence of peroxy radicals also alters the daytime partitioning of $NO_x$. In comparison to Eq. (7.35), the partitioning is shifted to favor larger $NO_2$ for a given $O_3$ concentration and $j_4$. Equation (7.35) has to be modified to

$$[NO_2]/[NO] = (k_3[O_3] + k_{12}[HO_2] + k_{40}[CH_3O_2])/j_4 \qquad (7.43)$$

Although the radical abundances (see Fig. 7.26) are orders of magnitude smaller than $O_3$, the terms involving the radical concentrations are not negligible because $k_{12}/k_3 \sim k_{40}/k_3 \sim 550$.

Laboratory measurements have shown that $k_{12}$ and $k_{40}$ are nearly equal. Thus, if measurements of temperature, $j_4$, $O_3$, and the $NO_2/NO$ ratio are made, it is possible to estimate the total peroxy radical concentration, $HO_2 + CH_3O_2$, from Eq. (7.43). An example from the MLOPEX experiment is given in Figure 7.26. The experimental determination is larger than predicted by the photochemical model, but considering that the result is very sensitive to the accuracy of the measurement of the $NO_2/NO$ ratio, the overall agreement is reasonable. The model also shows that $HO_2$ is the dominant peroxy radical at high sun and is therefore dominant in the rate of formation of $O_3$ (cf. Table 7.1). Also, from a comparison of Figures 7.25 and 7.26, it is noted that although one tends to think that the peroxy radical mixing ratios are small relative to more stable species, their predicted mixing ratio near noon exceeds that of NO by a factor of 5-6.

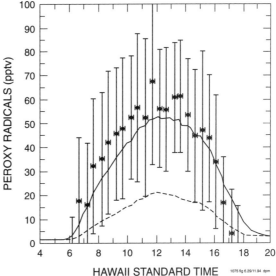

**Figure 7.26.** *The peroxy radical concentrations inferred from measurements of $NO_2$, NO, $O_3$, and $j(NO_2)$ during the MLOPEX study (dots plus standard deviation). The solid curve is the model prediction of the sum of $HO_2$ and $CH_3O_2$. The dashed curve is the model estimate of $CH_3O_2$ alone (from Ridley et al., 1992).*

Substitution of Eq. (7.43) into Eq. (7.42) yields

$$P(O_3) = j_4[NO_2] - k_3[NO][O_3] \qquad (7.44)$$

Thus the gross rate of formation of $O_3$ can be determined directly from daytime measurements of $NO_x$, $j_4$, $O_3$, and temperature, measurements that are currently simpler than techniques required for measurement of $HO_2$ and $RO_2$ species. Figure 7.27 shows that the average diurnal production rate of $O_3$ determined from Eq. (7.44) for the Mauna Loa experiment is in good agreement with a photochemical model.

## Table 7.1
### Ozone Production and Loss Rates for MLOPEX

| Reaction | 24-h Average Rate[a] | % of Total | Solar Noon Rate[a] | % of Total |
|---|---|---|---|---|
| *Production* | | | | |
| $HO_2 + NO \rightarrow OH + NO_2$ (7.12) | 2.79 | 69.9 | 8.15 | 69.5 |
| $CH_3O_2 + NO \rightarrow CH_3O + NO_2$ (7.40) | 1.20 | 30.1 | 3.58 | 30.5 |
| | | | | |
| *Destruction* | | | | |
| $O(^1D) + H_2O \rightarrow 2\ OH$ (7.53) | 2.36 | 48.2 | 9.44 | 56.4 |
| $O_3 + HO_2 \rightarrow OH + 2O_2$ (7.49) | 1.98 | 40.4 | 5.45 | 32.6 |
| $O_3 + OH \rightarrow HO_2 + O_2$ (7.50) | 0.51 | 10.4 | 1.68 | 10.0 |
| Other minor terms | 0.05 | 1.0 | 0.16 | 1.0 |
| | | | | |
| Net production[b] | −0.91 | −0.47 | −5.00 | −2.59 |

[a] Rates are in units of $10^5$ molecules cm$^{-3}$ s$^{-1}$.
[b] Rate in units of ppbv day$^{-1}$.

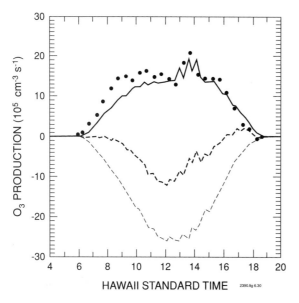

**Figure 7.27.** Components of the net rate of production of $O_3$ for clear-sky conditions during MLOPEX. The circles are the estimates of the production rate from Eq. (7.44), while the solid curve is the model production rate. The thin dashed line is the model prediction of the total rate of $O_3$ loss, $L(O_3)$. The thick dashed line gives the model prediction of the net rate of $O_3$ production, $P(O_3)-L(O_3)$ (from Ridley et al., 1992).

Just as in the $O_3$ catalytic destruction cycle in the stratosphere, the production cycles in the troposphere compete with reactions that deplete $HO_x$ and $CH_3O_2$ radicals or deplete $NO_x$. $NO_x$ removal may be permanent through physical processes such as deposition or incorporation in particles or temporary through chain-termination reactions that produce reservoir constituents. Being longer-lived than their precursors, reservoir species are subject to large-scale redistribution through transport processes. Reactions (7.18), (7.19), and (7.45)-(7.48) are some examples of the formation of reservoir constituents nitric acid, peroxynitric acid, and methyl hydroperoxide. Figure 7.28 shows that $HNO_3$ is the major $NO_x$ reservoir species at the remote MLOPEX site

$$OH + NO_2 + M \rightarrow HNO_3 + M \tag{7.18}$$

$$HO_2 + NO_2 + M \rightarrow HO_2NO_2 + M \tag{7.19}$$

$$N_2O_5 + H_2O/\text{aerosol} \rightarrow 2\ HNO_3 \tag{7.45}$$

$$HO_2 + HO_2 \rightarrow H_2O_2 + O_2 \quad (7.46)$$
$$HO_2 + OH \rightarrow H_2O + O_2 \quad (7.47)$$
$$HO_2 + CH_3O_2 \rightarrow CH_3OOH + O_2 \quad (7.48)$$

Some of the odd-nitrogen reservoirs are soluble and can be removed from the atmosphere through precipitation and/or surface deposition, processes that can shorten their atmospheric residence time considerably. Although precipitation events are highly localized and episodic, they are more frequent at lower altitudes so that formation of $HNO_3$ in the lower atmosphere can be an effective sink of active nitrogen. Conversely, the lifetime of soluble reservoirs like $HNO_3$ will be much longer in the upper troposphere (~1 month) so that the return of active nitrogen from $HNO_3$ or $HO_2NO_2$ (or other reservoirs such as PAN) through photolysis, reaction with OH, heterogeneous reactions, or unimolecular decomposition is important to $O_3$ production on the global scale.

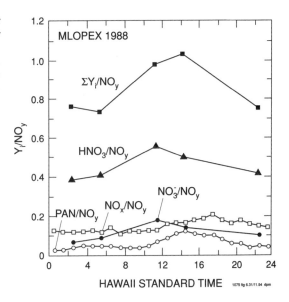

**Figure 7.28.** *Diurnal variation of $NO_y$ partitioning and the sum of all measured reactive nitrogen species during MLOPEX. Median ratios are shown. The results suggest a missing component of $NO_y$ during the night (Atlas et al., 1992).*

It is clear from Eq. (7.42) that in the limit of zero NO, methane and carbon monoxide oxidation cycles do not form $O_3$. In fact, at near zero $NO_x$ concentrations, radical chain termination reactions compete with the reactions

$$HO_2 + O_3 \rightarrow OH + 2\,O_2 \quad (7.49)$$
$$OH + O_3 \rightarrow HO_2 + O_2 \quad (7.50)$$

that interconvert $HO_x$ species and actually lead to catalytic $O_3$ loss (see Chapter 6 also). In the absence of $NO_x$, CO oxidation can be an efficient $O_3$ destruction process; that is,

$$OH + CO \rightarrow CO_2 + H \quad (7.36)$$
$$H + O_2 + M \rightarrow HO_2 + M \quad (7.37)$$
$$\underline{HO_2 + O_3 \rightarrow OH + 2O_2} \quad (7.49)$$

Overall :  $CO + O_3 \rightarrow CO_2 + O_2$

If $HO_2$ reacts with another peroxy radical [*e.g.*, (7.46) or (7.48)] instead of with $O_3$ (7.49), then there would be no direct change in $O_3$. Similarly methane oxidation could occur without changing $O_3$ directly, as follows

$$OH + CH_4 \rightarrow CH_3 + H_2O \qquad (7.38)$$

$$CH_3 + O_2 + M \rightarrow CH_3O_2 + M \qquad (7.39)$$

$$CH_3O_2 + HO_2 \rightarrow CH_3OOH + O_2 \qquad (7.48)$$

Another important loss of $O_3$, in fact the dominant one in remote regions at lower altitudes where the $H_2O$ concentration can be high, is photolysis of $O_3$ — the very process that leads to OH formation

$$O_3 + h\nu \rightarrow O(^1D) + O_2, \quad < 320\ nm \qquad (7.51)$$

$$O(^1D) + M \rightarrow O + M \qquad (7.52)$$

$$O(^1D) + H_2O \rightarrow 2\ OH \qquad (7.53)$$

Most of the electronically excited $O(^1D)$ atoms are quenched to ground-state O atoms, which immediately undergo recombination (7.34) to re-form $O_3$, but a few react with water vapor (7.53) to produce OH. Based on (7.49-7.53), the local or in situ gross rate of photochemical destruction of $O_3$ can be written approximately as

$$L(O_3) = (fj_{51} + k_{49}[HO_2] + k_{50}[OH])\,[O_3] \qquad (7.54)$$

Equation (7.54) is a lower limit to the overall gross destruction rate since we have neglected removal processes that can occur in cloud droplets, surface deposition (which can be very efficient depending upon location and the type of surface vegetation), and other minor processes such as direct consumption by reaction with unsaturated hydrocarbons. In Eq. (7.54), $f$ is just the fraction of excited $O(^1D)$ atoms that react with water, or the fraction of the photolytic process (7.51) that leads to an immediate loss of $O_3$:

$$f = k_{53}[H_2O]/\,(k_{52}[M] + k_{53}[H_2O]) \qquad (7.55)$$

Thus the gross photochemical loss rate of $O_3$ can increase with increasing local $O_3$ concentration and increase with decreasing stratospheric $O_3$ because $j_{51}$ is strongly affected by the overhead column abundance of $O_3$. Furthermore, it is important to note that while $HO_2$ radicals readily react with $O_3$, organic peroxy radicals (generally labeled $RO_2$, where $CH_3O_2$ is the smallest carbon number example) do not.

In sharp contrast to Eq. (7.42), which gives the gross rate of production of $O_3$, the gross loss rate given by Eq. (7.54) is independent of the $NO_x$ concentration. Based upon our current understanding of photochemistry, this is approximately true, since only a minor loss involving the gas-phase formation of $NO_3$ has been neglected. However, just as heterogeneous processes have revolutionized stratospheric photochemistry, it is possible that heterogeneous processes likely involving $NO_3$ or $N_2O_5$ formation can alter current projections of possible trends in tropospheric $O_3$. This is an active research area in tropospheric chemistry today. For example, heterogeneous processes may be an effective sink of active $NO_x$ (as well as radical species). Such processes may also have a substantial seasonal variation. Clearly, "new" chemistry that affects either $P(O_3)$ or $L(O_3)$ will alter current predictions of trends in $O_3$ or in the oxidizing capacity of the troposphere.

The overall or net rate of change of local $O_3$ due to in situ photochemistry is, of course, given by the difference $P(O_3)$-$L(O_3)$. Calculations of this net rate can be based on measurement of $NO_x$, peroxy radicals, hydroxyl radical, photolysis rates, etc., as described by Eqs. (7.42) and (7.54). Measurements of $HO_2$ and in particular OH are difficult, and are currently undergoing development and intercomparison. Thus in general, local $O_3$ loss rates are most frequently determined from model calculations. Table 7.1 and Figure 7.27 summarize both $P(O_3)$ and $L(O_3)$ determined for the Mauna

Loa experiment. From the table, the largest contribution to the production rate is from $HO_2$ radicals. However, the largest contribution (nearly 50%) to the loss rate is the $fj_{51}[O_3]$ term. The gross production and loss rates are nearly equal, but the average net rate of change of $O_3$, $P(O_3)$-$L(O_3)$, is a small destruction rate equivalent to 0.5 ppbv per day. Figure 7.29 shows the gross production and loss rates, averaged over a day, for a range of $NO_x$ mixing ratios using a photochemical model appropriate to the solar insolation, $CH_4$ and $CO$ concentrations of the MLOPEX experiment. Over this small range of $NO_x$ mixing ratio (1-100 pptv), the photochemical loss rate is nearly constant while the production rate is nearly linear in $NO_x$. Thus, over this small range of $NO_x$, the $HO_2$ and $CH_3O_2$ concentrations must be substantially independent of $NO_x$. Indeed, this near independence is a simplifying feature of the low-$NO_x$ atmosphere. In contrast, we shall see that in the high-$NO_x$ atmosphere typical of the continental boundary layer near source regions, the peroxy radical concentrations have a strong dependence on the $NO_x$ level.

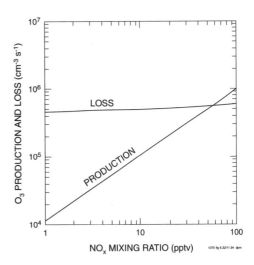

**Figure 7.29.** Model calculated 24-h average production and loss rates of $O_3$ versus the mixing ratio of $NO_x$ (Liu et al., 1992).

### 7.5.3.2 "Critical" $NO_x$

We have seen that whether the gross $O_3$ formation rate $P(O_3)$ or the gross destruction rate $L(O_3)$ predominates in $CO$ or $CH_4$ oxidation in the remote atmosphere depends sensitively on the $NO_x$ concentration. Although transport of $O_3$ into or out of a given remote region may well be the most important factor in determining the temporal behavior of $O_3$, the net local rate of photochemical production is just $P(O_3)$–$L(O_3)$. Within the approximations discussed for Eqs. (7.42) and (7.54), this difference is zero at a daytime NO concentration (sometimes called the critical concentration, $[NO]^*$) when

$$[NO]^* = \{(fj_{51} + k_{49}[HO_2] + k_{50}[OH])[O_3]\} / (k_{12}[HO_2] + k_{40}[CH_3O_2]) \qquad (7.56)$$

Since daytime $NO_x$ and NO are related through Eq. (7.43), the equivalent critical $NO_x^*$ can be determined. For example, in Figure 7.29 it is found at the intersection of the two curves, or at ≈55 pptv of $NO_x$ (≈20 pptv of NO) for a 24-h average. In regions where other $O_3$ loss rates might be important, the critical mixing ratio would be shifted to a larger value. When the NO mixing ratio is below (above) the critical value, photochemical processes give a net destruction (production) of $O_3$. $NO^*$ depends on all the factors that control radical production and loss and thus on the distribution of UV, water, and $O_3$ in the remote atmosphere. $NO^*$ usually falls within the range of 5-30 pptv of NO near noon in the lower remote troposphere. Thus it is clear why the small abundances of $NO_x$ or NO seen in Figures 7.19-7.25 are critically important to understanding the tropospheric budget of $O_3$ on the global scale. Although many of the remote atmosphere observations summarized in these figures fall within the low-$NO_x$ regime, many are close to the very low-$NO_x$ regime where small changes in active nitrogen can be the difference between a net photochemical production or a net photochemical destruction of $O_3$ in an air mass. To maintain steady-state conditions, regions of net destruction must import $O_3$ from net production regions and vice versa.

At present, a number of field studies have identified regions where net destruction is sustained. Largest net destruction rates have been found in the moist tropical to midlatitude remote marine boundary layer. Net destruction rates equivalent to 10-20% of $O_3$ per day have been reported. Indeed, in the equatorial Pacific marine boundary layer $O_3$ mixing ratios are low, typically 5-15 ppbv, mainly due to net photochemical destruction processes. Even over the Arctic/Subarctic regions of continental North America in summer, $NO_x$ concentrations in the lower and middle troposphere were found to be sufficiently small to allow net photochemical destruction of $O_3$. Table 7.1 has shown that a small net destruction of $\sim$0.5 ppbv/day occurs in the lower free troposphere in the North Pacific. Thus these regions characterized by net destruction currently act as a "buffer" against increases in global tropospheric $O_3$ proportionate to increases in active nitrogen through anthropogenic activity (WMO, 1995). Periodic import of higher concentrations of $O_3$ or its precursors to such regions will over a relatively short time lead to consumption of $O_3$. However, any increase in $NO_x$ globally increases the rate of production of $O_3$. How long this buffering action can last is unknown due to the limited knowledge of the global distribution of $NO_x$ over the seasons and due to uncertainties related to heterogeneous chemistry.

The regions of net $O_3$ destruction are mostly confined to the remote boundary layer and lower free troposphere. In the upper troposphere, at least in the Northern Hemisphere, available observations suggest that net $O_3$ production is usual. On the global scale both $P(O_3)$ and $L(O_3)$ are large. For example, Fehsenfeld and Liu (1993) have estimated $P(O_3)$ at $\sim 96 \times 10^{28}$ molecules s$^{-1}$ and $L(O_3)$ at $\sim 89 \times 10^{28}$ molecules s$^{-1}$ for the 1-12 km altitude region in the Northern Hemisphere summer season. The net production rate of $\sim 7 \times 10^{28}$ molecules s$^{-1}$ must be considered to have much greater uncertainty, since it is the difference between two large rates (see Chapter 13 for a more detailed discussion of these numbers).

### 7.5.3.3 Role of Active Nitrogen in Odd-Hydrogen Partitioning

A second major role for $NO_x$ in the atmosphere is in determining the partitioning of $HO_x$ between its most reactive (OH) and less reactive ($HO_2$) constituents. Indeed, the roles of $NO_x$ in $O_3$ production and in $HO_x$ partitioning are inseparable in the low-$NO_x$ atmosphere. $HO_2$ dominates $RO_2$ in the remote atmosphere, first because CO is more reactive than $CH_4$ and second because $CH_4$ oxidation also generates $HO_2$. At low altitude $HO_2$ also exceeds OH by a large factor so that the concentration of odd hydrogen, $HO_x$ = OH + $HO_2$, is very nearly that of $HO_2$. Figure 7.30 shows this more clearly and gives the variation of odd hydrogen radicals with increasing NO determined from a model where CO and $CH_4$ were the oxidizable fuels. In tropical or midlatitude near-surface regions in summer there is remarkably little change in $HO_2$ with increasing NO until the latter reaches mixing ratios of 200-300 pptv (or $\sim$400-600 pptv of $NO_x$); this result was noted indirectly earlier in that $P(O_3)$ was nearly linear in NO or $NO_x$ (cf. Fig. 7.29). For example, if we ignore $RO_2$ or consider an atmosphere in which CO is the only fuel, then the change in the production rate of

$O_3$ with NO over this limited range of $NO_x$ is nearly constant:

$$dP(O_3)/d[NO] = k_{12}[HO_2] \qquad (7.57)$$

The behavior of OH is quite different. OH increases only slowly at first while NO is increased from near zero up to mixing ratios near the critical level (~10 pptv). In contrast, further increase in NO up to the 200-300 pptv level results in a strong increase in OH. However, throughout this range of $NO_x$, $HO_2$ remains much larger than OH. Thus $HO_x$ is essentially independent of NO, while OH is strongly dependent on NO over this mixing ratio range. An examination of Figures 7.19-7.25 shows that in the remote atmosphere observed mixing ratios of NO or $NO_x$ are typically in the range where the OH concentration is or is beginning to be strongly dependent on active nitrogen.

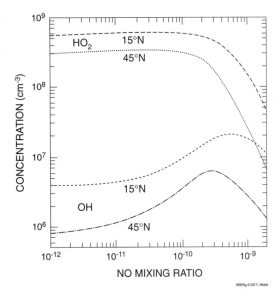

**Figure 7.30.** *The dependence of OH and $HO_2$ on the mixing ratio of NO calculated for conditions appropriate to the remote atmosphere (Logan et al., 1981).*

The explanation for the difference in behavior for the hydrogen radicals is qualitatively straightforward. At very low NO, for example, ~1 pptv in Figure 7.30, the $HO_x$ concentration is determined by its rate of production from photolysis and water vapor reaction (the $fj_{51}[O_3]$ term given by (7.51-7.53)), the fuel (CO, $CH_4$) loading, and its loss through radical termination reactions such as (7.46-7.48). For example, the larger $HO_x$ concentration at 15°N compared to 45°N in Figure 7.30 is mainly due to a greater production rate as a result of stronger photolysis and larger water concentration in the tropical region. With the scale of $HO_x$ set by these factors, increasing NO changes the partitioning between OH and $HO_2$, but has a near-negligible effect on production or loss of $HO_x$ until NO is increased beyond 200-300 pptv. Below approximately the critical level of NO, the partitioning of $HO_x$ is mainly governed by the interconversion reactions (7.49) and (7.50), or by the $O_3$ concentration. NO is so small that its reaction with $HO_2$ (7.12) does not compete successfully with the reaction of $O_3$ with $HO_2$ (7.49). As NO is increased, the rate of (7.12) overtakes and eventually dominates (7.49), leading to a larger $OH/HO_2$ ratio. It is also (7.12) [and (7.40)] that determines the $O_3$ production rate. Stronger increases in NO and, via Eq. (7.43), $NO_2$ eventually lead to an increased loss of OH and hence $HO_x$. These losses occur mainly by the termination reaction, which consumes OH and $NO_x$ to form long-lived $HNO_3$ (7.18). In the upper troposphere, the termination reaction forming peroxynitric acid (7.19) is also important.

The low-$NO_x$ regime has been defined by the constraint that the rate of production of $O_3$ be nearly linear in $NO_x$ or that the $HO_2$ concentration be nearly independent of $NO_x$. In the lower troposphere, by comparison with Figure 7.30, this condition holds approximately for increasing $NO_x$ concentrations until $HO_2$ falls rapidly, or roughly to the maximum in the OH concentration. At concentrations of $NO_x$ larger than at the maximum in OH, the rate of production of $O_3$ decreases because the $HO_2$ concentration decreases more precipitously than the increase in $NO_x$. The neighborhood of the maximum in OH is the transition region between the low- and high-$NO_x$ regimes. It is also the region where the onset of nonlinearity in $O_3$ production begins and persists through the high-$NO_x$ regime.

In the upper troposphere, the qualitative relationship between $HO_x$ partitioning and $NO_x$ is similar to that of Figure 7.30, but there are some important differences in detail. First, because the water concentration is much lower, the rate of production of OH (and $HO_x$) by Reactions (7.51-7.53) is much slower. Other processes such as photolysis of $H_2O_2$ and reactions involving formaldehyde and maybe acetone become more important in determining the steady-state $HO_x$ concentration relative to the lower troposphere. Second, the $HO_2$/OH ratio is significantly lower. For example, at midlatitudes in summer near 10 km, the predicted ratio is ∼20 compared to values of 100 or more in the low-$NO_x$ regime of Figure 7.30. Third, reservoir formation reactions, in addition to generation of $HNO_3$, become relatively more important. For example, net formation of peroxynitric acid via (7.19) is much more important in the cold upper troposphere where its thermal stability is much higher. The overall result is that the maximum in the OH concentration, or the "falloff" in $HO_2$, occurs at a smaller concentration of $NO_x$. Consequently, in the upper troposphere, at least in the Northern Hemisphere, available data suggest that current levels of $NO_x$ are in the neighborhood of the transition region, where the net rate of production of $O_3$ is nonlinear. The expected differences at higher altitude in the troposphere are shown from one model in Figure 7.31. Note that at 45°N in summer the concentration of OH near 10 km is a factor of 4-7 times lower than shown for low altitudes in Figure 7.30. Also, at 10 km in summer a strong reduction in OH is predicted to occur at a $NO_x$ mixing ratio as low as ∼100 pptv. The net rate of production of $O_3$ is clearly nonlinear in the neighborhood of 100-400 pptv of $NO_x$. Thus global increases in $NO_x$ in the upper troposphere are not expected to result in directly proportional increases in upper tropospheric $O_3$.

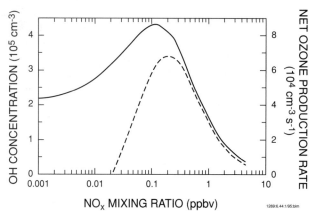

**Figure 7.31.** Calculated dependence of the OH concentration (solid line) and the net $O_3$ production rate (dashed line) on the $NO_x$ mixing ratio for an altitude of 10 km, 45°N, summer (from Ehhalt and Rohrer, 1994).

### 7.5.3.4 Catalytic Efficiency of $NO_x$ in $O_3$ Production

The coupling between the $HO_x$ and $NO_x$ cycles in the remote atmosphere can be summarized by recognizing that the catalytic cycling to produce $O_3$ continues until chain-termination reactions occur. One possible definition of the catalytic efficiency (CE) of $NO_x$ is the ratio of the gross rate of production of $O_3$ to the rate of loss of $NO_x$, $L(NO_x)$; that is,

$$CE \sim P(O_3)/L(NO_x) = P(O_3)\tau_{NO_x}/[NO_x] \qquad (7.58)$$

where $\tau_{NO_x}$ is the photochemical lifetime of $NO_x$, which is discussed in more detail in Section 7.5.5 (WMO, 1995). This definition is only approximate because it does not include terms for the release of $NO_x$ from reservoir constituents such as $HNO_3$. Such processes are most important in the middle and upper troposphere. If a 24-h average is considered, then in the low-$NO_x$ case where $P(O_3)$ is linear in $NO_x$ (cf. Fig. 7.31) and $HO_2$ is nearly constant, CE is proportional to the product of the $HO_2$ concentration and $\tau_{NO_x}$. Even in the low-$NO_x$ regime, CE will decrease with increasing $NO_x$ because $\tau_{NO_x}$ will usually decrease. For example, formation of nitric acid (7.18) leads to loss of $NO_x$. The rate of this reaction will increase as OH increases as a result of the increased $NO_x$.

Further insight into the gross variation of CE can be gained if CO is considered as the only fuel and formation of nitric acid is considered a permanent loss of $NO_x$, then:

$$CE \sim k_{12}[HO_2][NO]/k_{18}[OH][NO_2][M] \qquad (7.59)$$

The instantaneous catalytic efficiency then depends upon the product of the $HO_2/OH$ and $NO/NO_2$ ratios. At lower altitudes in summer, the $HO_2/OH$ ratio is large and the $NO/NO_2$ ratio via Eq. (7.43) is less than unity. However, from Figure 7.30, as $NO_x$ is increased beyond the critical concentration of NO, the $HO_2/OH$ ratio decreases strongly. Indeed, even beyond the maximum in OH, the $HO_2$ concentration decreases faster than does OH. Consequently, the $HO_2/OH$ ratio continues to decrease even at high $NO_x$ and the catalytic efficiency continues to decrease throughout the low-$NO_x$ to high-$NO_x$ regimes.

CE is thus usually highest in the remote atmosphere or in the low-$NO_x$ environment. In the upper troposphere, the $HO_2/OH$ ratio is expected to be smaller, as discussed previously, while the $NO/NO_2$ ratio is larger mostly because of the strong temperature dependence of (7.3). Thus there is some compensation in the change in CE with altitude in the remote atmosphere.

Additions of $NO_x$ to the remote atmosphere realistically occur in episodic transport events from continental source regions or from other processes including convection or stratospheric exchange, lightning, or aircraft exhausts. Since typical mixing ratios of $O_3$ in the free atmosphere are 30-80 ppbv (much higher than typical $NO_x$ levels), a high catalytic efficiency does not mean that the net increase in $O_3$ in such events is large because the integration is effectively only over the relatively short lifetime of $NO_x$. An extreme example would be small additions of $NO_x$ such that the increase keeps the mixing ratio less than the critical level. The integrated change in $O_3$ over time would still result in a decrease of $O_3$ compared to its starting mixing ratio, although the decrease would clearly not be as great as without the addition of $NO_x$. The overall effect of the additions of $NO_x$ to an air mass is governed by the net production rate, $P(O_3)$-$L(O_3)$, not just $P(O_3)$. Indeed, a net $O_3$ production efficiency could be defined analogous to Eq. (7.58).

Another example concerns the effects of current (and projected increases of) aircraft exhaust emissions of $NO_x$ on the $O_3$ budget of the upper troposphere. As discussed previously, at least in the Northern Hemisphere, the upper troposphere may be in the transition region where net $O_3$ production is beginning to be nonlinear in $NO_x$ and where CE decreases more rapidly. Most models run with and without current emissions from aircraft give an increase in $O_3$ of only $\sim 7\%$ in the 8-12 km region at midlatitudes, while the increase in $NO_x$ due to aircraft in this altitude region is estimated to be of the order of 25-40%. The projections are uncertain because the change in $O_3$ is also dependent upon the levels of $NO_x$ prescribed or adopted for the model without aircraft emissions, and observations in this altitude region are limited. This dependence on the adopted or calculated distribution of $NO_x$ without aircraft emissions is implied by Eq. (7.58), since the $HO_2/OH$ ratio depends upon the absolute mixing ratio of $NO_x$ even within the low-$NO_x$ regime (cf. Fig. 7.30).

In contrast, in the high-$NO_x$ case found near surface source regions, an even stronger decrease in CE can be outweighed by the high $NO_x$ abundances that are sustained by continual near-surface emissions. The result can be large net increases in $O_3$, that is, pollution episodes. It is fortunate that CE in such regions is low, otherwise pollution events would be more serious than they are currently.

### 7.5.4 Odd Nitrogen in the Continental Boundary Layer near Source Regions: The High-$NO_x$ Case

Odd-nitrogen chemistry within the continental boundary layer (CBL) near anthropogenic and natural emissions is not fundamentally different from its action in the remote troposphere. Oxidation processes are, however, much more complicated because of the large number of nonmethane hydrocarbons available as fuel (see Chapters 9 and 13). In addition, secondary VOCs produced in the oxidation processes are subject to further degradation. In or downwind of urban centers anthropogenic hydrocarbons (for which controls are possible) are often but not always dominant, while in rural regions natural emissions of hydrocarbons such as isoprene and terpenes (for which controls are impractical) are often most important.

With the usually much higher concentrations of primary and secondary constituents present in the CBL (cf. Figs. 7.18-7.19), processes that are negligible in the remote atmosphere become important and, typically, both the rate of production and the rate of loss of constituents are much greater. For example, the variety and concentrations of peroxy radicals ($RO_2$) and the $NO_x$ concentrations can be much greater in this regime. These conditions give an $O_3$ production rate much larger than observed in the remote atmosphere. However, constituent loss rates are also usually much larger. Formation of $HNO_3$ can become large enough that it becomes an important component of acidic precipitation. Near the surface, wet and dry deposition become very important. Furthermore, many unsaturated hydrocarbons react directly with $O_3$, which initially will increase the loss rate of $O_3$.

It is no surprise to anyone living in or near an urban center that photochemical smog generation over large scales is a key environmental concern. For example, $O_3$ often exceeds regulatory limits over large regions of the north and southeastern United States and over central Europe in summer. A large number of urban and rural field studies have shown conclusively that increased population and related human activities have increased boundary layer $O_3$ concentrations over large scales compared to preindustrial times. It was also shown that human activities resulting in the production and emission of active nitrogen are especially critical to this $O_3$ generation. Indeed, the total $O_3$ formed over these regions in summer is often limited by the availability of catalyst $NO_x$. This means that even more $O_3$ enhancement could occur with additional $NO_x$ available in the atmosphere. Thus it is likely that compliance with environmental standards for $O_3$ on regional scales will require greater emphasis on control of $NO_x$ emissions in addition to those that have focused on anthropogenic hydrocarbon emissions.

Pollution in the boundary layer is not limited to the environs of urban/industrialized regions. Biomass burning, whether attributable to natural or anthropogenic activity, can be a very important source of photochemical precursors and of $O_3$ production over large scales (see Chapters 5 and 13).

Transport of $O_3$ and especially of its precursors produced in the high-$NO_x$ boundary layer are surely responsible for the small increasing trend in tropospheric $O_3$ currently observed in the Northern Hemisphere. Since the recent trend in lower stratospheric $O_3$ is downward, it is doubtful that changes in the amplitude or frequency of stratospheric-to-tropospheric transport is responsible for the trend. Indeed, a good deal of experimental and modeling attention is being given to understanding the processes of transport from and to continental source regions in, for example, episodic convective cloud transport or in synoptic-scale frontal system movement and development.

### 7.5.4.1 Generalized NMHC Oxidation Cycle: $O_3$ Production

If NMHC are abbreviated further to $RH$, and carbonyl products (aldehyde or ketone) are denoted $R'CHO$, where $R'$ denotes an organic fragment having one fewer carbon atom than $R$, then the generalized cycle is

$$OH + RH \rightarrow R + H_2O \qquad (7.60)$$
$$R + O_2 + M \rightarrow RO_2 + M \qquad (7.61)$$
$$RO_2 + NO \rightarrow RO + NO_2 \qquad (7.40i)$$
$$RO + O_2 \rightarrow HO_2 + R'CHO \qquad (7.41i)$$
$$HO_2 + NO \rightarrow OH + NO_2 \qquad (7.12)$$
$$2\,(NO_2 + h\nu \rightarrow NO + O) \qquad (7.4)$$
$$\underline{2\,(O + O_2 + M \rightarrow O_3 + M)} \qquad (7.34)$$

$$\text{Overall}: \quad RH + 4O_2 + 2\,h\nu \rightarrow R'CHO + H_2O + 2\,O_3$$

This cycle is completely analogous to the one given earlier for methane oxidation, where in that case the carbonyl is formaldehyde. Furthermore, the carbonyl products can also be consumed in $O_3$ production cycles or in other processes such as photolysis. Consequently, there is an amplification in the rate of production of peroxy radicals ($HO_2 + RO_2$) and the variety of organic peroxy radicals ($RO_2$). As in the CO or $CH_4$ oxidation sequence, $NO_x$ is not consumed directly. However, alternate pathways for permanent or temporary removal of active $NO_x$, other than those considered for the remote atmosphere, become very important, resulting in a substantially decreased lifetime relative to the remote atmosphere. Some of these processes are discussed in Section 7.5.5. Thus other terms are required in the denominator of Eq. (7.59), and the catalytic efficiency can be reduced relative to the efficiency in the remote atmosphere. Nevertheless, $O_3$ production rates become so large that mixing ratios can build to levels exceeding standards set by environmental agencies.

The gross production rate of $O_3$ has the same form as Eq. (7.42), but the expression must now sum over all of the various organic peroxy radicals present. In addition, their individual rate coefficients may not be nearly identical to that for $HO_2$ or $CH_3O_2$:

$$P(O_3) = (k_{12}[HO_2] + \Sigma k_{40i}[RO_2]_i)\,[NO] \qquad (7.62)$$

A further and very important complication is that now the concentration of $NO_x$ is often in the region where the concentration of peroxy radicals may be strongly and nonlinearly dependent upon $NO_x$; that is, $NO_x$ is in the region of strong falloff of OH or $HO_2$ (Fig. 7.30). $NO_x$ remains a strong determinant in the partitioning of $HO_x$ into OH and $HO_2$, but $RO_2$ can produce both $O_3$ and $HO_2$ through (7.40$i$) and (7.41$i$). The $HO_x$ concentrations are also a strong function of the loading and variety of hydrocarbons, and the total organic peroxy radical concentration can exceed that of $HO_2$. Consequently, the formation of $O_3$ is strongly and nonlinearly affected by the $NMHC/NO_x$ ratio.

Figure 7.32 provides an example of this nonlinear chemistry where a model is used to illustrate the consequences of adding the natural unsaturated hydrocarbon isoprene (2-methyl-1,3-butadiene) to a rural continental atmosphere under July solar insolation conditions. Case (a) represents the case in which only typical concentrations of CO and $CH_4$ are the oxidizable fuels. The OH and $HO_2$ variation with $NO_x$ is similar to that from the model used to generate Figure 7.30. Note that in Figure 7.32 the ordinate scale is linear rather than logarithmic. Case (b) is run with the isoprene mixing ratio held constant at 400 pptv for various fixed mixing ratios of $NO_x$ ranging from 0.1 to 10 ppbv. Here, for $NO_x$ below ~1 ppbv, the OH concentration is reduced substantially relative to Case (a), mainly due to the reactivity of OH with isoprene compared to CO and $CH_4$. In contrast, peroxy radicals ($HO_2 + RO_2$) are enhanced considerably and thus so is the rate of production $O_3$ at an equivalent value of $NO_x$. The overall result in this example is a much larger abundance of $O_3$.

In urban centers, the $NO_x$ mixing ratio can be much larger than 10 ppbv. Consequently the peroxy radical concentration can be reduced to extremely small values, and the rate of $O_3$ production is then limited by their availability. Under these conditions, (or under conditions of low insolation due to clouds or wintertime at mid- and high latitudes), $O_3$ can be lost or "titrated" by these large $NO_x$ abundances through Reaction (7.3),

$$NO + O_3 \rightarrow NO_2 + O_2 \qquad (7.3)$$

provided $NO_2$ is lost or transformed to other reservoirs competitively with photolysis to O atoms. Extreme examples are found in measurements of strong reduction in $O_3$ on flights through point sources of strong emissions such as exhaust plumes from coal or gas-fired electrical utilities. However, $O_3$ in urban centers where $NO_x$ can be very high is also often subject to this effect. Consequently, controls on $NO_x$ alone

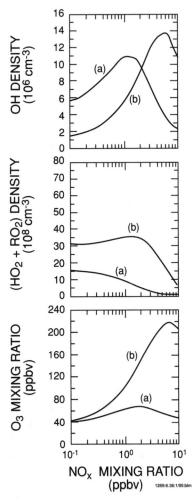

**Figure 7.32.** Model calculated OH, $HO_2 + RO_2$, and $O_3$ for July insolation at a rural continental surface site showing the effect of addition of 400 pptv of the natural hydrocarbon isoprene. (a) No isoprene, (b) with isoprene (courtesy of M. Trainer).

may actually lead to an increase in urban $O_3$, but a decrease in rural (or downwind) $O_3$.

### 7.5.5 Odd-Nitrogen Loss and Transformation Processes in the Continental Boundary Layer: Lifetime of $NO_x$

A strong determinant of the catalytic efficiency of $NO_x$ is its atmospheric lifetime [see Eq. (7.58)]. If we consider nitric acid formation (7.18) to be the only process by which $NO_2$ (or $NO_x$) is transformed to a reservoir constituent, then the lifetime of $NO_x$ may be expressed in terms of the time constant for $NO_2$ to $HNO_3$ conversion:

$$\tau_{NO_x} = \tau_{NO_2}(1 + [NO]/[NO_2]) \tag{7.63}$$

where $\tau_{NO_2} = 1/k_{18}[OH][M]$. Thus for this restricted case, the photochemical lifetime of $NO_x$ depends strongly on the OH concentration and the $[NO]/[NO_2]$ ratio. In the upper troposphere, where the OH concentration is usually smaller than near the surface, the smaller OH and large $NO/NO_2$ ratio [cf. Eq. (7.43)] yield $\tau_{NO_x} \sim$ 4-7 days depending upon the OH concentration. In contrast, in the lower troposphere, high OH concentrations and a small $NO/NO_2$ ratio yield a photochemical lifetime of the order of a day when only (7.18) is considered. However, with the high concentration and variety of hydrocarbons present in the continental boundary layer, a number of new $NO_x$ loss and transformation processes become important. These processes and wet and dry deposition conspire to keep the lifetime of $NO_x$ in the summer CBL as short as a few hours. Some of the transformation processes have been given in the previous discussion of the stratosphere and troposphere, but a more complete listing is given in the remainder of this chapter. Some of these new processes are also important in the remote atmosphere, whether it be in the boundary layer or free troposphere. However, many of the details are not well known, especially those concerning a variety of possible heterogeneous processes. Nevertheless, as research progresses, and as new processes are identified that could significantly reduce the lifetime of $NO_x$, the general effect will be to reduce the calculated efficiency of the $NO_x$ catalyst in $O_3$ production cycles. It must also be recalled that many of the transformation reactions are dynamic in that over time the reservoirs can undergo reactions that liberate $NO_x$. Thus with transport of air masses to locations distant from formation regions, these reservoirs can be important sources of active nitrogen to remote regions. However, it is fair to say that limitations on our understanding of the details of transport and loss processes, rather than the gas-phase photochemical interactions, cause the biggest uncertainty in large-scale model calculations.

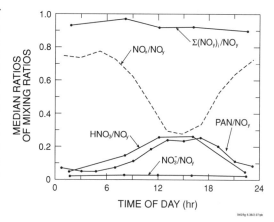

**Figure 7.33.** The diurnal variation of the partitioning of $NO_y$ observed at a continental U.S. surface site, Scotia PA, during summer 1987. In comparison to the observations made at a remote site shown in Figure 7.28, $NO_x$ and $HNO_3$ are larger and smaller components of $NO_y$, respectively. There is a suggestion of a small amount of missing reactive nitrogen, but when all of the measurement uncertainties are considered, it is not significant (Fehsenfeld et al., 1988; reprinted with kind permission from Kluwer Academic Publishers).

An example of the abundance of reservoir species is given by Figure 7.33, which illustrates the partitioning of total odd nitrogen, $NO_y$, from measurements at a rural continental site in the United States. The major reservoir species, in order of relative abundance, are $HNO_3$, PAN, other organic nitrates, and particulate nitrate. $HNO_3$ constitutes a smaller fraction of $NO_y$ at Scotia than in the remote Pacific (cf. Fig. 7.28), mostly due to larger removal rates to the surface in the CBL. Because of the proximity to sources at Scotia, the $NO_x$ fraction is much larger than at Mauna Loa.

## 7.6 Experimental Summary of the Influence of Odd Nitrogen in the Continental Boundary Layer

Although the oxidation chemistry in the CBL can be very complex, recent studies allow a convenient experimental summary of the sensitivity of $O_3$ to $NO_x$ emissions in regions influenced significantly by urban centers. Just as fuels are oxidized to produce $O_3$, $NO_x$ is transformed or oxidized to $NO_y$ reservoir constituents. The *overall* effect of all the production and loss processes can be determined approximately from observations of $O_3$ and $NO_y$ constituents other than active $NO_x$, that is, $NO_y - NO_x$. The latter difference represents the odd-nitrogen reservoirs and is more readily measured than are all of the individual odd-nitrogen species. Figure 7.34 shows $O_3 + NO_2$ versus $NO_y - NO_x$ from a long-term study in a rural region of Germany that can be influenced, depending upon the wind direction, by emissions from the heavily populated Rhine Valley region. The data show a significant correspondence between these two variables. Figure 7.35 shows results from a number of studies in rural areas of the northeastern United States and one from southern Ontario, Canada, in summer. These data have been filtered to include only measurements when the photochemical activity was high. The behavior and magnitude of the slope is similar to the studies made in Germany. Furthermore, the relationship is in good agreement with a regional model and is quite insensitive to the hydrocarbon loading, confirming that the availability of $NO_x$ is dominant in the extent of $O_3$ production. Because the rapid loss or transformation of $NO_x$ to reservoirs roughly balances the emissions of active nitrogen in these regions, the slopes of the curves in the figures are interpreted approximately as the amount of $O_3$ formed per unit of $NO_x$ emitted or consumed in the oxidation sequences. At reservoir mixing ratios of 5-10 ppbv, these studies, whether in Germany or the United States, give 5-8 ppbv of $O_3$ produced per ppbv of $NO_x$ emitted, and slightly larger production rates at lower values of $NO_y - NO_x$. The curvature is interpreted mainly as a result of the change in the catalytic efficiency of $NO_x$, although it can also be

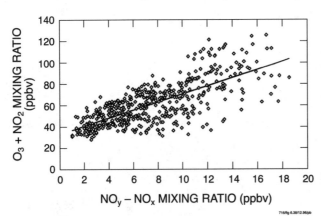

**Figure 7.34.** The relation between $O_3 + NO_2$ versus $NO_y - NO_x$ from a ground site in Schauinsland, Germany (Volz-Thomas et al., 1993). The slope of this plot is interpreted approximately as the ozone production efficiency, the number of $O_3$ molecules produced for each $NO_x$ molecule emitted or oxidized to reservoir components. In these studies the $NO_2$ mixing ratios were a significant fraction of the $O_3$ mixing ratios, and thus their sum is used as the ordinate scale.

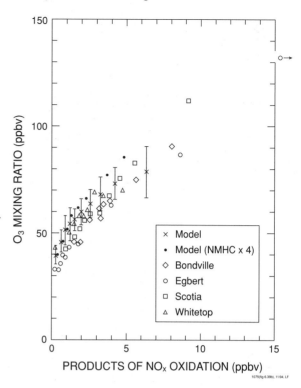

**Figure 7.35.** The average observed $O_3$ versus $NO_y$ - $NO_x$ or the products of $NO_x$ oxidation from four widely dispersed (~1600 km north to south) ground sites in the Eastern United States and Canada (Trainer et al., 1993). The data have been restricted to active photochemical periods. Model calculations for two different levels of NMHC are in good agreement with the observations. The slope of the correlation, the ozone production efficiency per $NO_x$ emitted, is slightly higher than observed in Germany. In these studies $NO_2$ was a small fraction of the $O_3$ mixing ratio.

affected by the NMHC/$NO_x$ ratio. These experimentally derived estimates are upper limits, since $O_3$ is likely not removed as efficiently as some of the $NO_y$ reservoirs by physical processes such as deposition. Nevertheless, they allow an upper-limit estimate of the $O_3$ abundance expected from anthropogenic and biogenic emission inventories in the CBL near source regions.

In winter, measurements made in Germany show that the association between $O_3$ and $NO_y$ - $NO_x$ is not nearly as robust. On average there is a decrease in $O_3$ with increasing odd-nitrogen reservoir concentrations. That large emissions of $NO_x$ can lead to a significant reduction of $O_3$ under conditions where photochemical activity is weak was noted earlier. The decrease may also have a component from heterogeneous processes involving the formation of $NO_3$ and $N_2O_5$, which are expected to be more important in the low insolation of winter.

## 7.7 $NO_3$ Chemistry

As described earlier in this chapter, the nitrate radical, $NO_3$, plays a role in both tropospheric and stratospheric chemistry. It is not usually regarded as a member of the $NO_x$ family, since its importance is mostly limited to nighttime, while NO-related chemistry can usually only be important during the day. Nonetheless, $NO_3$ is closely coupled with $NO_2$, and they are easily interconverted by reactions or photolysis. The major source of $NO_3$ in all regions of the atmosphere is the reaction of $NO_2$ with $O_3$

$$NO_2 + O_3 \rightarrow NO_3 + O_2 \qquad (7.15)$$

The reaction of $HNO_3$ with OH also provides a daytime source of $NO_3$

$$HNO_3 + OH \rightarrow H_2O + NO_3 \qquad (7.22)$$

During the day, NO$_3$ is photolyzed rapidly to NO or NO$_2$, with a lifetime of about 5 sec for overhead sun at the Earth's surface

$$NO_3 + h\nu \rightarrow NO + O_2 \qquad (7.7a)$$
$$NO_3 + h\nu \rightarrow NO_2 + O \qquad (7.7b)$$

The latter process dominates by about a factor of 12. The reason for the rapid photolysis rate is the intense absorption spectrum in the visible (Fig 7.36). The band at 662 nm is not dissociative, while absorption in the 623 nm band only leads to dissociation with an efficiency of 10-15%. However, absorption in the broad, weakly structured spectral region below 615 nm is sufficient to account for the short atmospheric lifetime of NO$_3$ during daytime.

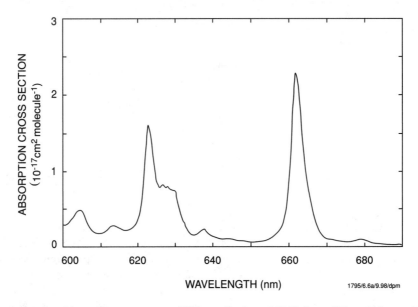

**Figure 7.36.** *Absorption spectrum of NO$_3$ radicals at 298 K from 600 to 690 nm. The strong bands at 623 and 662 nm have both been used to monitor NO$_3$ in the atmosphere by DOAS spectroscopy. At colder temperatures the bands become more intense (from Sander, 1986; reprinted with permission from J. Phys. Chem., copyright 1986, American Chemical Society).*

At night, on the other hand, NO$_3$ is quite long lived, since it does not react particularly rapidly with organic molecules, and the concentrations of free radicals, with which it does react, tend to be small at night. The predominant reaction of NO$_3$ in both the troposphere and stratosphere is with NO$_2$ to form N$_2$O$_5$; the consequences of this reaction are quite different in the two regions, however, due to the differences in temperature encountered.

### 7.7.1 NO$_3$ in the Troposphere

The reaction of NO$_3$ with NO$_2$ to form N$_2$O$_5$ is reversible, and at the temperatures encountered in the lower troposphere equilibrium is usually maintained

$$NO_3 + NO_2 + M \rightleftharpoons N_2O_5 + M \qquad (7.16)$$

The equilibrium constant, $K_c$, is given by:

$$K_c = [N_2O_5]/[NO_3][NO_2] = 2.7 \times 10^{-27} \exp\left(\frac{11,000}{T}\right) \text{ cm}^3 \text{ molecule}^{-1}$$

The lifetime of $N_2O_5$ with respect to thermal decomposition is about 50 sec at 298 K and 1 atmosphere pressure. Thus the three molecules are in dynamic equilibrium in the lower troposphere. Consequently, $NO_3$ and $N_2O_5$ show coupled behavior, and anything that removes one will feed back into the other.

The most important reactions of $NO_3$ are with organic compounds, particularly alkenes. (For a discussion of the reaction mechanism, see Chapter 9.) The rate coefficients of these reactions are fairly small, usually falling between $2 \times 10^{-16}$ and $6 \times 10^{-12}$ cm$^3$ molecule$^{-1}$ s$^{-1}$, although the rate coefficients for some terpenes with $NO_3$ approach $1 \times 10^{-10}$ cm$^3$ molecule$^{-1}$ s$^{-1}$ (Wayne et al., 1991). Consequently, the lifetime of $NO_3$ is often longer than 1000 s, for relatively unpolluted air. Abstraction reactions of $NO_3$ with alkanes are very slow, but reaction with aldehydes is more rapid

$$NO_3 + CH_3CHO \rightarrow HNO_3 + CH_3CO \tag{7.64}$$

$NO_3$ also reacts rapidly with organic sulfur species, such as DMS

$$NO_3 + CH_3SCH_3 \rightarrow HNO_3 + CH_3SCH_2 \tag{7.65}$$

The occurrence of these $NO_3$ reactions is important since they allow oxidation processes to continue at night, when the OH concentration is low. The reactions also provide nighttime sources for peroxy radicals, which react further with $NO_3$ or with each other.

The reaction of $NO_3$ with NO is very rapid, $k = 2.5 \times 10^{-11}$ cm$^3$ molecule$^{-1}$ s$^{-1}$, and $NO_3$ does not normally build up to measurable levels in regions of high NO, such as those impacted by fresh automobile exhaust

$$NO + NO_3 \rightarrow 2\, NO_2 \tag{7.66}$$

Heterogeneous uptake on aqueous droplets can also affect the $NO_3$ concentration in the troposphere. The uptake of $N_2O_5$ is known to be rapid, and since $N_2O_5$, $NO_3$, and $NO_2$ are usually in equilibrium, the presence of droplets often reduces the $NO_3$ concentration to levels below the detection limit of current instrumentation. Direct heterogeneous uptake of $NO_3$ may also be possible, although the reaction between $H_2O$ and $NO_3$ is thought to be slow. However, $NO_3$ is a strong oxidizing agent, and it reacts rapidly with many anions to give nitrate and radicals such as Cl atoms, which themselves may play a role in aqueous-phase chemistry

$$NO_3 + Cl^- \rightarrow NO_3^- + Cl \tag{7.67}$$

### 7.7.2 $NO_3$ in the Stratosphere

As a result of the strong temperature dependence of the equilibrium constant $K_c$, the formation of $N_2O_5$ tends to be irreversible in the stratosphere, and $N_2O_5$ becomes a very effective reservoir for $NO_3$. As shown earlier, the build up of $N_2O_5$ after dusk is linear with time. $N_2O_5$ can under these circumstances be transported over long distances before undergoing either thermal dissociation, Reaction (7.16), or photolysis to release the $NO_3$

$$N_2O_5 + h\nu \rightarrow NO_2 + NO_3 \tag{7.17a}$$
$$N_2O_5 + h\nu \rightarrow NO + NO_3 + O \tag{7.17b}$$

In the stratosphere, the photolysis of chlorine nitrate can also be a significant source of $NO_3$

$$ClONO_2 + h\nu \rightarrow Cl + NO_3 \tag{7.6}$$

### 7.7.3 Measurements of $NO_3$

The strong absorption spectrum of $NO_3$ in the visible (see Fig. 7.36) has been used to measure $NO_3$ over long, open paths in the atmosphere. Both horizontal paths using artificial light sources and upward-looking paths using the moon or starlight have been used in different situations (Solomon et al., 1989a, and references therein).

#### 7.7.3.1 Tropospheric Measurements of $NO_3$

The most frequently used method to measure $NO_3$ involves the DOAS technique (differential optical absorption spectrometry), in which a bright lamp, usually a high-power xenon lamp, is directed over a path of several kilometers (Platt et al., 1984; Carslaw et al., 1997). The beam can be either detected at the remote site, or reflected back to a detector near the source to double the pathlength of the measurement. The first tropospheric measurements of $NO_3$ were made in a fairly remote continental site (Noxon et al., 1980); soon afterwards $NO_3$ was detected in marine regions (Noxon, 1983). When the lifetime of $NO_3$ was calculated from its formation rate and steady-state concentration, it was found that the lifetime was usually less than about 1000 s. It was at first hypothesized that the short lifetime was an indication of a thermal decay of $NO_3$, or the presence of a fairly well-mixed scavenger throughout the troposphere

$$NO_3 + M \rightarrow NO + O_2 + M \tag{7.68}$$

However, it is now thought more likely that the reduction of the $NO_3$ steady-state concentration is a result of scavenging of $N_2O_5$ by aerosols, which also reduces the $NO_3$ by virtue of the rapid equilibrium described above. Figure 7.37 shows some of the first measurements of $NO_3$ and $NO_2$ made using this technique.

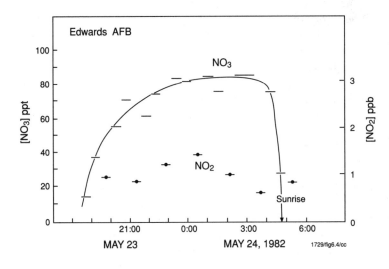

**Figure 7.37.** Nighttime concentration profiles for $NO_3$ and $NO_2$ at Edwards Air Force Base, California, May 24, 1982. The line shown is a smooth curve through the data, ignoring possible local maxima and minima in the concentration profiles (from Platt et al., 1984; reprinted with permission from Environ. Sci. Technol., copyright 1984, American Chemical Society).

#### 7.7.3.2 Stratospheric Measurements of $NO_3$

Measurements of $NO_3$ in the lower to mid-stratosphere have been made using the moon or stars as a light source and either ground-based (Noxon et al., 1978) or ballon-borne spectrometers (Naudet et al., 1981). In the case of ground-based measurements, the total $NO_3$ column above the spectrometer can be measured. However, by varying the slant angle of the measurement, information can be gained about the vertical distribution of the $NO_3$. Figure 7.38 shows measurements of the vertical column of $NO_3$ as a function of the solar zenith angle through sunrise at 40° North. As the sun rises, the $NO_3$ is photolyzed away irreversibly.

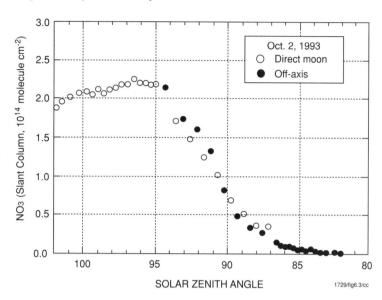

**Figure 7.38.** $NO_3$ slant column abundance versus solar zenith angle for the morning of October 2, 1993. Measurements were made alternating between direct Moon and the sky next to the Moon. Note that time increases from left to right, as the Sun rises (solar zenith angle decreases) (from Weaver et al., 1996).

## 7.8 Gaseous Acid and Particulate Nitrate Formation

As discussed previously, formation of $HNO_3$ is one of the dominant $NO_x$ transformations that occurs throughout the troposphere

$$OH + NO_2 + M \rightarrow HNO_3 + M \qquad (7.18)$$
$$HNO_3 + h\nu \rightarrow OH + NO_2 \qquad (7.21)$$
$$HNO_3 + OH \rightarrow H_2O + NO_3 \qquad (7.22)$$
$$HNO_3 + H_2O_{(l)}/\text{aerosol} \rightarrow \text{removal from gas phase} \qquad (7.69)$$

Being highly soluble, $HNO_3$ is readily incorporated in cloud or fog droplets, scavenged by aerosols, and it is therefore subject to wet and dry deposition. In the presence of gaseous ammonia it readily forms ammonium nitrate aerosol. In the marine boundary layer it is also readily scavenged by sea-salt aerosol. Thus, in the boundary layer and lower free troposphere, the lifetime of $HNO_3$ can be very short (<1 day) and is likely controlled more by physical or heterogeneous processes than by gas-phase reactions (7.21) and (7.22). Thus formation of $HNO_3$ can be an effective sink of active nitrogen in the lower atmosphere and boundary layer, where cloud formation is frequent and aerosol loading can be high. In the middle and upper troposphere, the lifetime can be much longer. For example, the lifetime against photolysis or reaction with OH is of the order of several weeks to a month so that transport is a major determinant of its large-scale distribution.

Other processes leading to the formation of $HNO_3$ and particulate nitrate involve the nighttime chemistry of $NO_x$, as was discussed in the previous section

$$NO_2 + O_3 \rightarrow NO_3 + O_2 \qquad (7.15)$$
$$NO_3 + NO_2 + M \rightleftharpoons N_2O_5 + M \qquad (7.16)$$

Nighttime loss or transformations of either $NO_3$ or $N_2O_5$ would be an important loss of $NO_x$ [and $O_3$ (7.3)]. Indeed these processes may effectively control the lifetime of

$NO_x$ in the long winter nights at middle and high latitudes. For example,

$$NO_3 + RH \rightarrow HNO_3 + R \tag{7.70}$$

$$NO_3 + R'H \rightarrow \text{organic nitrates} \tag{7.71}$$

$$NO_3 + \text{aerosols}/H_2O_{(l)} \rightarrow HNO_{3(l)} \tag{7.72}$$

where $R$H are organic compounds other than alkenes and $R'$H are alkenes. Hydrolysis of $NO_3$ may occur in fog and cloud droplets, although the heterogeneous reactions on, or scavenging by, tropospheric aerosols to produce $HNO_3$ in the aerosol phase or particulate nitrate have not been characterized in detail.

More detailed information is available for some of the heterogeneous processes involving $N_2O_5$, studies motivated largely by their importance to $O_3$ depletion in the lower stratosphere. Kinetic parameters for some of these reactions are available in Appendix H. In sulfuric acid aerosols (sa) the $HNO_3$ formed is mostly released to the gas phase due to the low solubility of $HNO_3$ in sulfuric acid solutions. In sea-salt aerosol or clouds, on the other hand, the $HNO_3$ (or $NaNO_3$) likely remains in the condensed phase

$$N_2O_5 + H_2O_{(l)} \rightarrow 2\, HNO_{3(l)} \tag{7.45}$$

$$N_2O_5 + NaCl_{(s)} \rightarrow ClNO_2 + NaNO_{3(s)} \tag{7.73}$$

$$N_2O_5 + H_2O_{(sa)} \rightarrow 2\, HNO_{3(g)} \tag{7.32}$$

Peroxynitric acid has not been measured in the troposphere, but model calculations suggest that its thermal instability limits its importance to the colder middle and upper troposphere (cf. Fig. 7.24)

$$HO_2 + NO_2 + M \rightleftharpoons HO_2NO_2 + M \tag{7.19}$$

$$HO_2NO_2 + OH \rightarrow H_2O + O_2 + NO_2 \tag{7.26}$$

$$HO_2NO_2 + h\nu \rightarrow HO_2 + NO_2 \tag{7.24}$$

$$\rightarrow OH + NO_3 \tag{7.25}$$

Nitrous acid has been measured in the boundary layer. It is readily photolyzed in the troposphere, and its formation can be a significant source of OH and NO in the early morning (7.75)

$$OH + NO + M \rightarrow HONO + M \tag{7.74}$$

$$HONO + h\nu \rightarrow OH + NO \tag{7.75}$$

Thus these processes will not lead to a significant loss of $NO_x$ unless the nitrous acid is removed by cloud or fog droplets or surfaces. There is also evidence that nitrous acid is formed predominantly through heterogeneous processes in the CBL rather than by the gas-phase reaction (7.74).

## 7.9 Chemistry of Organic Nitrates

Loss of $NO_x$ by reactions with the large variety of NMHC available in the CBL limits the production of $O_3$. Alkanes, alkenes, and other organic species can be oxidized in the presence of $NO_x$ to form organic nitrate compounds. This class of nitrogen-containing compounds is significant in the chemistry of reactive nitrogen in the atmosphere because they represent a temporary reservoir of active nitrogen, can act as a radical sink in much the same way as nitric acid, and thus can limit the production of $O_3$ in the CBL. Because some of the organic nitrates have lifetimes of weeks or more and can be transported to the more remote atmosphere, these compounds can be a source of $NO_x$, and hence can affect $O_3$ production in regions far from primary sources of $NO_x$ and NMHC. The full suite of different atmospheric organic nitrates is only beginning to be explored, and their role in atmospheric chemistry on regional and global scales is receiving more attention. Organic nitrate formation from $C_2$-$C_7$ alkanes and alkenes is used to illustrate the complex chemistry involved since both classes of hydrocarbons have important natural as well as anthropogenic sources.

### 7.9.1 Organic Nitrate Formation

Peroxyacetyl nitrate (PAN) is the predominant organic nitrate produced in the atmosphere, although higher carbon homologues, particularly peroxypropionyl nitrate (PPN), have been identified. PAN is formed from the reaction of acetyl peroxy radicals and $NO_2$ according to:

$$CH_3C(O) + O_2 + M \to CH_3C(O)O_2 + M \tag{7.76}$$

$$CH_3C(O)O_2 + NO_2 + M \rightleftharpoons CH_3C(O)O_2NO_2 + M \tag{7.77}$$

Acetyl radicals, $CH_3C(O)$, are produced during the oxidation of acetaldehyde, which is itself an oxidation product of $>C_2$ hydrocarbons. Thus PAN formation may be related to the oxidation chain of a variety of different hydrocarbons. In a similar way, higher peroxyacyl nitrates may be formed from peroxyacyl radicals with a longer carbon skeleton.

The primary mechanism for alkyl nitrate formation has been shown to be the reaction of NO with alkylperoxy radicals. The reaction sequence is initiated by OH radical reaction with an alkane ($RH$) followed by the oxidation of the alkyl radical

$$RH + OH \to R + H_2O \tag{7.60}$$

$$R + O_2 + M \to RO_2 + M \tag{7.61}$$

$$RO_2 + NO + M \to \{RO_2NO\} \to RONO_2 + M \tag{7.78i}$$

Reaction (7.78$i$) is a chain-terminating step and serves as a sink for both $NO_x$ and radicals. A competing reaction is the chain-propagating step:

$$RO_2 + NO \to RO + NO_2 \tag{7.40i}$$

Atkinson et al. (1982) demonstrated that $\alpha_i$ [$= k_{78i}/(k_{78i} + k_{40i})$] depended on the alkyl chain length and molecular configuration as well as the pressure. The fraction of molecules, $\alpha$, reacting via (7.78) increased from $\leq 0.014$ for ethane to 0.33 for octane, while the nitrate yield from methyl peroxy is thought to be $< 10^{-3}$.

It is interesting to note here the relationship between organic nitrate production and ozone formation. From the discussion of NMHC oxidation earlier, it was shown that Reaction (7.40$i$) leads directly to ozone production. Since a fraction of the time $RO_2$ radicals react to form $RONO_2$ rather than $NO_2$, there is a relationship between the rate of alkyl nitrate production and ozone production [Reaction (7.62)]. This relationship can be expressed as

$$P(RONO_2) = \overline{\alpha} \times P(O_3) \tag{7.79}$$

where, for small $\alpha_i$,

$$\overline{\alpha} \approx \frac{\Sigma \alpha_i \times (k_{40})_i \times [RO_2]_i}{\Sigma (1 - \alpha_i) \times (k_{40})_i \times [RO_2]_i} \tag{7.80}$$

This relationship has been used to evaluate the contribution of $RO_2$ to ozone production in a photochemically active air mass in Germany (Flocke et al., 1994). This analysis showed that alkyl-$RO_2$ contributed about 8.5% to the total ozone production during transport from an urban source to a mountain receptor site.

Alkene oxidation is also expected to produce organic nitrates. Both OH and $NO_3$ radicals add to the double bond in alkenes, followed by addition of $O_2$ to form a peroxy radical (see Section 9.3.1.1). In general, difunctional nitrates are produced — β-hydroxy nitrates from OH addition [via Reaction (7.78)] and dinitrates or carbonyl nitrates following $NO_3$ addition. For example, in the laboratory oxidation of propylene, the reported products are α-(nitrooxy)acetone, 1-nitrooxy-2-propyl-alcohol, and 2-nitroxy-1-propyl alcohol as major nitrate products, with minor amounts of propylene glycol dinitrate (Shepson et al., 1985).

Of potentially greater importance than the oxidation of simple alkenes is $NO_3$ reaction with unsaturated hydrocarbons, such as isoprene and others, from biogenic emissions. Laboratory studies have shown that $NO_3$-alkene reactions are rapid and can have a potentially significant impact on isoprene chemistry. Organic nitrate formation from the oxidation of biogenic hydrocarbons has not been studied in great detail, but its importance as a potential reservoir of atmospheric reactive nitrogen has been noted by several workers.

Other reactions of radical species with NO or $NO_2$ can also occur, but the significance of the products formed to odd-nitrogen chemistry has not been demonstrated. One reaction that may occur is the reaction of $RO_2$ radicals with $NO_2$ to form alkyl peroxynitrates:

$$RO_2 + NO_2 + M \rightarrow RO_2NO_2 + M \qquad (7.81)$$

These species are thermally much less stable than PAN (similar to $HO_2NO_2$) and may serve as $NO_x$ reservoirs only at the low temperatures of the upper troposphere.

Organic nitrites and nitrates may be formed at high $NO_x$ from reactions of alkoxy radicals with NO or $NO_2$

$$RO + NO + M \rightarrow RONO + M \qquad (7.82)$$
$$RO + NO_2 + M \rightarrow RONO_2 + M \qquad (7.83)$$

However, alkoxy radicals tend to react with $O_2$ (which is present in high abundance) or decompose, rather than reacting with $NO_x$ species (see Chapter 9). Thus Reaction (7.78) is the major source of alkyl nitrates in the troposphere.

### 7.9.2 Occurrence of Organic Nitrates in the Global Troposphere

The most abundant organic nitrate, PAN, has been measured extensively in urban regions, and less commonly in remote continental or marine atmospheres. Table 7.2 gives a summary of observations. Typically, urban centers have been found to produce the largest amounts of PAN. Mixing ratios of PAN of >10 ppbv have been reported in the Los Angeles basin, while values near 1 ppbv are not uncommon in many urban centers. The photochemical origin of PAN in these areas is usually reflected in a strong diurnal cycle. PAN mixing ratios tend to peak in the late afternoon and are well correlated with levels of other secondary photochemical products such as ozone.

Given the urban source of PAN from hydrocarbon and $NO_x$ precursors, it has been surprising that measurements of PAN from more remote areas showed that this compound is nearly ubiquitous throughout the troposphere (Table 7.2 and Figure 7.22b). As noted above, this finding has important implications on the long-range transport of reactive nitrogen stored in a mobile chemical reservoir species, PAN. The concentration of PAN generally decreases from urban centers ($\geq 1$ ppbv) to nonurban continental areas (0.1-1 ppbv) to the marine atmosphere (<0.2 ppbv). This decrease reflects, in part, the availability of precursor hydrocarbons and $NO_x$ in the more remote regions. Away from strong ground-level sources of PAN, the mixing ratio of PAN tends to increase with increasing altitude. Thus there is a potential upper-level source of reactive nitrogen throughout the troposphere. However, whether this PAN is formed in situ at altitude or is transported to the colder free troposphere where it is more stable and may accumulate is still undetermined.

**Table 7.2**
*Some Representative Measurements of Peroxyacetyl Nitrate (PAN) in the Troposphere; Mixing Ratios Given in ppbv*

| Site | Average | Maximum |
|---|---|---|
| *Urban* | | |
| Delft, The Netherlands | 0.42 | 0.94 |
| Munich, FRG (January, 1990) | 0.27 | 2.23 |
| (April, 1990) | 0.61 | 1.99 |
| Bonn, FRG | 0.08 | 0.17 |
| Philadelphia, PA | 1.0 | 3.7 |
| Denver, CO | 0.64 | 2.04 |
| Claremont, CA | 3.2 | 11.8 |
| Dübendorf, Switzerland | 0.29 | 4.40 |
| *Nonurban continental* | | |
| Badger Pass, CA | 0.13 | 0.22 |
| Simcoe, ONT | 0.4 | 1.7 |
| North Bay, ONT | 0.61 | 2.5 |
| Alert, NWT | 0.47 | — |
| Tanbark Flat, CA | 2.9 | >16 |
| Aircraft <2.6 km (N. America) | 0.172 | — |
| Aircraft 4.6-6.1 km (N. America) | 0.149 | — |
| Arctic/Subarctic 0-2 km | 0.025 | — |
| Arctic/Subarctic 2-4 km | 0.13 | — |
| Arctic/Subarctic 0-2 km | 0.26 | — |
| Niwot Ridge, CO (Easterly) | 0.810 | 4.02 |
| Niwot Ridge, CO (Westerly) | 0.205 | 1.47 |
| Scotia, PA | 1.000 | — |
| Davos, Switzerland | 0.11 | 0.98 |
| Lindau, FRG (April, 1989) | 1.14 | 4.7 |
| (December, 1988) | 0.16 | 1.14 |
| *Remote/marine* | | |
| N. Pacific Ocean | 0.074 | 0.4 |
| N. Hemisphere (Atlantic) | 0.05 | 0.16 |
| S. Hemisphere (Atlantic) | 0.01 | 0.024 |
| N.E. Atlantic, Boundary layer | 0.184 | — |
| N.E. Atlantic, Free troposphere | 0.094 | — |
| N. Pacific, 3.5 km | 0.017 | — |
| N. Pacific, 5.4-5.0 km | 0.137 | — |
| N. Pacific, 0.6 km | 0.007 | — |
| N. Atlantic, surface "background" | 0.005 | — |

Partially adapted from data in Roberts (1990)

Whether the PAN reservoir is increasing or decreasing has important implications on oxidant formation in remote areas. As discussed earlier, much of the remote troposphere is in near balance with respect to rates of ozone formation and destruction. An increase in $NO_x$ from an increasing reservoir of PAN (and other labile odd-nitrogen reservoirs) could theoretically drive the net balance toward ozone production in large areas of the troposphere that are currently sinks for tropospheric ozone.

Evidence for a general temporal trend in PAN is mixed. Long-term measurements of PAN in the Netherlands do suggest an increase of nearly 10% annually in the average concentration from 1973 to 1985. Over a similar time period, measurements in the Los Angeles area show no similar increase. Insufficient data are available to evaluate long-term trends in remote areas or in the free troposphere, though no apparent trend has been found over five years in the Canadian Arctic or Eastern Canada.

A limited database is currently available for other organic nitrates in the troposphere. Alkyl nitrates have been measured in different environments and have been found in low concentrations in virtually all regions of the Northern Hemisphere, with trace quantities (<1 pptv) found even in remote areas of the Equatorial and South Pacific Ocean. It is now thought that the oceans may provide a relatively large source of methyl nitrate and thereby a source of $NO_x$ to the remote troposphere. Multifunctional nitrates, such as hydroxy-alkyl nitrates, have been tentatively identified in air from Shenandoah National Park, Virginia, Boulder, Colorado, and rural regions of Canada (O'Brien et al., 1995), but reliable quantitative data are not yet available for these compounds. Similarly, complex peroxyacylnitrates (e.g., MPAN from methacrolein) have been tentatively identified in rural areas with large isoprene emissions, but quantitative data are still sparse.

### 7.9.3 Sinks for Organic Nitrates

The main sinks for organic nitrates discussed here include thermal degradation and photolysis as well as oxidation by OH radical. Thus there is a strong seasonal component as well as altitudinal dependence to the loss rate of organic nitrates from the atmosphere.

PAN can be slowly photolyzed, but its main loss process in the lower troposphere is via thermal degradation. The overall lifetime of PAN is dependent on the ratio of $NO/NO_2$ as well as the temperature. Acetyl peroxy radicals undergo competition between reaction with NO and $NO_2$

$$CH_3C(O)O_2 + NO_2 + M \rightarrow PAN + M \quad (7.77)$$

$$CH_3C(O)O_2 + NO \rightarrow CH_3CO_2 + NO_2 \quad (7.84)$$

Reaction with $NO_2$ (7.77) leads to no net loss of PAN, while the NO reaction leads to irreversible loss. Hence, the effective lifetime of PAN is given by

$$\tau_{\text{eff}} = \tau_d \left(1 + \frac{k_{77}\,[NO_2]}{k_{84}\,[NO]}\right)$$

where $\tau_d$ is the lifetime for decomposition. Singh et al. (1990) calculate that the lifetime of PAN (in the absence of new PAN synthesis) corresponds to 2, 24, and 600 sunlit hours for altitudes of 0 km (294 K), 3 km (279 K), and 6 km (261 K). This temperature-dependent stability is a major factor in producing the observed increase of PAN with altitude. A similar temperature sensitivity is expected for other acylperoxy nitrates.

Alkyl nitrates have a much higher thermal stability than the peroxyacyl (or peroxyalkyl) nitrates, and they are essentially stable at all temperatures found in the troposphere. Photolysis and oxidation by OH radical are more important loss processes for these compounds. With photolysis rates of $C_1$-$C_4$ alkyl nitrates on the order of 1 to $3 \times 10^{-6}$ sec$^{-1}$, the photolytic lifetime is on the order of 3-12 days, which is similar to the lifetime against OH oxidation. Thus photolysis may be the most important removal process for $C_1$-$C_4$ alkyl nitrates, while OH oxidation rates exceed photolytic degradation for >$C_5$ nitrates. The combined effects of OH oxidation, photolytic degradation, and transport lead to the observed seasonal signal in the abundance of alkyl nitrates in the troposphere. This is illustrated for the mid-Pacific troposphere with data from the MLOPEX II experiment (Fig. 7.39).

Loss rates for multifunctional and more complex organic nitrates are estimated for only a few of the organic nitrates. Recent studies show that photolysis and oxidation by OH are competitive processes for some difunctional organic nitrates. Photolysis lifetimes on the order of 1-6 days are estimated (Barnes *et al.*, 1993). As organic nitrates add functional groups, they tend to become less volatile and will likely become associated with aerosol particles. Also, increased solubility, for example, of hydroxynitrates, might increase the probability of precipitation scavenging of these compounds.

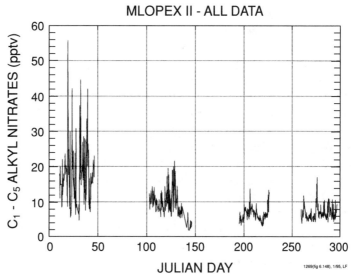

**Figure 7.39.** Alkyl nitrates measured during MLOPEX 2 (E. Atlas, private communication, 1997).

## Further Reading

Glueckauf, E. (1950) The composition of atmospheric air, in *Compendium of Meteorology*, American Meteorological Society, Washington.
Junge, C. E. (1963) *Air Chemistry and Radioactivity*, Academic Press, New York.
Leighton, P. A. (1961) *Photochemistry of Air Pollution*, Academic Press, New York.
Smith, R. A. (1872) *The Beginnings of a Chemical Climatology*, Longmans, Green and Co., London.

Ian Galbally

## Time's Arrow

It is perhaps interesting to look at the lineage of this book and the history of research concerning nitrogen oxides in the atmosphere.

The first book on atmospheric chemistry (known to but not seen by the writer) is Smith's *The Beginnings of a Chemical Climatology*, published in 1872. Perhaps the first clear indications of the presence of nitrogen oxides in the atmosphere came about by measuring nitrates and nitrites in rainfall in a similar time period. Prior to the discovery of nitrogen-fixing organisms, fixed nitrogen in rainfall was considered to be the source of nitrogen for plant growth; so measurements were made in many locations worldwide.

It wasn't until the late 1930s that nitrogen dioxide was reliably measured in the atmosphere (using a technique that involved preconcentration on silica gel) showing concentrations of 20 ppb in large towns decreasing to less than 500 ppt on most occasions in the countryside. These observations were summarized by Glückauf (a distinguished atmospheric chemist of that time) in his classic review of the composition of the atmosphere in 1950, which suggested, on the basis of these observations, that nitrogen dioxide might not be a normal constituent of air, but one that occurred due to combustion processes and was found primarily in the polluted atmospheres of large towns.

The 1950s brought about a revolution in the understanding of chemical kinetics and photochemistry. At the same time it was discovered that nitrogen oxides, hydrocarbons, and sunlight were the precursors for photochemical smog. Philip Leighton synthesized the results of this decade of discovery in his book, *Photochemistry of Air Pollution*. He identified the photostationary state that links nitric oxide, nitrogen dioxide, and ozone. In the same monograph he suggested that free radicals are probably formed in the air during daylight hours and that peroxy radicals could oxidize nitric oxide in polluted air. This foreshadowed the work of Hiram "Chip" Levy II, Paul Crutzen, and Harold Johnston on the photochemistry of atmospheric ozone and nitrogen oxides in the troposphere and stratosphere.

In 1963 in his book, *Air Chemistry and Radioactivity*, Christian Junge suggested the possible importance of biological processes and lightning as sources of these nitrogen oxides. He also summarized the more recent observations up to that time that indicated around 1 ppb of nitrogen dioxide in the background atmosphere.

The evolution of modern analytic measurement methods for nitrogen oxides was central to the next round of developments. The gas-phase chemiluminescent nitric oxide detector, the photolytic convertor for nitrogen dioxide and the long-path optical absorption methods for nitrogen dioxide and nitrogen trioxide, and a host of other methods allowed, for the first time, the direct determination of the various sources and sinks and atmospheric distribution of nitrogen oxides, and allowed the chemical kinetic schemes developed in laboratories to be tested in the atmosphere by direct observations.

In spite of this progress, it is evident that we have not well quantified the sources of tropospheric nitrogen oxides, particularly those of soils and lightning. We have yet to be able to represent the three-dimensional distribution of nitrogen oxides realistically in our models, and a significant amount of the total nitrogen oxides ($NO_y$) remains as unidentified species. Perhaps the greatest challenge facing us in this field is to build regional and global models that couple atmospheric chemistry and the basic physical, biogeochemical, and anthropogenic processes that act as sources and sinks of atmospheric trace gases.

As surely as time's arrow speeds on, we can expect these and other issues to be unraveled before the students of this book grow old.

*Ian Galbally is leader of the Global Atmospheric Change Programme at the Division of Atmospheric Research of CSIRO in Australia. He has made important contributions to the study of trace gas exchanges between the biosphere and the atmosphere. He convened the meeting at Dookie College in 1988 that led to the creation of the International Global Atmospheric Project (IGAC).*

# 8 Halogen Compounds

## 8.1 Introduction

Interest in the chemistry of halogen compounds in the atmosphere largely resulted from the proposal in the early 1970s that chlorine chemistry in the stratosphere could lead to catalytic ozone destruction, and that the chlorofluorocarbons (CFCs) could provide a significant stratospheric source of chlorine (Molina and Rowland, 1974; Stolarski and Cicerone, 1974). The CFCs are entirely of anthropogenic origin, and have found use as foam-blowing agents, as refrigerants, and as solvents. These compounds are quite stable in the atmosphere and have lifetimes of between 50 and 500 years or more. These long lifetimes, coupled with their strong infrared absorptions, also make the CFCs significant greenhouse gases.

Although the CFCs make up the greatest fraction of the halogen burden in the atmosphere, a number of other compounds must also be considered when compiling an atmospheric halogen budget (see Chapter 5). Hydrofluorocarbons (HFCs) and hydrochlorofluorocarbons (HCFCs) are currently being developed as replacements for the CFCs, and some of these compounds are already in use and have been detected in the atmosphere. Halons, which are fully halogenated compounds containing bromine, have been manufactured as fire suppressants. In addition, some natural sources of halogen exist in the form of partially halogenated methanes and ethanes. HCl is also an important contributor to the tropospheric chlorine budget, with a wide variety of natural and anthropogenic sources.

In this chapter, the sources, atmospheric trends, and losses of the halogenated source gases that are emitted to the atmosphere will be described, and the chemistry that ensues following their destruction, both in the stratosphere and troposphere, will be discussed. Special attention is drawn to heterogeneous processes that affect the stratospheric chemistry of halogens and are implicated in the formation of the Polar Ozone Hole in the Antarctic. Finally, the strategies currently being pursued to limit halogen levels in the atmosphere and expected future trends will be summarized.

## 8.2 Scope and Definitions

The halogens are found in the atmosphere in two general forms, organic and inorganic. The organic species are often referred to as "source gases," since these species are released at the Earth's surface (from a variety of sources, some natural, and some manmade). Destruction of the source gases in the atmosphere leads to the release of the halogens, which are then partitioned among a number of inorganic forms through chemical reactions.

The organic source gases are often identified according to the following short forms:

CFC, HCFC, HFC: Chlorofluorocarbons, hydrochlorofluorocarbons, and hydrofluorocarbons. These compounds are manufactured for use as refrigerants, foam-blowing agents, and solvents. The manufacture of CFCs is now banned, due to their impact on stratospheric ozone levels, and the HCFCs will progressively be phased out of production also. The HFCs have been shown to be "ozone friendly" and are replacing their chlorine-containing analogs in industrial applications.

Halons: These bromine- and fluorine-containing organic compounds have been used as fire-extinguishing compounds in the past. Their production is now banned owing to their harmful effects on stratospheric ozone. Examples include Halon 1301 ($CF_3Br$) and Halon 1211 ($CF_2ClBr$).

The major forms of inorganic halogens are as follows (where $X$ = F, Cl, Br, and I):

$X$: Halogen atoms are produced in the oxidation of halogen-containing organic source gases. The halogen atoms react with ozone to generate $XO$, or react with hydrogen-containing compounds to produce $HX$. Lifetimes for $X$ atoms in the stratosphere are typically very short, on the order of a second or less.

$XO$: Halogen monoxides are formed in the reaction of $X$ with ozone, or from the destruction of halogen nitrates. They are highly reactive, and in the case of ClO and BrO, are involved in numerous catalytic cycles that destroy stratospheric ozone.

$HX$: The hydrogen halides are the least reactive of the inorganic halogen species. They serve as important reservoir species for stratospheric halogens.

$HOX$: Hypohalous acids are formed in the reaction of $XO$ with $HO_2$ and from the hydrolysis of $XONO_2$, and are usually destroyed by photolysis in the stratosphere.

$XONO_2$: The halogen nitrates, formed in the reaction of $XO$ with $NO_2$, are important reservoir species for stratospheric halogens. They are destroyed primarily by photolysis.

## 8.3 Sources of Halogens

Following the initial proposals of the potential impacts of CFCs on stratospheric ozone, measurement programs were established to monitor the concentrations of organic halogen source gases in the atmosphere and to determine temporal and spatial trends in their abundances. Thus a global data network is now available from which the budget of these halogen source gases can be assessed. The atmospheric abundances, trends, and lifetimes of the primary organic halogen species are listed in Table 8.1. All the CFCs, HCFCs, Halons, $CH_3CCl_3$, and $CCl_4$ have purely anthropogenic sources, while $CH_3Cl$ is the major contributor to the natural halogen budget.

### 8.3.1 Sources and Trends of Atmospheric Chlorine and Fluorine

The total amount of organic chlorine in the atmosphere is calculated by multiplying the mixing ratio of each compound by the number of chlorine atoms present in that compound. The resulting values for all species measured in the troposphere are then added together to give the burden of organic chlorine. The 1990 value for total organic chlorine in the troposphere based on the compounds listed in Table 8.1 is 3.8 ppbv. Approximately 85% of the total is from compounds with solely anthropogenic sources.

## Table 8.1
*Abundances, Trends, and Lifetimes of Halogenated Organic Source Gases*

| Molecule[a] | Concentration (ppt) | | | Trend (%/year) | | | Lifetime[c] |
| --- | --- | --- | --- | --- | --- | --- | --- |
| | 1990[b] | 1992[c] | 1995[d] | e | c | d | (years) |
| $CF_2Cl_2$ (CFC 12) | 474-479 | 503 | 532 | 3.7-4.0 | 2.6 | 1.1 | 102 |
| $CF_3Cl$ (CFC 11) | 254-263 | 268 | 272 | 3.7-3.8 | 0.9 | -0.2 | 50±5 |
| $CF_2ClCFCl_2$ (CFC 113) | 71-75 | 82 | 84 | 9.1 | 3.1 | -0.8 | 85 |
| $CF_2ClCF_2Cl$ (CFC 114) | 15-20 | 20 | | 6 | | | 300 |
| $CF_3CFCl_2$ (CFC 114a) | 5 | | | 6 | | | ~400 |
| $CF_3Cl$ (CFC 13) | 5 | | | | | | 640 |
| $CCl_4$ | 107 | 106[d] | 103 | 1.2 | -0.8 | -0.8 | 42 |
| $CH_3CCl_3$ | 140[d] | 135[d] | 109 | 3.7 | 2.2 | -13.5 | 5.4±0.4 |
| $CHF_2Cl$ (HCFC 22) | 90[f] | 102[f] | 117 | 6-7 | 6.9 | 4.8 | 13.3 |
| $CH_3Cl$ | 600 | 600 | ~550 | | | | 1.5 |
| $CHCl_3$ | 15 | | | | | | 0.6 |
| $CH_2Cl_2$ | 30 | | | | | | 0.4 |
| $CCl_2CCl_2$ | 10 | | | | | | 0.3 |
| $CH_2ClCH_2Cl$ | 35 | | | | | | |
| $CHClCCl_2$ | 2-3 | | | | | | 0.01 |
| $CH_3Br$ | 10-15 | | | — | | | 1.3 |
| $CH_2Br_2$ | 0.5-3 | | | | | | 0.5 |
| $CHBr_3$ | 0.2-0.3 | | | | | | ~0.06 |
| $CH_2ClBr$ | 1-2 | | | | | | |
| $CHClBr_2$ | 1 | | | | | | |
| $CHCl_2Br$ | 1 | | | | | | |
| $C_2H_4Br_2$ | <1 | | | | | | |
| $C_2H_5Br$ | 2-3 | | | | | | |
| $CF_2ClBr$ (H 1211) | 1.2-2.2 | | | 15 | 3 | 3.2 | 20 |
| $CF_3Br$ (H 1301) | 1.7-2.5 | | | 20 | 8 | 3.0 | 65 |
| $CH_3I$ | ≈1-10 | | | — | | | 0.01 |
| $CH_3CFCl_2$ | | 0.3 | 3.5 | nd | ~200 | 54.0 | 9.4 |
| $CH_3CF_2Cl$ | | 3.5 | 7.0 | nd | ~30 | 22.2 | 19.5 |
| $CF_3CH_2F$ | | 0.3[g] | 1.73[g] | nd | nd | 100[g] | 14 |
| Total Cl/Br | 3800/30 | | | | | | |

[a] Nomenclature for some of these compounds are given in Appendix D.
[b] Kaye et al., 1994.
[c] WMO, 1995.
[d] Montzka et al., 1996a.
[e] WMO, 1992.
[f] Montzka et al., 1993.
[g] Montzka et al., 1996b.

These anthropogenic compounds are used primarily as refrigerants (CFCs-11, -12, and -114, HCFC-22), foam-blowing agents (HCFCs-142b, -141b, and -22, CFCs-11 and -12), solvents (CFC-113, HCFC-141b, methyl chloroform, carbon tetrachloride, chloroform, trichlorethylene, and perchlorethylene), and fire retardants (Halon-1211). Global production of CFC-11 and -12 began in the early 1960s (along with $CCl_4$) and increased rapidly through the mid-1970s. Maximum production occurred in 1988 and has decreased substantially since that time (Fig. 8.1). Production of CFCs, methyl chloroform, carbon tetrachloride, and the Halons has been phased out and production of the HCFC and HFC replacements has been phased in by developed countries as of

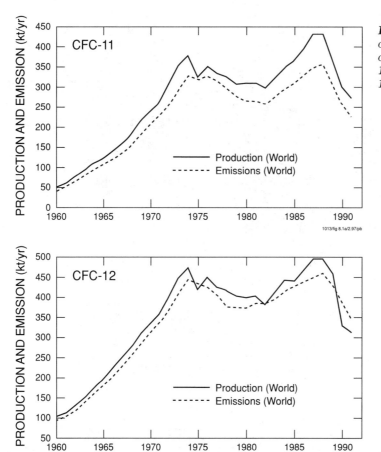

**Figure 8.1.** Estimates of global production of CFC-11 and CFC-12 (from Kaye et al., 1994).

1996, as called for by the 1987 Montreal Protocol report and subsequent amendments on regulation of substances that deplete the ozone layer.

However, emissions of the phased-out compounds are expected to continue for some time due to release of current inventories and use by developing countries. Figure 8.2 shows the emissions from 1972-1992 of the major CFCs, HCFCs, and Halons. The resulting effect of these emissions on the atmospheric mixing ratios for each compound depends on the individual lifetime (Table 8.1). Figure 8.3 shows the results of continuous monitoring of CFC mixing ratios at ground-based stations in the Northern and Southern Hemispheres since 1978. The monitoring efforts were conducted by two groups, NOAA Global Monitoring for Climate Change (NOAA/GMCC, now NOAA Climate Monitoring and Diagnostics Laboratory, NOAA/CMDL), and the Atmospheric Lifetime Experiment, now called the GAGE program. The mixing ratios of CFCs-11, -12, and -113 increased continuously until 1989, when their rate of increase began to decline. Dramatic changes have occurred in the mixing ratios and growth rates of a number of anthropogenic compounds since that time, as demonstrated in Figure 8.4. The observed changes are consistent with the changes in industrial production of halocarbons as called for by the Montreal Protocol and with the calculated lifetimes of each compound.

Methyl chloride is the most abundant halocarbon in the atmosphere and is the only natural organic chlorine compound that reaches the stratosphere in any substantial amount. Natural production of $CH_3Cl$ from the oceans and from biomass

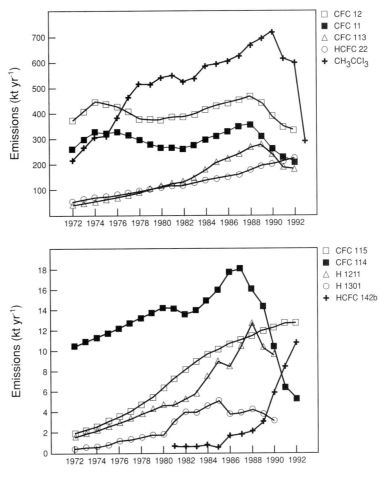

**Figure 8.2.** Annual emissions of halocarbons in kt yr$^{-1}$. The CFC-11, -12, and -113 data are estimates of global emissions, whereas the remaining estimates are based on data only from reporting companies (from WMO, 1995).

burning accounts for at least 75% of the global sources. The tropospheric mixing ratio of $CH_3Cl$ shows a seasonal cycle that has been attributed to seasonal variations in OH radical, which is the primary sink.

The sources of atmospheric fluorine are the CFCs and, to a lesser degree, $CF_4$, which is also anthropogenic. Therefore, the trends in CFCs result in the observed trends in both atmospheric chlorine and fluorine levels. Fluorine, however, is not an important part of stratospheric ozone destruction because of the high stability of the reservoir species HF. This is discussed in more detail in Section 8.5.

### 8.3.2 Sources and Trends of Atmospheric Bromine

The total amount of organic bromine in the atmosphere is calculated in the same manner as the total organic chlorine. The current value for tropospheric organic bromine is 25-30 pptv based on the compounds listed in Table 8.1. While this value is more than 100 times less than total organic chlorine, bromine is 40-100 times more effective (depending on location) at destroying ozone than chlorine; that is, one bromine atom in the stratosphere will potentially destroy 40-100 times as many ozone molecules as one chlorine atom. The primary organic bromine species measured in the stratosphere to date have been the Halons and $CH_3Br$. The other species listed in Table 8.1 that contribute to tropospheric bromine loading are mostly destroyed in the troposphere. Global production of the Halons became significant in the mid- to late

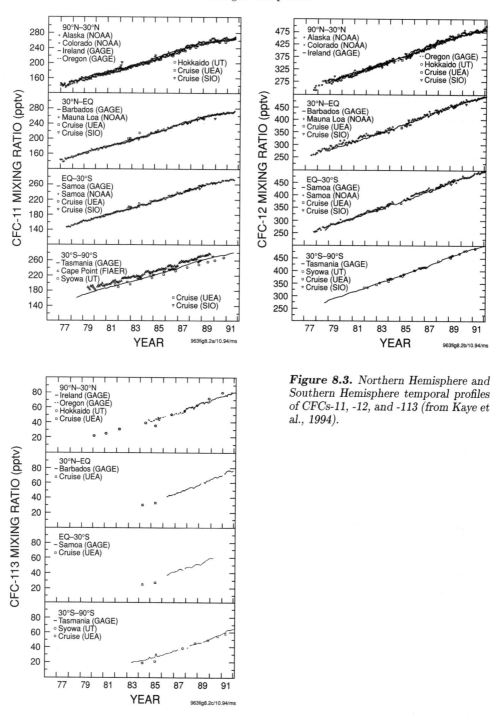

**Figure 8.3.** Northern Hemisphere and Southern Hemisphere temporal profiles of CFCs-11, -12, and -113 (from Kaye et al., 1994).

1970s. Annual emission rates for Halons 1301 ($CF_3Br$) and 1211 ($CF_2ClBr$) are shown in Figure 8.2. The resulting increase in atmospheric mixing ratios for both these compounds is presented in Figure 8.4. Methyl bromide is the primary naturally occurring organic bromine species that is present in the stratosphere. The four potentially important sources of $CH_3Br$ are biological activity in the ocean, agricultural usage, biomass burning (natural and anthropogenic), and exhaust from

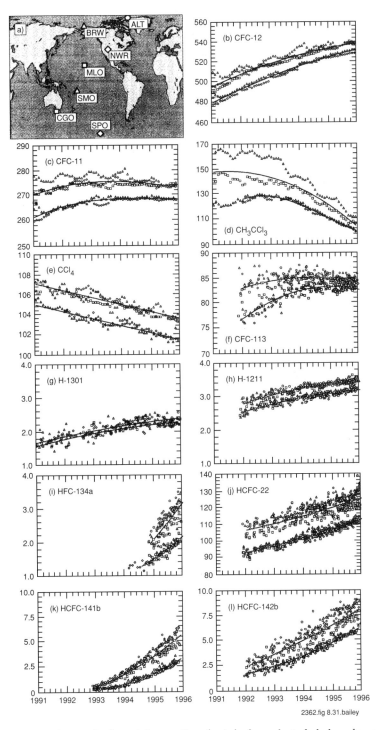

**Figure 8.4.** *Atmospheric mixing ratios (pptv) for selected halocarbons at Northern Hemispheric (upper) and Southern Hemispheric (lower) sites. Solid lines represent fits to the monthly hemispheric mean mixing ratios. Samples were collected at the locations shown in Panel A (reprinted with permission from Montzka, S. A., J. H. Butler, R. C. Myers, T. M. Thompson, T. H. Swanson, A. D. Clarke, L. T. Lock, and J. W. Elkins (1996a) A decline in the tropospheric abundance of ozone-depleting halogen, Science 272, 1318, ©1996, American Association for the Advancement of Science).*

automobiles using leaded gasoline. Approximately 40% of the total annual emissions of $CH_3Br$ are from anthropogenic sources, although uncertainties exist in estimates of the magnitudes of the sources. There is no clear trend in $CH_3Br$ mixing ratios over the time period 1978-1992. As a result of the greater efficiency of bromine for ozone destruction, anthropogenic production of $CH_3Br$ is currently under consideration for regulation.

### 8.3.3 Sources of Atmospheric Iodine

The most important source of iodine to the atmosphere is the oceans, with species such as $CH_3I$, $CH_2I_2$, and $C_2H_5I$ having been identified. Typical levels of $CH_3I$ in the marine boundary layer are less than 2 pptv, although values as high as 45 pptv (Oram and Penkett, 1994) have been observed and are probably associated with oceanic phytoplankton blooms. Biomass burning may also provide a source of $CH_3I$ and other iodocarbons, although this source is likely minor compared to the oceanic sources.

## 8.4 Loss Processes of Halogen Source Gases

The altitude at which a halogenated source gas is destroyed in the atmosphere will ultimately determine its potential to destroy ozone. If the compound is sufficiently long-lived to allow its dissociation to occur in the stratosphere, its potential for ozone depletion will obviously be greater than for a compound whose destruction occurs primarily in the troposphere. Photochemical destruction of all the halogenated source gases listed in Table 8.1 is initiated by one of three processes: photolysis, reaction with OH, or reaction with $O(^1D)$ atoms. The rate and altitude at which these processes occur for each molecule will determine its distribution (*i.e.*, its concentration profile versus altitude) in the atmosphere. In this section, the processes involved in the photochemical destruction of the various classes of halogenated source gases are examined.

### 8.4.1 Loss Processes for CFCs

One of the properties that led to the development of the CFCs as important industrial chemicals was their chemical inertness. However, it is this very property that leads to their harmful effects on the environment. Once emitted to the atmosphere from the surface, these compounds mix into the free troposphere, where there exists no known mechanism for their destruction or removal. This long tropospheric lifetime results in their eventual transport to the stratosphere, where photochemical destruction can occur.

The altitude at which a given CFC will be destroyed is dependent on its chemical composition. In general, more Cl atoms in the molecule shift its UV absorption spectrum to longer wavelengths, allowing photolysis to occur at lower altitudes and resulting in a decrease in the atmospheric lifetime of the compound. This effect can be seen within the series $CF_3Cl$, $CF_2Cl_2$, $CFCl_3$, and $CCl_4$ (see Table 8.1 and Fig. 8.5). A full latitude/altitude profile of $CF_2Cl_2$, measured from the Upper Atmosphere Research Satellite (UARS), is given in Plate 15.

The photochemical destruction of the CFCs leads to the release of free Cl atoms in the stratosphere:

$$CFCl_3 + h\nu \rightarrow CFCl_2 + Cl \qquad (8.1)$$
$$CFCl_2 + O_2 + M \rightarrow CFCl_2O_2 + M \qquad (8.2)$$
$$CFCl_2O_2 + NO \rightarrow CFCl_2O + NO_2 \qquad (8.3)$$

$$CFCl_2O + M \to COFCl + Cl + M \qquad (8.4)$$
$$COFCl + h\nu \to FCO + Cl \qquad (8.5)$$

$$CF_2Cl_2 + h\nu \to CF_2Cl + Cl \qquad (8.6)$$
$$CF_2Cl + O_2 + M \to CF_2ClO_2 + M \qquad (8.7)$$
$$CF_2ClO_2 + NO \to CF_2ClO + NO_2 \qquad (8.8)$$
$$CF_2ClO + M \to COF_2 + Cl + M \qquad (8.9)$$

Some destruction of CFCs also occurs as a result of reaction with $O(^1D)$, an excited electronic state of the oxygen atom that is produced in the short-wavelength ($\lambda \leq 320$ nm) photolysis of ozone. Reaction of $O(^1D)$ with CFCs may occur via a number of different channels, as illustrated here for CFC-12:

$$O(^1D) + CF_2Cl_2 \to CF_2Cl + ClO \qquad (8.10a)$$
$$\to O(^3P) + CF_2Cl_2 \qquad (8.10b)$$
$$\to CF_2ClO + Cl \qquad (8.10c)$$
$$\to COF_2 + Cl_2 \qquad (8.10d)$$

Production of ClO is probably dominant (with a yield near 60%), while quenching of $O(^1D)$ to $O(^3P)$ appears to be a minor channel (yield $\approx$ 15-20%). Occurrence of the other two channels has not been confirmed. Concentrations of $O(^1D)$, and hence the rate of CFC destruction via reaction with $O(^1D)$, increase rapidly with altitude (from about 10 atom cm$^{-3}$ at 15 km, to 100 atom cm$^{-3}$ at 25 km, and 1000 atom cm$^{-3}$ at 35 km). In fact, the increase in $O(^1D)$ concentrations with altitude roughly parallels the increase in the photolysis rate of CFC-11 and CFC-12 with altitude in the 20-35 km range, resulting in a nearly constant relative contribution of the two processes to CFC loss as a function of altitude. Reaction with $O(^1D)$ contributes of order 1% to the total loss of CFC-11, and about 30% to the total loss of the longer lived CFC-12.

Upper limits to the rate constants for OH reaction with CFCs ($k \leq 10^{-17}$ cm$^3$ molecule$^{-1}$ s$^{-1}$) obtained in laboratory studies rule out this process as a significant loss mechanism for CFCs.

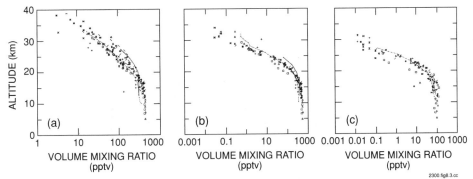

**Figure 8.5.** *Altitude profiles of various CFCs (from Kaye et al., 1994): (a) CFC-12; (b) CFC-11; (c) $CCl_4$.*

### 8.4.2 Loss Processes for the Halons

Qualitatively, the halons (mainly $CF_3Br$ and $CF_2BrCl$) behave similarly to the CFCs once emitted to the atmosphere, with photolysis being the dominant destruction pathway. In general, halons photolyze faster than CFCs because the presence of

Br in the molecule shifts the absorption spectrum to longer wavelengths and allows the photolysis of the halons to occur at lower altitudes than the analogous chlorinated species. For example, $CF_3Br$ is photolyzed at lower altitudes than $CF_3Cl$ and hence has a shorter atmospheric lifetime (see Table 8.1). In fact, all the halons except $CF_3Br$ are partially destroyed in the troposphere by photolysis. However, their tropospheric lifetimes are still sufficiently long to allow a significant fraction of these compounds to be transported to the stratosphere. Photolysis of the halons occurs via rupture of the C-Br bond,

$$CF_3Br + h\nu \to CF_3 + Br \tag{8.11}$$

(which is weaker than a C-Cl or C-F bond), releasing free Br atoms to the atmosphere.

Halons, like the CFCs, also are expected to react rapidly with $O(^1D)$ atoms, although no measurements of these rate constants have been reported. Nevertheless, even if it is assumed that these reactions occur at the maximum possible rate (*i.e.*, with a gas kinetic rate constant of $\approx 2 \times 10^{-10}$ cm$^3$ molecule$^{-1}$ s$^{-1}$), they do not make a significant contribution to the overall loss, because the photolysis rate of the halons is so rapid. Reaction with OH is very slow ($k \leq 5 \times 10^{-16}$ cm$^3$ molecule$^{-1}$ s$^{-1}$), and is not a significant sink for halons.

### 8.4.3 Loss Processes of Partially Hydrogenated Halogen Source Gases

Halogenated source gases that contain hydrogen atoms (such as the HCFCs, HFCs, $CH_3Cl$) possess an additional mechanism for their atmospheric destruction, *i.e.*, reaction with OH:

$$OH + CX_3H \to H_2O + CX_3 \tag{8.12}$$

where $X$ refers to H, F, Cl, Br, or I. Since OH is present throughout the atmosphere, destruction of partially hydrogenated species occurs in the troposphere, thereby reducing the amount available for transport to the stratosphere. It is this property that has led to the development of partially hydrogenated HCFCs and hydrofluorocarbons HFCs as replacements for the CFCs, an issue discussed in more detail in Section 8.6.

The rate at which these partially hydrogenated compounds are removed in the troposphere is then controlled by the rate of their reaction with OH. Rate coefficients for some of these hydrogenated species are presented in Table 8.2. In many cases, the rate coefficients are rather small so the tropospheric lifetimes are sufficiently long (a few months to a couple of years) that significant fractions of these compounds are transported to the stratosphere. Once in the stratosphere, reaction with OH, photolysis, and reaction with $O(^1D)$ lead to the continued removal of the source gas. Note that for compounds whose dominant loss mechanism is reaction with OH in both the troposphere and stratosphere, the stratospheric lifetime is considerably longer than the tropospheric lifetime owing to the rather large temperature dependence of the OH rate coefficients.

**Table 8.2**
*Rate Coefficients for Reaction of OH with Various Hydrogenated Halogen Source Gases*

| Molecule | \multicolumn{3}{c}{Rate Coefficient for Reaction with OH (cm$^3$ molecule$^{-1}$ s$^{-1}$)} | | |
|---|---|---|---|
| | A Factor | E/R | $k_{298}$ |
| $CH_3Cl$ | 4.0e-12 | 1400 | 3.6e-14 |
| $CH_2Cl_2$ | 3.8e-12 | 1050 | 1.1e-13 |
| $CHF_2Cl$ | 1.0e-12 | 1600 | 4.7e-15 |
| $CH_3CCl_3$ | 1.8e-12 | 1550 | 1.0e-14 |
| $CHCl_3$ | 2.0e-12 | 900 | 1.0e-13 |
| $CH_3Br$ | 4.0e-12 | 1470 | 2.9e-14 |
| $CH_2Br_2$ | 2.4e-12 | 900 | 1.2e-13 |

$^a k = A\exp(-E_a/RT)$, see Section 3.4.2.
From DeMore et al., 1997.

Tropospheric photolysis is a major sink for some bromine- and iodine-containing methanes, such as $CHBr_2Cl$, $CHBr_3$, and $CH_3I$. For example, $CH_3I$ (which is the dominant iodine source gas) has a tropospheric lifetime with respect to photolysis of only a few days.

### 8.4.4 Loss Processes for Perfluorinated Species

The perfluorinated species, the most abundant of which is $CF_4$, deserve special attention because the chemistry involved in their destruction is not at all like that of other halogen source gases. Indeed, the lifetimes of these compounds are very long as they are not susceptible to reaction with OH or $O(^1D)$, and photolyze only very slowly. As a result, different loss mechanisms (which occur on much longer timescales) need to be considered (Ravishankara et al., 1993; Morris et al., 1995). These include reaction with H atoms, destruction by Lyman-$\alpha$ radiation at 121.6 nm, uptake by oceans and soils, processing through combustion systems (such as power plants, engines, etc.), and reactions with electrons and ions in the mesosphere and thermosphere. For $CF_4$, a lifetime in excess of 50,000 years has been estimated, with the major loss probably being due to destruction in high-temperature combustion systems (which process approximately 0.01% of the air in the Earth's atmosphere per year). A lower limit of 2300 yrs to the atmospheric lifetime of $CF_4$ has been determined from measurements of its stratospheric abundance (Zander et al., 1996b).

### 8.4.5 Stratospheric Chlorine and Bromine Loading

As mentioned previously, not all the halogenated organic species contributing to tropospheric chlorine and bromine loading listed in Table 8.1 are present in the stratosphere. Measurements collected at or near the primary region of transport from the troposphere to the stratosphere, that is, the tropical tropopause, are necessary to define the species (and their respective mixing ratios) that are available for transport into the stratosphere. Measurements collected at the same time across the tropopause into the lower stratosphere allow calculations of the amount of Cl and Br released in the lower tropical stratosphere. Table 8.3 presents the mixing ratios and percent contributions to total organic chlorine and bromine from a comprehensive suite of simultaneously measured compounds at 24°N near the tropopause and in the lower stratosphere. Anthropogenically produced compounds contribute 85% of the total organic chlorine and 66% of the total organic bromine in the tropopause samples (assuming 50% of $CH_3Br$ and 100% of $CH_2Br_2$ is natural).

## 8.5 Inorganic Chemistry of Halogen Species

The destruction of halogen source gases in the atmosphere leads to the generation of free halogen atoms. It was shown in the previous section that the destruction of most fluorinated source gases occurs in the stratosphere, hence confining subsequent chemistry to that region of the atmosphere. Conversely, methyl iodide destruction and subsequent iodine chemistry occurs mainly in the troposphere. Destruction of chlorine and bromine source gases can occur throughout the troposphere and the stratosphere. In this section, a discussion of the inorganic chemistry of the halogens is presented, first in the stratosphere, where the chemistries of chlorine, bromine, iodine, and fluorine are considered and then in the troposphere, where the effects of chlorine, bromine, and iodine chemistry are discussed.

### Table 8.3a
Mixing Ratio (ppt) and Percent Contribution to Cl at the Tropopause and in the Lower Stratosphere

| Molecule | Mixing Ratio at Tropopause | % Contribution to Total Cl at Tropopause | Mixing Ratio in Lower Stratosphere | % Contribution to Cl in Lower Stratosphere |
|---|---|---|---|---|
| CFC 115 | 4.40 | 0.1 | 3.7 | 0.1 |
| HCFC 22 | 96.00 | 2.8 | 92.0 | 2.9 |
| $CH_3Cl$ | 531.00 | 15.1 | 469.0 | 14.9 |
| HCFC 142b | 2.80 | 0.1 | 2.1 | 0.1 |
| CFC 114 | 16.40 | 1.0 | 15.3 | 1.0 |
| $CH_2Cl_2$ | 14.90 | 0.9 | 11.1 | 0.7 |
| CFC 113 | 74.60 | 6.4 | 67.9 | 6.5 |
| $CHCl_3$ | 3.13 | 0.3 | 1.7 | 0.2 |
| $CH_3CCl_3$ | 117.00 | 10.0 | 100.0 | 9.5 |
| $CCl_4$ | 109.00 | 12.4 | 98.0 | 12.4 |
| CFC 12 | 494.00 | 28.2 | 465.0 | 29.5 |
| CFC 11 | 264.00 | 22.6 | 236.0 | 22.4 |

From Schauffler et al., 1993.

### Table 8.3b
Mixing Ratio (ppt) and Percent Contribution to Br at the Tropopause and in the Lower Stratosphere

| Molecule | Mixing Ratio at Tropopause | % Contribution to Total Br at Tropopause | Mixing Ratio in Lower Stratosphere | % Contribution to Br in Lower Stratosphere |
|---|---|---|---|---|
| Halon 1301 | 2.77 | 16.4 | 2.64 | 18.8 |
| Halon 1211 | 2.58 | 15.3 | 2.20 | 15.7 |
| Halon 2402 | 0.22 | 2.6 | 0.19 | 2.7 |
| $CH_3Br$ | 9.67 | 57.2 | 8.37 | 59.6 |
| $CH_2Br_2$ | 0.72 | 8.5 | 0.23 | 3.3 |

From Schauffler et al., 1993.

### 8.5.1 Halogen Species in the Stratosphere
#### 8.5.1.1 General Overview

In general, the destruction of halogenated source compounds leads to the rapid release of free halogen atoms. A complex set of chemical reactions then occurs, which leads to the partitioning of the halogen among a number of inorganic species, the major ones having the general formulae $X$, $XO$, $HOX$, $HX$, $XONO_2$ (where $X$ refers to F, Cl, Br, or I). These species are collectively referred to as $Cl_y$ for $X$ = Cl and $Br_y$ for $X$ = Br. The $X$ and $XO$ forms of inorganic halogen are often referred to as "active" forms, as it is these compounds that are involved in processes that catalytically destroy odd oxygen, such as the general scheme:

$$X + O_3 \rightarrow XO + O_2$$

$$\underline{XO + O \rightarrow X + O_2}$$
$$\text{Net: } O + O_3 \rightarrow O_2 + O_2$$

The $HX$ and $XONO_2$ compounds are often referred to as "inactive," as they tend to act as longer-lived reservoirs for halogens. It then follows that a key factor in the ability of a particular halogen to destroy ozone is the relative amount of time spent in

## Table 8.4a
*Rate Constants (where available) for Various Bimolecular Reactions Involving Inorganic Halogen Species*

| Reaction | X = Fluorine | | | X = Chlorine | | | X = Bromine | | |
|---|---|---|---|---|---|---|---|---|---|
| | A | E/R | $k_{298}$ | A | E/R | $k_{298}$ | A | E/R | $k_{298}$ |
| $X + CH_4 \to HX + CH_3$ | 1.6e-10 | 260 | 6.7e-11 | 1.1e-11 | 1400 | 1.0e-13 | No reaction - endothermic | | |
| $X + O_3 \to XO + O_2$ | 2.2e-11 | 230 | 1.0e-11 | 2.9e-11 | 260 | 1.2e-11 | 1.7e-11 | 800 | 1.2e-12 |
| $X + HO_2 \to HX + O_2$ | | | | 1.8e-11 | -170 | 3.2e-11 | 1.5e-11 | 600 | 2.0e-12 |
| $X + HO_2 \to XO + OH$ | | | | 4.1e-11 | 450 | 9.1e-12 | | | |
| $X + H_2O \to HX + OH$ | 1.4e-11 | 0 | 1.4e-11 | No reaction - endothermic | | | No reaction - endothermic | | |
| $XO + O \to X + O_2$ | 2.7e-11 | 0 | 2.7e-11 | 3.0e-11 | -70 | 3.8e-11 | 1.9e-11 | -230 | 4.1e-11 |
| $XO + NO \to X + NO_2$ | 8.2e-12 | -300 | 2.2e-11 | 6.4e-12 | -290 | 1.7e-11 | 8.8e-12 | -260 | 2.1e-11 |
| $XO + HO_2 \to HOX + O_2$ | | | | 4.8e-13 | -700 | 5.0e-12 | 3.4e-12 | -540 | 2.1e-11 |
| $XO + OH \to X + HO_2$ | | | | 1.1e-11 | -120 | 1.7e-11 | | | 7.5e-11 |
| $XO + XO \to$ products | 1.0e-11 | 0 | 1.0e-11 | 7.6e-12 | 1830 | 1.6e-14 | 1.5e-12 | -230 | 3.2e-12 |
| $OH + HX \to X + H_2O$ | No reaction - endothermic | | | 2.6e-12 | 350 | 8.0e-13 | 1.1e-11 | 0 | 1.1e-11 |
| $OH + HOX \to XO + H_2O$ | | | | 3.0e-12 | 500 | 5.0e-13 | | | |

From DeMore et al., 1997.

## Table 8.4b
*Rate Constants for Termolecular Reactions Involving Inorganic Halogen Species*
(see Appendix F for explanation of $k_0$, $k_\infty$, $n$, and $m$)

| Reaction | X = Fluorine | | | | X = Chlorine | | | | X = Bromine | | | |
|---|---|---|---|---|---|---|---|---|---|---|---|---|
| | $k_0$ | $n$ | $k_\infty$ | $m$ | $k_0$ | $n$ | $k_\infty$ | $m$ | $k_0$ | $n$ | $k_\infty$ | $m$ |
| $X + O_2 \to XO_2$ | 4.4e-33 | 1.2 | | | 2.7e-33 | 1.5 | | | | | | |
| $X + NO \to XNO$ | 1.8e-31 | 1.0 | 2.8e-10 | 0 | 9.0e-32 | 1.6 | | | | | | |
| $X + NO_2 \to XNO_2$ | 6.3e-32 | 2 | 2.6e-10 | 0 | 1.5e-30 | 2.0 | 2.0e-10 | 1.0 | 4.2e-31 | 2.4 | 2.7e-11 | 0 |
| $XO + NO_2 \to XONO_2$ | 2.6e-31 | 1.3 | 2.0e-11 | 1.5 | 1.8e-31 | 3.4 | 1.5e-11 | 1.9 | 5.2e-31 | 3.2 | 6.9e-12 | 2.9 |
| $XO + XO \to X_2O_2$ | | | | | 2.2e-32 | 3.1 | 3.5e-12 | 1.0 | | | | |

From DeMore et al., 1997.

the "active" versus "inactive" forms. Although the photochemical reactions involved in partitioning the halogens are qualitatively similar for fluorine, chlorine, and bromine (see Table 8.4), quantitative differences in the thermodynamics and reaction kinetics (as described in the subsequent sections) lead to substantial differences in the relative abundances of active and inactive species and hence in their abilities to destroy ozone.

### 8.5.1.2 Chlorine Chemistry in the Stratosphere

It has already been noted that destruction of CFCs and other chlorine source gases results in the production of free Cl atoms in the stratosphere. These Cl atoms react with $O_3$, leading to formation of ClO:

$$Cl + O_3 \to ClO + O_2 \tag{8.13}$$

ClO can then be converted back to Cl via a variety of pathways (see Table 8.4 and Fig. 8.6), many of which regenerate the ozone lost in Reaction (8.13):

$$ClO + NO \to Cl + NO_2 \tag{8.14}$$
$$NO_2 + h\nu \to NO + O \tag{8.15}$$
$$O + O_2 + M \to O_3 + M \tag{8.16}$$

$$ClO + h\nu \to Cl + O \tag{8.17}$$
$$O + O_2 + M \to O_3 + M \tag{8.16}$$

Cycles of this sort, in which all species consumed in the cycle are regenerated, are referred to as null cycles, since no net change in chemical composition occurs.

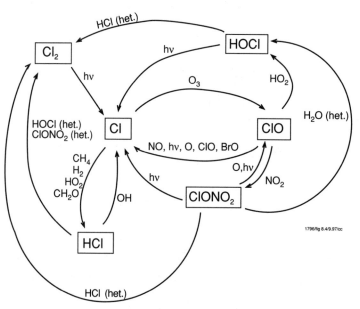

**Figure 8.6.** Inorganic chemistry involved in the interconversion of chlorine species in the stratosphere. (het.) refers to heterogeneous reactions that occur on particles.

However, some reactions that convert ClO back to Cl complete catalytic cycles that lead to loss of odd oxygen. The most important Cl cycle in the midstratosphere involves the reaction of ClO with O atoms:

$$Cl + O_3 \rightarrow ClO + O_2 \qquad (8.13)$$
$$\underline{ClO + O \rightarrow Cl + O_2} \qquad (8.18)$$

Net: $O + O_3 \rightarrow 2\, O_2$

Reaction of ClO with $HO_2$ also leads to net $O_3$ destruction:

$$Cl + O_3 \rightarrow ClO + O_2 \qquad (8.13)$$
$$ClO + HO_2 \rightarrow HOCl + O_2 \qquad (8.19)$$
$$HOCl + h\nu \rightarrow OH + Cl \qquad (8.20)$$
$$\underline{OH + O_3 \rightarrow HO_2 + O_2} \qquad (8.21)$$

Net: $2\, O_3 \rightarrow 3\, O_2$

Interconversion of Cl and ClO occurs on a time scale of one minute at 30 km, with ozone-destroying cycles (predominantly reaction of ClO with O) occurring with about 5% efficiency relative to null cycles (largely reaction of ClO with NO). The concentration ratio [ClO]:[Cl] is largely controlled by Reactions (8.13), (8.14), (8.17), and (8.18), and is approximately 5000 at 20 km, 500 at 30 km, and 100 at 40 km.

Were this the only chemistry involving the Cl and ClO species occurring in the stratosphere, the above cycles would continue unimpeded and stratospheric ozone levels would be greatly diminished. However, Cl and ClO can be converted to the reservoir species HCl and $ClONO_2$, respectively. Cl atoms react with $CH_4$, $H_2$, $HO_2$, and $H_2O_2$, to generate HCl

$$Cl + CH_4 \rightarrow CH_3 + HCl \qquad (8.22)$$
$$Cl + HO_2 \rightarrow HCl + O_2 \qquad (8.23)$$
$$Cl + H_2 \rightarrow HCl + H \qquad (8.24)$$
$$Cl + H_2O_2 \rightarrow HCl + HO_2 \qquad (8.25)$$

ClO reacts with $NO_2$ in a termolecular reaction to form chlorine nitrate

$$ClO + NO_2 + M \rightarrow ClONO_2 + M \qquad (8.26)$$

HCl is the longest-lived and most abundant Cl reservoir species. It has a lifetime of about one month in the lower stratosphere, and is returned to active Cl largely via reaction with OH

$$OH + HCl \rightarrow Cl + H_2O \qquad (8.27)$$

$ClONO_2$ is destroyed mainly by photolysis or via reaction with O atoms, leading to regeneration of active Cl species

$$ClONO_2 + h\nu \rightarrow Cl + NO_3 \qquad (8.28a)$$
$$\rightarrow ClO + NO_2 \qquad (8.28b)$$
$$O + ClONO_2 \rightarrow \text{products} \qquad (8.29)$$

The lifetime of $ClONO_2$ is approximately 6 hours in the midlatitude lower stratosphere (below 30 km) and decreases to about an hour at 40 km as the intensity of UV light increases.

Another shorter-lived chlorine reservoir is hypochlorous acid, HOCl, formed in the reaction of $HO_2$ with ClO:

$$HO_2 + ClO \rightarrow HOCl + O_2 \qquad (8.19)$$

The ability of HOCl to act as a reservoir is limited by its fairly rapid photolysis, which occurs on the time scale of less than one hour throughout the stratosphere.

It has recently become apparent that chemical reactions occurring on (or in) aerosol particles (commonly referred to as "heterogeneous chemistry"; see Chapters 3 and 4) can be of immense significance to stratospheric chemistry, and can lead to repartitioning of $NO_y$ and $Cl_y$ species in the stratosphere. Particles present in the stratosphere include polar stratospheric clouds (the effects of which will be discussed later in this chapter) and liquid sulfuric acid/water droplets (known as the Junge Layer), which are present throughout the lower stratosphere. The number density of these sulfate particles is maximum following large volcanic eruptions, such as El Chichón and Mount Pinatubo. The most important reaction occurring on sulfate aerosols is the conversion of $N_2O_5$ to $HNO_3$

$$N_2O_{5(g)} + H_2O_{(l)} \rightarrow 2\ HNO_{3(g)} \qquad (8.30)$$

While this reaction has no direct effect on the halogens, the conversion of $N_2O_5$ to $HNO_3$ (a more stable $NO_x$ reservoir) leads to a reduction in the abundance of $NO_2$. This in turn leads to an increase in the ratio of ClO to $ClONO_2$ in the stratosphere and amplifies the ability of Cl to destroy ozone. At low temperatures, when the water content of the sulfate aerosol is increased, the hydrolysis of $ClONO_2$ can occur

$$ClONO_{2(g)} + H_2O_{(l)} \rightarrow HOCl_{(g)} + HNO_{3(g)} \qquad (8.31)$$

Rapid photolysis of HOCl to produce Cl leads to the net conversion of inactive reservoir $ClONO_2$ to active Cl, and again heightens the effectiveness of Cl in the destruction of ozone. The increase in the ClO mixing ratio (and concomitant decrease in $O_3$ levels) as a result of these heterogeneous reactions on sulfate aerosols following the eruption of Mount Pinatubo has been modeled, for example, by Brasseur and Granier (1992) (see Fig. 7.17), and has been confirmed by observation (see, for example, Avallone et al., 1993).

The abundances of the chlorine species have been determined by numerous experimental techniques, some of which are described in Chapter 11. Organic source gases are determined spectroscopically from a variety of remote sensing platforms, or via whole air sampling techniques in which stratospheric air is collected from balloon, airplane, or rocket platforms and returned to laboratories for analysis by gas chromatographic techniques. Inorganic species are measured by remote sensing techniques employing microwave, millimeter-wave, or infrared spectroscopy or via in situ analysis (HCl by IR absorption, and ClO by conversion to Cl and subsequent detection by resonance fluorescence). The most complete, simultaneous measurement of all chlorine compounds to date comes from the analysis of high-resolution infrared solar spectra recorded by the Atmospheric Trace Molecule Spectroscopy (ATMOS) Fourier transform spectrometer operated onboard Spacelab 3. This instrument provided numerous sunrise and sunset measurements of stratospheric trace gases in the spring of 1985 (Zander et al., 1992). In addition to the chlorine species determined by the ATMOS spectrometer (HCl, $ClONO_2$, F-11, F-12, $CCl_4$, $CH_3Cl$, and HCFC-22), measurements from other instruments for methyl chloroform, CFC-113, COFCl, ClO, and HOCl were incorporated into an analysis of the chlorine budget between 12.5 and 55 km. These results are summarized in Figure 8.7. The mixing ratio of the CFCs and HCFCs can be seen to decrease as a function of altitude as these compounds are destroyed. Concomitant with the decrease in organic source gas concentration is the increase in levels of inorganic chlorine, mostly in the form of HCl (see Color Plate 20). The total Cl mixing ratio is seen to be almost constant with altitude at a level of about 2.6 ppbv. From a more recent ATMOS mission, the total stratospheric Cl mixing ratio for 1994 was found to be 3.53 ppbv (Zander et al., 1996a), indicating that between the two ATMOS missions of spring 1984 and fall 1994, the chlorine loading in the stratosphere increased by 0.10 ppbv/yr. The total atmospheric budget of Cl is expected to peak at about 4.0 ppbv near the turn of this century.

*Figure 8.7.* Stratospheric altitude profiles of chlorine and fluorine species as measured by instruments aboard Spacelab 3 (from Zander et al., 1992).

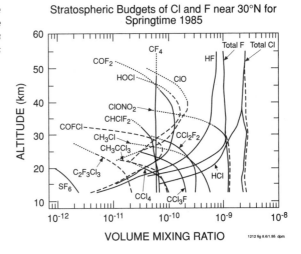

The relative abundances of the inorganic chlorine species will now be considered. HCl is the most abundant inorganic species at all altitudes due to its long lifetime. Indeed, HCl makes up over 95% of the chlorine budget at altitudes above 45 km. ClONO$_2$ is a significant reservoir below about 30 km, where its formation rate is most rapid. As its rate of formation decreases [due to the decreased efficiency of the termolecular reaction (8.26) as the pressure decreases] and its photolysis rate increases at higher altitudes, its mixing ratio decreases, leading to a rise in the abundance of ClO. The mixing ratio of ClO reaches a maximum of about 200 ppt at altitudes between 35 and 40 km. Mixing ratios of ClO throughout the stratosphere, measured from the UARS, are shown in Color Plate 22.

As noted earlier, it is of particular interest to examine the ratio of the concentrations of active (Cl + ClO) to inactive (ClONO$_2$ and HCl) chlorine, since this parameter is useful in determining the efficiency of ozone destruction. At low altitudes (15-20 km), active forms of Cl contribute of order 2% or less to the total inorganic chlorine, while this value increases to about 3% at 25 km, 10% at 30 km, and 35% at 40 km. This increase is primarily due to the aforementioned decrease in stability of ClONO$_2$ with increasing altitude. As will be presented in more detail in Chapter 14, Cl-based chemistry makes its largest fractional contribution to ozone destruction in the 35-50 km range.

It is important to point out that there still exist significant gaps in our understanding of the partitioning of chlorine in the stratosphere, as is demonstrated by some persistent quantitative disagreements between models and observations. For example, as pointed out by many authors (see Minschwaner et al., 1993, and references therein), modeled ClO/HCl concentration ratios consistently exceed those measured at altitudes above about 30 km. The possibility of a minor channel in the reaction of OH with ClO to form HCl,

$$OH + ClO \rightarrow HO_2 + Cl \qquad (8.32a)$$
$$\rightarrow HCl + O_2 \qquad (8.32b)$$

could alleviate this discrepancy, and requires further laboratory study. In contrast, in situ measurements of ClO/HCl ratios in the lower stratosphere (below about 20 km) (Avallone et al., 1993; Webster et al., 1993b) exceed predictions by a factor of two or more. This may be due to heterogeneous chemistry involving bromine species (see the following section; Hanson and Ravishankara, 1995; Hanson et al., 1996; Tie and Brasseur, 1996).

## 8.5.1.3 Bromine Chemistry in the Stratosphere

Although the chemistry of bromine in the stratosphere is not as well understood as that of chlorine, it is quite clear that the potential for ozone depletion by bromine is significantly larger than for an equivalent amount of chlorine. The reason for this difference (probably about a factor of 40-100) is that the reservoir species HBr and $BrONO_2$ are less abundant than their chlorine counterparts. While HCl is formed largely from the reaction of Cl with methane and $H_2$, the corresponding reactions involving Br are endothermic and hence cannot occur. Thus HBr formation can only occur via reaction of Br with $HO_2$, $CH_2O$, or possibly $H_2O_2$, species that are far less abundant than methane. In addition, the rate constant for reaction of OH with HBr is about 12-15 times larger than that of OH with HCl, lowering the atmospheric lifetime of HBr to a couple of days as compared to one month for HCl. While $BrONO_2$ is formed in an analogous way to $ClONO_2$,

$$BrO + NO_2 + M \rightarrow BrONO_2 + M \qquad (8.33)$$

it, too, is significantly less stable than its Cl analog (lifetime with respect to photolysis of about 10 min in the lower stratosphere compared to 6 h for $ClONO_2$).

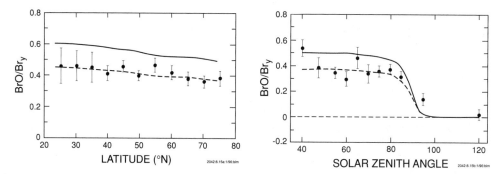

**Figure 8.8.** (a) $BrO/Br_y$ versus latitude. (b) $BrO/Br_y$ versus solar zenith angle; Sept. and Oct. 1991. Error bars are $2\sigma$ precision for BrO. Measurements were made aboard the NASA ER-2 at an altitude of about 20 km. The solid line is a model calculation; the dashed line is the model calculation scaled by 0.75 (from Avallone et al., 1995).

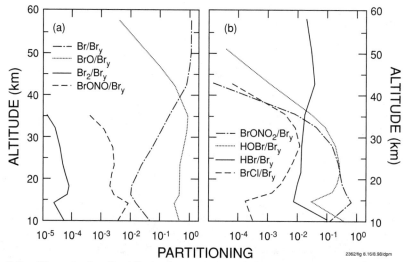

**Figure 8.9.** The calculated midlatitude partitioning of reactive bromine as a function of altitude for local noon at equinox (from Lary, 1996).

Measurements of inorganic bromine in the stratosphere are more difficult than for chlorine owing to its lower abundance. At present, the most measured form of inorganic bromine is BrO (see Fig. 8.8), which has been measured either via in situ (Brune et al., 1989; Avallone et al., 1995) or ground-based techniques (Solomon et al., 1989b). Recently, measurements of HBr and HOBr (Johnson et al., 1995b; Carlotti et al., 1995) have also been made. Owing to the relative paucity of measurement data, a full picture of the partitioning of inorganic bromine is best obtained from models. The results of one such model are shown in Figure 8.9 (Lary, 1996). Generally, during the daytime it is believed that about 50% of inorganic bromine is present in the form of BrO, with about 10-20% contributions from each of HOBr and $BrONO_2$. HBr makes up only a few percent of the total. Of great significance is the large abundance of BrO, making bromine catalytic cycles such as those listed below more efficient than their Cl counterparts.

$$Br + O_3 \rightarrow BrO + O_2 \qquad (8.34)$$
$$BrO + HO_2 \rightarrow HOBr + O_2 \qquad (8.35)$$
$$HOBr + h\nu \rightarrow OH + Br \qquad (8.36)$$
$$\underline{OH + O_3 \rightarrow HO_2 + O_2} \qquad (8.21)$$

Net: $2\ O_3 \rightarrow 3\ O_2$

$$BrO + NO_2 + M \rightarrow BrONO_2 + M \qquad (8.33)$$
$$BrONO_2 + h\nu \rightarrow Br + NO_3 \qquad (8.37)$$
$$NO_3 + h\nu \rightarrow NO + O_2 \qquad (8.38)$$
$$NO + O_3 \rightarrow NO_2 + O_2 \qquad (8.39)$$
$$\underline{Br + O_3 \rightarrow BrO + O_2} \qquad (8.34)$$

Net: $2\ O_3 \rightarrow 3\ O_2$

In addition, a cycle that couples the chemistry of chlorine and bromine through the reaction of ClO with BrO is also of importance in the lower stratosphere:

$$ClO + BrO \rightarrow Br + OClO \qquad (8.40a)$$
$$\rightarrow Br + ClOO \qquad (8.40b)$$
$$\rightarrow BrCl + O_2 \qquad (8.40c)$$

Production of OClO in this reaction results in a null cycle, since OClO photolysis regenerates odd oxygen,

$$OClO + h\nu \rightarrow O + ClO \qquad (8.41)$$

However, subsequent reactions of BrCl and ClOO,

$$ClOO + M \rightarrow Cl + O_2 \qquad (8.42)$$
$$BrCl + h\nu \rightarrow Br + Cl \qquad (8.43)$$

regenerate halogen atoms and lead to catalytic destruction of odd oxygen.

Heterogeneous reactions on sulfate aerosols [such as (8.44) and (8.45)] have been shown to be faster than the analogous chlorine reactions,

$$BrONO_{2(g)} + H_2O_{(l)} \rightarrow HOBr_{(g)} + HNO_{3(g)} \qquad (8.44)$$
$$HOBr_{(g)} + HCl_{(l)} \rightarrow BrCl_{(g)} + H_2O_{(l)} \qquad (8.45)$$

and may have some important effects on lower stratospheric chemistry (Hanson and Ravishankara, 1995; Hanson et al., 1996; Tie and Brasseur, 1996). Reactions (8.33) and (8.44) result in the conversion of $NO_x$ to the more stable $HNO_3$. Reaction

(8.44), followed by photolysis of HOBr [Reaction (8.36)], can be an important source of OH radicals, and increases the rate of conversion of HCl to active chlorine [via Reaction (8.27)].

### 8.5.1.4 Fluorine Chemistry in the Stratosphere

As was discussed above, the major sources of fluorine in the stratosphere are CFCs-11 and -12. Photolysis of these compounds leads to release of Cl atoms, while the fluorine is initially present in the form of the carbonyl compounds COFCl and $COF_2$. These carbonyl compounds are sufficiently long-lived to be detectable by column FTIR measurements, but are themselves photolyzed, releasing free fluorine:

$$COF_2 + h\nu \rightarrow FCO + F \tag{8.46}$$

$$COFCl + h\nu \rightarrow FCO + Cl \tag{8.5}$$

$$FCO + O_2 + M \rightarrow FC(O)O_2 + M \tag{8.47}$$

$$FC(O)O_2 + NO \rightarrow FCO_2 + NO_2 \tag{8.48}$$

$$FCO_2 + h\nu \rightarrow F + CO_2 \tag{8.49}$$

The free atoms produced in this chemistry mostly react with $O_2$ and participate in the following null cycle:

$$F + O_2 + M \rightarrow FO_2 + M \tag{8.50}$$

$$FO_2 + NO \rightarrow FNO + O_2 \tag{8.51}$$

$$FNO + h\nu \rightarrow F + NO \tag{8.52}$$

The F atoms eventually react with $CH_4$, $H_2O$, $H_2$, or with $O_3$:

$$F + CH_4 \rightarrow CH_3 + HF \tag{8.53}$$

$$F + H_2O \rightarrow OH + HF \tag{8.54}$$

$$F + H_2 \rightarrow HF + H \tag{8.55}$$

$$F + O_3 \rightarrow FO + O_2 \tag{8.56}$$

The formation of HF in these reactions is significant because it is a very stable reservoir species. Due to the very strong bond, HF cannot react with OH (the reaction is endothermic). In addition, photolysis cannot occur to any appreciable extent in the stratosphere, making HF an essentially permanent stratospheric reservoir for fluorine (see Color Plate 20). FO will react with NO or with O atoms,

$$FO + NO \rightarrow F + NO_2 \tag{8.57}$$

$$FO + O \rightarrow F + O_2 \tag{8.58}$$

regenerating F atoms that will rapidly be converted to HF, hence terminating any F/FO cycles that could potentially destroy odd oxygen.

A fairly simple picture then emerges for the overall budget of fluorine in the atmosphere, as presented in Figure 8.7 (Zander et al., 1992; Russell et al., 1996). Fluorine enters the stratosphere mainly in the form of CFCs and $CF_4$. The CFCs are photolyzed in the lower stratosphere (CFC-11 near 20 km and CFC-12 near 25 km) and converted to the temporary reservoirs $COF_2$ and COFCl. Subsequent photolysis of these compounds liberates free F atoms, which are converted to HF. The extreme stability of HF then ensures that virtually no fluorine is present in the form of F or FO. Due to its long lifetime, $CF_4$ mixing ratios remain essentially constant throughout the stratosphere (Zander et al., 1996b).

### 8.5.1.5 Iodine Chemistry in the Stratosphere

As discussed earlier, $CH_3I$ is the dominant source of iodine to the atmosphere. $CH_2I_2$, $CH_2ClI$, and $C_2H_5I$ are among the other iodinated source gases. Because all these compounds photolyze very rapidly (lifetimes of a few days to a few weeks), it was generally considered that the transport of these species to the stratosphere would be negligible. However, Solomon et al. (1994b) suggested that very rapid vertical transport during convective events could transport sufficient quantities of these short-lived species to the lower stratosphere, where their subsequent destruction would lead to significant ozone depletion. Because no significant reservoir species exists for iodine (HI, HOI, and $IONO_2$ would all be rapidly photolyzed), almost all inorganic iodine would exist as active IO and I (Solomon et al., 1994b). Coupling of the chemistry of iodine to that of bromine and chlorine through Reactions (8.59) and (8.60),

$$IO + ClO \rightarrow I + Cl + O_2 \tag{8.59}$$

$$IO + BrO \rightarrow I + Br + O_2 \tag{8.60}$$

followed by reaction of the free halogen atoms with ozone, could then lead to ozone depletion. At present, no iodine-containing species have been detected in the stratosphere, but the potential effects of iodine chemistry in the stratosphere are currently a topic of considerable research.

### 8.5.2 Special Chemistry of Halogens in the Polar Stratosphere

One of the most dramatic results of man's influence on the composition of the atmosphere is the recent discovery of near complete destruction of ozone in the lower stratosphere above Antarctica, commonly referred to as the Antarctic "ozone hole." It is now firmly established that heterogeneous reactions occurring on the surfaces of polar stratospheric clouds (PSCs, solid particles composed of $HNO_3$ and $H_2O$; see Chapter 4) are responsible for grossly altering the chemical composition of the Antarctic Polar Vortex during winter. The occurrence of these heterogeneous reactions sets the stage for ozone depletion upon the return of sunlight following the polar night, as the result of gas-phase chemical reactions involving chlorine and bromine species. The heterogeneous reactions currently believed to be important on the surfaces of PSCs are as follows:

$$N_2O_{5(g)} + H_2O_{(s)} \rightarrow 2\,HNO_{3(s)} \tag{8.30}$$

$$ClONO_{2(g)} + H_2O_{(s)} \rightarrow HOCl + HNO_{3(s)} \tag{8.31}$$

$$ClONO_{2(g)} + HCl_{(s)} \rightarrow Cl_2 + HNO_{3(s)} \tag{8.61}$$

$$N_2O_{5(g)} + HCl_{(s)} \rightarrow ClNO_{2(g)} + HNO_{3(s)} \tag{8.62}$$

$$HOCl_{(g)} + HCl_{(s)} \rightarrow Cl_{2(g)} + H_2O_{(s)} \tag{8.63}$$

These chemical reactions greatly alter the partitioning of inorganic chlorine within the Antarctic vortex. The reservoir species $ClONO_2$ and HCl are removed from the atmosphere via these processes, and are converted to species such as $Cl_2$, HOCl, and $ClNO_2$, all of which rapidly photolyze in the Antarctic spring to generate free Cl atoms:

$$Cl_2 + h\nu \rightarrow Cl + Cl \tag{8.64}$$

$$HOCl + h\nu \rightarrow OH + Cl \tag{8.20}$$

$$ClNO_2 + h\nu \rightarrow Cl + NO_2 \tag{8.65}$$

Another major effect of these reactions is to effectively remove odd nitrogen from the gas phase via its conversion to $HNO_3$, which remains in the PSC particles. Hence, the conversion of ClO to $ClONO_2$ via reaction with $NO_2$ no longer occurs to

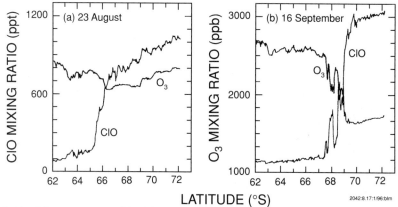

**Figure 8.10.** Measurements of highly enhanced ClO concentrations in the Antarctic Polar Vortex in 1987. The destruction of ozone during the period Aug. 23-Sept. 16 is evident, as is the anticorrelation of $O_3$ and ClO concentrations (adapted with permission from Anderson, J. G., D. W. Toohey, and W. H. Brune (1991) Free radicals within the Antarctic vortex: The role of CFCs in Antarctic ozone loss, *Science* **251**, 39, ©1991, American Association for the Advancement of Science).

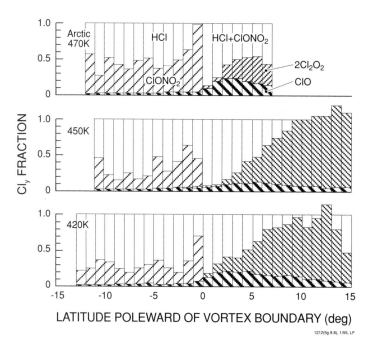

**Figure 8.11.** ClO/HCl/ClONO$_2$ levels in the Polar Vortex (from Kawa et al., 1992) at different altitudes (isentropic surfaces of 470, 450, and 420 K). The enhanced levels of ClO and $Cl_2O_2$ in the vortex are evident.

an appreciable extent. Similar heterogeneous processes probably occur for brominated compounds also.

Measurements made inside the polar vortex show a dramatic shift in concentrations of inorganic chlorine species, with levels of ClO enhanced by a factor of 100 and corresponding decreases in ClONO$_2$ and HCl (see Figs. 8.10 and 8.11). This drastic repartitioning of chlorine leads to the occurrence of a totally different chemistry not previously considered. The elevated concentrations of ClO allow these radicals to undergo self-reaction, which at low temperatures leads to formation of the chlorine peroxide (ClOOCl) molecule, the so-called ClO dimer, whose subsequent photolysis regenerates Cl atoms. The resulting catalytic $O_3$-destroying cycle, first proposed by Molina and Molina (1987),

$$2\,(Cl + O_3 \rightarrow ClO + O_2) \qquad (8.13)$$
$$ClO + ClO + M \rightarrow ClOOCl + M \qquad (8.66)$$
$$ClOOCl + h\nu \rightarrow Cl + ClOO \qquad (8.67)$$
$$\underline{ClOO + M \rightarrow Cl + O_2 \qquad (8.42)}$$
$$\text{Net: } 2\,O_3 + h\nu \rightarrow 3\,O_2$$

is believed to be the dominant ozone-destroying cycle in the Antarctic Polar Ozone Hole.

In addition, the enhanced mixing ratios of ClO (and probably to a lesser extent, BrO) in the vortex increase the importance of the previously discussed cycle involving reaction of ClO with BrO:

$$Cl + O_3 \rightarrow ClO + O_2 \qquad (8.13)$$
$$Br + O_3 \rightarrow BrO + O_2 \qquad (8.34)$$
$$ClO + BrO \rightarrow Br + OClO \qquad (8.40a)$$
$$\rightarrow Br + ClOO \qquad (8.40b)$$
$$\rightarrow BrCl + O_2 \qquad (8.40c)$$

The cycle involving the reaction of ClO with $HO_2$ radicals is also of increased importance owing to the increased mixing ratio of ClO:

$$Cl + O_3 \rightarrow ClO + O_2 \qquad (8.13)$$
$$ClO + HO_2 \rightarrow HOCl + O_2 \qquad (8.19)$$
$$HOCl + h\nu \rightarrow OH + Cl \qquad (8.20)$$
$$\underline{OH + O_3 \rightarrow HO_2 + O_2 \qquad (8.21)}$$
$$\text{Net: } 2\,O_3 \rightarrow 3\,O_2$$

Recent efforts designed to study the Arctic Polar Vortex also show the effects of heterogeneous chemistry and chlorine repartitioning. For example, Mankin et al. (1990) have observed a reduction in the ratio of HCl/HF column abundances in the Arctic, indicating the possible chemical removal of HCl via heterogeneous processing. More recently, dramatic increases in ClO abundances (mixing ratios in excess of 1 ppbv) and near total conversion of $Cl_y$ to ClO in the Arctic Winter Vortex have been reported, for example, by Waters et al. (1993), Webster et al. (1993a), and Toohey et al. (1993).

### 8.5.3 Inorganic Chemistry of Halogens in the Troposphere
#### 8.5.3.1 Chlorine Chemistry in the Troposphere

The tropospheric chemistry of chlorine has recently been reviewed by Graedel and Keene (1995). Basically, the compounds listed in Table 8.1 along with HCl and other inorganic compounds (HOCl, $Cl_2$, and $ClONO_2$) comprise the tropospheric Cl burden. The CFCs and other long-lived species will not be considered in this section because of their lack of reactivity in the troposphere. As shown in Section 8.3 and Table 8.2, the organic species containing hydrogen ($CH_3Cl$, etc.), as well as $C_2Cl_4$, will be destroyed in the troposphere via reaction with OH. This leads to the production of a variety of partially oxidized species (HCOCl, $COCl_2$, $CCl_3CHO$), as shown for $CH_3Cl$:

$$OH + CH_3Cl \rightarrow CH_2Cl + H_2O \qquad (8.68)$$
$$CH_2Cl + O_2 + M \rightarrow CH_2ClO_2 + M \qquad (8.69)$$

$$CH_2ClO_2 + NO \rightarrow CH_2ClO + NO_2 \tag{8.70}$$

$$CH_2ClO + O_2 \rightarrow HCOCl + HO_2 \tag{8.71}$$

These partially oxidized species are then either removed by rainout, or undergo further oxidation leading to free Cl-atom production:

$$HCOCl + h\nu \rightarrow HCO + Cl \tag{8.72}$$

Direct emissions of inorganic chlorine, mostly in the form of HCl, occur from a variety of sources. The largest of these sources is direct volatilization from seasalt aerosol:

$$HNO_{3(g)} + NaCl_{(s)} \rightarrow HCl_{(g)} + NaNO_{3(s)} \tag{8.73}$$

$$H_2SO_{4(s)} + 2NaCl_{(s)} \rightarrow 2HCl_{(g)} + Na_2SO_{4(s)} \tag{8.74}$$

Emission of HCl from volcanoes and from a variety of anthropogenic sources (coal combustion, refuse incineration, etc.) have also been identified (Graedel and Keene, 1995). However, due to the high solubility of HCl in rainwater, its tropospheric distribution is not expected to be very uniform. Typical HCl concentrations are 100-300 pptv in the marine boundary layer, with somewhat lower values over land and in the free troposphere. Elevated levels of HCl (a few ppb) can be found in urban areas. Production of other gas-phase chlorine species from reactions involving seasalt have recently been proposed (Graedel and Keene, 1995):

$$N_2O_{5(g)} + NaCl_{(s)} \rightarrow ClNO_{2(g)} + NaNO_{3(s)} \tag{8.75}$$

$$2\ Cl^- + O_{3(g)} + H_2O \rightarrow Cl_{2(g)} + 2\ OH^- + O_{2(g)} \tag{8.76}$$

The tropospheric inorganic chemistry of chlorine is largely the same as that of the stratosphere, involving the species Cl, ClO, HOCl, HCl, and ClONO$_2$, although the effects of this chemistry are generally of little importance. The only clearly observed influence is in ozone-depletion events that occur at the Earth's surface at high northern latitudes during spring (Barrie *et al.*, 1994, and references therein). Here, episodic events occur in which ozone levels near the surface drop rapidly from normal levels of about 40 ppb to near zero. Enhanced levels of reactive chlorine and bromine are thought to be present in the gas phase, presumably obtained from reactions involving seasalt (NaCl and NaBr). Although it is bromine chemistry that is thought to be responsible for the ozone destruction (see following section), measurements of the distribution of hydrocarbons (Jobson *et al.*, 1994) during these events indicate that their concentrations are being influenced by reaction with Cl atoms:

$$Cl + R\text{-}H \rightarrow R + HCl \tag{8.77}$$

Enhanced levels of reactive chlorine, again resulting from reactions on seasalt, may also play a role on a global scale in the marine boundary layer (see, for example, Singh *et al.*, 1996). Initiation of the oxidation of hydrocarbons and dimethyl sulfide via reaction with Cl atoms has been proposed (Graedel and Keene, 1995).

### 8.5.3.2 Bromine Chemistry in the Troposphere

The production of Br atoms in the troposphere from the destruction of source compounds, such as CH$_3$Br, leads to the formation of HBr, BrONO$_2$, HOBr, and BrO via the same reactions as are involved in the stratospheric chemistry. While this chemistry appears to have little consequence in most regions of the troposphere, it has been implicated in the rapid destruction of ozone at the Earth's surface that occurs in the Arctic springtime (Fig. 8.12). Measurements conducted at Alert, North West Territories, and Barrow, Alaska, often indicate very rapid and near-total loss of ozone, concurrent with increases in so-called "filterable" bromine (bromine compounds such

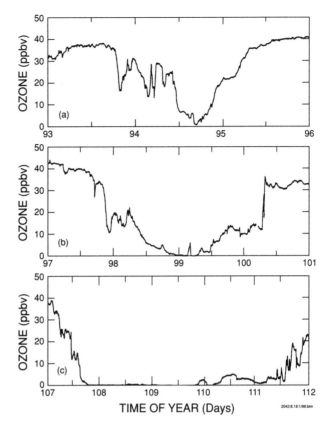

**Figure 8.12.** Hourly mixing ratios of surface-level ozone as measured near Alert, NWT, Canada, during three major ozone-depletion events: (a) April 3-5; (b) April 6-10; and (c) April 17-21, 1993 (from Anlauf et al., 1994).

as HBr, BrONO$_2$, Br, BrO, HOBr, and Br contained in particles, which are collected on cellulose filters and subsequently detected as Br$^-$; see McConnell et al., 1992, and references therein).

Despite the apparent involvement of Br in these ozone-depletion events, a quantitative mechanism has not yet been put forward. Ozone depletion most likely occurs through catalytic cycles involving BrO, analogous to those involved in stratospheric ozone depletion:

$$2 \ (Br + O_3 \rightarrow BrO + O_2) \tag{8.34}$$

$$BrO + BrO \rightarrow 2 \ Br + O_2 \tag{8.78a}$$

$$\underline{BrO + BrO \rightarrow Br_2 + O_2} \tag{8.78b}$$

$$\underline{Br_2 + h\nu \rightarrow 2 \ Br}$$

Net: $2 \ O_3 \rightarrow 3 \ O_2$

However, it is not clear how sufficiently high levels of active BrO and Br can be maintained to destroy ozone on the short timescales (a few days) that are observed. Models of the gas-phase chemistry occurring in these regions show that Br and BrO will rapidly be converted to reservoir species, mainly HBr and BrONO$_2$:

$$BrO + NO_2 + M \rightarrow BrONO_2 + M \tag{8.33}$$

$$Br + CH_2O \rightarrow HBr + HCO \tag{8.79}$$

$$Br + HO_2 \rightarrow HBr + O_2 \tag{8.80}$$

Although elevated concentrations of source gases (like CHBr$_3$ and CH$_2$Br$_2$) are present in the late winter and early spring over the Arctic Ocean (probably due to emissions by sea ice algae), their concentrations appear to be too low and their photolysis rates too slow to constantly regenerate supplies of free bromine necessary to deplete ozone at a sufficient rate. Hence, other sources of active bromine have been proposed. The first of these involves the heterogeneous reaction of $N_2O_5$ on Br-containing sea-salt particles (Finlayson-Pitts et al., 1990),

$$N_2O_5 + NaBr \rightarrow NaNO_3 + BrNO_2 \qquad (8.81)$$

$$BrNO_2 + h\nu \rightarrow Br + NO_2 \qquad (8.82)$$

leading to the production of free Br atoms. Another hypothesis involves the heterogeneous reaction of inorganic bromine reservoir species on sulfate aerosols (Fan and Jacob, 1992):

$$BrONO_{2(g)} + H_2O_{(l)} \rightarrow HOBr_{(aq)} + HNO_{3(aq)} \qquad (8.44)$$

$$HOBr + HBr \rightarrow Br_{2(g)} + H_2O \qquad (8.83)$$

Evaluations of these hypotheses await more detailed measurements and, in particular, require speciation of the bromine compounds present during these ozone-depletion episodes.

### 8.5.3.3 Iodine Chemistry in the Troposphere

Photolysis of $CH_3I$, the only significant source of atmospheric iodine, liberates free I atoms to the troposphere, which react with ozone to produce IO:

$$I + O_3 \rightarrow IO + O_2 \qquad (8.84)$$

Subsequent photolysis of IO or reaction with NO regenerates I in null cycles with respect to ozone destruction:

$$IO + h\nu \rightarrow I + O \qquad (8.85)$$

$$IO + NO \rightarrow I + NO_2 \qquad (8.86)$$

Major reservoir species for iodine include $IONO_2$ (formed from reaction of IO with $NO_2$), HI (formed from the reaction of I with $HO_2$), HOI (formed in the reaction of IO with $HO_2$), and $I_2O_2$ (formed via the self-reaction of IO). Although very little is known about the fate of these reservoir species, they are expected to photolyze rapidly. In addition, all these reservoir species are also likely to be subject to rapid deposition, leading to the removal of iodine from the atmosphere.

## 8.6 Controlling the Detrimental Effects of Halogens on the Atmosphere: Future Outlook

*8.6.1 Introduction*

With the conclusive evidence of the link between CFC destruction in the stratosphere and ozone depletion, the global community has responded in an effort to curtail and eventually eliminate anthropogenic sources of CFCs and halons. The Montreal Protocol on Substances that Deplete the Ozone Layer was brought into existence in 1987 in an effort to control and reduce the production and use of halocarbons. Subsequent amendments to the Protocol (London, 1990; Copenhagen, 1992; Vienna, 1995) served to increase the number of compounds regulated by the Protocol, and to accelerate the timetable for the phase-out of their production. Concomitantly, intense efforts have been focused on the development of the HCFCs and HFCs as replacements for the CFCs, and on the assessment of the environmental acceptability of these replacements with respect to ozone depletion, global warming, and the toxicity of their oxidation products.

## 8.6.2 Development of CFC Replacement Compounds

The rationale behind the development of CFC replacements was to find compounds that possessed similar physical properties (so that they can be used for industrial application), but that possessed significant tropospheric removal processes to allow the rapid removal of these compounds from the atmosphere. The general idea was to develop molecules possessing hydrogen atoms, which would be susceptible to reaction with OH in the troposphere (see Section 8.4.3). A whole suite of molecules, some of which are listed in Table 8.5, have now been proposed for industrial use. The tropospheric lifetimes of these molecules (also given in Table 8.5) are solely determined by the rate of their reaction with OH.

**Table 8.5**
*Atmospheric Lifetimes and Ozone Depletion Potentials (where available) for Various HCFCs and HFCs*

| Molecule | Rate Coefficient for Reaction with OH | | | Tropospheric Lifetime (yrs) | ODP |
|---|---|---|---|---|---|
| | A Factor | E/R | $k_{298}$ | | |
| HCFC-123 | 7.7e-13 | 900 | 3.8e-14 | 1.6 | 0.02 |
| HCFC-124 | 6.6e-13 | 1250 | 1.0e-14 | 6.7 | 0.022 |
| HCFC-141b | 1.7e-12 | 1700 | 5.7e-15 | 12 | 0.11 |
| HCFC-142b | 1.3e-12 | 1800 | 3.1e-15 | 23 | 0.065 |
| HCFC-225ca | 1.0e-12 | 1100 | 2.5e-14 | 3 | 0.025 |
| HCFC-225cb | 5.5e-13 | 1250 | 8.3e-15 | 8.1 | 0.033 |
| HCFC-31 | 2.8e-12 | 1270 | 3.9e-14 | 1.5 | |
| HCFC-132b | 3.6e-12 | 1600 | 1.7e-14 | 4.3 | |
| HCFC-133a | 5.2e-13 | 1100 | 1.3e-14 | 5.6 | |
| HCFC-243cc | 7.7e-13 | 1700 | 2.6e-15 | 30 | |
| HFC-23 | 1.0e-12 | 2440 | 2.8e-16 | 411 | 0 |
| HFC-32 | 1.9e-12 | 1550 | 1.0e-14 | 6.7 | 0 |
| HFC-41 | 3.0e-12 | 1500 | 2.0e-14 | ~4 | 0 |
| HFC-125 | 5.6e-13 | 1700 | 1.9e-15 | 26 | 0 |
| HFC-134 | 1.6e-12 | 1680 | 5.7e-15 | ~13 | 0 |
| HFC-134a | 1.5e-12 | 1750 | 4.2e-15 | 16 | 0 |
| HFC-143 | 4.0e-12 | 1650 | 1.6e-14 | ~4 | 0 |
| HFC-143a | 1.8e-12 | 2170 | 1.2e-15 | ~70 | 0 |
| HFC-152 | 1.7e-11 | 1500 | 1.1e-13 | 0.4 | 0 |
| HFC-152a | 6.0e-12 | 1530 | 3.5e-14 | 1.7 | 0 |
| HFC-161 | 7.0e-12 | 1100 | 1.7e-13 | 0.2 | 0 |

Adapted from Kaye *et al.*, 1994; WMO, 1992; and DeMore *et al.*, 1997.

Criteria were required on which to base the environmental acceptability of the newly developed molecules. To this end, the concepts of the Ozone Depletion Potential (ODP) and Global Warming Potential (GWP) were developed. These parameters were determined through the use of computer models; a known amount of the replacement compound is added to the model atmosphere, and its effects on either total stratospheric ozone (ODP) or radiative forcing (GWP) relative to an equivalent amount of CFC-11 are determined. The use of CFC-11 as a reference compound in these studies helps to remove uncertainties in the parameterizations of transport and chemistry in the models used. A large number of models, developed by groups around the world, have been employed for the evaluation of ODPs of the proposed substitutes; average values are presented in Table 8.5 (WMO, 1992).

### 8.6.3 Chemistry of HCFCs and HFCs

The mechanism of the OH-initiated photo-oxidation of HFCs and HCFCs is quite complex, as is shown schematically in Figure 8.13. This process leads to the production of a number of partially oxidized compounds, some of which still contain halogens. Table 8.6 lists the major halogen-containing products obtained from OH attack on the various substitutes; not shown are the hydroperoxides ($ROOH$) and alkyl peroxynitrates ($ROONO_2$), which are only temporary reservoirs (see Fig. 8.13).

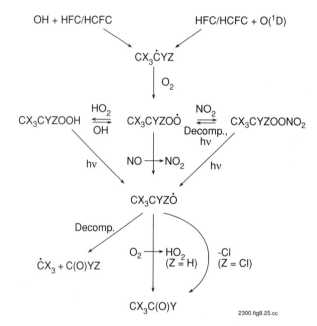

**Figure 8.13.** Degradation scheme for HFC/HCFC initiated by the reactions with OH and $O(^1D)$ (from WMO, 1992).

**Table 8.6**
*Gas-Phase Atmospheric Degradation Products of HFCs and HCFCs*

| Compound | Major Carbon-Containing Degradation Products[a] |
|---|---|
| HFC-32 ($CH_2F_2$) | $COF_2$ |
| HFC-125 ($CF_3CF_2H$) | $COF_2$, $CF_3OH$ |
| HFC-134a ($CF_3CFH_2$) | $HCOF$, $CF_3OH$, $COF_2$, $CF_3COF$ |
| HFC-143a ($CF_3CH_3$) | $CF_3CHO$, $CF_3OH$, $COF_2$, $CO_2$ |
| HCFC-22 ($CHF_2Cl$) | $COF_2$ |
| HCFC-123 ($CF_3CHCl_2$) | $CF_3COCl$, $CF_3OH$, $COF_2$, $CO$ |
| HCFC-124 ($CF_3CHFCl$) | $CF_3COF$ |
| HCFC-141b ($CH_3CFCl_2$) | $CFCl_2CHO$, $COFCl$, $CO$, $CO_2$ |
| HCFC-142b ($CH_3CF_2Cl$) | $CF_2ClCHO$, $COF_2$, $CO$, $CO_2$ |
| HCFC-21 ($CHFCl_2$) | $COFCl$ |
| HCFC-31 ($CH_2FCl$) | $HCOF$ |
| HFC-41 ($CH_3F$) | $HCOF$ |
| HFC-152a ($CH_3CHF_2$) | $COF_2$, $CH_2O$, $CO$ |
| HFC-152 ($CH_2FCH_2F$) | $HCOF$ |

[a] Temporary reservoirs, such as hydroperoxides and alkyl peroxynitrates, are also expected to be formed, but are not listed in the table.
Adapted from Wallington et al., 1994a,b; and Tuazon and Atkinson, 1993.

**Figure 8.14.** Destruction rates of HCFCs via various processes as a function of altitude (from Orlando et al., 1991). Left panel: HCFC-123; right panel: HCFC-124.

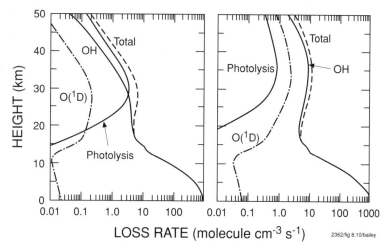

Table 8.7
*Atmospheric Degradation Mechanisms for HFC and HCFC Byproducts*

|  | Uptake into Clouds | Reaction with OH | Photolysis | Approx. Trop. Lifetime (days) |
|---|---|---|---|---|
| $C(O)F_2$ | ✓ |  |  | 5-10 |
| $C(O)FCl$ | ✓ |  |  | 5-20 |
| $CF_3C(O)F$ | ✓ |  |  | 5-15 |
| $HC(O)F$ | ✓ |  |  | 150-1500 |
| $CF_3OH$ | ✓ |  |  | 2 |
| $CF_3CHO$ | ✓ | ✓ | ✓ | ~10 |
| $CF_2ClCHO$ | ✓ | ✓ | ✓ | ~10 |
| $CFCl_2CHO$ | ✓ | ✓ | ✓ | ~10 |
| $CF_3C(O)Cl$ | ✓ |  | ✓ | ~30 |

Adapted from Wallington et al., 1994b.

Although OH-initiated HCFC oxidation does occur in the troposphere, the rate of this process is not sufficient for complete destruction to occur before HCFC transport to the stratosphere occurs (see tropospheric lifetimes of Table 8.5). Once in the stratosphere, photolysis or reaction with $O(^1D)$ may compete with the OH reaction as important degradation products, as shown in Figure 8.14.

Some oxidation products still contain chlorine and could, if sufficiently long-lived, transport this chlorine to the stratosphere. Long-lived oxidation products also need to be assessed in terms of their contributions to global warming, and their potential toxicity to human and plant life.

Products obtained in HFC and HCFC oxidation are, in general, destroyed by heterogeneous uptake into clouds, reaction with OH, direct photolysis, or thermal decomposition (Wallington et al., 1994b). The important destruction pathways for these compounds and their approximate tropospheric lifetimes are given in Table 8.7. There are apparently no chlorine-containing compounds produced in sufficient quantities or with sufficiently long lifetimes that large amounts of chlorine can be transported to the stratosphere. The eventual breakdown products are, for the most part, compounds found naturally in the atmosphere and thus pose no concerns regarding their toxicity. One possible exception is trifluoroacetic acid (TFA, $CF_3COOH$), formed by the hydrolysis of $CF_3C(O)F$ in cloud water, which has been shown to have adverse effects on plant growth at high concentrations. However, concentrations of TFA in nonstagnant water are not likely to reach harmful levels, given that microbial activity is known to degrade TFA.

Although it was initially believed that the HFCs would have ODPs of zero (due to the fact that they contain no chlorine), it was suggested that the chemistry involved in the oxidation of $CF_3$ radicals could lead to some ozone depletion. $CF_3$ is obtained as a product of the oxidation of HFCs 23, 125, 134a, and 143a, as well as from HCFCs 123, 124, and 133a. In the atmosphere, $CF_3$ will be rapidly converted to $CF_3O$:

$$CF_3 + O_2 + M \rightarrow CF_3O_2 + M \qquad (8.87)$$

$$CF_3O_2 + NO \rightarrow CF_3O + NO_2 \qquad (8.88)$$

However, until recently the atmospheric fate of $CF_3O$ has been something of a mystery. Reaction with $O_2$ and thermal decomposition are endothermic, and hence do not occur:

$$CF_3O + O_2 \rightarrow CF_2O + FO_2 \qquad (8.89)$$

$$CF_3O + M \rightarrow CF_2O + F + M \qquad (8.90)$$

Since no loss mechanism was readily apparent for $CF_3O$, discovery of a reaction with $O_3$ spawned the idea that catalytic cycles involving $CF_3O$ could lead to significant ozone depletion in the stratosphere:

$$CF_3O + O_3 \rightarrow CF_3O_2 + O_2 \qquad (8.91)$$

$$\underline{CF_3O_2 + O_3 \rightarrow CF_3O + 2\,O_2} \qquad (8.92)$$

Net: $2\,O_3 \rightarrow 3\,O_2$

$$CF_3O + O_3 \rightarrow CF_3O_2 + O_2 \qquad (8.91)$$

$$\underline{CF_3O_2 + O \rightarrow CF_3O + O_2} \qquad (8.93)$$

Net: $O + O_3 \rightarrow 2\,O_2$

The reaction rate constant for $CF_3O$ with $O_3$ has now been determined to be relatively small, $k \approx 10^{-14}$ cm$^3$ molecule$^{-1}$ s$^{-1}$, as has the rate coefficient for reaction of $CF_3O_2$ with $O_3$. In addition, other reactions occur that permanently remove $CF_3O$ from the atmosphere and limit its ability to participate in catalytic ozone-destruction cycles:

$$CF_3O + NO \rightarrow CF_2O + FNO \qquad (8.94)$$

$$CF_3O + RH \rightarrow CF_3OH + R \qquad (8.95)$$

$$CF_3O + H_2O \rightarrow CF_3OH + OH \qquad (8.96)$$

where RH is a hydrocarbon species. Incorporation of this newly discovered chemistry into stratospheric models has led to the conclusion that no significant $O_3$ depletion will occur as a result of $CF_3O$ (Ravishankara et al., 1994).

### 8.6.4 Future Trends in Atmospheric Chlorine and Bromine

The timetable for the reduction in halogen source gas production and consumption, as outlined by the Copenhagen Amendment, calls for the complete phase-out of the production and consumption of CFCs 11, 12, 113, 114, and 115 by the year 1996; the phase-out of halon production and consumption by 1994; and the phase-out of most HCFCs by the year 2040. Projections made following the Copenhagen Amendment (see Fig. 8.15) indicated that the peak chlorine burden would occur in the late 1990s, at a value of about 4 ppbv. The total chlorine burden should then decrease, reaching

3 ppbv in about 2040 and 2 ppbv in about 2060. Recent measurements (Montzka et al., 1996a), illustrated in Figure 8.4, have indeed shown that the mean global tropospheric chlorine abundance reached a maximum in 1994 and is now decreasing at a rate of about 25 pptv per year. Trends in stratospheric chlorine loading will mirror those observed in the troposphere, after accounting for the 4-6 year time lag associated with the transport of air to the stratosphere. It is important to note that the onset of the appearance of the Antarctic ozone hole coincided with chlorine levels of approximately 2 ppbv. Hence, even with current regulations on future emissions of chlorine source gases, ozone depletion in the Antarctic vortex can be anticipated to continue for another 60 years or so.

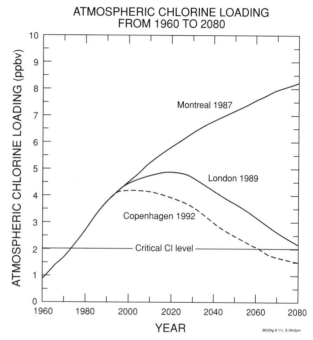

**Figure 8.15.** Predicted future atmospheric burdens of chlorine and bromine.

The atmospheric bromine burden (when considering only compounds that have the potential to reach the stratosphere) is expected to peak near 25 pptv near the turn of the century, and to decrease to a value of about 18 pptv by the year 2050. The smaller change in Br concentrations in the future results from the larger contribution of natural sources (such as $CH_3Br$ and $CH_2Br_2$) to the bromine budget. The implications of these predicted Cl and Br atmospheric burdens for future stratospheric ozone depletion and global warming are the subject of subsequent chapters.

## Further Reading

Brasseur, G. and S. Solomon (1986) *Aeronomy of the Middle Atmosphere*, 2nd ed., D. Reidel, The Netherlands.

Calvert, J., ed. (1994) *The Chemistry of the Atmosphere: Its Impact on Global Change*, Blackwell Scientific Publications, Oxford, UK.

Warneck, P. (1988) *Chemistry of the Natural Atmosphere*, Academic Press, San Diego.

Wayne, R. P. (1991) *Chemistry of Atmospheres*, 2nd ed., Clarendon Press, Oxford, UK.

Mario Molina

## CFCs and Stratospheric Ozone Depletion

Volatile organic compounds containing halogens are used in a wide variety of applications that result in their release to the atmosphere. Concerns about the environmental consequences of such release were originally focussed only on local effects related to their potential toxicity or to their contribution to urban smog.

Fully halogenated hydrocarbons are, however, nontoxic and are chemically too inert to contribute to local air pollution. In the early 1970s atmospheric measurements indicated that some chlorofluorocarbons (CFCs) were present throughout the globe in amounts corresponding to their total industrial production to that date. In 1973, F. Sherwood Rowland and I decided to investigate the ultimate fate of those CFCs; we concluded that they would be destroyed predominantly in the upper stratosphere by solar photodissociation in the ultraviolet. This is a slow process that leads to overall residence times in the environment for these compounds of the order of a century.

It was clear early on that the CFCs would mix throughout the atmosphere before decomposing; it then became apparent that their decomposition products would reach levels of at most a few parts per billion, and hence, it appeared unlikely that any significant environmental effects would ensue. As I started to learn more about the atmosphere, I realized that even trace amounts of certain species may have global consequences.

There are only a few parts per million of ozone present in the stratosphere, and yet this compound serves an extremely important function, namely, to absorb solar ultraviolet radiation, preventing it from reaching the earth's surface. In the process, ozone heats the atmosphere and thus gives rise to the inverted temperature profile that defines the stratosphere. On the other hand, the amount of solar energy absorbed by ozone — in spite of its relative scarcity — is many times larger than the total amount of energy consumed by human society. Thus at first sight it seemed that human activities should hardly affect large natural systems such as the ozone layer.

I first became aware of the potential global impact of CFCs when I compared their industrial production rates with the natural production rates of nitrogen oxides (NO and $NO_2$) in the stratosphere. As first pointed out by Paul J. Crutzen, these compounds are present there at parts per billion level, and play a major role in controlling ozone abundances by means of catalytic cycles: Each nitric oxide molecule can destroy thousands of ozone molecules, as it is continuously being regenerated. The amplification factor related to such cycles explains how it is that relatively small amounts of chlorine-containing chemicals of industrial origin can have global consequences. The CFC-ozone problem is truly global: The largest effects are found over Antarctica, as far away as possible from the industrial sources. The reason is not that the CFCs or their decomposition products concentrate at those latitudes, but rather that it is over the South Pole that the lowest temperatures are reached, leading to the formation of polar stratospheric clouds: The cloud particles activate chlorine and deactivate nitrogen oxides, by scavenging nitric acid. Chlorine free radicals are particularly efficient at destroying ozone in the absence of high levels of nitrogen oxides.

Another factor to consider in connection with global issues is the complexity of the atmosphere and the many feedbacks at play in such a large system. Some of these feedbacks act as powerful stabilizers: For example, if some ozone is destroyed at high altitudes, more solar radiation penetrates to lower altitudes, where it increases the ozone production rate, partially compensating for the high-altitude loss. Other effects, however, can be more complicated. Consider, for example, those related to increases in greenhouse gases: While the Earth's surface might warm up, the stratosphere will cool. On the one hand, lower stratospheric temperatures would lead to less efficient ozone-destroying catalytic cycles at tropical latitudes, because gas-phase reaction rates would slow down. On the other hand, the lower temperatures would enhance ozone depletion at high latitudes, because the efficiency of the heterogeneous reactions responsible for chlorine activation would increase. Similarly, increases in the concentration of methane would lead to less efficient chlorine catalytic cycles as a consequence of a faster HCl formation rate; however, here again the effect is opposite at high latitudes: The chlorine activation rate would increase, since methane is oxidized in the stratosphere yielding significant amounts of water, which would in turn lead to increased surface areas for heterogeneous reactions. The picture that evolves is that while some components of the atmospheric system may be robust, others may be quite vulnerable, and hence susceptible to damage by human activities.

*Mario Molina is a Professor of Atmospheric Chemistry at the Massachusetts Institute of Technology in Cambridge, MA. Jointly with F. S. Rowland, he suggested that ozone depletion could be expected from the release in the atmosphere of industrially manufactured chlorofluorocarbons. Dr. Molina is a member of the U.S. National Academy of Sciences. In 1995, he shared the Nobel Prize for Chemistry with P. J. Crutzen and F. S. Rowland.*

# 9 Carbon-Containing Compounds

## 9.1 Introduction

No atmospheric element is more versatile than carbon. The four electrons in outer orbitals of carbon atoms can form single, double, and triple bonds with many other elements: hydrogen, oxygen, nitrogen, sulfur, and halogens, among others. Carbon-carbon bonds are also very common, and can lead to the formation of branched chain and ring molecules. Literally thousands of organic species are known; many of these are volatile and some can play an important role in atmospheric chemistry. Carbon-containing compounds in the atmosphere include carbon monoxide (CO), carbon dioxide ($CO_2$), hydrocarbons having the general molecular formula $C_nH_m$, and substituted hydrocarbons most often containing oxygen, nitrogen, halogens, or sulfur.

The atmospheric budget of hydrocarbons is controlled by emissions that can be either anthropogenic or natural (frequently biogenic), and atmospheric destructions with a lifetime set, in most cases, by photochemical reactions with atmospheric oxidants, such as OH, $NO_3$, and $O_3$. Local atmospheric concentrations can adversely affect air quality, while global amounts contribute to the radiative budget of the atmosphere. At all geographic scales, hydrocarbon reactivity is coupled to the $O_x$-$HO_x$-$NO_x$ chemistry that controls the oxidizing capacity of the troposphere.

The total global emissions of methane and other hydrocarbons were estimated in Chapter 5 (see Tables 5.6-5.8). Methane, the simplest hydrocarbon, has the highest atmospheric abundance, about 1.8 ppmv in 1994. Because of its long lifetime, methane is distributed throughout the lower atmosphere, and its chemistry is important globally. Similarly, CO is globally distributed and impacts the atmosphere on a planetary scale.

Other hydrocarbons (often called nonmethane hydrocarbons, or NMHCs) are more reactive, and have smaller emission rates, making their atmospheric concentrations much smaller (except at locations very near emission sources). The chemistry of these NMHCs plays an important role in many regions of the troposphere. As examples, emissions of hydrocarbons from fossil fuel combustion are a key ingredient in photochemical smog production in urban areas. Large emissions of isoprene from forests in the southeastern United States are a major local contribution to recurring high-ozone episodes in the region. Also, NMHC emissions from biomass burning contribute to ozone production in the tropics.

In this chapter, the chemistry involved in the oxidation of CO, $CH_4$, and the NMHCs will be elucidated. Further discussions on the impacts of this chemistry on the regional and global scales are presented in other chapters (see Chapters 7 and 13).

## 9.2 Scope and Definitions

Because of the large number of different organic species present in the atmosphere, it is convenient to group them into classes of compounds, based on their molecular structure, with each class behaving in a similar fashion in the atmosphere. Some of the common classes of compounds present in the atmosphere are given in Table 9.1.

**Table 9.1**
*Types of Organic Species Present in the Atmosphere*

| Type of Compound | General Chemical Formula | Examples |
|---|---|---|
| Alkanes | R-H | $CH_4$, methane |
| | | $CH_3CH_3$, ethane |
| Alkenes | $R_2C=CR_2$ | $CH_2=CH_2$, ethene or ethylene |
| | | $CH_3\text{-}CH=CH_2$, propene |
| Alkynes | $RC\equiv CR$ | $HC\equiv CH$, acetylene |
| Aromatics | $C_6R_6$ (cyclic) | $C_6H_6$, benzene |
| | | $C_6H_5(CH_3)$, toluene |
| Alcohols | R-OH | $CH_3OH$, methanol |
| | | $CH_3CH_2OH$, ethanol |
| Aldehydes | R-CHO | $CH_2O$, formaldehyde |
| | | $CH_3CHO$, acetaldehyde |
| Ketones | RCOR | $CH_3C(O)CH_3$, acetone |
| Peroxides | R-OOH | $CH_3OOH$, methylhydroperoxide |
| Organic acids | R-COOH | $HC(O)OH$, formic acid |
| | | $CH_3C(O)OH$, acetic acid |
| Organic nitrates | $R\text{-}ONO_2$ | $CH_3ONO_2$, methyl nitrate |
| | | $CH_3CH_2ONO_2$, ethyl nitrate |
| Alkyl peroxy nitrates | $RO_2NO_2$ | $CH_3O_2NO_2$, methyl peroxynitrate |
| Acylperoxy nitrates | $R\text{-}C(O)OONO_2$ | $CH_3C(O)O_2NO_2$, |
| | | peroxyacetyl nitrate (PAN) |
| Alkyl radicals | $R^\cdot$ | $CH_3^\cdot$, methyl radical |
| | | $CH_3CH_2^\cdot$, ethyl radical |
| Acyl radicals | $RC^\cdot O$ | $CH_3C^\cdot=O$, acetyl radical |
| Alkoxy radicals | $RO^\cdot$ | $CH_3O^\cdot$, methoxy radical |
| Peroxy radicals | $ROO^\cdot$ | $CH_3O_2^\cdot$, methylperoxy radical |
| | | $CH_2(OH)CH_2O_2^\cdot$ |
| | | 2-hydroxyethylperoxy radical |
| Biogenic compounds | $C_5H_8$ | $CH_2=C(CH_3)\text{-}CH=CH_2$, isoprene |
| | $C_{10}H_{16}$ | $\alpha$-pinene, $\beta$-pinene |
| Multifunctional species | | $CH_3C(O)CHO$, methylglyoxal |
| | | $CH_2(OH)CHO$, glycolaldehyde |

## 9.3 Atmospheric Photochemistry of Hydrocarbons
### 9.3.1 General Features

The overall photo-oxidation sequence for a generic hydrocarbon is illustrated in Figure 9.1 and discussed in more detail below. The atmosphere is an oxidizing environment, and the hydrocarbons are gradually degraded toward their oxidative endpoints, $CO_2$ and $H_2O$, through a sequence of radical and nonradical intermediates (although these endpoints are seldom reached in the atmosphere for individual volatile organic compounds). In the process, hydrocarbons fuel the photochemistry of the troposphere, with their rich supply of hydrogen atoms.

Given that hundreds of different hydrocarbons can be present in the atmosphere, the task of deciding all the relevant reactions and their respective rate coefficients may at first appear nearly impossible. Fortunately, there are definite patterns and semiempirical relationships that make the problem considerably easier. For example,

**Figure 9.1.** General oxidation sequence of hydrocarbons (RH).

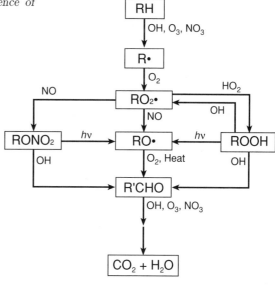

the reactions of OH radicals with $n$-butane ($CH_3CH_2CH_2CH_3$) are fairly similar to the reactions with $n$-pentane ($CH_3CH_2CH_2CH_2CH_3$) after some straightforward consideration of the effect of the additional $-CH_2-$ group.

### 9.3.1.1 Initial Oxidation

The atmospheric reactions of a hydrocarbon usually begin with attack by a strong atmospheric oxidizer such as the hydroxyl radical (OH), ozone ($O_3$), or the nitrate radical ($NO_3$). Rate coefficients for reactions of these oxidants with representative organic species at 298 K are given in Table 9.2.

Reaction with OH is generally the most rapid process, and controls the lifetimes of most organic species. The large variation in the rate coefficients for reaction of the organic species with OH (over four orders of magnitude) thus leads to widely varying lifetimes of these species. For an approximate OH concentration of $10^6$ molecule cm$^{-3}$ throughout the lower atmosphere, lifetimes for organic species range from a few hours for the biogenic NMHCs (isoprene, pinenes) to a few days for compounds such as ethane and toluene, to a couple of months for ethane and acetone, to about ten years for methane.

Hydroxyl radicals react with alkanes via hydrogen abstraction, as illustrated for ethane:

$$OH + CH_3CH_3 \rightarrow CH_3CH_2^{\cdot} + H_2O \qquad (9.1)$$

This reaction is favored because the new bond formed in water is stronger than the C-H bond broken in ethane. For longer alkanes, OH will abstract the most weakly bound hydrogen preferentially, and the rate constant for the overall reaction will reflect the bond strengths of the parent hydrocarbon as well as the number of available hydrogen atoms. These initial reactions have been studied in the laboratory in considerable detail, and semiempirical rules (Atkinson, 1987; Kwok and Atkinson, 1995) have been developed for computing the rate coefficients for many cases. Reactions for OH abstracting an H atom are generally faster for tertiary (>CH–) groups than for secondary (–CH$_2$–) groups, and slowest for primary (–CH$_3$) groups. For example, application of these rules to the reaction of OH with propane ($CH_3CH_2CH_3$) predicts a rate coefficient of $1.2 \times 10^{-12}$ cm$^3$ molecule$^{-1}$ s$^{-1}$, in good agreement with the

## Table 9.2
*Rate Coefficients for Reactions of Several NMHCs with OH, $O_3$, and $NO_3$ ($cm^3$ $molecule^{-1}$ $s^{-1}$)*

| Molecule | $10^{12}$ $k_{OH}$ | $10^{18}$ $k_{O_3}$ | $10^{16}$ $k_{NO_3}$ |
|---|---|---|---|
| methane | 0.0063 | — | ≤0.004 |
| ethane | 0.24 | — | 0.08 |
| n-butane | 2.4 | — | 0.55 |
| n-hexane | 5.5 | — | 1.05 |
| ethene | 8.5 | 1.6 | 2.0 |
| propene | 26 | 10 | 95 |
| 1-butene | 31 | 9.6 | 135 |
| isoprene | 100 | 12.8 | 6800 |
| β-pinene | 79 | 15 | 25000 |
| α-pinene | 54 | 87 | 62000 |
| d-limonene | 170 | 200 | 120000 |
| acetylene | 0.8 | 0.008 | 0.5 |
| formaldehyde | 10 | — | 5.8 |
| acetaldehyde | 14 | — | 24 |
| acetone | 0.22 | — | ≤0.1 |
| methanol | 0.9 | — | 2.1 |
| dimethyl ether | 3.0 | — | ≤30 |
| benzene | 1.2 | ≤0.01 | <0.3 |
| toluene | 6.0 | ≤0.01 | 0.7 |

Data from DeMore et al., 1997; Atkinson, 1997; and Atkinson et al., 1997b.

measured value of $(1.1 \pm 0.2) \times 10^{-12}$ (DeMore et al., 1997), with 70% of the reaction occurring via H-atom abstraction from the secondary carbon (i.e., from the $-CH_2-$ group), and 15% from each primary ($-CH_3$) group.

The carbon-centered radicals produced in these reactions [e.g., the ethyl radical $CH_3CH_2^\cdot$ in Reaction (9.1)] will, under atmospheric conditions, attach rapidly (usually in less than a ms) to $O_2$ to form organic peroxy radicals,

$$CH_3CH_2^\cdot + O_2 + M$$
$$\to CH_3CH_2O_2^\cdot + M \qquad (9.2)$$

the chemistry of which is discussed below.

While reaction of OH with saturated hydrocarbons occurs via abstraction, reaction with unsaturated compounds (alkenes and alkynes) occurs predominantly via addition of OH to the double bond:

$$OH + CH_2=CH_2 + M \to HOCH_2-CH_2^\cdot + M \qquad (9.3)$$

which is quickly followed by addition of $O_2$ to form a substituted peroxy radical:

$$HOCH_2-CH_2 + O_2 + M \to HOCH_2-CH_2O_2^\cdot + M \qquad (9.4)$$

Addition of OH to alkenes is in general quite rapid (rate constants range from about $9 \times 10^{-12}$ $cm^3$ $molecule^{-1}$ $s^{-1}$ for ethene to $1 \times 10^{-10}$ $cm^3$ $molecule^{-1}$ $s^{-1}$ for isoprene), thus limiting the atmospheric lifetimes of alkenes to a couple of days or less. Methods for calculating rate coefficients for OH with various alkenes are summarized by Atkinson (1987).

Initiation of the oxidation of unsaturated species can also occur via reaction with ozone. Though the rate coefficients for these reactions are quite small, $k = (1-1000) \times 10^{-18}$ $cm^3$ $molecule^{-1}$ $s^{-1}$ (Atkinson, 1990; Atkinson, 1997b), the ubiquity of ozone makes these reactions significant in some instances, particularly at night when OH levels are significantly reduced. The mechanism of $O_3$-alkene reactions is not completely understood. The initial reaction is believed to proceed via the formation of a carbonyl species and the so-called Criegee biradical, as shown for ethylene:

$$O_3 + CH_2=CH_2 \to CH_2O + (\cdot CH_2OO \cdot)^* \qquad (9.5)$$

where the (*) indicates that the biradical is initially excited. This excited Criegee biradical can decompose or be stabilized by collisions

$$(\cdot CH_2OO\cdot)^* \rightarrow CO + H_2O \qquad (9.6a)$$
$$\rightarrow CO_2 + H_2 \qquad (9.6b)$$
$$\rightarrow 2\,H + CO_2 \qquad (9.6c)$$
$$+M \rightarrow \cdot CH_2OO\cdot + M \qquad (9.6d)$$

The stabilized biradical is then thought to react with ambient species such as $H_2O$, NO, $SO_2$, and $CH_2O$. However, the rates and mechanisms of these reactions are not well understood.

A small amount of $HO_x$ radicals ($HO_x$ = OH, $HO_2$, and organic peroxy radicals; see below) arises from the ozone-ethene reaction (9.6c and d). There is also a growing body of evidence that OH is produced directly (along with $RO_2$ radicals) in high yield from $O_3$ reactions with larger alkenes. The reaction of $O_3$ with anthropogenic and biogenic alkenes appears to be an important source of $HO_x$ radicals in urban and forested regions, respectively.

$NO_3$ radicals can also be involved in the initiation of the oxidation of organic species. They can add to double bonds

$$CH_2 = CH_2 + NO_3 + M + O_2 \rightarrow CH_2(ONO_2)CH_2O_2\dot{} + M \qquad (9.7)$$

or abstract hydrogen atoms from relatively weak carbon-hydrogen bonds, for example, in aldehydes

$$CH_3CHO + NO_3 + O_2 \rightarrow CH_3C(O)O_2\dot{} + HNO_3 \qquad (9.8)$$

The $NO_3$ reactions are usually most important at night, since during the day the $NO_3$ concentration is suppressed by its rapid photolysis. The rate and mechanism of the reactions of $NO_3$ with organic species have been reviewed by Wayne et al. (1991).

### 9.3.1.2 Peroxy Radicals

The initial hydrocarbon oxidation step almost always leads to the formation of organic peroxy radicals. The lifetime of these radicals is usually relatively short (seconds), but can extend to minutes under relatively clean conditions since the peroxy radicals do not react appreciably with the most common atmospheric gases such as $N_2$, $O_2$, $H_2O$, $O_3$, hydrocarbons, $SO_2$, or $NH_3$. They do, however, react with nitrogen oxides (NO and $NO_2$) and with other peroxy radicals (both organic peroxy radicals, and their inorganic analog $HO_2$). These two classes of reactions represent a critically important fork in the hydrocarbon oxidation path. In environments with sufficiently high $NO_x$ concentrations (30 ppt or more), the chemistry of a generic peroxy radical is dominated by its reaction with NO

$$RO_2\dot{} + NO \rightarrow RO\dot{} + NO_2 \qquad (9.9a)$$

This results in extremely active photochemistry, as the corresponding alkoxy radical, $RO\dot{}$, is highly reactive, and the NO $\rightarrow$ $NO_2$ conversion leads to ozone production (as discussed in Chapters 7 and 13).

This radical-propagating reaction is in competition with a number of reactions that tend to dampen the reactivity. For many hydrocarbons, an alternate pathway is known, leading to the formation of relatively inert organic nitrates,

$$RO_2\dot{} + NO + M \rightarrow RONO_2 + M \qquad (9.9b)$$

The organic nitrate yields are known for many alkanes, and are generally larger for longer and more complex carbon chains.

Peroxy radicals also react with $NO_2$ to form complex species, known as peroxynitrates:

$$RO_2^{\cdot} + NO_2 + M \rightarrow RO_2NO_2 + M \tag{9.10}$$

When the $R$ group is a simple alkyl group (e.g., $\cdot CH_3$), the peroxynitrate is thermally unstable [that is, Reaction (9.10) is reversible] and the alkyl peroxynitrate concentration is small. However, more stable peroxynitrates can be formed in certain instances, such as in the reaction of peroxyacetyl radicals with $NO_2$:

$$CH_3C(O)O_2 + NO_2 + M \rightarrow CH_3C(O)O_2NO_2 + M \tag{9.11}$$

The resulting compound (known as PAN, peroxyacetylnitrate), as well as the organic nitrates formed in Reaction (9.9$b$), can act as signficant reservoirs for reactive nitrogen, and may play a role in its long-range transport (see Chapter 7 for more details).

The other class of reaction important for the peroxy radicals is their reaction with other peroxy radicals (either $HO_2$ or $RO_2$).

Reaction with $HO_2$,

$$RO_2^{\cdot} + HO_2 \rightarrow ROOH + O_2 \tag{9.12}$$

dampens reactivity by forming the longer-lived organic peroxides, $ROOH$. Reactions of the peroxy radicals with other organic peroxy radicals typically have more than one possible set of products. Nonradical products include ketones, alcohols, and organic acids, while production of alkoxy radicals is also possible. For example,

$$CH_3CH_2O_2^{\cdot} + CH_3CH_2O_2^{\cdot} \rightarrow CH_3CH_2OH + CH_3CHO + O_2 \tag{9.13a}$$
$$\rightarrow CH_3CH_2O^{\cdot} + CH_3CH_2O^{\cdot} + O_2 \tag{9.13b}$$

The first channel is an effective sink for the radicals, while the second channel propagates the radical chain. Under some conditions (low $NO_x$, high hydrocarbons), these reactions have a significant effect on the total radical concentrations and therefore on the oxidizing capacity of the atmosphere.

### 9.3.1.3 Alkoxy Radicals

The alkoxy radicals, formed when peroxy radicals lose one oxygen through self-reaction or reaction with NO, are highly reactive and therefore short-lived (less than 1 sec). Thermal decomposition, reaction with $O_2$, and (for the longer-chain radicals) isomerization are all possible, and often can compete with one another. Their relative rates or branching ratios, more than the absolute rates or rate coefficients, are the critical values of importance to the atmosphere. Considerable uncertainty still exists about these reactions. The rate coefficients for the $O_2$ reactions are known for several different alkoxy radicals, but much less information is available on the rates of decomposition and isomerization. In general, smaller alkoxy radicals tend to react with $O_2$,

$$CH_3CH_2O^{\cdot} + O_2 \rightarrow CH_3CHO + HO_2 \tag{9.14}$$
$$CH_3CH(O^{\cdot})CH_3 + O_2 \rightarrow CH_3C(O)CH_3 + HO_2 \tag{9.15}$$

while larger and more complex radicals are more prone to dissociation and isomerization:

$$CH_3CH(O^{\cdot})CH_2OH + M \rightarrow CH_3CHO + CH_2OH + M \tag{9.16}$$
$$CH_3CH(O^{\cdot})CH_2CH_2CH_3 \rightarrow CH_3CH(OH)CH_2CH_2CH^{\cdot}CH_3 \tag{9.17}$$

Estimation of decomposition rates are often made using the so-called Evans-Polanyi relationship, in which the activation energy of a unimolecular process is estimated from the enthalpy of the reaction. This process is complicated by substantial uncertainties in the heats of formation (especially of the radical species) that translate into larger uncertainties in the rate coefficients and branching ratios. The chemistry of alkoxy radicals has been reviewed by Atkinson (1997a).

The fate of alkoxy radicals, while a major point of uncertainty in the gas-phase hydrocarbon oxidation scheme, is of key importance. Decomposition of alkoxys breaks the longer hydrocarbon chains and speeds the oxidation toward their ultimate oxidation products ($CO_2$ and $H_2O$). In contrast, isomerization and $O_2$ reactions preserve the original carbon chain, and lead to the formation and subsequent chemistry of new long-chain partially oxidized intermediate species.

### 9.3.1.4 Chemistry of Partially Oxidized Species

As shown above, the oxidation of hydrocarbons leads to the production of partially oxidized species, including aldehydes, ketones, alcohols, organic acids, etc. Rate coefficients for reactions of a representative set of these compounds are given in Table 9.2. Comprehensive listings of rate parameters are available in Atkinson *et al.* (1997).

Most of these compounds react with OH, and abstraction of a hydrogen atom is most common. Abstraction from aldehydes,

$$OH + RCHO + O_2 \rightarrow H_2O + RC(O)O_2 \cdot \qquad (9.18)$$

should be recognized as particularly fast, and can ultimately form PAN and its analogous compounds, $RC(O)OONO_2$. Other reactions can also be important, such as addition of ozone and $NO_3$ to double bonds (if present), and $NO_3$ abstraction of hydrogen from an aldehyde group.

In addition to reaction with OH, $NO_3$, and $O_3$, organic intermediates with characteristic UV-absorbing groups (chromophores) also photolyze. Examples include ketones and aldehydes [$RC(O)R'$, $RCHO$], peroxides and peroxy acids [$RCH_2OOH$ and $RC(O)OOH$], and nitrates ($RCH_2ONO_2$). Photolytic lifetimes range from days to weeks for ketones and aldehydes (except formaldehyde), organic peroxides, peroxy acids, and organic nitrates, to only a few hours for formaldehyde and multifunctional compounds having several chromophores such as glyoxal [$CH(O)CHO$] and methyl glyoxal [$CH_3C(O)CHO$]. Photolysis of these species leads to the net production of odd hydrogen radicals ($HO_x$). Photolysis of formaldehyde in particular needs to be considered in determining the budget of $HO_2$ and OH radicals (see Chapter 6)

$$CH_2O + h\nu \rightarrow CO + H_2 \qquad (9.19a)$$
$$\rightarrow HCO + H \qquad (9.19b)$$
$$HCO + O_2 \rightarrow HO_2 + CO \qquad (9.20)$$
$$H + O_2 + M \rightarrow HO_2 + M \qquad (9.21)$$

### 9.3.1.5 Heterogeneous Reactions

Some hydrocarbons and their oxidation products can also react at liquid and solid surfaces and thus be removed from the gas phase. Surfaces in contact with the atmosphere include cloud droplets and ice particles, aerosols, the ground (including vegetation and other structures), and surface water. The importance of these heterogeneous reactions is probably greater for oxygenated hydrocarbons, and especially those with multiple functional groups. Radicals may also react rapidly on surfaces, but fast gaseous reactions usually dominate their lifetime.

The fate of the organic species after such heterogeneous reactions is of some interest. Often the reaction represents a final removal from the atmosphere (surface deposition, rain out), but in other cases (clouds, aerosols) further chemical transformation and subsequent re-entry into the gas phase are possible. An example of this is the production of formic acid in clouds,

$$CH_2O_{(g)} + H_2O_{(aq)} \rightarrow CH_2(OH)_{2\,(aq)} \qquad (9.22)$$

$$CH_2(OH)_{2\,(aq)} + OH_{(aq)} + O_2 \rightarrow HC(O)O^-_{\,(aq)} + H^+ + H_2O + HO_2 \qquad (9.23)$$

These reactions affect the pH of cloud water, and evaporation of the cloud can introduce new compounds (formic acid, HCOOH, in the above example) to the gas phase.

By their nature, these heterogeneous reactions are complex and not as well understood as gas-phase hydrocarbon oxidation. The rates are difficult to measure even under well-controlled laboratory conditions, and the physical and chemical variety of natural surfaces does not lend itself to generalizations. The parameterization of these reactions remains a major contributor to the uncertainties about the cycling of hydrocarbons through the atmosphere.

### 9.3.2 Chemistry of Some Representative Carbon-Containing Species
#### 9.3.2.1 Carbon Monoxide

While CO is not really a hydrocarbon, its atmospheric chemistry illustrates the basic principles involved in hydrocarbon oxidation. As presented in Chapter 5, CO is globally distributed with a wide variety of sources. It is produced from the burning of biomass and fossil fuel, is emitted from vegetation and the oceans, and is a product of the oxidation of methane and other hydrocarbons (see following sections on methane and isoprene oxidation).

The dominant sink for CO, as is the case for most hydrocarbons, is its reaction with OH:

$$OH + CO \rightarrow H + CO_2 \qquad (9.24)$$

In fact, this reaction provides the major sink for OH in the remote troposphere. The rate coefficient for Reaction (9.24) is a complex function of pressure. Near the Earth's surface the rate coefficient is $2.4 \times 10^{-13}$ cm$^3$ molecule$^{-1}$ s$^{-1}$, leading to a CO lifetime of about 1-2 months. Subsequent reaction of H with $O_2$ results in $HO_2$ production (Reaction 9.21). When sufficient NO is available, $HO_2$ reacts with NO, regenerating OH and leading to net ozone production

$$HO_2 + NO \rightarrow OH + NO_2 \qquad (9.25)$$
$$NO_2 + h\nu\,(+O_2) \rightarrow NO + O_3 \qquad (9.26)$$

Under cleaner conditions, where $NO_x$ concentrations are low, $HO_2$ radicals react with themselves, with other peroxy radicals, or with $O_3$ (resulting in net loss of ozone).

#### 9.3.2.2 Methane

Methane is of great importance to global tropospheric chemistry, as it is usually the most abundant and certainly the most ubiquitous hydrocarbon. Its chemistry is more complex than that of CO, but still relatively simple compared to that of other hydrocarbons, and illustrates some of the general features described in the earlier sections. Laboratory studies have elucidated most of the reaction sequence and kinetics from the initial OH attack to the ultimate production of $CO_2$, but even for this relatively simple molecule there are several points of uncertainty. The reactions describing methane oxidation are shown in Table 9.3, together with appropriate kinetic data.

The reaction between OH and methane [Reaction (9.27) in Table 9.3] is one of the best-studied hydrocarbon reactions and is known accurately over a wide range of temperatures (200-1500 K). The kinetic data given in the table are valid near room temperature (298 K), but more accurate values at different temperatures can be obtained by the modified Arrhenius expression

$$k_{27}(T) = 1.59 \times 10^{-20} T^{2.84} \exp\left(-\frac{978}{T}\right) \text{ cm}^3 \text{ molecule}^{-1} \text{ s}^{-1}$$

The methyl radical formed in this reaction is extremely short-lived in the lower atmosphere (ca. $10^{-7}$ s at sea level) due to the rapid occurrence of Reaction (9.28), which is therefore frequently combined with Reaction (9.27) to write a single net reaction leading to the production of the methyl peroxy radical,

$$CH_4 + OH\,(+ O_2) \rightarrow CH_3O_2^{\cdot} + H_2O$$

with the rate coefficient of Reaction (9.27).

Table 9.3
Methane Oxidation Scheme

| Reaction | | $k$ (298 K)[a] | $E/R$[b] |
|---|---|---|---|
| (9.27) | $CH_4 + OH \rightarrow CH_3^{\cdot} + H_2O$ | $6.3 \times 10^{-15}$ | $1.8 \times 10^3$ |
| (9.28) | $CH_3^{\cdot} + O_2 + M \rightarrow CH_3O_2^{\cdot} + M$ | $1.1 \times 10^{-12}$ | $-1.2 \times 10^3$ |
| (9.29) | $CH_3O_2^{\cdot} + NO \rightarrow CH_3O^{\cdot} + NO_2$ | $7.6 \times 10^{-12}$ | $-2.8 \times 10^2$ |
| (9.30) | $CH_3O_2^{\cdot} + HO_2 \rightarrow CH_3OOH + O_2$ | $5.6 \times 10^{-12}$ | $-8.0 \times 10^2$ |
| (9.31) | $CH_3O_2^{\cdot} + CH_3O_2^{\cdot} \rightarrow CH_3O^{\cdot} + CH_3O^{\cdot} + O_2$ | $1.4 \times 10^{-13}$ | $7.5 \times 10^2$ |
| (9.32) | $CH_3O_2^{\cdot} + CH_3O_2^{\cdot} \rightarrow CH_2O + CH_3OH + O_2$ | $2.6 \times 10^{-13}$ | $-3.8 \times 10^2$ |
| (9.33) | $CH_3O^{\cdot} + O_2 \rightarrow CH_2O + HO_2$ | $1.9 \times 10^{-15}$ | $9.0 \times 10^2$ |
| (9.34) | $CH_3O^{\cdot} + NO_2 + M \rightarrow CH_3ONO_2 + M$ | $1.5 \times 10^{-11}$ | 0 |
| (9.35) | $CH_3OOH + h\nu \rightarrow CH_3O^{\cdot} + OH$ | $6.5 \times 10^{-6}$ [c] | |
| (9.36a) | $CH_3OOH + OH \rightarrow CH_3O_2^{\cdot} + H_2O$ | $5.2 \times 10^{-12}$ | $-2.0 \times 10^2$ |
| (9.36b) | $CH_3OOH + OH \rightarrow CH_2O + OH + H_2O$ | $2.2 \times 10^{-12}$ | $-2.0 \times 10^2$ |
| (9.37a) | $CH_3OH + OH \rightarrow CH_3O^{\cdot} + H_2O$ | $1.3 \times 10^{-13}$ | $6.0 \times 10^2$ |
| (9.37b) | $CH_3OH + OH\,(+ O_2) \rightarrow CH_2O + HO_2 + H_2O$ | $7.7 \times 10^{-13}$ | $6.0 \times 10^2$ |
| (9.38) | $CH_3ONO_2 + h\nu \rightarrow CH_3O^{\cdot} + NO_2$ | $1.6 \times 10^{-6}$ [c] | |
| (9.19a) | $CH_2O + h\nu \rightarrow H_2 + CO$ | $4.6 \times 10^{-5}$ [c] | |
| (9.19b) | $CH_2O + h\nu \rightarrow HCO + H$ | $3.5 \times 10^{-5}$ [c] | |
| (9.39) | $CH_2O + OH \rightarrow HCO + H_2O$ | $1.0 \times 10^{-11}$ | 0 |
| (9.40) | $CH_2O + NO_3 \rightarrow HCO + HNO_3$ | $5.8 \times 10^{-16}$ | $2.9 \times 10^3$ |
| (9.41) | $CH_2O + HO_2 \rightarrow CH_2(OH)O_2^{\cdot}$ | $5.0 \times 10^{-14}$ | $-6.0 \times 10^2$ |
| (9.42) | $CH_2(OH)O_2^{\cdot} \rightarrow CH_2O + HO_2$ | $1.3 \times 10^{2}$ | $7.0 \times 10^3$ |
| (9.43) | $CH_2(OH)O_2^{\cdot} + NO\,(+O_2) \rightarrow HCOOH + HO_2 + NO_2$ | $7.6 \times 10^{-12}$ | $-1.8 \times 10^2$ |
| (9.20) | $HCO + O_2 \rightarrow HO_2 + CO$ | $5.5 \times 10^{-12}$ | $-1.4 \times 10^2$ |
| (9.24) | $CO + OH\,(+O_2) \rightarrow CO_2 + HO_2$ | $2.4 \times 10^{-13}$ | 0 |

[a] Rate coefficient at 298 K and 1 atm pressure; units are $\text{cm}^3$ $\text{molecule}^{-1}$ $\text{s}^{-1}$ for bimolecular reactions and $\text{s}^{-1}$ for photolysis reactions.
[b] Activation energy at 298 K, divided by gas constant $R$; units are K.
[c] Photodissociation rate coefficient for typical high sun conditions.
Rate coefficient data from DeMore et al., 1997, and Atkinson et al., 1997.

The reactions of $CH_3O_2^{\cdot}$ follow the general pattern described in Section 9.3.1: Reaction (9.29) converts NO to $NO_2$ and propagates the radical chain by producing methoxy radicals, $CH_3O^{\cdot}$, while Reaction (9.30) destroys two radicals and produces the longer-lived intermediate methyl hydroperoxide, $CH_3OOH$. The yield of $CH_3ONO_2$ from Reaction (9.29) is negligible, in contrast to the peroxy radicals of higher alkanes. Reactions among methyl peroxy radicals have several possible outcomes, the main ones being the production of two methoxy radicals [radical propagation, Reaction (9.31)] and the production of formaldehyde and methanol [radical termination, Reaction (9.32)]. Reactions of $CH_3O_2^{\cdot}$ with peroxy radicals derived from other hydrocarbons are also possible (though not shown in the table), and can lead to both radical and nonradical products.

Methoxy radicals react almost exclusively with $O_2$ (Reaction 9.33), even in polluted urban environments. Reaction (9.34) is not an important sink of $CH_3O^{\cdot}$, but is of interest as a possible source of methyl nitrate, $CH_3ONO_2$, under very high $NO_x$ conditions. The reactions of the radicals $CH_3O_2^{\cdot}$ and $CH_3O^{\cdot}$ occur only a few seconds or, at most, minutes after the initial OH attack on $CH_4$, and yield the intermediates $CH_2O$, $CH_3OOH$, $CH_3OH$, and $CH_3ONO_2$. These longer-lived nonradicals have atmospheric lifetimes ranging from several hours to months, and have their own chemical photo-oxidation sequences. The production of numerous oxygen- and nitrogen-containing intermediate species is a general characteristic of the atmospheric chemistry of all hydrocarbons, and the four new species produced from methane are a modest example compared to the complex speciation that arises from the photo-oxidation of larger hydrocarbons.

The fate of methyl hydroperoxide can have a significant impact on the budget of atmospheric radicals. It should be recalled that the formation of $CH_3OOH$ occurred at the expense of two radicals, $CH_3O_2^{\cdot}$ and $HO_2$, in Reaction (9.30). The photolysis of $CH_3OOH$, Reaction (9.35), offsets this radical loss by producing $CH_3O^{\cdot}$ and OH. The reactions of $CH_3OOH$ with OH (9.36a and 9.36b), however, cause an overall loss of radicals. Abstraction of a hydrogen from the $CH_3$ group [Reaction (9.36b)], followed by breaking of the weak O-O bond, leads to no new radicals and therefore makes irreversible the initial loss of two radicals incurred in Reaction (9.30). Abstraction of the hydrogen atom from the –OOH group (9.36a) destroys one OH radical while regenerating $CH_3O_2^{\cdot}$, which can then react again with $HO_2$. Reactions (9.36a) and (9.30) may in fact be viewed as a catalytic cycle for the destruction of the radicals OH and $HO_2$.

Formaldehyde is a major intermediate in the degradation of methane (and many other hydrocarbons). In the absence of heterogeneous losses, essentially every methane molecule is converted to $CH_2O$ [Reactions (9.27-9.38)]. The lifetime of formaldehyde is relatively short, and both photolysis and OH reactions result in the formation of CO. Note that two photolysis channels are available, and while their probabilities are comparable, they differ importantly in their effect on $HO_x$ radicals. Reaction (9.19b), in particular, is a source of two $HO_2$ radicals, since the photolysis products H and HCO react rapidly with $O_2$,

$$CH_2O + h\nu \to H + HCO \qquad (9.19b)$$

$$H + O_2 + M \to HO_2 + M \qquad (9.21)$$

$$HCO + O_2 \to HO_2 + CO \qquad (9.20)$$

Other reactions of $CH_2O$ include abstraction of a hydrogen atom by $NO_3$ [Reaction (9.40)], which may be significant at night, and addition of $HO_2$ to form the hydroxy methyl peroxy radical [Reaction (9.41)]. Although the rate coefficient for this reaction is reasonably fast, its reversal by thermal decomposition of the radical [Reaction (9.42)] is also fast at the temperatures of the lower troposphere, thus preventing significant buildup of the $CH_2(OH)O_2^{\cdot}$ radical. However, other reactions of this radical can be competitive under some conditions, and may be among the few gas-phase sources of formic acid [*e.g.*, Reaction (9.43)]. Aqueous conversion of formaldehyde to formic acid is also possible, as already discussed in Section 9.3.1.5.

Methanol and methyl nitrate are produced from methane in small amounts. Their destruction is relatively slow, and eventually results in the production of formaldehyde, with little direct impact on photochemical radical chains. However, the photolysis of methyl nitrate [Reaction (9.38)] may contribute to $NO_x$ sources in remote regions of the atmosphere.

### 9.3.2.3 Isoprene

Isoprene constitutes a non-negligible source of carbon to the atmosphere (see Tables 5.7 and 5.8), and its high reactivity makes it a very important contributor to the chemistry of the troposphere. For example, high ozone levels observed over large regions of the southeastern United States have been attributed, at least in part, to chemistry involving the oxidation of isoprene in the presence of nitrogen oxides from nearby anthropogenic sources. In addition, isoprene oxidation provides a globally significant source of CO.

Although it behaves in general as an alkene, the chemistry of isoprene is highly complex, and even a simplified explicit description of its oxidation would involve about 35 species and 100 reactions. Hence, only the major pathways involved will be outlined here, and the reader is referred to the original literature (Tuazon and Atkinson, 1990; Paulson *et al.*, 1992; Paulson and Seinfeld, 1992) for a more complete treatment of the chemistry. Since isoprene is emitted by plants during photosynthesis (*i.e.*, when light is present), its emissions coincide with high OH radical concentrations, making the OH reaction its dominant loss pathway. Isoprene is also reactive towards $NO_3$ and $O_3$, but these reactions are usually of lesser atmospheric significance.

As in the general case of an alkene, OH adds to the doubly bonded carbon atoms of isoprene. The radicals formed from addition to the terminal carbon atoms [compounds (a) and (d) below] are more probable than the radicals obtained from internal addition [(b) and (c)].

$$\begin{matrix} H_2C=C(CH_3)-CH=CH_2 + OH \longrightarrow 0.4 \{ \text{(a) } HOCH_2-C(CH_3)=CH-\cdot CH_2 + \text{(b) } HOCH-C(CH_3)=CH-\cdot CH_2 \} \\ + 0.6 \{ \text{(c) } H_2C=C(CH_3)-\cdot COH-CH_2 + \text{(d) } H_2C=C(CH_3)-\cdot C-H_2COH \} \end{matrix} \quad (9.44)$$

However, the major products of the reaction depend mainly on which bond OH attacks, and not on whether addition takes place at the terminal or internal carbon. The branching ratio, inferred from product yields, is about 60:40 in favor of addition to the double bond containing the methyl substitution.

Each of the substituted alkyl radicals that result from OH addition will add $O_2$ in the usual way, to form peroxy radicals:

$$\text{(a) } HOCH_2-\cdot CH-C(CH_3)=CH_2 + O_2 \xrightarrow{[M]} HOCH_2-CH(OO\cdot)-C(CH_3)=CH_2 \quad (9.45)$$

When the NO concentration is greater than about 30 ppt, the peroxy radicals react predominantly with NO. At lower NO levels, the peroxy radicals can also react with $HO_2$ and other organic peroxy radicals ($RO_2$) to form a variety of products. For simplicity, the discussion that follows corresponds to conditions of high $NO_x$. Reaction of the radicals (a)-(d) (after addition of $O_2$) with NO results mainly in $\beta$-hydroxyalkoxy radicals; that is, the hydroxy (–OH) group and the alkoxy group (–O·) are on adjacent carbons. Alkyl nitrates are formed with a yield of about 13% ($x=0.13$),

$$\text{·O-O}\diagdown\text{CH}_3 \atop \text{HOCH}_2\diagup\text{CH-C}\diagdown\text{CH}_2 + \text{NO} \longrightarrow (x)\ \text{O}_2\text{N-O}\diagdown\text{CH}_3 \atop \text{HOCH}_2\diagup\text{CH-C}\diagdown\text{CH}_2 + (1-x)\ \text{·O}\diagdown\text{CH}_3 \atop \text{HOCH}_2\diagup\text{CH-C}\diagdown\text{CH}_2 + (1-x)\ \text{NO}_2 \quad (9.46)$$

While most alkoxy radicals have several reaction pathways available, including decomposition, isomerization, and reaction with $O_2$, the $\beta$-hydroxy alkoxy radicals are known to decompose. Decomposition of the alkoxy radicals from both (a) and (b) leads to the formation of methacrolein and formaldehyde:

$$(9.47)$$

(The small half-arrows represent rearrangements of single electrons.) The $CH_2OH$ radical reacts rapidly with $O_2$ to yield formaldehyde,

$$\cdot CH_2OH + O_2 \rightarrow CH_2O + HO_2 \quad (9.48)$$

The alkyl peroxy radicals from (c) and (d) react in analogous fashion to produce methyl vinyl ketone and formaldehyde:

$$(9.49)$$

$$(9.50)$$

Methylvinyl ketone (MVK) and methacrolein (MACR), along with their co-product $CH_2O$, account for about 60% of the products obtained in the initial oxidation of isoprene under high-$NO_x$ conditions. The remaining 40% of the products are believed to include a number of organic nitrates, carbonyl, and hydroxycarbonyl compounds (Kwok et al., 1995). MVK and MACR are themselves very reactive and have atmospheric lifetimes of less than one day. Both compounds are unsaturated; thus reaction with OH is a facile process. In addition, methacrolein possesses an aldehyde functionality, and abstraction of the aldehydic H by OH occurs in competition with reaction at the double bond. There is also the possibility that these compounds could undergo photolysis. A wide variety of products are obtained from the oxidation of MVK and MACR in the presence of $NO_x$, including PAN analogues [e.g., $CH_2=C(CH_3)C(O)OONO_2$], CO, $CO_2$, $CH_2O$, and a number of multifunctional compounds — hydroxyacetaldehyde ($HOCH_2CHO$) also

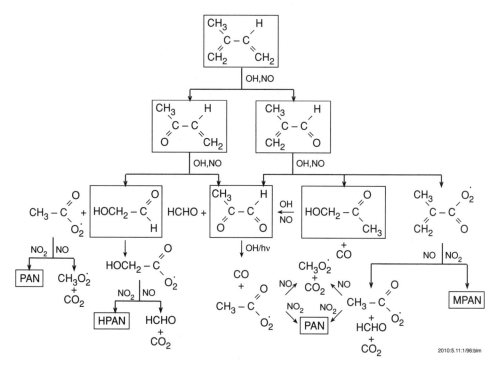

**Figure 9.2.** Simplified schematic of isoprene oxidation in the presence of $NO_x$. The following abbreviations are used: PAN, $CH_3C(O)O_2NO_2$; MPAN, $CH_2=C(CH_3)C(O)O_2NO_2$; HPAN, $HOCH_2C(O)O_2NO_2$.

known as glycolaldehyde, hydroxyacetone [$HOCH_2C(O)CH_3$], and methylglyoxal [$CH_3C(O)CHO$]. These multifunctional species further react with OH or photolyze, leading to the eventual production of CO, $CO_2$, and $CH_2O$. The overall yield of CO in high-$NO_x$ conditions has been estimated to be about three molecules per molecule of isoprene consumed (Miyoshi et al., 1994). Figure 9.2 is a schematic of isoprene oxidation in the presence of $NO_x$. The species in boxes are sufficiently long-lived to be measured in the atmosphere by conventional techniques.

The OH-initiated oxidation of isoprene under low-$NO_x$ conditions has been less intensely studied. In general, yields of MACR, MVK, and CO are greatly reduced, and hydroperoxide species, produced in the reaction of $RO_2$ radicals with $HO_2$, are observed (Miyoshi et al., 1994).

$NO_3$ and $O_3$ also react with isoprene. The $O_3$ reaction is rather slow compared to the OH reaction; hence $O_3$ is not expected to be a major daytime loss path for isoprene, although it can contribute to its chemistry at night. As with the OH-initiated oxidation of isoprene, methacrolein, methyl vinyl ketone, and formaldehyde are important products. The $NO_3$ reaction is fast, but because $NO_3$ photolyzes rapidly in sunlight, appreciable concentrations of $NO_3$ do not coincide with peak isoprene emissions.

The chemistry of monoterpenes (general formula, $C_{10}H_{16}$), also emitted by vegetation, is even more complex than that of isoprene. The initial step in their oxidation again occurs via reaction with OH, $O_3$, or $NO_3$. While some of the initial products formed in the oxidation process have been identified, the full oxidation process for the terpenes is far from established. The low vapor pressures of some of the oxygenated products likely leads to their incorporation into organic aerosol.

## 9.4 Distribution of Hydrocarbons

Hundreds of different hydrocarbons have been identified in the atmosphere. The distribution of these species depends on the strength and location of their sources (see Chapter 5), on their reactivity (as described above), and on the processes that distribute them. The sources and distribution of NMHCs have recently been reviewed by Singh and Zimmerman (1992).

Many of the more reactive NMHCs are photochemically oxidized in the planetary boundary layer (PBL), which typically extends up to 2000 m over terrestrial areas in the daytime (see Chapter 2). At night, the thickness of this layer may be less than 100 m. Transport of emissions out of the PBL is slow, since the layer is usually capped by strong temperature inversions. However, instabilities, such as cloud processes and other forms of convection, may rapidly mix NMHCs throughout the PBL and through the inversion, into the free troposphere. In the case of convective events, NMHCs and other trace gas emissions may be transported rapidly to the upper troposphere. Once out of the planetary boundary layer, stronger winds and larger-scale processes may disperse the NMHCs regionally or globally, depending on the individual rates of chemical destruction.

Table 9.4
*Median Mixing Ratios from Urban Areas (ppbv)*

| Hydrocarbon | US[a] | Japan[b] | Germany[c] |
|---|---|---|---|
| ethane | 11.5 | 2.7 | 2.8 |
| ethylene | 10.7 | 4.0 | 1.0 |
| acetylene | 6.5 | 2.0 | 0.89 |
| propane | 7.7 | 2.6 | 1.35 |
| propene | 2.7 | 0.7 | 0.17 |
| $i$-butane | 3.8 | 1.1 | 0.27 |
| $n$-butane | 10.0 | 1.9 | 0.78 |
| 1-butene + 2-butene | 1.5 | 0.4 | 0.024 |
| $i$-pentane | 9.0 | 1.6 | 0.43 |
| $n$-pentane | 4.4 | 1.1 | 0.15 |
| 2-methylpentane | 2.5 | 0.8 | |
| 3-methylpentane | 1.8 | 0.4 | |
| $n$-hexane | 1.8 | 1.0 | |
| benzene | 2.2 | 1.2 | |
| toluene | 4.9 | 4.9 | |
| ethylbenzene | 0.8 | 1.0 | |
| $m+p$-xylene | 2.3 | 1.2 | |
| $o$-xylene | 0.9 | 0.5 | |

[a] Seila *et al.*, 1989, ground-based samples, average of 39 cities.
[b] Uno *et al.*, 1985, aircraft samples at 350-650 m.
[c] Geiss and Volz-Thomas, 1992 (Freiburg, 10 km downwind of urban center).

### 9.4.1 Planetary Boundary Layer
#### 9.4.1.1 Urban Atmospheres

Urban environments contain a wide variety of NMHCs, with total mixing ratios usually in the range of 100-1000 ppbC. Typical concentrations of major hydrocarbons in selected cities are given in Table 9.4. Differences in absolute concentrations of

NMHCs among the U.S., Japanese, and European urban areas are due to differences in source character and strength, as well as dilution processes. The ratios of individual hydrocarbons to each other may give the relative abundance in the averaged emission source.

There are strong diurnal and seasonal variations in the distribution of urban hydrocarbons. Afternoon minima in the mixing ratios of many NMHCs are due to deeper daytime atmospheric mixed layers and higher photochemical losses. Higher mixing ratios at night often reflect a shallow nighttime boundary layer with continued emissions of NMHC into this layer. Higher values of anthropogenic NMHCs are usually found in winter compared to summer as a result of frequent atmospheric stagnation and lower removal rates.

### 9.4.1.2 Rural and Remote Continental Boundary Layer

Measurements from remote continental sites show significantly lower average concentrations compared with urban sites. Although many of the hydrocarbons reported are clearly of anthropogenic origin (from urban or industrial areas or areas where biomass burning is occurring, etc.), a larger contribution from biogenic sources is usually seen.

Biogenic hydrocarbons, such as isoprene and terpenes, are highly reactive and have atmospheric lifetimes of only a few hours. Consequently, they are mostly detected near their sources in the boundary layer, not in the mid- to upper troposphere. In addition, these emissions are seasonal in midlatitudes. Most biogenic emissions occur in the warmer spring and summer periods. In the tropics, the dry season varies with location and biomass burning occurs mostly during the dry season. In rural and, in some cases, urban areas where biogenic emissions are moderately high, these species can compete with the anthropogenic hydrocarbon burden in $O_3$ formation. Typical mixing ratios for isoprene, $\alpha$-pinene, and several NMHCs from some remote areas are shown in Table 9.5.

**Table 9.5**
*NMHC Mixing Ratios (ppbv) in
Several Rural and Remote Terrestrial Ecosystems
(Surface and Boundary Layer)*

| Ecosystem | Ethane | Ethene | Propane | Benzene | Isoprene | $\alpha$-Pinene |
|---|---|---|---|---|---|---|
| Sub-Alpine forest[a] | 2.2 | 0.46 | 1.27 | 0.24 | 0.63 | 0.14 |
| Tropical rainforest[b] | 0.98 | 0.97 | 0.37 | 0.08 | 2.04 | 0.10 |
| Tropical rainforest[c] | 1.10 | 0.70 | 0.16 | 0.17 | 5.45 | 0.20 |
| Tropical rainforest[d] | 0.73 | 0.29 | 0.10 | 0.07 | 1.21 | 0.06 |
| Wooded savanna[e] | 0.65 | 0.33 | 0.11 | 0.16 | 0.04 | <0.01 |
| Wooded savanna[f] | 0.26 | 0.09 | 0.05 | 0.04 | <0.01 | <0.01 |
| Mixed temperate forest[g] | x | x | 0.88 | 0.13 | 4.33 | 0.28 |

[a] Niwot Ridge, CO, USA, August/September, surface (Greenberg and Zimmerman, 1984).
[b] Amazon Tropical Forest, Brazil, dry season, boundary layer (Zimmerman et al., 1988).
[c] Amazon Tropical Forest, Brazil, wet season, boundary layer (Zimmerman et al., 1988).
[d] Nigerian Tropical Forest, wet season, surface (Zimmerman et al., 1988).
[e] Kenya, dry season, surface (Greenberg et al., 1985).
[f] Kenya, wet season, surface (Greenberg et al., 1985).
[g] Jacquin, Alabama, July 1990, boundary layer (Guenther et al., 1995).

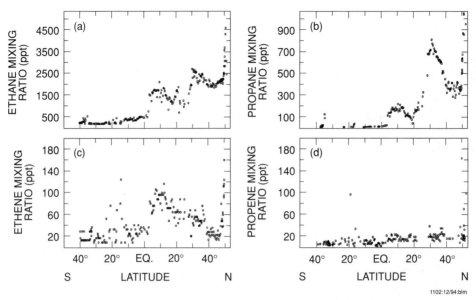

**Figure 9.3.** Latitudinal profiles of $C_2$ and $C_3$ hydrocarbons over the Atlantic Ocean during March and April 1987 (data taken from Rudolph and Johnen, 1990).

#### 9.4.1.3 Marine Boundary Layer

Generally, lower concentrations of saturated hydrocarbons are found in the marine atmosphere compared to continental sites because the ocean is not a significant source of most hydrocarbons. However, marine sources of reactive alkenes—particularly ethylene and propylene—have been measured. Latitudinal profiles of NMHCs over oceans (Fig. 9.3) can reflect transport of anthropogenic alkanes, aromatics, and acetylene, from the northern midlatitude anthropogenic sources to remote regions. Ethylene and propylene do not show the latitudinal gradient of alkanes because of their oceanic emissions.

Measurements of various NMHCs made in the marine boundary layer show strong annual cycles, with mixing ratios peaking in the winter months. This is largely due to the seasonality in the abundance of OH—increased photochemical activity leads to higher OH concentrations in the summer months.

### 9.4.2 Latitudinal and Vertical Distribution of CO and Hydrocarbons

$CH_4$ and CO, the longest-lived organic species ($\simeq$10 years and 2 months, respectively), are the most abundant in the background atmosphere and are, therefore, the most important to global atmospheric chemistry. The annual patterns of surface concentrations for $CH_4$ and CO [Figs. 9.4a ($CH_4$) and 9.4b (CO)] show marked seasonality and interhemispheric differences. Both sources and atmospheric chemical losses of $CH_4$ and CO are generally greater in the warmer months; the net effect, however, is that lowest concentrations are observed in the summer period and largest concentrations in the winter. Consequently, the seasonal cycles for the Northern and Southern Hemispheres are approximately 6 months out of phase. The overall mixing ratios and magnitudes of the seasonal cycle for both $CH_4$ and CO are greater in the Northern Hemisphere, where total atmospheric inputs are larger. Similar patterns have been measured for other hydrocarbons at several locations; the latitudinal gradient for some NMHCs in the mid-Atlantic has already been shown in Figure 9.3.

**Figure 9.4(a).** Three-dimensional representation of the global distribution of background concentrations of methane (from Steele et al., 1987).

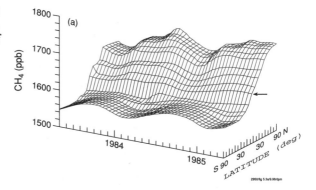

**Figure 9.4(b).** Three-dimensional representation of the global distribution of background concentrations of CO (adapted with permission from Novelli, P. C., K. A. Masarie, P. Tans, and P. M. Lang (1994) Recent changes in atmospheric carbon monoxide, *Science* **263**, 1587, ©1994, American Association for the Advancement of Science).

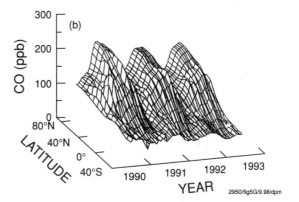

The nearly global distribution of carbon monoxide in the free troposphere has been observed from space by the MAPS (Measurement of Air Pollution from Satellites) instrument on board the Space Shuttle. These observations (see Color Plate 8) reveal strong hemispheric and seasonal differences in the CO abundance as well as a large impact of biomass burning sources in the tropics during the dry season.

The vertical distribution of $CH_4$ and CO has been observed during aircraft campaigns (Figs. 9.5a-d and Color Plate 7). These figures represent a snapshot of the atmosphere over limited time periods. However, these snapshots reveal several important features:

1. Strong surface concentrations of $CH_4$ and CO appear near many terrestrial (source) areas;

2. Concentrations are generally highest in the planetary boundary layer and markedly lower above this surface layer;

3. Gradients of decreasing concentration are exhibited going from midlatitudes of the Northern Hemisphere to midlatitudes of the Southern Hemisphere;

4. Above the planetary boundary layer, concentrations do not decrease consistently with altitude up to the tropopause. Concentrations drop sharply above the tropopause (approximately 15 km at tropical and 10 km at polar latitudes).

Similar vertical structure is seen for other hydrocarbons. Because all hydrocarbons are released from the Earth's surface, decreasing concentrations of these compounds are usually found with increasing altitude. However, a monotonic decrease of concentration with altitude is often not seen in individual profiles, since

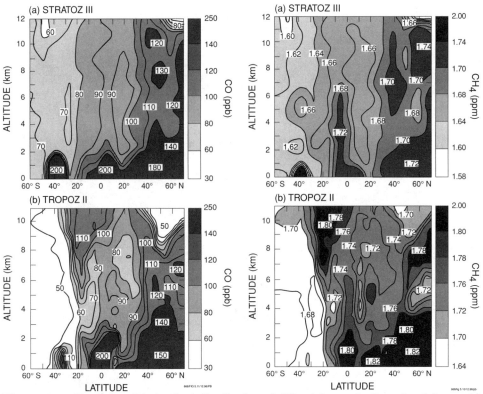

**Figure 9.5.** Vertical and latitudinal distribution of CO mixing ratio (in ppbv, left panels) and $CH_4$ (in ppmv, right panels) observed during the (a) STRATOZ III and (b) TROPOZ II campaigns (A. Marenco, private communication, 1990).

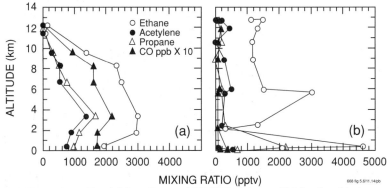

**Figure 9.6.** Vertical profile of CO and several NMHCs (a) near Fairbanks, Alaska, and (b) over the Pacific Ocean near San Diego, California (Greenberg et al., 1990).

various mixing processes of different scales distribute hydrocarbons throughout the troposphere. Figure 9.6 gives vertical profiles for CO and several hydrocarbons from the surface up to the tropopause.

The characteristics of hydrocarbon sources may also affect their distribution in the atmosphere. The emissions of isoprene, a biogenic hydrocarbon emitted from vegetation, occur only during the daytime when sunlight is above a certain minimum intensity. $\alpha$-pinene, another biogenic compound, is emitted throughout the day and night. Figure 9.7 illustrates the vertical distribution of these two biogenic trace gases

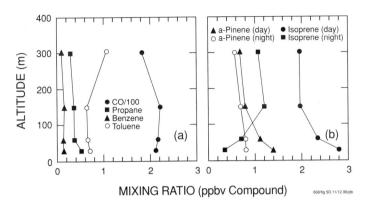

Figure 9.7. Average profiles of CO and NMHC over the Amazon tropical forest near Manaus, Brazil (Zimmerman et al., 1988).

in the boundary layer over the Amazon tropical forest. Also shown are the vertical distributions for CO and several other hydrocarbons.

It has been shown earlier in this chapter that the oxidation of hydrocarbons in the atmosphere leads to the production of a whole suite of partially oxidized intermediate species. A great deal of progress is currently being made on the development of analytical tools for the measurement of these oxygenated species, and measurements of a number of these compounds are now available.

Two of the more important and ubiquitous oxygenated species for which measurements are available are formaldehyde, $CH_2O$, and acetone, $CH_3C(O)CH_3$. Formaldehyde is produced in the oxidation of methane and a wide variety of NMHCs, and is destroyed via photolysis and reaction with OH. Largely due to the fact that it is a byproduct of methane oxidation, formaldehyde is found throughout the troposphere. Latitude-altitude profiles for formaldehyde, obtained during the TROPOZ II campaign (Arlander et al., 1995), are shown in Figure 9.8.

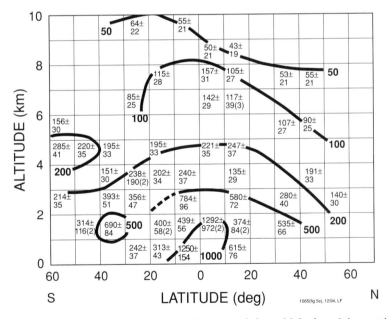

Figure 9.8. Altitude and latitude distribution of formaldehyde mixing ratio measured during the southbound flights during the TROPOZ II field campaign. Mixing ratios are ppt formaldehyde. Actual measurements and approximate concentration contours are indicated in latitude-altitude bins (from Arlander et al., 1995).

Acetone in the atmosphere originates from a number of diverse regional sources (Singh et al., 1994, 1995). These include both direct emissions (from biogenic and anthropogenic sources) and production from the oxidation of numerous NMHCs (including propane, alkenes, and possibly biogenic compounds). Because of its long chemical lifetime (of order 100 days), acetone is found throughout the free troposphere, with typical concentrations of order 1 ppbv (Singh et al., 1994, 1995). As is the case for most oxygenated organics, destruction of acetone occurs via reaction with OH and via photolysis. Although some uncertainty exists in the exact rate of its photolysis, this process may represent an important source of $HO_x$ radicals in the free troposphere.

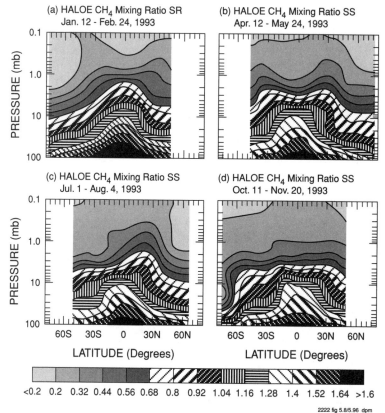

**Figure 9.9.** $CH_4$ mixing ratios as measured by HALOE in 1993. (a) Measurements taken at sunrise; (b)-(d) measurements taken at sunset (from Luo et al., 1995).

## 9.4.3 Stratospheric Methane

The only organic compound sufficiently long-lived to be transported to the stratosphere in significant amounts is methane (atmospheric lifetime about 10 years). As discussed earlier, the mixing ratio of methane is quite uniform in the troposphere, with a present-day value of about 1.8 ppm. Thus air entering the stratosphere from the troposphere contains approximately this amount of methane. The destruction of methane in the stratosphere is via reaction with OH, just as in the troposphere, and via reaction with $O(^1D)$ and with Cl atoms:

$$O(^1D) + CH_4 \rightarrow OH + CH_3 \tag{9.51}$$
$$Cl + CH_4 \rightarrow HCl + CH_3 \tag{9.52}$$

Because of the strong temperature dependence of the reaction of OH with methane and the low concentration of $O(^1D)$ and Cl atoms in the lower stratosphere, chemical destruction of methane in the lower stratosphere is quite slow. Thus its mixing ratio can be used as a tracer for dynamical processes (its chemical lifetime is longer than the timescale for the dynamical processes).

Methane mixing ratio profiles have recently been obtained from the Halogen Occultation Experiment (HALOE) aboard the Upper Atmosphere Research Satellite (UARS) (Luo et al., 1995), as is shown in Figure 9.9 and Color Plate 20. As discussed in Chapter 6, destruction of $CH_4$ results in near-quantitative production of water. Thus, "total hydrogen," defined as $2\,[CH_4] + [H_2O]$, is essentially constant throughout the stratosphere, young stratospheric air containing about 1.8 ppm $CH_4$ and 3 ppm $H_2O$ and older air containing less methane and more water.

## Further Reading

Finlayson-Pitts, B. J. and J. N. Pitts (1986) *Atmospheric Chemistry*, Wiley & Sons, New York.
Warneck, P. (1988) *Chemistry of the Natural Atmosphere*, Academic Press, San Diego.
Wayne, R. P. (1991) *Chemistry of Atmospheres*, 2d ed., Clarendon Press, Oxford, UK.

## Hydrocarbons

Hydrocarbons (HCs) are ubiquitous components of the Earth's atmosphere with large natural and anthropogenic sources. They are intricately linked with many aspects of tropospheric chemistry and are themselves key tools in studies of atmospheric transport and chemistry. The most abundant HC in the global atmosphere is methane (1.8 ppm), and its oxidation forms the heart of what we nominally call tropospheric chemistry. The carbon flux associated with the release of nonmethane HCs or NMHCs ($\approx 10^3$ Tg yr$^{-1}$) is comparable to that of methane, but NMHCs are far more reactive and can influence atmospheric composition over a short period of time. Atmospheric lifetime of NMHCs can vary from a few hours to several months, while methane can persist for years. The first impetus to study NMHCs came in the 1950s following the recognition by Haagen-Smit that Los Angeles smog was produced as a result of reactions between NMHCs and nitrogen oxides (NO$_x$), and ozone was a major component of smog. In parallel with these developments were suggestions by Went that plant biomass was a major source of terpenoid hydrocarbon emissions. Emission rates of biogenic HCs vary greatly among the plant species and are often a strong function of temperature and sunlight. The most important of these is perhaps isoprene, but others (*e.g.*, terpenes, aldehydes, ketones, alcohols) have also been identified. Oceans have also been found to be sources of light (C$_2$-C$_5$) and very heavy (C$_9$-C$_{28}$) HCs. In most cases the mechanisms that produce these HCs in the plant/oceanic biological system are not understood. Once in the atmosphere, HCs decompose to produce a complex array of oxygenated products.

In the polluted atmosphere, NMHCs dominate the oxidation chemistry and are primarily responsible for urban/rural smog. Ozone, a key component of smog, must be controlled in the atmosphere because of its many deleterious health and welfare effects. In the presence of NO$_x$, catalytic cycles are formed that can lead to the formation and accumulation of ozone. While NO$_x$ is necessary for ozone formation, HCs are essential for its buildup. The photochemical oxidation cycle of a simple alkane ($R$H) can generate carbonyls (aldehydes and ketones) as well as ozone

$$\begin{aligned}
R\text{H} + \text{OH} \,(+\,\text{O}_2) &\rightarrow R\text{O}_2 + \text{H}_2\text{O} \\
R\text{O}_2 + \text{NO} \,(+\,\text{O}_2) &\rightarrow (\text{carbonyls}) + \text{NO}_2 + \text{HO}_2 \\
\text{NO}_2 + h\nu \,(+\,\text{O}_2) &\rightarrow \text{NO} + \text{O}_3 \\
\hline
R\text{H} + \text{OH} \,(+\,3\,\text{O}_2) &\rightarrow (\text{carbonyls}) + \text{O}_3 + \text{HO}_2 + \text{H}_2\text{O}
\end{aligned}$$

Hydrocarbons such as methane, ethane, and propane can produce formaldehyde, acetaldehyde, and acetone, respectively, as intermediates. In the United States, the strategy to control ozone in the air we breathe primarily depends on the control of NMHC emissions through methods such as the use of catalytic converters in cars. Since 1996 in California, reformulated gasoline has been used to further reduce emissions of all NMHCs, but especially of certain highly toxic species such as benzene. Fossil fuel combustion is a main source of anthropogenic NMHCs in the Northern Hemisphere, while other sources such as biomass burning are far more important in the Southern Hemisphere. It has become increasingly clear that natural (*e.g.*, isoprene) as well as anthropogenic HCs can produce ozone in any environment where abundant NO$_x$ is available, generally from anthropogenic emissions.

In remote atmospheres the dominant HC is methane, except in forested areas, where biogenic emissions (*e.g.*, isoprene) may be important. It is believed that methane (and CO) oxidation provides a key mechanism for ozone formation in the global free troposphere.

NMHC abundance in remote atmospheres is generally quite low, but can still have a profound influence that is only now being understood. In addition to their role in ozone formation, NMHCs have also been used as tracers of atmospheric motions and chemistry. Following are some of the key conclusions that have resulted from studies of NMHCs:

- Carbonyls, which are key products of NMHC oxidation, can sequester reactive nitrogen in the form of PAN.
$$CH_3CHO + OH(+O_2) \rightarrow CH_3C(O)O_2 + H_2O$$
$$CH_3C(O)O_2 + NO_2 \leftrightarrow CH_3C(O)O_2NO_2 (PAN)$$
PAN (and like compounds) is long-lived at cold temperatures and can transport reactive nitrogen over long distances around the globe. It can also release $NO_x$ under warm conditions, allowing the processes of ozone formation to occur in remote places. In polluted atmospheres, the same PAN is phytotoxic and is associated with both eye irritation and plant damage.
- Oxygenated HCs are just beginning to be measured, but some of them (e.g., acetone) may provide an important source of $HO_x$ radicals in the upper troposphere, influencing ozone chemistry in this region. Alcohols (e.g., $CH_3OH$) are both primary biogenic and anthropogenic emissions as well as secondary products of hydrocarbon oxidation, but few atmospheric measurements are available. In general our knowledge of the distribution and sources of oxygenated HCs is quite rudimentary.
- Because some NMHCs react with chlorine atoms at a rate that is $10^2$-$10^3$ times faster than OH (e.g., $C_2H_6$), these have been used to infer the presence of Cl atoms in the marine boundary layer. How these Cl atoms are produced is not known, but they can oxidize many species such as dimethyl sulfide far more efficiently than OH.
- NMHC concentration ratios (e.g., $C_4H_{10}/C_2H_2$) have been used as indicators of OH radicals in urban/rural atmosphere. Other ratios such as that of $C_2H_6/C_3H_8$ provide qualitative measures of the age of an air mass. Seasonal cycles of NMHC provide a means of validating the chemistry and dynamics of global photochemical models.
- Concentration ratios of peroxides of widely different solubility (e.g., $HOOH/CH_3OOH$, both products of HC oxidation) provide a convenient measure of the degree of precipitation scavenging encountered by an air mass during convective transport.

The chapter on HCs does much to inform the reader about the complexities of HC oxidation in the atmosphere. Much future progress is needed before we can fully understand the distribution, sources, and sinks of this complex group of species. Accurate identification and measurement of atmospheric NMHCs are required both in polluted environments, where the presence of anthropogenic and biogenic mixtures increases complexity, and in remote environments where the concentrations are extremely low. A more accurate knowledge of the sources of NMHCs, especially biogenic sources from plants and oceans, is required. Little attention has been devoted to the role of heavy ($>C_{10}$) as well as oxygenated HCs, in large part due to the difficulties in their measurements. While methane, a climatologically and chemically important gas, has increased at a rate of nearly 1%/yr, little is known about the atmospheric trends of NMHCs. Significant progress has been made in recent decades, and the effort to put our knowledge on a solid quantitative footing is well on its way.

*Hanwant Singh is a senior scientist in the Earth Sciences Division of the NASA Ames Research Center. He has made a number of important scientific contributions in several aspects of atmospheric chemistry. Among these are novel approaches to the elucidation of $NO_x$, $HO_x$, and $ClO_x$ cycles in the troposphere. He is the recipient of the Frank A. Chambers award of the Air and Waste Management Association for "outstanding achievements in the art and science of air pollution."*

# 10 Sulfur Compounds

## 10.1 Introduction

Sulfur is the fourteenth most abundant element in the Earth's crust and is one of the essential elements to all life on Earth. As shown in Chapter 5, sulfur is present in, and constantly exchanged between, the lithosphere (the regime made up of rocks and clays), the biosphere, the hydrosphere, and the atmosphere. In the atmosphere, sulfur in both the gaseous and aerosol forms impacts regional and global chemistry, climate change, as well as the health of various living organisms. Sulfur dioxide ($SO_2$), released from fossil fuel combustion, is converted to sulfuric acid ($H_2SO_4$). The resulting acidic precipitation in the form of rain, fogs, and mists has long been implicated in playing some role in forest decline and foliar damage (Schindler, 1988; Binkley et al., 1989; Linthurst, 1984). Other deleterious effects include soil acidification and the resultant release of toxic metals like aluminum into lakes and streams (Schindler, 1988), root damage, and plant and nutrient leaching (Schindler, 1988; Binkley et al., 1989). Moreover, the reaction of $H_2SO_4$ on mineral dust particles, primarily aluminosilicate clays, is responsible for the increased solubility of aluminum (Winchester, 1988). It has been suggested that prolonged exposure to the aluminum thus solubilized could lead to detectable respiratory effects in humans.

It has been hypothesized by Charlson et al. (1992b) that sulfuric acid aerosols may affect the radiation balance in the troposphere on a regional scale by reflecting incoming solar radiation back toward space. Charlson et al. (1987) also postulate that dimethyl sulfide (DMS) released by aquatic plankton is oxidized to form $H_2SO_4$ aerosols, which may serve as cloud condensation nuclei in remote marine areas. The increased particle number density thus generated leads to a decrease in the mean particle radius, which enhances cloud reflectance. This can possibly be involved in a climate feedback loop.

Finally, sulfur gases can affect the chemistry of the stratosphere. The relatively unreactive and insoluble sulfur compound carbonyl sulfide (OCS), which is produced in the troposphere from both biogenic and anthropogenic sources, can be transported into the stratosphere, where it is converted to $SO_2$ and ultimately to fine $H_2SO_4$ aerosols. This process was first postulated by Crutzen (1976) to explain the existence of the background aerosol sulfate layer known as the Junge layer. With the exception of carbon disulfide ($CS_2$), which is one of the main precursors of OCS, other sulfur gases are too reactive to affect the Junge layer by direct transport from the troposphere. However, violent volcanic emissions, such as those from Mt. Pinatubo and El Chichón, inject large quantities of $SO_2$ directly into the stratosphere, and these emissions significantly enhance the Junge layer. It has been postulated that long-term enhancements to this aerosol layer may influence the Earth's radiation budget and climate through increased solar scattering. More recently, Hofmann and Solomon (1989) and Granier and Brasseur (1992) have shown the potential for enhanced stratospheric ozone destruction on a global scale due to heterogeneous reactions on sulfate aerosols.

Thus sulfur chemistry has influenced and will continue to influence atmospheric composition and climate in a significant way. This chapter presents the key sulfur species of atmospheric importance followed by a discussion of the important chemical reactions involving these species, including aqueous-phase chemistry. Next, there follows a discussion of measurements of sulfur gas abundances, distributions, potential atmospheric trends, key unresolved issues, and existing uncertainties. The reader is referred to Chapter 5 for an inventory of the relevant sulfur gases, which includes their sources and sinks. The role of sulfur in aerosol formation and heterogeneous chemistry is presented in Chapter 4.

## 10.2 Scope and Definitions

$H_2S$: Hydrogen sulfide, the major reduced sulfur gas released from terrestrial ecosystems.

$CH_3SCH_3$: Dimethyl sulfide, DMS. Reduced sulfur gas released by marine algae; constitutes the largest single reduced sulfur emission.

$CH_3SH$: Methane thiol, methyl mercaptan; reduced sulfur gas.

$CH_3SSCH_3$: Dimethyl disulfide; reduced sulfur gas.

$CS_2$: Carbon disulfide; reduced sulfur gas; released by both natural and industrial processes.

OCS: Carbonyl sulfide; the most abundant atmospheric sulfur gas. Both released and taken up by vegetation; also produced in the oxidation of $CS_2$.

$SO_2$: Sulfur dioxide; formed in large quantities in fossil fuel combustion. Also formed in the oxidation of most reduced sulfur gases.

$SO_3$: Sulfur trioxide; intermediate in the oxidation of $SO_2$; precursor to $H_2SO_4$.

$H_2SO_4$: Sulfuric acid; acidic gas; stable end product of sulfur oxidation; condenses to form particles that can affect climate.

$CH_3SO_3H$: Methane sulfonic acid, MSA; minor product in the oxidation of DMS; taken up by particles.

$CH_3S(O)CH_3$: Dimethyl sulfoxide, DMSO; minor product in the oxidation of DMS.

$HSO_3^-$: Bisulfite anion; dissolved form of $SO_2$.

$SO_3^{2-}$: Sulfite anion; doubly ionized form of dissolved $SO_2$.

$HSO_4^-$: Bisulfate anion; ionized form of $H_2SO_4$.

$SO_4^{2-}$: Sulfate anion; doubly ionized form of $H_2SO_4$.

nss-$SO_4$: Non-sea-salt sulfate; the concentration of sulfate found in marine aerosols in excess of that found in seawater; nss-$SO_4$ is normally attributed to uptake of the products of DMS oxidation.

## 10.3 Sulfur Compounds

Sulfur compounds exist in both oxidized and reduced forms, with oxidation states ranging from −2 to +6. Once emitted into the oxidizing environment of the atmosphere, the general tendency is for reduced sulfur compounds, primarily biogenic in origin, to be oxidized to the +4 oxidation states of $SO_2$ and methane sulfonic acid ($CH_3SO_3H$, MSA) and ultimately to the +6 state of $H_2SO_4$. The +6 state is the thermodynamically stable form in the presence of oxygen. The total concentration of sulfur species present in solution in the +4 and +6 states is usually referred to as S(IV) and S(VI), respectively.

Generally, biogenic emissions of sulfur consist of the following reduced organic and inorganic compounds: methane thiol ($CH_3SH$, also known as methyl mercaptan), dimethyl sulfide ($CH_3SCH_3$, DMS), dimethyl disulfide ($CH_3SSCH_3$, DMDS), carbonyl sulfide (OCS), carbon disulfide ($CS_2$), hydrogen sulfide ($H_2S$). Sulfur dioxide emissions, which are mainly anthropogenic in nature, arise from the combustion of fossil fuels, which can contain between 0.05 and 0.14 sulfur by weight. There also exist many reactive sulfur intermediates at much lower concentrations such as $CH_3S$, HS, SO, $CH_3SO_x$(x=1,2,3), $HSO_3$, and $SO_3$. In the liquid phase the sulfur ions $SO_4^{2-}$, $SO_3^{2-}$, $HSO_4^-$, and $HSO_3^-$ play a role.

The major gas-phase loss process for all the reduced sulfur compounds in the atmosphere is reaction with the hydroxyl radical (OH) (Tyndall and Ravishankara, 1991). At night, reaction of DMS with the nitrate radical ($NO_3$) in coastal and continental regions represents an additional chemical loss mechanism. Aqueous-phase loss processes for both $SO_2$ and DMS can also be important. The chemistry associated with these various loss processes will now be discussed.

## 10.4 Tropospheric Chemistry of Sulfur Compounds

As mentioned previously, the atmospheric chemistry of sulfur species is driven by the oxidation of sulfur to forms containing several S-O bonds. These species tend to be more acidic than their precursors, and thus contribute to local and regional acidification of precipitation. The chemistry of the reduced biogenic compounds is first discussed, followed by the oxidized forms ($SO_2$, MSA, $H_2SO_4$). The detailed pathways by which $SO_2$ and MSA are formed from the reduced sulfur compounds have not yet been described completely. The $CH_3S$ radical is thought to be a common intermediate in the oxidation of several of the reduced compounds, and the critical branching step to $SO_2$ or MSA is thought to involve one of the reactive free radicals ($CH_3SO_x$, where x could be 1, 2, or 3) produced from $CH_3S$ oxidation. Although the exact details of the free radical chemistry are still to be determined, the slowest step in the oxidation of sulfur compounds is always the initial attack on the parent compound. In the following discussion an OH concentration of $8 \times 10^5$ molecule cm$^{-3}$ is used to calculate lifetimes.

### 10.4.1 Oxidation of Sulfur Gases

Hydrogen sulfide reacts predominantly with OH, to give the HS radical

$$OH + H_2S \rightarrow H_2O + HS \tag{10.1}$$

HS then reacts with $O_3$ or $NO_2$, to form HSO

$$HS + O_3 \rightarrow HSO + O_2 \tag{10.2}$$

$$HS + NO_2 \rightarrow HSO + NO \tag{10.3}$$

and the HSO is converted rapidly to $SO_2$ according to:

$$HSO + O_3 \to HSO_2 + O_2 \tag{10.4}$$

$$HSO_2 + O_2 \to HO_2 + SO_2 \tag{10.5}$$

The reaction of $H_2S$ with $NO_3$ is slow enough to be neglected in the atmosphere. Using a rate coefficient of $4.7 \times 10^{-12}$ cm$^3$ molecule$^{-1}$ s$^{-1}$ for OH attack leads to an atmospheric lifetime of $\approx 70$ hours for $H_2S$.

The OH reaction with carbon disulfide, $CS_2$, is complex, involving reversible addition of OH to the $CS_2$, followed by reaction of the adduct with $O_2$ (Lovejoy et al., 1990; Stickel et al., 1993). The ultimate products have been shown to be $SO_2$ (115-120% per $CS_2$), OCS (80-85%), and CO ($\approx 15\%$). These observations can be rationalized as follows:

$$OH + CS_2 \rightleftharpoons CS_2OH \tag{10.6}$$

$$CS_2OH + 2O_2 \to HO_2 + SO_2 + OCS, \quad 80-85\% \tag{10.7a}$$

$$\to HCO + 2SO_2, \quad 15-20\% \tag{10.7b}$$

$$HCO + O_2 \to HO_2 + CO \tag{10.8}$$

The oxidation of $CS_2$ is thus a large source of OCS in the troposphere. Since the overall reaction with OH depends on the stability of the adduct $CS_2OH$, and on its reaction with $O_2$, no simple expression is possible for the rate coefficient. However, based on typical tropospheric conditions for $O_2$ and temperature, a value of $3 \times 10^{-12}$ cm$^3$ molecule$^{-1}$ s$^{-1}$ is appropriate, giving a lifetime of around 120 h. Thus $CS_2$ will be converted to $SO_2$ and OCS relatively close to its point of emission. The reaction of $CS_2$ with $NO_3$ is negligibly slow.

The reaction of OCS with OH is very slow ($k \sim 2 \times 10^{-15}$ cm$^3$ molecule$^{-1}$ s$^{-1}$, leading to a tropospheric chemical lifetime of about 10 yrs). Consequently, a fraction of the tropospheric OCS is transported to the stratosphere, where more rapid destruction via photolysis and reaction with O atoms can occur.

As shown in Chapter 5, DMS is the biogenic sulfur compound with the highest total emission rate. It is removed mainly by reaction with the hydroxyl radical, but the reaction with the nitrate radical is fast enough to be potentially important at night in coastal regions which are impacted by anthropogenic $NO_x$. Both reactions are initiated predominantly by the abstraction of a hydrogen atom:

$$OH + CH_3SCH_3 \to H_2O + CH_3SCH_2 \tag{10.9a}$$

$$OH + CH_3SCH_3 \to CH_3S(OH)CH_3 \text{ (adduct)} \tag{10.9b}$$

$$NO_3 + CH_3SCH_3 \to HNO_3 + CH_3SCH_2 \tag{10.10}$$

The OH reaction pathways vary as a function of temperature; the abstraction reaction (10.9a) has a weak, positive temperature dependence, while the addition reaction (10.9b) has a fairly strong, negative temperature dependence. At 298 K, 80% of the total reaction proceeds by abstraction, while at 273 K the abstraction and addition channels are roughly equal. The $CH_3SCH_2$ radical behaves as other alkyl radicals

$$CH_3SCH_2 + O_2 + M \to CH_3SCH_2O_2 + M \tag{10.11}$$

$$CH_3SCH_2O_2 + NO \to CH_3SCH_2O + NO_2 \tag{10.12}$$

$$CH_3SCH_2O \to CH_3S + CH_2O \tag{10.13}$$

$$CH_3SCH_2O_2 + HO_2 \to CH_3SCH_2OOH + O_2 \tag{10.14}$$

The $CH_3S$ radical, which is a common breakdown product of many organic sulfur compounds, will be discussed separately. The existence of the DMS-OH adduct has been inferred from laboratory studies of the OH radical (Hynes et al., 1986; 1995). It is possible that adduct formation allows a facile route to MSA without going through the CH$_3$S radical. DMSO is also a potential product. For atmospheric conditions, a rate coefficient of about $6 \times 10^{-12}$ cm$^3$ molecule$^{-1}$ s$^{-1}$ is appropriate for reaction of OH with DMS, leading to a DMS lifetime of about 60 h. Thus one would expect DMS to be oxidized mostly in the lower troposphere. However, both DMS and SO$_2$ have been found in the upper troposphere. This implies that rapid vertical transport must have occurred, probably through entrainment into convective clouds. There even exists the possibility that DMS is injected directly into the stratosphere via such a mechanism, which would further contribute to the formation of sulfuric acid in the Junge layer. The DMS-NO$_3$ reaction, which forms nitric acid, can provide an efficient sink for both DMS and NO$_x$ in coastal regions. A further potential loss for DMS is dissolution in aqueous particles followed by reaction with OH or O$_3$ (Lee and Zhou, 1994). DMS, however, is not very soluble in aqueous particles; so the contribution of aqueous oxidation to the DMS lifetime will normally be small.

The reactions of methane thiol with OH and NO$_3$ are both thought to proceed by abstraction, leading to the CH$_3$S radical. The rate coefficient for OH attack is $3.3 \times 10^{-11}$ cm$^3$ molecule$^{-1}$ s$^{-1}$, while that for the NO$_3$ reaction is $9 \times 10^{-13}$ cm$^3$ molecule$^{-1}$ s$^{-1}$

$$OH + CH_3SH \rightarrow H_2O + CH_3S \qquad (10.15)$$

$$NO_3 + CH_3SH \rightarrow HNO_3 + CH_3S \qquad (10.16)$$

In the case of dimethyl disulfide, the reaction mechanism is not known with certainty. Part of the time addition of OH to one of the sulfur atoms precedes breaking of the S-S bond, with a rate coefficient of $2 \times 10^{-10}$ cm$^3$ molecule$^{-1}$ s$^{-1}$

$$OH + CH_3SSCH_3 \rightarrow CH_3S + CH_3SOH \qquad (10.17)$$

The CH$_3$S radical is then oxidized as described in Section 10.4.2. The CH$_3$SOH probably leads directly to MSA formation (Barnes et al., 1994)

$$CH_3SOH + O_2 \rightarrow CH_3SO_3H \qquad (10.18)$$

## 10.4.2 Formation of SO$_2$ and MSA from CH$_3$S Radicals

As noted above, the CH$_3$S radical is a key intermediate in the oxidation of the methylated sulfur compounds. Early chamber studies of DMS and CH$_3$SH oxidation showed high yields of MSA relative to SO$_2$, while studies that used very low NO$_x$ found almost quantitative conversion of the sulfur to SO$_2$. This has led to an intense period of study on the CH$_3$S radical, using both direct spectroscopic studies and final product measurements. These experiments show that the yield of MSA is very dependent on the NO$_x$ concentration, and predict that very low MSA levels should be found in the troposphere (Berreshiem et al., 1995, and references therein). This prediction is borne out by most measurements at midlatitudes, which favor an MSA yield of roughly 5% relative to nss-SO$_4$ (the amount of sulfate present in aerosols in excess of that expected from seawater). At higher latitudes, larger ratios of MSA to SO$_2$ are found. It is not known, however, whether this reflects the NO$_x$ dependence of the radical chemistry or the temperature dependence of the OH reaction that initiates the oxidation.

Laboratory studies have suggested that the CH$_3$S radicals should react with O$_3$ or NO$_2$ to give mainly CH$_3$SO (Dominé et al., 1992; Turnipseed et al., 1993)

$$CH_3S + NO_2 \rightarrow CH_3SO + NO \qquad (10.19)$$

$$CH_3S + O_3 \rightarrow CH_3SO + O_2 \qquad (10.20)$$

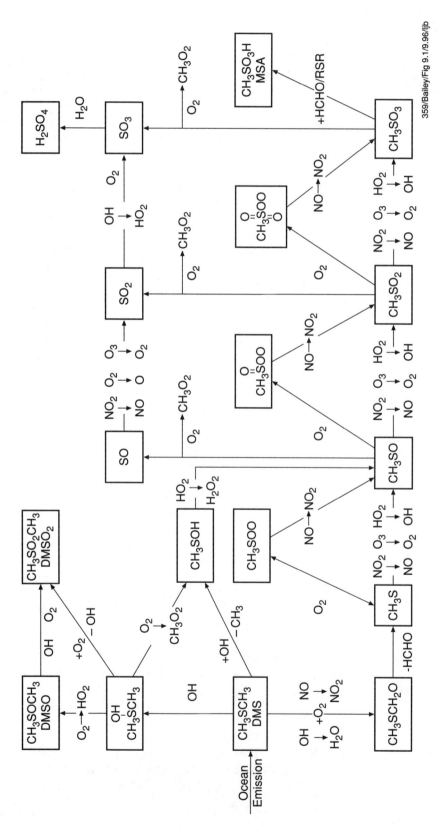

**Figure 10.1.** Schematic of the oxidation of DMS by OH to final products. Note that many branched pathways exist and that the exact details are speculative (adapted from Barnes, 1993).

The reaction of $CH_3S$ with $O_2$ has been the subject of some speculation. While direct measurements of $CH_3S$ loss in the presence of $O_2$ suggest only a very slow reaction, it has been shown that $O_2$ adds reversibly to $CH_3S$ (Turnipseed et al., 1993). While it is unlikely that this process leads to a change in the mechanism of DMS oxidation, it should be investigated further

$$CH_3S + O_2 \rightleftharpoons CH_3SOO \tag{10.21}$$

The detailed chemistry of the $CH_3SO$ radical has not yet been evaluated. Rate coefficients have been measured for its reactions with $O_3$ and $NO_2$, which are assumed mostly to give $CH_3SO_2$. An addition reaction with $O_2$ is also probable

$$CH_3SO + NO_2(O_3) \rightarrow CH_3SO_2 + NO\,(O_2) \tag{10.22}$$

$$CH_3SO + O_2 \rightarrow CH_3S(O)O_2 \tag{10.23}$$

Figure 10.1 summarizes some of the known and suggested pathways in the atmospheric oxidation of DMS. Although it is not yet established at which point the production of MSA becomes inevitable, the scheme contains the overall features.

Chamber studies of the $NO_3$-initiated oxidation of DMS (which necessarily use high $NO_x$ concentrations) have indicated the formation of a species $CH_3SO_xO_2NO_2$ (where x = 1 or 2), which is a sulfonyl analogue of PAN

$$CH_3S(O)_xO_2 + NO_2 \rightleftharpoons CH_3S(O)_xO_2NO_2 \tag{10.24}$$

This peroxynitrate is thermally unstable, like its alkyl and acyl equivalents. If it is allowed to decompose in the presence of NO, to scavenge the peroxy radical, large amounts of MSA and aerosol particles are formed (van Dingenen et al., 1994). While $CH_3SO_xO_2NO_2$ probably does not play a role in the atmosphere, its chemistry does indicate one route to the formation of MSA

$$CH_3S(O)_2O_2 + NO \rightarrow CH_3SO_3 + NO_2 \tag{10.25}$$

$$CH_3SO_3 + RH \rightarrow CH_3SO_3H + R \tag{10.26}$$

It has recently been questioned whether DMS oxidation has to actually go through $SO_2$ in order to reach $H_2SO_4$. Levels of nss-$SO_4$ appear to be higher than can be explained by the gas-phase oxidation of ambient levels of $SO_2$, which proceeds rather slowly (Huebert et al., 1993). This has led to the conjecture that $CH_3SO_3$ decomposes to $SO_3 + CH_3$, which produces sulfuric acid aerosol directly (Bandy et al., 1992a). This has repercussions for the rate of formation of aerosols following DMS oxidation

$$CH_3SO_3 \rightarrow CH_3 + SO_3 \tag{10.27}$$

Although many experiments have been carried out on the overall mechanism of sulfide oxidation, the details are still not finalized. Many of the experiments used $NO_x$ levels much higher than those found in the marine troposphere, and this may affect the results, particularly with regard to the quantitative formation of MSA. At present, two potential pathways to the formation of MSA have been identified — the reaction of $CH_3SOH$ with $O_2$ (10.18) and the abstraction of a hydrogen atom by $CH_3SO_3$ radicals (10.26). The reactions of $CH_3S$ and $CH_3SO$ with $NO_2$ appear to leave the C-S bond intact, while the corresponding reactions with $O_3$ lead to some fragmentation (Dominé et al., 1992; Barnes et al., 1994). Thus substantial formation of MSA from $CH_3S$ radicals will probably only occur under high-$NO_x$ conditions, that is, coastal regions. Future studies using the direct detection of the $CH_3SO_x$ (x=1,2,3) intermediates are badly needed, in order to put the mechanism on a quantitative basis, and to be able to predict the amounts of $SO_2$, $SO_3$, and MSA in the atmosphere. In addition, the participation of oxygenated species such as $CH_3SCH_2OOH$ and $CH_3S(O)O_2NO_2$ should be investigated. Finally, it should be pointed out that, due to the variable

### 10.4.3 Larger Oxygenated Compounds

It is possible that the oxidation of DMS leads to the production of such molecules as dimethyl sulfoxide (DMSO) and dimethyl sulfone (DMSO$_2$), although no gas-phase reactions have been unambiguously identified that produce these molecules from DMS. DMSO has been detected at very low levels in ambient marine air, while DMSO$_2$ has been reported in one study as a product of DMS oxidation in the absence of NO$_x$. Hynes et al. (1995) have indirect evidence that the reaction of the DMS-OH adduct with O$_2$ leads, at least part of the time, to the formation of DMSO. Both DMSO and DMSO$_2$ have very low vapor pressures, and may be expected to deposit onto particles. Both react with OH, and the ultimate product formed in the gas phase will probably be SO$_2$. In the liquid phase dimethyl sulfoxide and dimethyl sulfone could form from DMS oxidation by O$_3$ or OH more readily than in the gas phase.

### 10.4.4 Oxidation of SO$_2$

According to our current level of understanding, SO$_2$ is a common intermediate in the oxidation of all of the reduced sulfur compounds. It is also a major direct emission from combustion sources. The oxidation of SO$_2$ can take place in the gas or liquid phase. The gas-phase mechanism was determined by Stockwell and Calvert (1983); it takes place by addition of OH to SO$_2$, with a rate coefficient of $9 \times 10^{-13}$ cm$^3$ molecule$^{-1}$ s$^{-1}$ at atmospheric pressure:

$$\text{OH} + \text{SO}_2 + M \rightarrow \text{HOSO}_2 + M \tag{10.28}$$

$$\text{HOSO}_2 + \text{O}_2 \rightarrow \text{HO}_2 + \text{SO}_3 \tag{10.29}$$

Sulfur trioxide, SO$_3$, is thought to react with water, either in the gas phase, or following uptake into droplets via:

$$\text{SO}_3 + \text{H}_2\text{O} \rightarrow \text{H}_2\text{SO}_4 \tag{10.30}$$

The gas phase reaction is thought to involve the consecutive reaction of two water molecules with SO$_3$ (Lovejoy et al., 1996). No other reactions of SO$_3$ are expected to occur, due to the preponderance of gas- and liquid-phase water in the atmosphere, and the expected high sticking coefficient of SO$_3$ on aqueous surfaces. Sulfuric acid likewise is taken up either into existing droplets or by the creation of new ones (see Chapter 4).

Sulfur dioxide is moderately soluble, and can also be taken up into aqueous droplets, resulting in homogeneous liquid-phase oxidation and in situ acidification of the droplets. The rate of such oxidation in many circumstances surpasses the gas-phase pathway (Hegg, 1985). Although aqueous-phase oxidation can be quite complicated (see, for example, the review by Hoffmann and Jacob, 1984), involving a large number of species including H$_2$O$_2$, O$_3$, CH$_2$O, and various transition metal ions such as Co, Cu, Fe, Mn, Ni, and V, H$_2$O$_2$ and O$_3$ are recognized to be the two most important oxidants. The bisulfite ion HSO$_3^-$ ion is thought to play a major role in this oxidation.

In discussing aqueous-phase reactions of SO$_2$, one must first consider the partitioning of SO$_2$ between the gas and liquid phases. For dilute solutions, as is certainly the case in the atmosphere, this distribution is given by Henry's Law (Section 4.8.1):

$$[\text{SO}_2(aq)] = \mathcal{H}_{\text{SO2}} P_{\text{SO2}} \tag{10.31}$$

where [SO$_2$(aq)] is the aqueous concentration of SO$_2$ in M, and $\mathcal{H}_{SO2}$ is the Henry's Law constant in M atm$^{-1}$ when the gas-phase partial pressure, $P_{SO2}$, is expressed in atmospheres. The Henry's Law coefficient for SO$_2$ in liquid water is 1.3 M atm$^{-1}$ at 25°C. However, the solubility of SO$_2$ is modified by the formation of the bisulfite ion according to the following equilibria and associated equilibrium constants $K_i$:

$$SO_2(aq) \stackrel{K_1}{\rightleftharpoons} H^+ + HSO_3^- \tag{10.32}$$

$$HSO_3^- \stackrel{K_2}{\rightleftharpoons} H^+ + SO_3^{2-} \tag{10.33}$$

For atmospheric droplets, $HSO_3^-$ is the dominant form. In this case it is more useful to define an effective Henry's Law coefficient, $\mathcal{H}^*_{S(IV)}$, in terms of the total concentration of dissolved species, [S(IV)], and $P_{SO_2}$:

$$\begin{aligned}\mathcal{H}^*_{S(IV)} &= [S(IV)]/P_{SO_2} \\ &= \left([SO_2(aq)] + [HSO_3^-] + [SO_3^{2-}]\right)/P_{SO_2}\end{aligned} \tag{10.34}$$

$$\mathcal{H}^*_{S(IV)} = \mathcal{H}_{SO_2} \left(1 + \frac{K_1}{[H^+]} + \frac{K_1 K_2}{[H^+]^2}\right) \tag{10.35}$$

Thus the overall SO$_2$ solubility is shown to be a function of the solution pH, a particularly important aspect when considering the overall rate of SO$_2$ oxidation in the aqueous phase. The overall SO$_2$ solubility decreases at low solution pH, and the effective Henry's Law coefficient then approaches the Henry's Law coefficient for physical solubility.

The aqueous-phase oxidation of SO$_2$ is dominated by H$_2$O$_2$ for pH values less than 5. This reaction, which has been studied quite extensively (see, for example, Schwartz, 1984b, and references therein), is believed to follow the mechanism:

$$HSO_3^- + H_2O_2 \rightarrow A^- + H_2O \tag{10.36}$$
$$A^- + H^+ \rightarrow H_2SO_4 \tag{10.37}$$

where the anion $A^-$ may possibly be peroxymonosulfurous acid ion (see Hoffmann and Jacob, 1984, for more details). The reaction rate, $R$, for SO$_2$ oxidation to H$_2$SO$_4$ can be written as:

$$R = k^{(2)} [S(IV)][H_2O_2] \tag{10.38}$$

Here the empirical second-order coefficient, $k^{(2)}$, was found to depend on the solution pH. As shown in Figure 10.2 (taken from Schwartz, 1984b), the rate constant falls off strongly with pH. Over the pH range 2-8, Schwartz found a first-order dependence of $k^{(2)}$ on [H$^+$]. Because this dependence is offset by the reverse dependence for the SO$_2$ solubility discussed above, the net overall pH dependence of the rate of this reaction is essentially flat, as shown in Figure 10.3 (taken from Martin, 1984). Since the Henry's Law coefficient for H$_2$O$_2$ is very high ($7 \times 10^4$ M atm$^{-1}$ at 25°C) and the reaction rate scales linearly with [H$_2$O$_2$], very high oxidation rates result from this reaction.

Ozone is not very soluble in water, but oxidizes S(IV) rapidly, particularly in alkaline solution, indicating the participation of both $HSO_3^-$ and $SO_3^{2-}$:

$$HSO_3^- + O_3 \rightarrow HSO_4^- + O_2 \tag{10.39}$$
$$SO_3^{2-} + O_3 \rightarrow SO_4^{2-} + O_2 \tag{10.40}$$

$HSO_4^-$ is the ionized form of sulfuric acid:

$$H_2SO_4 \rightleftharpoons HSO_4^- + H^+ \tag{10.41}$$

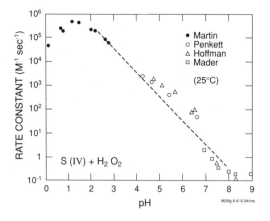

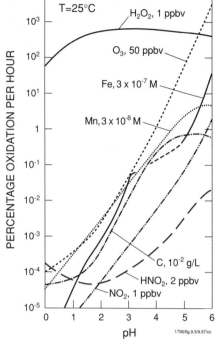

**Figure 10.2.** Second-order rate constant $k^{(2)}$ for oxidation of sulfur (IV) by hydrogen peroxide according to $d[S(VI)]/dt = k\,[H_2O_2]\,[S(IV)]$, as a function of solution pH. Dashed line represents slope of $-1$ (i.e., $k \propto [H^+]$) arbitrarily drawn through the data (from Schwartz, 1984b; citations for datasets shown are given there).

**Figure 10.3.** (right) Percent per hour oxidation of total sulfur (IV) in a cloud containing 5 ppb (volume) of $SO_2$ and 1 mL of liquid water m$^{-3}$ for various reactants. The oxidation rate on the ordinate is expressed in terms of the equivalent gas-phase oxidation rate in percent per hour oxidation of total S(IV) at a temperature of $25°C$. It is assumed that there are no rate limitations due to mass transport; temperature = $25°C$ (from Martin, 1984).

The following second-order rate expression can be written for aqueous-phase reactions of $SO_2$ with ozone:

$$R = k^{(2)}\,[O_3(aq)][S(IV)] \tag{10.42}$$

The pH dependence of the rate constant, which is shown in Figure 10.4 taken from Schwartz (1984b), is opposite to that of $H_2O_2$. Both the rate constant and $SO_2$ solubility increase with increasing pH, resulting in the steep increase in overall rate with pH shown in Figure 10.3.

In addition to $H_2O_2$ and $O_3$, Figure 10.3 displays the pH dependence of the $SO_2$ oxidation rate by oxygen catalyzed by Fe, Mn, and C. Although different profiles can be obtained using different assumptions, particularly temperature and solar flux (see for example, the more detailed profiles of Calvert et al., 1985), attention should be focused on the comparative rates for the two most important oxidants, $H_2O_2$ and $O_3$. As can be seen, $H_2O_2$ is the dominant oxidant below pH values $\approx 5$, and $O_3$ dominates above this. Since sea water has a pH of 8, the oxidation of $SO_2$ in sea-salt aerosol initially proceeds by reaction with $O_3$. However, as the pH drops, $H_2O_2$ becomes the major oxidant (Sievering et al., 1992; Chameides and Stelson, 1992).

For the assumptions given (in Fig. 10.3), the aqueous-phase oxidation rate can surpass the gas-phase rate with OH. However, it must be noted that ozone photolysis, which produces $O(^1D)$ and initiates the production of OH, is also responsible for producing the oxidizing agents (OH, $H_2O_2$, and $CH_3O_2H$) necessary for aqueous-phase $SO_2$ oxidation. The production of $H_2O_2$, for example, arises from the hydroperoxy

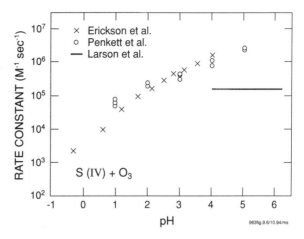

**Figure 10.4.** Second-order rate constant $k^{(2)}$ for oxidation of sulfur (IV) by ozone according to $d[S(VI)]/dt = k[O_3(aq)][S(IV)]$, as a function of solution pH. Temperature = 25°C (from Schwartz, 1984b; citations for the datasets shown are given there).

radical $HO_2$:

$$HO_2 + HO_2 \rightarrow H_2O_2 + O_2 \qquad (10.43)$$

Hydrogen peroxide can also be produced directly in the aqueous phase via reactions of $HO_2$ with itself or $Cu^{2+}$ ions, for example (Mozurkewich *et al.*, 1987).

In addition, the free radicals OH and $HO_2$ can react with S(IV). These species can be produced in the liquid phase via photolytic reactions or scavenged out of the gas phase. The relative importance of these two mechanisms is not yet fully understood, but as the aqueous chemistry of OH and $HO_2$ becomes better understood, the mechanisms will be evaluated more fully. The following reactions are thought to occur:

$$HSO_3^- + HO_2 \rightarrow HSO_4^- + OH \qquad (10.44)$$

$$HSO_3^- + OH \rightarrow HSO_3 + OH^- \qquad (10.45)$$

$$HSO_3 + O_2 \rightarrow HSO_5 \qquad (10.46)$$

$$HSO_5 + HSO_3^- \rightarrow SO_4^- + SO_4^{2-} + 2H^+ \qquad (10.47)$$

The liquid-phase addition reaction of $HSO_3$ with $O_2$ should be contrasted with the gas-phase reaction (10.29), which proceeds by abstraction.

The presence of formaldehyde in solution affects the oxidation of $HSO_3^-$, through the formation of the relatively stable and unreactive hydroxymethanesulfonate ion

$$HSO_3^- + CH_2O \rightleftharpoons HOCH_2SO_2O^- \qquad (10.48)$$

If sufficient $CH_2O$ is present, a large part of the S(IV) can thus be tied up in a form that is not easily oxidized. This clearly will affect the rate of $SO_2$ oxidation in polluted areas of the troposphere.

It is likely that many other soluble organic compounds contribute to the overall rate of $SO_2$ oxidation in solution. Compounds containing peroxide linkages, such as methyl hydroperoxide ($CH_3OOH$) and peroxyacetic acid [$CH_3C(O)O_2H$], have been considered, and it is probable that many more will be discovered as the links between droplet-phase chemistry and the gas-phase oxidation of organic compounds are clarified.

As discussed by Calvert *et al.* (1985), the general shape of the yearly sulfate ion distribution matches well with the $j_{O_3}[O_3]$ values; namely, the highest rates occur during the summer months when the OH production rate from reaction of $O(^1D)$ with $H_2O$ is the highest. This is in contrast to nitric acid deposition, which shows no

### 10.4.5 Particulate MSA and Sulfate in the Remote Atmosphere

Methane sulfonic acid and sulfuric acid, the observed end products of DMS oxidation, both have fairly low vapor pressures over aqueous solutions, and will tend to be lost by condensation into the liquid phase. The vapor pressure of MSA is sufficiently high that it is not expected to form new particles by condensation. Very few measurements of gas-phase MSA are available; however, MSA has long been known to be a component of marine aerosol, presumably due to its deposition onto existing particles (Saltzman et al., 1983). A correlation has been found between the gas-phase DMS and the particulate MSA in several remote locations. Figure 10.5 shows measurements of MSA in rain water along with gas-phase DMS, as measured at Amsterdam Island in the Southern Indian Ocean over a 2-year period. The high levels of DMS found in the Southern Hemisphere summer are accompanied by clear increases in the MSA.

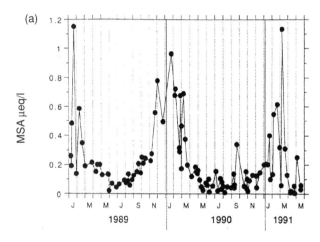
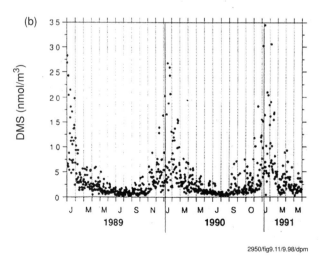

**Figure 10.5.** (a) MSA concentrations in rain water from January 1989 to May 1991 and (b) daily variations of atmospheric DMS concentrations at Amsterdam Island during 1989-1991 (Mihalopoulos et al., 1993).

Sulfuric acid, on the other hand, is capable of forming new aerosol particles by condensation, as well as contributing to the growth of existing particles. Since many marine particles are derived from sea water, which itself contains sulfate, it is difficult to determine the exact amount of DMS-derived sulfate in a given aerosol, and more so whether the particle was formed as a result of DMS oxidation. The amount of sulfate in a particle above that found in sea water is referred to as non-sea-salt sulfate (nss-$SO_4$) and is usually calculated according to nss-$SO_4 = (SO_{4_{total}} - 0.0603\,[Na^+])$. Good correlations between nss-$SO_4$ and DMS have been reported, but the interpretation of the measurements is made difficult by the need to subtract off the amount of sea-salt-derived sulfate.

The factors governing aerosol formation and growth are complicated, and include the vapor pressures of the condensing gases, temperature, degree of water vapor supersaturation, and the presence of existing droplets. The consequences of new particle formation are far-reaching, due to the climatic importance of the formation of particles and cloud condensation nuclei (CCN). Observation of new particle growth in the atmosphere is hampered by the scavenging of newly created nuclei by existing particles. However, new particles have been observed in several instances (Covert et al., 1992; Weber et al., 1995), notably after a precipitation event has scavenged most of the pre-existing nuclei.

## 10.5 Measurements of Sulfur Gas Abundances and Distributions

Despite numerous research studies focusing on sulfur chemistry, our knowledge about the biogenic and anthropogenic sources and sinks as well as the global distributions of volatile sulfur compounds is still very limited. This paucity in our understanding is due to several reasons: (1) the diversity of the sulfur species emitted; (2) the diversity of the ecosystems needed to be studied; (3) the spatial and temporal spottiness of sulfur gas fluxes from various ecosystems; (4) problems with obtaining representative measurements and extrapolations of limited data over a few localities to global scales; and (5) analytical problems in accurately measuring background concentrations of various sulfur compounds, which as a class are very reactive. As a result, the global database for volatile sulfur compounds is, at best, very limited and will no doubt change in the future. The current assessment of the abundances and distributions of these compounds is presented here. A comprehensive discussion of the budgets of these species is presented in Chapter 5, and where appropriate is briefly summarized here.

### 10.5.1 OCS Distributions

Carbonyl sulfide is the most abundant gaseous sulfur species in the unpolluted atmosphere and is thus the major form of gas-phase atmospheric sulfur on a global scale. Because of this and its role in producing stratospheric sulfate aerosols, ambient distributions of OCS, including possible anthropogenic perturbations, are of great interest.

The atmospheric concentration of OCS is perhaps the quantity in its global budget with the least associated uncertainty. Measurements of OCS in the troposphere have consistently resulted in mixing ratios around $500 \pm 70$ pptv (see, for example, Fried et al., 1991, and references therein), with no apparent statistically significant variation with altitude or between continental and maritime locations. This ubiquitous nature is due in part to its relatively low water solubility, relatively low chemical reactivity with OH and other species, and in part to the rather diffuse nature of its sources and sinks. In addition, most ambient measurements of OCS have been carried out with a precision of 5-15% using a variety of chromatographic methods. This is much too coarse to discern subtle spatial and temporal variations.

From the few time series measurements available (Bandy et al., 1992b; Rinsland et al., 1992), there have been no observed trends in the ambient OCS concentration over the past decade. The Rinsland et al. study is particularly significant in this regard since it is based on total column abundances of OCS using infrared spectroscopy with known spectroscopic parameters. This approach avoids the uncertainties associated with calibration standards and their potential drift with time. Latitudinal measurements by Bingemer et al. (1990) and Johnson et al. (1993) over the Atlantic Ocean in the boreal spring and summer show large interhemispheric gradients of 15-25% (north > south). However, more recent measurements by Weiss et al. (1995) over the Pacific Ocean reveal a much smaller OCS gradient of 3% during the spring and −3.8% during the boreal winter. Although anthropogenic OCS sources from Europe and North America may play some role in the differences between the springtime Atlantic and Pacific profiles, the cause of the discrepancy remains largely unknown. The reverse gradient (south > north) in the boreal winter reflects the strong dependence of the OCS oceanic flux on available sunlight and precursor concentrations and suggests an oceanic sink in the Northern Hemisphere. Such a sink suggests that present estimates of the global OCS flux from the oceans are too high. The most recent OCS budget (Weiss et al., 1995) shows an excess of sources over sinks by about a factor of 2. This is, however, inconsistent with the time series measurements above.

### 10.5.2 $CS_2$ Distributions

Because of the very high efficiency of converting $CS_2$ to OCS via its reaction with OH (80-85% of $CS_2$ molecules produce OCS), atmospheric $CS_2$ is also important in determining the stratospheric sulfur balance during nonvolcanic periods. This is further shown in Figure 5.29.

In contrast to OCS, much lower background concentrations and higher variability have been observed in ambient $CS_2$ measurements. Such variance in the background $CS_2$ concentrations no doubt reflects both the short atmospheric residence time of this gas and the inhomogeneous nature of $CS_2$ sources. Background free tropospheric and predominantly maritime measurements revealed $CS_2$ concentrations primarily in the 2-6 pptv range (see references in Bandy et al., 1993b). This is very similar to shipboard measurements of 2.2 pptv by Saltzman and Cooper (1988) in the vicinity of the Bahamas, and aircraft measurements of 0.5 to 5 pptv in the remote marine atmosphere during the CITE-3 experiment (Bandy et al., 1993b). Although these measurements are in line with $CS_2$ flux estimates from the oceans (Toon et al., 1987), additional background $CS_2$ measurements are needed to characterize its global spatial distribution more fully. In air influenced by $CS_2$ sources from anthropogenic activities and biomass burning, much higher concentrations in the 11 to 339 pptv range have been measured by many investigators (see Bandy et al., 1993b). The specific identity of these sources, however, has not been firmly established. As discussed in Chapter 5, more work in this area is needed. In addition, because of its importance as an OCS source, it would be very desirable to carry out time series measurements of $CS_2$.

### 10.5.3 DMS Distributions

Although DMS is found widely over the Earth's oceans, the measured distributions show great variability as a result of variations in the source strengths and its short atmospheric lifetime. As stated in Chapter 5, DMS in the oceans is in general supersaturated, and the flux across the ocean surface reflects not only the rate of production, but also the rate of transfer. Mixing ratios found in temperate regions are typically in the range 10-100 pptv, with a fairly strong seasonal cycle

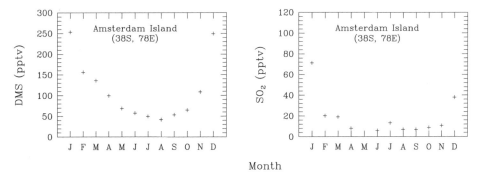

**Figure 10.6.** Seasonal variation of (left) atmospheric DMS and (right) SO$_2$ measured at Amsterdam Island in the Southern Indian Ocean. The SO$_2$ is thought to arise largely from DMS oxidation (courtesy of M. Barth; adapted from Putaud et al., 1992).

**Figure 10.7.** Schematic representation of the distribution of atmospheric DMS in the marine boundary layer measured during different cruises (from Staubes and Georgii, 1993).

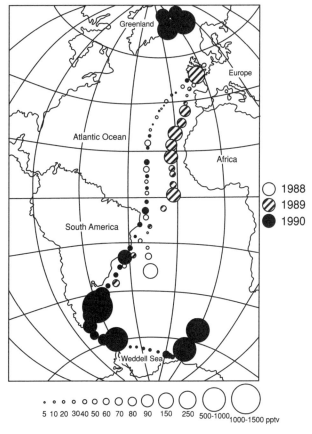

due to an enhanced production rate in summer (Fig. 10.6a). In the polar regions the local concentrations can be much higher during the phytoplankton "blooms," and concentrations of up to 3 ppbv have been reported, due to the high production rate and low photochemical removal. Values in excess of 10 ppbv have been reported at low tide over coastal algae fields, for instance, on the coast of Brittany, France. Figure 10.7 depicts the latitudinal variation of DMS schematically.

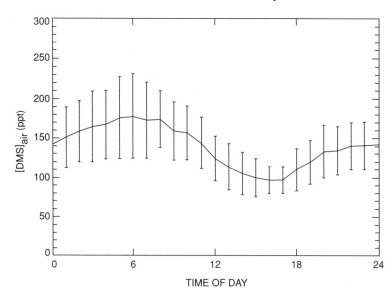

**Figure 10.8.** The diurnal variation of DMS averaged over a nine-day cruise in the south Atlantic in March, 1991. The measurements were taken at a latitude of roughly 20° S. The variation is a result of daytime oxidation by OH superimposed on an essentially steady flux from the ocean (from Suhre et al., 1995).

The mixing ratio of DMS also exhibits a diurnal variation due to its gas-phase oxidation. In the remote marine boundary layer, where its oxidation is controlled by OH radicals, a minimum is found in the late afternoon or early nighttime (Fig. 10.8). In coastal areas, the opposite may occur, since the oxidation by $NO_3$ radicals at nighttime can outweigh daytime OH oxidation if anthropogenic $NO_x$ predominates.

### 10.5.4 $H_2S$ Distributions

Hydrogen sulfide mixing ratios in the troposphere are highly variable. As discussed in Chapter 5, the sources of $H_2S$ are minor in comparison to DMS, and the mechanisms for its production are less well understood. Mean daily mixing ratios of between 5 and 100 pptv have been reported for marine air, with diurnal cycles observed due to its rapid reaction with OH. Coastal and wetland regions often show much higher $H_2S$ mixing ratios, and values of several hundred ppt are typical, with measurements possibly as high as 1 ppb.

### 10.5.5 $SO_2$ Distributions

Sulfur dioxide is perhaps the most difficult of the sulfur gases discussed in this chapter to measure with any reliability in the remote troposphere, where concentrations are typically less than 100 pptv. This difficulty arises in part from the reactivity of $SO_2$, particularly on moist inlet surfaces present when sampling in the marine atmosphere, and the necessity of most techniques to employ either cryotrapping or some collection/trapping technique. Such techniques raise questions about $SO_2$ reactions during collection. Aircraft measurements, furthermore, require sampling times on the order of 3 minutes or less to achieve good spatial resolution. Much progress has been made in the past few years on the measurement of gas-phase $SO_2$. In a ground-based intercomparison (Stecher et al., 1997), it was found that six independent techniques gave reasonable agreement (±25%) down to about 50 ppt $SO_2$.

Despite these problems, measurements of $SO_2$ in the remote marine atmosphere are extremely important for elucidating the long-range transport of sulfur compounds far removed from source regions, and our understanding of atmospheric sulfur oxidation and deposition. Acidic deposition in Bermuda, for example, is thought to occur by long-range transport of sulfur and nitrogen compounds from North America. Long-range transport of sulfur compounds, including $SO_2$ and acidic aerosols into the Arctic, may also play a significant role in climate, by affecting the radiation balance in the polar regions. Long-term measurements of $SO_2$ have been carried out at Amsterdam Island in the Southern Indian Ocean. They are shown in Figure 10.6b.

Aircraft flights over the remote marine atmosphere of the Western Atlantic Ocean during the WATOX-86 compaign (Thornton et al., 1987) resulted in a median $SO_2$ concentration of 0.065 ppbv in the free troposphere compared with a value of 1.02 ppbv in the boundary layer. More recent measurements by Bandy et al. (1993a) during CITE-3 revealed much lower background $SO_2$ concentrations ranging from 0.015 to 0.060 ppbv at 4 km in the free troposphere over the Northeast coast of Brazil. Such low $SO_2$ concentrations are more in line with the modeling results (0.045 ppbv at 6 km) of Toon et al. (1987). However, DMS to $SO_2$ ratios measured during the SAGA-3 program (Huebert et al., 1993) were significantly higher than can be explained by production of $SO_2$ from DMS. This implies that $SO_2$ is not the principal product of DMS oxidation. As previously mentioned in Section 10.3.6, one possibility is that DMS oxidation forms sulfate directly without the need for an $SO_2$ intermediate. Clearly, additional background measurements of $SO_2$, DMS, nss-$SO_4$, as well as other gases are required to understand the DMS oxidation process.

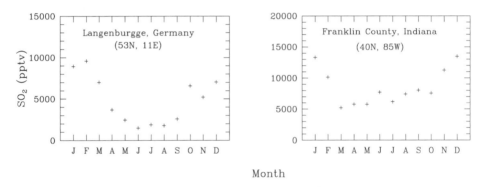

**Figure 10.9.** Annual variation of atmospheric $SO_2$ measurements at industrialized sites in Germany and North America (courtesy of M. Barth).

The majority of ambient $SO_2$ measurements are centered around highly polluted regions of Europe and North America, where the anthropogenic emissions are high. Figure 10.9 depicts the annual variation of $SO_2$ at sites in North America and Germany. Both sites show a wintertime maximum associated with increased fossil fuel burning. In the absence of measurements in many remote stations, true global coverage must rely primarily on tropospheric chemistry-transport models. Langner and Rodhe (1991) present such a model in three dimensions that takes into account processes governing the transport, transformation, and deposition of DMS, $SO_2$, and sulfate ($SO_4^{2-}$) aerosol. This model is based on emission inventories where available, and characteristic $SO_2$-$CO_2$ emission ratios using known $CO_2$ fossil fuel emission patterns where direct emission rates were not available.

Figure 10.10a shows the resulting global simulation for the annual average $SO_2$ distribution in units of pptv for the lowest model layer. The major source regions of Europe, North America, China, and Japan are very apparent. Maximum $SO_2$

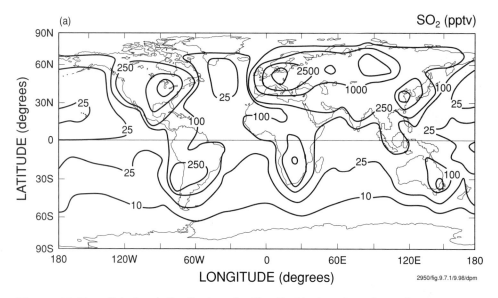

**Figure 10.10a.** *Calculated distribution of sulfur dioxide (pptv) at the surface. Isolines are: 10, 25, 100, 250, 1000, 2500, 5000 (from Langner and Rodhe, 1991).*

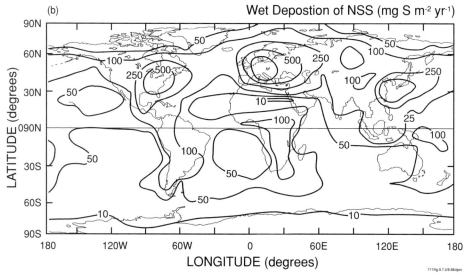

**Figure 10.10b.** *Calculated distribution of surface wet deposition of sulfate. Isolines are: 10, 50, 100, 250, 500, 1000 (from Langner and Rodhe, 1991).*

concentrations attain values as high as 2.5 ppbv over parts of Europe. Emissions over southern regions of Africa and South America, caused by smelters, are also clearly evident. Figure 10.10b shows the resulting annual wet deposition initiated by both gas and aqueous-phase $SO_2$ oxidation. The major source regions, eastern United States, parts of Europe, China, and Japan, again show the highest values.

The model of Langner and Rodhe (1991) is generally within a factor of 2 to 3 of ambient $SO_2$ measurements carried out at the surface. The model does a reasonable job of following the general shape of observed $SO_2$ distributions; in accordance with observations, high $SO_2$ concentrations in the ppbv range are predicted in the boundary layer over major anthropogenic source regions of Europe and the United States and much lower values over the North Atlantic. Model predictions fall within a factor of 2 to 3 of the sub-100 pptv values observed over remote marine regions.

## 10.6 $SO_2$ and Acid Precipitation

Anthropogenic emissions of $SO_2$ and nitrogen oxides are the dominant precursors responsible for acid precipitation. The principal sources of $SO_2$ are coal and oil combustion and smelting, while internal combustion engines, residential and commercial furnaces, industrial and electric utility boilers are the major sources of nitrogen oxide emissions.

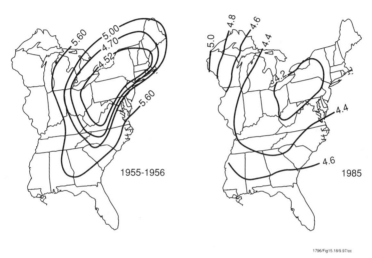

**Figure 10.11.** *Isopleths (lines of equal value) of the average annual pH of precipitation over the eastern United States for different years. Notice the broadening of the region of pH values lower than five from 1955 to 1987 (after Laws, 1993).*

It has long been recognized that the industrial emissions above were responsible for dense acidic $H_2SO_4$ and $HNO_3$ "smogs" observed in Europe and North America. It soon became apparent that such events were causing considerable health effects. To address these concerns in the United States, emissions of $SO_2$ were reduced under the 1970 Clean Air Act. Air quality standards for $SO_2$ were set and maintained by individual states, resulting in approximately a 20% reduction in annual $SO_2$ emissions. However, throughout the 1970s and 1980s, over 50 areas exceeded the $SO_2$ standards, and acidic pollutants in North America continued to be transported over long distances far removed from their sources with consequent damage to lakes, streams, and forests. Figure 10.11 shows the change in rainfall pH between the 1950s and the 1980s. The National Acid Precipitation Assessment Program (NAPAP) was first authorized in 1980 and reauthorized under the 1990 Clean Air Act to further strengthen the clean air initiatives established earlier. This program not only developed a comprehensive and coordinated research and monitoring approach, but also legislative mandates to reduce further the total atmospheric loading of $SO_2$ and nitrogen oxides (refer to the NAPAP report to Congress, 1992). Innovative marketing approaches for controlling the release of these pollutants from electric utilities, the major emitters, were also adopted. The first phase of the control program was implemented in January 1995, and when fully implemented in the year 2010 will place annual emission caps allowable from electric utilities and other industrial sources in the United States.

Canada has also been actively involved in reducing $SO_2$ and nitrogen oxide emissions. In addition to a landmark agreement with the United States in 1991, Canada has established a permanent national cap for $SO_2$ emissions by the year 2000 and committed to reducing nitrogen oxide emissions in the same time frame. Beginning in the mid-1980s European countries developed strong interest in revising international $SO_2$ and nitrogen oxide emission protocols. The first Sulfur Protocol was signed in 1985 under the United Nations' Economic Commission for Europe Convention on Long Range Transboundary Air Pollution. At the time of this writing, discussions related to a Second Sulfur Protocol are underway.

During the first ten years of NAPAP, a series of Congressional reports were published that summarized the major conclusions related to air concentrations and deposition of pollutants. Some of the key findings of these reports are:

- Anthropogenic emissions of sulfur and nitrogen oxides are the greatest sources of sulfur and nitrogen deposition over most of eastern North America.

- For regionally representative sites in the eastern United States, dry deposition of sulfur and nitrogen generally accounts for one-third to two-thirds of the total deposition (wet plus dry).

- Concentrations and deposition of pollutants are generally higher in the vicinity of major point sources and in urban areas than at regionally representative sites. The area of influence varies with season and with prevailing meteorological conditions.

- The aqueous chemistry associated with clouds and precipitation scavenging tends to be nonlinear; changes in wet deposition over eastern North America may not be proportional to emissions changes. Changes in total sulfur deposition are more likely to be proportional to changes in emissions.

- Except for areas impacted by local sources, air concentrations and deposition of pollutants are highest at high elevation (locations above 1400 m) in the Appalachians.

- The acidic content of aerosols has not been well characterized over large geographic regions. There are no reliable estimates for the uncertainties in measuring total sulfur and total nitrogen deposition.

In the United States, the above protocols have resulted in $SO_2$ and nitrogen oxide emission reductions of 9% and 6%, respectively, over the years 1980 to 1990 (1992 NAPAP Report to Congress). Even larger reductions of 39% and 16%, respectively, are projected (relative to 1980 levels) in the year 2010. In Canada, the corresponding figures are: 21% reduction in $SO_2$ emissions and a 5% increase in nitrogen oxide emissions. Although the emission reductions due to the abatement programs are clear, detecting the specific benefits to soil and water chemistry as well as forest health will be significantly more challenging.

## 10.7 Stratospheric Sulfur Chemistry

The chemistry of sulfur in the stratosphere is much simplified due to the fact that most of the reduced gases are removed in the troposphere. In general, the only reduced gas reaching the stratosphere is OCS, although others may occasionally be injected during convective events. The stratospheric oxidation of OCS occurs by photolysis or reaction with O atoms

$$OCS + h\nu \to CO + S \quad (10.49)$$
$$O + OCS \to CO + SO \quad (10.50)$$
$$S + O_2 \to SO + O \quad (10.51)$$
$$SO + O_2 \to SO_2 + O \quad (10.52)$$
$$SO + NO_2 \to SO_2 + NO \quad (10.53)$$

The $SO_2$ formed is oxidized to $SO_3$ and $H_2SO_4$ by gas-phase OH, in a mechanism analogous to that occurring in the troposphere. Photolysis of $SO_2$ is also possible in the stratosphere above 20 km. Since both $SO_2$ photolysis and the reaction of SO with $O_2$ result in the production of O atoms, the overall effect is the catalytic production of two $O_3$ molecules, and Bekki and co-workers have shown that a small production of $O_3$ may occur at midlatitudes following a volcanic eruption of sufficient strength to eject sulfur gases into the stratosphere (Bekki et al., 1993).

Although OCS accounts for most of the sulfur reaching the stratosphere, concerns have recently been raised that a large fleet of supersonic aircraft might deposit considerable amounts of sulfur directly into the lower stratosphere. The impacts of such an increased aerosol burden have been calculated to be nontrivial, both from chemical and climatic standpoints (Bekki and Pyle, 1992; 1993).

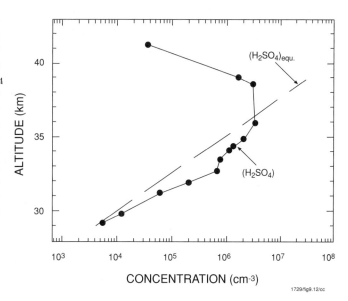

*Figure 10.12.* Concentration of $H_2SO_4$ derived from ion measurements (Reiner and Arnold, 1997). The dashed line shows the $H_2SO_4$ saturation vapor pressure calculated from the temperature at each altitude. Below about 35 km, the gas-phase $H_2SO_4$ is in equilibrium with aerosol particles. Above 35 km the concentration falls below the equilibrium amount as the source becomes negligible.

No direct measurements of $H_2SO_4$ have been made in the stratosphere, although its concentration has been inferred from measurements of cluster ions (see the following section) using mass spectrometry. The distribution of gas-phase $H_2SO_4$ is controlled by its rate of production and by its equilibrium with liquid sulfate aerosol. Throughout most of the stratosphere the temperature is sufficiently low that most of the $H_2SO_4$ resides in the aerosol. However, between about 30 and 40 km the temperature rises to levels where release of $H_2SO_4$ into the gas phase occurs. Above about 40 km the rate of production from OCS oxidation is no longer significant, resulting in a sharply peaked $H_2SO_4$ distribution. Figure 10.12 shows the most recent $H_2SO_4$ observations, along with the expected vertical profile. Above 40 km photolysis of $H_2SO_4$ returns $SO_2$ to the gas phase, as recently demonstrated in satellite measurements (Rinsland et al., 1995).

## 10.8 Gas-Phase Ionic Chemistry in the Stratosphere

In addition to the neutral species already considered there is also a rich ionic chemistry in the stratosphere that is closely tied to the presence of $H_2SO_4$. Ions are produced by the impact of cosmic rays on the major gases, notably $O_2$. After a series of charge-transfer reactions the negative ion distribution is governed by clusters containing the thermodynamically most stable ions, $HSO_4^-$ and $NO_3^-$. The clustering reactions are of the form:

$$NO_3^- + HNO_3 \rightarrow [NO_3(HNO_3)]^- \qquad (10.54)$$

$$[NO_3(HNO_3)]^- + (n-1)HNO_3 \rightarrow [NO_3(HNO_3)_n]^- \qquad (10.55)$$

$$[NO_3(HNO_3)_n]^- + H_2SO_4 \rightarrow [HSO_4(HNO_3)_n]^- + HNO_3 \qquad (10.56)$$

By further exchange reactions, clusters consisting only of $H_2SO_4$ and $HSO_4^-$ can be obtained. In situ mass spectrometric measurements show that below 30 km and above 40 km $NO_3^-$ core ions are dominant over $HSO_4^-$ core ions, a conclusion supported by the 2D modeling studies of Beig et al. (1993).

A complementary positive ion chemistry also exists in the stratosphere. In this case the initially formed ions are proton hydrates $[H(H_2O)_n]^+$. However, one of the water molecules can be replaced by a molecule with a high proton affinity such as $CH_3CN$ to form a so-called nonproton hydrate, for example,

$$[H(H_2O)_n]^+ + CH_3CN \rightarrow [H(H_2O)_n(CH_3CN)]^+ \qquad (10.57)$$

Acetonitrile is released in the troposphere, and a fraction is transported to the stratosphere. Thus stratospheric ion chemistry is linked intimately to the chemical and transport processes occurring at lower altitudes.

## Further Reading

Baker, D. J., D. Bear, C. M. Browner, J. R. Lyons, E. A. Rieke, and S. F. Tierney (1993) *National Acid Precipitation Assessment Program 1992 Report to Congress.*

Special issue of the *Journal of Geophysical Research,* Volume 98 (1993) on NASA's GTE/CITE-3 Experiment.

Saltzman, E. S. and W. J. Cooper, eds. (1989) *Biogenic Sulfur in the Environment,* ACS Symposium Series 393.

Warneck, P. (1988) *Chemistry of the Natural Atmosphere,* Academic Press, San Diego.

Robert J. Charlson

## Sulfur, Aerosols, Clouds, and Rain

Despite its low abundance, the element sulfur plays several essential roles here on Earth — roles that are made possible by its wide range of oxidation states and its middling location in the periodic table. Notably, it forms gaseous compounds, liquid and highly water-soluble ones, as well as solid forms, allowing it to be transported globally by the atmosphere and hydrosphere. Because of a multitude of chemical reaction pathways, physical phase changes occur frequently in the global sulfur cycle. Importantly, sulfur has two abundant and two not so abundant stable isotopes ($^{32}$S, 95% natural occurrence, $^{34}$S, 4.2%, $^{33}$S, 0.76%, and $^{36}$S, 0.014%) that are fractionated strongly in both chemical reactions and phase changes, allowing tracking of sulfur from place to place and form to form by study of ratios, usually $^{34}$S/$^{32}$S. There are no radioactive isotopes with long half-lives ($^{35}$S has $t_{1/2} = 87$ d) such that use of radioactive tracing is limited but possible. While major features of its isotopic distribution are fairly well established, there are many interesting puzzles still available for study.

The presence of sulfide-sulfur (−2) in two essential amino acids (cysteine and methionine) allows cross-linking of protein structures, such as cell walls, and thereby dictates that biological processing of sulfur is a major feature of the sulfur cycle. Notably, biological processing reduces sulfur from the +6 oxidation state of the aqueous sulfate ion to the −2 state in amino acids. There, depending on the local presence or absence of oxygen, reduced gaseous forms can emerge, notably $H_2S$ from anaerobic processes and $(CH_3)_2S$ from aerobic ones. Thus reduced gaseous forms of sulfur are generated naturally and provide reactive sulfur compounds to the atmosphere, where they undergo rapid (few day) oxidation in the atmosphere back to the +4 and +6 forms. $SO_2$ can react further or be removed directly to surfaces, such as biota. Given that the dominant oxidation products in the air, $H_2SO_4$ and its ammonium sulfate salts, have very low vapor pressures, physical condensation occurs, forming an absolutely essential entity within the atmosphere-aerosol particles. These submicrometer entities are essential to the functioning of the atmosphere in that they provide sites for condensation of liquid water as so-called *cloud condensation nuclei* or CCN. Without the abundant number of such sulfate particles (typically a few hundred per cm$^3$ in unpolluted air), cloud droplets could only form on much rarer particles such as sea-salt or soil dust (one or less per cm$^3$), making clouds entirely different than we know them. Thus the entire physical functioning of clouds and the hydrologic cycle is dependent on the biological production of reduced sulfur compounds.

In addition to this unique role as the dominant form of CCN, the acidity of $H_2SO_4$ provides the dominant strong acid in natural/unpolluted as well as polluted rain water. While many textbooks have described the pH of "natural" cloud and rain water as being controlled by dissolution of $CO_2$, and a pH $= 5.6$, the presence of pH values around 4.8 to 5 in the cleanest rain water on Earth (*e.g.*, at Cape Grim, Tasmania) along with sulfate molarities of ca. $10^{-5}$ illustrate control by the natural sulfur cycle. Interestingly, the naturally occurring molar ratio of $NH_4^+$ to $SO_4^{2-}$ in the aerosol is of order unity, such that the pH of the system is very sensitive to perturbation by the addition of more acids. Close to large sources of $SO_2$ (within a few hundred to ca. 1000 km), pH values as low as 3.5 can be observed, showing a clear excess of $H_2SO_4$ over the available base.

Continuing with this global perspective, the natural cycle of sulfur results in a more or less steady flux of reduced S gases to the atmosphere. The balance of this system depends greatly on the oxidative state of the atmosphere, surface, and soil water. In fact, sulfate ion in water is the dominant oxidizing agent in anaerobic systems, producing $H_2S$, so there are global redox aspects to the sulfur cycle as well as the above role in governing the acid-base balance of rainwater.

Yet another global feature of the natural sulfur cycle is the influence of sulfate aerosols on the optical properties of both clear and cloudy air. These submicrometer particles have large scattering efficiencies for sunlight, making sulfates an important factor in determining the visual clarity of air and the amount of sunlight reflected by the atmosphere. When acting as CCN, these same particles can influence the number population of cloud droplets and hence the albedo of clouds. Whether the natural biological sulfur cycle is actually coupled into feedbacks in the climate system is still only a hypothesis; however, there are no doubts that important connections exist that need to be understood (Charlson et al., 1987).

Finally, the picture of a steady-state functioning of the natural sulfur cycle has been radically disturbed by the production of large amounts of $SO_2$, largely from burning of fossil fuels and from smelting of metal ores. Figure 5.31 shows the development of the Northern Hemisphere $SO_2$-S fluxes compared to the natural flux to the atmosphere. While a degree of uncertainty exists about the exact magnitude of the natural flux, it is clear that the anthropogenic fluxes exceed the natural ones by several fold — probably a factor of 2 to 5. This makes the atmospheric sulfur cycle one of the most perturbed elemental cycles of Earth (cf. the carbon and nitrogen cycles) and connects to several of the functions of the natural S cycle that are described above:

- The acid-base balance of rainwater over large regions of industrial activity is radically altered from the natural state to a much more acidic condition. pH values of 4 or less are commonly observed. Most industrial activity is in the NH.

- Clouds in polluted locales have larger numbers of droplets than in clear marine clouds. In fact, clouds over the entire NH are slightly more reflective than in the SH, allegedly due to anthropogenically enhanced sulfate CCN (Han et al., 1994).

- Air in the NH is hazier than in the SH, particularly in and near industrial activity, largely due to sulfate aerosol. The amount of the sulfate aerosol is calculated to reflect 0.4-0.8 W m$^{-2}$ of sunlight away from the whole globe, and up to 5 or 10 W m$^{-2}$ from the industrial areas. Because this is comparable in magnitude but opposite in sign to climate forcing by greenhouse gases (global mean increase of heating of $+2$ W m$^{-2}$), it is necessary to consider sulfate aerosols as a factor in climate change.

It is clearly useful to view the sulfur cycle from a global perspective even though the processes are not at all uniform over the globe. Figure 5.32 is a schematic diagram of the natural and perturbed sulfur cycle of the atmosphere and is intended to provide a context into which the details of sulfur chemistry logically fit.

*Robert J. Charlson is a Professor of Atmospheric Chemistry in the Department of Atmospheric Sciences at the University of Washington. He has made major contributions regarding the global biogeochemical cycles of sulfur and nitrogen, the optical role of aerosols, and the importance of sulfate particles for the Earth's climate. Dr. Charlson received an honorary doctorate from Stockholm University in 1993.*

# Part 3

# Tools

# 11 Observational Methods: Instruments and Platforms

## 11.1 Introduction

Measurements form the base on which our understanding of the atmosphere is built. New or improved measurement capability or application of measurement techniques in new ways or new situations has often resulted in a significant expansion or modification of our understanding. Measurements provide the information for testing models and validating theoretical concepts. In atmospheric chemistry, no measurements are more fundamental than measurements of the composition of the atmosphere. The range of concentrations that must be measured is tremendous. Diatomic nitrogen constitutes 78% of the atmosphere, while radical species such as OH can be less than ten parts per quadrillion; this implies a range of concentration of $10^{14}$ to be measured. It presents a difficult challenge for instrumentation and requires the development of many types of instruments. This chapter will survey some of the major types of instruments for measuring concentrations of atmospheric gases and related parameters that are important for atmospheric chemistry. It cannot cover all methods, nor treat most in depth.

## 11.2 Instrumentation for Constituent Measurements

Measurements of concentration of atmospheric constituents may be made in situ (on the portion of the atmosphere that is located at the instrument) or remotely (that is, on atmosphere located at a distance from the instrument). The two types of measurements involve different techniques, although some principles may be common.

### *11.2.1 In Situ Measurements*
11.2.1.1 Absorption Measurements

Spectroscopic methods for the measurement of constituents are based on the fact that different chemical compounds absorb electromagnetic radiation differently at specific wavelengths (see Section 3.2.2). The pattern of absorption with wavelength provides a "fingerprint" for the chemical compound, and the intensity of absorption depends upon the amount of absorber. This simple approach has been developed into a wide variety of spectroscopic and radiometric instruments used in atmospheric chemistry. Such instruments can be used for in situ measurements to determine the concentration of compounds in air located at or within the instrument or for remote sensing measurements where the air sampled may be located at great distances from the instrument. Spectroscopic instruments can work at widely different wavelengths from the ultraviolet to radio waves and can vary from extremely narrow-band high-resolution spectrometers to broad-band radiometers.

Ultraviolet absorption is used to measure ozone in situ (Proffitt and McLaughlin, 1983). The ozone absorption cross section is very large, about $1.2 \times 10^{-17}$ cm$^2$ molecule$^{-1}$ at its peak (see Fig. 3.7), and is relatively insensitive to pressure and temperature. This peak nearly coincides with the resonance line of mercury at 253 nm. A low-pressure mercury emission lamp, filtered to produce only this line, is a convenient source for measuring the monochromatic absorption. Air containing the ozone to be measured flows through a tube around 50 cm in length, light is sent along the tube, and the intensity at the end of the path is detected. A similar tube with its own detector from which ozone has been removed is used to monitor varying brightness of the source. A typical ozone concentration of 1 ppm in the lower stratosphere produces an absorption of around 0.1%. To remove imbalance effects, the sample and zero tubes are interchanged. The air can be passed first through a scrubber that destroys the ozone in order to determine the instrument zero. Figure 11.1 shows a schematic of an instrument that has been used on a high altitude aircraft.

A much more difficult UV absorption experiment has been used to measure OH radicals in the A-X band near 308 nm (Mount and Harder, 1995). Because the OH concentration is perhaps a million times less than that of ozone, a long path is needed. The path used is between two mountains 10 km apart. The source and detector are located on one mountain, and an array of cube-corner retro-reflectors on the other mountain reflects the beam back to the detector, with absorption along the intervening path. A monochromator is used to disperse the laser radiation and resolve the OH absorption lines.

This is one example of differential optical absorption spectroscopy (DOAS), in which the radiation from a continuous UV or visible light source located a great distance from the spectrometer is absorbed by trace gases in the intervening atmosphere (Platt *et al.*, 1979). Because very small absorptions can be measured with sensitive detectors at the short wavelengths involved, this is a good technique for those species that have structured UV absorption, such as $SO_2$. The use of DOAS to measure $NO_3$ radicals is discussed in Section 7.7. UV absorptions are usually not extremely narrow spectrally because many rotation lines are smeared together, so high spectral resolution is not required. The long path involved limits the spatial resolution; this technique could be considered either in situ or remote sensing.

Since virtually all molecules (except noble gases and homonuclear diatomics such as $H_2$ and $N_2$) have absorptions at infrared wavelengths, infrared spectroscopy provides a versatile technique for measuring chemical species. It is, however, relatively insensitive and requires long absorption paths to achieve substantial absorption. Typical molecules require a path length-concentration product around 10 ppm-m (*i.e.*, a path length of 10 m at a mixing ratio of 1 ppm or a path length of 10 km at a mixing ratio of 1 ppb) to produce a 1% absorption. With reasonable path lengths achievable in an instrument (0.1 to 100 m), the sensitivity for straight absorption is tens of ppb. While this is useful for measuring minor gases such as $N_2O$ or CO, many of the important species that absorb in the infrared cannot be measured in this way. Lasers offer the possibility of measuring much smaller absorption.

Tunable diode laser absorption spectroscopy is a highly selective and sensitive technique for in situ measurement of trace gases (Reid *et al.*, 1978). It has been used most commonly on the ground or on airplanes, with the sample air drawn into a cell at low pressure (to reduce the absorption linewidth), but has also been used with open paths on balloons or on the ground.

The tunable diode laser emits in the mid infrared region, where many atmospheric molecules have strong absorptions. Lightweight molecules have sharp absorption lines from vibrational-rotational transitions. At low pressures, the linewidth is limited

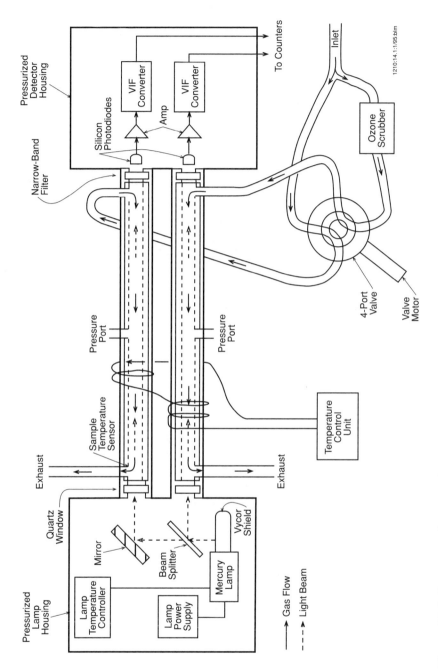

**Figure 11.1.** Instrument for measuring ozone concentration by ultraviolet absorption. This instrument has flown on a high-altitude aircraft to measure ozone in the lower stratosphere. The sample air is alternated in the two absorption tubes with air from which the ozone has been chemically removed. The difference in the transmission of the two cells in the two modes supplies the ozone absorption by a sample independently of the lamp intensity or cell transmission (after Proffitt and MacLaughlin, 1983).

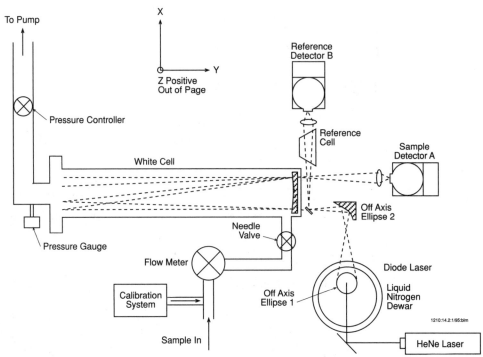

**Figure 11.2.** Schematic of tunable diode laser instrument for the measurement of atmospheric trace gases. The laser is cooled by liquid nitrogen, and its temperature and current determine the wavelength emitted. The radiation passes through a multiple reflection cell to achieve long path length. Absorptions less than $10^{-5}$ can be measured (from Fried et al., 1991).

by the Doppler shift of the individual molecules, and the lines do not overlap. By adjusting the laser temperature and current, the laser output can be produced at a single frequency, which can be scanned continuously over bandwidths of several $cm^{-1}$.

Figure 11.2 shows a typical tunable laser instrument employed for atmospheric measurements. The diode laser is cryogenically cooled; its operating temperature is stabilized in the 80-120 K range. The laser current, typically ~100 mA, is increased in a sawtooth fashion to scan the laser wavelength. The laser radiation is directed through an absorption cell through which the sample flows continuously at low pressure. Multipass absorption cells allow absorption paths of 10-200 m. The resulting spectrum is recorded by a photodetector as the laser scans. Many rapid scans can be averaged.

Two different data acquisition approaches are commonly used. In direct absorption, the transmitted intensity at line center is used with measurements of pressure, temperature, and path length, along with the experimentally determined line intensity to obtain the concentration of absorber from the Beer-Lambert Law (Section 3.2.2). This method requires no gas calibration standards when the atmospheric measurement is made.

Alternatively, an external modulation waveform at high frequency can be added to the scanning laser current. A lock-in amplifier is used for synchronous detection of the measured signal. In most instances, the second harmonic of the modulation frequency is used for detection. In effect, this compares the signal at the line center with the average of the signal at two points a small distance on either side. When the modulation amplitude is chosen to be near the linewidth, at line center this method measures the depth of the absorption line, removing the constant and linearly varying terms from the background. Figure 11.3 shows an example of a second harmonic signal. For weak absorptions in direct absorption, it is necessary to measure the small difference in two large signals to obtain the absorption, whereas these large values are eliminated in harmonic absorption measurements. Harmonic absorption is usually used for greatest sensitivity and can measure absorptions of $10^{-5}$ to $10^{-6}$. With a path length of 100 m, this corresponds to minimum detectable concentrations from a few pptv to ppbv, depending upon the strength of the observed line. Additional advantages of harmonic detection are reduced susceptibility to low-frequency noise, because of the higher detection frequency, and enhanced discrimination against signals without a rapid wavelength dependence, such as the broad absorption tails of nearby water lines or aerosol absorptions. Second-harmonic detection systems are calibrated by replacing the sample by a calibration gas with known concentration, or by measuring the ratio of the observed signal to the signal from a reference cell containing the gas in question.

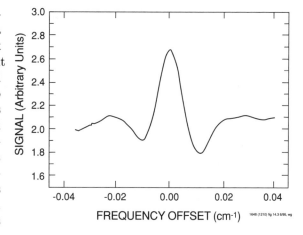

**Figure 11.3.** The instrument response at a frequency twice the modulation frequency; the signal is almost proportional to the second derivative of the gaseous absorption spectrum. This spectrum corresponds to 550 ppt of OCS in air (from Fried et al., 1991).

A different approach to achieving high spectral resolution with high signal in the infrared is provided by gas correlation radiometers (Sebacher, 1978). The key to this technique is the use of a sample of the target gas itself to provide the high spectral resolution. Because the absorption spectrum of the reference is perfectly correlated with the absorption spectrum of the target gas in the atmospheric sample, and usually very poorly correlated with the spectrum of interfering species, an instrument whose spectral sensitivity is given by the spectrum of the reference produces a measurement with high sensitivity (because all of the lines are used simultaneously, not just a single line) and with minimal interference. This is arranged by alternately passing radiation that has been emitted or transmitted by the sample through or around the reference sample. For radiation of those wavelengths for which the reference absorbs, this produces an alternating signal, while other wavelengths are unaffected and produce only a dc signal. Thus detection of the ac signal results in measurement of radiation only of wavelengths at which the reference sample absorbs. This can be effected by moving an absorption cell containing the reference optically or mechanically into the beam (Houghton and Smith, 1970; Abel et al., 1970), or by alternating the signal radiation to two detectors, one of which has the reference sample in its optical path, and measuring the difference in output of the two detectors (Russell et al., 1993). The most challenging part of the measurement is in maintaining the balance in the two beams (ensuring that the signal is zero when the sample does not contain the target gas). Figure 11.4 shows a gas correlation instrument used for measuring CO in situ.

### 11.2.1.2 Resonance Fluorescence

Radicals are particularly difficult to measure because of their high reactivity and low concentration, yet they are key to much of the chemistry of the atmosphere.

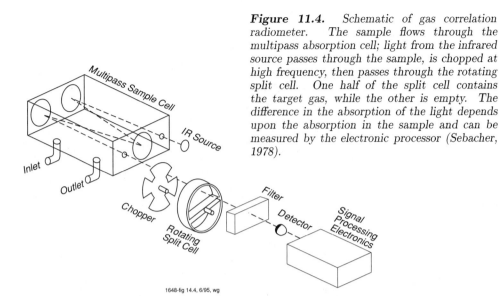

**Figure 11.4.** Schematic of gas correlation radiometer. The sample flows through the multipass absorption cell; light from the infrared source passes through the sample, is chopped at high frequency, then passes through the rotating split cell. One half of the split cell contains the target gas, while the other is empty. The difference in the absorption of the light depends upon the absorption in the sample and can be measured by the electronic processor (Sebacher, 1978).

Measurement techniques for radicals must be sensitive at the parts-per-trillion range. One technique that has been used for such measurements in situ is resonance fluorescence (Anderson, 1978). A beam of photons is produced at a wavelength that corresponds to an electronic transition from the ground state of the species in question. The beam passes through the sample, and the sample molecules absorb photons from the source beam and re-emit them in all directions. The resonantly scattered photons are detected at right angles to the incident photons and the flow of the sample through the cell. Photons may be scattered elastically by all molecules (Rayleigh scattering), as well as resonantly (when the energy of the photon just corresponds to the energy of an excited state of the target atom or molecule). The cross section for resonant scattering, however, is many orders of magnitude larger than for Rayleigh scattering (a factor of $10^{11}$, for example, for Cl atoms at 119 nm); this is the basis for the selectivity of the device. The signal from Rayleigh scattering is calibrated out by replacing the sample with air not containing the target radical. The match between the wavelength of the source lamp and the radical to be detected may be achieved by using a microwave discharge lamp with the discharge in the target species, for example, a discharge in Cl to detect atomic Cl. Lines other than the desired resonance line are filtered out.

A variety of light sources has been used, but the resonance lamp is the most useful. The lamp must be designed to produce high-intensity, spectrally pure light that is not self-absorbed within the lamp. The signal produced is proportional to the flux of photons from the lamp and the concentration of the target species. The flux is measured by monitoring the intensity of the direct beam from the lamp and obtaining the ratio of the scattered beam to the first measurement. Calibration is performed by inserting a sample containing a known concentration of the radical at a temperature, pressure, and flow rate similar to the measurement sample or by chemical titration of the sample.

The flow of the sample through the instrument must be considered carefully because the radicals are so reactive that a single wall collision will usually destroy them. Laminar flow is used with a geometry that makes the time for diffusion to the walls large compared to the transit time through the instrument. Because of the viewing geometry, only the center of the beam is used.

This technique has been used principally for the measurements of OH, ClO, and BrO in the stratosphere (see examples in Chapters 6 and 8). Most of the radical chlorine in the stratosphere is in the form of ClO, but the resonance fluorescence technique measures atomic Cl. To measure ClO, the incoming air sample is mixed with an excess of NO, which quickly reacts with ClO to produce atomic Cl and $NO_2$. The resonance fluorescence then is used to detect the Cl that is produced. Turning off the flow of the NO permits the background to be measured. Figure 11.5 illustrates the ClO instrument flown on a balloon.

A variant of the resonance fluorescent technique uses a laser rather

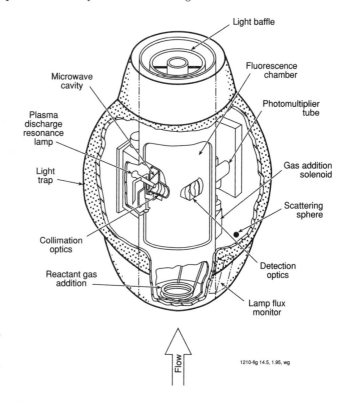

**Figure 11.5.** Balloon-borne resonance fluoresence instrument for the measurement of atomic chlorine and ClO in the stratosphere (Anderson, 1978).

than an atomic resonance lamp to provide the radiation that lifts the target molecules from their ground state to an excited state, from which they decay radiatively (Davis et al., 1979; Wang, 1981). Again, the photon emitted when the excited molecule decays is detected optically. Laser-induced fluorescence (LIF) can be used for molecular species as well as atomic species, whereas resonance lamps generally cannot be constructed for molecules because the molecules dissociate in the electrical discharge in the lamp. One form of LIF that is useful for trace molecules is two-photon LIF. In the technique, two lasers are used to excite the molecule to the radiating upper state in a two-step process; the first laser excites the molecule to a quantum state at lower energy than the fluorescent state, and a second laser excites it from this state to the fluorescent state before it can decay from the first intermediate state. The advantage of this rather complicated technique is that the photons from the exciting lasers are of lower energy, and hence longer wavelength, than the fluorescent photon that is detected; thus any fluorescence from contaminants, which will be of longer wavelength than the exciting radiation, can be ignored in the detection process.

### 11.2.1.3 Chemiluminescence

Chemiluminescence occurs when light is emitted from a molecule during a transition from an excited electronic or vibrational state produced by an exothermic chemical reaction. For example, the reaction of ozone with nitric oxide is used as the basis for chemiluminescent detection of nitric oxide in the atmosphere (Ridley and Howlett,

1974; Drummond *et al.*, 1985). The reactions for this measurement are:

$$O_3 + NO \rightarrow NO_2^* + O_2 \tag{11.1}$$

$$NO_2^* \rightarrow NO_2 + h\nu, \qquad 590 < \lambda < 3000 \text{nm} \tag{11.2}$$

The asterisk indicates a molecule in an electronically excited state; the excited molecule decays to the ground state with the emission of a photon. The emitted light can be detected using conventional photomultipliers with suitable wavelength response. Under constant conditions in the reaction volume, the intensity of the emitted light is proportional to the mixing ratio of the emitting molecule.

The chemiluminescent reactions (11.1) and (11.2) have been widely used as the basis of detectors for different atmospheric odd nitrogen species (see Chapter 7 for several examples). Because the reaction is specific for NO molecules, other nitrogen species, such as $NO_2$, need to be converted to NO prior to detection. Currently, the most specific conversion technique for $NO_2$ involves the photolysis of $NO_2$ to NO. To measure total reactive oxidized nitrogen in the atmosphere, all odd-nitrogen species are catalytically reduced to NO using a CO reagent gas in the presence of gold at 300°C. The quantity measured with this technique is commonly called $NO_y$, and it is the sum of all oxidized nitrogen species, including NO, $NO_2$, $HNO_3$, HONO, $N_2O_5$, and organic nitrates. The instrument shown in Figure 11.6 has been used for measuring NO in the stratosphere.

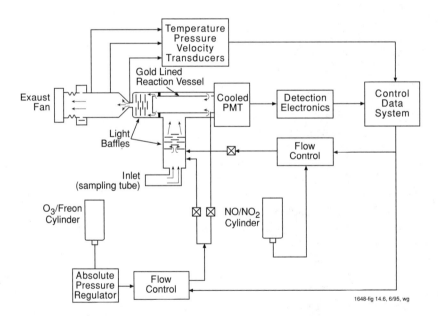

**Figure 11.6.** *Chemiluminescent detector for NO. The sample air is mixed with ozone in the reaction vessel, and the photons from the resulting chemiluminescence are detected by the photomultiplier tube. If the ozone is added before the reaction chamber, the luminescence takes place outside the view of the PMT and gives a background level (Ridley and Howlett, 1974).*

While the measurement of $NO_x$ is probably the primary use of chemiluminescence detection in atmospheric chemistry, other uses have been reported. For example, the same set of Reactions (11.1) and (11.2) have been used to measure ozone. In this case, an excess of NO, rather than $O_3$, is added for the reaction. The chemiluminescent reaction of alkenes and $O_3$ also has been described to monitor a major biogenic hydrocarbon (isoprene) in plant emissions (Hills and Zimmerman, 1990), or to measure $O_3$ by the addition of excess ethene.

One novel use of the technique of chemiluminescence has been the measurement of radicals with chemical amplification (Cantrell *et al.*, 1993b). In this approach, ambient peroxy radicals ($HO_2$ and $RO_2$) are converted through a chain reaction to a larger amount of $NO_2$, which is detected by chemiluminescent reaction with an organic dye. The conversion occurs through the following chain reaction sequence:

$$HO_2 + NO \rightarrow OH + NO_2 \quad (11.3)$$
$$OH + CO \rightarrow H + CO_2 \quad (11.4)$$
$$H + O_2 + M \rightarrow HO_2 + M \quad (11.5)$$

The reagent gases NO and CO are added to ambient air so that the concentrations are around 3 ppmv and 10%, respectively. For these conditions, with a reaction time of 10 s, up to 200 $NO_2$ molecules are produced for each $HO_2$. The $NO_2$ is then measured by an appropriate technique such as chemiluminescence. Organic peroxy radicals are measured if they are converted to $HO_2$ in the NO-rich environment of the inlet by, for example:

$$CH_3O_2 + NO \rightarrow CH_3O + NO_2 \quad (11.6)$$
$$CH_3O + O_2 \rightarrow CH_2O + HO_2 \quad (11.7)$$

The radical signal must be modulated because of ambient ozone that converts NO to $NO_2$. Nitrogen is substituted for CO to measure this radical background. The difference between the background and the signal with CO present is the amplified radical signal. Detection limits for this technique depend on the variability of ambient ozone, but often approach 2-3 pptv.

## *11.2.2 Measurements on Atmospheric Samples*
### 11.2.2.1 Sampling Methods

When analyses of atmospheric constituents cannot be performed by direct techniques, samples can be collected for later analysis using procedures that are optimized for the specific chemicals to be analyzed. Among the techniques used are whole air sampling, filter collection, and matrix isolation.

Whole air sampling involves filling a suitable container with air without any separation of major and minor components of the atmosphere. Most often, the containers used are specially cleaned and treated (electropolished) stainless steel vessels, or glass flasks, with volumes in the range of 0.5 to 5 L, but Teflon and Tedlar bags are also commonly used for some analyses. While the simplicity of this technique is appealing, successful application of whole air sampling relies on the stability of the analyte gases in the sample container. Thus whole air sampling is most often applied to stable gases with little affinity for surface adsorption or reaction. Compounds such as fluorocarbons and many nonmethane hydrocarbons are suitably sampled in stainless steel canisters, while reactive gases such as ozone are unstable in these containers. Also, gases with a high affinity for surface adsorption, such as nitric acid, are not suitable for storage in whole air containers.

Whole air samples may be collected by simply opening a valve on an evacuated container and then closing it after the container has filled to ambient pressure, or by pumping the air into the container with a noncontaminating pump. Data collected using this technique are presented in Table 8.3. High-altitude samples have been collected by immersing the sample container in liquid neon, which is cold enough to condense essentially all the air (Lueb et al., 1975). When the valve is opened, sample air rushes in and condenses, reducing the pressure and collecting larger samples.

For trace gases, samples may be preconcentrated cryogenically by cooling the air until the desired compounds have condensed and then by pumping the abundant oxygen and nitrogen away (see, for example, Goldan et al., 1995, or Greenberg et al., 1996, and references therein). This increases the fractional concentration of the analyte gases in the sample by a large factor and eases the analysis.

An alternate set of sampling techniques for chemicals that cannot be suitably stored in canisters is based on specific adsorption or reaction on a solid substrate. Of course, aerosols are collected on inert (glass and/or Teflon) filters for analysis of organic and inorganic species, but other filter materials are useful for collecting gas-phase components of the sample. For example, nylon filters have been shown to collect acidic gases, such as nitric acid, very efficiently (Norton et al., 1992). Filters treated with different chemicals, too, are used for reaction with the atmospheric species of interest. Studies have shown that formaldehyde, for example, can be collected by reaction with bisulfite-treated filter media. Formaldehyde forms a stable adduct, hydroxymethanesulfonic acid (HMSA), on the filter surface, and this compound is subsequently treated for analysis (Dasgupta et al., 1980). While suitable precautions must be taken to avoid artifacts during collection, treated filters have been widely used for atmospheric chemistry studies. Dinitrophenylhydrazine (DNPH) has also been used to "derivatize" aldehydes. A summary of current methods of detecting formaldehyde is given in Gilpin et al. (1997).

One problem associated with filtration is the potential for ambiguous separation of particulate and gas-phase components of compound. Volatilization of a compound from a filter surface or adsorption onto the filter surface will produce artifacts that confuse the measured partition of a compound between the gas and aerosol phases. One technique that has been developed to minimize this problem is sample collection with a diffusion denuder (Ferm, 1986). This technique takes advantage of the different rates of diffusion of gaseous and aerosol compounds from a sample air stream to a treated surface. In this mode of sample collection, air first flows through an open tube, a series of concentric tubes, or across parallel plates. During passage through the inlet, gas-phase species are removed by diffusion and reaction on the treated solid surfaces. Some treatments that have been described are citric acid to collect ammonia and bicarbonate solution to collect organic acids (Norton, 1992). Aerosols pass through the inlet and are collected by a filter or series of filters downstream of the denuder inlet. This process has been reasonably successful in accurately characterizing the partitioning of chemical species between gas and aerosol phases.

Variations of the filtration and denuder techniques involve collection of compounds on other solid or liquid substrates. Use of solid adsorbent cartridges containing porous organic polymers or carbon molecular sieves are increasingly being used to collect a wide variety of different organic compounds (O'Brien et al., 1995). A range of techniques also has been described that use water or aqueous solutions to collect soluble gases, such as formaldehyde, hydrogen peroxide, and nitric acid (e.g., Lee and Zhou, 1993). An air stream mixed with a small volume of water provides a large concentration factor of the soluble gas, and the water can be analyzed using different wet chemical techniques.

Techniques that collect a sample in a container for later analysis are not suitable for collection of species that react on the walls of the container. In particular, most free radicals are so reactive that they will not survive in a container. Matrix isolation can be used to sample such species, as well as many stable species. The reactive molecules and radicals are trapped in an inert crystalline matrix formed by condensing the sample onto a very cold metal surface, usually at liquid nitrogen temperature (77 K). By condensing the sample at this temperature, oxygen and nitrogen may be pumped away, leaving the sample highly concentrated in the trace species in a matrix of $CO_2$ and water. The reactive molecules are frozen in place and cannot find other molecules with which to react and so survive as long as they are kept cold. The analysis method must operate on the frozen sample. Infrared spectroscopy has been used for measurement of trace gases (Griffith and Schuster, 1987) and electron paramagnetic spectroscopy for the detection of peroxy radicals such as $HO_2$ (Mihelcic et al., 1985).

### 11.2.2.2 Analysis Methods

The measurement methods applied to air samples collected by the techniques just described encompass a wide range of analytical techniques. In this section we survey a limited number of techniques that include chromatography, isotopic measurements, and some specialized techniques.

In a broad sense, the term *chromatography* describes a wide range of techniques that separate mixtures of chemicals into their component parts. Given that air is a complex chemical mixture, there is immediate appeal to a technique that can deal with the complex composition of the atmosphere. The detector, then, does not need to be specific to each compound. Chromatographic techniques used in atmospheric chemistry include gas, liquid, and ion chromatography. A number of different detectors are used to detect the sample compounds in gas chromatography. Any technique that responds differently to the target molecules than it does to the carrier gas can be used as a detector. In essence, the detector senses the presence of target gas; the identity of the target gas is determined by the length of time it was retained in the chromatographic column. Some of the detectors that have been commonly used are thermal conductivity detectors, which measure the change in cooling of a heated filament when the target gas flows over it, flame ionization detectors, which detect the change in the number of ions produced when the gas stream is burned in a hydrogen flame, flame photometric detectors (used especially for sulfur compounds), which detect the characteristic atomic emission in a flame of an atom in the target species, and electron capture detectors, which measure the very high affinity of halogen compounds for capturing an electron from a small beta-emitting radioactive source.

Because of its versatility and the nature of the medium to be analyzed, gas chromatography is the most widely used technique for measuring complex mixtures. Analysis of trace atmospheric components (ppbv and less) often involves the pre-concentration of these compounds from the air ($N_2/O_2$) matrix. Given that the sensitivity of the different detectors is in the nanogram ($10^{-9}$) to picogram ($10^{-12}$ g) range, sample volumes of 0.1-1 L are used to measure compounds in the pptv range. A gas chromatograph is shown schematically in Figure 11.7. For many applications to ambient samples, high-resolution capillary columns up to 100 m in length are required to achieve the necessary separation between closely related compounds. The inside of the capillary column is coated with a substance that attracts the molecules in the sample and slows their passage through the column, relative to the carrier gas. Substances that are bound more tightly to the wall coating require longer to pass through the column. Frequently the temperature of the column is varied in a precise way during the collecting of the chromatogram in order to improve the speed and resolution of the results. Figure 11.8 illustrates the chromatogram of a suite of hydrocarbons in an atmospheric sample. Even with the high resolution supplied by capillary gas chromatography, increased specificity of compound detection is required in many cases. The use of a mass spectrometer (MS) as a chromatographic

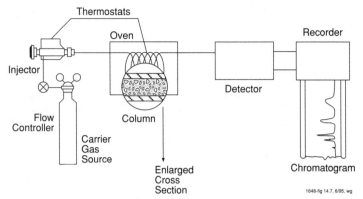

**Figure 11.7.** Schematic of a gas chromatograph. The sample is injected into the carrier gas; different components pass through the heated column at different rates and thus are detected by the detector at different times (from Okamura and Sawyer, 1978).

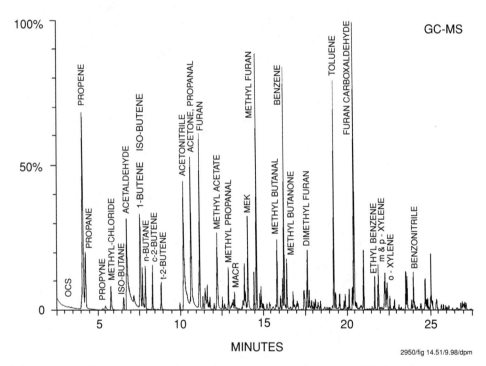

**Figure 11.8.** Chromatograph of organics in ambient air (courtesy of J. Greenberg).

detector provides enhanced detection and confirmation of compound identities. Ion fragments produced in the MS ion source can be monitored individually to obtain high-sensitivity measurements of target compounds whose fragmentation is known. Alternately, unknown compounds may be identified in a sample from the characteristic "fingerprint" of mass fragments produced in the MS.

While mass spectrometers have most commonly been used in the laboratory for analyzing collected samples, mass spectroscopy using highly specialized ionization techniques, such as chemical ionization or selected ion chemical ionization, prior to ion mass analysis and detection, has been used in the field for very sensitive detection of specific compounds. The use of chemical ionization improves mass spectrometric detection limits from the parts per million or billion range to the low parts per quadrillion range. At the same time, this technique adds an additional level of chemical specificity, but at the expense of measurement diversity. The number of compounds to which this technique can be optimally applied is quite limited, but where it is well suited, it typically offers far more sensitivity and faster time response than other techniques.

Chemical ionization mass spectrometry (CIMS) takes advantage of several chemical and physical processes (Eisele and Tanner, 1991; Eisele and Berresheim, 1992). Ionizing radiation creates ions that react with other molecules in the sample, eventually producing the most stable ions, negative ions of sulfuric acid, for example. In selective chemical ionization mass spectrometry, specific ions are created that are introduced into the sample to react only with certain specific compounds. The ions produced for analysis have an even higher proton affinity or acidity than the initial reactant ions, thus allowing them to be formed readily by proton exchange reactions with the initial reactant ions and be measured in the positive or negative ion spectrum, respectively. Since the initial reactant ion will react with only a few trace atmospheric constituents, the charge associated with these ions can be transferred almost exclusively to the compounds of interest, even though these compounds may constitute only one part in $10^8$-$10^{11}$ of the atmospheric molecules present. This provides a high degree of chemical selectivity in addition to the mass selectivity. Since only ions and not neutral compounds are analyzed and detected by the mass spectrometer system, the detection sensitivity of the mass spectrometer has also been enhanced by the same 8 to 11 orders of magnitude.

While chemical ionization sources operate at ambient pressure, mass spectrometers operate in a high vacuum. The chemical ionization source is coupled to a mass spectrometer vacuum system by a small aperture, which allows both ions and neutral gas to enter the vacuum region. Electrostatic lenses focus the ions into a narrow beam on the entrance to the analysis system, while neutral gas molecules are pumped away.

Quadrupole or tandem quadrupole mass spectrometers are the most common instruments used in conjunction with chemical ionization, although magnetic-sector and time-of-flight mass spectrometers are also used. Quadrupole spectrometers use a combination of static and radio frequency electric fields to define a stable trajectory for an ion with a specific charge-to-mass ratio along the central axis of four parallel conducting poles. An electron multiplier (ion detector) at the exit end of this set of four poles then measures the number of ions detected as a function of the ion mass setting of the quadrupole.

One important measurement for which chemical ionization mass spectroscopy is used is the measurement of hydroxyl radicals (Eisele and Tanner, 1991). OH reacts rapidly with $SO_2$ to form $HSO_3$, which is converted through a chain of reactions to $H_2SO_4$, which can be measured with high sensitivity by CIMS. Isotopically labeled $SO_2$ is used to react with the ambient OH, and the labeled $H_2SO_4$ is detected by CIMS. Such data are shown in Figure 6.15.

Measurement of isotopic composition of specific gases has found several uses in atmospheric chemistry. The two principal applications are in the determination of age of samples from the decay of radioactive isotopes or the determination of physical conditions under which the molecules have existed. Radon, produced in tests of nuclear weapons in the atmosphere, has also been used as a tracer of atmospheric motions, since it can be measured with very high sensitivity; it has also been used to determine how long an air parcel has been over the ocean since the land surface is a source of radon, but ocean surfaces are not. Measurements of $^{14}C$ in methane have been used to indicate how recently the methane was formed, and hence whether it is of biological or fossil origin. $^{14}$C is continually produced in the high atmosphere by the action of cosmic rays, and incorporated into biological systems. It then decays with its characteristic half-life of 5700 years. By measuring the ratio of $^{14}$C to $^{12}$C, then, it is possible to determine how much of the $^{14}$C has decayed since the methane was formed, and thus the age of the methane. This helps to determine its source, since methane from current biological activity has the standard amount of $^{14}$C, while methane from natural gas, for instance, has virtually no $^{14}$C left.

The isotopic composition of molecules affects the molecular mass. This, then, is related to vibrational energy in the ground state and can affect the rate of chemical reactions slightly (see Section 3.4.9). Further, fractionation can occur in physical processes, as the slightly greater mass of HDO compared to $H_2O$ results in a slightly different vapor pressure, and thus the isotopic species fractionate in phase changes. These differences can help describe the history of such processes. For instance, oxygen isotopes in ice cores are related to the vapor pressure of water and thus the temperature of the global atmosphere. These can be used to help establish past climate variations (Nairn and Thorley, 1961; Jouzel et al., 1993; see Fig. 15.5). Information on the sources of hydrocarbons can be obtained from measurements of the $^{12}C/^{13}C$ ratio (see Chapter 5).

Isotopic measurements are usually made by mass spectrometers. Infrared spectroscopic methods can also be used to measure isotopic composition of gaseous samples, although usually with somewhat less precision than mass spectrometers. IR methods have the additional advantage of indicating which atom has been isotopically substituted; for example, it can distinguish $^{14}N^{15}N^{16}O$ from $^{15}N^{14}N^{16}O$. Diode laser measurements, similar to those described for measuring trace gases, are promising in this application. In this case, two almost coincident lines having similar intensities, one from each isotopic species, are compared.

A novel use of isotopic measurements is the use of radioactive $^{14}$C for the measurement of OH (Felton et al., 1992). Excess $^{14}$CO is added to the sample containing OH. The OH quickly reacts with the $^{14}$CO to produce $^{14}CO_2$. All the $CO_2$ in the sample, including the isotopically labeled $^{14}CO_2$ produced by the reaction with OH, is chemically separated from the sample. The amount of $^{14}CO_2$ is determined from the rate of radioactive decay of the $^{14}$C, and gives directly the amount of OH in the sample, since each OH produces one $^{14}CO_2$ molecule.

## 11.2.3 Remote Sensing of Composition
### 11.2.3.1 Passive Remote Sensing

Remote sensing is characterized by the fact that the location of the measurement is different from that of the instrument. This requires that information be propagated by electromagnetic (or occasionally acoustic) radiation from the atmosphere to the instrument. The instrument measures the radiation and from these measurements deduces characteristics of the atmosphere, a process known as retrieval. The information may be contained in the intensity, spectral distribution, arrival time, or polarization of the received radiation. Since atoms and molecules interact with radiation at specific electromagnetic frequencies determined by the quantum mechanical energy levels of the molecule, a large class of remote sensing instruments includes spectrometers, which measure the intensity of radiation as a function of frequency. Such instruments measure the radiation emitted by the portion of the atmosphere in question or they receive radiation from some other source that has been modified by absorption or scattering by the atmosphere. The external radiation source is usually the Sun, but the moon and stars have also been used, as well as light scattered or emitted by the Earth's surface.

The interaction between the molecule and the radiation may change the rotational energy of the molecule, the vibrational energy, or the electronic energy. These correspond to the far-infrared and microwave regions of the electromagnetic spectrum, the middle- and near-infrared regions, and the visible and ultraviolet regions, respectively (see Section 3.1). All these spectral regions have been used for chemical analysis of the atmosphere. The technology for spectroscopy of these regions is different, depending substantially upon the detection method.

Ozone has strong absorption in the ultraviolet region of the spectrum; indeed this is one of the principal reasons for great interest in ozone. In the Hartley bands, between 200 and 310 nm, the absorption is very strong, with cross sections reaching about $1.2 \times 10^{-17}$ cm$^2$ molecule$^{-1}$ at the peak around 255 nm. With this strength of absorption, no sunlight reaches the surface. At longer wavelengths, ozone absorbs in the Huggins bands, but not nearly so strongly, and the fraction of sunlight that reaches the surface increases rapidly with wavelength, from nearly complete absorption at 290 nm to little ozone opacity at 340 nm. This rapid variation in ozone opacity causes the UV cutoff of the solar spectrum; this cutoff may be used to measure the ozone column from the surface of the Earth.

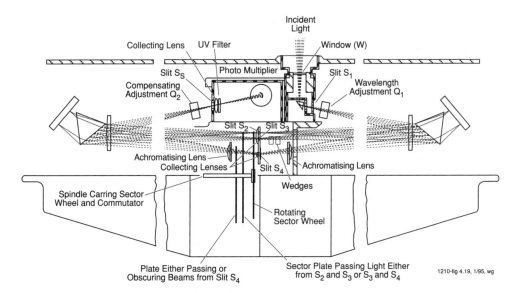

*Figure 11.9.* Optical schematic of Dobson photoelectric ultraviolet spectrometer for measurement of ozone. The rotating sector wheel alternately allows radiation from two nearby wavelengths to fall on the detector. The intensity of longer wavelength is attenuated by a calibrated absorber until the alternating signal is null; the required attenuation provides a measure of the ozone (Dobson, 1968).

A spectrophotometer, shown in Figure 11.9, was developed to measure ozone by its UV spectrum by G. M. B. Dobson, one of the pioneers of ozone research, in the 1920s (Dobson, 1968). This Dobson spectrometer has been used for over 60 years and is the basis of a network of ground-based stations that have given the longest record of ozone measurements. The instrument uses direct or scattered sunlight in the Huggins bands. The instrument operates at pairs of wavelengths, about 20 nm apart, at which ozone absorption differs markedly. Using one pair of wavelengths, a measurement is made that includes differential absorption by ozone, but also differential scattering by molecules and aerosols. Measurements at two or more pairs of wavelengths permit the separation of ozone from other effects and the determination of the column of ozone. The spectrometer is a double quartz prism instrument to reduce scattered light. The first prism separates radiation of the two wavelengths onto two fixed slits, while the second, with reversed dispersion, recombines these two wavelengths onto the same spot on a detector, a photomultiplier in modern instruments. A chopper alternates radiation of the two wavelengths onto the detector, and the radiation of the longer wavelength (which is more intense) is attenuated by a variable wedge until there is no alternating current in the output of the detector. This null measurement technique was necessitated by the limited electronic technology available at the time of the development. The attenuation of the wedge then determines the ratio of signal at this pair of wavelengths. The value outside the atmosphere is determined by plotting the log of the signal versus the path length through the atmosphere and extrapolating to zero atmosphere. Knowledge of the ozone absorption coefficients allows extraction of the ozone column, which is usually reported in Dobson units, one Dobson unit being equivalent to a layer of pure ozone 0.001 cm thick at STP, or $2.69 \times 10^{16}$ molecule cm$^{-2}$.

In the visible and ultraviolet, detection sensitivity is generally limited by the fluctuation in the number of photons received from the source; the inherent instrument noise is relatively small. UV/visible spectrometers usually use photomultipliers for detectors at a single wavelength, or arrays of silicon photodiodes for detection at many wavelengths simultaneously. There are relatively few species that can be measured in this wavelength region, primarily because ozone is such a strong absorber at wavelengths shorter than 300 nm. This limits measurements using astronomical sources to longer wavelengths. In the troposphere, long path measurements can be made using artificial sources; these have been discussed above as in situ measurements. Only a few molecules have substantial absorption in the visible and near UV, but several of these are important in the chemistry of the atmosphere. Probably the most important of these is $NO_2$, which has significant absorption in the blue region of the spectrum, near 430 nm. This absorption has been extensively used for measurements of the stratospheric $NO_2$ from the ground, because radiation in the strongest infrared band of $NO_2$ is completely absorbed by water vapor in the troposphere, although infrared measurements have been used from aircraft and balloons. Measurements are made in direct absorption using the sun or moon as a source, or more commonly, using scattered light at twilight. Looking vertically when the solar zenith angle is slightly greater than 90°, the scattered photons come from the upper atmosphere, which is still illuminated by sunlight rather than from the lower atmosphere, which is in the dark. The horizontal path length through the stratosphere is much longer than the vertical path from the scattering altitude to the ground; so the method is relatively insensitive to tropospheric contamination. Successive spectra as the solar zenith angle increases, that is, as the sun sets, give some information on vertical distribution, although the inversion problem is complicated by multiple scattering and by temporal variation in constituent mixing ratios, especially for $NO_2$, which has a rapid diurnal variation at sunset. The measured atmospheric absorptions are small, and the solar spectrum is not smooth, having strong Fraunhofer lines formed in the solar atmosphere. The solar lines are removed by comparing a solar spectrum taken at high sun, in which the atmospheric lines are negligible, with the twilight spectrum. There is a complication, however, in that the Fraunhofer lines in the scattered spectrum are not as deep as in the direct solar spectrum, a phenomenon known as the Ring effect, which is apparently due to Raman scattering from distant wavelengths and to the fluorescence of aerosols.

## Box 11.1 Measurement of the Vertical Distribution of Ozone

Several methods are available to measure the vertical profile of the ozone concentration. The very first was a small photographic spectrometer that measured the cutoff of the solar spectrum at different altitudes as a balloon ascended. The first widely used technique was a variation of the Dobson method called the Umkehr (a German word meaning "turning back") method (Götz et al., 1934). The UV brightness of the zenith sky is measured as a function of the solar zenith angle. When the solar zenith angle exceeds 90°, the intensity increases because the sunlight is going through the ozone layer at a steeper angle. This curve can be inverted to give a profile, but the vertical resolution is limited to about a scale height. The technique requires rapid, accurate measurements of the brightness at solar zenith angles greater than 80°.

The most common method of measuring the ozone profile is with an ozonesonde, a lightweight ozone-measuring instrument carried aloft on a small balloon, much like a radiosonde. The measuring instrument is an electrochemical device using iodine-iodide redox electrodes to measure the ozone. The sondes contain electrodes in an aqueous solution of potassium iodide. When air containing ozone is pumped through the cell, each ozone molecule releases an $I_2$ molecule from the electrolyte, which is reconverted by the cell to iodide causing two electrons to flow through the cell's external circuit. Thus a measurement of current combined with a knowledge of the amount of air pumped through the cell provides a measure of the ozone concentration.

Two types of ozonesondes, the electrochemical concentration cell (ECC) developed by Komhyr (1969) and the Brewer-Mast sonde (Brewer and Milford, 1960), which differ mostly in the detail of the electrodes, have been used for several decades as part of an international network. The cells must be calibrated and corrections made for temperature and the efficiency of the pump at different pressures.

Optical sondes, which measure the decrease of solar UV radiation due to absorption by ozone, have also been used, primarily on rockets, for investigation of higher altitudes. The faster time response of these devices makes them appropriate for rocket measurements.

Measurements may be made from the ground by means of differential absorption lidar, as described in Section 11.2.3.2. Lidars give the highest vertical resolution, limited by the length of the laser pulse and the integration required to get sufficient signal-to-noise ratios. Lidar measurements of ozone can reach altitudes around 50 km, but these are large, expensive instruments with optical apertures of typically a meter.

Microwave spectrometer measurements can invert the shape of the atmospheric emission line to obtain a vertical profile with resolution 0.5-1 times the scale height (Parrish et al., 1992). Their measurements go to altitudes around 80 km. High-resolution infrared absorption measurements are beginning to be used in a similar way, but because of the larger Doppler linewidth in the infrared, the upper altitude limit is around 30 km.

Finally, measurements of stratospheric ozone profiles are made from satellites with solar occultation or infrared or microwave emission in a limb viewing mode. These methods have good vertical resolution, and the emission experiments get daily global coverage. Clouds prevent measurements much below the tropopause. Measurements of the profile in the troposphere with nadir viewers are difficult because of the necessity of looking through a much larger stratospheric ozone amount.

With care, it is possible to retrieve good values for the stratospheric column of $NO_2$, $O_3$, $NO_3$ (at night, using the moon), and, in the winter polar stratosphere, OClO and BrO (Noxon, 1975; Solomon et al., 1989a,b).

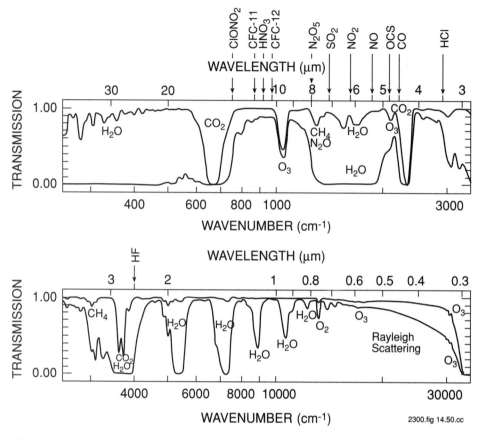

**Figure 11.10.** Infrared spectrum of the atmosphere, showing regions of strong absorption by minor gases in the atmosphere and location of bands of trace gases used for analysis of their atmospheric concentration. The upper and lower lines represent the transmission at altitudes of 11 and 0 km, respectively (courtesy of D. P. Edwards).

Infrared spectroscopy provides a particularly valuable method for atmospheric chemistry because, as noted earlier, almost all molecules have absorptions in this region (Farmer, 1974; Mankin, 1978). Figure 11.10 illustrates the infrared spectral bands with strong atmospheric absorption and the location of absorption regions for a number of trace gases. When used in a limb viewing mode, that is, looking almost horizontally through the atmosphere, the effective path length is large, providing adequate sensitivity for most species with concentration greater than around 100 ppt. In favorable cases, even greater sensitivity is achieved. Infrared spectroscopy can be used in emission or in absorption, using the sun as a source. The latter has greater sensitivity, but is only available at limited times and is not suitable for measuring diurnal variations. Infrared spectroscopy has been used from the ground, from aircraft, from balloons, and from satellites.

The spectral resolving power, $\lambda/\delta\lambda$, required for the observation depends upon the nature of the observation. If the species that one wants to measure is the dominant species in the given spectral region, generally modest resolving power is adequate. But in a spectral region where weak lines of the target species are obscured by much stronger lines of more abundant interfering species, then the spectral resolution $\delta\lambda$ needs to be comparable to the width of the spectral line. In the lower atmosphere where pressure broadening produces lines with a width around 0.1 cm$^{-1}$, a resolving power of 10,000 or so is adequate. But in the upper atmosphere, resolving power of $10^6$ or more may be required for maximum sensitivity.

To obtain adequate signal-to-noise ratios at high spectral resolution, most atmospheric observations are taken using Fourier transform spectrometers. These devices, based on a Michelson interferometer, produce a signal, the interferogram, which is the Fourier transform of the desired spectrum (see Box 11.6). The measured interferogram is transformed numerically on a computer to produce the spectrum. The Fourier transform spectrometer has the advantage over a classical prism or grating spectrometer of observing all the wavelengths simultaneously for the entire period of observation rather than sequentially. The longer observation at a particular wavelength gives an increase in the signal-to-noise ratio; the increase is proportional to the square root of the number of spectral wavelengths measured. For high resolution this increase can be quite large, justifying the additional complexity of the interferometric approach.

Inverting a series of spectra made at different angles through the atmosphere (or different tangent heights, viewed from outside the atmosphere) leads to a profile of the absorber. A great deal of work has gone into the optimum way to do this, but the most common approach used with spectra is the so-called "onion peeling" method. The spectrum at the highest angle or tangent height is used to obtain the absorption in the highest layers. Then, accounting for the changing geometry, the absorption produced by this amount of absorber is computed for the geometry of the next lower observation. Additional absorption is calculated to match the observation and attributed to the next layer down. In this way the retrieval proceeds downward, obtaining the concentration in successively lower layers. More modern methods use maximum likelihood inversions to retrieve the entire profile simultaneously from a series of spectra.

Three different wavelength regions have been used for infrared atmospheric spectroscopy. The near-infrared region, 1-2 $\mu$m, contains overtone vibrational bands and weak electronic bands of many molecules. Because these absorptions violate the first-order selection rules, the bands are generally much weaker than in the middle infrared; for this reason, the near-infrared is seldom used in atmospheric spectroscopy except to measure abundant species such as water, methane, and carbon dioxide.

The middle-infrared region, roughly 2-20 $\mu$m, has been the most useful for atmospheric infrared spectroscopy. Here molecules absorb by changing both their rotational and vibrational energy. The differing vibrational energy moves the absorptions of different species to different wavelengths so that there is less of an interference problem than in the far-infrared. The signals are larger, except in emission at short wavelengths, due to the larger Planck function. It is difficult to use this region in emission at wavelengths short of about 6 $\mu$m because the Planck function falls off exponentially at shorter wavelengths.

Gases such as $CO_2$, $O_3$, $H_2O$, and $HNO_3$ have strong spectral bands that dominate over other species' spectral features for several portions of the infrared spectrum. Thus it is possible to observe, for example, $HNO_3$, by employing an optical filter that isolates a $HNO_3$ spectral band from the rest of the infrared spectrum. An instrument using such a filter is called a *filter radiometer*. It has advantages of simplicity and sensitivity, particularly in an emission instrument, because it observes radiation from more wavelengths simultaneously.

In the far-infrared, wavelengths from around 20 to 1000 $\mu$m, absorptions come from changes in the rotational energy of the molecules. Only molecules with a permanent dipole moment can be observed. Obtaining a high signal-to-noise ratio and resolution in this spectral region is difficult because detectors are less sensitive than at shorter wavelengths, but it is easier to achieve adequate resolution. Absorption by water vapor is strong throughout this region, and it is useful only in the stratosphere where the water vapor concentration is low. Even here, ozone lines interfere with many weaker species. This wavelength region is particularly useful for light molecules such as HCl and OH, which have large permanent dipole moments and rotational constants that produce their absorptions at the higher-frequency end of the region (Johnson et al., 1995a).

The microwave spectral region (0.1-10 cm) has emission and absorption lines similar to the far-infrared in that they arise in pure rotation transitions in polar molecules. The technology of the measurement, however, is entirely different (Parrish et al., 1988). Instead of a wide-band spectrometer, the instrument is a radio receiver that measures frequencies covering only a narrow spectral region, typically encompassing only a single spectral line. The shape of this line is measured with very high spectral resolution. Normally the receiver accepts radiation from a narrow cone a few degrees above the horizon. The acceptance cone is determined by the antenna, a waveguide, which is frequently coupled to a parabolic reflector. The Planck function is so small at these wavelengths and the spectral bandpass is so narrow that signals are minute, but receivers can be made very sensitive. The limiting noise is usually thermal in the mixer, which combines the atmospheric radiation with radiation from a local oscillator to produce a radio frequency signal at the difference in the two frequencies. By cooling the preamplifier and mixer to cryogenic temperatures, the sensitivity can be increased. Nevertheless, it is frequently necessary to integrate the signal for hours to obtain an adequate signal-to-noise ratio.

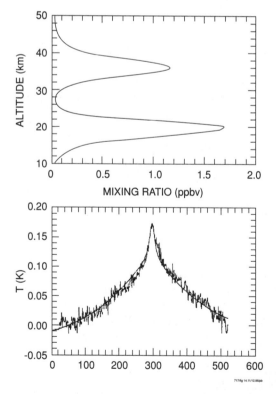

**Figure 11.11.** Microwave atmospheric emission spectrum of chlorine monoxide measured in Antarctica, shown along with the stratospheric ClO profile derived from the spectrum displaying the two-peaked structure in the lower and upper stratosphere. The ordinate of the spectrum is the brightness temperature; if divided by the actual atmospheric temperature, this gives the emissivity of the atmosphere (de Zafra, 1989).

The microwave spectrometer obtains the shape of the spectral line, producing data that can be inverted to obtain a vertical profile. Since the width of the line is proportional to pressure, the wings of the line arise at higher pressures and lower altitude than the central regions of the line. Fitting the observed shape with the shape calculated for a given distribution of emitters enables one to determine the profile in the stratosphere and mesosphere with a resolution of around a scale height (5-10 km). Figure 11.11 illustrates an observed microwave emission line with the fitted line profile (de Zafra, 1989). The peak in ClO mixing ratio at higher altitude comes from gas-phase chemistry of chlorine; the peak at lower altitude is due to the ClO produced by heterogeneous chemistry in the polar winter that produces the Antarctic ozone hole, as described in Chapters 8 and 14.

### 11.2.3.2 Active Remote Sensing

Lidar (an acronym for LIght Detection And Ranging, by analogy with radar) is an optical measurement technique in which a pulse of laser light is transmitted into the atmosphere, usually vertically; the atmosphere scatters a small portion of the light back to a receiver. From measurement of the intensity of the returned light the amount of scattering can be determined, and from the time delay between the transmission of the pulse and the receipt of the scattered return, the distance to the scattering location can be determined. Continuous measurement of the returned scattered intensity permits the scattering amount to be determined as a function of the distance from the lidar. The interpretation of the scattered intensity in terms of the atmospheric information desired may be done in a variety of ways depending upon the nature of the light transmitted and its interaction with the atmosphere.

The most common type of lidar utilizes scattering from aerosol particles in the atmosphere and Rayleigh scattering from atmospheric molecules (Carswell, 1983). Particles have a much larger scattering cross section than molecules, and thus if there are a substantial number of particles, then the scattered return is much larger than the return from the gases alone. In regions with substantial aerosol content, the quantity that is usually determined from the lidar is the "backscatter ratio," or the ratio of the observed backscattered signal to what would have been obtained from molecular Rayleigh scattering alone. To determine the Rayleigh scattering component, it is necessary to know the atmospheric density profile, which can be obtained from the temperature profile and the surface pressure via the hydrostatic equation. Conversely, at altitudes at which the aerosol component is negligible, the variation of backscatter can be used to determine the density profile and hence the temperature profile. This is particularly used for measurements of temperatures in the upper stratosphere and mesosphere, where conventional weather balloons cannot reach. Measurement of the Doppler shift between the transmitted and received signals permits determination of the component of the wind along the line of sight at the scattering volume.

A typical lidar consists of a transmitter with a solid-state laser such as ruby or Nd:YAG (neodymium-doped yttrium aluminum garnet) and a small telescope to project the laser beam upward into the atmosphere, and a receiver with a large telescope (typically 0.3 to 1.5 m in diameter) to collect the return beam, optical filters to reduce background light (vital for use in daytime and useful even in the more common nighttime operation), a photomultiplier detector, and electronics for recording the return signal as a function of time from the transmission of the laser pulse. Usually the transmitter and receiver telescopes are coaxial for good alignment at all ranges. A monitor on the transmitter laser measures the time the pulse is transmitted and the transmitted power. The duration of the laser pulse must be small

## Box 11.2 Analysis of Lidar Signals

The signal returned from the atmosphere when a lidar pulse is transmitted can be written as

$$S(R)\,\Delta R = \frac{E\lambda}{hc}\,\xi(R)\,\frac{A}{4\pi R^2}\,\eta(\lambda)\beta(R)\Delta R \exp\left(-2\int_0^R \epsilon(r)dr\right)$$

where $S(R)\,\Delta R$ is the number of photoelectrons generated from returns at a range (distance) $R = ct/2$, where $c$ is the velocity of light and $t$ is the time since the transmission; $E$ is the energy of the pulse transmitted at wavelength $\lambda$; $\xi(R)$ is the degree of overlap of the transmitted beam and the receiver field of view; $A$ is the collection area of the receiver; $\eta(\lambda)$ is the product of the quantum efficiency of the detector (number of photoelectrons generated per photon striking the detector) and the overall transmission of the optical system; $\beta(R)$ is the backscatter coefficient, equal to the number density of scatterers times the total scattering cross section times an efficiency factor for scattering in the backward direction; $\Delta R$ is the range integrated over, proportional to the integration time; and $\epsilon(r)$ is the extinction of radiation by absorption and scattering along the line of sight. It is seen that the signal falls very rapidly with increasing range, becoming, for reasonable estimates of the parameters, only a few photoelectrons per pulse at a range of 50 km; to get satisfactory signals at such ranges, many thousands of pulses must be averaged.

The pulse energy transmitted is measured for each pulse; the other instrument parameters are determined by calibration. If the extinction is negligible (or can be estimated) the backscattering coefficient can be determined. If only molecular Rayleigh scattering is important, then the scattering cross sections are known and the molecular density determined. If aerosol scattering is important, $\beta$ is usually reported as the ratio of total scattering to molecular scattering, indicating the product of the number density of aerosol particles and their backscatter cross-section.

In the case of a differential absorption lidar or DIAL, measurements are made at two wavelengths, usually close together, at which the target molecule has strongly different absorptions. Using subscripts to indicate parameters for the two wavelengths, the ratio of the two returned signals is

$$\frac{S_2}{S_1} = \frac{E_2}{E_1}\,\frac{\lambda_2}{\lambda_1}\,\frac{\eta_2}{\eta_1}\,\frac{\beta_2}{\beta_1}\,\exp\left(-2\int_0^R [(\sigma_{s_2} - \sigma_{s_1})\,N_t + (\sigma_{a_2} - \sigma_{a_1})\,N_a]\right)dr$$

where the extinction has been written as the sum of a scattering term, $\sigma_s N_t$, and an absorption term, $\sigma_a N_a$. All molecules and aerosols contribute to the scattering but only the target molecule to the absorption. Then

$$\frac{d}{dR}\ln\left(\frac{S_2}{S_1}\right) = \frac{d}{dR}\ln\left(\frac{\beta_2}{\beta_1}\right) - 2(\sigma_{s_2} - \sigma_{s_1})\,N_t - 2(\sigma_{a_2} - \sigma_{a_1})\,N_a$$

Assuming that the ratio of backscatter $\beta_2/\beta_1$ at the two wavelengths does not vary with range (which it may as the ratio of aerosol to molecular scattering changes with altitude), then we can solve for the absorber density

$$N_1 = \frac{-1}{2(\sigma_{a_2} - \sigma_{a_1})}\frac{d}{dR}\ln\left(\frac{S_2}{S_1}\right) - \frac{(\sigma_{s_2} - \sigma_{s_1})}{(\sigma_{a_2} - \sigma_{a_1})}\,N_t$$

The second term can be made small by choosing wavelengths close together so that $\sigma_{s_1}$ and $\sigma_{s_2}$ are nearly equal, but at which $\sigma_{a_1}$ and $\sigma_{a_2}$ are quite different.

enough to obtain the desired range resolution; a 1 μs pulse gives a resolution of 150 m. The pulses are repeated as rapidly as the laser can be fired; 10-100 pulses per second is common. The electronics to record the signal must be able to record the value with a large dynamic range at a very high data rate, determined by the desired range resolution. Signals from many individual laser shots, sometimes millions, are averaged to produce a usable signal-to-noise ratio in the final result. Because the magnitude of the returned signal varies so much with range, the photomultiplier may saturate at small ranges. This may be dealt with in several ways. Sometimes the transmitter and receiver telescopes are offset from each other so that their fields of view do not overlap until the range is large enough to avoid saturation. A mechanical shutter can be phased to the pulse transmit time so that it reduces the signal for small range. Sometimes the return signal is split between two detectors with different sensitivities; the less sensitive detector is used for small ranges and the more sensitive for large ranges.

Starting from a Nd:YAG laser, pulses may be generated at several wavelengths by frequency multiplication and using the pulses to pump dye lasers or Raman scattering cells. The output of the dye lasers may again be doubled. In this way, it is possible to generate pulses at several wavelengths in the visible and ultraviolet in order to obtain a wavelength at which the scattering cross sections are larger. If two pulses have wavelengths at which aerosol and molecular scattering are similar, but at which some gas such as ozone has a markedly different absorption coefficient, the returns will be modified by the absorption by the gas. This technique, called DIAL (DIfferential Absorption Lidar), permits measurement of the density of the absorbing gas along the path, and thus vertical profiles of mixing ratio of the gas may be obtained from an instrument on the ground or in an aircraft. This technique has been applied principally to measurements of stratospheric ozone, but may also be used for other gases. An airborne ozone DIAL is shown schematically in Figure 11.12.

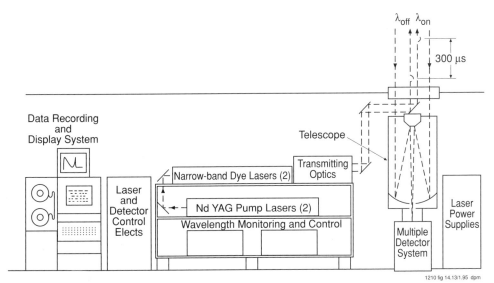

**Figure 11.12.** *Airborne DIAL system used for measuring scattering by molecules and aerosols in the stratosphere and attenuation of scattered light by ozone (after Browell, 1989).*

DIAL has been used widely both from the ground and from aircraft for measurements of the vertical profile of ozone. DIAL is particularly valuable in obtaining profiles with good vertical resolution.

## Table 11.1
*In Situ and Remote Sensing Methods for Measuring Important Atmospheric Species.*

| Compound | In Situ Methods | Remote Sensing Methods |
|---|---|---|
| $H_2O$ | Frost point hygrometer, Lyman-$\alpha$ absorption, TDL | IR, microwave spectroscopy, Raman lidar, filter radiometry |
| $CO_2$ | GC, IR gas correlation | IR spectroscopy, filter radiometry |
| CO | Gas correlation, chemical conversion | IR spectroscopy, gas correlation radiometry |
| $CH_4$ | GC, TDL, gas correlation | IR spectroscopy, filter radiometry |
| Nonmethane hydrocarbons | GC-MS | IR spectroscopy (limited number of compounds) |
| $O_3$ | UV absorption, chemiluminescence | UV, IR, microwave spectroscopy, lidar |
| $N_2O$ | GC, TDL | IR spectroscopy, radiometry |
| NO | Chemiluminescence | IR spectroscopy |
| $NO_2$ | Photolysis + chemiluminescence | visible spectroscopy, IR spectroscopy |
| $HNO_3$ | TDL, ion chromatography, filter and wet chemistry | IR spectroscopy, filter radiometry |
| PAN and organic nitrates | GC | IR spectroscopy (PAN) |
| $N_2O_5$ | | IR spectroscopy |
| Halogenated hydrocarbons | GC-electron capture detection | IR (more abundant species) |
| HCl, HF | TDL | IR spectroscopy |
| $ClONO_2$ | | IR spectroscopy |
| Cl, ClO | Resonance fluorescence | Microwave spectroscopy |
| OCS | GC, TDL | IR spectroscopy |
| $SO_2$ | Ion chromatography, chemiluminescence | UV, IR spectroscopy |
| DMS, $CS_2$ $H_2S$ | GC | |
| OH | Resonance fluorescence, laser-induced fluorescence, chemical ionization mass spectroscopy, radioisotope chemistry | Lidar, UV spectroscopy, DOAS, far-IR spectroscopy |
| $HO_2$ | Radical amplifier, laser-induced fluorescence | Far-IR spectroscopy |
| $CH_2O$ | GC, TDL | IR spectroscopy |
| $H_2O_2$ | High-performance liquid chromatography | Far-IR spectroscopy |
| $O_2$, $N_2$, $H_2$ Ar, Ne, He | Mass spectroscopy | |

Note: The list above is not intended to be exhaustive, but illustrates the more common methods of measuring the species in question. Other methods may be more relevant in some circumstances.

Scattering from the atmosphere is not necessarily elastic; that is, the scattered photon may have a different energy and wavelength than the incident photon. This may be used to distinguish the molecule that scattered the photon, another lidar technique for chemical analysis. One such mechanism is Raman scattering, where the scattered photon is at a frequency lower than the incident frequency by the vibrational frequency of the scattering molecule. Since this Raman shift is characteristic of the scattering molecule, the lidar signal is proportional to the concentration of the target molecule. In this way, one can obtain the density of $N_2$ independent of aerosol, and, via the hydrostatic equation, the temperature. It can also be used to determine the profile of a specific chemical. The Raman scattering cross section, however, is much smaller than the Rayleigh cross section; so the signals are much weaker. This limits the usefulness of the chemical Raman lidar to fairly abundant species such as water and methane and relatively small ranges. A variant of this technique that improves the sensitivity is to transmit at a frequency at which the absorption by the target molecule is resonant and to look for resonant scattering or fluorescence of the molecule. Resonant scattering cross sections are so large that very small concentrations can be detected. This approach has been applied to observe the sodium layer in the mesosphere (Mégie and Blamont, 1977; Gardner, 1986).

In the preceding discussion, a number of experimental methods were described for measuring the concentration of specific chemical compounds. Most of these methods can be used for a variety of compounds, although some are quite specific. Most chemicals can be measured in more than one way. Table 11.1 summarizes the more important methods of measuring various compounds or classes of compounds, both in situ and by remote sensing.

## 11.3 Flux Measurements

In many cases, not only is the concentration of a chemical species in the atmosphere of interest, but also the rate with which it is changing. This depends upon both the reactions that produce or remove the chemical within the volume of air and upon the transfer of the substance into or out of the atmospheric volume. This flux could, for example, be produced by emission of methane by a swamp, emission of terpenes by leaves of plants, or destruction of ozone on contact with the ground. Fluxes are measured by a variety of techniques (Dabberdt et al., 1993); the relation of the measured quantity to the desired flux depends upon the structure of atmospheric turbulence, and so an understanding of turbulence structure is necessary for some techniques.

The simplest method of measuring flux is the enclosure method. The surface where the chemical species is produced is covered with a chamber. As the compound is released at the surface, the concentration of the chemical in the chamber will increase with time. Hence by measuring the concentration at different times the flux can be estimated. Of course, one must make sure that the chemical species in question is not destroyed on the walls of the enclosure. Any of the methods described earlier can be used for the concentration measurement as appropriate. The enclosure can be a bag placed over a leaf or branch of a tree or a box covering a known area of the ground. Figure 11.13 shows a flux enclosure on a plant leaf. The greatest difficulty with the enclosure method is that the measurement may itself affect the measured quantity, so that the results are not representative of the situation in the free atmosphere. For instance, the buildup of the concentration of the chemical in the enclosure may inhibit the emission process. This problem is reduced by using a flowing enclosure, where ambient air flows through the chamber and the concentration of the target species is measured at the input and at the output. The analytical technique must be sensitive if the difference in the input and output concentration is small, as is desirable for minimizing feedback effects. The enclosure may also absorb some of the incident radiation, changing the radiation environment of the surfaces; this can be especially serious in the case of photosynthesizing plants, where chemical activity is sensitive to radiation. Temperature rise in the enclosure due to the suppression of convection can also modify fluxes. There is also a sampling problem: How can one be sure that the branch or surface sampled is representative of the entire system studied? Chambers can vary from simple bags enclosing a branch to flowing chambers with radiation and temperature control. Teflon film is often used as the enclosing material for plant

**Figure 11.13.** Flux chamber for measuring the emission of nonmethane hydrocarbons from leaves (courtesy of A. Guenther).

studies because it is inert, has little out-gassing, and is transparent to photosynthetic radiation. For trace gases such as methane where radiation effects are not important, stainless steel chambers can be used. Many chambers incorporate a fan to keep the air well mixed.

Large-scale flux measurements (measurements averaged over a large area) are best done with micrometeorological techniques. The vertical flux is the average over time and area of the product of the vertical velocity and the mixing ratio of the species. In other words, there is an upward flux if the air parcels moving upward have a higher concentration, on average, of the species in question than those moving

**Figure 11.14.** The ASTER facility for measuring fluxes of heat, moisture, and chemical species by micrometeorological effects. Meteorological data come from sonic anemometers, propeller anemometers, fast response resistance temperature sensors; chemical data come from a variety of sensors. Auxiliary data on radiation, precipitation, and other factors are also measured (courtesy of T. Horst).

downward. With fast instruments for measuring both vertical wind velocity and chemical concentration, the flux may be calculated directly. The instruments must be fast enough to respond to the smallest turbulent eddies that carry significant flux, while being sensitive enough to measure the small differences in concentration in the upward and downward flows. This is a challenging combination. Figure 11.14 shows a field deployment of the Atmosphere-Surface Turbulent Exchange Research (ASTER) facility (Businger et al., 1990) for measuring fluxes by eddy correlation. The velocity is usually measured with a sonic anemometer, which measures the difference in the velocity of sound upward and downward. Instruments that can measure the concentration of a chemical species very rapidly and precisely are not common; they tend to be large and expensive. It is possible to take a sample in a very short period of time, use a larger time to analyze it, and multiply it by the instantaneous value of the vertical velocity and obtain the flux, provided the measurements are made at intervals greater than the time over which a variable is correlated with itself.

The eddy correlation method just described is the most direct way to measure fluxes, but is not always feasible. Other techniques have been developed to circumvent the requirement for rapid response chemical instruments. One method is conditional sampling. Samples are collected over an extended period of time and analyzed by a conventional method. A particular sample represents air determined by some property of the vertical velocity, for instance, in eddy accumulation, one sample contains only air that had a positive vertical velocity; a second sample contains only air that had a negative vertical velocity. By measuring the difference in the concentration of the target species in the two samples, it is possible to estimate the flux.

Fluxes are also estimated by the gradient method; in the surface layer, micrometeorological theory predicts a unique relation between the flux and the vertical gradient of species concentration when the source or sink is at the surface. The relationship depends upon meteorological conditions. By measuring the concentration at two heights, say, 1 and 10 m, the flux may be determined. For example, the gradient of two species may be determined, the flux of one of which can be determined directly. The ratio of the fluxes of the two species is the same as the ratio of their gradients (with certain assumptions). For species with long lifetimes, the gradient can be quite small, and high precision is needed in the measurement.

For point measurements of fluxes to be meaningful in terms of ecosystems, the area where the measurements are made must be uniform over a significant region upstream, since the turbulent eddies that transport the flux also bring in quantities of the chemical species from upwind by horizontal advection. A rule of thumb is that the area should be uniform upwind for a distance of 100 times the height of the measurement, although under stable conditions, a larger fetch may be required.

### Box 11.3 Measurement of Water Vapor

Water vapor absorbs or releases large quantities of heat when it evaporates or condenses; this heat is important in driving motions of the atmosphere. For this reason, water in its various phases is an important meteorological quantity. It is also the most important greenhouse gas. Water vapor, however, is also an important chemical constituent, as it serves as the source of OH radicals.

The chief difficulty in measuring water vapor is the large range of densities encountered. Near the surface in the tropics, the mixing ratio of water can be up to 5%, whereas in the coldest part of the tropical tropopause, it may be as low as 3 ppm. In situ instruments, in particular, are subject to errors caused by contamination of the instrument by water vapor at low altitudes.

Capacitance or resistance sensors are used at the surface and on radiosondes for measurement of water vapor in the troposphere. A hygroscopic material is used as the dielectric in a capacitor of the conducting material in a resistor; the material absorbs water, the amount of which is dependent on humidity and temperature. The dielectric constant or conductivity varies with the amount of absorbed water, and hence an electrical measurement of the capacitance or resistance indicates the humidity.

A method commonly used for measurement of water vapor in the lower atmosphere is the determination of the temperature at which the water vapor present in the atmosphere would represent the saturation vapor pressure. One method of making this measurement is with a psychrometer, a thermometer covered by a piece of cloth wet with water. The water evaporates, cooling the thermometer. When the cloth-covered thermometer reaches the temperature at which saturation occurs, there is no net evaporation and no further drop in temperature occurs. From a table of vapor pressure, one computes the vapor pressure of water in the atmosphere and hence the water mixing ratio.

More precise measurements are made optically. A mirror is cooled until dew or frost condenses on its surface. The presence of the dew or frost is accurately detected by a drop in the apparent reflectivity of the mirror, due to scattering by the layer of water. The temperature of the mirror when the vapor first condenses is the temperature at which water vapor in the atmosphere is saturated. Again, tables of vapor pressure versus temperature give the absolute humidity.

Optical absorption measurements are also used to measure water vapor. Water vapor has numerous absorption bands in the near-infrared; tunable diode lasers may be used to measure individual absorption lines in these bands. By choosing lines of different strength and by using different path lengths, the sensitivity of the instrument may be chosen to match the concentration of water vapor to be measured.

The far-ultraviolet region is also used in absorption. Specifically the Lyman-$\alpha$ line of hydrogen is used as the radiation wavelength because, although oxygen is a strong absorber at such short wavelengths, the Lyman-$\alpha$ line occurs in a small window in the oxygen spectrum where oxygen absorption is low and water absorption is high. Generation and detection of this wavelength is relatively easy, and usually only a short path through the atmosphere suffices for the measurement, resulting in an instrument that is compact and fast.

Absorption of Lyman-$\alpha$ radiation results in the dissociation of the water and forms OH radicals in an electronically excited state. The water vapor may be measured with high specificity by measuring the fluorescence emitted by these radicals at 308 nm. While some of the radicals decay radiatively, others are quenched by collisions with other molecules, a process proportional to the atmospheric density. Since the number of radicals produced is proportional to the water vapor density, the quenching results in a signal proportional to water vapor mixing ratio rather than to density; this is an advantage in an instrument designed to operate over a large range of ambient pressure.

Remote sensing of water vapor most commonly uses the infrared absorption bands, especially the $\nu_2$ band near 6 $\mu$m and the far-infrared pure rotation band, or microwave spectroscopy using one of the pure rotation lines. Some satellite experiments also use absorption of sunlight in one of the near-infrared bands. Microwave spectroscopy operating from the ground utilizes the core of the atmospheric emission line originating above about 20 km, because the pressure-broadened line from lower in the atmosphere is too wide for the receiver of the spectrometer. The emission from the lower atmosphere merely provides a broad continuum on which the observations rest.

## 11.4 Measurements of Atmospheric Radiation

Atmospheric radiation from all directions contributes to the photolysis rate of a molecule; so all downwelling (direct, diffuse, and reflected) as well as upwelling (albedo) radiation needs to be considered. Measurements of photolysis rates must be carefully planned to be representative of the true atmospheric photolysis rates. The two principal experimental techniques used to determine atmospheric photolysis rates are chemical actinometers, which measure the actual photolysis, and radiometers, which measure the incident radiation from which the photolysis rate may be calculated.

Chemical actinometers have been used to measure photolysis rates of $NO_2$ and $O_3$. Some photolysis rates of organic nitrates have also been measured. A schematic diagram of a flowing chemical actinometer is shown in Figure 11.15. It consists of an apparatus to generate a known flow, pressure, and concentration of the gas of interest, a photolysis cell of known volume for exposing the gas to ambient sunlight, and a way of detecting one of the products of the photolysis. Post-photolysis reactions in the gas stream of the species to be detected must be minimized or accounted for.

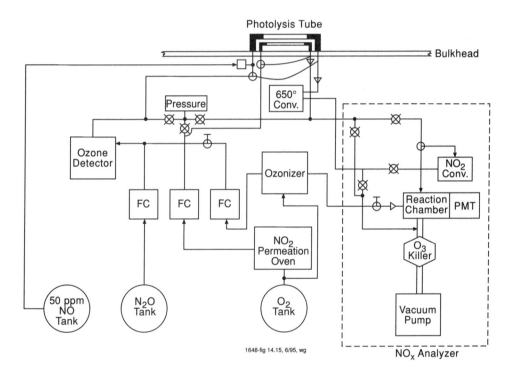

**Figure 11.15.** Schematic of an airborne chemical actinometer for measuring rate of photolysis of $NO_2$ and $O_3$. The photolysis tube is outside the airplane, receiving radiation from the whole sky; the NO produced is measured by the $NO_x$ analyzer. FC indicates flow controller and PMT is the photomultiplier tube (Dickerson and Stedman, 1980).

Photolysis cells are generally cylindrical quartz tubes mounted over a non-reflecting surface, giving a hemispherical field of view. With cancellation of internal and external reflections of incident light and a small correction for tube end cross section, cylindrical tubes can closely approximate true hemispherical light collectors if situated above objects that could occlude or reflect ambient light. In an actinometer with a properly designed photolysis cell, a molecule of gas will respond exactly the same to a photon of light from 0 or 90° zenith angle, as would a molecule in the atmosphere. This is particularly important for measurement of photolysis rates because, depending on the wavelength range of the photodissociation, a large portion of the effective radiation can come from diffuse or scattered radiation.

In $NO_2$ photolysis measurements

$$NO_2 + h\nu \rightarrow NO + O(^3P) \qquad (11.8)$$

knowing the starting $NO_2$ concentration, gas flow, pressure, cell volume, and determining the concentration of the NO produced, one can calculate the photolysis rate by Eq. (11.9), where [NO] is the concentration of the NO produced, $[NO_2]_0$ is the initial $NO_2$ concentration, and $\Delta t$ is the photolysis time calculated from the flow rate, cell volume, and cell pressure (Shetter et al., 1992).

$$j_{NO_2} = [NO]/([NO_2]_0 \Delta t) \qquad (11.9)$$

The sensitivity of the $j_{NO_2}$ actinometer is calibrated directly by the addition of NO calibration gas to the photolysis cell gas stream. Background signal levels are determined by covering the photolysis cell to occlude any ambient radiation.

In the measurement of $O_3$ photolysis to $O(^1D)$, it has not proven feasible to measure $O(^1D)$ directly; so $O(^1D)$ is detected indirectly by its reaction with $N_2O$ and subsequent detection of one of the products of the sequence of reactions (Shetter et al., 1996):

$$O(^1D) + N_2O \rightarrow N_2 + O_2 \qquad (11.10)$$
$$\rightarrow NO + NO \qquad (11.11)$$
$$NO + O_3 \rightarrow NO_2 + O_2 \qquad (11.1)$$
$$NO_2 + O_3 \rightarrow NO_3 + O_2 \qquad (11.12)$$
$$NO_2 + NO_3 + M \rightarrow N_2O_5 + M \qquad (11.13)$$

The photolysis rate of ozone is more complicated than $j_{NO_2}$ because UV radiation is scattered more strongly than visible radiation; 60% of the effective radiation for $j_{O_3}$ comes from diffuse light. In addition, the overhead ozone column attenuates the UV light, which reduces $j_{O_3}$, while not affecting visible light. In order to compare $j_{O_3}$ measurements from different locations, ozone column measurements are required.

Measurements of the flux of radiation can be made with a spectroradiometer. These measurements, combined with laboratory measurements of the absorption cross section of the molecule and the quantum efficiency for producing the desired product, can be used to compute photolysis rates, from the defining equation [Eq. (3.16)]. The spectroradiometer must be calibrated to produce absolute flux measurements. Filter radiometers have been used to determine $j_{NO_2}$ and $j_{O_3}$. A typical instrument is composed of three parts, a light receiver, a bandpass filter, and a phototube. Ambient radiation in the wavelength interval admitted by the filter is proportional to the photolysis process at those wavelengths. Spectroradiometers must be absolutely calibrated with an actinometer to determine the proportionality constant. Some instruments use a scanning double monochromator instead of a filter to provide detailed information about the wavelength dependence of the radiation. Most commercial units employ a flat plate receiver, which has a response that is proportional to the cosine of the solar zenith angle. Under clear sky conditions, the photolysis will depend on the cosine, but when clouds or aerosols are present, the angular distribution of the ambient radiation is a function of local cloud and aerosol conditions. Some spectroradiometers attempt to correct for this by using a hemispherical dome to collect radiation equally from the upper $2\pi$ steradians.

## 11.5 Instrumentation for Aerosol and Cloud Measurements

Aerosols have many effects upon atmospheric chemistry. They absorb soluble chemicals, provide surfaces for the occurrence of heterogeneous reactions, and alter the radiation environment. There are two main classes of measurements of aerosols, the size distribution and number density of the aerosols, their chemical composition and physical state. The former are usually determined from the optical scattering by the aerosols. Extremely small particles, those with radii less than 0.01 $\mu$m, are called condensation nuclei (CN) because they have little direct effect except to serve as a nucleus on which gases condense to form larger particles; CN are measured by exposing the sample to a supersaturated vapor. The vapor deposits on the CN, causing them to become large enough to be detected optically.

Larger aerosols are usually measured optically, by the amount of light that they scatter and the directionality of the scattering for smaller particles (0.1-1 $\mu$m radius) or by measuring the size of their shadow for larger particles (Dye and Baumgardner, 1984). Figure 11.16 illustrates the principle of an aerosol spectrometer based on measurement of optical scattering. The airstream containing the aerosols flows through a region with a focussed laser beam. Optical detectors measure the scattered radiation or the absence of radiation in the shadow region to size individual particles. The results are summed into size bins over the observation time to give a particle size spectrum (the number of particles per unit volume in each size range).

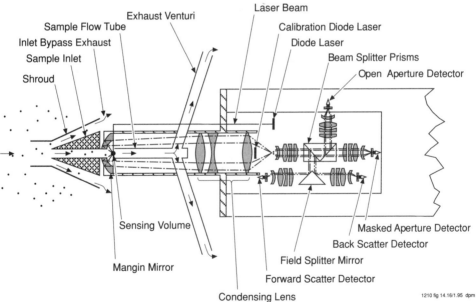

**Figure 11.16.** Aerosol scattering spectrometer for measuring the size distribution of atmospheric aerosol particles by their differential scattering cross sections (Baumgardner et al., 1994).

Particles are also collected to determine their composition by chemical analysis or to determine their shape by electron microscopy. Filters collect particles of all sizes above some minimum. The collected particles may all be lumped together to get a sample large enough for chemical analysis. Individual particles are sometimes collected on wires or grids for microprobe analysis. Particles may be sorted into size ranges and only particles in a given range subjected to an analysis. This is done using the fact that small particles can be accelerated by the airflow more readily. Passing the airflow along a curving path, the larger particles impact the walls and are collected, while the smaller ones pass on with the air. Varying the speed and curvature of the airflow allows collection of different size ranges.

## 11.6 Observing Platforms
### 11.6.1 Ground-Based Measurements

Instruments require some sort of platform on which to work, a support that supplies physical mounting, environment control, and electrical power. For atmospheric measurements, the platform may be a building, a truck or trailer, or a site in the open air sitting on the ground itself. A building is the easiest platform from which to work. A building can provide protection of the instruments from the elements and a stable environment with temperature, humidity, and conditions under which the instrument works best. Ample electrical power is usually available; air or water cooling can be provided. Normally, however, the interest is in the atmosphere at some specific location, usually not at the investigator's laboratory. Thus most measurements are made "in the field." A high-altitude site is frequently useful for the reduction in water vapor above the instrument and for avoiding pollution in the boundary layer. Even if a building is not available, some temporary shelter is usually used. To sample the atmosphere at some small distance above the surface, a tower may be erected; these are frequently 5-15 m high. Depending upon the type of instrument, the entire instrument may be placed on the tower, or just an inlet that samples the atmosphere and carries it to an instrument at the base of the tower. Instruments or samplers may also be used at higher altitude above a fixed location by the use of a tethered balloon. These are usually small, aerodynamically shaped balloons that can lift a few kilograms. The altitude is limited by the balance between the lift of the balloon and the weight of the line to hold the balloon in place, as well as potential interference with air traffic. Tethered balloons have been used to loft small payloads as high as 1000 m, although 100 m is more typical.

### 11.6.2 Aircraft and Balloon Measurements

The atmosphere is three dimensional and in situ instruments located at the surface cannot sample the vertical dimension. Even remote sensing instruments frequently need to be moved above the surface for optimum operation. It is also very important in many situations to sample over a horizontal extent. In all these cases, the instrument must be mounted on a mobile platform that can take the instrument to the location where the measurements must be made. One of the most versatile of these mobile platforms is the airplane. Airplanes can cover large areas fairly quickly, can provide a height range through most of the troposphere or sometimes higher, and can carry loads ranging from fairly small, perhaps 100 kg, up to large coordinated instrument packages weighing several tons. They can provide modest to large amounts of electrical power. Light planes can carry small packages at altitudes up to about 6 km, and can cruise at 250 km/h for several hours. Turboprop planes come in a wide variety of sizes from small two-engine aircraft designed for business use, which can carry 100-200 kg up to 7 km with a range of 1000-4000 km, to large four-engine transports that can carry many tons and can cruise for up to 10 h at altitudes from just above the surface to ceilings of typically 8 km. Turbine engines offer an additional advantage for many chemical experiments; the compressed gas can be used to pump large flows through instruments. Again, jet aircraft range in size from small business jets with payloads of a few hundred kilograms to large airliners or cargo aircraft with payloads of many tons. Typically, they have ceilings of 10-12 km and ranges of 2000 to 10000 km. Higher ceilings can be achieved by specialized aircraft such as the ER-2 or B-57 with a ceiling of around 20 km. For example, many of the critical observations made in the Antarctic and Arctic polar vortices were made from the ER-2 (see Fig. 8.10). The instruments must be automated. During the flight the experimenter cannot have direct hands on access to an instrument located in a cargo bay. Remotely piloted vehicles (drones) that can be operated from the ground are being developed for atmospheric research. These may be able to carry small payloads to even higher altitudes or to achieve longer endurance.

Airplanes make it easy to move the instrument over large horizontal distances, but are limited in their vertical range. They can cover the entire troposphere and parts of the lower stratosphere, but aircraft instruments for the middle or upper stratosphere or higher regions must be remote sensing instruments. Even remote sensing instruments in many cases find it difficult to sound the upper atmosphere because of the interposed, much denser, lower atmosphere.

To achieve greater altitude capability, experimenters may use balloons or rockets. Scientific balloons are typically large, zero-pressure (pressure inside the balloon is equal to the ambient pressure outside) balloons made of thin plastic films. The payload capacity is the product of the volume of the balloon and the difference between the density of the helium in the balloon and the air outside, minus the weight of the balloon itself and its associated equipment. Large scientific balloons are usually $10^5$-$10^6$ m$^3$ in volume and can lift payloads of around 1000 kg to altitudes of 30-45 km. The instruments, of course, have to be fully automatic and not sensitive to low pressure. The temperature of the payload usually is quite cold when passing through the tropopause, but at altitude may become quite warm in daylight, but cold at night. Thermal control has to be designed carefully. Launch procedures for large balloons can be elaborate; good weather, especially low winds, is needed. Such balloons cannot be launched everywhere. Most are launched in the southwestern United States, southern France, or northern Sweden. Other sites are possible, but are more difficult. Consideration must be given to avoiding interference with aircraft and also to recovery locations for the payload, which is returned to Earth by parachute. Flight durations of 24 h or so are possible only when the stratospheric winds are low, because the balloon may drift too far to recover when the winds are 100 km/h or more. The low winds occur (over the southern U.S.) in May, when the winter westerlies give way to summer easterlies and in September when the reverse occurs.

Smaller balloons also have a role in atmospheric chemistry. They are used to launch small payloads, frequently disposable, to measure profiles with in situ instrumentation on ascent or descent. They usually do not float at altitude for long periods. These are used for measuring temperature and humidity routinely (radiosondes) and with electrochemical cells for measuring ozone concentration (ozonesondes). They have also been used for measuring stratospheric aerosol profiles of number density and size distribution, and for collecting whole air samples.

Balloons are seldom used above 45 km, because the low air density limits the payloads even for large balloons. For higher-altitude sampling, sounding rockets may be used. These rockets usually carry payloads of a hundred kilograms or so to heights of a few hundred kilometers. They may be used for remote sounding instruments for the few minutes they are above the atmosphere, or they may have an in situ instrument that samples the atmosphere as the payload descends through the atmosphere on a parachute.

**Box 11.4 Networks for Global Change Observation**

To detect and determine the nature and magnitude of changes in the chemistry of the atmosphere that produce global change requires careful measurement over a long period of time, as many of the rates of change are small and it is important to determine the change before it has accumulated to a large value. Measurements must be made at widely varying places around the globe, away from local influences. Good accuracy is important in the measurements, but good precision (the ability to get the same answer from a constant input day after day) is even more important. Two of the networks that have been important in tracking the concentration of trace gases near the Earth's surface are the ALE/GAGE (Atmospheric Lifetime Experiment/ Global Atmospheric Gases Experiment) and the NOAA CMDL (Climate Monitoring and Diagnostics Laboratory) network.

Anthropogenic and biogenic trace gases have been measured since 1978 by the ALE/GAGE network (Prinn et al., 1983). The global network program provides gas chromatographic measurements of atmospheric concentrations of methane, nitrous oxide, several chlorofluorocarbons, and other chlorine-containing compounds. These gases are important both as greenhouse gases and as contributors to the stratospheric chemistry, especially the chlorine budget. The recently initiated Advanced GAGE uses a new fully automated system designed by the Scripps Institution of Oceanography containing a custom-designed sample module and gas chromatographic instruments.

Data are collected at monitoring sites at Cape Grim, Tasmania; Port Matatula, American Samoa; Ragged Point, Barbados; Mace Head, Ireland; and Cape Meares, Oregon. The sites are chosen to represent both the Northern and Southern Hemispheres, and to have clean, maritime air with minimal influence of local sources. While sharing similar goals with ALE/GAGE, the CMDL conducts research particularly related to atmospheric constituents that are capable of forcing change in the climate of the Earth through modification of the atmospheric radiative environment, and those that may cause depletion of the global ozone layer. CMDL accomplishes this goal primarily through long-term measurements by a variety of techniques of key atmospheric properties such as the concentrations of carbon dioxide, carbon monoxide, methane, nitrous oxide, surface and stratospheric ozone, halogenated compounds including CFC replacements, aerosols, and solar and infrared radiation at sites spanning the globe. Through these measurements, CMDL documents global changes in the key atmospheric species, which are all affected by mankind, identifying sources, sinks, and interannual variability. CMDL provides data that are used to assess climate forcing and ozone depletion, to develop and test predictive models, and to inform policy makers and the public.

Measurements of stratospheric change are made by the Network for Detection of Stratospheric Change using a variety of remote sensing instruments. The major goal of the network is to provide the earliest possible detection of changes in the stratosphere, and the means to understand them. In the shorter term, it provides data on temporal and spatial variability of stratospheric composition and structure as well as providing ground-truth and complementary measurements for satellite systems. Five primary stations spanning the latitude range from the Arctic to the Antarctic use a common set of instruments to measure profiles and column amounts of ozone and a number of key species involved in the ozone chemistry. Techniques used include lidars for measuring temperature, pressure, aerosol content, and tropospheric and stratospheric ozone, UV-visible spectrometers and Fourier transform infrared spectrometers for column measurements of numerous species, and microwave spectrometers for measuring the vertical profiles of a few species including ClO. Additionally, UV spectroradiometers measure the ultraviolet irradiance at the ground. These primary stations are supplemented by a number of complementary sites with smaller sets of instruments of less continuous coverage. The goal is to maintain the dataset continuously for two solar cycles and to use careful calibrations and instrument and data intercomparisons to ensure that any trends detected are real. The stations are supported and operated by scientists from over a dozen nations under the guidance of an international Steering Committee.

Information about observing networks and data from many of them, as well as other observations, can be obtained from the Internet.

## 11.6.3 Satellite Measurements

Satellites are used to obtain global coverage of observations. Obviously, satellite experiments must use remote sensing techniques. To date, satellite experiments for atmospheric chemistry have been passive remote sensing, using radiation emitted or scattered by the atmosphere, or measuring radiation from the Sun or a star that has traversed various layers of the atmosphere. Radars have been used in space, and lidars are currently being developed for satellites, but these are primarily for fields other than atmospheric chemistry.

A variety of techniques have been used to sense the Earth's atmosphere from space. Instruments look along limb (horizontal) viewing paths, nadir (downward) viewing paths, or a hybrid combination of viewing geometry, for both occultation and emission experiments. In an *occultation experiment*, the sun or a bright star is observed as the line of sight from the object to the viewing instrument passes through the atmosphere. In an *emission experiment*, the radiation emitted by the atmosphere is observed. In a scattering experiment, solar radiation scattered within the atmosphere is observed. Instruments have probed the atmosphere using ultraviolet, visible, infrared, and microwave wavelengths.

The geographic coverage of a satellite instrument depends upon the orbit as well as upon the orientation of the instrument with respect to the orbital plane. For satellites in low Earth orbits, the orbital period is typically 90-100 min; so the rotation of the Earth means that successive orbits move the observation point around 24° of longitude. The plane of the orbit is not fixed in inertial space, but precesses due to the gravitational force from the nonspherical Earth and from effects of the Sun and moon. By adjusting the plane of the orbit, it can be arranged so that the period of precession is one year; in this way the precession just keeps up with the movement of the Earth in its orbit around the Sun and the orbit maintains a constant relation to the Sun. In this Sun-synchronous mode, the observation point crosses a given longitude at the same two local times each day. Observations can be made at, for example, local noon and local midnight around the orbit. Many of the atmospheric experiments use this mode so that geographical variation can be distinguished from diurnal variation.

Since the variety of satellite instruments that have been used or are under development for the study of atmospheric chemistry is too large to discuss comprehensively, we will illustrate the range of types of instrumentation by four diverse examples of instruments for remote sensing of atmospheric chemistry that have operated successfully in space. They are chosen to illustrate the use of a variety of wavelengths and sounding techniques for measuring a variety of species in different parts of the atmosphere. Box 11.5 describes the Upper Atmosphere Research Satellite (UARS), a satellite with nine instruments for a coordinated study of the chemistry, dynamics, and energetics of the upper atmosphere.

The first of these is ATMOS, the Atmospheric Trace Molecular Spectroscopy Experiment (Farmer, 1987). This is a high-resolution Fourier transform spectrometer (see Box 11.6), shown in Figure 11.17, that has flown on four missions on the Space Shuttle between 1985 and 1994. It is used to measure the atmospheric transmission

**Box 11.5 The Upper Atmosphere Research Satellite (UARS)**

One of the premier examples of the application of satellites to the study of atmospheric chemistry is the Upper Atmosphere Research Satellite (UARS) (Reber et al., 1993). This large satellite carried nine instruments for a comprehensive study of the chemistry and dynamics of the upper atmosphere, primarily the stratosphere, but also the mesosphere and lower thermosphere. It has produced one of the most complete data sets gathered on the energy inputs to the atmosphere, winds, temperature, and chemical composition. The instruments form a tightly integrated set that, combined with correlative data and theoretical studies, provides a systematic, unified research approach to this region of the atmosphere. UARS was launched by the Space Shuttle in June 1991 into an orbit 600 km above the Earth's surface, inclined at 57°. Because the orbit precesses relative to the Sun, the spacecraft is rotated every 36 days to keep the same side facing the Sun. The limb viewing instruments in alternate periods look from 34°S to 80°N and from 80°S to 34°N.

Of the nine instruments, three are designed to measure the energy input in the form of electromagnetic radiation and particle precipitation from the Sun to the upper atmosphere, two measure winds directly, and six measure temperature and a large number of chemical species. All the instruments for measurement of the atmospheric composition observe the upper atmosphere in a limb view. They use a variety of wavelengths and observe in absorption or emission.

The Halogen Occultation Experiment (HALOE) uses the Sun as a source of radiation to measure the absorption by the atmosphere (Russell et al., 1993). By using the Sun, HALOE can observe at shorter infrared wavelengths than instruments that observe atmospheric emission; in particular, it can observe the vertical profiles of HCl and HF, two very important species for understanding the role of halogens in stratospheric chemistry, as described in Chapter 8 (see also Color Plate 20). In addition, it measures methane, carbon dioxide, ozone, water vapor, nitric oxide, and nitrogen dioxide as well as extinction by aerosols. For HCl and HF, as well as $CH_4$ and NO, a gas correlation cell is used to provide the spectral discrimination. Other channels, which measure more abundant species, use broad-band radiation to measure the total absorption in a spectral band; in these channels, the instrument signal is a direct measurement of the energy absorbed during occultation.

The Cryogenic Limb Array Etalon Spectrometer (CLAES) measured temperature and the concentrations of ozone, NO, $NO_2$, $N_2O$, $HNO_3$, $N_2O_5$, $ClONO_2$, water, methane, and CFCs by measuring the infrared emission of these species along a limb path (Roche et al., 1993) (see Color Plates 15 and 16). It achieved good spectral resolution by using a solid etalon Fabry-Perot interferometer that was tuned by tilting it with respect to the incoming radiation. This gave sufficient resolution for discriminating the species even in spectral regions where the bands overlapped. The higher resolution required very sensitive detectors. To enhance the sensitivity, not only were the detectors cooled to 16 K by solid neon, but the entire optical instrument was cooled to 150 K to reduce background radiation. Because the supply of cryogen was limited by the size of the instrument, it ran out of cryogen after 20 months, as planned, and stopped operating.

The Microwave Limb Sounder (MLS) also measures the thermal emission along a limb path; however, it uses millimeter wavelength radiation (Barath et al., 1993). Profiles of geophysical parameters are inferred from the intensity and spectral characteristics of this emission, and from its variation as the MLS line of sight is scanned vertically through the atmospheric limb. The weakness of the Planck function in the millimeter region and the diffraction of the relatively long-wavelength radiation requires a large antenna to collect the radiation.

However, heterodyne receivers at these wavelengths are very sensitive, much more so than infrared detectors. MLS measures temperature and a number of chemical species including ozone, water, and nitric acid, but the most important is ClO, which is a key component of the catalytic chemistry that destroys ozone at high altitude as well as in the lower stratosphere in polar spring time (See Chapter 14 and Plates 19, 21, and 22).

The Improved Stratospheric and Mesospheric Sounder (ISAMS), like CLAES, measured the infrared emission of the atmosphere along a limb path (Taylor et al., 1993). Like HALOE, it used a gas correlation technique to achieve high spectral resolution in a filter radiometer. In ISAMS, however, a pressure modulated radiometer is used to perform the correlation, instead of sending the radiation along two paths, with and without a gas cell, as in HALOE. In the PMR, a single cell is used in the optical path, but the pressure in the cell, and hence the absorption by the cell gas, is varied at a rate of several Hz, and the intensity of the atmospheric radiation passing through the cell is measured as a function of the cell pressure. Conventional filter bands are used for more abundant gases and aerosols. In addition to several gases of the nitrogen family, the instrument measured temperature, water, ozone, methane, and carbon monoxide as well as aerosols.

In addition to these four instruments for measuring composition and temperature, two instruments measure winds by measuring the Doppler shift of radiation emitted or scattered by the atmosphere. The High Resolution Doppler Imager (HRDI) (Hays et al., 1993) and the Wind Imaging Interferometer (WINDII) (Shepherd et al., 1993) use high-resolution interferometers to observe Doppler shifts of atomic and molecular emission lines in the lower thermosphere. HRDI also measures winds in the stratosphere using oxygen lines in the spectrum of scattered sunlight. To measure velocities to an accuracy of the order of 10 m s$^{-1}$, the shift must be measured to a small fraction of the Doppler linewidth, and care must be used in subtracting any effect of the spacecraft motion.

Two instruments, the Solar Ultraviolet Spectra Irradiance Monitor (SUSIM) (Brueckner et al., 1993) and the Solar/Stellar Irradiance Comparison Experiment (SOLSTICE) (Rottman et al., 1993), measure the solar ultraviolet radiation, which drives the ozone photochemistry. In particular, they examine the temporal change of this radiation and its correlation with solar activity. The difficulty is maintaining a highly stable intensity calibration over a long period in space. The two experiments use radically different approaches. SUSIM carries a bank of calibration lamps. Some are used frequently to calibrate the instrument; others are used rarely to check the stability of the calibration lamps themselves. SOLSTICE, on the other hand, compares the Sun to the average of a group of stars, using an innovative technique for handling the huge range of brightnesses.

The Particle Environment Monitor (PEM) determines the type, amount, energy, and distribution of charged particles injected into the Earth's atmosphere (Winningham et al., 1993). In addition, it measures the X-ray flux from the Sun. These measurements provide information on the deposition of high-energy radiation into the upper atmosphere.

Collectively, the UARS instruments are providing the most comprehensive picture yet obtained of the chemistry, dynamics, and energy balance on a global scale in the upper atmosphere. As of 1997, all the instruments except CLAES and ISAMS are still providing data routinely.

in occultation (measuring the radiation from the Sun after it passes through the atmosphere at sunrise or sunset) as seen from space. The sunlight is directed into the instrument by a two-axis photo-guided suntracker. The signal is detected by a HgCdTe photoconductor cooled by a mechanical refrigerator for optimum sensitivity. The apodized spectral resolution is 0.015 cm$^{-1}$. Almost a million sample points are recorded in the two second scan and telemetered to the ground, where a high-speed computer performs the Fourier transform. The spectra of sunlight after passing through the atmosphere are divided by the spectrum taken above the atmosphere so

**Box 11.6 Fourier Transform Spectroscopy**

In a Michelson interferometer, a beam of radiation is divided by a partially reflecting beamsplitter into two parts, which are reflected from two mirrors back to the beamsplitter, where they recombine. If the light is monochromatic of intensity $B$, the beams will interefere constructively and produce a maximum intensity if the difference in the round trip paths to the two mirrors is a multiple of the wavelength. If the path difference is an additional half-wavelength, the waves of the two beams will interfere destructively and cancel, producing a zero intensity. Thus, if $x$ is the difference in the round trip distance from the beamsplitter to the mirrors and back, the intensity at the output of the interferometer will be

$$I(x) \propto \frac{B}{2}\left[1 + \cos\left(2\pi x/\lambda\right)\right]$$

Subtracting the constant term, which is easily done electronically, and considering a source of radiation at many wavelengths, it can be seen that the signal, called an interferogram [assuming $B(\sigma)$ is an even function], is

$$I(x) \propto \int_{-\infty}^{\infty} B(\sigma)\cos\left(2\pi\sigma x\right) d\sigma$$

where $\sigma = 1/\lambda$ is the wavenumber of a particular component of the radiation. It is seen that the interferogram is now the cosine Fourier transform of the source. Since the Fourier transform is an invertible operation, the spectrum can be recovered by inverse Fourier transform

$$B(\sigma) \propto \int_{-\infty}^{\infty} I(x)\cos\left(2\pi\sigma x\right) dx$$

Of course, since the interferogram is not an analytic function, but rather is the measured output of a radiation detector, the Fourier transform has to be taken by a numerical approximation. There are two shortcomings to this approximation: The interferogram that goes into the calculation is no longer continuous, but is sampled at fixed spacings $\delta x$, and there are only a finite number of points, so the integral does not extend to infinity. Each of these introduces certain elements into the computed spectrum. The sampling makes it impossible to separate all different frequencies, a problem known as aliasing; if the samples are exactly one wavelength apart for a certain wavelength, the sampling will also produce the same values for a wavelength one-third as great. This problem can be eliminated by band limiting the spectrum so that there is a minimum of two samples per wavelength of the shortest wavelength present. Shorter wavelengths must be eliminated from the input radiation.

The finite value of the maximum optical path difference limits the spectral resolution. To separate two different wavelengths in the spectrum, one must have one-half more wave than the other in the maximum path difference $L$,

$$L = n\lambda_1 = \left(n + \frac{1}{2}\right)\lambda_2$$

$$\frac{\delta\lambda}{\lambda} = \frac{\lambda_1 - \lambda_2}{\lambda} = \frac{L}{2\lambda}\frac{1}{(n)\left(n+\frac{1}{2}\right)} = \frac{\lambda}{2L}$$

We see that the resolving power $\lambda/\delta\lambda$ is $2L/\lambda$. Thus high-resolution interferometers must have large path difference. The number of points to be recorded in the interferogram is $L/\delta x = 2L/\lambda_{\min}$. This number can become quite large, a million or more in modern high-resolution instruments.

Why would anyone use an instrument that requires so much computation? The answer lies in the comparison of signal-to-noise ratios achievable compared with a grating spectrometer. In a grating spectrometer, the entrance slit allows radiation from only a small portion of the source to enter the spectrometer, whereas the circular aperture of an interferometer can admit much more radiation; this signal increase is called the *throughput advantage*.

In a grating spectrometer, the exit slit allows radiation from only one spectral element at a time to strike the detector; in an interferometer, the detector sees radiation from all wavelengths simultaneously, each encoded with its specific modulation frequency. If detector noise, independent of the signal, is the principal source of noise, as is commonly the case in the infrared, this additional measurement time allows the signal-to-noise ratio to be increased by $\sqrt{N}$, where $N$ is the number of wavelengths observed. This *multiplex advantage* can be very significant in high-resolution spectroscopy. It should be noted that if the noise is due to photon noise of the source, which is proportional to the square root of the signal, the multiplex advantage disappears, and if there is noise proportional to the signal (such as fluctuations due to atmospheric turbulence between the source and the instrument), there is actually a multiplex *disadvantage*.

The development of the computer and particularly the fast Fourier transform (FFT) algorithm allowed the computation of high-resolution spectra from interferograms to become practical. In the past 20 years, Fourier transform infrared (FT-IR) spectroscopy has become the standard method of high-resolution broadband spectroscopy in the infrared.

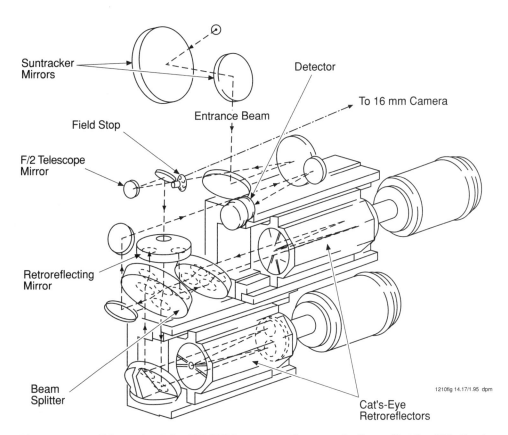

**Figure 11.17.** *Schematic of the ATMOS instrument flown on the Space Shuttle. This Fourier transform spectrometer measures high-resolution atmospheric transmission spectra by observing sunlight that has passed through the atmosphere (Zander et al., 1989).*

the solar spectrum and instrumental response effects are eliminated. The brightness of the Sun allows sufficient signal to use high spectral resolution; this enables many chemical species to be measured. To date, profiles of more than three dozen species have been measured by ATMOS, ranging in concentration from $N_2$ to $SF_6$, with a mixing ratio of less than 2 ppt.

The geometry of the occultation gives high vertical resolution. The tangent point is the lowest point in the ray path from the Sun through the atmosphere to the instrument. No absorption is contributed by the atmosphere below the tangent point, since the rays do not pass through this part of the atmosphere. Above the tangent point, the density of the atmosphere falls off rapidly with altitude and the path length through a given altitude range decreases; so relatively little absorption arises from heights much greater than the tangent altitude. Because the tangent altitude decreases rapidly, about 2 km per second, during sunset, the instrument must record a complete spectrum in about a second.

The first step in the analysis is the determination of the pressure and temperature of the atmosphere (at the tangent point) associated with the spectrum. The density is determined from the absorption of $CO_2$ and the known mixing ratio of this stable species. The temperature can be obtained from the height variation of the density, via the hydrostatic equation. Synthethic spectra are then calculated for this model atmosphere for various densities of other absorbers, and the amount adjusted to obtain the best fit to the observed spectrum. In this way profiles of many trace gases are built up.

The principal disadvantage of ATMOS for the study of stratospheric chemistry is the fact that it only observes at sunrise and sunset. This means that at most 24 sunrises and as many sunsets are generated each day and the coverage is not global. Also, any species that have diurnal variation because of the variation of sunlight for photolysis show rapid variation at sunrise and sunset, making interpretation of the measurements over a long horizontal path difficult. These objections may be overcome by measuring atmospheric emission rather than absorption. Emission measurements may be made at any point on the Earth, regardless of local time. In this way, a polar orbiting satellite can build up a picture of the entire stratosphere in one day. The emission depends upon the atmospheric temperature as well as composition. The temperature is determined by measurement of the emission from a gas of known mixing ratio such as carbon dioxide. The temperature profile can be used with the hydrostatic equation to obtain a vertical profile of pressure. It is necessary to know the pressure at one viewing angle to provide the constant of integration for the hydrostatic equation. This can be obtained by making measurements in two parts of the $CO_2$ band, permitting the determination of pressure and temperature simultaneously at that height. The known pressure-temperature profile is used to convert measurements of the radiance in spectral bands emitted by other trace gases to profiles of the mixing ratio of those gases. Of course, the radiance of the thermal emission of the Earth's atmosphere is much less than that of the Sun; therefore, it is necessary to use broad spectral bands for the measurement rather than a high-resolution spectrum to obtain sufficient signal. It is also necessary to use cryogenically cooled detectors to minimize the noise. These ideas have been used in the Limb Infrared Monitor of the Stratosphere (LIMS) flown on Nimbus 7 in 1978 (Gille and Russell, 1984).

LIMS is a six-channel filter radiometer, illustrated in Figure 11.18. It uses a telescope of 15 cm aperture to collect infrared radiation in the 6-15 $\mu$m wavelength range from a limb view of the stratosphere. The radiation from six spectral bands within this wavelength range is selected by interference filters, and an optical system relays the radiation to six cooled HgCdTe photoconductive detectors. The detectors

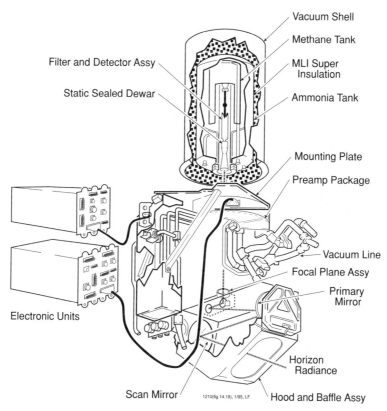

**Figure 11.18.** *Schematic of the LIMS instrument flown on the Nimbus 7 satellite. This filter radiometer measured the emission from the atmosphere in six infrared bands. High sensitivity was achieved by cooling the detectors with solid methane; the cryogen lasted for seven months in orbit. Limb viewing geometry enhanced the vertical resolution and the sensitivity (Gille and Russell, 1984).*

were cooled to 65 K by a solid methane cryogen. The methane was shielded from external heat sources by a solid ammonia cryogen. These cryogens gradually sublimed as they absorbed heat, and were exhausted after about seven months, limiting the lifetime of the observation.

The field of view of each detector is limited by a mask to a height of 2 km at the limb. The fields are scanned across the limb by a rotating plane mirror, building up a profile in the atmosphere every 6 s. These narrow fields of view, combined with the inherent geometry of the limb viewing, gave profiles with high vertical resolution. Limb viewing has the additional advantages that the horizontal path is long, giving a substantial amount of absorber in the path even for trace gases at the low pressures of the stratosphere. Additionally, the black background of space implies that all the signal comes from the atmosphere. This sensitivity enabled LIMS to measure global distributions of trace gases with concentrations in the ppb range for the first time. The six channels measured $CO_2$ (two channels) for determination of the temperature-pressure profile, ozone, water vapor, nitric acid, and nitrogen dioxide. A planned instrument HIRDLS (High Resolution Dynamics Limb Sounder) to be launched as part of the Earth Observing System will build on these principles and obtain distributions of a dozen trace gases with high vertical and horizontal resolution.

Another satellite experiment concerned with stratospheric chemistry is TOMS (Total Ozone Mapping Spectrometer) (Heath *et al.*, 1975). Unlike the previous two

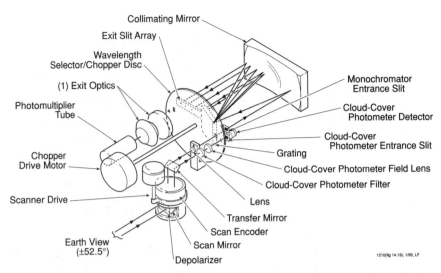

**Figure 11.19.** *The TOMS ultraviolet photometer used to measure global distributions to total ozone columns from Nimbus 7 (Heath et al., 1975).*

instruments, which operate in the infrared, TOMS senses the total atmospheric ozone column, utilizing the strong absorption by ozone in the ultraviolet region in a wavelength range in which few other atmospheric molecules absorb strongly. The atmosphere scatters the ultraviolet radiation from the Sun strongly; the combination of scattering and absorption allows the determination of ozone amount from the spectrum of the radiation returned to space. The TOMS instrument, shown in Figure 11.19, is an ultraviolet grating spectrometer that measures the albedo of the Earth at six wavelengths between 213 and 380 nm; comparison of the radiation scattered back from the atmosphere with the incident solar radiation permits measurement of the ozone column. The spectrometer has a fixed grating and an array of exit slits, one for each wavelength. An external scan mirror scans the 3° field of view across the satellite ground track to build up a global picture of the entire illuminated part of the atmosphere; there are, unfortunately, no measurements in the polar night. The ozone absorption coefficient varies so strongly with wavelength that the albedo decreases by about a factor of 100 with a 12 nm change in wavelength. Thus the need for spectral purity in the measurements is a challenge. A rotating disk successively selects radiation from one of the target wavelengths to pass to the photomultiplier detector. A diffuser plate of roughened aluminum is illuminated by the direct sunlight; the spectrometer periodically views this diffuser plate. The comparison of the brightness of the atmosphere with the brightness of the diffuser yields the albedo of the atmosphere. The wavelength dependence of this albedo is determined by the ratio of scattering to absorption in the atmosphere, and so reflects the strongly varying ozone absorption coefficient and the total amount of ozone.

The original TOMS gave the most complete global picture of the distribution of ozone and its change in the years from its 1979 launch until its failure in 1992; subsequent versions are maintaining the long-term ozone record. Because $SO_2$ also has strong absorption features in this spectral region (but with a different wavelength dependence from ozone), TOMS has been able to track the stratospheric cloud from several volcanic eruptions and to estimate volcanic sulfur inputs to the stratosphere.

One difficulty with the use of TOMS data for determination of long-term trends in ozone has been uncertainty in the degradation of the calibration plate. Inevitably the surface receives some contamination in space from out-gassing of various surfaces

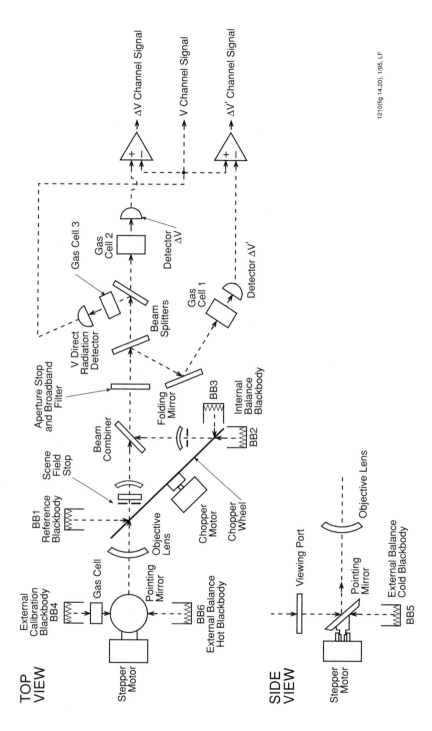

**Figure 11.20.** *The optical layout of the MAPS instrument used to measure the distribution of tropospheric CO from the Space Shuttle. Gas cells containing CO at different pressures give the instrument optimum sensitivity for measuring CO emission from various heights in the troposphere (Reichle et al., 1986).*

in the spacecraft and instruments. Hydrocarbons in this contamination darken when exposed to the hard ultraviolet from the Sun. This gradually reduces the reflectivity of the diffuser, introducing uncertainty in the absolute measurements. One of the advances made in recent years has been the determination of this degradation by examining the wavelength dependence of the changing signal, so that ozone trends of a few tenths of a percent per year can be determined. Comparison of the TOMS measurements with ground-based Dobson measurements has also been useful.

Relatively few spaceborne experiments have been used for tropospheric chemistry. There are several reasons for this. The prevalence of clouds in the troposphere requires downward viewing instead of limb viewing. This geometry has poorer vertical resolution. The lack of temperature contrast between the surface and the lowest part of the atmosphere makes it impossible to measure the boundary layer in thermal emission, despite the chemical interest in this region. One example of a tropospheric experiment is the MAPS (Measurement of Air Pollution from Satellites) flown on the space shuttle (Reichle et al., 1986); this experiment is illustrated in Figure 11.20. This nadir viewing experiment was designed to measure the global distribution of carbon monoxide in the middle and upper troposphere (see Color Plate 8). It operates in the 4.8 $\mu$m region, in the fundamental vibration region of the CO molecule. At this wavelength, most of the radiation is thermal emission from the surface and the atmosphere, although reflected sunlight also must be considered. The CO signal consists of widely spaced narrow absorption lines. The instrument uses gas filter correlation techniques to obtain a high throughput and high effective spectral resolution. The scene is viewed by three detectors, one of which looks directly at the Earth and atmosphere, while the other two view it through cells containing CO at different pressures. The difference between the signals from the broad-band detector and one in which the gas in the cell absorbs radiation just at the wavelengths of the CO lines indicates the amount of radiation in the CO lines, and hence the amount of CO in the atmosphere. The cell with the higher pressure affects primarily the radiation arising in the middle troposphere where the pressure and therefore the width of the absorption lines is larger, while the cell with the lower pressure produces a signal dependent upon CO higher in the troposphere. The spectral difference in absorption coefficient between the vacuum cell and the gas cell, multiplied by the interference filter defining the overall response band of the instrument, gives the wavelength dependence of the instrument response, which just matches the CO spectrum. The biggest experimental difficulty is balancing the sensitivity of the two channels so that the difference in their response is due to the atmospheric emission in the CO lines and not to a gain difference in the response to the much larger broad-band blackbody radiation. This balance condition is established in flight by viewing two onboard black-bodies at different temperatures and adjusting the gain of the gas detector to eliminate the response to blackbody radiation (i.e., with no CO signal) for both the low- and high-temperature sources, spanning the range of terrestrial sources. Viewing an onboard blackbody that has a cell in front of it containing a known amount of CO allows measurement of the scale factor of the response. The emissivity of the terrain and the solar zenith angle have to be included in the analysis. The instrument has very little sensitivity to CO in the boundary layer, because it is at very nearly the same temperature as the ground and so emits almost as much radiation as it absorbs.

This instrument had a limited observing time. It produced valuable data on the CO in the middle and upper troposphere in geographical regions where other measurements of this quantity were not available. It should be noted that MOPITT (Measurements of Pollution in the Troposphere) (Drummond and Mand, 1996), an instrument operating on similar principles, will produce global measurements of CO and $CH_4$ from the EOS satellites. MOPITT operates both at short wavelengths, where absorption of reflected sunlight provides the total CO column, and in the fundamental band where, like MAPS, it determines the middle and upper tropospheric amounts. Subtraction allows determination of concentration near the surface.

## Further Reading

Beer, R. (1992) *Remote Sensing by Fourier Transform Spectrometry*, Wiley, New York.

Dieminger, W., G. K. Hartmann, and R. Leitinger, eds. (1996) *The Upper Atmosphere*, Springer-Verlag, Berlin, Sec. II.4.

Grant, W. G., ed. (1989) *Ozone Measuring Instruments for the Stratosphere*, Optical Society of America, Washington, D.C..

Killinger, D. K., and A. Mooradian, eds. (1983) *Optical and Laser Remote Sensing*, Springer-Verlag, Berlin.

Kuwana, T., ed. (1978) *Physical Methods in Modern Chemical Analysis*, Academic Press, New York.

Roscoe, H. K. and K. C. Clemitshaw (1997) Measurement techniques in gas-phase tropospheric chemistry: A selective view of the past, present, and future, *Science* 276, 1065.

Gérard Mégie

## From Individual Measurements to Scale Integration Strategies

Chemical, dynamical, and radiative processes take place in the Earth's atmosphere at spatial scales ranging from the microscopic scale of turbulence and elementary collision processes to the planetary scale of long-range transport, and at temporal scales extending from the microsecond typical of fast chemical processes to the millennial scales characteristic of long-term evolutions linked to astronomical cycles. Reproducing the physical conditions of the atmosphere in the laboratory, with temperatures varying from +40°C at the surface to −100°C at the summer mesopause, a pressure decrease from 1 atm at the surface to $10^{-6}$ atm at 100 km, and a large variation with altitude of the spectrally resolved solar radiation, is almost impossible. In addition, one has to take into account the large dynamical range of relative abundances of species in the atmosphere, from 78% for molecular nitrogen down to $10^{-14}$ for the OH radical (a key species in tropospheric and stratospheric chemistry) or even less when considering the excited oxygen atom $O(^1D)$ (which produces OH in the troposphere and destroys stable species in the lower stratosphere), initiating the fast chemistry component. There is, therefore, no other possibility than to consider the atmosphere itself as the laboratory in which experiments have to be performed. Different measurement strategies are then conducted, using surface-based, airborne, and spaceborne observational techniques, which range from individual measurements to coordinated campaigns measuring multiple species.

Individual measurements of species concentrations often require pushing experimental methods to their limits. Scientists involved in the development of instruments for atmospheric studies thus participate in the experimental research on the most sensitive techniques conducted in fundamental physics and chemistry, such as spectroscopy, mass spectrometry, gas and liquid chromatography, and chemical kinetics. As a permanent challenge, the discovery of new processes implies the development of new techniques to measure not only gaseous components of the atmosphere, but also composition and physical properties of various types of particles ranging from cloud droplets and ice particles to natural and anthropogenic aerosols. Similarly, an innovative measurement often opens a new field of research, as was the case in the past for active chlorine measurements in the stratosphere, background OH concentration in the troposphere, or chemical analysis of aerosol particles.

A single measurement, however, is certainly not sufficient to constrain the elaborate chemical transport models that are currently being developed, or to evaluate long-term trends in atmospheric variables. We have learned over the past thirty years that any type of instrument has its own limits in terms of precision, accuracy, spatial, and temporal resolution. Only combined measurements using both in situ and remote sensing techniques from various platforms can be considered appropriate to quantify atmospheric processes. This required measurement strategy opens a new type of international collaboration, defining coordinated campaigns or satellite payloads to use the best available techniques developed over the world. In this respect, all instruments find their place. The Antarctic ozone hole was not discovered by an up-to-date spaceborne instrument, but by the time-validated Dobson spectrometer. Measurements performed in the past century on surface ozone, or in the 1930s on stratospheric ozone, are still being used to assess long-term trends. The combination of ground-based and spaceborne observation is now considered the only way to detect atmospheric changes precisely at various scales, as illustrated by the implementation of science-driven networks with a world wide extension.

Finally, it is increasingly obvious that atmospheric measurements have to be planned and integrated in strong interaction with model assessments. One will never be in a position to measure every variable of interest, everywhere, at any time, and with an infinite resolution. One of the hardest challenges that the atmospheric science community currently faces is certainly the scale integration problem: Processes at the local scale directly influence large-scale effects. This applies to such problems as, for example, the evaluation of regional or global budgets of species, including interaction between the atmosphere and the terrestrial and marine biospheres, the quantification of the influence of local pollution at the regional scale, and the dislocation of the polar vortex in shorter-scale features such as laminae or cutoff lows. New concepts have to be developed there, based on innovative instruments and measurement methodologies, making use of data assimilation techniques. New researchers in atmospheric science will have to be trained to address all these aspects simultaneously. Ten years from now, the existing distinction between experimentalists and modellers might only be a reminder of the past.

*Gérard Mégie is Professor at the University Pierre-et-Marie-Curie in Paris, France, and Director of the Service d'Aéronomie CNRS and the Institut Pierre-Simon-Laplace for Global Environmental Sciences. His main research interests concern stratospheric ozone, tropospheric chemistry, and climate chemistry interactions. Gérard Mégie has been involved in several instrumental developments, which include lidar remote sensing, tunable diode laser spectrometry, spaceborne, and airborne experiments.*

# 12 Modeling

## 12.1 Introduction

Theoretical models are developed to study chemical, physical, and biological processes in the global environment and to describe the complex nonlinear interactions and feedbacks that affect the Earth system. In particular, such models simulate the spatial distribution and temporal evolution of chemical compounds in the atmosphere and their exchanges with the ocean and the terrestrial biosphere. By definition, models are developed to test hypotheses that need to be verified by observations or experiments. Models can be viewed as mathematical representations of the fundamental laws that govern the fate of the atmosphere and other components of the geosphere, and attempt to replicate the complex processes occurring in this natural system. As *diagnostic* tools, they assist in the interpretation of field observations and in the identification of key variables and processes. They are also used to test the sensitivity of calculated quantities to physical or chemical processes that are not always well understood. Deviation of model results from observations generally signals insufficient understanding (or omission) of key processes and generates new experimental programs. As *prognostic* tools, models provide information on the future state of the Earth system, but such predictions are regarded as reliable only if sufficient confidence has been acquired in the model. As a basis for regulatory measures, several model assessments have been performed in recent years to predict, for example, future changes in the chemical composition of the atmosphere or in the climate system in response to expected anthropogenic perturbations (*e.g.*, release of chlorofluorocarbons or carbon dioxide in the atmosphere).

Over the past two decades, atmospheric models have been widely used to provide weather forecasts and climate predictions. They are now also regarded as powerful tools to study the transport and chemical conversions of trace gases and aerosols in the troposphere and in the middle atmosphere. Their development is necessarily an iterative process involving the design of complex codes, the validation of results against available observations, and the completion of sensitivity tests. Models can become sufficiently complicated that, even with state of the art supercomputers, the chemical and transport equations cannot be solved in a reasonable amount of computer time. Therefore, compromises have to be made concerning the degree of details included in the models and the spatial resolution used in the numerical treatment of the equations. The type of simplifications introduced in the models is usually a function of the scientific question to be addressed. Because models differ greatly in their degree of complexity, this chapter will focus primarily on the basic methods and numerical techniques used to develop such models. The concepts will, however, be illustrated through several examples highlighting the results provided by existing zero-, one-, two-, and three-dimensional models.

## 12.2 Model Equations

At the heart of chemical models is the continuity equation, which is written as

$$\frac{\partial \rho}{\partial t} + \vec{\nabla} \cdot (\rho \vec{v}) = \widetilde{\mathbf{Q}} \qquad (12.1)$$

where $\boldsymbol{\rho} = (\rho_1, \rho_2, \ldots, \rho_i, \ldots, \rho_N)^T$ represents the mass (or number) density, and $\widetilde{\mathbf{Q}} = (\widetilde{Q}_1, \widetilde{Q}_2, \ldots, \widetilde{Q}_i, \ldots, \widetilde{Q}_N)^T$ the net chemical source term of $N$ chemical constituents, $\vec{v} = (u, v, w)$ is the transport velocity vector, and $t$ is time. The mass (or number) density $\boldsymbol{\rho}$ and the wind $\vec{v}$ are explicit functions of time $t$ and spatial coordinates $\vec{x} = (x, y, z)$. The chemical forcing $\widetilde{\mathbf{Q}}$ also explicitly depends on $t$ and $\vec{x}$, and in general also depends explicitly on constituent densities $\rho_j$. In Eq. (12.1), if $\rho$ is expressed in kg m$^{-3}$ (particles cm$^{-3}$), $\widetilde{\mathbf{Q}}$ is expressed in kg m$^{-3}$ s$^{-1}$ (particles cm$^{-3}$ s$^{-1}$).

Equation (12.1) is the so-called flux form of the continuity equation. Several chemistry models use the advection form of the continuity equation

$$\frac{\partial \boldsymbol{\mu}}{\partial t} + \vec{v} \cdot \vec{\nabla} \boldsymbol{\mu} = \mathbf{S} \qquad (12.2)$$

wherein $\boldsymbol{\mu} = (\mu_1, \mu_2, \ldots, \mu_i, \ldots, \mu_N)^T$ is the mass (or volumetric) mixing ratio vector defined by $\boldsymbol{\mu} = \boldsymbol{\rho}/\rho_a$, $\mathbf{S} = \widetilde{\mathbf{Q}}/\rho_a$ is the source term (expressed here in s$^{-1}$), and $\rho_a$ is the total atmospheric mass (or number) density. Transforming Eq. (12.1) into Eq. (12.2) is accomplished by expanding $\boldsymbol{\rho}$ in terms of $\boldsymbol{\mu}$ and utilizing overall fluid conservation

$$\frac{\partial \rho_a}{\partial t} + \vec{\nabla} \cdot (\rho_a \vec{v}) = 0 \qquad (12.3)$$

It should be noted that Eq. (12.2) is often rewritten as

$$\frac{d\boldsymbol{\mu}}{dt} = \mathbf{S} \qquad (12.4)$$

where

$$\frac{d}{dt} \simeq \frac{\partial}{\partial t} + \vec{v} \cdot \vec{\nabla}$$

is the "total" derivative. The terms $\vec{\nabla} \cdot (\rho \vec{v})$ and $\vec{v} \vec{\nabla} \boldsymbol{\mu}$ are interchangeably referred to as the "transport operator" appropriate to the particular representation of the continuity equation. Continuity equations (12.1) or (12.2), together with initial and boundary conditions, form the basic analytical system for describing the spatial and temporal behavior of atmospheric constituents. Unfortunately, in all but the simplest scenarios, this system does not yield an analytic solution. Thus the only recourse is to numerical methods, by which the differential equations (12.1) or (12.2) are replaced by discrete analogues. The solutions of these discrete analogues are derived at given times and discrete locations (called grid points). Figure 12.1 shows a three-dimensional grid that is typical of global models. The advective process can be divided into (spatial and temporal) scales that can be resolved explicitly by the "numerics" of the model, and those that occur on smaller scales. So, typically the transport operator is replaced by a term representing the *resolved* transport processes, and a set of terms representing the *unresolved* transport processes. These latter terms include a representation of turbulent diffusion, and (at least) one representing other organized dynamical processes such as convection. The net source term is often replaced by production and loss terms for the constituents due to chemical conversion mechanisms, and includes other processes such as wet scavenging. The evolution

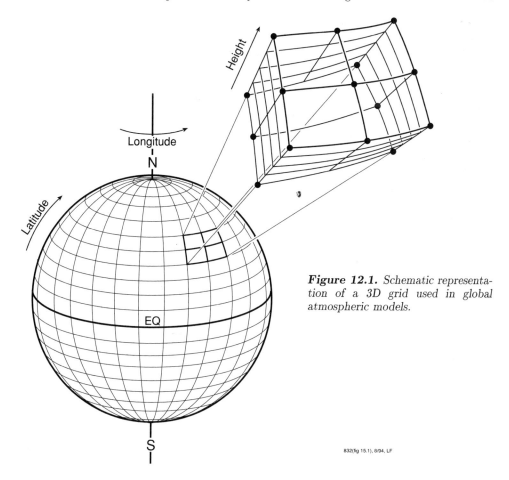

**Figure 12.1.** Schematic representation of a 3D grid used in global atmospheric models.

equation (advective form) for the mixing ratio $\mu$ can thus be explained as

$$\frac{\partial \mu}{\partial t} = -\vec{v} \cdot \vec{\nabla} \mu + \left(\frac{\partial \mu}{\partial t}\right)_{\text{diff}} + \left(\frac{\partial \mu}{\partial t}\right)_{\text{conv}} + \mathbf{S} \qquad (12.5)$$

There is an equivalent flux form equation as well.

Typically Eq. (12.5) is solved numerically by using an "operation split" formulation. In this case, the different component processes (resolved advection, diffusion, convection, chemistry) are isolated and treated sequentially for each time step $\Delta t$, at the cost of some accuracy in the solution $\mu$. A discussion of the implications of this operator splitting is beyond the scope of this chapter. The following sections will deal with the solution to the component equations, rather than Eq. (12.5) itself.

The chemical forcing $\mathbf{S}$ can be represented as

$$\mathbf{S}(\mu, \mathbf{x}, t) = \mathbf{e}(\vec{x}, t) - A(\vec{x}, t) \cdot \mu + B(\mu, \vec{x}, t) \cdot \mu \qquad (12.6)$$

where $\mathbf{e}(\vec{x}, t)$ is the zeroth-order or "external" forcing (independent of system variables $\mu$), $A(\vec{x}, t)$ is an $N \times N$ matrix for first-order forcing (typically photolysis, thermal decay, radioactive decay), and $B(\mu, \vec{x}, t)$ is an $N \times N$ matrix for nonlinear forcing (typically bi- and trimolecular reactions). Note that individual elements of $\mathbf{e}$, $A$, and $B$ may be positive or negative, denoting production or loss, respectively. Also note that the $B$ matrix entries explicitly depend on the solution set $\mu$ and, as such, represent *nonlinear forcing*. Typically terms in $B(\mu, \vec{x}, t) \cdot \mu$ represent a reaction between $\mu_i$ and $\mu_j$, where $\mu_i + \mu_j \xrightarrow{r} \mu_k$, and thus the term $r\mu_i\mu_j$ appears as a loss in the $ith$ and $jth$ constituent equations and as a production in the $kth$ equation.

The lifetime of the various chemical constituents belonging to a given chemical system can be calculated from the Jacobian matrix $\partial \mathbf{S}/\partial \boldsymbol{\mu}$. Because the rates of the different reactions involved can vary by many orders of magnitude, the lifetimes can range from perhaps microseconds to centuries. In this case, the system is said to be *stiff*. More specifically, System (12.6) is stiff when the ratio between the largest and the smallest eigenvalues of the Jacobian matrix is much larger than unity.

The integration of chemical-transport equations requires that initial conditions be prescribed for the abundance of the trace constituents and that conditions be specified at the boundaries of the spatial domain over which the model is applied. Initial conditions may be set with preliminary estimates of the solution or with observed concentrations. Boundary constraints represent the influence of the external environment on the chemical system. In most cases, these conditions are not easily established, since they reflect complex processes at the interface between different layers of the atmosphere, or between the atmosphere and other components of the Earth system. For example, emissions or deposition of trace gases at the Earth surface, which are frequently used as boundary conditions for chemical-transport models, are dependent on several factors, including biological processes and anthropogenic activity (Chapter 5). Emission inventories are established based on comprehensive information concerning a variety of ecosystems as well as industrial and agricultural activities. The oceans, which cover 70 percent of the Earth's surface, provide a large area for exchange of chemical compounds (see Chapter 5). The flux $\Phi$ representing gas transfer from the ocean to the atmosphere is generally expressed by

$$F = K(v, T)\,\Delta n \qquad (12.7a)$$

where $K(v, T)(> 0)$ is the transfer velocity and

$$\Delta n = n_w - n_a \mathcal{H} \qquad (12.7b)$$

where $\mathcal{H}\,(M\,\text{atm}^{-1})$ is the Henry's law constant, $n_a$ (atm) is the concentration of the gas in air, and $n_w\,(M)$ its concentration in sea water. Water is said to be supersaturated if $\Delta n$ is higher than zero, so that the flux is from the ocean to the atmosphere; otherwise ($\Delta n < 0$) it is subsaturated, and transfer occurs from air to water. The transfer velocity, which can be expressed in terms of resistance to transfer in the water and in the air, respectively, varies greatly with the intensity of the wind above the surface of the sea and the chemical reactivity of the gas.

Over continents, in addition to direct emissions, the transfer of gases and aerosols from the atmosphere to land surfaces occurs through wet and dry deposition. The flux for dry deposition is expressed by

$$F_D = -n_a v_D \qquad (12.8)$$

where the species-dependent deposition velocity $v_D(> 0)$ varies with the nature of the surface (*e.g.*, canopy) and the aerodynamic conditions in the surface layer of the atmosphere. Table 5.1 provides typical values of deposition velocities for several species.

From a mathematical point of view, two types of boundary conditions need to be considered. The simplest of them (Dirichlet condition) consists of a prescribed concentration, based, for example, on observations. When a flux is prescribed at the boundary, the concentration gradient normal to the boundary (Neumann condition) is specified.

## 12.3 Modeling Chemical Processes

Chemical "numerics" embodies the full range of numerical algorithms applicable to solving the following system

$$\frac{d\boldsymbol{\mu}}{dt} = \mathbf{S} \qquad (12.9)$$

where, as in Eq. (12.2), $\boldsymbol{\mu} = (\mu_1, \ldots, \mu_N)^T$, and $\mathbf{S} = (S_1, \ldots, S_N)^T$. Further, $\mathbf{S}$, the forcing vector, takes the form (12.6) expressed in the previous section. Eq. (12.9) is a system of nonlinear ordinary differential equations. In the following, common temporal discretizations for solving Eq. (12.9) will be discussed.

The simplest and computationally cheapest method is the Euler forward or *fully explicit* method given by

$$\boldsymbol{\mu}^{n+1} = \boldsymbol{\mu}^n + \Delta t \cdot \mathbf{S}(t_n, \boldsymbol{\mu}^n), \qquad t \in [t_0, T] \qquad (12.10)$$

where $t_n = t_o + n\Delta t\, (n = 0, 1, \ldots)$, $\Delta t = t_{n+1} - t_n$ is the time step (assumed here to be uniform), $\boldsymbol{\mu}^n = \boldsymbol{\mu}(t_n)$ and $\boldsymbol{\mu}^{n+1} = \boldsymbol{\mu}(t_{n+1})$ are approximations to the analytical solution at $t = t_n$ and $t = t_{n+1}$, respectively, and $\mathbf{S}(t_n, \boldsymbol{\mu}^n)$ is the forcing evaluated at $t_n$ and for the known approximation $\boldsymbol{\mu}^n$. Equation (12.10) is a single-step method; only the times $t_{n+1}$ and $t_n$ and the variables at these two times are involved. Note that in Eq. (12.10) the unknown $\boldsymbol{\mu}^{n+1}$ is represented strictly in terms of *known* quantities at $t_n$ and thus is denoted "fully explicit."

Although Eq. (12.10) is appealing in that the solution is easily computed, the forward Euler method must be utilized carefully, if at all, for realistic chemical systems due to numerical stability restrictions (see Box 12.1).

---

**Box 12.1 Stability of Fully Explicit Euler Forward Scheme**

A very simple, linear example will reveal the basic shortcoming of the fully explicit method: numerical stability restrictions. For the single-component system

$$\frac{d\mu}{dt} = -r\mu \qquad (1)$$

with $r > 0$, we have via Eq. (12.10) that

$$\mu^{n+1} = \mu^n (1 - r\Delta t) \qquad (2)$$

From the analytic solution to (1) we know that

$$\mu(t_n + \Delta t) = \mu(t_n)\, e^{-r\Delta t} \qquad (3)$$

Basic stability analysis requires that $|\mu^{n+1}/\mu^n| < 1$ for all $n$. Equation (2) will only meet this stability criterion if the following condition holds

$$\Delta t < \max \left| \frac{2}{r} \right|$$

For large values of $|r|$, $\Delta t$ is severely restricted.

---

The problem of a restrictive time step to ensure numerical stability, a universal constraint for explicit methods, is overcome with the Euler backward or *fully implicit* scheme given by

$$\boldsymbol{\mu}^{n+1} = \boldsymbol{\mu}^n + \Delta t \cdot \mathbf{S}\left(t_{n+1}, \boldsymbol{\mu}^{n+1}\right) \qquad (12.11)$$

This algorithm is unconditionally stable for any $\Delta t$, and thus the time step may be selected purely for accuracy considerations (see Box 12.2). As is almost always the case in practical simulations, when the forcing function **S** includes nonlinear terms, a direct, single-step solution for Eq. (12.11) does not exist in general. The system, which can be written as

$$\mathbf{G}\left(\boldsymbol{\mu}^{n+1}\right) \equiv \boldsymbol{\mu}^{n+1} - \boldsymbol{\mu}^n - \Delta t \cdot \mathbf{S}\left(t_{n+1}, \boldsymbol{\mu}^{n+1}\right) = 0 \qquad (12.12)$$

where **G** is an $N$-valued, nonlinear vector function, is most frequently solved by iterative methods or linearization. Prominent among the iterative algorithms is the Newton-Raphson iteration,

$$\boldsymbol{\mu}^{n+1}_{(m+1)} = \boldsymbol{\mu}^{n+1}_{(m)} - J^{-1} \cdot \mathbf{G}\left(\boldsymbol{\mu}^{n+1}_{(m)}\right), \qquad m = 0, 1, 2, \ldots \qquad (12.13)$$

with

$$\boldsymbol{\mu}^{n+1}_{(0)} = \boldsymbol{\mu}^n$$

where $J$ is the Jacobian matrix whose elements $J_{ij} = \partial G_i / \partial \mu_j$. This iteration converges to the solution for time steps smaller than a problem-dependent upper bound. Thus, even though the fully implicit method places no theoretical time step limitations, the algebraic system solution via Newton-Raphson iteration reimposes a time step limitation. Nonetheless, for real world problems the fully implicit scheme with Newton-Raphson iteration proves to be vastly superior to the fully explicit approach.

---

**Box 12.2 Stability of Fully Implicit Euler Backward Scheme**

For Eq. (1) in Box 12.1 the Euler backward scheme (12.11) leads to the following linear algebraic equation

$$\mu^{n+1} = \frac{\mu^n}{1 + r\Delta t}$$

Clearly for any $\Delta t \in (0, \infty)$ $\left|\mu^{n+1}/\mu^n\right| < 1$ so that the fully implicit scheme is unconditionally stable. Additionally, as $\Delta t \to \infty$ the above ratio tends to zero, mirroring the analytic solutions. Comparing the numerical solutions with the analytic solutions [(3) in Box 12.1] we see that the crux of numerical accuracy involves comparing $e^{-r\Delta t}$ with $1/1 + r\Delta t$.

---

Another popular iterative solution to Eq. (12.12) used by Shimazaki (1985) derives from rewriting the forcing function **S** as

$$\mathbf{S}(t, \boldsymbol{\mu}) = \mathbf{P}(t, \boldsymbol{\mu}) - \boldsymbol{\beta}(t, \boldsymbol{\mu}) \cdot \boldsymbol{\mu} \qquad (12.14)$$

and "linearizing" equation (12.11) as

$$\boldsymbol{\mu}^{n+1}_{(m+1)} = \boldsymbol{\mu}^n + \Delta t \left[ \mathbf{P}\left(t_{n+1}, \boldsymbol{\mu}^{n+1}_{(m)}\right) - \boldsymbol{\beta}\left(t_{n+1}, \boldsymbol{\mu}^{n+1}_{(m)}\right) \boldsymbol{\mu}^{n+1}_{(m+1)} \right], \qquad m = 0, 1, 2, \ldots \qquad (12.15)$$

with

$$\boldsymbol{\mu}^{n+1}_{(0)} = \boldsymbol{\mu}^n$$

Equation (12.15) is computationally far less expensive than Eq. (12.13) in that only a diagonal rather than a full matrix system needs be solved at each iteration. Again, there are theoretical convergence restrictions on $\Delta t$ that depend on $\boldsymbol{\mu}^{n+1}_{(0)}$, the initial iterate, and the exact functional forms of **P** and $\boldsymbol{\beta}$.

So far, we have examined two single step methods (backward and forward Euler) that can be regarded as particular cases of multistep methods represented by (Byrne and Hindmarsh, 1975)

$$\boldsymbol{\mu}^{n+1} = \sum_{k=0}^{K} \alpha_k \boldsymbol{\mu}^{n-k} + \Delta t \sum_{k=-1}^{K} \gamma_k \mathbf{S}\left(t^{n-k}, \boldsymbol{\mu}^{n-k}\right) \qquad (12.16)$$

where $\alpha_k$ and $\gamma_k$ are method-specific constants selected to ensure stability and $K$ is the overall order of the method. For the forward Euler method, $K = 0$, and only $\alpha_0 = \gamma_{-1} = 1$ are nonzero. A well-known solution that is particularly adequate for "stiff" systems is due to Gear (1971). This algorithm is composed of so-called backward difference formulas (BDF) up to order six, wherein only $\gamma_{-1}$ is nonzero and $\alpha_k$ are selected for stability and accuracy. Gear's BDF family is extremely robust and stable but does, like backward Euler methods, require solving nonlinear algebraic systems. Elegant solver codes based on Gear's method continuously and automatically vary the time step and order to meet user-specified solution error tolerances. Unfortunately, such codes require much computer memory and time, and are computationally impractical for most multidimensional simulations. They are often used, however, in zero- and one-dimensional models. Efficient coding of the Gear method developed specifically for vectorized machines (Jacobson and Turco, 1994) is now available.

In addition to the computationally tractable backward Euler or Shimazaki methods, there is the unique "semianalytic" linearization due to Hesstvedt et al. (1978), also called the quasi-steady-state approximation (QSSA). Rephrasing Eq. (12.9) as in Eq. (12.14) gives

$$\frac{d\boldsymbol{\mu}}{dt} = \mathbf{S}\left(\boldsymbol{\mu}, t\right) = \mathbf{P}\left(\boldsymbol{\mu}, t\right) - \boldsymbol{\beta}\left(\boldsymbol{\mu}, t\right) \cdot \boldsymbol{\mu} \qquad (12.17)$$

The "semianalytic" analogue for the $i$th component of Eq. (12.17) is

$$\mu_i^{n+1} = (\mu_i)_{ss} + [\mu_i^n - (\mu_i)_{ss}] \exp\left(-\beta_i\left(\mu^n, t_{n+1}\right) \Delta t\right) \qquad (12.18)$$

where $(\mu_i)_{ss} = P_i\left(\mu^n, t_{n+1}\right)/\beta_i\left(\mu^n, t_{n+1}\right)$ is the value of the mixing ratio at steady state. This method is stable. Clearly for a scalar, linear equation with constant coefficients, Eq. (12.18) is the analytic solution. The semianalytic scheme represented in Eq. (12.17) can itself be iterated.

While this semianalytic method is relatively straightforward and computationally efficient, it is not conservative (see Box 12.3). Within the framework of chemical modeling, a system of equations [such as (12.18)] is said to be conservative if

$$\mathbf{W}\frac{d\boldsymbol{\mu}}{dt} = \mathbf{W} \cdot \mathbf{S}\left(\boldsymbol{\mu}, t\right) = 0 \qquad (12.19)$$

for a nonzero weight vector $\mathbf{W} = (w_1, w_2, \ldots, w_N)^{\mathrm{T}}$.

The methods described above bring out the difficult tradeoffs in selecting a numerical algorithm. Implicit algorithms are in theory stable and conservative, but require iterative solutions for the general nonlinear case. Explicit techniques, although conservative and inexpensive on a per time step basis, generally impose unacceptably stringent time step limits. Linearizations such as the QSSA method are relatively efficient and stable, but are not inherently conservative. In a practical chemical simulation, one often divides an overall system into groups that are treated with a combination of explicit, linearized, and implicit techniques. There is no set of "golden" rules to guide the actual solution technique partitioning of a system. However, in such a procedure one ideally would like to minimize the size of the implicit group while maximizing the number of constituents in the explicit group.

**Box 12.3 Mass Conservation in Chemical Schemes**

It is highly desirable that any numerical scheme for solving Eq. (12.9) retains the inherent conservation in the sense of Eq. (12.19). For example, suppose we have the following two-component chemical system

$$\mu_1 \xrightarrow{r_1} \mu_2$$

or

$$\mu_2 \xrightarrow{r_2} \mu_1$$

The corresponding kinetic equations are

$$\frac{d\mu_1}{dt} = -r_1\mu_1 + r_2\mu_2 \tag{1a}$$

$$\frac{d\mu_2}{dt} = +r_1\mu_1 - r_2\mu_2 \tag{1b}$$

so that

$$\frac{d}{dt}(\mu_1 + \mu_2) = 0 \tag{2}$$

In this case, if we add the two equations together, the sum of $\mu_1$ and $\mu_2$ is time invariant or "conserved." In terms of Eq. (12.19), the weighting vector $w = (1,1)$. This analytic conservation implies that for any numerical analogue $\mu_1^{n+1} + \mu_2^{n+1} = \mu_1^n + \mu_2^n$ or that from time step to time step the sum of the numerical solutions should be constant. Examining the forward Euler method (12.10) applied to system (1) yields that

$$\mu_1^{n+1} = \mu_1^n - r\Delta t \mu_1^n + r\Delta t \mu_2^n \tag{3a}$$

$$\mu_2^{n+1} = \mu_2^n + r\Delta t \mu_1^n - r\Delta t \mu_2^n \tag{3b}$$

or that $\left(\mu_1^{n+1} + \mu_2^{n+1}\right) = \left(\mu_1^n + \mu_2^n\right)$ and thus forward Euler is conservative. Similarly, backward Euler (12.11) is conservative. In the case of the Hesstvedt method (12.17), Eqs. (1a) and (1b) become

$$\mu_1^{n+1} = \frac{r_2}{r_1}\mu_2^n + \left(\mu_1^n - \frac{r_2}{r_1}\mu_2^n\right)e^{-r_1 \Delta t} \tag{4a}$$

$$\mu_2^{n+1} = \frac{r_1}{r_2}\mu_1^n + \left(\mu_2^n - \frac{r_1}{r_2}\mu_1^n\right)e^{-r_2 \Delta t} \tag{4b}$$

Clearly Eq. (4a-b) does not obey the conservation constraint in that $\mu_1^{n+1} + \mu_2^{n+1}$ does not equal $\mu_1^n + \mu_2^n$. It should be pointed out that the Shimazaki method (12.15), if iterated only once, is not conservative either. In practice, very few real chemical systems are conservative in the sense of Eq. (12.19), but it is still important to determine whether a numerical algorithm is inherently conservative.

---

To maximize the potential candidates for explicit treatment and/or to minimize the stiffness of the implicit grouping, chemical modelers often invoke the technique of "family" grouping. The mathematical formalism of family grouping is completely analogous to the discussion on conservation [Eq. (12.19) and Box (12.3)]. In this case, however, the weighting factor is a matrix instead of a vector

$$W\frac{d\boldsymbol{\mu}}{dt} = W\mathbf{S} \tag{12.20a}$$

or

$$\frac{d\widetilde{\boldsymbol{\mu}}}{dt} = \widetilde{\mathbf{S}} \tag{12.20b}$$

with $\widetilde{\boldsymbol{\mu}} = W\boldsymbol{\mu}$ and $\widetilde{\mathbf{S}} = W\mathbf{S}$, where $W = N \times N$ matrix of weights.

Furthermore, the weight matrix $W$ is chosen not necessarily to yield a null forcing function $\widetilde{\mathbf{S}}$ for specific rows, although that is ideal, but rather to reduce the stiffness in the forcing term $\widetilde{\mathbf{S}}$. Again, there are no universal rules for this grouping process. With experience, chemists have formed groups of constituents with simple components such as the $O_x$ group composed of O and $O_3$ or the $NO_x$ group composed of NO and $NO_2$.

Accompanying the chemical "family" grouping approach is the technique of photochemical equilibrium. In this technique the family equation is formed as discussed, and the remaining individual family members are computed by assuming $d\mu_i/dt \simeq 0$ or $S_i(\mu,t) \simeq 0$. Thus not only is the original system [Eq. (12.9)] transformed to a less stiff, more smoothly time varying counterpart, but the actual number of fully time-dependent equations in the system is reduced. Although this may seem like a panacea, care must be exercised in choosing the groups. Many times the equations for photochemical equilibrium within a group can be nonlinear and difficult to solve, or worse, the assumption that $d\mu_i/dt \approx 0$ may not hold (*e.g.*, during nighttime).

In principle, *aerosols* can be modeled in the same way as molecules, by writing a balance equation for classes of particles with a given size and chemical composition (see Box 12.4). However, this rapidly leads to an inflation of the number of equations. For instance, an accurate description of nucleation, condensation, and coagulation of a single aerosol species in the size range between 1 nm and 1 $\mu$m would require at

---

**Box 12.4 General Equation for Aerosol Dynamics**

State-of-the-art aerosol models provide estimates of the size distribution and chemical composition of particles in the atmosphere, and, in most cases, are coupled with a gas-phase chemical model. The general equation for aerosol dynamics (see, *e.g.*, Gelbard and Seinfeld, 1979; Seinfeld and Pandis, 1998) accounts for nucleation, condensation, coagulation, transport, and surface deposition (see Chapter 4). Neglecting the effects of atmospheric transport (described in Section 12.4) and adopting the mass $m$ of the aerosol particles as the independent variable, the *number density* of particles $n(m,t)\,dm$ at time $t$ having a mass in the range $[m, m+dm]$, can be derived from the following conservation equation (Meng et al., 1998)

$$\begin{aligned}\frac{\partial n(m,t)}{\partial t} = &-\frac{\partial}{\partial m}\left[I(m,t)\,n(m,t)\right] \\ &+\frac{1}{2}\int_0^m \beta(m',m-m',t)\,n(m',t)\,n(m-m',t)\,dm' \\ &-n(m,t)\int_0^\infty \beta(m,m',t)\,n(m',t)\,dm' \\ &+\widetilde{S}(m,t)\end{aligned} \qquad (1)$$

where $I(m,t)$ represents the rate of change in the total mass of a particle of mass $m$ as a result of condensation and evaporation, $\beta(m,m')$ is the coefficient accounting for coagulation between particles of masses $m$ and $m'$, and $\widetilde{S}(m,t)$ is a net source term that accounts for processes such as nucleation, gravitational settling, and wet scavenging. Theoretical expressions for each of these quantities in relation to different physical processes are given, for example, in Seinfeld and Pandis (1998).

Note that the aerosol distribution, and hence the general equation for aerosol dynamics, can also be written in terms of aerosol volume or, if the particles are assumed to be spherical, as a function of aerosol diameter or radius.

A general dynamic equation analogous to (1) provides the size distribution of the aerosol *mass density* for each chemical species $i$ included in the model. If transport (advection and diffusion) and particle settling are now taken into account, this equation can be written (Wexler *et al.*, 1994; Lurmann *et al.*, 1997)

$$\frac{\partial \rho_i(m,\mathbf{x},t)}{\partial t} + (\mathbf{v}(\mathbf{x},t) - v_s(m)\mathbf{k})\boldsymbol{\nabla}\rho_i(m,\mathbf{x},t)$$
$$- \boldsymbol{\nabla}(\mathbf{K}(\mathbf{x},t)\boldsymbol{\nabla}\rho_i(m,\mathbf{x},t))$$
$$= H_i(m,\mathbf{x},t)\rho(m,\mathbf{x},t) - \frac{\partial(m\rho_i H)}{\partial m}$$
$$+ \int_0^m \beta(m', m-m', \mathbf{x},t)\rho_i(m',\mathbf{x},t)\frac{\rho(m-m',\mathbf{x},t)}{m-m'}\,dm'$$
$$- \rho_i(m,\mathbf{x},t)\int_0^\infty \beta(m',m,\mathbf{x},t)\frac{\rho(m',\mathbf{x},t)}{m'}\,dm'$$
$$+ S_i(m,\mathbf{x},t) \qquad (2)$$

where $\rho(m,\mathbf{x},t)$ is the total mass distribution at location $\mathbf{x}$ and time $t$, $\rho_i(m,\mathbf{x},t)\,dm$ is the mass concentration of species $i$ in the mass range $[m, m+dm]$ with

$$\rho = \sum_{i=1}^N \rho_i$$

if $N$ is the number of chemical species, $m_i$ is the mass of species $i$ in a particle of total mass

$$m = \sum_{i=1}^N m_i$$

$H_i = (1/m)\,dm_i/dt$ is the inverse of the characteristic time for particle size change from condensation or evaporation, with

$$H = \sum_{i=1}^N H_i$$

$\beta(m,m') = \beta(m',m)$ is the binary coagulation coefficient, $\mathbf{v}$ is the wind velocity vector, $v_s$ is the particle settling velocity, $\mathbf{k}$ is the unit vector in the vertical direction, $\mathbf{K}(\mathbf{x},t)$ is the turbulent diffusion tensor, and $S_i$ is the net source term of species $i$ that accounts for nucleation, emission and mass change due to chemical reactions.

Note that coagulation reduces the total particle number density $n$, but does not modify the total aerosol mass density $\rho$. On the other hand, condensation increases total particle mass, but does not affect the total number of particles.

Numerical techniques used to solve the general dynamic equation have been reviewed by Seigneur *et al.* (1986). In the so-called *sectional representation*, the continuous aerosol mass or size distribution is approximated by a series of step functions (Gelbard and Seinfeld, 1980; Warren and Seinfeld, 1985) covering a finite number of sections or "bins". It is assumed that the aerosol density is uniformly distributed within each section, and, for each of the $N$ chemical species, the general dynamic equation is replaced by a system of $K$ differential equations, where $K$ is the number of bins used in the model.

least 30 size bins. Most approaches to simplify the description of the size distribution are based on the notion that any distribution $n(Dp)$ can be accurately reconstructed from its moments $M_0, M_1, M_2, \ldots, M_\infty$, with (see Section 4.4.2)

$$M_k = \int_0^\infty D_p^k \, n\,(Dp) \, dDp \tag{12.21}$$

where $Dp$ is the particle diameter.

Rather than tracking the change of the full-size distribution, simplified models track the lower-order moments in time, which have a physical meaning ($M_0$ = total number concentration, $M_1$ = mean diameter, $M_2$ = total surface, ...). For certain problems in aerosol dynamics, a closed set of balance equations can be obtained, in terms of only a few moments (Friedlander, 1983; McGraw, 1997). In atmospheric applications, this is not possible, and an assumption regarding the form of the size distribution needs to be made. One sensible assumption is that the distribution is lognormal (Section 4.4.2) or a sum of lognormals (Whitby, 1978). In the case of one lognormal, the distribution is determined by the total number concentration $N$ [see Eq. (4.6)], the geometric mean diameter, and geometric standard deviation. The latter two can be described in terms of $M_0$ and two arbitrary other moments (Whitby et al., 1991). The most drastic assumption is to use monodisperse distributions, in which case only $M_0$ and $M_1$ need to be tracked.

Whatever method is used, care must be taken that simplified models still capture the nonlinearity inherent in aerosol dynamics (Raes and Van Dingenen, 1995).

## 12.4 Modeling Atmospheric Transport

In atmospheric models, the transport of mass can be represented by advection, diffusion, and convection (Chapter 2). Advective transport of chemical species is driven by air motion or winds. Air motions with spatial scales smaller than the resolution of the grid used in the model, and which contribute significantly to the transport of constituents, must be parameterized. Turbulence, for example, in the planetary boundary layer, an important process that can rapidly transfer gases released at the surface to the free troposphere, is parameterized by an eddy diffusion coefficient along each spatial dimension. Deep convection by cumulo-nimbus clouds is another important subscale process that rapidly displaces air from the planetary boundary layer to the upper troposphere. Convection is parameterized by a variety of methods.

In atmospheric models, transport terms are frequently approximated with finite-difference analogues due to their efficiency and simplicity. Another discretization technique, the finite-element method, while computationally more expensive, has been successfully implemented in three-dimensional chemistry-transport models (*e.g.*, CALGRID model, Yamartino *et al.*, 1992), as well as in weather forecasting models (*e.g.*, Canadian regional weather forecasting model, Tanguay *et al.*, 1989). Spectral formulations of the transport equations (truncated series expansions) are sometimes adopted in spectral general circulation models (especially for water vapor), but tend to produce oscillations in the vicinity of sharp gradients, and are therefore being progressively abandoned.

### *12.4.1 Advection*

There is no single universal method that can be recommended to approximate advection. A numerical method is selected that meets a subset of the following criteria simultaneously:

- *Accuracy*, which for a simple problem can be estimated by comparing the numerical solution with its analytical counterpart;
- *Stability*, which often imposes a restriction on the time step;
- *Transportivity*, which requires that any perturbation is advected downwind;
- *Locality*, such that the solution of the advection problem at a given point is not significantly influenced by the field far from that point;
- *Conservation*, which requires that no gain nor loss of mass occurs during the transport;
- *Monotonicity (shape preserving)*, through which the occurrence of new extrema is prohibited; these extrema (noise) are characterized by undershoots and overshoots near regions of strong gradients;
- *Efficiency*, such that the computer time consumed is not prohibitive, and the storage requirement does not exceed computer capacity.

A review of many of the methods is given, for example, by Oran and Boris (1987) and by Rood (1987). Performance comparisons have been reported by Høv et al. (1989) or Chock (1991).

Modern transport models generally rely on numerical algorithms that are too complex to be described here in any detail. As an introduction to these advanced techniques, it is useful to consider simple algorithms for the one-dimensional advection equation

$$\frac{\partial \psi}{\partial t} + u \frac{\partial \psi}{\partial x} = 0 \qquad (12.22)$$

where $\psi$ is the advected quantity (*e.g.*, density), $t$ is time, and $u$ is the velocity in the $x$ direction. Depending on the choice of the method, the numerical solution can exhibit significant artificial diffusion and/or dispersion. Figure 12.2 shows the behavior of diffusive and dispersive schemes compared to the exact solution (superimposed in bold) when a triangular pattern is advected. Many numerical methods have been proposed in an attempt to minimize this behavior.

Three groups of advective schemes can be distinguished: the Eulerian, Lagrangian, and semi-Lagrangian techniques. The Eulerian formulation is referred to an observer (fixed at the spatial grid points), and Eq. (12.22) is its mathematical expression. The Lagrangian and semi-Lagrangian approaches consider the observer fixed on air parcels displaced along trajectories, and their mathematical expression is the corollary of Eq. (12.22)

$$\frac{d\psi}{dt} = 0 \quad \text{with} \quad \frac{d}{dt} = \frac{\partial}{\partial t} + u \frac{\partial}{\partial x} \qquad (12.23)$$

*(a) Eulerian algorithms.* In Eulerian schemes, the classical choice for many applications, the spatial grid points are fixed and the flux of air mass passing through them is computed. The general limitation of these methods comes from the stability criteria. They are stable provided that the Courant-Friedricks-Lewy (CFL) condition

$$\frac{|u|\Delta t}{\Delta x} \leq C \qquad (12.24)$$

where $C$ is a constant of order unity, is satisfied ($\Delta t = t_{n+1} - t_n$ is the time step and $\Delta x = x_{j+1} - x_j$ is the space interval between grid points $j$ and $j+1$). This relationship bounds the maximum time step by the ratio of the spatial resolution and the wind speed. As the spatial resolution of the model increases, the time step must decrease proportionally. Table 12.1 presents popular algorithms of this group. The Euler forward scheme is unconditionally unstable and must be rejected. The Chapeau and upwind methods are most efficient, but the first is dispersive and the second is diffusive. The leapfrog scheme is shape preserving, but is not monotonic (noise near strong gradients). Bott (1989) has presented a positive definite, second-order algorithm that offers a good compromise between accuracy and computational cost. A very accurate algorithm, which is recommended for coarse resolution grids, has been proposed by Prather (1986). This algorithm represents the tracer concentration inside a grid box by second-order polynomials and its spatial distribution by second-order moments. The method is characterized by small numerical diffusion, but is computationally expensive since, at each grid point, 10 variables (in the three-dimensional case) need to be stored. A comparison of the positive-definite scheme of Bott (1989) and the method of Prather (1986) for a 1D advective procedure is shown in Figure 12.3 (Müller, 1992a). Schemes with higher-order accuracy (>1) preserve shape, but are not monotonic. The solution must then be filtered to eliminate noise, which, for nonlinear chemical systems, can lead to a rapid divergence of the solution. The filter can be linear (*e.g.*, Shapiro, 1971) or nonlinear (*e.g.*, Forester, 1977), but in any case must remove noise without significantly reducing the amplitude of the resolved waves. The only class of Eulerian scheme that produces stable solutions, maintains steep gradients, and preserves monotonicity is that of the nonlinear algorithms. These

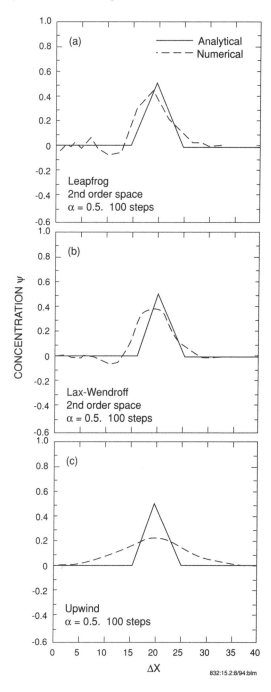

**Figure 12.2.** Performance of three numerical schemes for the one-dimensional advection of a triangular function: (a) leap frog, (b) Lax-Wendroff, and (c) upwind methods (100 time steps with Courant number $\alpha = 0.5$). The exact (analytical) solution is also shown (adapted from Rood, 1987).

### Table 12.1
*Algebraic (Discretized) Schemes for the Advection Equation*

| Scheme | Algebraic Form | Stability | Accuracy |
|---|---|---|---|
| Euler forward | $\Psi_j^{n+1} = \Psi_j^n + \frac{u\Delta t}{2\Delta x}\left(\Psi_{j-1}^n - \Psi_{j+1}^n\right)$ | Unstable | $O\left(\Delta t, \Delta x^2\right)$ |
| Upwind | $\Psi_j^{n+1} = \Psi_j^n + \frac{u\Delta t}{\Delta x}\left(\Psi_{j-1}^n - \Psi_j^n\right)$ | $\frac{u\Delta t}{\Delta x} \leq 1$ | $O\left(\Delta t, \Delta x\right)$ |
| Leapfrog | $\Psi_j^{n+1} = \Psi_j^{n-1} + \frac{u\Delta t}{\Delta x}\left(\Psi_{j-1}^n - \Psi_{j+1}^n\right)$ | $\frac{u\Delta t}{\Delta x} \leq 1$ | $O\left(\Delta t^2, \Delta x^2\right)$ |
| Lax-Wendroff | $\Psi_j^{n+1} = \Psi_j^n + \frac{u\Delta t}{2\Delta x}\left(\Psi_{j-1}^n - \Psi_{j+1}^n\right) +$ $\frac{1}{2}\left(\frac{u\Delta t}{\Delta x}\right)^2 \left(\Psi_{j-1}^n - 2\Psi_j^n + \Psi_{j+1}^n\right)$ | $\frac{u\Delta t}{\Delta x} \leq 1$ | $O\left(\Delta t^2, \Delta x^2\right)$ |
| Crank-Nicholson | $-\frac{u\Delta t}{2\Delta x}\Psi_{j-1}^{n+1} + 2\Psi_j^{n+1} + \frac{u\Delta t}{2\Delta x}\Psi_{j+1}^{n+1} =$ $\frac{u\Delta t}{2\Delta x}\left(\Psi_{j-1}^n - \Psi_{j+1}^n\right)$ | Stable | $O\left(\Delta t^2, \Delta x^2\right)$ |
| Chapeau (FEM) | $\left(\frac{1}{3} - \frac{u\Delta t}{2\Delta x}\right)\Psi_{j-1}^{n+1} + \frac{4}{3}\Psi_j^{n+1} + \left(\frac{1}{3} + \frac{u\Delta t}{2\Delta x}\right)\Psi_{j+1}^{n+1} =$ $\left(+\frac{1}{3} - \frac{u\Delta t}{2\Delta x}\right)\Psi_{j-1}^n + \frac{4}{3}\Psi_j^n + \left(\frac{1}{3} - \frac{u\Delta t}{2\Delta x}\right)\Psi_{j+1}^n$ | Stable | $O\left(\Delta t^2, \Delta x^4\right)$ |

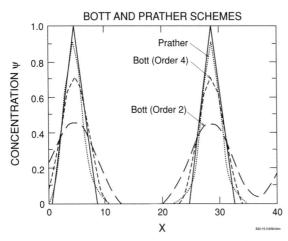

**Figure 12.3.** Comparison of the Prather scheme and Bott algorithms (order 2 and 4) for the one-dimensional advection of two triangular functions (adapted from Müller, 1992a).

schemes are in general complex and time consuming. Smolarkiewicz (1984), for example, has introduced in the upwind scheme an appropriate antidiffusive velocity that reduces the numerical diffusion characteristic of this algorithm.

(b) *Lagrangian algorithms.* In the Lagrangian schemes (*e.g.*, Walton et al., 1988; Taylor et al., 1991), distinct air parcels, in which the tracers are assumed to be homogeneously mixed, are followed as they are displaced by the winds. Lagrangian schemes are simple in concept and are not subject to spurious diffusion. However, the accumulation of errors in the determination of the parcel location introduces a limitation when applied to global chemistry-transport models. Nevertheless, in spite of its shortcomings, the Lagrangian method is used to determine, for example, the history, over several days, of an air parcel reaching a given location in the atmosphere (trajectory calculations). Lagrangian algorithms are frequently used in regional pollution studies, that is, to compute the dispersion of pollutants in industrialized areas. A review of these models is given by Pasquill and Smith (1983).

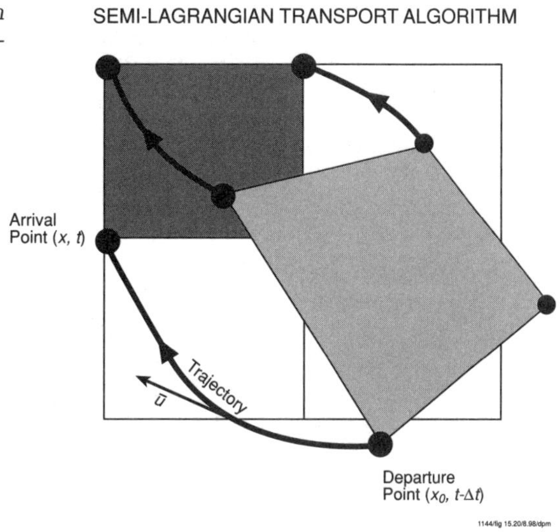

**Figure 12.4.** *Schematic description of the semi-Lagrangian transport algorithm.*

(c) *Semi-Lagrangian algorithms.* In semi-Lagrangian formulations (Robert, 1981), the solution on prescribed grid points is derived at each time step on the basis of a Lagrangian backward calculation (Fig. 12.4). In the absence of dissipative processes (*e.g.*, chemical reactions), the tracer value (*e.g.*, mixing ratio) is conserved along the trajectory. The initial position ($x_0$) of an air parcel, which after one time step arrives at the mesh point ($x$), is calculated by

$$x_0 = x + \int_t^{t-\Delta t} u(x,t)\, dt \qquad (12.25)$$

In general, the point at position $x_0$ does not coincide with a grid point, so that the velocity $u(x,t)$ has to be estimated by interpolation. The success of the method (accuracy, monotonicity, shape preserving) is greatly dependent on the interpolation scheme used (see Staniforth and Côté, 1991, for a review). Linear interpolations, while simple and monotonic, introduce strong numerical diffusion. This drawback can be avoided with higher-order methods (Williamson and Rasch, 1989), but, once again, their implementation is complex. Intrinsically, interpolation methods are not conservative. Therefore, most chemical-transport models include a "fixer" applied after the advection calculation to correct for deviations from exact mass conservation.

Semi-Lagrangian advection schemes are now commonly used in general circulation models (*e.g.*, to simulate the transport of water vapor) and in global tracer models.

### 12.4.2 Subgrid Transport

Transport processes that occur at a geometric scale smaller than the grid size of the model (typically several hundreds of kilometers in the case of a global chemical transport model) or at a temporal scale smaller than the time step of the model (typically 30 min in these models) cannot be represented explicitly by the advective scheme. Several formulations with different levels of complexity have been used to account for subgrid boundary layer exchanges and convective transport in the free troposphere.

### 12.4.2.1 Boundary Layer Exchanges

Mixing processes in the planetary boundary layer occur mainly during the day when the solar heating induces buoyancy. The air is rapidly transported from the surface up to approximately one kilometer above the ground ("bottom-up" motion). At the same time, there is large, slow, downward transport to preserve mass conservation ("top-down" transport). Turbulence is expressed, mathematically, by a number of unknowns in the sets of equations for turbulent flow greater than the number of equations (closure problem). A good compromise is first-order closure employing a vertical diffusion coefficient that is a function of temperature and velocity gradients at the surface and incorporates the asymmetry between "top-down" and "bottom-up" motions (Wyngaard and Brost, 1984). A more sophisticated model, such as the large eddy simulation model of Moeng (1984), resolves the equations of the turbulent part of the variables and can be used for validation of the vertical diffusion parameterization (Holtslag and Moeng, 1991).

The "diffusive transport" of a quantity $\psi$ in direction $x$ (one-dimensional problem) can be formally expressed by the following equation

$$\frac{\partial \psi}{\partial t} = \frac{\partial}{\partial x}\left(K \frac{\partial \psi}{\partial x}\right) \quad (12.26)$$

in which $K$ is the so-called diffusion coefficient and $t$ is time. Table 12.2 (in which $K$ is assumed to be a constant) presents several algorithms that are available to solve this equation. A more general analysis can be found in Fletcher (1988). The explicit (Euler) scheme is stable only if

$$\frac{2K\Delta t}{(\Delta x)^2} \leq 1 \quad (12.27)$$

while the Crank-Nicholson and the fully implicit scheme are unconditionally stable and are used in most applications. These two schemes result in tridiagonal algebraic systems that can easily be solved.

Multidimensional problems can be efficiently solved by considering each spatial direction separately. For example, the alternating direction method (Peaceman and Rachford, 1955) is second-order accurate and unconditionally stable.

### 12.4.2.2 Convective Transport

Transport in deep convective clouds (*e.g.*, cumulonimbus clouds) greatly changes the distribution of chemical species in the troposphere. Again, convective transport is very localized: In a region typically covering the area of a grid cell in a global model (*e.g.*, 100 km × 100 km), perhaps 99.9% of the area is not affected by convective updraft; in the remaining 0.1%, the vertical exchanges between the boundary layer and the upper troposphere take place in approximately 1 h. As in general circulation models, parameterizations must be introduced to estimate the effect of these localized, but intense, processes on the global distribution of trace constituents. The simplest of these parameterizations is based on a vertical eddy diffusion, which is increased by several orders of magnitude in convectively unstable regions (see, *e.g.*, Mahlman and Moxim, 1978; Levy *et al.*, 1982, 1985). Although such a method of local exchanges might improve the calculated distribution of long-lived species, it is generally regarded

## Table 12.2
### Algebraic (Discretized) Schemes for the Diffusion Equations

| Scheme | Algebraic Form | Stability | Accuracy |
|---|---|---|---|
| Euler forward | $\Psi_j^{n+1} = \Psi_j^n + \frac{K\Delta t}{\Delta x^2}\left[\Psi_{j-1}^n - 2\Psi_j + \Psi_{j+1}^n\right]$ | $\frac{K\Delta t}{\Delta x^2} \leq 0.5$ | $O(\Delta t, \Delta x^2)$ |
| Crank-Nicholson | $-\frac{K\Delta t}{2\Delta x^2}\Psi_{j-1}^{n+1} + \left(1 + \frac{K\Delta t}{\Delta x^2}\right)\Psi_j^{n+1} - \frac{K\Delta t}{2\Delta x^2}\Psi_{j+1}^{n+1} =$ $\frac{K\Delta t}{2\Delta x^2}\Psi_{j-1}^n + \left(1 - \frac{K\Delta t}{\Delta x^2}\right)\Psi_j^n + \frac{K\Delta t}{2\Delta x^2}\Psi_{j+1}^n$ | Stable | $O(\Delta t, \Delta x^2)$ |
| Fully Implicit | $-\frac{K\Delta t}{2\Delta x^2}\Psi_{j-1}^{n+1} + \left(1 + \frac{K\Delta t}{\Delta x^2}\right)\Psi_j^{n+1}$ $-\frac{K\Delta t}{2\Delta x^2}\Psi_j^{n+1} = \Psi_j^n$ | Stable | $O(\Delta t, \Delta x^2)$ |
| Chapeau (FEM) | $\left(\frac{1}{6} - \frac{K\Delta t}{2\Delta x^2}\right)\Psi_{j-1}^{n+1} + \left(\frac{2}{3} + \frac{K\Delta t}{\Delta x^2}\right)\Psi_j^{n+1}$ $+ \left(\frac{1}{6} - \frac{K\Delta t}{2\Delta x^2}\right)\Psi_{j+1}^{n+1} = \left(\frac{1}{6} + \frac{K\Delta t}{2\Delta x^2}\right)\Psi_{j-1}^n$ $+ \left(\frac{2}{3} - \frac{K\Delta t}{\Delta x^2}\right)\Psi_j^n + \left(\frac{1}{6} - \frac{K\Delta t}{2\Delta x^2}\right)\Psi_{j+1}^n$ | Stable | $O(\Delta t, \Delta x^2)$ |

as inappropriate for species with lifetimes of less than a month. In addition, because the rapid updraft inside clouds is accompanied by slow downward motion in the surrounding area, and even more complex processes such as side entrainment and detrainment, more detailed parameterization schemes must be implemented (see Fig. 12.5). These range from "quasidiffusive" schemes (such as the scheme in Hack (1994), which, in regions of vertical conditional instability, redistributes heat, moisture, and chemical compounds of a given model layer into the two neighboring layers during a given time step) to so-called "entraining plume" models. Among these latter formulations, which can move air parcels over the entire troposphere during a given time step, are the parameterizations proposed by Tiedtke (1989), Feichter and Crutzen (1990), and Emanuel (1991). The performance of these and other convective transport schemes, which are used in different chemical-transport models, is analyzed in one dimension by Mahowald et al. (1995). These schemes account easily for the asymmetry between upward and downward transport and are mass conservative.

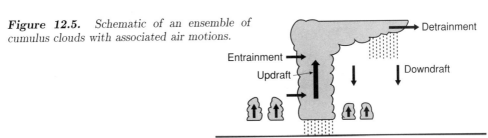

**Figure 12.5.** Schematic of an ensemble of cumulus clouds with associated air motions.

Convective transport parameterization is a major source of uncertainty in chemical transport models. Schemes have to be evaluated by comparing the calculated concentration of gases such as $^{222}$Rn with observations (see, e.g., Brost and Chatfield, 1989; Jacob and Prather, 1990). $^{222}$Rn is emitted by soils with a well-established mean flux value (see, e.g., Turekian et al., 1977; Lambert et al., 1982) and is only removed by radioactive decay (with a half-life of 3.824 days, corresponding to an e-folding time of 5.52 days). Figure 12.6 shows the vertical profile of $^{222}$Rn calculated with different one-dimensional convective schemes. The profiles are compared to the average of 16 vertical profiles from the western U.S. summer. An intercomparison and evaluation of several global 3D atmospheric transport models using $^{222}$Rn and other short-lived tracers has recently been presented by Jacob et al. (1997). Finally, it is worth noting that cloud convection acts very differently on soluble and non-soluble species. Convective precipitation is a very efficient mechanism for removing soluble compounds from the atmosphere. The parameterization of wet deposition is not straightforward, since it has to account for the physical state of the clouds (water droplets, ice, snow) and for complex processes such as below-cloud evaporation (Section 4.7.2).

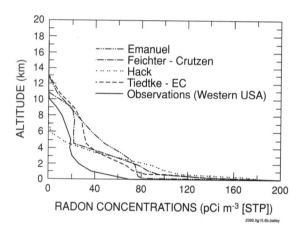

*Figure 12.6.* Vertical distribution of $^{222}$Rn (expressed in $pCi\ m^{-3}$) calculated by different one-dimensional moist convective schemes and a simple adiabatic mixing boundary layer scheme. A profile representative of 16 vertical summer profiles in the western U.S. (taken from Liu et al., 1984) is also shown (adapted from Mahowald et al., 1995). Advection of air from oceanic as well as continental regions, which affect the observed vertical profile is ignored in this comparison of cumulus parameterizations.

## 12.5 Examples and Illustrations

Existing models of the atmosphere vary greatly in their complexity and their computational requirements. Zero-dimensional models (which do not explicitly account for transport processes) are best suited for studying complex chemical systems at a given location, or for following an air parcel along its trajectory in the atmosphere. One-dimensional models provide vertical distributions of chemical compounds, but do not account for horizontal inhomogeneities associated with solar insolation and transport processes; they offer the advantage of rapid calculations even with fairly detailed chemistry. Two-dimensional models (latitude, altitude) are often used for the stratosphere where longitudinal exchanges are rapid. Finally, three-dimensional chemical transport models provide the most realistic representation, but are usually computer intensive. Especially in the case of global models, their spatial resolution must remain limited to 100 km or more. Regional three-dimensional models can use much higher resolution and formulate more explicitly some of the physical processes over a limited geographical domain, as long as suitable lateral boundary conditions are prescribed.

**Box 12.5 Behavior of Nonlinear Chemical Systems**

Chemical systems in the atmosphere are generally nonlinear. In certain cases, they exhibit different regimes, and transitions between these regimes. This is the case, for example, for the tropospheric $O_x$, $HO_x$, $NO_x$, CO, $CH_4$ system which exhibits a turnaround in the ozone and OH concentrations as $NO_x$ increases through a critical level of approximately 0.5-1 ppbv (White and Dietz, 1984; Kasting and Ackerman, 1985; Stewart, 1995). Transition zones between two chemical regimes are often characterized by multiple and/or unstable solutions (Stewart, 1993; Hess and Madronich, 1997). Figure 12.7 represents the number density of ozone as the external NO source increases, with the stable and unstable portions of the solution. Note the sudden and discontinuous change in the steady-state concentration of ozone that occurs when the NO source passes through a critical level. For low CO sources, the jump with decreasing NO sources occurs at a lower source value than does the opposite jump with increased sources. Under specific conditions, the solution can become oscillatory (Hess and Madronich, 1997) with periods that can vary from years to thousands of years.

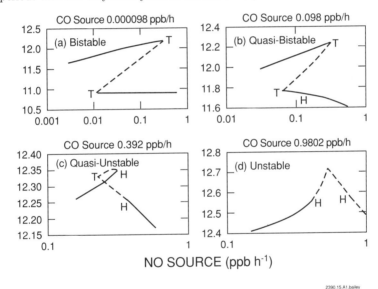

*Figure 12.7.* Logarithm of the ozone number density as a function of the NO source for given CO sources (ppb $h^{-1}$). Stable and unstable portions of the solutions are represented by solid and dashed lines, respectively. The labels T and H refer to turning points and Hopf bifurcation points, respectively (see Stewart, 1995, for more details).

## 12.5.1 Zero-Dimensional Models

Zero-dimensional chemical models simulate the evolution of a chemical system inside a single, well-mixed air parcel and neglect any coupling between chemistry and transport. The system is represented by $N$ differential equations ($N$ being the number of chemical species involved), which are generally nonlinear and numerically stiff (see Section 12.2). Time-dependent factors such as the photolytic radiation and temperature affect the production and destruction of the chemical species through chemical and photochemical reaction rates. These models are used mostly to analyze complex chemical systems (Box 12.5) that encompass a large number of reactions. As

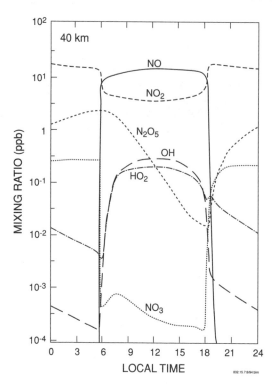

**Figure 12.8.** Diurnal variation in the mixing ratio of several chemical species at 40 km altitude calculated by the Gear method (Ramaroson et al., 1992).

shown by Figure 12.8, they are also used to calculate the diurnal evolution of the fast-reacting species belonging to a given chemical family, and to perform sensitivity analyses (Box 12.6).

### 12.5.2 One-Dimensional Models

Much of the pioneering work in atmospheric chemistry and aeronomy has been performed using simple one-dimensional models. In these models, which ignore all variations with longitude and latitude and hence treat the atmosphere as a vertical column, the exchanges of mass along the vertical are represented by an eddy diffusion formulation

$$\frac{\partial \mu_i}{\partial t} = \frac{1}{\rho_a} \frac{\partial}{\partial z} \left( K_z \rho_a \frac{\partial \mu_i}{\partial z} \right) + S_i \qquad (12.28)$$

where the eddy diffusion coefficient $K_z$ is determined empirically, for example, from measurements of long-lived tracers and $z$ is the altitude. Typical values for $K_z$ are 10 m$^2$ s$^{-1}$ in the troposphere, where vertical mixing is rapid, 0.5 m$^2$ s$^{-1}$ in the lower stratosphere, where vertical exchanges are slow due to the temperature inversion, and 100 m$^2$ s$^{-1}$ in the mesosphere, where gravity wave breaking associated with the negative vertical temperature gradient produces intense mixing. One-dimensional models provide a very crude representation of dynamical exchanges, but, because they are computationally inexpensive, are used mainly to make sensitivity studies or to perform long-term integrations. Figure 12.9 shows typical output provided by a one-dimensional model of the troposphere (Kasting and Singh, 1986).

### 12.5.3 Two-Dimensional Models

Two-dimensional models are widely used to simulate dynamical (temperature, winds, etc.) and chemical processes in atmospheric regions such as the stratosphere and

> **Box 12.6 Sensitivity Analyses with Chemical Models**
>
> When calculated in a numerical model, the atmospheric concentration of species $i$ ($i = 1, N$) depends on various input parameters $\alpha_j$ (e.g., chemical rate constants, boundary conditions) whose values are known with a given uncertainty. The impact of such uncertainties of input parameters on model outputs needs to be estimated. It is also useful to identify which parameters most affect solutions of the model equations. Methods used for sensitivity analyses are described, for example, by Rabitz et al. (1983).
>
> Introducing the sensitivity coefficient $s_{ij}$ of concentration $\mu_i$ to input parameter $\alpha_j$
>
> $$s_{ij} = \frac{\partial \mu_i}{\partial \alpha_j}$$
>
> and differentiating the kinetics equation with respect to $\alpha_j$ (12.9)
>
> $$\frac{\partial \mu_i}{\partial t} = S_i\left(\boldsymbol{\mu}, \boldsymbol{\alpha}\right) \quad (1)$$
>
> where $\boldsymbol{\mu}$ is the concentration vector, $\boldsymbol{\alpha}$ the input parameter vector, and $S_i$ the net source term of species $i$, the following equations are obtained ($i = 1, N$)
>
> $$\frac{\partial s_{ij}}{\partial t} = \sum_{m=1}^{N} \frac{\partial S_i}{\partial \mu_m} s_{mj} + \frac{\partial S_i}{\partial \alpha_j} \quad (2)$$
>
> These equations, linear in the sensitivity coefficients $s_{ij}$, are coupled to the kinetics equations since the Jacobian $\partial S_i/\partial \mu_m$ and the inhomogeneity term $\partial S_i/\partial \alpha_j$ are generally dependent on the concentrations determined from Eq. (1). Several methods are available to solve this coupled system. The direct method can easily become burdensome if the number of chemical constituents ($N$) is large, and if one desires to test the sensitivity of the chemical system to a large number of parameters. Other methods, such as those based on the Green's function (see, e.g., Dougherty et al., 1979) or adjoint functions (e.g., Hall and Cacuci, 1983), are receiving increasing attention. Monte Carlo techniques are also used to assess the effect of chemical kinetics uncertainties on calculated concentrations (Thompson and Stewart, 1991). An emerging approach to derive the sensitivity of model output to model input is to use automated computer programs to differentiate the original model source code.

mesosphere, where the conditions are more homogeneous than in the troposphere. In these models, each three-dimensional field $\psi(\lambda, \phi, z)$ is longitudinally averaged, resulting in the zonal mean variable $\overline{\psi}(\phi, z)$ defined by

$$\overline{\psi}(\phi, z) = \frac{1}{2\pi} \int_{2\pi} \psi(\lambda, \phi, z)\, d\lambda \quad (12.29)$$

where $\lambda$ is the longitude and $\phi$ is the latitude (see Section 2.4). The average defined by Eq. (12.29) and noted by an overbar is calculated for fixed values of coordinates $\phi, z$, and $t$, and is therefore referred to as an Eulerian mean. It provides a simple framework in which the atmospheric quantities are separated into mean and deviation quantities. The primitive equations for the Eulerian-mean flow and tracer distribution can be easily deduced from Eqs. (2.92) to (2.97). One difficulty with this approach, however, is that the zonally averaged flux of any quantity $\psi$ cannot be derived from

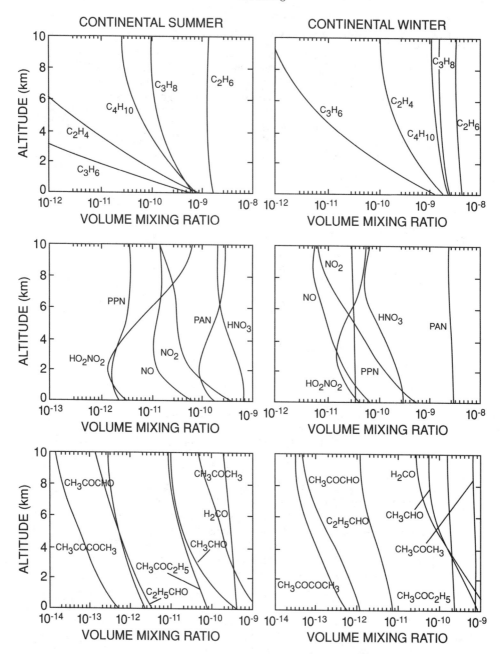

**Figure 12.9.** Vertical distribution of several hydrocarbons (upper panels), nitrogen compounds (middle panels), and oxygenated organic compounds (lower panels) calculated by the 1D model of Kasting and Singh (1986) at tropospheric heights for continental summer and winter conditions.

the product of the mean quantity $\overline{\psi}$ and the mean velocity $\overline{v}$, but is expressed by

$$\overline{\psi v} = \overline{\psi} \cdot \overline{v} + \overline{\psi' v'} \qquad (12.30)$$

if the variables denoted by a prime represent the departure from the mean

$$\psi'(\lambda, \phi, z) = \psi(\lambda, \phi, z) - \overline{\psi}(\phi, z) \qquad (12.31)$$

The eddy flux term $\overline{\psi'v'}$ accounts for the flux associated with atmospheric disturbances (*e.g.*, Rossby waves). Its value is generally provided by a closure relation expressed, for example, by an eddy diffusion formulation. Because of the large degree of cancellation between the poorly quantified eddy and mean transport terms in the basic equations (see Chapter 2), Eulerian two-dimensional models are relatively inaccurate and are often replaced by transformed Eulerian mean (TEM) models (see, *e.g.*, Section 2.4.2 and Garcia and Solomon, 1983; 1994; Brasseur *et al.*, 1990). These models are formulated as a function of the residual mean meridional circulation [see definition and Eq. (2.110) in Chapter 2].

Although the TEM formulation is simpler and more meaningful than the classic Eulerian approach, closure equations have to be added to the primitive equations to account for chemical eddy transport and subgrid effects. In prognostic models that derive the mean meridional circulation and temperature consistently with the zonal mean distribution of chemical constituents, the forcing of planetary and gravity waves is expressed in the zonal mean equation (2.105) through the Eliassen-Palm (EP) flux divergence. Several parameterizations of this latter quantity as a function of the mean dynamical state of the atmosphere have been proposed (Hitchman and Brasseur, 1988; Garcia *et al.*, 1992). The parameterization proposed by Garcia (1991) is based on the solution of the quasigeostrophic potential vorticity equation with a dissipative term accounting for damping by Rayleigh friction, Newtonian cooling, and wave breaking.

Forcing resulting from gravity wave breaking also affects the momentum budget of the middle atmosphere and is believed to be responsible for the observed deceleration of the zonal winds in the mesosphere and the summer cold mesopause temperatures. Vertical mixing of trace constituents also occurs in relation with gravity wave breaking. Parameterizations have been proposed, for example, by Lindzen (1981) and by Fritts and Lu (1993).

Other sources of momentum, especially in the tropics, may be responsible for the occurrence of the semiannual and quasibiennial oscillations. Several attempts have been made to simulate these effects in two-dimensional models (Garcia *et al.*, 1992; Gray and Pyle, 1986; 1987), but more work is clearly needed.

The meridional distribution of nitrous oxide derived by Garcia *et al.* (1992) for January is shown in Figure 12.10. Many of the features appearing in the observations by the space-borne Stratospheric and Mesospheric Sounder Instrument (Jones and

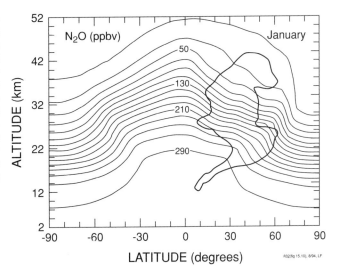

**Figure 12.10.** Global distribution of nitrous oxide (ppbv) calculated by the model of Garcia et al. (1992) on January 30. The heavy contour superimposed on the $N_2O$ mixing ratio lines encloses the region where the eddy diffusion coefficient associated with planetary wave breaking exceeds $10^6$ $m^2$ $s^{-1}$ and where, consequently, the meridional gradient of $N_2O$ is relatively weak.

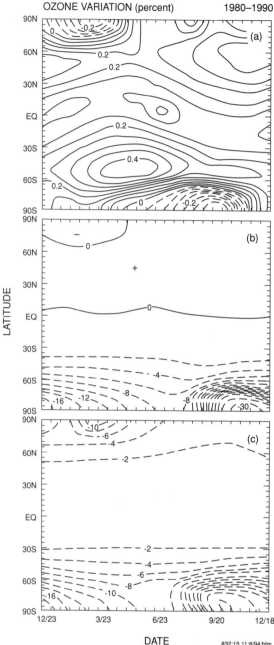

**Figure 12.11.** Relative change (percent) in the ozone column abundance represented as a function of latitude and season, from 1980 to 1990, as calculated by the model of Granier and Brasseur (1992). The model accounts for the observed trends in $CO_2$, $CH_4$, $N_2O$, and the chlorofluorocarbons. The top panel refers to a case in which only gas-phase chemistry is taken into account, while panel (b) includes the effects of heterogeneous chemistry on the surface of particles in polar stratospheric clouds. In panel (c), additional heterogeneous processes on the surface of background aerosol particles are taken into account.

Pyle, 1984) are reproduced by the model: maximum mixing ratios in the tropics, deep minima at both poles, and meridional gradients at midlatitudes that differ with latitude and season. The subtropical and midlatitude gradients are most shallow during winter when large-scale mixing associated with planetary wave dissipation occurs.

Another illustration of a two-dimensional model is provided by Figure 12.11a-c, which shows as a function of latitude and season the percentage change in the ozone column abundance calculated by Granier and Brasseur (1992) for different chemical schemes. In the first case, only the effect of gas-phase chemistry is taken into account, while in the second and third cases, the contribution of heterogeneous chemical conversions due to the presence of polar stratospheric clouds and background aerosol, respectively, are also considered. The results shown in the third case (Fig. 12.11c) are closest (although not in perfect agreement) to the observed ozone trend during the 1980s and 1990s. The respective roles of these additional processes are discussed in Chapters 4 and 14.

### 12.5.4 Three-Dimensional Global Models

Global chemical transport models (CTM) are being developed to simulate the distribution and the evolution of trace constituents in the troposphere and/or stratosphere. These models are often very complex and computationally expensive. They include an advective scheme driven by dynamical variables (*e.g.*, wind components) produced either by a general circulation model or by the analysis of observed meteorological variables. The transport code also includes parameterizations for unresolved (subgrid) exchanges associated with convection and mixing. Surface emissions and deposition of trace gases, as well as chemical conversions in the atmosphere, are simulated at levels of details that depend on the complexity of the model. Typical resolutions for global CTMs are 1 to 5 degrees in longitude and latitude and 15 to 60 min in time.

Pioneering work in the modeling of atmospheric tracers on the global scale was begun in the 1970s by Mahlman and Moxim (1978). Chemical transport models first focussed on long-lived species such as nitrous oxide (Levy et al., 1982) or some halocarbons (Prather et al., 1987). These compounds were chosen because their chemical destruction is linear and easy to introduce in atmospheric transport models. More recently, relatively detailed chemical schemes have been implemented in global transport models. These models have been first applied to the stratosphere, and used to simulate the behavior of ozone and related species (see, e.g., Rose and Brasseur, 1989; Kaye and Rood, 1989; Grose et al., 1989; Rasch et al., 1995), and to study the role played by heterogeneous reactions in polar regions (Chipperfield et al., 1993; Lefèvre et al., 1994; Brasseur et al., 1997). When applied to the troposphere, the formulation of CTMs is more complex since these models must account for the effects of clouds (including aqueous-phase chemistry, wet deposition, convective exchanges, light scattering), the influence of surface sources and deposition, and the exchange between the boundary layer and the free troposphere. Such models, although still in their infancy, have already been used to investigate several scientific questions such as the impact of regional pollution sources on the global atmosphere, the natural and anthropogenic processes affecting the oxidizing capacity of the atmosphere, and the role of sulfate aerosols on climate forcing.

Different categories of CTMs can been distinguished. Some of them are linked to existing general circulation models (GCMs) and account for a large fraction of the atmospheric variability. Simpler models using mean wind/temperature climatologies and relatively long time steps reproduce only the seasonal features, and are used primarily to do sensitivity tests at an acceptable computer cost.

Very often, CTMs are run "offline." In this case, the dynamical variables needed to drive atmospheric transport are precalculated in a GCM or in an assimilation model (see Section 12.7), and stored periodically (e.g., every 3 to 12 h, depending on the model) on history tapes. The content of these tapes is then read, interpolated in time, and used in the CTM. Models can also be run "online." Under this configuration, the CTM is integrated into a GCM, and both models are integrated simultaneously. Although this approach may be computationally more expensive (since the dynamics have to be recalculated each time the model is executed), it has the advantage of providing variables instantaneously at each time step. In addition, dynamics and chemistry are fully coupled, which is particularly important for certain types of scientific questions. This is the case, for example, for studies of the stratosphere, where heating rates (which directly influence the atmospheric dynamics) are highly sensitive to the distribution of chemical species such as ozone and water vapor (which is directly affected by atmospheric dynamics). In the troposphere, this type of coupling is weaker, and "offline" calculations may be sufficient for most applications. However, to treat the coupling processes between chemistry and climate (e.g., between sulfate aerosols, cloudiness, and radiative forcing), "online" models are preferable.

A key aspect of model development is the evaluation of these models through comparison with available observations. For stratospheric studies, global measurements of chemical constituents are provided by satellites. In the case of the troposphere, observations are generally local, and based on a limited number of measurements by ground-based, airborne, or balloon-borne instruments (*e.g.*, ozone soundings). Model evaluation remains therefore very limited, which stresses the need for more global measurements of tropospheric chemical species. Individual aspects of CTMs can be evaluated through comparisons of specific model output with atmospheric observations of selected tracers. For example, $^{222}$Rn is a useful tracer to assess the performance of the model with regard to vertical convective exchanges (*e.g.*, Jacob and Prather, 1990; Feichter and Crutzen, 1990; Balkanski *et al.*, 1992; Mahowald *et al.*, 1995; Allen *et al.*, 1996). The $^{210}$Pb and $^{7}$Be isotopes are appropriate to evaluate the formulation of wet deposition, while $^{14}CO_2$ is used to analyze stratosphere/troposphere exchanges (*e.g.*, Feichter *et al.*, 1991; Brost *et al.*, 1991; Koch *et al.*, 1996). Chlorofluorocarbons and the $^{85}$Kr isotope are used to test the representation of interhemispheric transport (see, *e.g.*, Prather *et al.*, 1987; Zimmermann *et al.*, 1989). These tracers are released mostly in the Northern Hemisphere, so that the interhemispheric gradient determined from observations at selected stations located in both hemispheres provides some estimate of the transport time through the intertropical convergence zone.

Figure 12.12 shows the calculated distributions of carbon monoxide in January and ozone in July near the surface provided by MOZART, a CTM developed at the National Center for Atmospheric Research (NCAR). This model accounts for approximately 40 chemical constituents and 130 reactions. The transport is calculated using a semi-Lagrangian formulation (Williamson and Rasch, 1989). Surface sources associated with biological and anthropogenic activity, as well as dry deposition, are specified as lower boundary conditions. Rainout of soluble species by precipitation, as well as convective transport, are parameterized according to the schemes of Giorgi and Chameides (1985).

The distribution of carbon monoxide reflects the existence of large sources due to biomass burning in the tropics and human activity in industrialized regions. The transport of CO from its source regions toward more pristine areas is also visible. Ozone is most abundant over the eastern United States and Europe, where large amounts of precursors are released as a result of human activities. The effect of transport is also visible.

## 12.5.5 Three-Dimensional Regional Models

Regional models are developed to study in detail the formation and fate of chemical compounds over a limited spatial domain (typically 1000 by 1000 km). These models have been used extensively to determine the reductions in anthropogenic emissions needed to achieve air quality standards. Data that characterize the topography, land-use type, emissions, and three-dimensional wind fields for the episode being simulated over a region are input to these models. As output, the concentration of chemical species at high spatial resolution (typically 1-10 km) describes the air quality over a given region. One of these models, RADM (Regional Acid Deposition Model), was developed at NCAR to simulate acid deposition over the eastern United States. It is an Eulerian chemical-transport model (Chang *et al.*, 1987). Atmospheric transport is calculated using meteorological fields produced by the Mesoscale Meteorological Model (Anthes *et al.*, 1987), including diffusion (semi-implicit finite-differencing scheme), convection (Anthes-Kuo parameterization, Anthes *et al.*, 1987), and advection (nonlinear upstream finite difference; Smolarkiewicz, 1984). Chemical processes, including gas-phase reactions, heterogeneous reactions on wetted aerosols, and aqueous-phase reactions in clouds are resolved by a one-step, semi-iterative technique (Stockwell, 1986; Stockwell *et al.*, 1990). Scavenging processes, such as dry

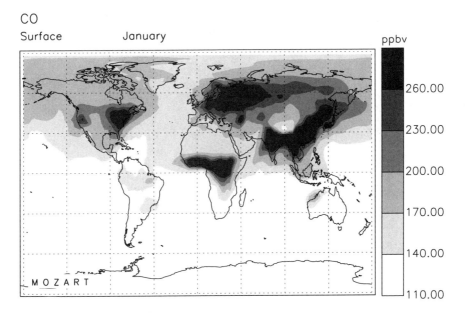

**Figure 12.12a.** Distribution of the carbon monoxide mixing ratio (ppbv) at the Earth's surface calculated (January average) by the NCAR global chemical transport model MOZART (Brasseur et al., 1998; Hauglustaine et al., 1998).

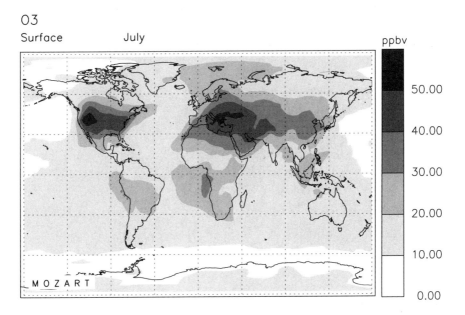

**Figure 12.12b.** Distribution of the ozone mixing ratio (ppbv) at the Earth's surface calculated (July average) by the NCAR global chemical transport model MOZART (Brasseur et al., 1998; Hauglustaine et al., 1998).

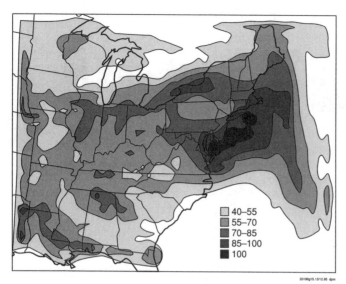

**Figure 12.13.** Simulation by the NOAA Aeronomy Laboratory Regional Model of the ozone episode observed in the eastern United States on July 7, 1986 (McKeen et al., 1991a). The figure shows the average ozone mixing ratio (1-5 PM) near the surface in ppbv.

deposition and wet deposition of soluble species, are also considered. The equations are discretized on a spatial domain covering the northeastern United States and southeastern Canada, which is subdivided into 35 × 38 horizontal grid meshes with six or fifteen vertical levels. Another three-dimensional Eulerian regional model, developed by McKeen *et al.* (1991a) at the NOAA/Aeronomy Laboratory, is used to study ozone buildup and other regional pollution problems over the eastern United States. The general approach is similar to that of the RADM model. Figure 12.13 shows a distribution of ozone calculated at the lowest model layer (averaged over 1-5 PM) for conditions representative of July 7, 1986. The situation is discussed more fully in Chapter 13.

One difficulty associated with regional CTMs is the specification of chemical concentrations or fluxes at the lateral boundaries of the model. Since measurements are generally not available, climatological values can be used, but these influence, sometimes excessively, the solutions over the model domain. An alternative is to "nest" the regional model into a global model, preferably a global CTM driven by observed (assimilated) dynamical quantities. The grid of a global model can also be designed so that the spatial resolution is dramatically increased in a chosen region of the world where detailed studies are requested.

## 12.6 Modeling Global Budgets and Biogeochemical Cycles

Because global Earth system models with detailed treatments of coupled physical, chemical, and biological processes are not yet available, biogeochemical cycles are often studied with simple box models. In these models, the mass $M$ of a given trace gas included in a rather homogeneous reservoir (*e.g.*, the atmosphere or the ocean) and the fluxes between these reservoirs are calculated as a function of time on the basis of global budget equations. Although this approach is based on empirical relations rather than on first principles and does not provide any insight on the detailed processes occurring inside a given reservoir, it is easily applicable to systems of connected reservoirs in which the material is transferred in a cyclic fashion.

Consider first a single reservoir whose mass (burden) is denoted $M$. If $P$ is the flux of incoming material (expressed in mass per unit time), and $L$ the flux of outgoing material (assumed to be proportional to the burden, or $L = kM$), the budget for the reservoir is expressed by

$$\frac{dM}{dt} = P - kM \qquad (12.32)$$

The turnover time associated with the reservoir is provided by the ratio of its burden at steady state ($M_{ss} = P/k$) to the sum of its sinks (or its sources)

$$\tau = \frac{M_{ss}}{L} = \frac{1}{k} \qquad (12.33)$$

and is thus the inverse of the loss coefficient $k$. This quantity is a measure of the average time spent by an individual particle (molecule or atom) in the reservoir. If $M_0$ is the burden at an initial time $t_0$, the system evolves toward the equilibrium solution $M_{ss}$ according to

$$M(t) = M_{ss} - (M_{ss} - M_o) \exp\left(-\frac{t - t_0}{\tau}\right) \qquad (12.34)$$

with a time scale given by the turnover time $\tau$ (see Fig. 12.14).

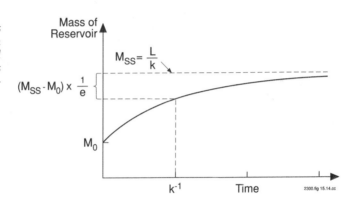

**Figure 12.14.** Adjustment of the mass $M$ in a given reservoir, from its initial value $M_0$ to its value at steady state $M_{ss}$. The response time is equal to $k^{-1}$ (Rodhe, 1992).

The same formalism can be applied when several reservoirs are coupled through specific physical or chemical exchanges. In the linear case, if the flux from reservoir $i$ to reservoir $j$ is expressed by

$$F_{ij} = k_{ij} M_i \qquad (12.35)$$

the budget equation can be written by the matrix equation

$$\frac{d\mathbf{M}}{dt} = \overline{\overline{\mathbf{K}}} \mathbf{M} + \mathbf{S} \qquad (12.36)$$

where the components $M_i$ and $S_i$ of vectors $\mathbf{M}$ and $\mathbf{S}$ represent, respectively, the burden and any external source for reservoir $i$, and $\overline{\overline{\mathbf{K}}}$ is the exchange matrix whose elements $K_{ij}$ are such that

$$K_{ij} = k_{ji}, \quad i \neq j \qquad (12.37a)$$

$$K_{ii} = -\sum_{j=1}^{N} k_{ij} \qquad (12.37b)$$

where $N$ is the number of reservoirs in the system. In the absence of external sources ($\mathbf{S} = 0$), the evolution of system (12.36) toward its equilibrium state can be expressed by a finite number of exponential decay functions whose corresponding time constants are the nonzero eigenvalues of matrix $\overline{\overline{\mathbf{K}}}$. The overall response time of the system is provided by the inverse of the smallest absolute value of these eigenvalues.

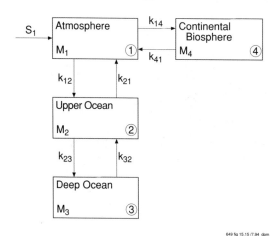

**Figure 12.15.** Linear system of four coupled reservoirs including (1) the atmosphere, (2) the surface layer of the ocean, (3) the deep ocean, and (4) the continental biosphere. Exchange rates of chemical compounds (e.g., carbon) between these reservoirs are represented by the exchange coefficients $k_{ij}$, and $S_1$ represents an external source (e.g., carbon release by fossil fuel burning; Machta, 1971).

The behavior of a linear system of several coupled reservoirs is illustrated by the case represented in Figure 12.15. The four reservoirs for carbon include the atmosphere (reservoir 1), the surface waters in the ocean (2), the deep ocean (3), and the continental biosphere (4). In addition, $S_1$ represents the emission into the atmosphere of carbon resulting from fossil fuel burning. The exchanges between reservoirs are first supposed to be represented by linear relations such as Eq. (12.36). The evolution of carbon inside the four reservoirs can be represented by the system

$$\frac{dM_1}{dt} = -(k_{12} + k_{14})M_1 + k_{21}M_2 + k_{41}M_4 + S_1 \quad (12.38a)$$

$$\frac{dM_2}{dt} = k_{12}M_1 - (k_{21} + k_{23})M_2 + k_{32}M_3 \quad (12.38b)$$

$$\frac{dM_3}{dt} = k_{23}M_2 - k_{32}M_3 \quad (12.38c)$$

$$\frac{dM_4}{dt} = k_{14}M_1 - k_{41}M_4 \quad (12.38d)$$

The state of the system prior to the perturbations associated with fossil fuel burning is found for $S_1 = 0$. Denoting by $M = M_1 + M_2 + M_3 + M_4$ the total burden of carbon in the system (which is constant under unperturbed conditions), the amount of carbon in one of the reservoirs, for example, the atmosphere ($i = 1$), can be expressed by

$$M_1 = \frac{M}{1 + \frac{k_{12}}{k_{21}}\left(1 + \frac{k_{23}}{k_{32}}\right) + \frac{k_{14}}{k_{41}}} \quad (12.39)$$

As the system becomes perturbed by the external fossil fuel source $S(t)$, the total burden in the system increases as

$$M(t) = M(t_0) + \int_{t_0}^{t} S(u)\, du \quad (12.40)$$

and the time-dependent solution of Eq. (12.38) needs to be determined. The time constants characterizing exchanges between the different reservoirs can vary substantially (see Fig. 12.16). Therefore, it is often possible to assume that equilibrium conditions exist between strongly coupled reservoirs, while time-dependent equations must be solved for determining the burden in reservoirs affected by the slowest exchanges.

**Figure 12.16.** Approximate characteristic time for exchange of air and water parcels, respectively, between different parts of the atmosphere and oceans (Rodhe, 1992; reprinted by permission of the publisher Academic Press).

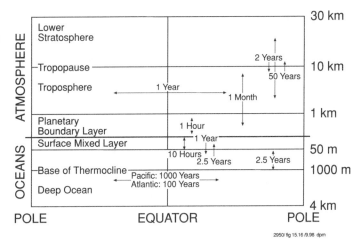

In many cases, the exchange between reservoirs cannot be represented by linear functions of the burden, as assumed so far in the discussion. The carbon cycle provides an important example of a nonlinear system. Let $M_i$ represent the total carbon content of each reservoir $i$, and assume specifically that $M_2$ and $M_3$ include all forms of carbon ($CO_2$, $H_2CO_3$, $HCO_3^-$, and $CO_3^{2-}$) dissolved in the upper and deep ocean, respectively. Since the carbon flux from the ocean to the atmosphere is assumed to be proportional to the burden of dissolved $CO_2$ in the surface waters (reservoir 2 in Fig. 12.15), and the partial pressure of $CO_2$ in water is related to the total carbon content by Expression (5.13) (see Chapter 5), the flux of carbon from the surface waters to the atmosphere can be expressed by

$$F_{21} = k_{21} M_2^{\mathcal{R}} \qquad (12.41)$$

where $\mathcal{R}$ is the so-called Revelle factor (Revelle and Suess, 1957), whose value is typically 10. Similarly, the flux from the atmosphere to the terrestrial ecosystems is often represented by

$$F_{14} = k_{14} M_1^{\alpha} \qquad (12.42)$$

with an exponent $\alpha$ considerably less than unity. Rodhe (1992) has estimated that if the known accessible reserves of fossil carbon (6000 Pg) were entirely released in the atmosphere, the new carbon burden at equilibrium would be higher than the original content by approximately 170%, 22%, 12%, and 12% in the case of the atmosphere, the continental biosphere, surface waters, and the deep ocean, respectively. If all exchange fluxes were proportional to the reservoir content (linear model), the enhancement would be equal to 15% in all reservoirs. Thus, although in absolute terms, most of the carbon released by fossil fuel burning is expected to be stored in the deep ocean, in relative terms, the atmosphere is expected to be the most perturbed reservoir.

Box models have a long tradition in atmospheric science and have been used extensively to study, for example, the global carbon cycle (Craig, 1957; Revelle and Suess, 1957; Bolin and Ericksson, 1959; Machta, 1971; Keeling, 1973). A more adequate treatment of carbon exchanges within the deep ocean has been achieved by replacing the well-mixed deep ocean reservoir with a series of layers in which vertical exchanges are represented by an eddy diffusion formulation. This type of box-diffusion model (Oeschger *et al.*, 1975), which has been improved to include polar outcropping areas (Siegenthaler, 1983) or a more detailed representation of carbon exchanges with the continental biosphere (Bolin, 1981), is currently in widespread use. It tends to be progressively replaced by higher-resolution models based on the fundamental laws of physics, biology, and chemistry.

## 12.7 Data Assimilation

In order to be able to characterize and predict the time evolution of the atmosphere, one has to know its present state. That state is characterized by a number of physical parameters such as wind vectors, temperatures, chemical concentrations, and radiation field. It is impossible to measure all the atmospheric state variables simultaneously in time and space. Thus, to describe the atmosphere some additional information has to be brought into consideration.

Such information can be introduced in two ways. It can be given by theoretical equations that allow one to deduce some components of the state from others or to derive them from a smaller set of model parameters. In this case the number of independent components of the state is reduced and the effective density of observations is increased. Alternatively, the additional information can be given by independent, a priori estimates of some components of the state and their corresponding uncertainties. These can be regarded as additional "virtual measurements" (Rodgers, 1977). Numerical models or climatologies can be used to obtain such estimates.

Objective approaches to combining our a priori knowledge about a physical system under consideration with available (usually sparse and irregular) observations are often referred to as *data assimilation*. "Geophysical data assimilation is a quantitative, objective method to infer the state of the earth-atmosphere-ocean system from heterogeneous, irregularly distributed, and temporally inconsistent observational data with different accuracies" (NRC, 1991).

The mathematical basis of data assimilation is estimation theory or inverse problem theory. Inverse theory is an organized set of mathematical techniques for reducing data to obtain useful information about the physical world on the basis of observations (Menke, 1984). In a conventional "forward" problem one uses a set of a priori parameters to predict the state of the physical system. In the "inverse" or estimation problem one attempts to use available observations of the state of the system to estimate poorly known model parameters and/or the state itself.

The meanings of the "model," "state," "model parameters," and "observations" depend on the particular application. In numerical weather prediction, for example, one attempts to estimate the initial conditions (model parameters) for a general circulation model from available past observations of winds, temperatures, and moisture. These initial conditions are then used to predict the time evolution of the atmosphere (the state). In remote optical sensing applications a radiative transfer model is used to deduce concentrations of chemical species (model parameters) from measured radiances (observations). In problems related to biogeochemical cycles, observations of long-lived tropospheric species such as carbon dioxide and methane are often used to estimate poorly known surface fluxes of these constituents (model parameters) using global tropospheric transport models. In turn, the estimated fluxes can be used to predict global distributions of these species in the atmosphere (the state). In the case of satellite observations measurements of atmospheric constitutents are often irregular and sparse in time and space, and it is highly desirable to have estimates of constituent concentrations at regular grid points (the state). This problem can be formalized as the inverse of an algebraic interpolation operator (the model) that interpolates species concentrations at the regular grid points to the location of the observations. Finally, in atmospheric photochemical modeling, measurements of some of the modeled species can be used to constrain certain model parameters (*e.g.*, winds, initial distributions of chemical species) and improve the model prediction of the state, including prediction of distributions of nonobserved species.

### 12.7.1 Mathematical Formalism

Some key concepts of the data assimilation methodology can be presented in a fairly concise mathematical form (*e.g.*, Lorenc, 1986). Consider a model of a physical system represented by an operator $M$ (generally nonlinear), and let the vector $\mathbf{x}$ with dimension $N_x$ be a set of input parameters for the model. These input parameters are used to predict the state of the system, the vector $\mathbf{y}$ with dimension $N_y$

$$\mathbf{y} = M(\mathbf{x}) \tag{12.43}$$

Let vector $\mathbf{y}_0$ be the actual observations of the state. Usually, the dimension of $\mathbf{y}_0$ is significantly less than $N_y$, so generally the operator $M$ also includes an interpolation of the model prediction to the observations $\mathbf{y}_0$. The analysis problem, then, is to find the "best" value of $\mathbf{x}$ that inverts Eq. (12.43) for a given $\mathbf{y}_0$, allowing for observation errors and other prior information (Lorenc, 1986).

"Best" here means that the errors of the final analysis are minimal. An exact value of a physical quantity can never be determined. One can only say that this value lies within a certain range with a certain probability. Therefore, all estimates of the "best" value of $\mathbf{x}$, or $\mathbf{x}_0$, obtained from the observed $\mathbf{y}_0$ are probabilistic in nature. A mathematically robust definition of the "best" or "optimal" value of $\mathbf{x}$ is, for instance, that the probability density for $\mathbf{x}$ is maximal given the available observations $\mathbf{y}_0$. This is the so-called *maximum likelihood* definition.

Probability density functions are usually described in terms of error covariances. A common assumption is that the probability density functions (PDFs) are Gaussian

$$\text{PDF}(\mathbf{x}) \sim \exp\left[-\frac{1}{2}(\mathbf{x}-\mathbf{x}_0)^T C^{-1} (\mathbf{x}-\mathbf{x}_0)\right] \tag{12.44}$$

where $C$ is the error covariance matrix for $\mathbf{x}$. Its diagonal elements are the uncertainties (standard deviations) of $\mathbf{x}$, and the off-diagonal elements represent correlation between uncertainties of different elements of vector $\mathbf{x}$. If $\mathbf{x}_t$ is the unknown true value of $\mathbf{x}$, then the covariance matrix $C$ is defined as

$$C = <(\mathbf{x}-\mathbf{x}_t)^T (\mathbf{x}-\mathbf{x}_t)>$$

where angular brackets represent averaging over all available realizations (measurements) of $\mathbf{x}$. Discussion of practical methods of error covariances estimation can be found, for instance, in Daley (1991).

Another definition of the optimal value of $\mathbf{x}$ is the so-called *minimum variance* estimate or the expected value estimate, which is defined as follows

$$\mathbf{x}_0 = \int \mathbf{x}\, \text{PDF}(\mathbf{x})\, d\mathbf{x}$$

It can be shown that in the case of Gaussian PDFs this definition is equivalent to the maximum likelihood definition. "Other prior information" in the above definition is given by an independent estimate of $\mathbf{x}$, or $\mathbf{x}_b$, often called the background, and the corresponding background error covariances.

If all probability density functions could be approximated by Gaussian functions, the solution minimizing the final analysis errors is given by a minimum of the following function (Lorenc, 1986)

$$J(\mathbf{x}) = [\mathbf{y}_0 - M(\mathbf{x})]^T (O + F)^{-1} [\mathbf{y}_0 - M(\mathbf{x})] + [\mathbf{x} - \mathbf{x}_b]^T B^{-1} [\mathbf{x} - \mathbf{x}_b] \qquad (12.45)$$

where $O$, $F$, and $B$ are the observational, model, and background error covariance matrices. They characterize, respectively, our confidence in the measurements, the model, and the a priori background estimate. $J(\mathbf{x})$ is often called the *misfit* or *cost* function.

Thus the solution of the optimal analysis problem can be written as

$$\mathbf{x}_0 \text{ such that } J(\mathbf{x}_0) = \min \{J(\mathbf{x})\} \qquad (12.46)$$

Once the optimal estimate of $\mathbf{x}$ is obtained, Eq. (12.43) can be used to derive the best estimate of $\mathbf{y}$. Note that both types of independent information mentioned earlier in this section are included in Eq. (12.45). The information given by theoretical equations is contained in operator $M$, and the a priori estimate of model parameters is given by $\mathbf{x}_b$. In the ultimate case when one measures the state directly, $M$ is the identity operator and $\mathbf{x} = \mathbf{y}$. If no a priori first guess is available, the background error covariances become infinity, and the second term in Eq. (12.45) vanishes.

Expressions (12.45) and (12.46) are simply a generalization of the well-known fact that independent measurements should be added with weights inversely proportional to the square of the measurement errors to form an optimal estimate. In practical applications one has to find an appropriate way to compute the error covariances and to minimize Eq. (12.45). Two techniques are often used: variational assimilation and the Kalman-Bucy filter.

To illustrate these techniques, assume that vector $\mathbf{x}$ represents initial conditions for a time-dependent model (*e.g.*, initial species concentrations for a numerical photochemical model). In the case of a box model that includes $N$ species, the dimension of vector $\mathbf{x}$ would be $N$, while in a multidimensional model that contains $M$ grid points, the dimension of $\mathbf{x}$ would be $M \times N$. These numbers can be quite large; for instance, a 3D model with 20 vertical levels and a $5° \times 5°$ horizontal resolution contains 54,020 grid points. If 20 chemical species are included, the dimension of $\mathbf{x}$ is 1,080,400. Both the Kalman-Bucy filter and variational technique require linearization of the nonlinear photochemical model.

### 12.7.1.1 Variational Technique

In the variational method a minimization algorithm is used to find the initial conditions for the model that minimize the misfit between model results (obtained using these initial conditions) and observations for the whole analysis period. The analysis period is usually much longer than the model time step. The misfit function is given by Eq. (12.45). Operator $M$ represents both the photochemical model and the subsequent interpolation of the model results (in time and space) to the location of the observations. The model, observations, and the background error covariances, as well as the first guess estimate $\mathbf{x}_b$ are considered to be known a priori, and the background error covariances do not change during the assimilation procedure.

The variational data assimilation technique can be thought of as a constrained least-squares fit to a set of observations distributed over some period of time. The constraints are given by the model equations. The choice of the analysis period is somewhat arbitrary and is dictated by the frequency of the observations and the characteristic time scales of the modeled system. For example, in most cases it would be impractical to search for the initial conditions of a photochemical model that fit observations available over a year.

Most minimization algorithms for Eq. (12.46) require the gradient of the cost function $J(\mathbf{x})$ with respect to $\mathbf{x}$, $dJ(\mathbf{x})/d\mathbf{x}$. The so-called adjoint method (*e.g.*, Talagrand and Courtier, 1987) is often used to compute $dJ(\mathbf{x})/d\mathbf{x}$. The adjoint method relies on the linearization of the photochemical model. This imposes additional requirements on the length of the analysis period; it should be short enough for the linear approximation to be valid.

### 12.7.1.2 Kalman-Bucy Filter

In the Kalman-Bucy method observations are "blended" with model simulations with certain weights as they become available to form new initial conditions for the model for the next time step (*e.g.*, Lorenc, 1986; Lyster *et al.*, 1997). The model simulations are considered to be the a priori background estimate, $\mathbf{x}_b$. The operator $M$ simply represents an interpolation of the model results, $\mathbf{x}_b$, to the locations of observations, $\mathbf{y}_0$. The solution of Eq. (12.46) is computed explicitly for each new observation by analytically solving $dJ(\mathbf{x})/d\mathbf{x} = 0$. If the interpolation operator $M$ is nonlinear, the solution is found iteratively.

The observational error covariance matrix $O$ and the error covariances of the interpolation operator $F$ are assumed to be known. The initial background error covariance is assumed to be known a priori, and all the subsequent (in time) background error covariances $B$ are recomputed for each assimilated observation using linearized photochemical model equations.

In the case of the Kalman-Bucy filter, the analysis period is essentially the time interval between two consecutive observations. The four-dimensional analysis (used in the variational technique when the analysis is done simultaneously in space and time) is replaced by a time series of three-dimensional analyses.

Mathematical methods of estimation and inverse theory have long been used in numerical weather prediction, data retrievals from remote sensing experiments (particularly satellite measurements), and so-called inverse modeling. Inverse modeling is usually used for determination of surface fluxes of various gases, and a number of empirical and more formal methods have been developed for this purpose. An overview of inverse modeling applications is given in the following section.

Recently, attempts have been made to apply estimation theory methods to assimilate ozone measured by satellites (Riishøjgaard, 1996; Levelt *et al.*, 1996) as well as other compounds such as methane using the Kalman filter (Lyster *et al.*, 1997) and $O_3$, $NO_2$, and $HNO_3$ with the variational technique (Fisher and Lary, 1995).

Figure 12.17 shows an example of assimilation of the UARS (Upper Atmosphere Research Satellite) measurements of $O_3$, $NO_2$, $HNO_3$, $ClO$ and $ClONO_2$ into a photochemical box model with the variational method. A trajectory model was used to calculate movements of a parcel of air for 48 h at 840 K potential temperature surface (approximately 10 mb). UARS observations (see Box 11.5) located within certain spatial and temporal limits from the parcel's trajectory were collected, and the adjoint of the photochemical box model was employed to calculate the gradient of the cost function (a sum of weighted squared differences between model results and data) with respect to the initial conditions. The final model run was then made with the optimized initial conditions for the whole 48 h interval. The approach described was suggested and implemented for the first time by Fisher and Lary (1995).

Recent studies concentrating on satellite data assimilation and analysis have proven interesting. The global coverage of satellite measurements makes them very attractive for detailing global distributions of chemical species. Such distributions are needed for comparisons with models, independent observations, and as climatological

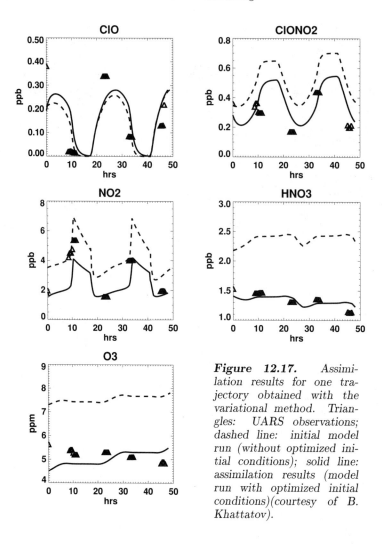

*Figure 12.17.* Assimilation results for one trajectory obtained with the variational method. Triangles: UARS observations; dashed line: initial model run (without optimized initial conditions); solid line: assimilation results (model run with optimized initial conditions)(courtesy of B. Khattatov).

data sets. However, orbital geometry imposes strict limitation on the space and time sampling. The asynoptic nature of most satellite observations in general does not allow the resolution of temporal scales of less than two days and zonal wave numbers larger than half the number of satellite orbits per day (Salby, 1987). The data assimilation methodology makes use of the additional constraints given by a priori knowledge and thus makes it possible to overcome some of these limitations.

## 12.8 Inverse Modeling

Inverse modeling of atmospheric trace gases is a common application of the estimation (inverse) theory techniques. Atmospheric models provide estimates of the distribution of chemical constituents for specified surface sources and sinks, and for given chemical scheme and transport formulation. Surface fluxes (*e.g.*, emission and deposition) are usually poorly quantified, as they are determined by complex biological and anthropogenic processes. Concentrations of atmospheric trace gases measured at different locations across the world provide information about the location of the sources/sinks of these gases. By comparing the observed concentrations with concentrations provided by numerical models, it is in principle possible to obtain information about the magnitude and the geographical distribution of the net surface flux that best reproduce the observations. There are different methods to address this inverse problem. In one of them, named the synthesis method (see, *e.g.*, Fung *et al.*, 1991), the surface fluxes are adjusted until the calculated concentrations match the observations. Such an empirical approach can eliminate some possible source/sink scenarios and illustrate possible distributions of surface fluxes; however, it does not produce an optimal estimate of the error. More formal methods have been proposed (*e.g.*, Enting, 1985) and used to estimate the surface fluxes of long-lived species such as $CO_2$ (Enting and Mansbridge, 1989; Enting *et al.*, 1995) and chlorofluorocarbons (Hartley and Prinn, 1993).

If the relation between the source $S(x,t)$ and the spatially varying part of the mixing ratio $\Delta\mu(x,t)$ at location $x$ and time $t$ is expressed by

$$\Delta\mu(x,t) = \int\int G(x,x',t,t')\, S(x',t')\, dx'dt' \qquad (12.47)$$

where $G$ is the Green's function of the problem, the inverse problem can be solved by inverting this equation, given a finite number of observed mixing ratios $\Delta\mu_k$ at $K$ sites $(k=1,\ldots,K)$.

One method to solve the inverse problem (especially in the case of slowly varying sources) relies on calculations by a global chemical transport model of the influence of each source $(S_i)$ on each observation site $(\mu_i)$. These can be calculated in a transport model to obtain a Jacobian $(J)$.

$$J_{ij} = \frac{\partial \mu_i}{\partial S_j} \qquad (12.48)$$

Changes in the sources are then determined by

$$\Delta S = J^{-1}\left(\mu^{obs} - \mu^{model}\right) \qquad (12.49)$$

There are more general forms of this equation such as the Kalman filter

$$\Delta S = CJ^T \left(JCJ^T + N\right)^{-1} \Delta\mu \qquad (12.50)$$

where $C$ is the estimated error in the sources and $N$ is the error in observations. Thus the simple inverse in Eq. (12.49) is now weighted by the error in observations.

In general, because the number of observation sites that provide time series of trace concentrations is limited, the inverse problem suffers from being underdetermined. In fact, only as many source characteristics as the number of observations can be determined. This means that some a priori information is often required.

## 12.9 Chemical-Transport Models in the Future

The trend to global chemical-transport models incorporating increasingly complex chemical systems is clear. The near term goal is an "Earth" system model with interacting global dynamics and chemistry in the atmosphere and ocean. Present-day three-dimensional models with restricted chemistry strain the best supercomputers, which sustain about 1 to 20 gigaflops (billions of floating point operations per second). Even at this performance level a detailed real world chemical-transport simulation (with typically 50 chemical constituents and 200,000 gridpoints) can take one or two machine days to model an annual cycle. Doubtless there will be basic algorithmic enhancements and perhaps entirely new numerical schemes that will increase overall computational efficiency. However, the best hope for significantly enhanced modeling lies in the increasing computational power of leading-edge supercomputers.

At this time two distinct architectures are rapidly evolving. The traditional architecture is represented by machines with a smaller number of very powerful vector processing elements all sharing a common memory. To date, these machines have been the "state of the art" supercomputing platforms. The newer architecture, referred to as massively parallel, is characterized by a large number (often in the hundreds) of standard microprocessing elements with distributed and/or shared memory. Central to fully exploiting these newer, highly parallel machines is the ability to divide and synchronize the computational burden conveniently among the individual processing elements. Parallel environment tools such as compilers and code analyzers, even though primitive at present, are the focus of great effort, and we can look forward to continued improvement in the near future. Given that supercomputer computational power has historically increased an order of magnitude every four years, we can realistically look to the immediate future when very rigorous global chemical-transport models will be common and valuable assessment and analysis tools.

## Further Reading

Aiken, R., ed. (1985) *Stiff Computation*, Oxford University Press, New York.

Gear, C. W. (1971) *Numerical Initial Value Problems in Ordinary Differential Equations*, Prentice Hall, New York.

Ojima, D., ed. (1992) *Modeling of the Earth System*, University Corporation for Atmospheric Research, Office for Interdisciplinary Earth Studies, Boulder, Colorado.

Press, W., S. A. Teukolsky, W. T. Vetterling, and B. P. Flannery (1992) *Numerical Recipes*, 2nd ed., Cambridge University Press.

Seinfeld, J. H. and S. N. Pandis (1998) *Atmospheric Chemistry and Physics*, Wiley, New York.

Trenberth, K. E., ed., (1992) *Climate System Modeling*, Cambridge University Press.

Washington, W., and C. L. Parkinson (1986) *An Introduction to Three-Dimensional Climate Modeling*, University Science Books, Mill Valley, California, and Oxford University Press, Oxford.

Henning Rodhe

## How Complex Do Models Need to Be?

During the past 30 years computers have made it possible to use two- and three-dimensional models to study transport, transformation, and removal of chemical species in the atmosphere. Until a few years ago, three-dimensional models of the global atmospheric chemistry system were severely limited by the capacity of the computers, such that either the number of chemical reactions or the degree of sophistication of the meteorological part had to be strongly restricted. Today (1996) we are quickly moving to a situation where computers can handle atmospheric chemistry systems that are quite complex from both a chemical and meteorological point of view. This opens up fascinating possibilities, but it also introduces considerable risks.

On the positive side: Increased spatial and temporal resolution enables an explicit description of a larger part of the spectrum of meteorological processes such as winds, clouds, etc. This will make the transport calculations more realistic, and it will make it possible to improve the parameterization of, for example, precipitation scavenging and chemical transformations occurring in clouds. The introduction of more chemical species and reactions will also enable the study of more sophisticated interdependences in the chemical web.

On the other hand: The more complex the model, the more difficult it is to understand the results completely; to see where the key uncertainties lie and to pinpoint the weak parts that need to be improved. With a complex model, the risk for "bugs" also increases at the same time as it becomes extremely difficult to identify such "bugs." With powerful computers now being readily available, it is tempting even for small research groups, or individual scientists, to perform simulations with complex global 3D models. What might not be appreciated from the outset is the very considerable effort needed to test the various aspects of the model before credible results can be obtained and published.

Another risk with the current rapid development of complex atmospheric chemistry models is the mismatch between the degree of detail of the model output and the observations needed for validation. We will soon reach — in some cases we have already reached — the point where the improvement in our understanding of global distributions of chemical species is limited by lack of observations. Further development of sophisticated models will then be of no use until more data have become available.

Any attempt to perform model simulations should start by considering the data available for model validation and the specific scientific questions to be addressed. A general rule should then be to choose the simplest possible model that is still adequate for the given purposes.

The chapter on modeling in this textbook emphasizes the need for careful consideration of numerical methods for solving the differential equations governing transport and transformation processes. This is a reasonable approach, since inadequate numerical schemes may make model results meaningless even if the physical and chemical processes are well described. It is worth pointing out that one has to be careful with numerical techniques even with very simple models (like systems of a few connected boxes), especially when nonlinear relations are involved. Obviously, it is not enough for a good model to be based on adequate numerical schemes. The equations representing the physical and chemical processes also have to be correct and relevant.

A particularly important, but difficult, problem is the description of clouds processes — physical and chemical transformation processes inside cloud droplets and on ice surfaces, precipitation scavenging, and rapid vertical mixing in convective clouds — in atmospheric chemistry models. The intermittent nature of cloud encounters and the small spatial scale of most clouds call for a statistical treatment. We are still very far from having an ideal parameterization of such cloud processes. Here is a challenge for students entering this field.

*Henning Rodhe is a professor of Chemical Meteorology at Stockholm University. His main research interest is modeling of the global distribution of sulfur and nitrogen compounds and the impact of anthropogenic emissions on terrestrial systems (acidification) and on climate.*

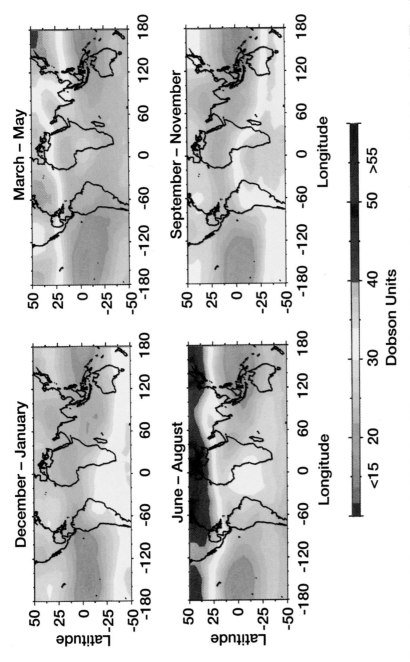

*Plate 1:* Seasonal climatological distribution of the tropospheric residual ozone abundance (expressed in Dobson Units). The values are obtained from the difference between total ozone columns derived by the Total Ozone Mapping Spectrometer (TOMS) (Version 7 data) and stratospheric abundances measured during the Stratospheric Aerosol and Gas Experiment (SAGE) (from Fishman and Brackett, 1997).

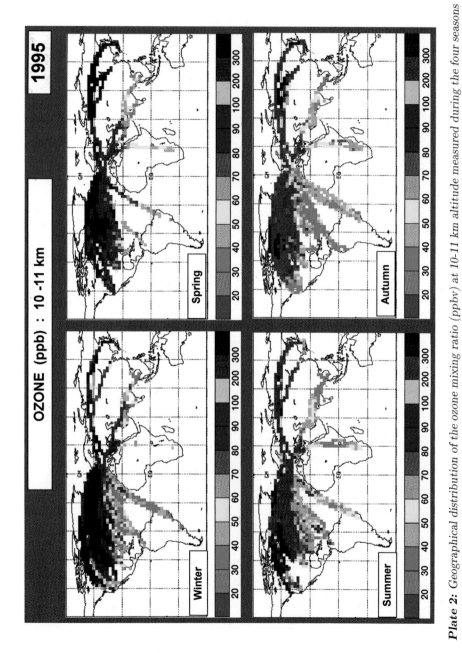

*Plate 2:* Geographical distribution of the ozone mixing ratio (ppbv) at 10-11 km altitude measured during the four seasons of 1995 by five commercial Airbus aircraft as part of the EC-funded MOZAIC (Measurement of OZone by Airbus In-service airCraft) Program. Values are integrated in 2.5° × 2.5° cells, and are represented for Northern Hemisphere winter, spring, summer, and fall, respectively (courtesy of A. Marenco, Laboratoire d'Aérologie, OMP, Toulouse, France).

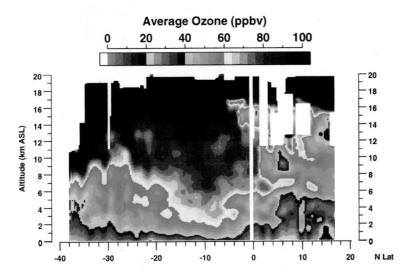

**Plate 3:** Average ozone distribution over the South Atlantic basin obtained from ozone lidar measurements during the NASA Global Tropospheric Experiment (GTE)/Transport and Atmospheric Chemistry Near the Equator-Atlantic (TRACE-A) field experiment conducted in September-October, 1992. Enhanced ozone associated with biomass burning in Africa and South America can be readily seen across the entire troposphere between 0 and 25°S (from Browell et al., 1996).

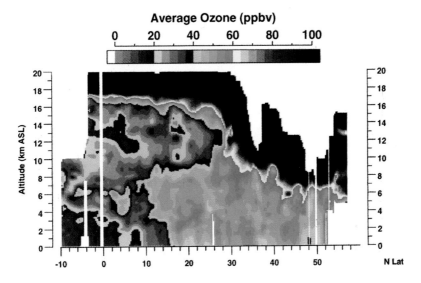

**Plate 4:** Average ozone distribution over the Western Pacific obtained from airborne lidar measurements during the NASA GTE/Pacific Exploratory Mission-West (PEM-West B) conducted during February-March, 1994. Ozone destruction near the surface in the tropics produces the observed low ozone levels in the lower troposphere, and deep cloud convection in the tropics transports low ozone from near the surface to the upper troposphere (courtesy of E.V. Browell, NASA Langley Research Center, Hampton, Virginia).

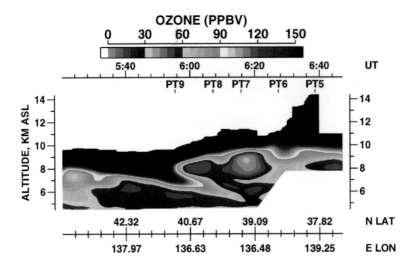

**Plate 5:** Example of tropopause fold event observed on 11 March, 1994 over Western Pacific during the NASA/GTE/PEM-West-B field experiment. The intrusion of stratospheric air (as indicated in this case by ozone levels greater than 90 ppbv) into the troposphere can be seen in the figure. The intrusion occurs around the jet stream. The tropopause height changes from about 10 km to the south of the jet to about 8 km to the north. Stratospheric ozone in the intrusion extends down to about 5 km (courtesy of E. V. Browell, NASA Langley Research Center, Hampton, Virginia).

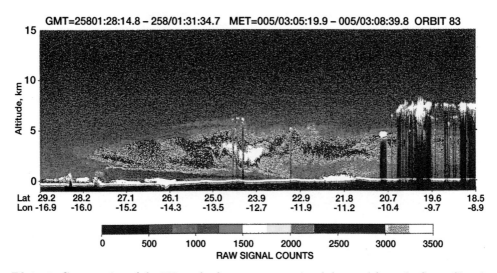

**Plate 6:** Cross section of the 532 nm backscatter return signal detected from the Space Shuttle during the NASA's Lidar In-space Technology Experiment (LITE, see Winker et al., 1996) showing an elevated layer of Saharan Dust extending from Mauritania (right) to the Canary Islands (left) (courtesy of NASA).

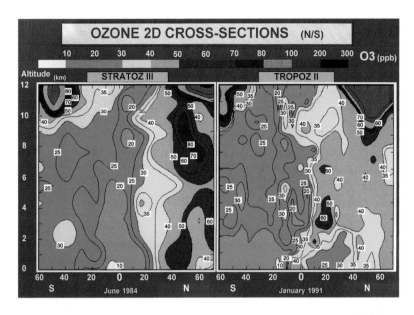
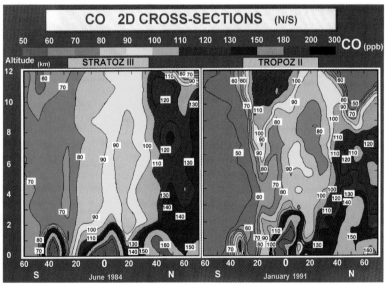

**Plate 7:** Latitude-altitude cross-section of the ozone (upper panel) and CO (lower panel) mixing ratio (ppbv) measured during the southbound flights (Greenland to Patagonia along the east coast of North America and the west coast of South America) during the STRATOZ III (June, 1984) and TROPOZ II (January, 1991) airborne campaigns (courtesy of A. Marenco, Laboratoire d'Aérologie, OMP, Toulouse, France).

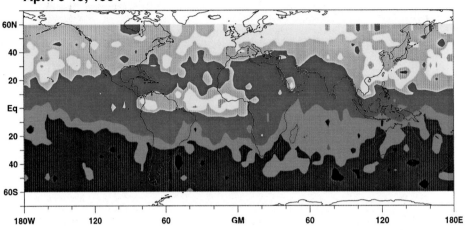

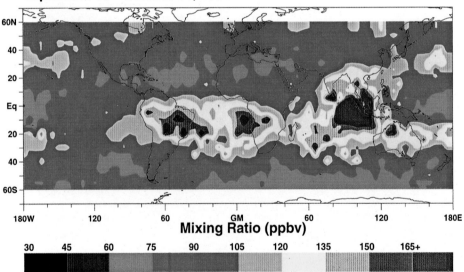

**Plate 8:** Geographical distribution of carbon monoxide mixing ratio (ppbv) in the mid-troposphere derived from the MAPS instrument onboard the Space Shuttle during two periods of time (upper panel: April 9-19, 1994; lower panel: September 30-October 11, 1994). These global observations reveal a strong hemispheric difference in the CO abundance during April (typical mixing ratios of 120 ppbv in the NH and 60 ppbv in the SH), and a substantial impact of tropical biomass burning during October (mixing ratios reaching more than 150 ppbv in the tropical region south of the equator) (courtesy of V. Connors, NASA Langley Research Center, and H. Reichle, North Carolina State University).

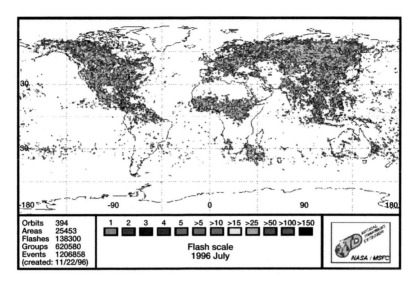

**Plate 9:** Global distribution of lightning flashes detected by the Optical Transent Detector (OTD) during July, 1996. Lightning represents a major source of nitrogen oxides in the free troposphere (courtesy of NASA Marshall Space Flight Center).

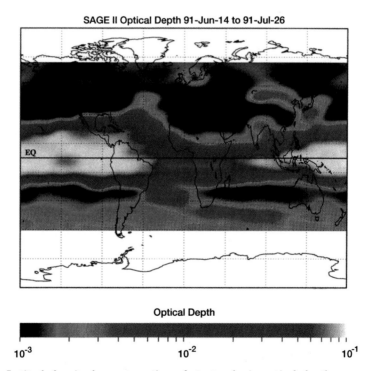

**Plate 10:** Latitude-longitude cross section of stratospheric optical depth measured by the Stratospheric Aerosol and Gas Experiment II (SAGE II) between June 14 and July 26, 1991. The large values in the tropics are the result of sulfur dioxide and some crustal material injected into the stratosphere by the eruption of Mt Pinatubo in mid-June, 1991 (from McCormick and Veiga, 1992).

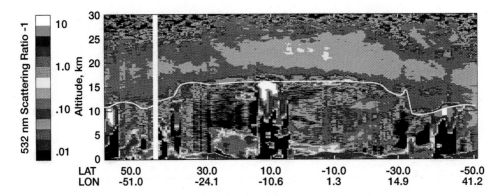

*Plate 11:* Intensity plot of lidar scattering ratio at 532 nm versus latitude and altitude, showing the horizontal and vertical distribution of the stratospheric aerosol along LITE orbit 115 on September 17, 1994. The tropopause heights are indicated by the white line (from Kent et al., 1997).

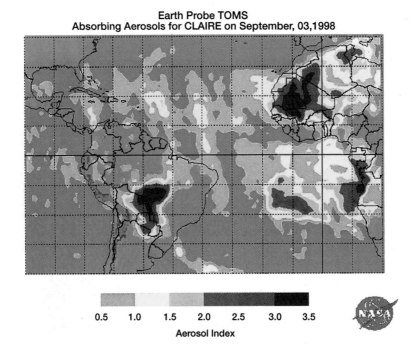

*Plate 12:* Absorbing aerosols observed by the Earth probe Total Ozone Mapping Spectrometer (TOMS) on September 3, 1998 (during the LBA/CLAIRE field campaign in Amazonia), highlighting the contribution of mineral dust from the Sahara and of soot from biomass burning in Africa and South America (courtesy of J. Herman, NASA/Goddard Space Flight Center).

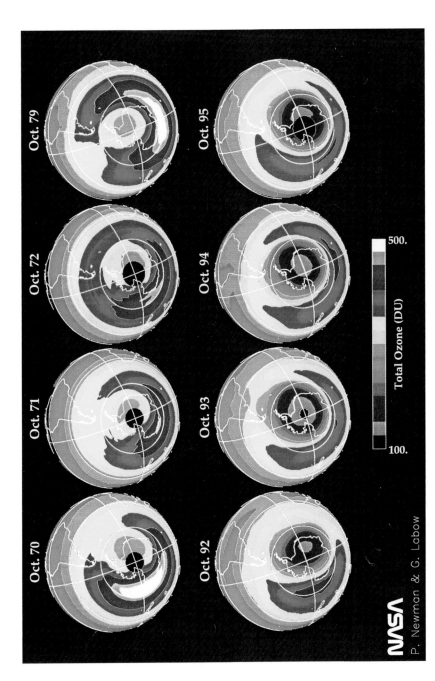

**Plate 13:** *Evolution between 1970 and 1995 of the October mean total ozone column abundance measured by the Backscatter Ultraviolet (BUV) and the Total Ozone Mapping Spectrometer (TOMS) space instruments (Dobson Units). The figure clearly shows the gradual reduction in the springtime Antarctic ozone (ozone hole) (courtesy of NASA/Goddard Space Flight Center).*

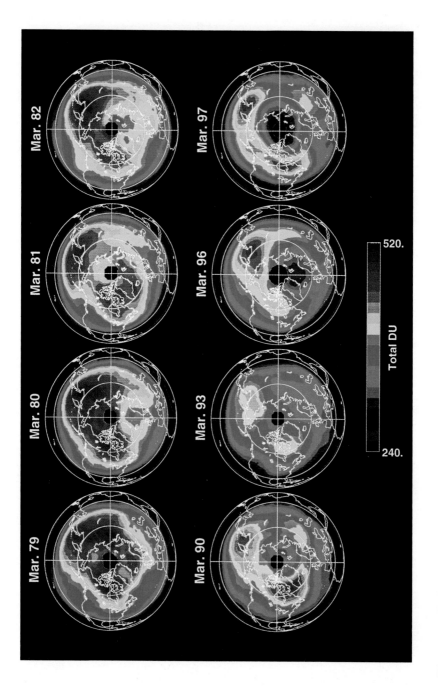

**Plate 14:** Evolution between 1979 and 1997 of the total ozone column abundance (March average) measured by the Total Ozone Mapping Spectrometer (TOMS) (Dobson Units). The figure highlights the substantial reduction in Arctic ozone during the 1990's (courtesy of P. Newman, NASA/Goddard Space Flight Center).

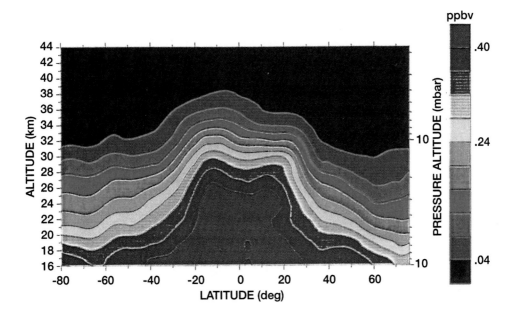

*Plate 15:* Zonally average distribution of the Chlorofluorocarbon-12 mixing ratio (in ppbv) measured by the CLAES instrument on board the Upper Atmosphere Research Satellite (UARS), March 20-27, 1992 (from Nightingale et al., 1996).

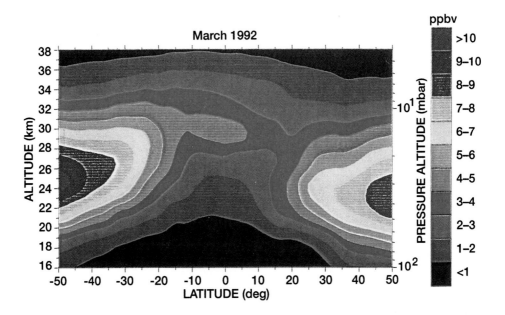

*Plate 16:* Zonally averaged distribution of the nitric acid mixing ratio (in ppbv) measured by the CLAES instrument on board the Upper Atmosphere Research Satellite (UARS) during March, 1992 (courtesy of NASA. See also Roche et al., 1994).

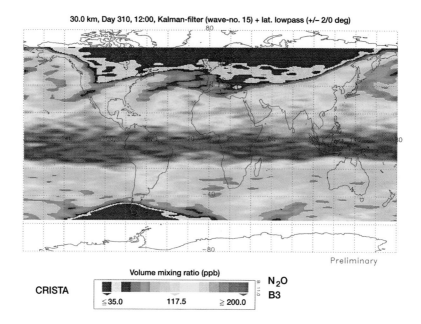

**Plate 17:** Horizontal distribution of the nitrous oxide mixing ratio (ppbv) at 30 km altitude measured in November, 1994 by the CRISTA instrument (University of Wuppertal, Germany) during a Space Shuttle flight. Intrusion of tropical air into mid- and high-latitude regions are visible in the vicinity of East Asia and near the eastern coast of the United States (courtesy of D. Offermann, University of Wuppertal, Germany).

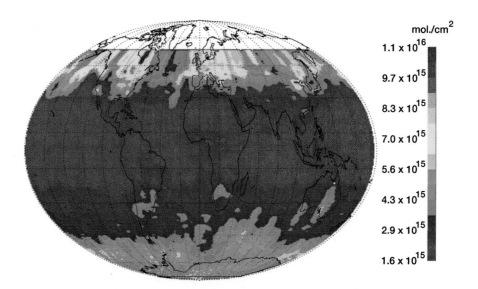

**Plate 18:** Vertical column of $NO_2$ (molecules $cm^{-3}$) measured by the ESA spaceborne GOME instrument on December 13-15, 1996 (courtesy of J. Burrows, University of Bremen, Germany).

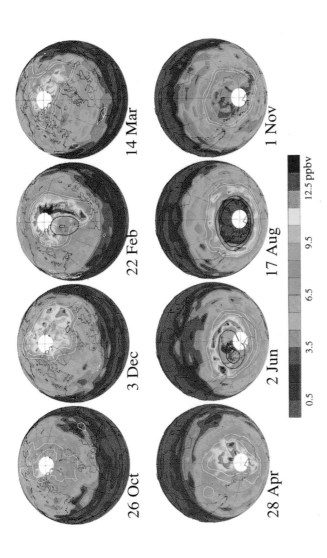

**Plate 19:** Observations of nitric acid in the lower stratosphere during the 1992/93 northern winter (top row) and during the 1992 southern winter (bottom row) from the Microwave Limb Sounder (MLS) instrument on board the Upper Atmosphere Research Satellite (UARS). The color bar gives the mixing ratio of nitric acid. Measurements were not possible within the white area poleward of 80°. The white lines are contours of potential vorticity and define the boundary of the vortex. The inner and outer black contours enclose areas where the temperature was 190 K or lower and 195 K or lower (outer), respectively. October 26 and December 3 in the north and April 18 in the south illustrate periods before temperatures fell low enough for PSC formation. February 22 in the north and June 2 and August 17 in the south illustrate times when temperatures were low enough for PSC formation. March 14 in the north and November 1 in the south illustrate times after temperatures had increased to above the threshold for PSC formation (Santee et al., 1994; adapted from WMO, 1995).

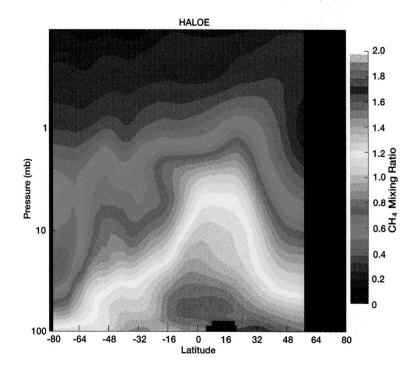
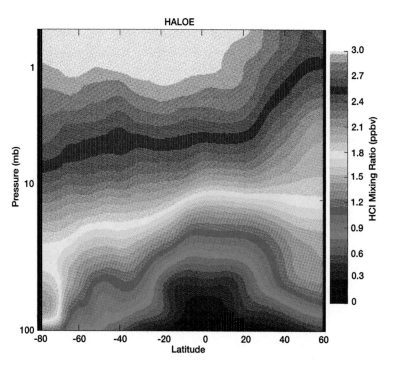

**Plate 20a, b:** Observed zonally averaged distribution (from 80° S to 60° N) of methane (above) and hydrogen chloride (below) by the HALOE instrument on board the Upper Atmosphere Research Satellite (UARS) from September 21 to October 15, 1992 (courtesy of J. M. Russell III, Hampton University, Virginia).

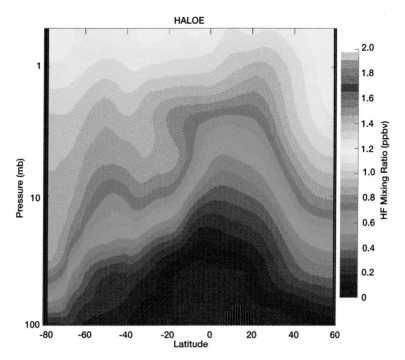

**Plate 20c:** *Observed zonally averaged distribution (from 80° S to 60° N) of hydrogen fluoride by the HALOE instrument on board the Upper Atmosphere Research Satellite (UARS) from September 21 to October 15, 1992 (courtesy of J.M. Russell III, Hampton University, Virginia).*

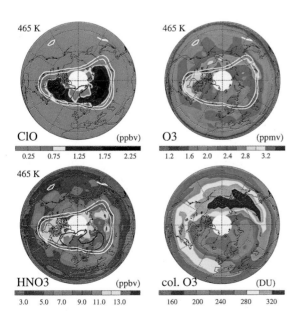

**Plate 21:** *Lower stratospheric mixing ratio of ClO (ppbv), $HNO_3$ (ppbv), and ozone (ppmv) on the 465 K isentropic surface, as well as ozone column abundance (Dobson Units) measured by the UARS Microwave Limb Sounder (MLS) on January 30, 1996 in the Northern Hemisphere. The location of the polar vortex is indicated by white lines. Chlorine activation is visible in the polar region where the temperature is low, and polar stratospheric clouds (PSCs) are formed (within the black contour inside the polar vortex). Denitrification is visible within the PSC region. The ozone column is low inside the polar vortex (courtesy of J. Waters, Jet Propulsion Laboratory, Pasadena, California).*

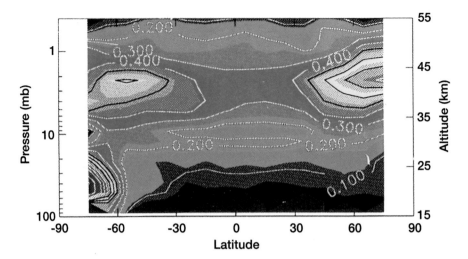

**Plate 22:** Zonally averaged daytime ClO mixing ratio (ppbv) between 15 and 55 km for August-October, 1992 observed by the MLS instrument on board UARS. The maximum concentrations of less than 1 ppbv are visible in the upper stratosphere (high latitudes). High ClO abundances seen south of 60° S in the lower stratosphere result from chlorine activation on polar stratospheric clouds (from Jackman et al., 1996).

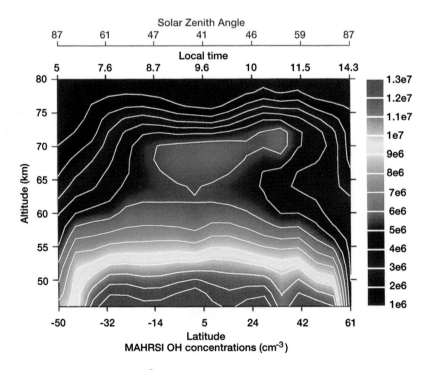

**Plate 23:** Number density ($cm^{-3}$) of the hydroxyl radical (OH) observed as a function of latitude and height on November 5, 1994 by the Middle High Resolution Spectrograph Investigation (MAHRSI) instrument on board the Space Shuttle during a single orbit. The solar zenith angle and local solar time corresponding to the observation are also indicated (from Summers et al., 1996).

# Part 4

## Ozone, Climate, and Global Change

# 13 Tropospheric Ozone

## 13.1 Introduction

Although tropospheric ozone is only a trace gas, it plays a controlling role in the oxidation capacity of the atmosphere. Ozone and its photochemical derivative, OH, are the major oxidants for most reduced gases. Without ozone, reduced gases such as CO, hydrocarbons, and most of the sulfur and reactive nitrogen compounds would accumulate to levels substantially above those in the present atmosphere. Environmentally, high levels of ozone at the surface are a major pollutant (see Box 13.1) because of their detrimental effects on human health and plants. Elevated and potentially harmful levels of ozone have been observed extensively in industrialized as well as developing countries. Radiatively, tropospheric ozone also plays an important role because its strong absorption band, centered at 9.6 $\mu$m, is within the so-called atmospheric window and thus makes it an effective greenhouse gas, particularly in the upper troposphere where the temperature is low (Fishman *et al.*, 1979; Wang *et al.*, 1986).

In the second half of the nineteenth century, surface ozone was a major topic of research because of its (erroneously) assumed role in checking the spread of epidemics. Most measurements were made then using Schönbein test paper, which gave only semiquantitative information. It was not until 1876 that the Paris Municipal Observatory at Montsouris started 34 years of continuous quantitative measurements (Bojkov, 1986).

The classical view of the origin of tropospheric ozone is that ozone is transported from the stratosphere and destroyed at the surface (Regener, 1949). Photochemical production as the source of ozone in the ambient atmosphere outside polluted areas

*Figure 13.1.* Major processes affecting tropospheric ozone on a global scale (courtesy of F. Stordal, NILU, Norway).

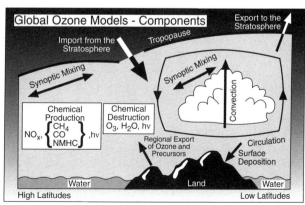

> **Box 13.1 Urban and Industrial Air Pollution: An Old Problem**
>
> Throughout antiquity, air pollution in the form of smoke was identified as the cause for the observed blackening of religious temples in large cities like ancient Rome, but it was in European cities much later that the problem of urban pollution received wide attention. During the twelfth century, philosopher Moses Maimonides (1135-1204) noted that urban air was often "turbid, thick, misty, and foggy." In the 1300s as the shortage of wood led industries in Great Britain to adopt coal as a fuel, King Edward I prohibited London merchants from selling coal during the sessions of Parliament. In 1661 John Evelyn, in his small book entitled *Fumifugium*, reported the presence in the same city of a "horrid smoake" that polluted the rain and the air, and started to damage materials and buildings. Episodes of stagnant smoke and dense fog, causing severe health problems (pulmonary diseases and heart failure) became relatively frequent in London after the industrial revolution and were called London *smog*, a term introduced by British physician Harold Des Voeux, coined from the words smoke and fog.
>
> During the nineteenth century, concern was expressed about the acidification of precipitation that resulted from increasing emissions of sulfur and nitrogen oxides produced by industrial and domestic combustion. The term *acid rain* (*pluie acide*) was introduced by French scientist M. Ducros in 1845, and the impact of air pollution on rainfall acidity was further considered by Angus Smith in his book *Air and Rain*, published in 1872. The problem of environmental acidification received a lot of attention in the 1980s when it became evident that aquatic life and forests had been severely damaged in Europe and North America. Legislative measures were taken to reduce industrial emissions of acid rain precursors.
>
> The twentieth century has also seen several major pollution incidents, such as the deadly events caused by smelters in the Meuse Valley (Belgium, 1929) and in Donora, Pennsylvania (1948), the industrial accidents in Seveso, Italy (1976) and in Bhopal, India (1984, causing the death of more than 2000 people), the nuclear reactor accident of Chernobyl in 1986, and the Kuwait Oil Fires during the Iraqi war in 1990.
>
> Measures to reduce the emissions of primary pollutants, such as sulfur from coal, have reduced the frequency and intensity of London smog events, but have not prevented the occurrence of the dense haze often reported during summertime in large urbanized areas such as Los Angeles, Mexico City, or Tokyo. In the early 1950s research conducted in California suggested that the necrotic patterns observed on plant leaves, and health symptoms commonly reported during haze events in Los Angeles (eye irritation, sore throat, asthma, etc.), were not caused by primary pollutants like $SO_2$, but by enhanced levels of ozone and other oxidants. The biochemist Arie Jan Haagen-Smit suggested that the formation of urban ozone results from the action of sunlight on reactive hydrocarbons and nitrogen oxides released, for example, by oil refineries and automobiles. Several decades after the discovery by Haagen-Smit, and in spite of regulatory measures such as the Clean Air Act (1970) in the United States and its subsequent amendments, *photochemical air pollution* remains a major problem in many areas of the United States and Europe, despite improvements in urban air quality.

was not recognized until Crutzen (1973) and Chameides and Walker (1973) proposed that oxidation of $CH_4$ and CO in the presence of $NO_x$ would lead to a significant amount of ozone production. This is shown schematically in Figure 13.1. Substantial advances have been made in the understanding of the budget of tropospheric ozone in the past two decades. Nevertheless, the subject remains a major focus of intensive research.

## 13.2 Distribution and Trends

Until recently, our knowledge of the distribution of tropospheric ozone (see Color Plates 2-5) has come primarily from ozonesonde, surface, and aircraft observations. Surface ozone observations have shown that elevated and potentially harmful levels of ozone are commonly observed in industrialized as well as developing countries. An example is illustrated in Figure 13.2. This figure depicts the elevated ozone mixing ratios (ppbv) that occurred during a pollution episode that covered a vast area of

**Figure 13.2.** Plot of elevated ozone levels (ppbv) that were observed over the eastern United States during a pollution event that occurred during the summer of 1986. These results were obtained from the Environmental Protection Agency (EPA) ozone monitoring network (Fehsenfeld and Liu, 1993).

the eastern United States and southern Canada. Although the highest levels were recorded in and near urban areas, ozone levels recorded over rural areas were high enough to do significant damage to sensitive vegetation. The photochemical production of ozone also contributes to a relatively large diurnal variation in its mixing ratio. Figure 13.3 shows a typical diurnal variation observed in rural areas of industrialized countries. The decrease of ozone near sunset is the result of surface deposition and the formation of a stable surface inversion layer that prevents ozone from being transported from aloft. This mechanism and the breakup of the surface inversion layer in the morning can also lead to a relatively smaller diurnal variation in clean continental areas, where the ozone production is small.

Figures 13.4a-d show the pressure-latitude distributions of the long-term average ozone mixing ratio (in ppmv) for four seasons as derived from observations at 24 ozonesonde stations (London and Liu, 1992). Most of the stations are located

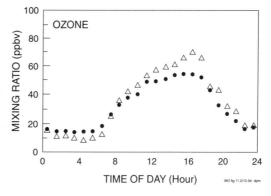

**Figure 13.3.** The average ozone diurnal mixing ratio at Scotia, PA, for 25 June to 26 July 1986 (circles) and the average diurnal ozone mixing ratio at this site during high-pressure events (i.e., when the atmospheric pressure exceeded the period average atmospheric pressure, 970 mb) (triangles). The sampling height was 5 m above the ground (from Trainer et al., 1987b; reprinted with permission from Nature, copyright 1987, Macmillan Magazines Limited).

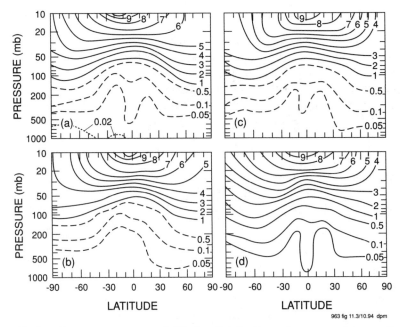

**Figure 13.4.** *Long-term pressure-latitude distribution of ozone mixing ratio (ppmv). (a) December, January, February, (b) March, April, May, (c) June, July, August, (d) September, October, November. Values less than 1 ppmv are shown as dashed lines, and values less than 50 ppbv as dotted lines (from London and Liu, 1992).*

over industrialized Northern Hemisphere (NH) midlatitude regions. Only five stations are in the Southern Hemisphere (SH). Ozone mixing ratios in general increase with altitude, indicating a downward flux throughout the troposphere. In the middle and upper troposphere (500-200 mb), the NH midlatitude mixing ratios are almost twice those in the SH. Since surface deposition is significantly faster over land than ocean, the hemispheric difference in ozone mixing ratio suggests a greater photochemical source and/or a greater stratospheric flux in the NH. Ozone is transported from the lower stratosphere into the upper troposphere through tropopause folding and other processes at mid- and high latitudes (Danielsen, 1985; Holton, 1990; see also Color Plate 5). This transport flux in the NH (about $5 \times 10^{10}$ molecule cm$^{-2}$ s$^{-1}$) is estimated to be about double that in the SH (Gidel and Shapiro, 1980; Mahlman et al., 1980). On the other hand, as discussed in Chapter 7, the rate-limiting ozone precursor, $NO_x$, is significantly more abundant in the NH than the SH. This will lead to a significantly higher photochemical production rate in the NH than in the SH. In the tropics the mixing ratio is low compared to higher latitudes and is nearly constant with height up to about 200 mb, apparently the result of photochemical destruction, efficient vertical transport, and small stratospheric influence. Although the tropical ozone distribution is based on only two stations (Samoa, 14°S, and Natal, 6°S), the general features of the distribution is substantiated by aircraft observations.

There is a relatively clear seasonal variation at midlatitudes in the NH. A representative station is Hohenpeissenberg (48°N), which has used the same type of instrument since November 1966 and has more frequent observations than other locations. Figure 13.5 shows the long-term (22 years) seasonal variation in the monthly average mixing ratio observed at the station. In the low to middle troposphere (surface to 500 mb) the mixing ratio during late spring and summer is from 25% to 50% higher than in late fall and winter. This feature can also be seen in Figures 13.4a-d. The

**Figure 13.5.** Long-term (1967-1988) pressure-month distribution of ozone mixing ratio (ppmv) for Hohenpeissenberg (48° N) (from London and Liu, 1992).

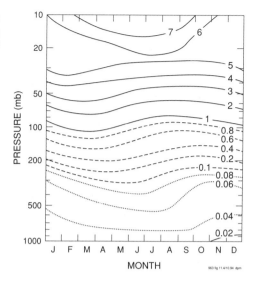

summer maximum is obviously a result of photochemical production of ozone due to near-surface emissions of ozone precursors, consistent with surface observations over industrialized regions. Above 500 mb the seasonal maximum is shifted to April at 100 mb and to January-February at about 30 mb. This is likely to be a result of the strong downward ozone transport from the stratosphere and the long photochemical lifetime (a few months) of ozone in the winter.

A technique that derives the tropospheric column density of ozone from the difference between results of two satellite instruments, namely, TOMS and SAGE, which measure total column ozone and stratospheric ozone, respectively, has been developed by Fishman et al. (1992). Color Plate 1 depicts the latitudinal and longitudinal distributions of tropospheric ozone column density in Dobson Units (1 DU = $2.69 \times 10^{16}$ molecule cm$^{-2}$; see Box 1.4) for four seasons. Two major features may have important implications to the ozone budget. First, the ozone column density in the tropics is about a factor of two lower than values at NH midlatitudes and a factor of 1.4 lower than SH midlatitude values, consistent with the ozonesonde observations discussed above. The second and more intriguing feature is the seasonal variation. At midlatitudes in both hemispheres there is a clear increase in ozone column density in the respective springs. At NH midlatitudes there is an additional increase in the summer. This summertime increase is consistent with the seasonal variation observed by midlatitude ozonesonde stations and is likely due to ozone production from ozone precursors of anthropogenic origin. The spring ozone increase that has been observed over many nonurban stations has been traditionally attributed to enhanced stratospheric flux. However, recently it was hypothesized that accumulation of anthropogenic ozone precursors in the winter coupled with increased ozone photochemical lifetime in winter and spring could lead to the spring buildup of ozone in the NH (Penkett and Brice, 1986; Liu et al., 1987). This hypothesis seems to be contradicted by the existence of the austral spring ozone maximum at SH midlatitudes, where industrial emissions are negligibly small. However, Fishman et al. (1992) proposed that there was evidence supporting an anthropogenic origin for the austral spring ozone maximum at SH midlatitudes. The anthropogenic pollution source is the widespread biomass burning in Africa and probably South America during the austral winter, that is, the dry season (see Box 13.2).

**Box 13.2 Ozone in the Tropical Troposphere**

As large amounts of the savanna are burned each year during the dry season in the tropics, large amounts of ozone precursors, such as $NO_x$ and CO, are released into the boundary layer of the atmosphere. With intense solar radiation available at these latitudes, photochemical processes are efficient and large amounts of ozone are produced (see Fig. 13.6). Observations made over Africa, for example, have revealed that the ozone density at 2-4 km

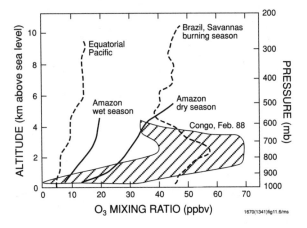

*Figure 13.6.* Vertical distribution of the ozone mixing ratio at different locations in the troposphere (from Andreae et al., 1992).

altitude during the dry season can reach values that are typical of polluted days (60-100 ppbv) in industrialized regions (Andreae *et al.*, 1992; Fishman *et al.*, 1990). Visibility is also reduced by the haze produced by the fires. Finally, substantial quantities of ozone and ozone precursors are probably transported away from the biomass-burning areas, and when entrained into tropical convective systems reach relatively high altitudes in the free troposphere. It is now realized that the numerous fires, which support agricultural activities in Africa, South America, and Asia, contribute substantially to the global pollution of the troposphere.

Using the Andes mountains to compute tropospheric ozone columns from the difference between TOMS-measured ozone columns over the mountains and nearby sea-level columns, Kim and Newchurch (1996) show clear evidence of a significant seasonal cycle in lower tropospheric ozone. This seasonal variation peaks between September and November, the height of the biomass-burning period in South America. Kim and Newchurch also computed increasing trends in lower tropospheric ozone between latitudes 12°S and 23°S over the period 1979 to 1993, which are consistent with the trends computed by Jiang and Yung (1996). Applying the same technique over New Guinea, Kim and Newchurch (1998) again find a significant seasonal cycle in lower tropospheric ozone exactly out of phase with the rainfall seasonality (*i.e.*, positively correlated with the biomass-burning cycle) downwind of the biomass burning region. Upwind of the biomass burning region, no seasonal cycle in the ozone exists. Consistent with a biomass-burning source of ozone, the trends upwind of the source are essentially zero, while the downwind trends are approximately 1% per year, similar to the increasing trends west of South America.

## Table 13.1
### Long-Term Ozone Trend

| Station | Lat. | Long. | 500 mb | | | 50 mb | | |
|---|---|---|---|---|---|---|---|---|
| | | | Mean (ppmv) | Slope (%yr$^{-1}$) | $\sigma$ (%yr$^{-1}$) | Mean (ppmv) | Slope (%yr$^{-1}$) | $\sigma$ (%yr$^{-1}$) |
| Resolute (1966 - 1988) | 75°N | 95°W | 0.052 | 0.75 | 0.26 | 3.42 | −0.64 | 0.17 |
| Churchill (1974 - 1988) | 59°N | 94°W | 0.053 | 1.89 | 0.69 | 3.22 | −0.74 | 0.32 |
| Edmonton (1973 - 1988) | 54°N | 114°W | 0.051 | 1.50 | 0.47 | 3.11 | −0.50 | 0.30 |
| Goose Bay (1970 - 1988) | 53°N | 60°W | 0.055 | 0.67 | 0.36 | 3.22 | −0.80 | 0.23 |
| Legionowo (1979 - 1988) | 52°N | 21°E | 0.067 | 0.65 | 0.74 | 2.94 | −0.89 | 0.50 |
| Lindenberg (1975 - 1988) | 52°N | 14°E | 0.073 | 0.26 | 0.46 | 2.93 | −0.26 | 0.24 |
| Hohenpeissenberg (1967 - 1988) | 48°N | 11°E | 0.052 | 2.42 | 0.22 | 2.88 | −0.63 | 0.17 |
| Payerne (1969 - 1988) | 47°N | 7°E | 0.050 | 1.18 | 0.39 | 2.71 | −0.73 | 0.14 |
| Wallops Island (1971 - 1988) | 38°N | 75°W | 0.063 | 0.32 | 0.35 | 2.62 | 0.30 | 0.29 |
| Aspendale (1966 - 1982) | 38°S | 145°E | 0.039 | 0.12 | 0.46 | 2.64 | −0.27 | 0.22 |

The identification of the origin of the spring ozone increase is still a subject of active research. The outcome will have a profound implication on understanding the impact of anthropogenic activities on the tropospheric ozone distribution.

Table 13.1 shows the long-term trends in tropospheric ozone at 500 mb at 10 stations compiled from available data by London and Liu (1992). For comparison, the trends at 50 mb are also shown. All stations are located in North America and Northern Europe except the one at Aspendale, Australia. Before discussing the trends, it is important to note that changes in instrument types (e.g., the four Canadian stations) and calibrations could introduce a bias in the calculated trends (Logan, 1994). This is evident in the large scattering of the trends among these stations, some within a few hundred kilometers of each other. The scattering is even greater when year-to-year changes are examined. Therefore, the trends probably should be regarded as being qualitative rather than quantitative. In this context, Table 13.1 indicates that there is probably an increase in midtropospheric ozone over Northern Europe. The increase over North America is probably not significant due to change of instruments. The decreasing trends at 50 mb are significant and are much greater than the tropospheric trends in absolute value.

In regard to the long-term trends of ozone at surface stations, there were some interesting studies of the data taken at the Paris Municipal Observatory at Montsouris from 1876 to 1909 (Bojkov, 1986; Volz and Kley, 1988). They found that the ozone concentrations at the station, which at that time was in a rural environment, were significantly lower than the present-day concentrations of surface ozone in rural Europe throughout the year. In particular, Volz and Kley, who reconstructed a similar instrument to those used at Montsouris and calibrated it using modern methods, found that the present-day ozone at a rural station in Europe showed a significantly greater seasonal variation than that at Montsouris between 1876 and 1886, where there was almost no seasonal variation. The present-day ozone had a late spring and summer maximum that was about three to four times the value at Montsouris, while the ratio of the winter minima was about two.

> **Box 13.3 Ozone in the Arctic Troposphere**
>
> The Arctic is surrounded by industrialized continents (Siberia, Europe, and North America), which, particularly in spring, contribute to substantial gaseous and particulate pollution (including the so-called Arctic haze). A remarkable phenomenon observed in spring is the quasidisappearance of ozone near the surface during a succession of episodes. These episodes are not associated with the Arctic haze, but are characterized by enhanced concentrations of particulate and gaseous bromine. The duration of these events ranges from several hours to several days. The depletion of surface ozone (typically from 40 ppbv to less than 0.5 ppbv), first reported by Barrie *et al.* (1988) and observed at polar sunrise, is believed to be due to catalytic reactions involving bromine and perhaps other halogens of marine origin (probably liberated from sea-salt and aerosol particles). The exact mechanisms are not yet fully understood (see Section 8.4.3.2), but are believed to involve heterogeneous reactions similar to those involved in the stratospheric ozone hole which generate halogen atoms and hence lead to rapid ozone destruction.

## 13.3 Production and Loss of Ozone

*13.3.1 Formulation*

In the early 1950s, Haagen-Smit correctly proposed that the ozone and other components of photochemical smog (see Box 13.4) observed in urban areas could be formed in the atmosphere as a result of photochemical reactions involving $NO_x$ and nonmethane hydrocarbons (NMHC) present in automobile exhaust. Since there are several hundreds of different hydrocarbons emitted by automobiles, the photochemistry involved is extremely complex. However, extensive research on the reactions of hydrocarbons (described in Chapter 9) has indicated that the following major reactions are responsible for the formation of ozone in the urban atmosphere. For clarity, the reaction scheme is simplified and NMHC abbreviated as $RH$. In addition, carbonyl products (aldehyde or ketone) are denoted $R'CHO$, where $R'$ denotes an organic fragment having one carbon atom fewer than $R$

$$OH + RH \to R + H_2O \tag{13.1}$$

$$R + O_2 + M \to RO_2 + M \tag{13.2}$$

$$RO_2 + NO \to RO + NO_2 \tag{13.3}$$

$$RO + O_2 \to HO_2 + R'CHO \tag{13.4}$$

$$HO_2 + NO \to OH + NO_2 \tag{13.5}$$

$$2\,(NO_2 + h\nu \to NO + O) \tag{13.6}$$

$$\underline{2\,(O + O_2 + M \to O_3 + M)} \tag{13.7}$$

$$\text{Overall:} \quad RH + 4O_2 + h\nu \to R'CHO + H_2O + 2\,O_3 \tag{13.8}$$

**Box 13.4 Photochemical Smog**

Nitric oxide and volatile organic compounds are common products of human activities, and are released in large quantities by automobile engines. In urban areas, under stable meteorological conditions (with almost no vertical exchanges and little ventilation), these pollutants tend to accumulate in the boundary layer, and during summer months, when solar ultraviolet radiation is intense, to undergo photochemical transformations. Deleterious products such as ozone, aldehydes, and peroxyacetyl nitrate (PAN) are formed [see Chapters 7 and 9, and Reactions (13.1-13.8)]. Figure 13.7 shows a typical sequence of a smog event in Los Angeles. Ozone precursors ($NO_x$, hydrocarbons) begin to build up during the morning rush hour, while the abundance of ozone reaches its maximum in the early afternoon. These pollutants are detrimental to the biosphere (plant damage, reduction in crop productivity) and cause health problems (eye and nose irritation, respiratory problems). They can be transported downwind and affect rural environments in the vicinity of metropolitan areas.

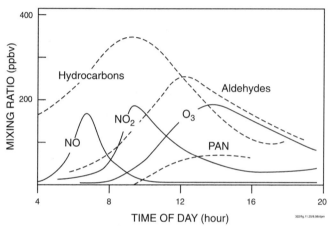

*Figure 13.7. Evolution of the chemical composition of the lower atmosphere during a smog event (Goody, 1995).*

The carbonyl compounds can undergo further photochemical reactions that will result in additional production of organic and hydrogen radicals and, in turn, produce more ozone. NMHC and $NO_x$ are referred to as ozone precursors. In the above reactions, NMHC are consumed, while $NO_x$ acts as a catalyst. However, there are simultaneous reactions in the atmosphere that consume $NO_x$ by converting it to $HNO_3$ or nitrate, which are readily removed from the atmosphere through wet and dry deposition.

In the free troposphere and remote oceanic boundary layer, NMHC concentrations are relatively small. The ozone production is dominated by oxidation of CO and $CH_4$ instead of NMHC. The reaction schemes for CO and $CH_4$ are similar to the generalized scheme for NMHC (see also Chapter 9):

$$OH + CO \rightarrow CO_2 + H \tag{13.9}$$

$$H + O_2 + M \rightarrow HO_2 + M \tag{13.10}$$

$$HO_2 + NO \rightarrow OH + NO_2 \tag{13.5}$$

$$NO_2 + h\nu \rightarrow NO + O \tag{13.6}$$

$$O + O_2 + M \rightarrow O_3 + M \tag{13.7}$$

$$\text{Overall}: \quad CO + 2\, O_2 + h\nu \rightarrow CO_2 + O_3 \tag{13.11}$$

$$OH + CH_4 \rightarrow CH_3 + H_2O \tag{13.12}$$
$$CH_3 + O_2 + M \rightarrow CH_3O_2 + M \tag{13.13}$$
$$CH_3O_2 + NO \rightarrow CH_3O + NO_2 \tag{13.14}$$
$$CH_3O + O_2 \rightarrow HO_2 + CH_2O \tag{13.15}$$
$$HO_2 + NO \rightarrow OH + NO_2 \tag{13.5}$$
$$2\,(NO_2 + h\nu \rightarrow NO + O) \tag{13.6}$$
$$\underline{2\,(O + O_2 + M \rightarrow O_3 + M)} \tag{13.7}$$
$$\text{Overall}: \quad CH_4 + 4O_2 + h\nu \rightarrow CH_2O + H_2O + 2\,O_3 \tag{13.16}$$

Because CO and $CH_4$ are readily available, $NO_x$ is usually the rate-limiting precursor of ozone production, as described in Section 7.5.2. In this context, additional insight can be gained by examining the mathematical formulation of the ozone photochemical production and loss. The formulation is defined somewhat differently by researchers according to a number of factors, including conceptual clarity, ease of formulation, and computational convenience. Of course, as long as it is done self-consistently, different formulations will result in the same ozone level because the level is determined by the net ozone production (*i.e.*, production minus loss), which is independent of the formulation used. In the following, the concept of rate-limiting ozone production is chosen, since it provides conceptual clarity.

Consider the CO oxidation reactions (13.9), (13.10), (13.5), (13.6), and (13.7), along with the two following reactions

$$NO + O_3 \rightarrow NO_2 + O_2 \tag{13.17}$$
$$O_3 + h\nu \rightarrow O_2 + O \tag{13.18}$$

When these last two reactions are combined with Reactions (13.6) and (13.7), null cycles are formed in which $O_3$ is not produced or destroyed

$$NO + O_3 \rightarrow NO_2 + O_2 \tag{13.17}$$
$$NO_2 + h\nu \rightarrow NO + O \tag{13.6}$$
$$\underline{O + O_2 + M \rightarrow O_3 + M} \tag{13.7}$$
$$\text{Overall}: \quad \text{No change in ozone.}$$

$$O_3 + h\nu \rightarrow O_2 + O \tag{13.18}$$
$$\underline{O + O_2 + M \rightarrow O_3 + M} \tag{13.7}$$
$$\text{Overall}: \quad \text{No change in ozone.}$$

Reactions (13.6) and (13.7) both have greater rates than (13.5), which thus can be regarded as the rate-limiting reaction of ozone production. A similar argument can be made for the oxidation of $CH_4$ and NMHC to obtain the following rate-limiting ozone production formulation

$$P(O_3) = k_3\,[RO_2]\,[NO] + k_5\,[HO_2]\,[NO] \tag{13.19}$$

An empirical extension of Eq. (13.19) is to define a new species, odd oxygen ($O_x$), consisting of molecules equivalent to ozone, as:

$$O_x = O_3 + O + O(^1D) + NO_2 + 2\,NO_3 + 3\,N_2O_5 + 1.5\,HNO_3 + 0.5\,OH$$
$$+ 0.5\,HO_2 + H_2O_2 + 0.5\,RO + 0.5\,RO_2 + ROOH + ROOR$$
$$+ 1.5\,\text{PAN (peroxyacetylnitrate)} + \text{other equivalent species} \tag{13.20}$$

The criterion of including a species in $O_x$ is that the species interchange readily with $O_3$ through photochemical reactions that do not result in actual production of $O_3$.

The coefficients in (13.20) are self-evident for species such as O and $NO_2$, which are readily converted to $O_3$ on a one-to-one basis. For $NO_3$, the coefficient is determined under the assumption that the photodissociation of $NO_3$ produces O and $NO_2$ only. Thus each $NO_3$ is equivalent to two $O_3$ molecules. For $N_2O_5$, the photodissociation and thermal dissociation of $N_2O_5$ are assumed to produce $NO_2$ and $NO_3$ only. For hydrogen and organic radicals, each species is assigned a coefficient of 0.5 because Reaction (13.23) shows that two OH molecules are produced for each $O(^1D)$ reaction with $H_2O$. Summing over all reactions,

$$P(O_x) = k_3 [RO_2][NO] + k_5 [HO_2][NO] \qquad (13.21)$$

and

$$L(O_x) = k_{23} [O(^1D)][H_2O] + k_{24} [HO_2][O_3] + L(NO_y) \qquad (13.22)$$

where $L(NO_y)$ denotes wet and dry deposition of odd nitrogen species, and $k_{23}$ and $k_{24}$ are the reaction rate coefficients for the reactions

$$O(^1D) + H_2O \to 2\, OH \qquad (13.23)$$

and

$$HO_2 + O_3 \to OH + 2\, O_2 \qquad (13.24)$$

While Eq. (13.21) can be understood from earlier discussion, Eq. (13.22) needs some explanation. A simple way to understand Eq. (13.22) is to consider loss of $O_x$ as the sum of loss of odd hydrogen, odd nitrogen, and $O_y$ [$= O_3 + O + O(^1D)$]. Then, the second term of $L(O_x)$ can be understood as the major loss of $O_y$, and the third term as the loss of $NO_y$. The first term appears to be erroneous because it is a loss of $O_y$, but also is a production of odd hydrogen. It is included as an approximation for the loss of odd hydrogen because the production of odd hydrogen is usually equal to its loss; that is, odd hydrogen is in photochemical stationary state. For details of the formulation, the reader is referred to Levy et al. (1985).

It should be noted that Eqs. (13.21) and (13.22) are defined for $O_x$, not ozone. However, since the ozone concentration is much greater than those of other $O_x$ species, the concentration of $O_x$ is a good approximation for ozone except in some urban centers. Thus, Eq. (13.19) is actually an approximation for Eq. (13.21) under nonurban conditions. In urban areas, Eq. (13.21) should be used and $O_3$ should be calculated from Eq. (13.20) by subtracting other species from $O_x$. The major advantage of the above formulation is that, under most tropospheric conditions, Eqs. (13.21) and (13.22) are good approximations to the actual photochemical production and loss of ozone, respectively; they include little contribution from the null cycles. To illustrate this point, assume that $O_x$ does not include O, then the reaction

$$O_3 + h\nu \to O_2 + O \qquad (13.18)$$

has to be regarded as a photochemical loss for ozone. This will lead to the obviously false conclusion that the lifetime of ozone is less than two hours. Another advantage is to realize that $O_x$ includes species that have similar photochemical characteristics to ozone, that is, oxidants. In addition, some of the species in $O_x$ interconvert so quickly that their mass continuity equations can be more efficiently treated numerically when they are grouped together as $O_x$.

### 13.3.2 Lifetime of Ozone

Equation (13.21) can be used to calculate the photochemical lifetime of ozone. Table 13.2 shows the lifetime calculated at two latitudes for various altitudes and seasons. In the summer, the ozone lifetime ranges from five days to a few weeks. In the boundary layer, surface deposition further reduces the lifetime, especially over the continent. Deposition velocities as high as 1 cm s$^{-1}$ have been observed over growing vegetation. As a result, the ozone distribution tends to be regional in the boundary layer, while in the free troposphere it tends to be influenced by synoptic or larger-scale transport processes. In the winter, the ozone lifetime at midlatitudes is more than three months. In addition, the surface deposition on dormant vegetation is reduced substantially. Therefore, in the winter, ozone can be transported efficiently over most of the mid- and high latitudes. In this context, note that even small net production or loss can lead to a significant seasonal variation.

### Table 13.2
*Model-Calculated Photochemical Lifetime (in days) of Ozone at Various Altitudes, Seasons, and Latitudes*

| Altitude (km) | 40°N | | 20°N | |
|---|---|---|---|---|
| | Summer | Winter | Summer | Winter |
| 0 | 8 | 100 | 5 | 17 |
| 5 | 15 | 160 | 10 | 35 |
| 10 | 40 | 300 | 30 | 90 |

The lifetime of ozone discussed above is defined in the conventional fashion. However, it is usually not very useful for analyzing observed distributions of ozone because in most of the troposphere there is enough $NO_x$ present to lead to a significant in situ production of ozone. As a result, the ozone distribution is usually governed by an apparent lifetime that is substantially greater than the lifetime defined above. This apparent lifetime is defined by ozone concentration divided by the net ozone loss rate, that is, $[O_3]$/net loss, where

$$\text{net loss} = L(O_3) - P(O_3) = -\text{Net Production} = -\text{Flux Divergence}$$

### 13.3.3 Global Production and Loss Rates of Ozone

Despite the simple expression of Eq. (13.21), the ozone production rate in the troposphere is difficult to calculate accurately. The major difficulty lies in the evaluation of the distribution of $NO_x$, which has a short photochemical lifetime and widely scattered sources. Furthermore, the oxidation products of $NO_x$, such as $HNO_3$ and $N_2O_5$, are readily removed by wet and dry deposition, the rates of which are difficult to calculate accurately. In addition, the ozone production rate has a highly nonlinear dependence on the $NO_x$ concentration. In comparison, the loss rate of ozone is simpler to evaluate with reasonable confidence. The first term of Eq. (13.22), which is the easiest one to calculate, happens to be the predominant term. With a typical NH midlatitude summer distribution of ozone, one can calculate a clear sky diurnal average tropospheric column ozone loss rate of about $50 \times 10^{10}$ molecule cm$^{-2}$ s$^{-1}$ with about 50% uncertainty. This is about ten times the average NH ozone flux from the stratosphere. In order to maintain the observed ozone level, there must be an ozone production rate on the same order of magnitude as the column ozone loss rate. This is probably the strongest argument for a significant photochemical ozone production in the troposphere.

By fixing the $NO_x$, CO, $CH_4$, and NMHC concentrations in a box model, the ozone production rate can be calculated. The dashed line in Figure 13.8 depicts the results as a function of $NO_x$ for the boundary layer under summer conditions (Liu et al., 1987). Since $NO_x$ is the rate-limiting precursor, the ozone production rate is divided by the primary loss rate of $NO_x$ to $HNO_3$. This gives the amount of ozone produced for each $NO_x$ molecule emitted into the atmosphere because in stationary state the $NO_x$ emission rate equals its loss rate. For comparison, results from a three-dimensional regional model for the boundary layer in the eastern United States are shown in open squares. Considering differences in ambient conditions and model structures between the two models, the results are in good agreement. Both models show a very nonlinear relationship; that is, for each $NO_x$ molecule emitted into the atmosphere, more ozone molecules are produced at low $NO_x$ mixing ratio than at high $NO_x$. This has a profound implication for ozone production in the troposphere. It implies that for a constant emission of $NO_x$, a spatially diffuse source will lead to a significantly greater ozone production rate than a point source. A good practical example is an aircraft source that is dispersed rapidly in the upper troposphere versus an urban combustion source that tends to build up over a limited area. Another ramification of this nonlinear effect is that the ozone production tends to be increased by transport processes because transport is almost always dispersive and thus dilutes the $NO_x$ mixing ratio (Liu et al., 1987). A remarkable example is convective transport of boundary layer $NO_x$ to the free troposphere that results in substantial enhancement of ozone production (Pickering et al., 1990).

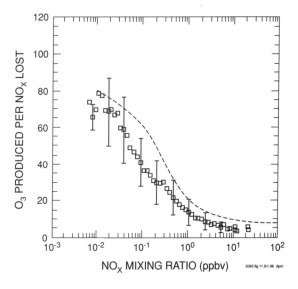

*Figure 13.8.* The number of ozone molecules produced for each $NO_x$ molecule lost plotted as a function of $NO_x$ mixing ratio. Open squares are boundary layer average values calculated from a three-dimensional model. The bars represent two standard deviations. The broken line results from a box model with fixed $NO_x$ and HC (3 p.m. EST, 0–1800 m, clear sky values; Liu et al., 1987).

It is important to note that Figure 13.8 implicitly assumes that $NO_x$ is lost from the atmosphere after it is oxidized to $HNO_3$. This is likely a good assumption in the lower troposphere, where wet and dry deposition of $HNO_3$ is fast compared to its conversion back to $NO_x$. In the upper troposphere, there may be significant recycling of $NO_x$ through $HNO_3$. If this happens, additional ozone can be produced. On the other hand, Figure 13.8 does not include losses of $NO_x$ other than the reaction of $NO_2$ with OH. Reactions such as $N_2O_5$ with aerosols and other heterogeneous reactions that remove $NO_x$ from the atmosphere can substantially reduce the ozone production efficiency shown in Figure 13.8, particularly in winter, when the OH concentration is low at midlatitudes (Ehhalt and Drummond, 1982; Liu et al., 1987; Dentener and Crutzen, 1993).

Using the values in Figure 13.8, the ozone production rate in the boundary layer of industrialized areas in the summer can be estimated by multiplying the ozone production efficiency by the emission flux of $NO_x$. The value calculated is about $26 \times 10^{28}$ molecule s$^{-1}$, which is shown in Table 13.3 (Fehsenfeld and Liu, 1993). Estimates by other methods such as a one-dimensional model calculation give consistent values (Fishman et al., 1985). The loss rate of ozone in the boundary layer of industrialized areas in summer shown in the table is estimated from the results calculated using a three-dimensional regional model (McKeen et al., 1991a). Thus the net production or flux divergence is about $8 \times 10^{28}$ molecule s$^{-1}$. This is the flux of ozone transported to other parts of the troposphere.

Table 13.3 also includes estimated rates of ozone production, loss, and net production for various other regimes of the troposphere for the summer season. In addition, the ozone burden of each regime and the stratospheric flux are included. For the free troposphere, it is not practical to estimate the ozone production from the emission flux of $NO_x$ because the sources of $NO_x$ in the free troposphere are poorly understood and there may be significant recycling of $NO_x$. As an alternative, ozone production rates for the free troposphere and the oceanic boundary layer can be calculated by using median values of the observed NO concentration. For details of the calculation, the reader is referred to Fehsenfeld and Liu (1993).

**Table 13.3**
*Global Ozone Production, Loss, and Net Production[a] in Important Regions in Summer.*

|  | Production | Loss | Net Production | $O_3$ Burden |
|---|---|---|---|---|
| Industrial boundary layer (0 to 1.5 km) | 26 | 18 | 8 | 1% |
| Oceanic boundary layer (0 to 0.5 km) | 0 | 20 | −20 | <5% |
| Clean continental boundary layer (0 to 1.5 km) | ? | ? | ? | <5% |
| Free troposphere (1 to 12 km) | 96 | 89 | 7 | >90% |
| Stratospheric flux | NA | NA | 20 | NA |

[a] All expressed in units of $10^{28}$ molecule s$^{-1}$.
The average stratospheric flux is shown under the net production column.
The percent ozone burden of each region is also listed.

The most important feature in Table 13.3 is the large ozone production and loss rates in the free troposphere. This, coupled with its greater than 90% burden of total tropospheric ozone, means that the budget of tropospheric ozone is controlled by its production in the free troposphere. Since $NO_x$ is the rate-limiting ozone precursor, the critical question becomes the identification and quantification of sources of free tropospheric $NO_x$. If the ozone production in the free troposphere is entirely from natural $NO_x$ sources, anthropogenic impact such as export of ozone produced in the industrial boundary layer would contribute less than 10% of the total ozone burden. On the other hand, if most of the free tropospheric $NO_x$ is from anthropogenic sources, then the tropospheric ozone would be controlled by human activities.

Despite the overwhelmingly large emissions of $NO_x$ in the industrial boundary layer, the ozone production is only about a quarter of that in the free troposphere.

Although this seems to be counterintuitive, it is the result of the combination of the following factors: nonlinearity of ozone production efficiency, recycling of $NO_x$ in the free troposphere, various diffuse $NO_x$ sources such as lightning, and transport of a significant amount of $NO_x$, including that from combustion and biomass burning, from the boundary layer to the free troposphere.

Another significant feature of Table 13.3 is that there is net production in the free troposphere and net destruction in the oceanic boundary layer (Liu et al., 1980; Chameides et al., 1988; Davis et al., 1996; Jacob et al., 1996). This is the result of higher $NO_x$ concentrations in the free troposphere compared to those in the oceanic boundary layer (see Chapter 7). However, it is not clear what sources and/or processes are responsible for the higher $NO_x$ concentrations in the free troposphere. $NO_x$ from lightning, convective transport of $NO_x$ emitted near the surface, aircraft emissions, and transport from the stratosphere have been proposed as the likely contributors. In this context, recycling of $NO_x$ has to be considered even if it is not a primary source of $NO_x$.

### 13.3.4 Production of Ozone in Industrialized Areas and Control Strategy

As discussed earlier, high levels of ozone are produced in the urban centers and over vast rural areas of the industrialized countries (Fig. 13.2). The precursors contributing to the ozone production are anthropogenically emitted $NO_x$ and NMHC. In addition, it has been shown that natural NMHC also play an important role in ozone production (Trainer et al., 1987a; Chameides et al., 1988).

A large amount of effort has been devoted to developing an effective control strategy for ozone by reducing the emissions of ozone precursors. The key process in developing an effective ozone control strategy is to understand the relationship between ozone levels and the ozone precursors. An approach commonly used to examine the relationship in urban air is depicted in Figure 13.9, which shows the isopleths of the noontime ozone mixing ratio and its dependence on the initial (at 5 a.m.) anthropogenic $NO_x$ and NMHC mixing ratios (note that the NMHC mixing ratio has been multiplied by the carbon number of each hydrocarbon) calculated from a time-dependent box model (Lin et al., 1988). One can see that

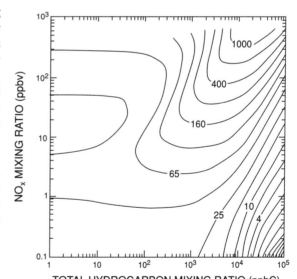

**Figure 13.9.** Isopleths of the noontime $O_3$ mixing ratio and its dependence on the initial (5 a.m.) $NO_x$ and hydrocarbon mixing ratios determined from a box model (Lin et al., 1988).

the ozone mixing ratio depends not only on the absolute levels of precursors, but also on their relative abundance. As a result, the inferred ozone control strategy can be very different for different mixtures of $NO_x$ and NMHC. For example, the typical initial mixing ratio of $NO_x$ in an urban center ranges from a few tens to a few hundreds ppbv, and the ratio of NMHC/$NO_x$ is on the order of 10:1. This will put the urban center in the upper right area of Figure 13.9, where the ozone mixing ratio is most sensitive to change of the initial NMHC mixing ratio. In other words, the ozone production is limited by the availability of NMHC. In this case, reduction of NMHC emissions is obviously more effective than the reduction of $NO_x$ emissions in controlling ozone levels. In fact, the figure indicates that ozone level can even increase when $NO_x$ emissions are reduced.

Figure 13.9 has some major shortcomings when it is applied to the real atmosphere. First, it does not include the effect of atmospheric transport that is a major process in controlling the concentrations of ozone and its precursors. Second, it does not account for the photochemistry and the ozone production of previous days. Finally, it does not include natural sources of NMHC. As a result, Figure 13.9 is not useful for characterizing the ozone buildup in the nonurban atmosphere. Even in the urban atmosphere, the figure can only be used as a qualitative indicator for the relationship between ozone levels and the precursors. Obviously, a three-dimensional model that incorporates realistic transport parameterization is needed. In the following, results from three-dimensional regional model studies of ozone formation in the eastern United States (McKeen et al., 1991a,b) are reviewed and their implications for the ozone control strategy discussed.

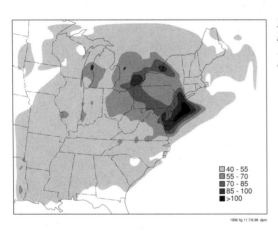

**Figure 13.10.** Model-calculated surface layer ozone mixing ratio (ppbv) for 6 July, 1986. The values are averaged over 1 to 5 p.m. EST (from McKeen et al., 1991a).

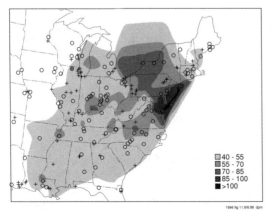

**Figure 13.11.** Observed surface ozone mixing ratios (ppbv) for rural and suburban stations over the same time period as Fig. 13.10 (from McKeen et al., 1991a).

A four-day period, 4-7 July 1986, was chosen for the model study. It was chosen because the synoptic conditions over the eastern United States were dominated by a high-pressure system that is conducive to high ozone buildup. Figures 13.10 and

13.11 give a comparison between the model-calculated ozone distribution in the surface layer (Fig. 13.10), averaged over 1 to 5 pm Eastern Standard Time (EST) of the third day (6 July 1986), and the corresponding observed values (Fig. 13.11). The reason for picking the afternoon period for comparison is to focus on the photochemical processes and avoid the effect of some fine structures of nocturnal inversion that do not have significant impact on the ozone distribution, other than in the shallow nocturnal surface layer. The observed values include only suburban and rural data, but exclude urban stations, which are not resolved in the model.

In the region where the influence of side-boundary conditions is small, most of the features are reproduced well. For example, the southern extent of the contours from the main body of high ozone is fairly well simulated east of Illinois and north of Alabama. Agreement between the position and magnitude of the highest ozone values is also apparent. The one region strongly influenced by coastal inflow (the Carolinas, Georgia, and Alabama) that is characterized by relatively low ozone is also well simulated by the model. The generally good agreement is reflected in the scatter plot of model-predicted versus observed ozone averaged from 1 to 5 pm EST (Fig. 13.12). However, the figure also shows a significant discrepancy between the model and observations. The model shows a distinct tendency to overpredict at the low ozone level and underpredict at the high end. The discrepancy is also a common feature of other three-dimensional models (Lamb, 1988; Schere and Wayland, 1989). The underprediction of high ozone is characteristic of those stations near urban centers, probably because of the inability of the model to simulate subgrid-scale urban plumes

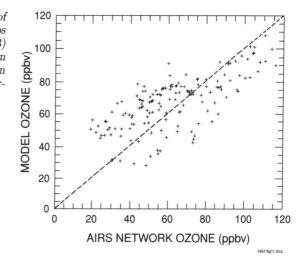

**Figure 13.12.** Scatter plot of model-calculated ozone mixing ratios averaged over 1 to 5 p.m. EST (day 3) versus observed values. AIRS is an acronym for Aerometric Information Retrieval System (adapted from McKeen et al., 1991a).

(Schere and Wayland, 1989). The overprediction at low ozone is probably the result of a combination of instantaneous mixing within the large grid (due to low model resolution) and the nonlinearity in ozone production (see Fig. 13.8). However, the fundamental problem is probably associated with a tendency for models to overpredict the OH concentration at low $NO_x$ levels (Eisele et al., 1994; Poppe et al., 1994; McKeen et al., 1997). The reason for the OH overprediction is not well understood. While the discrepancy does not appear to be serious, it contributes to a significant uncertainty in the modeled ozone budget, particularly at low $NO_x$ levels.

The sensitivity of ozone buildup to changes in the ozone precursors can be studied by uniformly reducing the emission of anthropogenic $NO_x$ or NMHC by 50%. Results are presented as the percentage of ozone changes in the surface layer. Figure 13.13

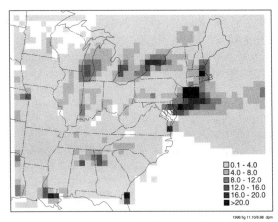

*Figure 13.13.* The percentage decrease of ozone in the model lowest layer, averaged from 1 to 5 p.m. EST, due to 50% reduction in anthropogenic NMHC emission (from McKeen et al., 1991b).

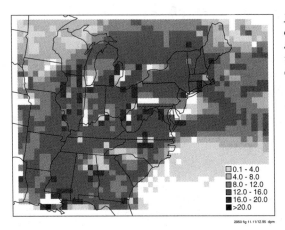

*Figure 13.14.* The percentage decrease of ozone in the model lowest layer, averaged from 1 to 5 p.m. EST, due to 50% reduction in anthropogenic $NO_x$ emission (from McKeen et al., 1991b).

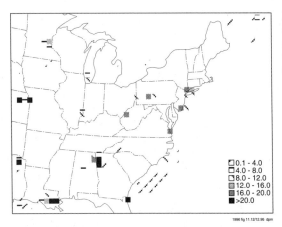

*Figure 13.15.* The percentage increase of ozone in the model lowest layer, averaged from 1 to 5 p.m. EST, due to 50% reduction in anthropogenic $NO_x$ emission (from McKeen et al., 1991b).

shows the percentage decrease of ozone due to a 50% reduction in anthropogenic NMHC emissions. Areas with less than 0.1% decrease should be ignored because they are either behind the cold fronts in Canada or associated with the onshore flow in the southeast, where little photochemical production of ozone has occurred. The average decrease over the continental region is only about 3%. However, decreases as large as

20% occur near certain urban areas: Boston and the New York-New Jersey corridor. Decreases in the Toronto, Chicago, and Atlanta areas are also relatively large. These percentage decreases are also significant because the absolute concentrations of ozone in these areas are high. The results for the northeastern United States agree very well with an earlier study that emphasizes this area (Possiel et al., 1989).

When the anthropogenic emission of $NO_x$ is reduced, the change in ozone is very different from the case when NMHC emissions are reduced. Figures 13.14 and 13.15 show, respectively, the percentage decrease and increase (at different locations) in ozone concentrations for the case when anthropogenic emission of $NO_x$ is reduced by 50%. The domain average decrease is about four times the decrease due to 50% NMHC reduction. However, near power plants and urban centers where $NO_x$ concentrations are high, Figure 13.15 shows that reducing $NO_x$ emissions actually leads to an increase in ozone concentration. This reflects the highly nonlinear nature of the relationship between ozone and $NO_x$.

The model results show that near urban and some suburban areas reduction of NMHC is a more effective strategy than reduction of $NO_x$ for controlling ozone levels. This is consistent with previous three-dimensional investigations (Milford et al., 1989; Possiel et al., 1989). However, beyond urban areas, $NO_x$ reduction is about a factor of four more effective than NMHC reduction in reducing ozone levels. Analysis of the ozone budget indicates that the limiting precursor of photochemical ozone production gradually shifts from NMHC in urban areas to $NO_x$ in the rural areas (McKeen et al., 1991b). Many factors contribute to the shift. First, the $NO_x$ concentration decreases faster with distance away from the urban centers because the $NO_x$ photochemical lifetime is shorter than the average lifetime of anthropogenic NMHC. In addition, there is a relatively large emission of natural NMHC from forests. Finally, in rural areas the background concentrations of CO and $CH_4$ become more important in producing ozone.

### 13.3.5 The Role of Biogenic Hydrocarbons in Ozone Production

It was shown in the previous section that a reduction of 50% in anthropogenic hydrocarbons would result in a decrease in ozone of only a few percent over much of the southeastern United States. This is due to the participation of natural hydrocarbons in the production of ozone over the region. Not only are hydrocarbons emitted strongly by vegetation in that area, but biogenic hydrocarbons tend to be very reactive with OH radicals, and hence potent sources of ozone. The secondary oxygenated organics formed as products from the hydrocarbons also may play a role in inducing ozone formation.

The dominant biogenic emission over much of the southeastern United States is isoprene, from oak and sweetgum trees, and related species (see Chapter 5). In other regions, monoterpenes produced by evergreens, particularly $\alpha$- and $\beta$-pinene, can play a role in affecting air quality. Trainer et al. (1987a) and Chameides et al. (1988) pointed out that biogenic hydrocarbons could play a major, if not dominant, role in ozone formation in rural and urban areas. These studies showed conclusively that while emissions from urban areas could affect rural regions, the converse was also true, and that the problem of ozone abatement is a complicated intertwining of meteorology, chemistry, and biology.

In the intervening years, much research has been carried out both in the lab and the field in an attempt to elucidate the mechanism behind ozone formation, and the best strategy for controlling it. A major effort was provided by the Southern Oxidants Study (SOS), a program sponsored by the United States Environmental

Protection Agency. SOS attempted to cover all aspects of the regional ozone problem, including studies of the physiology of isoprene emission, standardization of methods for measurement of hydrocarbons, and regional chemical/transport models of the southeastern United States. Several field studies were also initiated in the Southeast with the intent of characterizing the chemical and meteorological environments, and how these affected the rate and extent of ozone formation. The studies indicated that over much of the southeastern United States odd nitrogen is the limiting reagent for ozone formation, not hydrocarbons, and that the same may actually be true in some urban areas. The findings have initiated a new way of looking at regional pollution and form the basis of proposed changes in legislature that more accurately reflect regional and longer-term incidences of ozone exposure.

Research is still continuing on the relation between biogenic hydrocarbons and ozone production. One product of the activity has been the Biogenic Emissions Inventory System (BEIS), which can be used to estimate the emissions from any county in the United States as a function of season and meteorological conditions. The study of volatile organic compounds, particularly the oxygenated derivatives, is also continuing and promises to be a fruitful area of research for many years to come. Emissions inventories applicable to wider regions are described in Guenther et al. (1995) and Simpson et al. (1995).

## 13.4 Major Uncertainties and Research Needs

Since $NO_x$ is the rate-limiting precursor of ozone everywhere except in the urban atmosphere, any uncertainty in the budget and distribution of $NO_x$ impacts significantly on that of ozone. As discussed earlier, there are large uncertainties in identifying and quantifying the sources of $NO_x$ in the free troposphere. This means that the origin of free tropospheric ozone is poorly understood, and thus it is difficult to assess the anthropogenic impact on the global ozone distribution. Furthermore, because the free troposphere contains the great majority of the tropospheric ozone pool, the budget and distribution of free tropospheric $NO_x$ must have the highest priority among the research topics related to tropospheric ozone. Specifically, the budget and distribution of free tropospheric $NO_x$ involves a large number of important research topics, such as production rate and distribution of $NO_x$ produced from lightning, convective transport of $NO_x$ emitted in the boundary layer to the upper troposphere, recycling of $NO_x$, and heterogeneous reactions and scavenging of odd-nitrogen species. Well-coordinated field observations, laboratory measurements, and modeling studies are needed to address these research topics.

Another important area of research is the photochemistry of $HO_x$ (OH, $HO_2$, $H_2O_2$) and $RO_2$ since they impact ozone production directly. The discrepancy between observed OH concentrations and model-calculated values mentioned in Section 13.3.4 must be resolved in order to make a reliable evaluation of the ozone budget. The roles of NMHC as well as various heterogeneous processes in the budget of $HO_x$ and $RO_2$ need to be understood. In addition, the $HO_x$ photochemistry in the upper troposphere, where water vapor concentrations are low, has not been systematically studied. Uncertainties in the $HO_x$ concentrations in the upper troposphere could have a profound impact on the calculated global ozone production.

Finally, the global spatial and temporal distributions of ozone are not sufficiently well known to allow a comprehensive understanding of its budget and trends. Both in situ and satellite measurements are needed. Satellite measurements have the advantage of global coverage, while well-calibrated in situ measurements are essential for monitoring the long-term trends.

## Further Reading

Calvert, J., ed. (1994) *The Chemistry of the Atmosphere: Its Impact on Global Change*, Blackwell Scientific Publications, Oxford.

Finlayson-Pitts, B. J. and J. N. Pitts (1986) *Atmospheric Chemistry*, Wiley & Sons, New York.

National Research Council (1991) *Rethinking the Ozone Problem in Urban and Regional Air Pollution*, National Academy Press, Washington, D.C.

Paul Crutzen

## Tropospheric Ozone

The research of the past 25 years has clearly shown the great importance of ozone in tropospheric chemistry. Starting with a highly important paper by Levy (1971), it was realized that the photolysis of ozone by ultraviolet solar radiation, leading to the production of electronically excited $O(^1D)$ atoms

$$O_3 + h\nu \rightarrow O(^1D) + O_2, \quad \lambda \leq 325 \text{ nm}$$

followed by

$$O(^1D) + H_2O \rightarrow 2\,OH$$

would lead to the production of highly reactive OH radicals. This "detergent" of the atmosphere reacts with almost all compounds that are released into the atmosphere by natural processes and anthropogenic activities, and the lifetime of most gases in the atmosphere is consequently determined by their reaction coefficients with OH radicals. Following this discovery, interest in tropospheric ozone greatly increased. Although one would think that by now, some 20 years later, one should have good knowledge about the global distribution and the sources and sinks (the budget) of tropospheric ozone, this is far from the truth. First, especially in the subtropical and tropical atmosphere, there is a severe lack of ozone measurements. Second, its budget is not well known, which in turn goes back to a severe lack of knowledge of the distributions and sources of NO and $NO_2$, the catalysts that strongly promote ozone formation in the troposphere.

There is little doubt that the individual production and destruction terms in the tropospheric budget of ozone are dominated by in situ tropospheric reactions and not by the downward flux of ozone from the stratosphere. As Table 13.3 in the accompanying chapter shows, the gross production or destruction terms of ozone during summer time are about six times larger than the estimated influx from the stratosphere. I obtained very similar numerical results, albeit for the entire year (Crutzen, 1995). Another important feature of Table 13.3 is the great importance of the free troposphere for the ozone budget. This also means that $CH_4$ and CO oxidation processes are most important in the tropospheric ozone budget. Because we only have a very limited amount of $NO_x$ measurements in the free troposphere, this term in the ozone budget is very uncertain, due to a lack of knowledge about the degree of NO transfer from the boundary layer to the free troposphere as well as NO production rates by lightning.

Therefore, in the future much more has to be learned about the behavior of ozone in the troposphere. Although major advances have been made in realizing the great importance of ozone, much of this knowledge is still far from quantitative. We will especially have to improve our knowledge about the ozone chemistry of the tropics and subtropics. This is even more important because of the expected large growths in human populations, as well as industrial and agricultural developments in these regions. Without a greatly improved database, any projections of ozone and hydroxyl concentration distributions will be very uncertain. This in turn will affect our predictions of the future trends of a great number of trace gases that are removed by reactions with OH.

There is no doubt that the newcomers in atmospheric chemistry will find a highly interesting research field to be explored with great societal implications. It is hopeful that the chapters that are collected in this book will be a good start.

*Paul Crutzen is the Director of the Atmospheric Chemistry Department at the Max-Planck-Institute for Chemistry in Mainz, Germany. He has made a large number of important scientific contributions in several aspects of stratospheric and tropospheric chemistry. Jointly with M. Molina and F. S. Rowland, he is the laureate of the 1995 Nobel Prize for Chemistry.*

# 14 Middle Atmospheric Ozone

## 14.1 Introduction

Middle atmospheric ozone is important both radiatively and chemically. Early observations showed that ozone was concentrated in a broad layer of the atmosphere between about 20 and 40 km (see Box 14.1). S. Chapman (1930) proposed the first photochemical theory for the presence of ozone in the stratosphere and suggested that ozone is formed by UV photolysis of molecular oxygen ($O_2$), predominantly in the tropics or midlatitudes during summer where solar radiation is most intense. However, relatively simple photochemical models showed that his proposed mechanism alone gave higher than observed $O_3$ mixing ratios. In the early 1970s, P. Crutzen (1970) and H. Johnston (1971) identified the important photochemical role played by odd-nitrogen constituents in determining the balance of ozone in the stratosphere. Their publications inspired an exciting era of interdisciplinary research focused on understanding the chemical, physical, and dynamical interactions occurring in the stratosphere. Much of this activity occurred in response to the thesis that a significant reduction of stratospheric ozone would result from exhaust emissions (especially emissions of $NO_x$, see Chapter 7) from a then-proposed large fleet of supersonic transport aircraft (SST) that would fly within the lower stratosphere. Research programs were international. In the United States much of the effort involved in laboratory measurements of chemical and photolysis rates, in measurements of stratospheric concentrations of trace species by ground-based, high-altitude balloon, rocket, aircraft, and satellite instrumentation, and in chemical and dynamical modeling came under the umbrella of the Climatic Impact Assessment Program (CIAP, 1975).

Over roughly a decade, progress in understanding the interactions of the odd-nitrogen, hydrogen, oxygen, and halogen families and their effects on ozone was based almost entirely on homogeneous gas-phase processes. Heterogeneous processes were proposed, but were largely dismissed as being relatively unimportant. By the beginning of the 1980s, economic realities had limited the size of the SST fleet to a few aircraft (Concorde) and initial large projections of ozone reduction were decreased to a relatively small amount as progress in understanding occurred. Although many interesting questions remained and databases for some stratospheric constituents were becoming available from satellite instrumentation, stratospheric research settled into a more relaxed atmosphere.

In the spring of 1985, J. C. Farman, B. G. Gardiner, and J. D. Shanklin (1985) published their observations of a dramatic decline in the column abundance of ozone above Antarctica during austral spring, an observation now colloquially called the Antarctic "ozone hole" (see Plate 13). Over the period of 1979 to 1985 the springtime

### Box 14.1 Atmospheric Ozone

The German chemist Christian Friedrich Schönbein identified ozone by its particular odor following electrical discharges in air. Schönbein named this gas "ozone" from the Greek word όζειν (or ozein), which means "to smell." In a letter written in 1840 to Arago and submitted to the French Academy of Sciences, he suggested that this compound could belong to the chemical group of chlorine and bromine. In 1845 C. Marignac and M. de la Rive in Geneva, Switzerland, suggested that ozone was produced by the transformation of oxygen, while 20 years later J. L. Soret in Basel identified ozone as an allotropic form of oxygen with the chemical formula: OOO or $O_3$. The first measurements of ozone were performed in 1858 by André Houzeau in Rouen, France. Later, French chemist Albert-Levy observed the abundance of ozone almost continuously from 1877 to 1907 at the municipal observatory of Parc Montsouris in Paris.

*Schönbein*

Early in its history, the optical properties of ozone were discovered. The measurement of the ozone spectrum in the laboratory by Walter Noel Hartley in 1881 led him to suggest that the observed cutoff of UV radiation ($\lambda < 300$ nm), discovered by French physicist Alfred Cornu in 1878, was due to absorption by atmospheric ozone. In 1880, eleven ozone absorption bands were observed by J. Chappuis in the visible between 500 and 700 nm, but these appeared to be several orders of magnitude weaker than the band discovered by Hartley in the ultraviolet near 250 nm. In 1890, William Huggins, who was observing the spectrum of the star Sirius, reported several absorption bands between 320 and 360 nm. In 1917, A. Fowler and R. J. Strutt attributed these bands to the presence of ozone in the atmosphere.

In 1920, Charles Fabry and Henri Buisson, two scientists working in Marseille, France, used UV measurements to make the first quantitative measurements of the ozone column above the ground. They estimated that the thickness of the ozone layer was of the order of 3 mm (under STP conditions), although substantial day-to-day variations were noticeable. In the 1920s the British scientist G. M. B. Dobson, who worked at Oxford University, developed the technique of Fabry and Buisson into a spectrophotometer with a double UV monochromator that for many years remained the only method for accurately measuring the ozone column. It is still in widespread use today and remains the basis for the current network of surface observing stations. Before the end of the 1920s, Dobson had installed his instruments at different locations, which led him to establish the latitudinal variation and seasonal evolution of the ozone overburden. He also discovered strong correlations between the ozone column abundance and meteorological conditions in the lower stratosphere.

*Dobson*

As the first observations of ozone became available, progress was also made in understanding the chemical processes affecting this gas in the atmosphere. At an ozone conference organized at Paris in 1929, British mathematician and geophysicist Sidney Chapman suggested that the formation of ozone resulted from the photolysis of molecular oxygen in the upper atmosphere.

As the absorption by $O_2$ at wavelengths beyond 200 nm had not yet been discovered, the mechanism proposed by Chapman led him to suggest that the ozone layer was probably located near 45 km altitude.

Measurements made by Swiss scientist Paul Götz in 1929 during a Spitsbergen expedition (by inverting solar intensities measured by the Dobson spectrophotometer for high solar zenith angles), and by German scientists Erich Regener and his son Viktor in 1934 (by analyzing solar spectra recorded from meteorological balloons) showed, however, that the maximum in the ozone density was located much lower, typically at 20-25 km altitude. It was also soon recognized that the photochemical mechanism introduced by Chapman was insufficient to reproduce the observations. Additional chemical reactions were in-

*Chapman*

troduced (see this chapter for details) and involved the presence of mesospheric water vapor (proposed in 1950 by David Bates in the United Kingdom and Marcel Nicolet in Belgium), of stratospheric $HO_x$ radicals (proposed in 1964 by J. Hampson in Canada), of nitrogen compounds (proposed in 1970 by the Dutchman Paul Crutzen), and of chlorine compounds (proposed in 1974 by American scientists Ralph Cicerone and Richard Stolarski). It soon became obvious, after the publication of a paper by Mario Molina and Sherwood Rowland in 1974, that industrially manufactured chlorofluorocarbons, a major source of stratospheric chlorine, would become a major threat to the ozone layer. In 1979 William Chameides and James Walker at the University of Michigan, as well as Crutzen, proposed a photochemical production mechanism for ozone in the troposphere, in the presence of methane, carbon monoxide, and nitrogen oxides. For their seminal discoveries concerning the chemistry of ozone Crutzen, Molina and Rowland were awarded the 1995 Nobel Prize in Chemistry.

*Crutzen*   *Rowland*   *Molina*

Important new discoveries took place in the 1980s. S. Chubachi, for example, observed very low ozone abundances at the Japanese Antarctic Station of Syowa during September and October, 1982. In 1985, a team of British scientists led by Joseph Farman reported alarming ozone depletion above the station at Halley Bay, Antarctica, during austral springtime. The mechanism proposed by U.S. scientist Susan Solomon and colleagues to explain the formation of this ozone hole, which involves the activation of anthropogenic chlorine on the surface of ice particles (polar stratospheric clouds), was later confirmed by laboratory studies as well as by several field campaigns conducted in Antarctica. In September 1987, an international protocol was signed in Montreal, Canada, to limit the production of chlorofluorocarbons. The phaseout of these compounds was decided in the early 1990s.

abundance showed a steady decrease from about 320 DU (Dobson Units) to less than 200 DU. (By 1992-1993 the decrease had reached record lows, below 100 DU.) The authors also noted a correspondence between the reduction in ozone and the growth of chlorofluorocarbons in the stratosphere. To say the least, these observations shocked the stratospheric research community, as models thought to have advanced well in their ability to reproduce the essentials of stratospheric transport and chemistry, which were based on a wide variety of observations, were unable to account for this dramatic signal of large-scale ozone depletion. Furthermore, the depletion of ozone occurred in the lower stratosphere, not in the 35-40 km region where the depletion of ozone by active chlorine would first be recognized according to current model scenarios then. Because the human consequences of a large ozone reduction were known to be profound from the CIAP and other programs, this failing sparked a renewed and vigorous experimental and theoretical investigation, which continues today. As a result of these studies, over the past 10 years remarkable changes have occurred in our understanding of stratospheric processes due to laboratory studies and to remarkably successful and comprehensive observational programs that have deployed instruments on high-altitude aircraft, the Space Shuttle, satellites, balloons, and the ground. The explanations for the "ozone hole" soon centered chiefly on the unexpected role of heterogeneous processes and the continued growth of the chlorine and bromine loading in the stratosphere. The accomplishment during this period is certainly one of the most remarkable examples of the unraveling of a global change issue over a short period of time. It is also ironic that the environmental concern about aircraft exhaust emissions, which gave birth to the CIAP program, and which really founded a new discipline in geoscience, has returned as a theme of upper tropospheric and lower stratospheric research initiatives today.

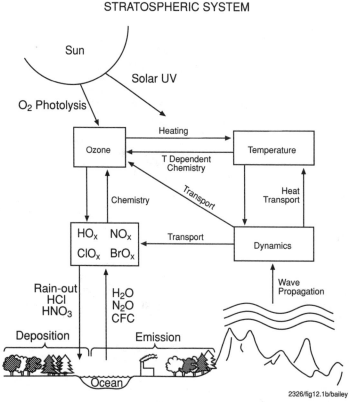

*Figure 14.1.* Representation of the coupled chemical, radiative, and dynamical processes in the middle atmosphere. Ozone, which is produced by $O_2$ photolysis, absorbs solar ultraviolet radiation, which heats the stratosphere and hence affects the atmospheric temperature. Ozone is destroyed by photochemical reactions involving radicals whose concentrations depend on the ozone abundance. Ozone and other chemical constituents are transported by winds that are related to the temperature distribution and are driven by the absorption in the atmosphere of vertically propagating waves (see Chapter 2). Chemical reaction rates are also temperature dependent.

Ozone is central in the highly coupled chemical, radiative, and dynamical processes of the middle atmosphere, shown schematically in Figure 14.1. Absorption of solar ultraviolet radiation by ozone in the range of 240 to 320 nm protects the surface from this potentially harmful radiation. Heating of the middle atmosphere following the absorption by ozone of solar UV and visible radiation and of Earth-emitted IR radiation contribute to the characteristic temperature profile of the middle atmosphere. The stability of the stratosphere is due primarily to the increase in temperature with increasing altitude through that region. It was probably the initial occurrence of ozone, some 2 billion years ago, with the associated screening of the surface from lethal UV, that allowed life to emerge from the oceans.

## 14.2 The Ozone Distribution

Figure 14.2 shows a typical vertical profile of the ozone volume mixing ratio for a midlatitude stratosphere. The basic shape of this profile with a distinct maximum is characteristic of any gas whose concentration depends on the balance between photochemical production driven from above and a gaseous total density decreasing with increasing altitude. In the case of ozone, the mixing ratio is typically 100-500 ppbv at the tropopause, 3 ppmv at 20 km, 8-10 ppmv at 35 km, and 2 ppmv at the stratopause.

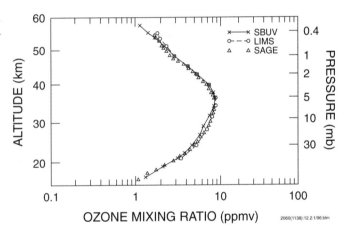

**Figure 14.2.** Monthly average ozone mixing ratio profile (ppmv) between 15 and 60 km altitude for April 1979 at 45°N from three satellite-borne instruments (SBUV, LIMS and SAGE) (WMO, 1986).

Ozone is one of the most extensively measured trace gases in the atmosphere; it is also one of the most variable. The total column of ozone (which is the vertically integrated ozone amount), long recognized for its role in shielding the surface from harmful UV radiation, has been measured for many years. Measurements were first made, and continue to be made, by an extensive network of ground-based UV spectrometers (Dobson, 1930) and more recently by satellite-borne sensors [e.g., the solar backscatter ultraviolet spectrometer (SBUV) and the total ozone mapping spectrometer (TOMS), Heath et al., (1975)]. Figure 14.3 shows the observed monthly mean total column abundance of ozone as a function of latitude and time of year. Column amounts vary with season and latitude. At some latitudes changes of up to 50% may occur in a few days at a given location. Interestingly, the distribution presented in Figure 14.3 shows a maximum column abundance at the latitudes of least production.

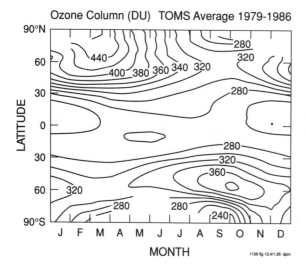

*Figure 14.3.* Total column ozone abundance (Dobson units) versus latitude and season from TOMS observations averaged from 1979-1986 (WMO, 1990a). Note the hemispheric asymmetry with ozone maxima near the North Pole in March (460 DU) and at $60°S$ in October (380 DU). The tropical ozone column exhibits little seasonal variability; its value is close to 260 DU.

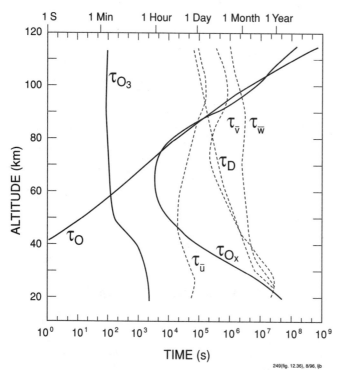

*Figure 14.4.* Chemical lifetimes of ozone ($O_3$), atomic oxygen (O), and odd oxygen ($O_x = O_3 + O$) compared to characteristic transport times. $\tau_{\bar{u}}$ denotes advection by the zonal wind, $\tau_{\bar{v}}$ by the meridional wind, $\tau_{\bar{w}}$ by the vertical wind, and $\tau_D$ the vertical eddy diffusion (Brasseur and Solomon, 1986; reprinted with kind permission from Kluwer Academic Publishers).

Ozone and atomic oxygen are both chemically highly reactive; they participate in reactions with many of the other trace gases in the middle atmosphere. As may be seen in Figure 14.4, the chemical lifetime of odd oxygen (defined as $O + O_3$, see below) in the lower stratosphere is comparable to the time constants for transport. This means that transport by the mean stratospheric circulation also plays a role in determining the ozone distribution.

As may be seen in Figure 14.3, the highest ozone column abundances are found at high latitudes in winter and early spring, while the lowest values are located in the tropics. Most ozone is produced at low latitudes, where the solar irradiance is highest. The observed distribution of total ozone is the result of the strong poleward and

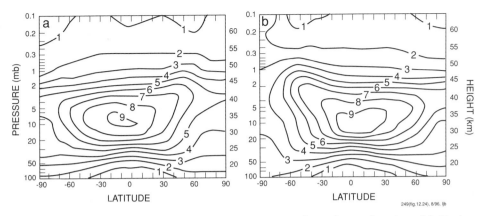

**Figure 14.5.** *Average zonal-mean ozone mixing ratios (ppmv) as a function of latitude and altitude (10-70 km) for (a) December 1978 and (b) May 1979. Data have been extended to the South Pole with the aid of SBUV observations, as described by Gille and Lyjak (1986).*

downward transport of mass in the stratosphere during winter. The hemispheric and seasonal differences in total ozone, illustrated in Figure 14.3, are attributed to differences in stratospheric dynamics associated with differences in surface orography. The very low ozone column observed in September and October over Antarctica is called the "ozone hole" and is known to be caused by chemical destruction (see below). In addition to the variations shown in Figure 14.3, at any given location there may be large day-to-day variations, which are more pronounced at higher latitudes. Figure 14.5 shows the LIMS observed zonal-mean cross section of ozone for two seasons. There are also significant interannual variations due to the influence of quasibiennial oscillations (QBO) and the 11-year cycle of solar output. Significant ozone reductions also have been reported after recent volcanic eruptions such as those of El Chichón (1982) and Mt. Pinatubo (1991).

## 14.3 Ozone Production

Chapman (1930) proposed that the reactions responsible for the formation (and destruction) of the stratospheric ozone layer were

$$O_2 + h\nu \rightarrow O + O, \qquad \lambda < 242 \text{ nm} \tag{14.1}$$

$$O + O + M \rightarrow O_2 + M \tag{14.2}$$

$$O + O_2 + M \rightarrow O_3 + M \tag{14.3}$$

$$O + O_3 \rightarrow 2\,O_2 \tag{14.4}$$

$$O_3 + h\nu \rightarrow O_2 + O \tag{14.5}$$

Laboratory studies have shown that Reaction (14.2) is too slow to be of importance in the stratosphere (but is important in the upper mesosphere and thermosphere). In the photolysis of $O_3$ (Reaction 14.5), atomic oxygen may be produced in two forms, in the ground state $O(^3P)$, mainly for absorbed wavelengths greater than 310 nm, or in the first excited state $O(^1D)$, for $\lambda < 310$ nm. In the upper part of the middle atmosphere $O(^1D)$ atoms also can be produced by molecular oxygen photolysis

$$O_2 + h\nu \rightarrow O(^1D) + O(^3P), \qquad \lambda < 175.9 \text{ nm} \tag{14.6}$$

$O(^1D)$ is rapidly quenched by collision with $N_2$ or $O_2$, both abundant in the atmosphere; thus nearly all the $O(^1D)$ produced in the stratosphere soon ends up as $O(^3P)$. Reaction (14.3) produces ozone, and Reaction (14.5) turns it back into O, the large reservoir of $O_2$ remaining essentially unchanged. Both these reactions are fast throughout the stratosphere (of the order of minutes); so there is a rapid exchange between $O_3$ and O. This interconversion allows us to define the sum of the concentrations of $O_3$ and O as $O_x$, commonly referred to as odd oxygen. Reaction (14.1) produces odd oxygen, Reaction (14.4) destroys it, and Reactions (14.3) and (14.5) have no effect on $O_x$, but determine the partitioning between O and $O_3$.

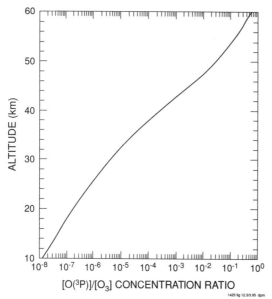

Figure 14.6 shows the daytime ratio $[O(^3P)]/[O_3]$ in the stratosphere. As may be seen, ozone accounts for more than 95% of the odd oxygen at all altitudes below 50 km. Above approximately 60 km altitude the density of atomic oxygen exceeds that of ozone during daytime.

Since the reactions forming atomic oxygen and ozone are initiated by sunlight, it might be expected that these constituents exhibit a diurnal variation. After sunset atomic oxygen is no longer produced by Reactions (14.1) and (14.5), but recombines with molecular oxygen to produce ozone and the ratio of [O] to $[O_3]$ becomes smaller. In the middle and lower stratosphere, where the ratio of O to $O_3$ is much less than unity, and the lifetime of ozone is greater than a day, little diurnal variation in the concentration of $O_3$ is observed. The photochemical lifetimes of ozone and atomic oxygen together with a typical transport lifetime are shown in Figure 14.4.

**Figure 14.6.** *Calculated daytime ratio of the concentration of $[O(^3P)]$ to $[O_3]$ in the stratosphere.*

## 14.4 Ozone Destruction

The general shape of the vertical distribution of ozone may be explained by the oxygen chemistry introduced by Chapman (Reactions 14.1-14.5). This chemical mechanism leads, however, to an overestimation of the ozone column abundance by almost a factor of two. Other mechanisms, therefore, need to be added to our representation of the ozone photochemistry, especially processes that destroy ozone. The direct loss mechanism for $O_x$ (Reaction 14.4), which is relatively slow, especially when the temperature is low, can be accelerated through several catalytic mechanisms (see Chapter 3), such as

$$X + O_3 \rightarrow XO + O_2$$
$$XO + O \rightarrow X + O_2$$
$$\text{Net}: O_3 + O \rightarrow 2O_2$$

even if the concentration of the catalyst $X$ is several orders of magnitude lower than that of ozone. Reactions (such as Reaction 14.4) or catalytic cycles that require atomic oxygen will only be effective above approximately 30 km altitude. Below that level, most of the ozone reduction occurs through cycles that do not require atomic oxygen.

These cycles generally lead to the conversion of two ozone molecules ($O_3$) into three oxygen molecules ($O_2$). Some of the important cycles which contribute to substantial destruction of ozone in the middle atmosphere are discussed in the next section.

### 14.4.1 $HO_x$ Chemistry

The first catalytic cycle to be identified as important to mesospheric ozone involves hydrogen (Bates and Nicolet, 1950). The hydroxyl radical, OH, is produced in the stratosphere by the oxidation of $H_2O$, $CH_4$, and $H_2$

$$H_2O + O(^1D) \to 2\ OH \tag{14.7}$$

$$CH_4 + O(^1D) \to CH_3 + OH \tag{14.8}$$

$$H_2 + O(^1D) \to H + OH \tag{14.9}$$

and, in the mesosphere (above 70 km), by photolysis of water vapor. OH (and $HO_2$) play critical roles in nearly all stratospheric and mesospheric ozone chemistry either as a direct reactant with odd oxygen or through control of the partitioning within other chemical families. Above about 40 km the major catalytic cycles involving OH (see Color Plate 23) that destroy odd oxygen are

Cycle 1
$$OH + O \to H + O_2 \tag{14.10}$$
$$H + O_2 + M \to HO_2 + M \tag{14.11}$$
$$\underline{HO_2 + O \to OH + O_2} \tag{14.12}$$
Net: $O + O \to O_2$

Cycle 2
$$OH + O \to H + O_2 \tag{14.10}$$
$$\underline{H + O_3 \to OH + O_2} \tag{14.13}$$
Net: $O + O_3 \to 2\ O_2$

As the ratio of [O] to [$O_3$] decreases, below 40 km, Cycle 3 becomes more important

Cycle 3
$$OH + O_3 \to HO_2 + O_2 \tag{14.14}$$
$$\underline{HO_2 + O \to OH + O_2} \tag{14.12}$$
Net: $O + O_3 \to 2\ O_2$

Below about 30 km, where there are very few oxygen atoms, Cycle 4 is the dominant hydrogen mechanism

Cycle 4
$$OH + O_3 \to HO_2 + O_2 \tag{14.14}$$
$$\underline{HO_2 + O_3 \to OH + 2\ O_2} \tag{14.15}$$
Net: $2\ O_3 \to 3\ O_2$

It is important to note that at various altitudes, as the concentrations of key reactants change and to a lesser extent as the temperature changes, the relative importance of the different catalytic cycles changes.

### 14.4.2 $NO_x$ Chemistry

The next catalytic cycle recognized as significant to $O_3$ involves nitrogen oxides (Crutzen, 1970; Johnston, 1971). Nitric oxide, NO, is produced in the stratosphere by the oxidation of nitrous oxide

$$N_2O + O(^1D) \to 2\ NO \tag{14.16}$$

There are two primary catalytic cycles initiated by NO which reform $O_2$ from odd oxygen

Cycle 5
$$NO + O_3 \rightarrow NO_2 + O_2 \qquad (14.17)$$
$$\underline{NO_2 + O \rightarrow NO + O_2} \qquad (14.18)$$
$$\text{Net: } O + O_3 \rightarrow 2\,O_2$$

Cycle 6
$$NO + O_3 \rightarrow NO_2 + O_2 \qquad (14.17)$$
$$NO_2 + O_3 \rightarrow NO_3 + O_2 \qquad (14.19)$$
$$\underline{NO_3 + h\nu \rightarrow NO + O_2} \qquad (14.20)$$
$$\text{Net: } 2\,O_3 \rightarrow 3\,O_2$$

Cycle 5 is the dominant cycle throughout the middle stratosphere. There are numerous reactions that control the amounts of NO and $NO_2$ at various altitudes, at night and in sunlight, and these are described in more detail in Chapter 7.

### 14.4.3 $Cl_x$ Chemistry

Cycles involving chlorine were the third to be recognized as significant to ozone destruction (Stolarski and Cicerone, 1974; Molina and Rowland, 1974). The natural source of reactive chlorine is from destruction of methyl chloride by photolysis (see Chapter 8)

$$CH_3Cl + h\nu \rightarrow CH_3 + Cl \qquad (14.21)$$

and by reaction with the OH radical

$$CH_3Cl + OH \rightarrow CH_2Cl + H_2O \qquad (14.22)$$

which is followed by a series of reactions eventually producing active chlorine. A large source (and one that increased steadily between the 1950s and the early 1990s) is provided by the manmade chlorofluorocarbon (CFC) gases, especially $CFCl_3$ and $CF_2Cl_2$ (see Chapter 8). These gases are photodissociated in the stratosphere to release active Cl

$$CF_2Cl_2 + h\nu \rightarrow CF_2Cl + Cl \qquad (14.23)$$
$$CFCl_3 + h\nu \rightarrow CFCl_2 + Cl \qquad (14.24)$$

The principal chlorine catalytic cycle destroying ozone in the upper stratosphere is

Cycle 7
$$Cl + O_3 \rightarrow ClO + O_2 \qquad (14.25)$$
$$\underline{ClO + O \rightarrow Cl + O_2} \qquad (14.26)$$
$$\text{Net: } O + O_3 \rightarrow 2\,O_2$$

When reactants from any of the three families, $HO_x$, $NO_x$, and $Cl_x$, are also taken into account, three other catalytic cycles are important

Cycle 8
$$Cl + O_3 \rightarrow ClO + O_2 \qquad (14.25)$$
$$OH + O_3 \rightarrow HO_2 + O_2 \qquad (14.14)$$
$$ClO + HO_2 \rightarrow HOCl + O_2 \qquad (14.27)$$
$$\underline{HOCl + h\nu \rightarrow OH + Cl} \qquad (14.28)$$
$$\text{Net: } 2\,O_3 \rightarrow 3\,O_2$$

Cycle 9
$$Cl + O_3 \rightarrow ClO + O_2 \qquad (14.25)$$
$$ClO + NO \rightarrow Cl + NO_2 \qquad (14.29)$$
$$\underline{NO_2 + O \rightarrow NO + O_2} \qquad (14.18)$$
$$\text{Net: } O + O_3 \rightarrow 2\,O_2$$

Cycle 10
$$Cl + O_3 \rightarrow ClO + O_2 \quad (14.25)$$
$$NO + O_3 \rightarrow NO_2 + O_2 \quad (14.17)$$
$$ClO + NO_2 + M \rightarrow ClONO_2 + M \quad (14.30)$$
$$ClONO_2 + h\nu \rightarrow Cl + NO_3 \quad (14.31)$$
$$\underline{NO_3 + h\nu \rightarrow NO + O_2} \quad (14.20)$$
$$\text{Net: } 2\,O_3 \rightarrow 3\,O_2$$

Cycle 9 is not a major destruction pathway for ozone. Cycle 7 is the major ozone destruction mechanism involving chlorine above 20 km; at lower altitudes Cycles 8 and 10 become dominant, along with Cycle 11 below.

### 14.4.4 $Br_x$ Chemistry

A fourth chemical family with potential for ozone destruction is bromine, and it has a cycle similar to that for chlorine. Sources of stratospheric bromine are given in Chapter 8. For example, methyl bromide, $CH_3Br$, from natural and anthropogenic surface sources, is destroyed in the stratosphere to produce Br and BrO. The effectiveness of bromine for destroying ozone is enhanced by reactions coupling the chlorine and bromine cycles

Cycle 11
$$BrO + ClO \rightarrow Br + Cl + O_2 \quad (14.32)$$
$$Br + O_3 \rightarrow BrO + O_2 \quad (14.33)$$
$$\underline{Cl + O_3 \rightarrow ClO + O_2} \quad (14.25)$$
$$\text{Net: } 2\,O_3 \rightarrow 3\,O_2$$

There are no stable reservoirs for Br or BrO as there are for Cl and ClO, since HBr and $BrONO_2$ are rapidly destroyed. Therefore, on a per molecule basis bromine is a more efficient catalyst for ozone destruction than is chlorine.

### 14.4.5 Summary of Ozone Destruction Mechanisms

The discussion above has considered each of the major ozone destruction mechanisms in turn and independently. The discussion has focused on the stratosphere, but a brief description of the chemistry above the stratosphere is presented in Box 14.2. Of course, these cycles are highly interrelated. Competition between cycles for key reactive species, such as OH, NO, or ClO, are very important. The cycles presented here represent possible reaction pathways, and consideration must be given to the actual concentrations of various gases and to the rates of reactions. The effects of various cycles are not additive. Only through the use of a comprehensive computer model of this complex chemistry, simultaneously solving for the kinetics of each reaction and incorporating variations with altitude and time, can an accurate picture emerge. What limits the efficiency of the catalytic cycles above is the chemical conversion of the reactive radicals involved into more stable "reservoir" compounds. The following gas-phase reactions are examples of such conversions

$$HO_2 + HO_2 \rightarrow H_2O_2 + O_2 \quad (14.34)$$
$$NO_2 + OH + M \rightarrow HNO_3 + M \quad (14.35)$$
$$Cl + CH_4 \rightarrow HCl + CH_3 \quad (14.36)$$
$$ClO + NO_2 + M \rightarrow ClONO_2 + M \quad (14.30)$$
$$BrO + NO_2 + M \rightarrow BrONO_2 + M \quad (14.37)$$

Reservoir species such as $HNO_3$ or HCl are slowly transported from the lower stratosphere into the troposphere, where they are removed in rain droplets. The

**Box 14.2 Chemistry of the Upper Atmosphere**

As the rate of collisional recombination decreases exponentially with height, and since photochemistry above approximately 70 km is affected primarily by short-wave radiation (solar extreme ultraviolet radiation including the Lyman-$\alpha$ line at 121.6 nm and X-rays), chemical elements tend to be more and more in the form of atoms (as opposed to molecules) and in the form of ions (as opposed to neutral constituents) with increasing altitude. For example, in the thermosphere the photolysis of molecular oxygen in the Schumann-Runge continuum becomes so rapid that the relative abundance of this molecule (21% in the lower and middle atmosphere) starts to decrease above approximatively 80 km, and becomes lower than the relative abundance of atomic oxygen above 120 km. At these heights, due to its slow recombination, atomic oxygen becomes a dominant species, while the abundance of ozone remains low. In addition, since the thermospheric lifetime of the oxygen atom is larger than a month, the spatial distribution of O is strongly affected by transport (including tides and gravity waves).

Similarly, the photolysis of water vapor by the solar Lyman-$\alpha$ line is a major source of atomic hydrogen. Again, because of its slow recombination, the hydrogen atom becomes a major long-lived constituent of the thermosphere. Its distribution is also strongly affected by transport.

Above 100 km altitude, where vertical mixing becomes weak, molecular diffusion tends to distribute chemical species according to their atomic or molecular weight: lighter gases are transported upward and become relatively more abundant at higher altitudes (see Fig. 14.7). Atomic hydrogen can even escape the terrestrial environment and be lost to space.

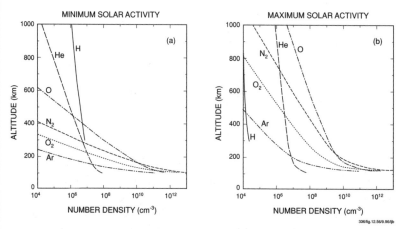

*Figure 14.7.* Vertical distribution of chemical species above 100 km altitude calculated for solar minimum and solar maximum conditions (US Standard Atmosphere, 1976).

In the mesosphere and lower thermosphere, the oxygen and hydrogen chemical families are strongly coupled. For example, the reaction between ozone and atomic hydrogen

$$O_3 + H \rightarrow OH \ (v < 9) + O_2$$

produces vibrationally excited hydroxyl radicals, which, during their deactivation, produce the Meinel bands associated with airglow.

The ionization above 50 km of neutral species (including $O_2$, O, $N_2$, and NO), by solar EUV and X-rays and by particle precipitation, leads to the formation of several electron layers, which affect radio-wave propagation. The region of the atmosphere in which a significant portion of the atoms and molecules are electrically charged is referred to as the *ionosphere*. The

> positive ion composition in the lower thermosphere is dominated by $NO^+$ and $O_2^+$ (during daytime). Below 90 km, proton hydrates $[H^+(H_2O)_n]$ are also present and become the dominant positive ion in the lower mesosphere. Above 90 km, negative charges are carried exclusively by electrons, while below that level, negative ions become relatively abundant. The abundance of ions is determined by chemical reactions and, at sufficiently high altitudes, by molecular diffusion and interactions with the magnetic field. The *magnetosphere* is the region of the atmosphere where magnetic phenomena and high atmospheric conductivity caused by ionization are important in determining the behavior of charged particles.

photolysis or oxidation of these species can, however, convert them back to reactive radicals. Additional conversion can also occur on the surfaces of particles such as ice crystals in polar stratospheric clouds. These situations lead to rapid ozone destruction and are observed in both polar regions during winter and spring (see below). Heterogeneous chemical reactions on sulfate aerosol particles present in the lower stratosphere also significantly affect the destruction mechanisms for ozone. One such reaction

$$N_2O_{5(g)} + H_2O_{(sa)} \rightarrow 2\,HNO_{3(g)} \qquad (14.38)$$

can convert nitrogen oxides into nitric acid without consuming OH as in Reaction (14.35). This reaction reduces the efficiency of the $NO_x$ catalytic cycles (Cycles 5 and 6) and enhances the cycles involving halogens, since the effects of Reactions (14.30) and (14.37) are reduced. Heterogeneous effects become especially important to ozone chemistry after large volcanic eruptions. These effects do not increase linearly with the aerosol load, since for high aerosol surface area densities, the heterogeneous conversion of NO and $NO_2$ into $HNO_3$ is limited by the rate at which $NO_2$ is converted (by ozone) into $N_2O_5$.

Figure 14.8 illustrates the relative contributions of the different chemical families to odd-oxygen destruction rates, as derived from a 2D model calculation for 1992 conditions. This figure shows that the destruction of $NO_x$ is dominant in the middle stratosphere, while the contribution of $HO_x$ reactions is largest in the lower stratosphere and in the upper stratosphere and mesosphere. The chlorine family contributes to more than 20-30 percent near 45 km altitude as well as between 20 and 30 km. Bromine plays an important role in the lower stratosphere, especially at high latitudes in the winter hemisphere. The odd-oxygen destruction rate by the Chapman mechanism never exceeds 25 percent of the total loss. Over the past decade, the contribution of halogen in the odd-oxygen destruction has increased dramatically, especially in the polar regions. At the same time, the importance of the $NO_x$ destruction cycle has decreased and the destruction by $HO_x$ has increased in the lower stratosphere.

The relative importance of the different chemical families in the destruction rates of odd oxygen depends upon (1) season and latitude through the variation in photolysis rates, (2) temporal trends in the source gases responsible for the active species, for example, the burden of stratospheric $Cl_y$ has increased much faster than that for $NO_y$ (from $N_2O$) over the past decade or so, (3) processes that determine the partitioning of the active species, for example, processes affecting the $NO_2/NO$ ratio, (4) homogeneous and heterogeneous processes that determine the transformation of the active radical and atomic species to longer-lived reservoir species of the $NO_y$ or other families, for example, processes that lower the $NO_x/NO_y$ ratio decrease the efficiency of $NO_x$ in $O_3$ destruction, and (5) physical processes, that is, particulate

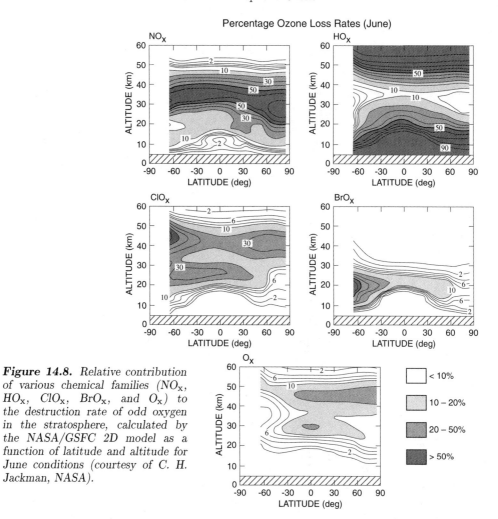

**Figure 14.8.** Relative contribution of various chemical families ($NO_x$, $HO_x$, $ClO_x$, $BrO_x$, and $O_x$) to the destruction rate of odd oxygen in the stratosphere, calculated by the NASA/GSFC 2D model as a function of latitude and altitude for June conditions (courtesy of C. H. Jackman, NASA).

sedimentation or transport processes that lead to the loss of active or reservoir species from the stratosphere. Over the past few years, advances in radical species measurements (e.g., ClO, BrO, HO, $HO_2$) in the lower stratosphere allow direct estimates of the relative efficiency of the different chemical families in odd-oxygen destruction. Figure 14.9 is an example for the ∼15 to ∼19 km altitude region of the lower stratosphere. In agreement with photochemical models, $O_3$ destruction in the lower stratosphere is currently dominated by the $HO_x$ family. The measurements also show a net photochemical production of $O_3$ at lower latitudes and a net destruction at northern latitudes.

## 14.5 Transport Effects

The global distribution of ozone depends not only on the local chemical production and loss but also on how ozone and the gases involved in its chemistry are transported by the meridional, zonal, and vertical winds (see Chapter 2 for more details on stratospheric transport). The significance of transport in determining constituent distributions depends on the relative magnitudes of chemical and transport lifetimes. Figure 14.4 shows vertical profiles of the chemical lifetime of odd oxygen along with typical transport timescales. Three cases may be distinguished: When the ozone

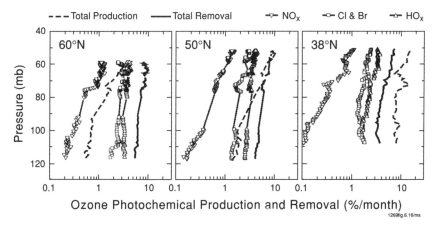

**Figure 14.9.** *The contributions of the $ClO_x + BrO_x$, $HO_x$, and $NO_x$ families to $O_3$ destruction in the lower stratosphere determined from in situ measurements of the radical constituents. Included are the total rates of production and loss of $O_3$ (adapated with permission from Wennberg, P. O., R. C. Cohen, R. M. Stimpfle, J. P. Koplow, J. G. Anderson, R. J. Salawitch, D. W. Fahey, E. L. Woodbridge, E. R. Keim, R. S. Gao, C. R. Webster, R. D. May, D. W. Toohey, L. M. Avallone, M. H. Proffitt, M. Loewenstein, J. R. Podolske, K. R. Chan, and S. C. Wofsy (1994) Removal of stratospheric $O_3$ by radicals: In situ measurements of OH, $HO_2$, NO, $NO_2$, ClO, and BrO, Science 266, 398, ⓒ1994, American Association for the Advancement of Science).*

chemical lifetime is much less than dynamical time constants, the ozone will be in photochemical equilibrium and the effects of transport will not be pronounced in the distribution. Gradients caused by the photochemistry will not be smoothed out by transport. If the chemical lifetime is much longer than the time required to transport ozone, the effect of the motions will be to reduce gradients. Note that some gases may exhibit a uniform distribution in one direction, but a significant gradient in another. When the time scales for chemical and dynamical processes are comparable, the distribution of ozone will depend on both mechanisms.

The effects of chemistry and transport are difficult to separate and a mathematical model must be used to disentangle the effects. Figure 14.10 shows a contour plot of the photochemical lifetime from 2D model calculations, which highlights the regions of chemical and dynamical control of the middle atmospheric odd oxygen. At altitudes above about 30 km photochemistry dominates the observed ozone distribution, while transport can play a significant role below 30 km and especially at high latitudes in winter and spring.

## 14.6 Polar Ozone

In polar regions during certain times of the year, a distinctive dynamical situation can develop that, when coupled with the recent (past 40 years) buildup of anthropogenic chlorine, can produce a dramatic, widespread destruction of ozone. The depletion is triggered by heterogeneous chemical reactions that occur on the surface of stratospheric aerosols and transform reservoir halogen species into active species that can catalytically destroy ozone with the return of solar insolation in spring.

Because of the rotation of the Earth and poleward flow, a pronounced circumpolar wind field is established in polar regions each winter (see Chapter 2). Figure 14.11 shows a schematic of the maximum potential vorticity gradient for a particular time in the Northern Hemisphere winter. The region of strong winds tends to confine the air

502    Middle Atmospheric Ozone

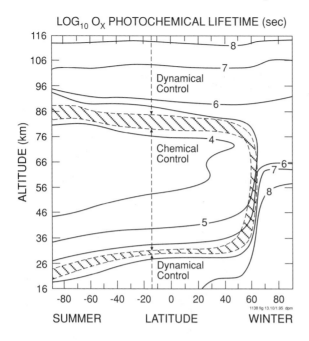

**Figure 14.10.** Photochemical lifetime of the $O_x$ family (logarithmic scale of values expressed in seconds) and the region of transition from photochemical to dynamical control (shaded) (WMO, 1986).

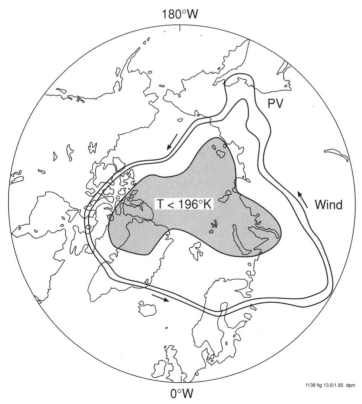

**Figure 14.11.** Typical Ertel potential vorticity and temperature on the 440 K surface (about 18 km) in a Northern Hemisphere winter. The contours marked "PV" indicate the maximum in the gradient of potential vorticity (usually near a value of $25 \times 10^{-6}$ K $m^2$ $kg^{-1}$ $s^{-1}$). The shaded area shows a region of cold temperature where PSCs are more likely to form.

poleward of the high winds and defines what is known as the polar vortex. Recall from the discussion in Chapter 2 that a large gradient in potential vorticity is an effective barrier to horizontal transport across that region. The maximum PV gradient may be regarded as the outer edge of the polar vortex.

Because of the lack of solar heating in winter, very cold temperatures are found within the vortices at both poles. The cold temperatures found in the lower stratosphere winter polar regions lead to the formation of chemically active aerosols and also enhance their reactivity. Depletion of ozone in the Southern Hemisphere polar region is severe and pervasive principally because temperatures in the lower stratosphere decline to values lower than required for abundant aerosol formation and persist for much of the winter season. Indeed, the Antarctic polar vortex is quite stable and nearly zonal, so that polar air is essentially isolated from midlatitude air masses over a lengthy period. In contrast, strong planetary wave activity induced by orography in the Northern Hemisphere polar region is more frequent (see Chapter 2). Furthermore, temperatures in the Arctic polar vortex do not decline to those required for formation of critical aerosols as frequently, nor do the low temperatures persist long enough for ozone depletion over large areas.

One class of aerosols found in the polar regions is commonly called polar stratospheric cloud particles (PSCs), which at times are sufficiently opaque to be visible from the ground. The other class is sulfuric acid aerosols, which are likely the seeds for formation of PSCs and are discussed in Chapter 4. The detailed composition of PSCs has not been determined. Nevertheless, much has been learned of their likely composition, formation, reactivity, and thermodynamic properties (WMO, 1992; 1995; also see Chapter 4). There is ample evidence that they are liquid or solid condensates of at least water, sulfuric acid, and nitric acid, with water being the major component. Type-I PSCs form at temperatures a few degrees above the frost point. When frozen they contain various forms of nitric acid hydrates. Typical sizes are 0.5-1.0 $\mu$m, but with continued cooling they may grow large enough to undergo some gravitational sedimentation. Because water vapor mixing ratios in the lower stratosphere are typically 3-5 ppmv and gaseous nitric acid is typically 1-10 ppbv, Type-I PSCs can sequester much of the available gaseous nitric acid without significantly reducing ambient water vapor. At temperatures below the frost point (about 190 K in the lower stratosphere), the aerosols continue to grow to 10 $\mu$m or larger size, large enough to undergo efficient sedimentation. These particles are composed chiefly of water ice and are labeled Type-II PSCs. The Antarctic winter vortex cools regularly and sufficiently to allow formation of both Type-I and Type-II PSCs. In contrast, the Arctic vortex is typically 5-15 K warmer and PSC formation is more episodic and localized.

There are two main effects of the heterogeneous reactions involving PSCs: the release of active chlorine from its usual stratospheric reservoirs and the removal of NO and $NO_2$ from the system by the production of $HNO_3$. The principal heterogeneous reactions involved are

$$ClONO_{2(g)} + HCl_{(s)} \to HNO_{3(s)} + Cl_{2(g)} \qquad (14.39)$$
$$ClONO_{2(g)} + H_2O_{(s)} \to HNO_{3(s)} + HOCl_{(g)} \qquad (14.40)$$
$$HOCl_{(g)} + HCl_{(s)} \to Cl_{2(g)} + H_2O_{(s)} \qquad (14.41)$$
$$N_2O_{5(g)} + H_2O_{(s)} \to 2HNO_{3(s)} \qquad (14.42)$$
$$N_2O_{5(g)} + HCl_{(s)} \to ClNO_{2(g)} + HNO_{3(s)} \qquad (14.43)$$

where $(s)$ and $(g)$ indicate that a reactant is in the solid or gas phase. Trapping of nitric acid in PSC particles through the above reactions, or the removal of $HNO_3$ if the particles settle out of the stratosphere, reduces the amount of NO and $NO_2$ available to return active chlorine or bromine to a less reactive reservoir via Reactions (14.30) and (14.37). Figure 14.12 shows a schematic of the four major processes leading to ozone depletion in the southern polar region: (1) formation of polar stratospheric

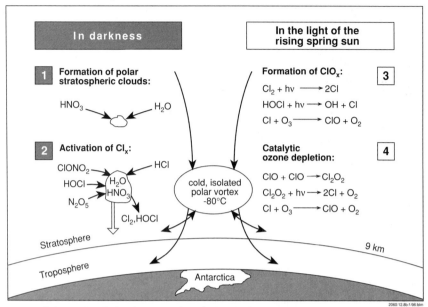

**Figure 14.12.** Schematic of the dynamical and chemical processes leading to ozone depletion within the Antarctic vortex. Processes (1) and (2) occur during the polar night, while processes (3) and (4) require the presence of sunlight.

clouds; (2) conversion of chlorine reservoirs in darkness; (3) formation of $ClO_x$; and (4) catalytic ozone depletion in the light of the spring sun.

In the darkness of autumn and early winter polar air cools and descends. The cold polar vortex isolates air and allows the formation of polar stratospheric clouds. Heterogeneous reactions on the surfaces of PSCs release chlorine from the usual reservoirs and convert NO and $NO_2$ into less reactive $HNO_3$. With the return of sunlight in the spring, $Cl_2$ and $HOCl$ (which have built up in the winter) are quickly photolyzed to Cl, which destroys ozone in Cycle 11 and, more important, in Cycle 12

| | | |
|---|---|---|
| Cycle 12 | $2\,(Cl + O_3 \rightarrow ClO + O_2)$ | (14.25) |
| | $ClO + ClO + M \rightarrow Cl_2O_2 + M$ | (14.44) |
| | $Cl_2O_2 + h\nu \rightarrow Cl + ClOO$ | (14.45) |
| | $ClOO + M \rightarrow Cl + O_2 + M$ | (14.46) |
| | Net: $2\,O_3 \rightarrow 3\,O_2$ | |

Figure 14.13 shows the dramatic erosion of ozone at altitudes between 10 and 20 km from observations within the Antarctic polar vortex in 1986 and 1987. Satellite-based observations of the total column of ozone have shown the full extent of depletion in the polar regions. Color Plate 13 shows contour maps of the Southern Hemisphere ozone column in October for a number of years. Note that the mean October $O_3$ amount within the vortex has decreased by as much as 50% and this depletion has been called the "ozone hole." Note also that the region of ozone destruction, comparable to the area of Antarctica, is 80% of the area of South America; so the fate of the large mass of ozone-poor air when the vortex finally breaks up is important to stratospheric chemistry in the entire hemisphere.

Figure 14.14 shows the change in polar inorganic chlorine as a function of the evolution of the polar vortex. Rapid conversion of $ClONO_2$ and HCl by heterogeneous reactions takes place as soon as PSCs are formed in early winter. The recovery of inactive chlorine reservoirs becomes efficient after the disappearance of cloud particles. The fastest process, in the presence of NO and $NO_2$, is

$$ClO + NO_2 + M \rightarrow ClONO_2 + M \qquad (14.30)$$

which produces a "peak" in $ClONO_2$. The formation of HCl is slower and results from

$$Cl + CH_4 \rightarrow HCl + CH_3 \qquad (14.36)$$

The balance between HCl and $ClONO_2$ is re-established later in spring.

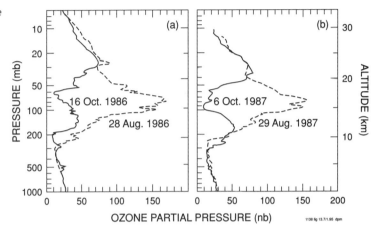

**Figure 14.13.** Ozone partial pressure versus altitude (surface to 35 km altitude) at 77°S in August and October of 1986 and 1987 (WMO, 1990a and Hofmann et al., 1989).

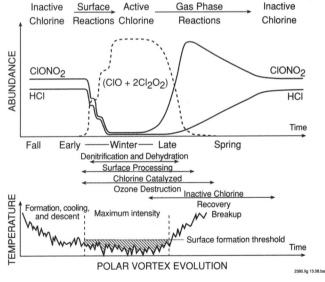

**Figure 14.14.** Schematic of chlorine photochemical and dynamical evolution of polar regions (WMO, 1995). Note the rapid conversion of $ClONO_2$ and HCl reservoirs into active chlorine as soon as polar stratospheric clouds are formed (air temperature reaching the PSC formation threshhold). The recovery of $ClONO_2$ in late spring is much faster than that of HCl.

Ozone destruction in the Arctic vortex has not been as dramatic as in Antarctica. Losses up to 35%, however, have been deduced from observations (see Color Plates 14 and 21). There are two major differences between the Arctic and Antarctic polar vortices, and the causes of these differences are related. First, the Arctic does not get as cold as the Antarctic, so PSCs are less likely to form and the processing associated with them less likely to occur. Second, the Arctic vortex is not as well formed or as persistent as in the Antarctic (largely due to Northern Hemispheric topography), and the breakup of the Arctic vortex comes two to three months earlier, with respect to the spring equinox, than the breakup in the south. This tends to cut short the processes that destroy ozone in the Arctic, but most of the mechanisms at work in the Antarctic are present in the Arctic, and the formation of an extensive Arctic ozone hole might require only a persistent cold winter. Satellite observations made since the 1970s (see Fig. 14.15 and Plate 14) suggest that total ozone values have been decreasing in the Arctic, especially since the late 1980s.

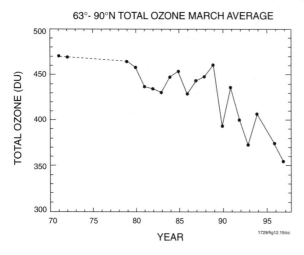

*Figure 14.15.* March average of total ozone from $63°N$ to $90°N$. Data are from Nimbus 4 BUV (1971-1972), Nimbus 7 TOMS (1979-1993), Meteor 3 TOMS (1994), NOAA 9 SBUV/2 (1996), and Earth Probe TOMS (1997) (from Newman et al., 1997). In March 1997 the average ozone column in the Arctic was more than 20% lower than normal, with a small region near the pole with values 40% lower than normal.

## 14.7 Ozone Perturbations
### 14.7.1 Anthropogenic Chlorine and Bromine

The important role of chlorine (and to a lesser extent bromine) in the control of ozone and especially in the chemistry in polar regions has been described above. Both chlorine and bromine have large anthropogenic components in their source budgets. At present some 85% of stratospheric chlorine and 60% of the stratospheric bromine are from manmade sources. Recognizing that human activity can have such an adverse effect on the ozone shield, the nations of the Earth have agreed to limit, and eventually cease, the production and release of compounds containing chlorine and bromine that can significantly affect stratospheric ozone. Even with the current restrictions, it is expected that the stratospheric abundances of chlorine and bromine will continue to increase until the end of the twentieth century. Efforts have been made to find substitutes for the major source gases $CFCl_3$ (CFC-11) and $CF_2Cl_2$ (CFC-12) (see Chapter 8). A number of hydrochlorofluorocarbons (HFCs and HCFCs) have been proposed as substitutes, each with some property, such as greater destruction in the troposphere, which makes it less likely to destroy stratospheric ozone. Numerical models, incorporating transport and chemistry, have been used to evaluate the potential of these replacement gases to destroy ozone. Given the emission of a certain gas at the Earth's surface, its effect on ozone depletion will depend on what fraction of the emission is delivered to the stratosphere, the mechanism controlling the breakdown of the halocarbon in the stratosphere, and the subsequent chemistry between the reactive halogen and ozone. For each of the halocarbons, an ozone depletion potential (ODP) is calculated, which is the amount of ozone destroyed

### Table 14.1
*Atmospheric Lifetimes and Ozone Depletion Potentials for Some Halogen-Containing Compounds*

|  |  | Stratospheric Lifetime (yr) | Tropospheric Lifetime (yr) | ODP |
|---|---|---|---|---|
| $CFCl_3$ | CFC-11 | 55 | $\infty$ | 1.0 |
| $CF_2Cl_2$ | CFC-12 | 116 | $\infty$ | 1.0 |
| $CF_2ClCFCl_2$ | CFC-113 | 110 | $\infty$ | 1.07 |
| $CF_2ClCF_2Cl$ | CFC-114 | 220 | $\infty$ | 0.8 |
| $CF_3CF_2Cl$ | CFC-115 | 550 | $\infty$ | 0.5 |
| $CHF_2Cl$ | HCFC-22 | 240 | 17 | 0.055 |
| $CF_3CHCl_2$ | HCFC-123 | 47 | 2 | 0.02 |
| $CF_3CHFCl$ | HCFC-124 | 129 | 7 | 0.022 |
| $CH_3CFCl_2$ | HCFC-141b | 76 | 13 | 0.11 |
| $CH_3CF_2Cl$ | HCFC-142b | 215 | 25 | 0.065 |
| $CF_3CF_2CHCl_2$ | HCFC-225 ca | 120 | 3 | 0.025 |
| $CF_2ClCF_2CHClF$ | HCFC-225 cb | 120 | 9 | 0.033 |
| $CCl_4$ | carbon tetrachloride | 47 | $\infty$ | 1.08 |
| $CH_3CCl_3$ | methyl chloroform | 6 | 7 | 0.012 |
| $CF_3Br$ | H-1301 | 77 | $\infty$ | 16 |
| $CF_2ClBr$ | H-1211 | 11 | 23 | 4 |
| $CF_2Br_2$ | H-1202 |  | 3.3 | 1.25 |
| $CF_2BrCF_2Br$ | H-2402 | 20 | 3.4 | 7 |
| $CF_2HBr$ | H-1201 |  | 5 | 1.4 |
| $CF_3CHFBr$ | H-2401 |  | 2.8 | 0.25 |
| $CF_3CHClBr$ | H-2311 |  | 0.8 | 0.14 |
| $CH_3Br$ | methyl bromide |  | 0.8 | 0.4 |

Approximate lifetimes and ozone depletion potentials (ODP) for CFC, HCFC and brominated compounds. ODP is a measure of the amount of ozone destroyed by the mass emission of a gas (expressed in tons $yr^{-1}$) over its entire atmospheric lifetime relative to that due to emission of the same mass of CFC-11.

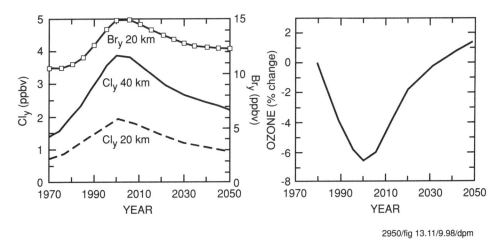

**Figure 14.16.** *Model calculations for chlorine, bromine, and ozone amounts at 40-50° N in March into the next century (WMO, 1992).*

by the gas over its entire atmospheric lifetime relative to that due to emission at the surface of the same mass of CFC-11. Estimates of ODP from a number of model calculations for a variety of CFC alternatives are contained in Table 14.1.

To evaluate the future effects of chlorine and bromine on ozone, projections for future surface emissions of these gases must be incorporated into numerical models. Figure 14.16 shows results at 20 km altitude from one such model for the projected chlorine and bromine amounts into the next century. Also shown in the figure is the predicted change in ozone column amount. These projections show that chlorine and bromine will continue to increase until about the year 2000, when the decrease in ozone will also be greatest. Values typical of the situation in 1980 will be attained sometime between 2030 and 2050.

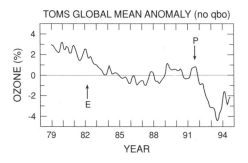

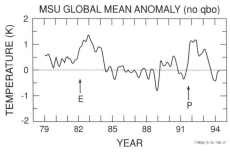

**Figure 14.17.** Globally integrated (65° N-S) and area-weighted TOMS ozone anomalies (differences from the long-term mean value) and MSU temperature anomalies over 1979-1994. QBO effects have been removed, and anomalies are calculated with respect to a 1987-1990 mean of zero. E and P denote the volcanic eruptions of El Chichón and Mt. Pinatubo, respectively (Randel et al., 1995).

### 14.7.2 Volcanic Effects

There are about 500 active terrestrial volcanoes on the Earth. Of these about 3% erupt each year and of that number about 10% have sufficient explosive power to transport gases and particles to the stratosphere. The composition of the gas injected into the stratosphere is extremely variable and composed mostly of $H_2O$, $CO_2$, and $SO_2$ (typically 70%, 20%, and 6%, respectively) and may contain up to 2% HCl. The HCl injected by volcanoes represents a small part of the global stratospheric load; studies after the eruptions of the El Chichón (1982) volcano showed HCl in the vicinity of the volcanic plume to be increased by up to 40%, which translates into approximately 9% of the global stratospheric HCl. The HCl injected into the stratosphere by the Mt. Pinatubo (1991) eruptions appeared to be substantially less than that from El Chichón.

From the perspective of ozone chemistry, more important than HCl emitted by volcanoes is the large amount of $SO_2$ that can be injected directly into the stratosphere. Within a few months of a volcanic injection, $SO_2$ is converted into $H_2SO_4$ aerosols (details of the aerosol formation are in Chapter 4). These aerosols provide sites for heterogeneous reactions that can affect ozone, similar to the heterogeneous reactions on PSC particles in polar regions. The reactions most likely to occur on volcanic sulfate aerosols, and which are also significant for background aerosol conditions, are

$$ClONO_{2(g)} + H_2O_{(l)} \rightarrow HNO_{3(g)} + HOCl_{(g)} \tag{14.47}$$

$$N_2O_{5(g)} + H_2O_{(l)} \rightarrow 2\ HNO_{3(g)} \tag{14.38}$$

in which chlorine is converted from its $ClONO_2$ reservoir into more reactive HOCl, and $NO_2$ (via $N_2O_5$) is converted into $HNO_3$. Model studies of the possible effects of these heterogeneous reactions, with volcanic loading similar to that after the Mt. Pinatubo eruptions, have predicted significant ozone reductions in midlatitudes. Analyses of long-term global observations have shown a reduction in global mean ozone amount of about 4% following the Mt. Pinatubo eruption, as shown in Figure 14.17. Volcanic effects may persist for 2-3 years after a major eruption, roughly the stratospheric residence time of the sulfate aerosols.

## 14.8 Impact of Ozone Depletion on UV Radiation

As mentioned earlier, atmospheric ozone protects the Earth's biosphere from harmful solar ultraviolet (UV) radiation. Such radiation has the potential to damage DNA in living cells, inhibit plant growth, damage animal and human skin, increase the likelihood of skin cancer, damage eye cornea and lenses (cataract), and affect the immune system. Living organisms are sensitive specifically to radiation at wavelengths less than 400 nm. Since wavelengths shorter than 280 nm are almost completely absorbed by the atmosphere, the radiation relevant to environmental biology is restricted to the UV-B (280-315 nm) and UV-A (315-400 nm) ranges.

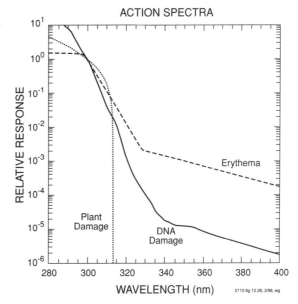

**Figure 14.18.** Action spectra for erythema induction, plant damage, and DNA damage (normalized to unity at 300 nm) between 280 and 400 nm (after Madronich, 1993; reprinted by permission of Oxford University Press).

Figure 14.18 shows the action spectra (*i.e.*, the effectiveness of the radiation to damage biological organisms as a function of wavelength) relevant to erythema induction (sunburn), DNA damage, and plant damage. These curves, normalized at 300 nm, are obtained by exposing biological targets to UV radiation at various isolated wavelengths. They show that biological processes are most sensitive to the UV-B region in which the transmission of solar radiation is strongly affected by atmospheric ozone (see Fig. 14.19, which represents the simultaneous observation of the ozone

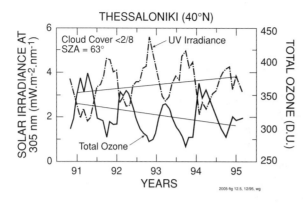

**Figure 14.19.** Solar irradiance ($mW\,m^{-2}\,nm^{-1}$) at 305 nm and total ozone abundance (DU) measured at Thessaloniki, Greece (40° N) between 1991 and 1995 for nearly clear sky conditions and a solar zenith angle of 63° (from C. S. Zerefos and A. F. Bais, private communication, 1995).

column abundance and UV-B radiation at Thessaloniki, Greece, between 1991 and 1995).

The biological effect of solar radiation is generally expressed in terms of the daily dose $D$ (J m$^{-2}$), which is defined as the integral over a full day (24 h) of the spectral irradiance $F_0(\lambda, t)$ reaching the surface multiplied by the biological action spectrum $A(\lambda)$, and integrated over wavelength $\lambda$. Thus

$$D = \int_{24h} \int_{\lambda} F_0(\lambda, t)\, A(\lambda)\, d\lambda\, dt \tag{14.48}$$

where $t$ is time. The dose is affected by the various parameters that determine the transmission of ultraviolet light in the atmosphere, that is, the ozone column abundance as well as the cloudiness, the aerosol load, and the brightness of the surface. The relative change $\Delta D/D$ in the biological dose produced by a change in the ozone column abundance $\Delta N(O_3)/N(O_3)$ is often expressed by a simple proportionality rule

$$\frac{\Delta D}{D} = -\text{RAF}\,\frac{\Delta N(O_3)}{N(O_3)} \tag{14.49}$$

where the constant RAF represents the so-called *radiation amplification factor*. Table 14.2 provides typical (experimental) values of this factor for various biological processes. Equation (14.49) is valid for small changes in the ozone column abundances. For example, for a 2% ozone depletion, the potential for photocarcinogenesis (skin cancer with RAF = 1.6) increases by 3.2%. For large changes in the ozone column, Eq. (14.49) should be integrated, leading to the power rule (Madronich, 1993)

$$\frac{D_2}{D_1} = \left[\frac{N_2(O_3)}{N_1(O_3)}\right]^{-\text{RAF}} \tag{14.50}$$

where $D_1$ and $D_2$ are the doses corresponding to the ozone column abundances $N_1(O_3)$ and $N_2(O_3)$, respectively.

Variations in the ozone column abundance observed during the past decades have produced changes in the average biological doses. Figure 14.20, for example, provides an estimate of the trend in the monthly mean DNA-damaging doses over the period 1979-1992 resulting from the global trends in total ozone observed during the same period (see Fig. 16.6) by the space-borne Total Ozone Mapping Spectrometer (TOMS). Large positive trends in these doses calculated in the polar region of the Southern Hemisphere during September and October are associated with the formation of the Antarctic ozone hole since the late 1970s. No significant trends are reported in the tropics, but an increase of up to 20% is seen at midlatitudes of the Northern Hemisphere during wintertime. These increases in UV doses may have been partly offset by enhanced levels of pollution, including changes in the aerosol load and tropospheric ozone in industrialized regions. Potential changes in cloudiness or in cloud reflectivity during the past decades could also have modified biological doses; the magnitude of this effect remains, however, highly uncertain.

**Table 14.2**
*Radiation Amplification Factors (RAFs) in January and July[a]*
*($30°N$ unless specified)*

|  | January | July |
|---|---|---|
| Erythema induction in humans | 1.1 | 1.2 |
| Skin cancer in SKH-1 hairless mice (Utrecht) | 1.5 | 1.4 |
| Skin elastosis | 1.1 | 1.2 |
| Generalized DNA damage | 2.2 | 2.1 |
| Damage to eye cornea | 1.2 | 1.1 |
| Damage to eye lens (cataract) | 0.8 | 0.7 |
| Generalized plant spectrum | 2.0 | 1.6 |
| DNA damage in alfalfa | 0.5 | 0.6 |
| Inhibition of photosynthesis, in Antarctic community | 0.8 | 0.8 |

[a] Adapted from Madronich et al. (1995).

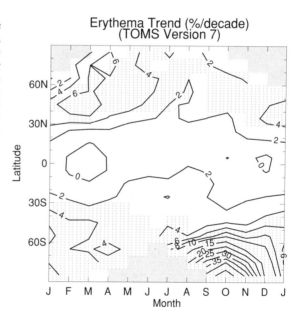

*Figure 14.20.* Calculated trends (in percent per decade) in erythema daily doses between the late 1970s and the early 1990s corresponding to measured changes in the ozone column during the same period (see Fig. 16.6). Shading denotes areas with significant trends. Grey areas correspond to regions with no ozone measurements available (S. Madronich and S. Flocke, private communication, 1997).

## Further Reading

Andrews, D. G., J. R. Holton, and C. B. Leovy, (1987) *Middle Atmosphere Dynamics*, Academic Press, Orlando, Florida.
Brasseur, G., ed. (1997) *The Stratosphere and Its Role in the Climate System*, NATO ASI Series I-54, Springer-Verlag, Berlin.
Brasseur, G., and S. Solomon (1986) *Aeronomy of the Middle Atmosphere*, 2nd ed., D. Reidel Publishing Co., Dordrecht, The Netherlands.
Calvert, J., ed. (1994) *The Chemistry of the Atmosphere: Its Impact on Global Change*, Blackwell Scientific Publications, Oxford, UK.

Gurney, R.J., J. L. Foster, and C. L. Parkinson (eds.) (1993) *Atlas of Satellite Observations Related to Global Change*, Cambridge University Press, Cambridge.

McEwan, M. J. and L. F. Phillips (1975) *Chemistry of the Atmosphere*, Edward Arnold Publishers, London, UK.

Shimazaki, T. (1985) *Minor Constituents in the Middle Atmosphere*, Terra Scientific Publishing Company, Tokyo, Japan.

Tervini, M., ed. (1993) *Environmental Effects of UV (Ultraviolet) Radiation*, Lewis Publisher, Boca Raton, Florida.

Wayne, R. P. (1991) *Chemistry of Atmospheres*, 2d ed., Clarendon Press, Oxford, UK.

World Meteorological Organization (WMO) (1991) *Scientific Assessment of Ozone Depletion, 1991*, Global Ozone Research and Monitoring Project Report No. 25.

World Meteorological Organization (WMO) (1994) *Scientific Assessment of Ozone Depletion, 1994*, Global Ozone Research and Monitoring Project Report No. 37.

Zerefos, C. S. and A. F. Bais, eds. (1997) *Solar Ultraviolet Radiation: Modeling, Measurements and Effects*, NATO ASI Series, Springer Verlag, Berlin, Germany.

Susan Solomon

## Ozone Depletion: From Pole to Pole

A distinguishing feature of the twentieth century has been the recognition of the fact that human activities are changing the Earth's atmosphere. Carbon dioxide, methane, and chlorofluorocarbon concentrations have increased, causing people around the world to come to a new realization: The atmosphere is vast but finite. There are now so many people on this planet that some of the gases we release are affecting the composition of our atmosphere. The most striking illustration of the concurrent development of scientific theory, observation, and societal implications of atmospheric change has been the depletion of the Earth's protective ozone layer, which is the subject of this chapter.

The questions being asked as a result of this understanding are very difficult ones: How do we know how large the changes are? What is natural and what is human-induced, and is there any coupling between them? Are the changes in our atmosphere affecting the quality of our environment? One reason that these are particularly challenging questions is that we cannot conduct a "controlled experiment" wherein chlorofluorocarbons are added to the atmosphere and then promptly removed, in order to discern their impact explicitly. The atmospheric composition changes are occurring on a global basis in what is clearly an "uncontrolled" manner, and science is faced with the difficult task of trying to detect possible results of those changes. The problem of the uncontrolled experiment is compounded by the fact that the atmosphere does vary - making it even tougher to determine whether changes are natural or not.

The first place where a human impact on the ozone layer was clearly distinguished was perhaps the most unexpected one: in the Antarctic. In 1985, scientists from the British Antarctic Survey reported that the springtime Antarctic ozone layer had decreased by almost half compared to measurements taken during the late 1950's, 1960's and early 1970's. This change was far greater than any natural variation observed in monthly averaged ozone there. Chlorofluorocarbons and bromocarbons were suspected as a possible cause.

Within a few years, aircraft and ground-based observations were carried out that measured not just ozone alone, but also the chemicals that can affect it. It was a privilege to be a scientist involved in the remarkable research of this period. As a result of the work of hundreds of researchers worldwide, it is now well known that the depletion is pronounced in Antarctica because it is indeed the coldest place on Earth. The extreme coldness of the Antarctic stratosphere allows chemical reactions to occur on surfaces that rapidly liberate reactive chlorine from chemically inert reservoirs, making the chlorine from chlofluorocarbons much more damaging to ozone than it would otherwise be. The most rapid ozone loss occurs in Antarctica during September, because both cold temperatures and sunlight are involved in the chemistry of Antarctic ozone depletion. Remarkably, at the time of this writing (about ten years after the discovery of the ozone hole), there are still key unanswered questions regarding the exact composition and phase of the surfaces (nitric acid/water, liquid, solid, etc.) involved in this essential chemical step in the chain that has created the ozone hole.

A logical next question is whether or not ozone depletion is also occurring in the Arctic. The answer is yes, but the changes are smaller there, primarily because the Arctic stratosphere is usually warmer than the Antarctic, particularly in spring. The Arctic stratosphere generally warms up sooner than the Antarctic does. This in turn means that the overlap between the cold temperatures that cause clouds to form and the sunlight that returns to the polar regions in spring is less effective in the North than in the South. However, many scientists are concerned by an important aspect of natural variability: The spring Arctic stratosphere can sometimes be very cold. In an unusually cold year, more Arctic ozone depletion is likely.

Ozone changes are happening at midlatitudes, too, but these are smaller still. Here again, there is suggestive evidence that these ozone changes are not natural, that mankind's use of chlorofluorocarbons is indeed the root cause of the changes. Observations also suggest that surface chemistry is involved in midlatitude ozone depletion, and that major volcanic eruptions can therefore lead to larger ozone depletions.

As a result of concerns about our changing ozone layer, governments worldwide have agreed to reduce emissions of chlorofluorocarbons to the atmosphere over the next decade or so. Eventually, the ozone layer can be expected to return to normal as a result, even in Antarctica. But it will take a long time. The chlorofluorocarbons remain in the Earth's atmosphere for time scales on the order of 50 to 500 years, depending on the compound. This means that the chlorofluorocarbon that is already present will continue to destroy ozone from one pole to the other well into the next century.

*Susan Solomon is a Senior Scientist at the Aeronomy Laboratory of the National Oceanic and Atmospheric Administration and an Affiliate Scientist at the National Center for Atmospheric Research in Boulder, Colorado. Her research has had a major impact on the understanding of stratospheric ozone depletion. Dr. Solomon led the first National Ozone Expedition in Antarctica, which showed that chlorofluorocarbons were the likely cause of the Antarctic ozone "hole." She is a member of the U.S. National Academy of Sciences and a Foreign Associate of the French Academy of Sciences.*

# 15 Atmospheric Chemistry and Climate

## 15.1 Introduction

Throughout the Earth's history, climate (defined as the long-term statistical behavior of the atmosphere) has been characterized by numerous cycles with successive fluctuations between colder and warmer periods. Even though the mean temperature of the Earth over the past geological periods has probably not varied by more than a few degrees Celsius, climate changes have produced dramatic variations: in the level of the oceans, in the geographical extent of the ice sheets, in water supply, and in the distribution of continental ecosystems, for example. Over the past centuries, perturbations associated with economic development driven by growing populations, and specifically with agricultural practices and industrial activities, have altered our chemical and physical environment with potential effects on the climate system.

An example is provided by the release to the atmosphere of increasing quantities of carbon dioxide ($CO_2$) as a result of fossil fuel consumption and biomass burning. As noticed many years ago (see Box 15.1), increases in the concentration of this gas tend to enhance the absorption by the atmosphere of outgoing terrestrial radiation from the surface, and at the same time to enhance the infrared radiation emitted at higher altitudes and colder temperatures. The net effect, often called the "greenhouse effect" is a positive radiative forcing, which tends to warm the lower atmosphere and the surface. Other radiatively active gases, such as methane ($CH_4$), nitrous oxide ($N_2O$), the chlorofluorocarbons (CFCs), and ozone ($O_3$), the atmospheric abundance of which has changed as a result of human activities, trap additional radiative energy in the Earth-atmosphere system.

A second example is provided by anthropogenic aerosols present in the troposphere. These small airborne particles reflect and absorb solar radiation. Through their effects on cloud properties, they can also alter cloud microphysical properties, including cloud reflectivity. In most cases the net effect is a negative forcing, which tends to cool the Earth's climate.

The response of the Earth system to these changes in the radiative balance is difficult to estimate. Atmospheric and oceanic temperatures as well as associated weather patterns are expected to be altered. Changes in the hydrological cycle, and specifically modifications in the precipitation and evaporation regimes (including drought and flood frequencies), and changes in cloudiness are also expected to result from modified radiative forcing.

Human-induced effects on climate are and will be superimposed on natural climate variability. It is therefore difficult to establish with a sufficient degree of confidence that the observed trend in the globally averaged temperature of the Earth (0.7 K since the year 1850) is the result of anthropogenic activity. However, on the basis of the geographical, seasonal, and vertical patterns in the observed temperature changes,

## Box 15.1 Climate Research: The Pioneers

The French mathematician Jean-Baptiste-Joseph de Fourier (1768-1830) suggested in 1827 that the atmosphere keeps the Earth warm by trapping heat as a plate of glass would. It is known today that the comparison of the atmosphere to a greenhouse is incorrect, but Fourier did suggest that the natural climate could be perturbed as a result of human activities. In 1859, the Irish physicist John Tyndall (1820-1893) measured the absorption of various gases, and established that water vapor, carbon dioxide, and methane were trapping infrared (terrestrial) radiation, while the most abundant atmospheric gases, nitrogen and oxygen, were not. He suggested that ice ages in the past could have been caused by changes in the atmospheric abundance of radiatively active gases.

*Tyndall*

*Arrhenius*

In 1896 the Swedish scientist Svante Arrhenius (1859-1927), who was awarded the Nobel Prize for Chemistry in 1903, theorized that a doubling in the natural concentration of carbon dioxide in the air would increase the Earth's global mean temperature by 5-6 degrees Celsius. He pointed out that such a change was likely due to the rapid expansion of industry, and because, as he stated, "we are evaporating our coal mines into the air." In 1938 Guy Stewart Callendar, an engineer working for British Electrical Industries, noted that the level of $CO_2$ had increased by some 10 percent since the 1890s and that it could explain the rise in temperature recorded over the same period of time.

It was not, however, until the early 1960s that the potential importance of global warming was recognized after the American geochemist Charles D. Keeling reported a continuous increase in the abundance of $CO_2$ observed in a region as remote as the Mauna Loa Observatory in Hawaii. In the past decades, the question of global warming has led to much debate and even to controversies. The complexity of the problem, including the role played by atmospheric constituents other than $CO_2$, is taken into account in current studies.

---

there is growing evidence of a human influence on the recent evolution of global and regional climate.

This chapter focuses on the interactions between atmospheric chemistry and climate and presents the different processes that contribute to the Earth's radiative balance. The relationship of past climate evolution to variations in the chemical composition of the atmosphere is also discussed. Finally, we examine the potential impact of anthropogenic trace gases and aerosols on the climate system.

## 15.2 Radiation in the Atmosphere
### 15.2.1 Solar Radiation

Solar radiation is the primary source of energy for the Earth system. This energy is provided mostly in the form of ultraviolet, visible, and near-infrared radiation (wavelength between 0.2 and 4 $\mu$m). At the top of the atmosphere, the shortwave energy flux intercepted by a surface normal to the direction of the Sun is approximately equal to 1370 W m$^{-2}$, and is called the solar constant. The corresponding energy captured by the Earth's system is on the average 342 W m$^{-2}$.

This energy is mainly absorbed in the atmosphere by molecular oxygen ($O_2$), ozone ($O_3$), and water vapor ($H_2O$), as described in Chapter 3. The absorption of solar radiation by ozone provides the energy that heats the stratosphere and mesosphere (see Chapter 2). The portion of solar radiation that is not absorbed in the atmosphere or backscattered to space reaches the Earth's surface. Figure 15.1 presents the spectrum of solar radiation outside the Earth's atmosphere and at sea level for clear sky conditions. Since the troposphere and the surface are coupled by convective exchanges, this energy almost simultaneously heats the soil, the vegetation, and the oceans as well as the entire troposphere, except in cases of temperature inversions near the surface (temperature increasing with height).

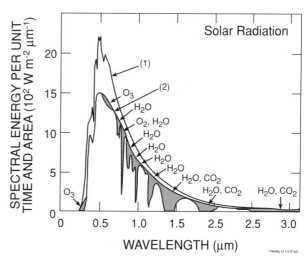

**Figure 15.1.** *Spectrum of solar radiation (1) outside the Earth's atmosphere and (2) at sea level for clear sky conditions (from Gast, 1961). The shaded area represents the energy absorbed by various gases in a clear atmosphere.*

The intensity of radiation emitted by the Sun is not entirely constant as a function of time. For example, variations in the "solar constant" of approximately a tenth of a percent are observed and linked to the 11-year solar cycle. Several attempts have been made to correlate the past evolution of climate with solar activity, but the subject remains controversial, since there are no reliable measurements of solar constant changes before the mid-1970s, and very little change has occurred since then.

## 15.2.2 Terrestrial Radiation

The energy provided by the Sun and absorbed by the Earth is reradiated as infrared radiation (see Box 15.2). This energy is absorbed by clouds as well as atmospheric molecules, the major absorbers being water vapor and carbon dioxide; these two gases are sufficiently abundant to trap a large fraction of the energy (mostly in the 12 to 20 $\mu$m spectral region) in the lowest layers of the atmosphere. In the 8 to 12 $\mu$m region, called the atmospheric window, terrestrial radiation propagates to space because of the relatively weak absorption in this region of the spectrum. Therefore, any gas with strong absorption properties in this spectral region is expected to be relatively efficient in trapping terrestrial radiation. The spectral locations of the absorption features of the main greenhouse gases in the atmospheric window are shown in Figure 15.2. An example of a terrestrial radiation spectrum measured at the top of the atmosphere by the Nimbus-3 IRIS instrument is shown in Figure 15.3. The absorbing bands such as the 9.6 $\mu$m band of $O_3$ and the 15 $\mu$m band of $CO_2$, as well as the atmospheric window and several other features ($H_2O$, $CH_4$), are noticeable. These radiatively active gases, also called greenhouse gases, absorb only a small fraction of solar energy, but they are very effective in absorbing as well as emitting longwave radiation. Their net effect is to reduce the amount of radiative energy emitted to space and to increase the radiative energy provided to the surface-troposphere system. The fundamental

**Box 15.2 The Radiative Transfer Equation**

The radiative field in a given medium is completely described by the intensity (radiance) $I(\mathbf{r}, \boldsymbol{\omega})$, which is defined at each point $\mathbf{r}$ as the flow of energy per unit area and unit solid angle propagating in direction $\boldsymbol{\omega}$. The net radiant flux (irradiance) $F(\mathbf{r},\boldsymbol{\omega})$ (W m$^{-2}$) traversing a plane surface, whose orientation in space is defined by direction $\boldsymbol{\omega}$, due to all light pencils of direction $\boldsymbol{\omega}'$ is given by the integral

$$F(\mathbf{r},\boldsymbol{\omega}) = \int_{4\pi} I(\mathbf{r},\boldsymbol{\omega}') \cos(\boldsymbol{\omega},\boldsymbol{\omega}') d\boldsymbol{\omega}'$$

The spherical radiant flux at point $\mathbf{r}$, $F_0(\mathbf{r})$ (W m$^{-2}$), is obtained by integrating the radiance over all solid angles

$$F_0(\mathbf{r}) = \int_{4\pi} I(\mathbf{r},\boldsymbol{\omega}') d\boldsymbol{\omega}'$$

The monochromatic radiance $I_\lambda = dI/d\lambda$ (expressed in W m$^{-2}$ sr$^{-1}$ nm$^{-1}$) represents the spectral density of the radiance at wavelength $\lambda$ (nm). Similarly, the monochromatic spherical radiant flux $F_{0,\lambda}$ (W m$^{-2}$ nm$^{-1}$) is defined as $F_{0,\lambda} = dF_0/d\lambda$. The actinic flux $q(\lambda)$ (photons m$^{-2}$ s$^{-1}$ nm$^{-1}$) is derived from

$$q(\lambda) = F_{0,\lambda} \frac{\lambda}{hc}$$

where $h = 6.6256 \times 10^{-34}$ J s (Planck's constant) and $c = 3 \times 10^8$ m s$^{-1}$ is the speed of light.

The value of the monochromatic radiance $I_\lambda$ is derived from the radiative transfer equation (see, *e.g.*, Lenoble, 1993), which expresses the energy balance in each unit volume of the atmosphere. Assuming that the atmosphere can be represented by a horizontally stratified medium (plane-parallel approximation), the radiative transfer equation can be expressed by

$$\mu \frac{dI_\lambda}{dz} = -k_\lambda(z)\left[I_\lambda(z;\mu,\varphi) - J_\lambda(z;\mu,\varphi)\right] \quad (A)$$

where $k_\lambda(z)$ is the extinction coefficient and $J_\lambda(z;\mu,\varphi)$ is the source function at altitude $z$ and for directions $(\mu,\varphi)$; $\mu$ is the cosine of the zenith angle and $\varphi$, the azimuthal angle, defines the direction of propagation of the beam under consideration.

*A. Solar Radiation*

Ignoring radiative emission by the medium (a good approximation for wavelengths less than 4 $\mu$m, where radiation of solar origin dominates), but considering scattering of light as well as absorption, the source function can be expressed by

$$J_\lambda = \frac{\Omega_\lambda(z)}{4\pi} \left[ \int_0^{2\pi} d\varphi' \int_{-1}^{+1} p_\lambda(z;\mu,\varphi;\mu',\varphi') I_\lambda(z,\mu',\varphi') d\mu' \right.$$
$$\left. + p_\lambda(z;\mu,\varphi;\mu_0,\varphi_0) \Phi_\lambda(\infty) \exp\left(-\frac{1}{\mu_0}\int_z^\infty k_\lambda(z') dz'\right) \right] \quad (B)$$

where $\Omega_\lambda(z)$ is the albedo for single scattering defined by the ratio

$$\Omega = \frac{k_{s,\lambda}}{k_\lambda}$$

where $k_{s,\lambda}$ and $k_\lambda$ are the scattering and total extinction coefficient at wavelength $\lambda$. $p_\lambda(z;\mu,\varphi;\mu',\varphi')$ is the phase function defining the probability that a photon originating from direction $(\mu',\varphi')$ be scattered in the direction $(\mu,\varphi)$. The direction $(\mu_0,\varphi_0)$ specifies the direction of the Sun, and $\Phi_\lambda(\infty)$ is the solar irradiance at the top of the atmosphere. Note that in

Eq. (B) the second term accounts for the scattering of the direct solar beam, while the first term accounts for multiple scattering.

The solution of the radiative transfer equation requires that the phase function $p_\lambda$ be represented by a mathematical expression. The simplest (although not very realistic) case is provided by the so-called isotropic scattering approximation, in which $p_\lambda = 1$. When scattering is due only to air molecules, the phase function is provided by the Rayleigh theory and

$$p_\lambda = \frac{3}{4}(1 + \cos^2 \theta)$$

where

$$\cos \theta = \mu \mu' + \left(1 - \mu^2\right)^{\frac{1}{2}} \left(1 - \mu'^2\right)^{\frac{1}{2}} \cos \left(\varphi - \varphi'\right)$$

In other cases (e.g., when solid and liquid particles contribute to the scattering of light), the phase function can be derived from the Mie theory (Mie, 1908) if the physical and chemical characteristics of the particles are known.

Numerous mathematical methods have been developed to solve the radiative transfer equation (A) in a scattering atmosphere (see, e.g., Lenoble, 1993). For example, in the discrete ordinate method, the radiance is calculated in a finite number of directions and the integrals in Eq. (B) are replaced by finite summation. In the two-stream approximations (see, e.g., Meador and Weaver, 1980; Toon et al., 1989), rather than calculating the fully directional radiance, the upwelling and downwelling diffusive fluxes are calculated. The various two-stream methods are distinguished by the amount of radiation scattered in the forward and backward directions and by the value chosen for the average direction of propagation of the diffuse light. An example of a popular two-stream method is provided by the $\delta$-Eddington approximation (Joseph et al., 1976).

### B. Terrestrial Radiation

At wavelengths larger than 4 $\mu$m, the only significant contribution to the radiation field in the atmosphere is provided by thermal emission in the Earth-atmosphere system. We will assume that scattering can be neglected. In the case of a plane-parallel atmosphere where local thermodynamic equilibrium conditions apply, the source function appearing in the radiative transfer equation (A) becomes

$$J_\lambda(z) = B_\lambda(T(z))$$

where the Planck function is

$$B_\lambda(T) = \frac{1.1911 \times 10^{-16} \left[\text{W m}^2 \text{ sr}^{-1}\right]}{\lambda^5 \left[\exp\left(1.438 \times 10^{-2} \,[\text{m K}] \Big/ \lambda\, T\right) - 1\right]}$$

if wavelength $\lambda$ is expressed in meters and temperature $T$ is in Kelvin.

Multiplying the radiative transfer equation

$$\mu \frac{dI_\lambda(z; \mu, \varphi)}{dz} = -k_\lambda(z) \left[I_\lambda(z; \mu, \varphi) - B_\lambda(T(z))\right]$$

by $d\mu\, d\varphi$ and integrating over all directions, it is easy to derive that

$$\frac{dF_\lambda(z)}{dz} = -k_\lambda(z) \left[F_{0,\lambda}(z) - 4\pi B_\lambda(T(z))\right]$$

where

$$F_\lambda(z) = \int_0^{2\pi} d\varphi \int_{-1}^{+1} \mu I(z; \mu, \varphi)\, d\mu$$

is the net flux of energy (irradiance) across a horizontal surface and

$$F_{0,\lambda}(z) = \int_0^{2\pi} d\varphi \int_{-1}^{+1} I(z;\mu,\varphi)\, d\mu$$

is the radiant spherical flux. Absorption (and emission) are not monochromatic, so that the irradiance must be integrated over a finite wavelength interval

$$F(z) = \int_\lambda F_\lambda(z)\, d\lambda$$

In most cases, the absorption coefficient $k_\lambda$ varies significantly with wavelength since absorption spectra of radiatively active molecules are composed of lines organized in bands. Band models have been developed to estimate the average absorbance, as well as the irradiance, over spectral intervals that are significantly broader than the width of individual lines. The radiative heating

$$\frac{\partial T}{\partial t} = -\frac{1}{\rho C_p} \frac{\partial F}{\partial z}$$

where $\rho$ is the air density and $C_p$ the specific heat at constant pressure, is provided by the difference between locally absorbed and emitted radiative energy. For more details, see Goody and Yung (1989) and Liou (1980).

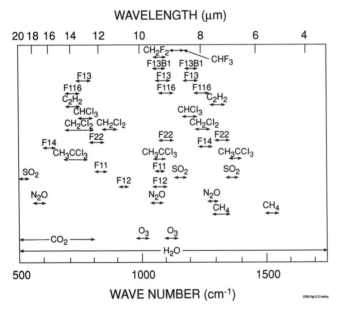

**Figure 15.2.** Spectral locations of the absorption features of several atmospheric gases (WMO, 1986).

reason for the existence of the "greenhouse effect" is that the temperature decreases with altitude in the troposphere. Radiatively active gases as well as clouds absorb the radiation emitted by the warmer surface, while their emission of radiation to space occurs at colder atmospheric temperatures. The trapping of the radiation by radiatively active molecules produces an increase in the surface temperature of about 33 degrees Celsius (assuming no change in albedo when atmosphere is removed). Without the "greenhouse effect" the average temperature at the surface would be only $-18°C$ and life would not be possible on Earth. At higher altitudes, the radiative

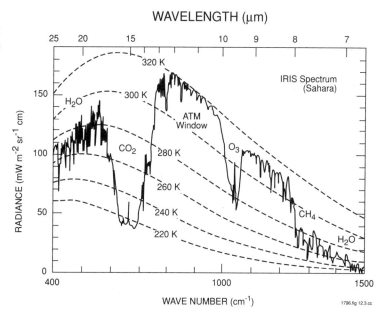

**Figure 15.3.** Example of terrestrial radiation spectrum obtained by the Nimbus 3 IRIS instrument for clear sky conditions (from Hanel et al., 1972).

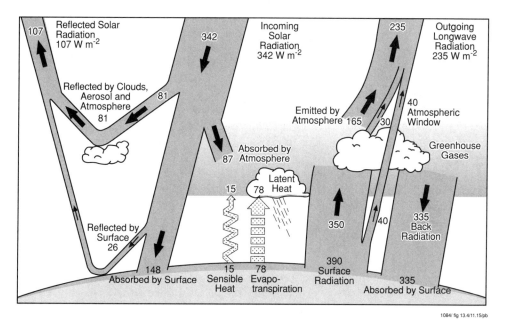

**Figure 15.4.** Schematic diagram of the global energy budget. The values are expressed in W m$^{-2}$ (IPCC, 1996).

emission to space in the 15 $\mu$m band of $CO_2$ contributes to a cooling in the stratosphere and mesosphere.

Surface climates are directly influenced by the radiation balance between incoming solar radiation and outgoing (reflected solar + infrared) radiation. A global energy budget of the Earth system can be approximately described as follows (Fig. 15.4): The solar energy penetrating into the Earth's system represents about 342 W m$^{-2}$, of which about 107 W m$^{-2}$ (or 31%) is returned to space (24% due to backscattering by clouds, air molecules, and particles, and 7% due to reflection at the Earth's surface),

87 W m$^{-2}$ (or 25%) is absorbed within the atmosphere, by ozone in the stratosphere, and by clouds and water in the troposphere. The remaining 148 W m$^{-2}$ (or 43%) is absorbed at the Earth's surface. From the terrestrial energy emitted by the surface (390 W m$^{-2}$ or 114%), only 40 W m$^{-2}$ (12% of the incoming solar radiation) escapes directly to space in the atmospheric window. The remaining 310 W m$^{-2}$ (or 90%) is absorbed within the troposphere by water vapor, $CO_2$, $O_3$, and the other greenhouse gases, as well as by clouds and aerosols. Finally, an energy of about 335 W m$^{-2}$ (or 98%) is emitted back to the surface, while 195 W m$^{-2}$ (or 57%) is emitted to space. The excess energy received by the surface is compensated by nonradiative processes such as evaporation (latent heat flux of 78 W m$^{-2}$, or 23%) and turbulence (sensible heat flux of 15 W m$^{-2}$, or 4%).

Note the difference between the radiative emission at the Earth's surface (390 W m$^{-2}$) and the total infrared emission to space (40 + 195 = 235 W m$^{-2}$). This energy trapped in the atmosphere (155 W m$^{-2}$) represents the greenhouse effect. With the exception of numbers given for the top of the atmosphere, these values are uncertain by approximately 10 to 20%.

Among the gases present in the atmosphere, the largest contribution to the greenhouse effect is provided by water vapor, followed by $CO_2$ and other trace gases such as $CH_4$, $N_2O$, $O_3$, CFCs, HCFCs, and HFCs.

Clouds also absorb and emit infrared radiation. In addition, they increase the planetary albedo (defined as the fraction of the incoming radiation that is reflected). Their net effect on the climate system is complex. It is believed that high-altitude cirrus clouds contribute to warming, while low-level stratus clouds contribute to cooling. Overall, the presence of clouds tends to cool the Earth's system. However, complex feedback effects could lead to warming or cooling effects, depending on specific cloud changes in response to changes in climate forcing.

As the concentration of several radiatively active gases are increasing as a result of human activities, there is great concern about a possible increase in the greenhouse forcing. When the concentration of a radiatively active gas increases, initially the longwave radiation to space is reduced. As a result, the energy budget is out of balance at the top of the atmosphere. At the same time, if this gas does not affect the absorption of solar radiation, the net radiative energy available in the lower atmosphere and at the Earth's surface increases and the energy balance is restored through a warming of the surface-troposphere system.

## 15.3 Natural Variations: Past Climates

One of the most fascinating confirmations of a link between atmospheric trace gas concentrations and climate is provided by the data obtained from ice cores collected in polar regions. Figure 15.5 shows the correlations between atmospheric trace gas content and temperature over the past 240,000 years. Clearly (assuming that the Vostok temperature series is representative of the global mean temperature), when the atmospheric abundances of $CO_2$ and $CH_4$ are low, the Earth is in a relatively cool climate state. To first order, when greenhouse gases are in relatively low abundance in the atmosphere, there is less infrared trapping of heat and the Earth surface is cool, perhaps cool enough to initiate glaciation. However, it should be noted that it is not clear if $CO_2$ was lower before or after the cooling occurred. Phase shifting of the temperature changes and atmospheric $CH_4$ and $CO_2$ concentrations is evident in some data sets. It is not straightforward to assign cause and effect relationships based on these time lags due to possible amplification effects (feedback mechanisms in the Earth system). Changes in surface temperatures and the areas covered by ice sheet have undoubtedly affected the exchange rates of greenhouse gases between the surface (continents, oceans) and the atmosphere. Simultaneously, the atmospheric abundance of these gases has affected the Earth's climate.

On even longer time scales, the atmospheric concentration of $CO_2$ may have been as high as 3000 ppmv during the Cretaceous period (90-65 million years ago). The Earth climate system was, based on geological evidence, very warm during this period. The area of the polar ice caps was substantially reduced, and probably absent, and the sea level was much higher than during present-day conditions. One of the most puzzling aspects of long-term changes in climate and the abundance of atmospheric greenhouse gases is what initiates a shift in dominant climate. A postulated cause is provided by periodic changes in the relative orientation of the Earth to the Sun. This effect is known as Milankovich forcing. Spectral analyses of geological records have highlighted typical periods of 20,000, 40,000, and 100,000 years in climate changes. The observed evolution in the temperature over the past millenia may have resulted from interactions between such external forcings (associated with changes in the orbital parameters of the Earth) and internal dynamics (*e.g.*, glacial feedback response).

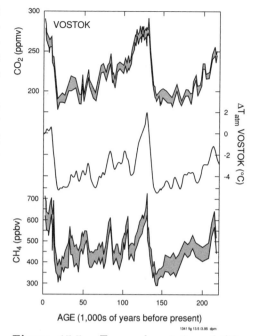

**Figure 15.5.** *Temporal variation of $CO_2$ and $CH_4$ concentrations and temperature in Antarctica over the past 240,000 years, based on information contained in an ice core collected at Vostok, Antarctica (Jouzel et al., 1993; reprinted with permission from* Nature, *copyright 1993, Macmillan Magazines Limited).*

It is interesting to note that the past 10,000 years in the history of our planet, during which civilization developed, were warm and marked by relatively consistent weather. The previous millenium, a long, cold period that ended about 11,000 years ago, is known as the "Younger Dryas" with low $CO_2$ and $CH_4$ concentrations. Recent analyses of ice cores extracted from the Greenland ice cap suggest that the winter temperature in Northern Europe has fluctuated by as much as 10°C over periods of time as short as a decade. These abrupt shifts, which may be related to reorganizations in the ocean's circulation, could be a manifestation of chaotic behavior of the climate system (Broecker, 1995). This situation probably limits our ability to predict future long-term climate evolution.

## 15.4 Impact of Anthropogenic Trace Gases on Climate

Although water vapor is the most important greenhouse gas, its distribution in the atmosphere is mainly driven by physical processes, but is only slightly affected by human activities (*e.g.*, deforestation on a large scale, which can matter regionally). Human activities are, however, responsible for significant changes in the abundance of other radiative gases, as seen in previous chapters.

Table 15.1 provides an estimate of the greenhouse gas concentrations and trends; the origin and magnitude of these trends are further discussed in Chapters 5, 8, and 9.

**Table 15.1**
*Assumed Evolution of Trace Gas Mixing Ratios at the Surface Level*

|  | Preindustrial | 1960 | 1980 | 1990 |
|---|---|---|---|---|
| $CO_2$ (ppmv) | 295 | 315 | 325 | 352 |
| $CH_4$ (ppbv) | 975 | 1270 | 1570 | 1675 |
| $N_2O$ (ppbv) | 290 | 300 | 303 | 310 |
| CFC-11 (pptv) | 0 | 11 | 173 | 275 |
| CFC-12 (pptv) | 0 | 33 | 297 | 468 |
| CFC-113 (pptv) | 0 | 0.2 | 15 | 51 |
| CFC-114 (pptv) | 0 | 0.2 | 4 | 7 |
| CFC-115 (pptv) | 0 | 0 | 2 | 5 |
| $CH_3CCl_3$ (pptv) | 0 | 5 | 105 | 150 |
| $CCl_4$ (pptv) | 0 | 75 | 95 | 105 |
| HCFC-22 (pptv) | 0 | 1 | 54 | 111 |
| Halon-1211 (pptv) | 0 | 0 | 0.5 | 1.5 |
| Halon-1301 (pptv) | 0 | 0 | 0.6 | 1.7 |
| $CH_3Cl$ (pptv) | 600 | 600 | 600 | 600 |
| Total $Cl_x$ (ppbv) | 0.6 | 1.0 | 2.5 | 3.5 |

From IPCC, 1990; WMO, 1990b.

Changes in the distributions of the trace gases are expected to affect the climate system through spatial and temporal changes in the flux of radiative energy into and out of the surface-troposphere system. This impact can, in principle, be quantified by calculating the induced change in the surface temperature; however, this quantity is largely dependent on complex feedback processes, which are not fully understood nor easily represented or verified in existing climate models. It is more straightforward to calculate the radiative forcing. This quantity is defined as the response in the net radiative energy flux at the tropopause to changes in the concentration of a given trace gas. Several factors determine the ability of an atmospheric gas to affect the radiative forcing: its atmospheric concentration, the strength and spectral position of its absorption bands, temperature, and pressure.

The radiative forcing associated with changes in the abundance of trace gases can be estimated by radiative models. The solution of the radiative transfer equations is complex since the absorption spectra of atmospheric molecules exhibit structures characterizing their numerous rotation and vibration-rotation lines (see Box 15.2). The absorption by these molecules varies considerably over small wavelength regions, and exact calculations require line-by-line integrations. However, such calculations require very large amounts of computer time. This direct approach is usually replaced by an approximate method in which the line or band characteristics are expressed in terms of global parameters. Radiative models have been developed to treat either sections of bands (narrow-band model) or entire bands (wide-band models) (Tiwari, 1978). The radiative codes included in multidimensional models (2D and 3D) use generally one of these two approaches.

*15.4.1 Direct Radiative Effects*

$CH_4$, $O_3$, $N_2O$, CFC-11, CFC-12, and various CFCs and HCFCs have strong absorption bands in the atmospheric window region. These trace gases absorb and emit radiation in bands composed of discrete lines with extended wings. For gases that are present in small quantities, such as the CFCs and HCFCs, the absorption increases quasilinearly with their atmospheric concentration. However, for gases with larger concentrations, such as methane and nitrous oxide, the absorption at the center of the bands is already saturated for present atmospheric abundances, and increasing absorption with increasing concentrations occurs mainly in the wings of the lines. In this case the radiative forcing is approximately proportional to the square root of the concentration change. For the most abundant radiatively active species such as $CO_2$, the atmosphere is almost entirely opaque in the center of the absorption lines, and the radiative effect of adding $CO_2$ is only noticeable in the wings of the lines. In this case, the absorption can be approximated by a logarithmic relationship with the $CO_2$ concentration. A doubling in the atmospheric abundance of $CO_2$ leads to an increase in the radiative forcing of about 4.6 W m$^{-2}$.

Simplified expressions providing the direct radiative forcing as a function of the greenhouse gases concentrations are given in Table 15.2 for both clear sky and average cloudiness conditions. These formulas are only approximate and should be applied only for small changes in the concentrations.

### Table 15.2
*Simplified Expressions to Calculate Radiative Forcing $\Delta F_R$ (W m$^{-2}$) Resulting from Changes in Atmospheric Concentrations*

1. $CO_2$: $\Delta F_R = a \ln(\mu/\mu_0)$ where $\mu$ and $\mu_0$ are new and initial $CO_2$ mixing ratios.

2. $CH_4$: $\Delta F_R = a \left[ \sqrt{\mu\,(CH_4)} - \sqrt{\mu_0\,(CH_4)} - f\left(\mu\,(CH_4), \mu_0\,(N_2O)\right) \right.$
   $\left. - f\left(\mu_0\,(CH_4), \mu_0\,(N_2O)\right) \right]$, where $\mu\,(CH_4)$ and $\mu_0\,(CH_4)$ are new and initial $CH_4$ mixing ratios (ppbv) and $\mu_0\,(N_2O)$ is the $N_2O$ mixing ratio (ppbv).

3. $N_2O$: $\Delta F_R = a \left[ \sqrt{\mu\,(N_2O)} - \sqrt{\mu_0\,(N_2O)} - f\left(\mu_0\,(CH_4), \mu\,(N_2O)\right) \right.$
   $\left. - f\left(\mu_0\,(N_2O), \mu_0\,(CH_4)\right) \right]$, where $\mu\,(N_2O)$ and $\mu_0\,(N_2O)$ are new and initial $N_2O$ mixing ratios (ppbv) and $\mu_0\,(CH_4)$ is the $CH_4$ mixing ratio (ppbv).

4. CFCs and HCFCs: $\Delta F_R = a\,(\mu - \mu_0)$, where $\mu$ and $\mu_0$ are new and initial mixing ratios (ppbv), respectively.

Function $f(x,y)$ is defined by (IPCC, 1990)
$$f(x,y) = 0.47 \ln\left[1 + 2.01 \times 10^{-5}\,(xy)^{0.75} + 5.31 \times 10^{-15} x\,(xy)^{1.52}\right]$$

*Values of factor a:*

|  | Clear sky | Cloudy conditions |
|---|---|---|
| $CO_2$ | 7.6 | 6.7 |
| $CH_4$ | 0.044 | 0.036 |
| $N_2O$ | 0.16 | 0.14 |
| CFC-11 | 0.27 | 0.21 |
| CFC-12 | 0.35 | 0.27 |
| CFC-113 | 0.38 | 0.29 |
| HCFC-22 | 0.20 | 0.15 |

Values of factor $a$ calculated by the 2D model of Brasseur et al. (1990). In the case of cloudy conditions, fractional cloud amounts are 0.23 for high clouds (10 km), 0.09 for middle clouds (4 km), and 0.31 for low clouds (2 km). Optical depths are 1.5, 6.1, and 16.3, respectively (from Manabe and Wetherald, 1967).

**Table 15.3**
*Direct and Indirect Radiative Forcing Relative to $CO_2$ per Unit Molecule Change and per Unit Mass Change for Current Concentrations (1990 Values)*

Direct radiative forcing relative to $CO_2$ per unit molecule change

|  | Clear sky | Clouds |
|---|---|---|
| $CO_2$ | 1 | 1 |
| $CH_4$ | 25 | 23 |
| $N_2O$ | 213 | 210 |
| CFC-11 | 13000 | 11700 |
| CFC-12 | 16800 | 14900 |
| CFC-113 | 18500 | 16000 |
| HCFC-22 | 9700 | 8300 |

Indirect radiative forcing relative to $CO_2$ per unit molecule change

|  | Clear sky | Clouds |
|---|---|---|
| $CO_2$ | 1 | 1 |
| $CH_4$ | 42 | 41 |
| $N_2O$ | 250 | 240 |
| CFC-11 | 8000 | 6900 |
| CFC-12 | 16600 | 15000 |

A useful index also used to evaluate the radiative impact of increasing greenhouse gases concentrations is provided by the relative radiative forcing. This index provides a direct comparison between the direct radiative forcing of a greenhouse gas and that of $CO_2$ (chosen here as the reference molecule). Values of the direct relative radiative forcing for several greenhouse gases are given in Table 15.3 for a 1990 reference atmosphere, for both clear sky and cloudy conditions, calculated for a 10% increase in the concentrations of the more abundant greenhouse gases. It should be emphasized that since radiative fluxes do not change linearly with the concentrations of trace gases, the calculated relative radiative forcing depends on the background concentrations for the greenhouse gases, as well as the magnitude of the changes applied to these concentrations. As an example, the relative forcing of methane as compared to $CO_2$ would be 24.4 for an increase of 1% in the concentrations of both $CO_2$ and $CH_4$, while it would be 27.6 for a doubling in the concentrations of both $CH_4$ and $CO_2$.

### 15.4.2 Indirect Effects: Chemical Feedbacks

Many climate models including general circulation models (GCMs), which are used to predict climate changes, use $CO_2$ as a proxy for other greenhouse gases and often estimate climate changes for a doubling in the equivalent $CO_2$ concentrations. Most of them currently do not explicitly account for the greenhouse effect of other trace gases. The use of a $CO_2$ proxy to represent the combined greenhouse forcing of $CO_2$ and the other radiatively active trace gases is questionable due to the differences in the spectral and chemical properties of all gases involved. For example, Figure 15.6 shows the difference in the heating rate calculated for a doubling of the $CO_2$ concentration (Fig. 15.6a) and an explicit treatment of the increase in the other trace gases (Fig. 15.6b). Compared with a doubling of the $CO_2$ concentration, the inclusion of the radiative effect of the other trace gases results in a value of the heating rate stronger in the lower stratosphere and much lower at the surface for high latitudes.

Gases such as $CH_4$, $N_2O$, and the CFCs are not only radiatively active, but they also produce chemical perturbations in the atmosphere and hence affect the abundance of other greenhouse gases. The oxidation of methane, for example, leads to an additional production of water vapor in the stratosphere and ozone in the troposphere. The breakdown of $N_2O$ and CFCs in the stratosphere leads to the production of active nitrogen or chlorine radicals that destroy ozone. $CO_2$ is chemically inactive in the atmosphere, but an increase in the $CO_2$ concentration and in the associated emission to space of the 15 $\mu$m radiation is expected to produce a cooling of the stratosphere and mesosphere. As the production and loss rates of ozone are strongly temperature

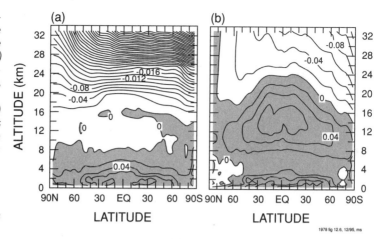

**Figure 15.6.** Variation in the longwave radiative heating rate (in degrees C per day) calculated in a climate model for (a) a doubling in the $CO_2$ concentration and (b) the explicit treatment of the increase of the other trace gases (from Wang et al., 1991; reprinted with permission from Nature, copyright 1991, Macmillan Magazines Limited).

dependent in the middle atmosphere, a $CO_2$ increase is expected to moderate the ozone destruction caused by chlorine and nitrogen compounds. Furthermore, a cooling of the winter polar stratosphere resulting from increasing $CO_2$ concentrations could lead to the formation of additional polar stratospheric clouds, which are associated with the observed dramatic destruction of ozone in the Antarctic polar stratosphere.

As a result of all these processes, stratospheric ozone could decrease globally, and more solar radiation could become available in the troposphere, leading to a warming of the surface. Less terrestrial radiation would be absorbed by ozone, leading in this case to a cooling of the surface. The net effect of stratospheric ozone changes on the climate system depends strongly on the magnitude and altitude of these changes. Moreover, all these changes could induce a modification of the circulation in the stratosphere, and thus affect the transport of other trace gases.

Human activities could also lead to an increase in the concentrations of ozone in the troposphere with potential impact (absorption of solar and terrestrial radiation) on the climate system. Such changes probably have a larger impact on the radiative forcing than those produced by ozone depletion in the stratosphere.

Increased emissions of trace gases at the Earth's surface could have a significant impact on the climate system (temperature, precipitation, frequency of extreme events), but the resulting effects are difficult to quantify because of strong non-linearities in the coupled chemical and climate systems. The available estimates of potential climate changes produced by perturbations in the chemical composition of the atmosphere are provided by interactive chemical-radiative-dynamical models.

The importance of these chemical feedbacks on the radiative forcing of the atmosphere is illustrated in Table 15.3. The indirect relative forcing of greenhouse gases has been derived by using an interactive two-dimensional model that is run to steady state. As for the calculations of the direct radiative forcing, a 1990 reference atmosphere is assumed; for each individual gas, a 10% increase in the background concentration is applied at the surface level.

Another illustration of the importance of chemical processes on the climate forcing is given in Figure 15.7. This figure represents the changes from 1900 to 1990 in the radiative forcing (1) when only considering direct radiative effects and (2) when chemical feedbacks are taken into account. The changes in the concentration of the greenhouse gases at the surface from 1900 to 1990 used in these calculations is given in Table 15.1. When chemical feedbacks are taken into account, the ozone produced as a result of the $CH_4$ release has a significant radiative effect, which is as strong as the direct radiative effect of methane.

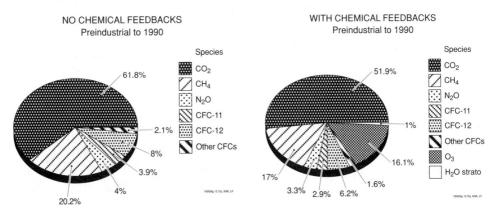

**Figure 15.7.** Contribution to the radiative forcing due to increases in greenhouse gas concentrations for the period 1900-1990. Note that these values, especially those including chemical feedbacks, are rather uncertain because of the strong nonlinearities in the coupled chemical and climate systems.

## 15.5 Global Warming Potentials (GWPs)

The radiative forcing provides an estimate of the change in the radiative flux at the tropopause in response to changes in the concentration of greenhouse gases. In order to take into account the lifetime of the gases in the atmosphere, and hence the period of time over which the climatic effect of a perturbation in their concentration is expected to be significant, an index called the Global Warming Potential (GWP) was defined. This concept was created in order to enable decision makers to evaluate options to regulate future emissions of various greenhouse gases without having to perform complex model calculations.

The GWP of a well-mixed gas is defined (IPCC, 1990) as the time-integrated change in the radiative forcing due to the instantaneous release of 1 kg of a trace gas $i$ expressed relative to that from the release of 1 kg of $CO_2$

$$\text{GWP} = \frac{\int_0^T \Delta F_{R,i}(t)\,dt}{\int_0^T \Delta F_{R,CO_2}(t)\,dt} \tag{15.1}$$

if $\Delta F_R$ represents the change in the forcing at the tropopause and $T$ is the time over which the integration is performed (time horizon).

Using a linear approximation,

$$\text{GWP} = \frac{\int_0^T a_i n_i(t)\,dt}{\int_0^T a_{CO_2} n_{CO_2}(t)\,dt} \tag{15.2}$$

where $a_i$ (expressed in W m$^{-2}$ kg$^{-1}$) is the instantaneous radiative forcing due to the increase in the concentration of trace gas $i$ and $n_i$ is the concentration of the gas $i$ remaining at time $t$ after the release (IPCC, 1990). $a_{CO_2}$ and $n_{CO_2}$ are the corresponding variables applied to $CO_2$, which is considered the reference gas. If $\tau_i$ is the lifetime of the molecule $i$ and $\tau$ an "effective" residence time for $CO_2$, the GWP of the gas $i$ can be approximated by

$$\text{GWP} = \frac{a_i \int_0^T e^{-t/\tau_i}\,dt}{a_{CO_2} \int_0^T e^{-t/\tau_{CO_2}}\,dt} = \frac{a_i \tau_i}{a_{CO_2} \tau_{CO_2}} \frac{1 - e^{-T/\tau_i}}{1 - e^{-T/\tau_{CO_2}}} \tag{15.3}$$

As indicated in the above expression, the estimation of the GWP for a trace gas requires estimates of the radiative forcing for the trace gas $i$ and for the reference gas $CO_2$ per unit of mass change, the lifetimes of species $i$ and of $CO_2$, and the definition of the time horizon $T$ over which the integration is performed. The indirect chemical effects resulting from the increase in the concentration of species $i$ also need to be evaluated.

The choice of the time horizon $T$ depends on the type of climate impact under consideration. As each response has its own characteristic time, there is no single universally accepted value of the time horizon $T$ that can be adopted. Table 15.4 illustrates the integration periods that are appropriate for different climatic responses.

**Table 15.4**
*Characteristic Integration Periods Appropriate for Different Climate Change Interactions*

| Climate Change Indicator | Integration Time Period |
|---|---|
| Maximum change in temperature | $\sim$ 100 years |
| Rate of change in temperature | $\sim$ 20-50 years |
| Maximum change in sea level | > 100 years |
| Rate of change in sea level | > 50 years |

From WMO, 1991.

**Table 15.5**

| Species | Formula | Lifetime (yrs) | Global Warming Potential | | |
|---|---|---|---|---|---|
| | | | $T$=20 years | 100 years | 500 years |
| Carbon dioxide | $CO_2$ | | 1 | 1 | 1 |
| Methane | $CH_4$ | 12-18[a] | 48-90 | 20-43 | 8-15 |
| Nitrous oxide | $N_2O$ | 121 | 290 | 330 | 180 |
| CFC-11 | $CFCl_3$ | 50 | 5000 | 4000 | 1400 |
| CFC-12 | $CF_2Cl_2$ | 102 | 7900 | 8500 | 4200 |
| CFC-113 | $CF_2ClCFCl_2$ | 85 | 5000 | 5000 | 2300 |
| HCFC-22 | $CF_2HCl$ | 13.3 | 4300 | 1700 | 520 |
| HCFC-142b | $CH_3CF_2Cl$ | 19.5 | 4200 | 2000 | 630 |
| Carbon tetrachloride | $CCl_4$ | 42 | 2000 | 1400 | 500 |
| Methyl chloroform | $CH_3CCl_3$ | 5.4 | 360 | 110 | 35 |
| HFC-134a | $CH_2FCF_3$ | 14 | 3300 | 1300 | 420 |
| Halon-1301 | $CF_3Br$ | 65 | 6200 | 5600 | 2200 |

[a]Includes the dependence of the methane abundance on the response time (indirect effect associated with changes in OH density).

As discussed in Chapter 5, the atmospheric abundance of $CO_2$ is regulated by the cycling of carbon between several biogeochemical reservoirs (atmosphere, ocean, biosphere). A single global residence time of $CO_2$ in the atmosphere cannot be derived. Carbon dioxide added to the atmosphere decays relatively rapidly over the first 10 years, with a more gradual decay over the next 100 years and a very slow decline over the 1000 year time scale. Expressions have been deduced from ocean-atmosphere-biosphere models that provide the decay of a perturbation in atmospheric $CO_2$ as a function of time. The study by Maier-Reimer and Hasselmann (1987), for example, provides an effective residence time of approximately 120 years for atmospheric $CO_2$. Typical global lifetimes of other trace gases are given in Table 15.5.

Table 15.5 presents a recent estimate of GWPs provided by IPCC (1995). GWPs were calculated by injecting a finite amount of trace gases to the abundance of the background atmosphere and by calculating the radiative response over several time horizons (20, 100, 500 years). The model of Siegenthaler and Joos (1992) was used to estimate the decay response of $CO_2$. For these calculations, the background atmospheric trace gas concentrations were held fixed (at current levels) and did not account for a possible future evolution of the atmospheric composition. Note that the decay time of a methane pulse (12-18 years) is higher than the global lifetime of this gas, since the concentration of OH decreases as $CH_4$ is added to the atmosphere, and the resulting GWP for $CH_4$ is higher than the direct GWP. The estimate of all indirect effects on the GWP (*e.g.*, changes in ozone, water vapor, temperature, etc.) is not straightforward, and is generally model dependent.

## 15.6 Radiative Effects of Aerosols
### 15.6.1 Direct Effects

Aerosols present in the atmosphere (including sulfate particles resulting from fossil fuel combustion and elemental carbon, EC, released by biomass burning; see Color Plate 12) absorb and scatter a significant fraction of incoming solar radiation back to space. The addition of anthropogenic (not EC) and volcanic aerosols to the atmosphere leads therefore to a reduction in the net radiation available at the surface and so to a cooling of the Earth's system. Aerosols also absorb terrestrial radiation and thereby produce a significant heating in dense aerosol layers.

An interesting modeling study of direct forcing by sulfate aerosols and comparison with greenhouse gas forcing has been presented by Kiehl and Briegleb (1993). Best estimates from observations and modeling studies of sulfate aerosol loadings and distributions were combined with imposed lognormal aerosol distributions (as discussed in Chapter 4) and derived optical properties, and the sensitivity of their findings to these aerosol parameters were examined (Fig. 15.8). They estimate that variations in size or chemical composition would alter the estimated forcing ($-0.3$ W m$^{-2}$, annually averaged) by $\pm 10\%$. The spatial distribution of aerosol properties may have a larger effect. This is because the greenhouse gas forcing occurs in different regions of the globe than does the anthropogenic aerosol forcing, which is strongest in the midlatitudes of the Northern Hemisphere, where most of the sources are located. In contrast, greenhouse gases, except ozone, generally become well mixed in the troposphere; their radiative effects are strongest in the region between $-30°$ and $+30°$ latitude. The combined effects of aerosols and greenhouse gases thus do not "cancel," but may change global temperature gradients.

It is known that organic matter can comprise a significant fraction of the tropospheric aerosol, and thus must also have a role in postulated climate effects. There is also substantial evidence that some of these species are hygroscopic and thus should contribute to indirect climate effects as well (Novakov and Penner, 1993). Soot has been detected in all regions of the globe, even in "remote" areas. Its strong solar radiation absorption characteristics suggest that climate forcing due to suspended soot aerosol will have a sign opposite that of sulfate. The net radiative forcing of a mixture of sulfate and soot could therefore be substantially smaller than the forcing calculated for sulfate only.

Widespread dust plumes are often detected in satellite images, and it might be expected that dust contributes to aerosol radiative forcing. Mineral dust aerosol may

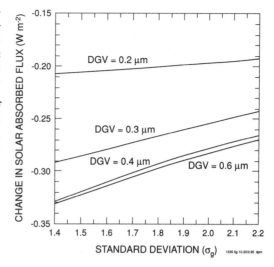

Figure 15.8. Variation of the globally averaged direct sulfate aerosol forcing (W m$^{-2}$) with geometric standard deviation ($\sigma_g$) for four different geometric mean diameters ($\mu$m). DGV = geometric mean diameter by volume (reprinted with permission from Kiehl, J. T. and B. P. Briegleb (1993) The relative roles of sulfate aerosols and greenhouse gases in climate forcing, Science 260, 311, ©1993, American Association for the Advancement of Science).

both scatter and absorb solar radiation, depending upon its composition and the wavelength of light considered. Sokolik et al. (1993) compared measurements of the complex refractive index for atmospheric dust aerosols and showed that the large range of values for the imaginary part of the refractive index leads to significant differences in estimates of radiative forcing. The uncertainty is magnified when one considers the effects of the presence of other suspended material (e.g., soot). Rather than dust inducing a significant effect on climate, the major impacts may follow in the opposite direction; that is, climate change may significantly affect dust production and transport. Changes in aridity in North Africa and shifts in large-scale atmospheric circulation patterns associated with climate change may alter the magnitude and pattern of Saharan dust transport to the North Atlantic (Arimoto et al., 1992).

Large volcanic eruptions, such as those of El Chichón (1982) and Mt. Pinatubo (1991), have substantially enhanced the aerosol load of the atmosphere for a few years, resulting in a noticeable cooling of the surface. One year after the eruption of Mt. Pinatubo the radiation forcing was estimated to have been –4 W m$^{-2}$ while an anomaly of –0.3 to –0.4°C in the global temperature was reported (Dutton and Christy, 1992; IPCC, 1995). Such volcanic perturbations are, however, transitory, with a typical time constant of 1-2 years.

The radiative forcing produced by the enhanced anthropogenic sulfate aerosol burden since preindustrial times is estimated (on the global scale) to be approximately –0.6±0.3 W m$^{-2}$ (IPCC, 1995). Because of their relatively limited lifetime (a few days), anthropogenic aerosols are mostly concentrated in industrialized regions (eastern United States, Europe, eastern Asia), where their radiative impact is believed to be significant. Although on the global scale their radiative impact is considerably smaller than the forcing caused by anthropogenic greenhouse gases (approximately 2.5 W m$^{-2}$), in industrialized areas the cooling caused by aerosols exceeds the warming produced by enhanced $CO_2$ and other radiatively active gases (Fig. 15.9).

Sulfate aerosols also serve as cloud condensation nuclei (CCN) and hence affect the formation and the radiative properties of clouds. This indirect climate impact of anthropogenic sulfur remains poorly quantified, but could be as large as or even larger than the direct forcing by aerosols of human origin.

An understanding of the role of aerosol mixtures in the past and for the present day, and extensions to predict climate change, is hindered by large uncertainties in many key quantities needed for such estimates. Quantification of major uncertainties and a proposal for strategies for minimizing them are presented in the review by Penner et al. (1993) (Table 15.6).

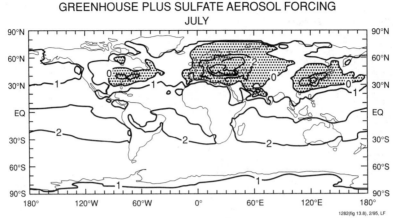

**Figure 15.9.** Radiative forcing in July (W m$^{-2}$) due to increases in greenhouse gases (preindustrial to present) and sulfate aerosols (Kiehl and Rodhe, 1995) since the preindustrial period. The greenhouse gases are $CO_2$, $CH_4$, $N_2O$, CFC-11, and CFC-12. Sulfate aerosol distribution from Pham et al. (1995). Contour interval 2 W m$^{-2}$ for negative values, 1 W m$^{-2}$ for positive values (reprinted with permission from Kiehl, J. T. and B. P. Briegleb (1993) The relative roles of sulfate aerosols and greenhouse gases in climate forcing, Science 260, 311, ©1993, American Association for the Advancement of Science).

**Table 15.6**
*Factors Contributing to an Estimate of the Direct Forcing by Anthropogenic Sulfate Aerosols and Their Estimated Range and Uncertainty Factors*

| Quantity | Central Value | Estimated Range | Uncertainty Factor |
|---|---|---|---|
| Aerosol mass scattering efficiency (m$^2$ g$^{-1}$) | 5 | 3.6–7 | 1.40 |
| Average lifetime of atmospheric $SO_4^{2-}$ (yr) | 0.016 | 0.012–0.022 | 1.38 |
| Aerosol hemispheric backscatter fraction | 0.15 | 0.12–0.19 | 1.27 |
| Fraction of $SO_2$ oxidized to $SO_4^{2-}$ aerosol | 0.5 | 0.4–0.6 | 1.20 |
| Proportionality coefficient (W m$^{-2}$) | 489 | 406–589 | 1.20 |
| Fractional increase in aerosol scattering efficiency due to hygroscopic growth | 1.7 | 1.4–2.0 | 1.20 |
| Source strength of anthropogenic S (Tg yr$^{-1}$) | 71 | 62–81 | 1.14 |
| Fraction of Earth not covered by cloud | 0.39 | 0.35–0.44 | 1.13 |
| Square of surface co-albedo | 0.72 | 0.65–0.80 | 1.11 |

Total uncertainty factor $= e^{[\text{sum squared log of uncertainty factor}]^{\frac{1}{2}}} = 1.89$

*Result:* If central value is 0.6 W m$^{-2}$, the range is from 0.3 to 1.1 W m$^{-2}$.
From Penner et al., 1993.

### 15.6.2 Indirect Effects

The subset of atmospheric aerosols active as CCN may have an "indirect effect" on climate by altering the albedo of clouds. Changes in the availability of CCN may change nucleated droplet number concentrations. As droplet number concentrations increase for fixed liquid water content, the mean droplet size decreases and the reflectivity of the cloud increases; the global energy balance is sensitive to such changes. Studies of potential "indirect" aerosol effects have focused upon the role of marine stratocumulus clouds, in part because of their ubiquity (they cover about 25% of the Earth's surface) and because the potential for perturbations to these clouds is high. Marine clouds generally have low droplet concentrations (on the order of 100 cm$^{-3}$), believed to be limited by the availability of CCN. Any process, then, that alters the relative amounts of CCN in marine regions may affect the albedo of these clouds. In contrast, clouds formed over continental regions are believed to have an excess of CCN available, and the number activated is most likely related to other factors such as the maximum supersaturation. However, there is observational evidence for a dependence of continental cloud drop concentration on aerosol loading (*e.g.*, Leaitch *et al.*, 1992). Han *et al.* (1994) derived effective cloud drop radii from satellite data and reported systematic differences in drop size between continental and marine water clouds and between marine clouds in the Northern and in the Southern Hemispheres. Smaller drop radii were found in those regions most affected by anthropogenic pollution, in support of the "indirect effect" hypothesis.

The importance of sulfate to aerosol and CCN concentrations led to the interesting hypothesis of Charlson *et al.* (1987) of a climate feedback loop involving marine phytoplankton, DMS emissions, CCN concentrations, and cloud radiative forcing. This work spurred much of the subsequent research and debate regarding indirect climate effects of aerosols. Some estimates suggest that a 30% change in CCN available to marine stratus will lead to a globally averaged forcing of 1 W m$^{-2}$. However, changes in CCN populations may have other effects that also influence climate. For example, enhanced droplet concentrations may reduce the likelihood of precipitation from clouds, altering cloud cover and cloud lifetime (Radke *et al.*, 1989). The response of cloud liquid water content to changes in CCN and climate is not well understood. Changes in precipitation would also change the atmospheric concentration of the most important greenhouse gas: water vapor.

Climate effects of aerosols are also postulated for polar regions. The climate of polar regions is of great interest in studies of global warming. As temperatures increase the extent of snow and ice is reduced, decreasing the surface albedo and further increasing the amount of sunlight that is absorbed by the Earth-atmosphere system. Conversely, a temperature decrease will increase the surface albedo and thus reinforce the cooling (*e.g.*, Curry *et al.*, 1993). This feedback mechanism results in the Arctic having an impact on the global climate as well as the local climate, since the ice-albedo feedback mechanism can result in substantial modification of the net energy retained by the Earth-atmosphere system.

Several types of polar aerosol effects are postulated. Soot aerosols deposited to snow and ice surfaces may alter their albedo in a "direct" effect. An "indirect effect" for polar ice-phase clouds (present even in the lower troposphere during the coldest months of the year) may also occur via the following mechanism. Polluted air has been typically shown to be deficient in ice-forming nuclei (IN). This relationship is believed to arise from an increased sulfate mass loading in polluted air; sulfate particles, which are poor IN, coagulate with potential IN and effectively deactivate them (Borys, 1989). If this hypothesis is correct, ice nucleation in the Arctic may be relatively enhanced during winter, if there is a decrease in the oxidation of $SO_2$ in the relative absence of sunlight and liquid water, which could result in a decreased amount of sulfate particles. Conversely, ice nucleation during "Arctic haze" events in the Spring would be suppressed. Thus, anthropogenic aerosol has the potential to impact the amount of condensed water and the total water budget in the Arctic by modifying the ice nucleation and the phase of condensed water.

## 15.7 Response of the Climate System to Radiative Forcing

The simplest model to estimate the response of the Earth's climate to radiative perturbations is based on the global balance between the incoming solar energy ($F_S$) that is absorbed by the Earth system and the outgoing terrestrial radiative energy ($F_T$) that is emitted to space. For the system to be at equilibrium

$$F_S = F_T \qquad (15.4)$$

If the system is perturbed by some radiative forcing $\Delta F_R$ (e.g., by an increase in the atmospheric abundance of carbon dioxide), the equilibrium between incoming and outgoing energy will be restored by a change ($\Delta$) in the initial fluxes ($F_S$ and $F_T$), such that

$$\Delta (F_T - F_S) = \Delta F_R \qquad (15.5)$$

Assuming that the balance is re-established by a change in surface temperature ($\Delta T_s$), the *climate sensitivity factor* $\lambda_c$, defined by

$$\Delta T_s = \lambda_c \Delta F_R \qquad (15.6)$$

is simply given by

$$\lambda_c = \left( \frac{\Delta F_T}{\Delta T_s} - \frac{\Delta F_S}{\Delta T_s} \right)^{-1} \qquad (15.7)$$

To make a first-order estimate of this factor, we assume that the globally averaged solar energy absorbed by the Earth system is given by

$$F_S = \frac{F_0}{4} (1 - \alpha) \qquad (15.8)$$

where $\alpha$ is the Earth's albedo (typically 0.3) and $F_0 = 1370$ W m$^{-2}$ is the solar constant (the averaged solar energy intercepted by a sphere is $F_0/4$). Similarly, we assume that the terrestrial energy radiated to space is expressed by the Stefan-Boltzmann law

$$F_T = \epsilon \sigma T_s^4 \qquad (15.9)$$

where $\sigma$ is the Stefan-Boltzmann constant, $\epsilon$ the emissivity of the atmosphere, and $T_s$ the surface temperature. Assuming that $\alpha$ and $\epsilon$ are constant, and neglecting all potential feedbacks in the climate system, one derives from Eqs. (15.7-15.9) that

$$\lambda_c = \left( 4\epsilon\sigma T_s^3 \right)^{-1} = \frac{T_s}{4F_T} \qquad (15.10)$$

Under these assumptions, the value of the climate sensitivity factor is approximately equal to 0.3 K (W m$^{-2}$)$^{-1}$ (Kiehl, 1992). Thus, for a perturbation associated with a doubling in the $CO_2$ abundance ($\Delta F_R = 4.6$ W m$^{-2}$), the increase $\Delta T_s$ in the mean Earth's temperature is 1.4 K, that is, significantly less than predicted by climate models or derived from satellite observations [$\lambda_c \simeq 0.6$ K (W m$^{-2}$)$^{-1}$].

The reason for this discrepancy is that important climate feedbacks have been ignored in this simple calculation. Examples of such feedbacks include those associated with the hydrological cycle. As a result of enhanced radiative forcing, the warmer atmosphere becomes more humid (Clausius-Clapeyron relation, e.g., see Hartmann, 1994), which produces an additional greenhouse forcing and hence a larger warming of the planet. Another positive feedback is produced by changes in the surface albedo [$\alpha$ in Eq. (15.8)] when the surface area covered by ice and snow varies in response to climate change. Feedbacks caused by changes in cloudiness in response to global warming/cooling are difficult to estimate since clouds reflect solar radiation back to space (cooling effect) and, at the same time, reduce the emission to space of terrestrial radiation (warming effect). Finally, the assessment of potential feedbacks involving the biosphere remains an important research topic.

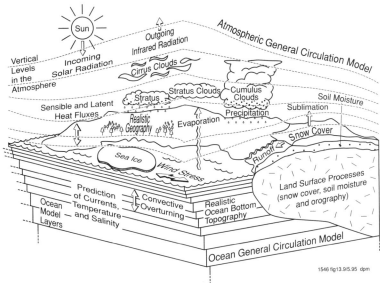

**Figure 15.10.** *Representation of physical processes in a coupled ocean-atmosphere climate model, including energy exchange, precipitation, and evaporation on land and oceans, ice and snow (courtesy of W. Washington).*

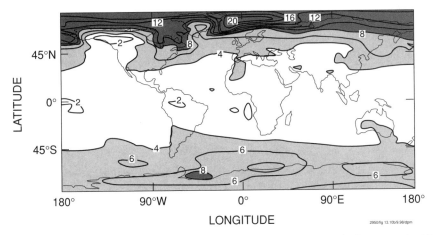

**Figure 15.11.** *Change in surface air temperature (10 year average) due to a doubling in the atmospheric concentration of $CO_2$ (mean values for December, January and February), calculated using the climate model of the Geophysical Fluids Dynamics Laboratory by Manabe and Wetherald (IPCC, 1990).*

Modern climate models account for many of the relevant feedback mechanisms. When used in a predictive mode, these models attempt to simulate the transient response of the climate system to changes in the radiative forcing. This transient response is determined by the thermal inertia of the system, that is, the effective heat capacity of the atmosphere, land, and ocean, as well as the radiative damping of the system (Schneider, 1992). The response of climate to a gradual increase in the atmospheric abundance of radiatively active gases can therefore be modeled accurately only with coupled ocean-atmosphere models. A representation of a complex climate model is shown in Figure 15.10. Such models show that the existence of feedbacks associated with water vapor, clouds, and ice tends to increase the value of the $\lambda_c$ factor to a value ranging between 0.4 and 1.25 K $(W\ m^{-2})^{-1}$ depending on the formulation in the model of the feedback processes. Thus, with these values of the feedback factor, the warming (at equilibrium) caused by a doubling in the atmospheric concentration of $CO_2$ should range from 1.8 to 5.8 K. The change in temperature for a $CO_2$ doubling, however, is not uniform in space and, as shown by Figure 15.11, is expected to be largest at high latitudes in winter. Improvements in climate predictions require, therefore, that physical, chemical, and biological processes be better understood and more accurately represented in the models.

## Further Reading

Brasseur, G., ed. (1997) *The Stratosphere and Its Role is the Climate System*, NATO ASI Series I-54, Springer-Verlag, Berlin.

Calvert, J., ed. (1994) *The Chemistry of the Atmosphere: Its Impact on Global Change*, Blackwell Scientific Publications, Oxford.

Goody, R. (1995) *Principles of Atmospheric Physics and Chemistry*, Oxford University Press, New York.

Goody, R. M., and Y. L. Yung (1989) *Atmospheric Radiation. Theoretical Basis*, Oxford University Press, Oxford and New York.

Hartmann, D. L. (1994) *Global Physical Climatology*, Academic Press, San Diego.

Intergovernmental Panel on Climate Change, IPCC (1990) *Climate Change*, J. T. Houghton, G. J. Jenkins, and J. J. Ephraums, eds., Cambridge University Press, Cambridge, UK.

Intergovernmental Panel on Climate Change, IPCC (1992) *Climate Change, 1992*, J. T. Houghton, B. A. Callander, and S. K. Varney, eds., Cambridge University Press, Cambridge, UK.

Intergovernmental Panel on Climate Change, IPCC (1995) *Climate Change: The IPCC Scientific Assessment*, J. T. Houghton, L. G. Meira Filho, J. Bruce, Hoesung Lee, B. A. Callander, E. Haites, N. Harris and K. Maskell, eds., Cambridge University Press, Cambridge, UK.

Intergovernmental Panel on Climate Change, IPCC (1996) *Climate Change, 1995*, J. T. Houghton, L. G. Meira Filho, B. A. Callander, N. Harris, A. Kattenberg, and K. Maskell, eds., Cambridge University Press, Cambridge, UK.

Kandel, R. (1990) *Our Changing Climate*, McGraw Hill, New York.

Peixoto, J. P. and A. H. Oort (1992) *Physics of Climate*, American Institute of Physics.

Ramanathan, V., L. Callis, R. Cess, J. Hansen, I. Isaksen, W. Kuhn, A. Lacis, F. Luther, J. Mahlman, R. Reck, and M. Schlesinger (1987) Climate-chemical interactions and effects of changing atmospheric trace gases, *Rev. Geophys.*, 25, 1441.

Schneider, S. H. (1989) *Global Warming*, Sierra Club Books, San Francisco.

Wuebbles, D. J. and J. Edmonds (1991) *Primer on Greenhouse Gases*, Lewis Publishers, Chelsea, Michigan.

Stephen H. Schneider

Can Climate Models Be Validated?

Chapter 15 provides an excellent introduction to the radiative forcing and feedback components of the climate change debate. Whether the sensitivity of the climate turns out closer to the higher or lower end of the often cited 1-5°C warming range for doubling of $CO_2$ is very critical for the impact and policy debates. Damages to natural or managed ecosystems, water supplies, or coastlines depend significantly on both the magnitude and rates of climate change. Since projection of such changes centers on climatic models of the main interacting subcomponents of the Earth system (*i.e.*, atmosphere, oceans, ice, surfaces, and biota), and since such coupled models contain many inherent uncertainties, evaluation of the reliability of these models becomes a research problem of prime importance.

"Validation" of climate model forecasts before the fact is not possible, strictly speaking, since the peculiar combination of rapidly increasing greenhouse gases and regionally heterogeneous aerosols has no known historical precedent. This is why simulation models must be used to project the detailed consequences. Fortunately, although no definitive validation of current forecasts is possible before the Earth system itself performs the ongoing global change "experiment," many subelements of the problem are well studied and we know from paleo changes much about natural rates of climate change. Therefore, subjective probabilities can be assigned to many aspects of climate change projections (*e.g.*, Morgan and Keith, 1995). Recently, there is an emerging consensus that increasing confidence can be expressed that real climatic changes are indeed happening (what is called the signal detection problem) and that at least some of this observed change can be linked (the so-called attribution problem) to an anthropogenic forcing (*e.g.*, see IPCC, WGI, 1995). What is the nature of this evidence?

First of all, the 1980s was the warmest decade in the instrumental record of surface thermometers, and there has been a 0.5±0.2°C century-long warming trend. But these facts were known in 1990, the previous record warm year (until 1995, an even warmer record year). The years 1992 and 1993 were substantially cooler, and ironically, this actually increased most scientists' confidence that human-induced global warming was being detected. The reason is that the explosive eruption of Mt. Pinatubo in the Philippines in 1991 spread a layer of sulfate dust particles in the stratosphere that filtered out a percent or two of the sun's heat. This, the computer models predicted, would for a few years cool the surface about a quarter of a degree C — very close to exactly what happened. Since the predicted cooling was made by the very same models that forecast global warming from enhancing the greenhouse effect, the credibility of the models increased as they fared well on this natural experimental test (primarily a test of the models' thermal response time).

Sulfate particles are not only a natural phenomena, but are generated by the people all over the industrialized and industrializing world, where high-sulfur coal and oil are burned. Up until 1994 general circulation models (GCMs) primarily considered only the effects of increased greenhouse gases in their predictions. Except for a factor of 2 crude correspondence with global-scale temperature rise, the results did not match up well with the regional patterns of climate change observed over the past 30 years. Critics charged that the models could not produce a "fingerprint" of climate change that looked like the observed patterns of changes of the past few decades and that the models were thus presumed suspect. Schneider (1994) responded that until the models are driven by the same time-evolving factors as the Earth (both the global warming from greenhouse gas increases and the regional cooling patterns from sulfate dust), "fingerprint" matching exercises between model-predicted patterns of change and observed changes do not prove anything. Since then, three such model calculations have been performed — at the U.K. Hadley Centre, the German Max-Planck-Institute, and the Lawrence Livermore National Laboratory (see Chapter 8, IPCC, 1995). All three studies produced patterns of change (*i.e.*, fingerprints) that are a much closer match to observed changes. So, perhaps ironically, it is the cooling effect of both natural and human-produced sulfate aerosols that has substantially increased IPCC scientists' confidence in the detection and attribution of observed climate changes known popularly as "global warming."

It is important to emphasize that increasing physical, chemical, or biological comprehensiveness or higher spatial and temporal resolution *by themselves* do not assure increasing simulation skill. Just because a process is observed to be important in nature at one scale does not imply that its *effects* are likewise important at the much larger scale of the lowest resolved model element (*e.g.*, the grid box). Processes may scale up from small scales to large according to laws observed at small scales, or they may interact with other processes such that their importance at large scale may be either negligible or of an entirely different character than observed at small scales (*e.g.*, see the discussion in Root and Schneider, 1995). Therefore, complex systems simulations should be evaluated primarily based on how well the model's simulations compare with appropriate observations *at the scale of the smallest resolved elements in the model* (*e.g.*, how well the model produces grid-box-averaged cloudiness, not necessarily how well it mimics the dynamics of individual clouds).

Evaluation at appropriate scales becomes a particular burden for testing the reliability of slowly changing factors such as deep oceans, glaciers, forests, or deserts. Thus, although many subcomponents of Earth systems models can be evaluated to a considerable degree, there will remain an uncomfortable degree of subjectivity in scientists' confidence in coupled models' projections. Furthermore, given both this subjectivity and the unprecedented nature of some anthropogenic forcings, it seems likely that current assessments have not yet anticipated all plausible outcomes of the current "global change experiment," let alone assigned them meaningful probabilities. In view of the possibility of such "surprises," the policy-making community has expressed interest in invoking the "precautionary principle" and is considering several concrete, albeit controversial, actions to mitigate anthropogenic forcings. In the meantime, it is incumbent on the scientific community both to improve its confidence in currently imagined projections of climate change and to take steps to address more explicitly the scientifically challenging problems of identifying surprises and bounding possible extreme outcomes. Chapter 15 has provided a useful foundation for those interested in both enhanced understanding of the climate system and improving the policy dialogue by helping to put decision making on a firmer factual basis.

*Stephen H. Schneider is a professor in the Department of Biological Sciences, a Senior Fellow at the Institute for International Studies and Professor by Courtesy in the Department of Civil Engineering at Stanford University. Dr. Schneider's current global change research interests include: climatic change; global warming; food/climate and other environmental/science public policy issues; ecological and economic implications of climatic change; climatic modeling of paleoclimates and of human impacts on climate. He is also active in advancing public understanding of science and in improving formal environmental education.*

# 16 Atmospheric Evolution and Global Perspective

## 16.1 Introduction

The Earth system has evolved constantly since its formation some 4.6 billion years (Ga) ago. In particular, the chemical composition of the atmosphere has changed as a result of a variety of natural processes, including the emergence of life. During the past century, major chemical perturbations in the lower and upper atmosphere have been caused by human activities. Although the focus of this book is on processes that lead to global changes in the Earth system on timescales of less than 100 years, it is important to examine these changes in the broader perspective of the Earth's history. The purpose of this last chapter is, therefore, to provide some insight on how the atmosphere has evolved over geological time scales and to discuss more recent changes that are largely of anthropogenic origin. Finally, recent model assessments will be used to examine how the chemical composition of the atmosphere could evolve in the decades to come if the population continues to grow, if agricultural activities develop further, and if pollution associated with industrial and domestic activities remains significant.

## 16.2 Atmospheric Evolution on Geological Timescales

It is generally believed that the universe was created some 15-20 billion years ago as a result of an enormous explosion (called the big bang) of a dense center of matter. Within the expanding gaseous mass produced by the explosion, clouds of hydrogen and helium condensed and, under gravitational forces, eventually contracted into swirling galaxies composed of billions of stars. At the outer rim of our galaxy, the Sun formed a little more than 4.6 billion years ago from a nebula consisting of a swirling cloud of gas and dust. The Earth was probably formed by accretion of cosmic dust and meteorites (planetisimals) that had condensed from the solar nebula. These grew in size as small bodies coalesced to produce a single planet. The gases and volatile compounds present at the early stage of the Earth's formation probably dissipated before the accumulation of the planetary mass became sufficient to produce a strong gravitational field able to retain the primordial atmosphere. Any primitive atmosphere was probably blown away by either giant impacts or strong solar winds. The large deficit in the present atmospheric concentration of noble gases (such as He, Ne, Ar) compared to their cosmic abundance provides evidence for the dissipation of gases during the accretion of the planet, since there is no known process that could have removed these unreactive gases from the atmosphere. Since the core of the emerging planet was extremely warm (due to radioactive decay of unstable isotopes and as a consequence of large impact events), volcanic and seismic activity as well as other outgassing mechanisms rapidly produced a secondary primitive atmosphere.

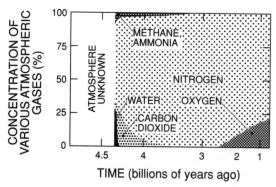

**Figure 16.1.** Probable evolution of the relative abundance (in percent) of chemical composition of the atmosphere during the Earth's history (Allègre and Schneider, 1994).

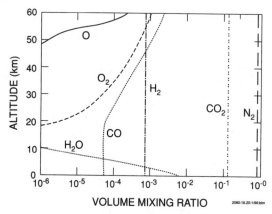

**Figure 16.2.** Vertical distribution of major atmospheric constituents in a weakly reduced, prebiotic atmosphere. The major gases are $N_2$ and $CO_2$. Photochemical destruction of $CO_2$ leads to the production of O and $O_2$ in the upper atmosphere (Kasting, 1990).

As the intense degassing of the planet liberated large quantities of water, a transient "steamy" atmosphere was formed. As soon as the temperature at the surface of the planet dwindled, however, the steam was rained out and formed oceans. The chemical composition of the remaining atmosphere (see Fig. 16.1) was probably dominated by nitrogen and carbon dioxide molecules, with traces of carbon monoxide, molecular hydrogen, methane, ammonia, sulfur dioxide, and hydrogen chloride (Fig. 16.2). No oxygen molecules were present at the surface in this weakly reducing medium, but the presence of methane and ammonia could have led to the formation of organic matter.

The historic experiment conducted in the 1950s by Miller (1953) and Miller and Urey (1959) tends to support this hypothesis: Biologically important molecules such as sugar and amino acids were formed by spark discharges (simulating lightning flashes) in a mixture of $CH_4$, $NH_3$, $H_2$, and $H_2O$ molecules contained in a sealed glass vessel. However, since volcanic gases might have been oxidized around 3.5-3.8 Ga, just before the first signs of life were recorded in sediments, ammonia and methane might not have been present in the primitive atmosphere. It is, however, possible that a *weakly* reduced atmosphere containing $CO_2$, $N_2$, and traces of CO and $H_2$ could have played a role in the emergence of life. Photochemical reactions could have generated formaldehyde ($CH_2O$ needed for the synthesis of sugars) and hydrogen cyanide (HCN used for the synthesis of amino acids). The pathway for the formation of HCN remains unclear, however, so that other mechanisms for the emergence of life have been proposed. For example, organic molecules could have been introduced into the Earth system by the bombardment of micrometeorites. Spectroscopic observations of interstellar space reveal the presence of molecules containing carbon, hydrogen, nitrogen, oxygen, etc., which are probably the source of the organic molecules detected in comets and observed in large meteorites that have penetrated the Earth's atmosphere and have reached the Earth's surface.

Despite the successful synthesis of primitive biomolecules under laboratory conditions representative of the early atmosphere, it now seems more likely that life began in the ocean (probably 3.8-4.2 billion years ago) in restricted, specialized environments such as volcanic vents. The high concentrations of sulfide found in the vicinity of such hot springs may have provided a mechanism for the formation of organic material, and these hot springs, even at present, support a remarkable biological system. The first evidence of life is provided by the fossilized imprints of a microbial community found in a 3.8 billion year old sedimentary rock.

At the time of the Earth's formation, the energy provided by the Sun (at that stage of its evolution) was probably 25-30 percent lower than current amounts. Degassing water, however, did not freeze, which suggests that the mean Earth temperature was maintained at a relatively high value by some terrestrial mechanism. This question, called the "faint young sun paradox," can be answered by assuming that either the Earth's albedo was lower in the past or that greenhouse warming was more pronounced. The concentration of $CO_2$ required to compensate for reduced solar luminosity in the past remains speculative, but could have been 600 times higher than the present atmospheric level. As the oceans were formed, large amounts of atmospheric $CO_2$ progressively dissolved in water.

One mechanism responsible for the decay of atmospheric $CO_2$ during the Archean era (2.5 Ga) is provided by weathering of rocks, when carbonic acid carried by rainwater (and produced when atmospheric $CO_2$ is dissolved in rain drops; see Section 5.4.1) reacts with silicate contained in the minerals. $CO_2$ was perhaps also removed from the atmosphere by photosynthesizing microorganisms present in the ocean, before being converted into calcium carbonate ($CaCO_3$) sediments. Carbon burial has been essential for allowing the buildup of oxygen and hence the development of certain forms of life on the planet. Over geological times, the abundance of $CO_2$ in the atmosphere has been regulated by weathering removal and release processes associated with volcanism and metamorphism (changes in rock structure or constitution due to pressure and temperature).

The transition from an oxygen-free, primitive reducing atmosphere to an oxidizing atmosphere that can sustain life for higher organisms was unquestionably the most important stage in the evolution of the Earth's atmosphere. This evolution has been very specific to the Earth. The contemporary atmospheres of neighboring planets such as Venus and Mars (see Table 16.1) are characterized by chemical compositions and climates that are very different from those encountered on Earth.

The initial mechanism that led to the formation of molecular oxygen was probably the photolysis of water vapor in the high atmosphere followed by the escape of hydrogen atoms to space. The abundance of $O_2$ remained low in the atmosphere since capture of this element by iron ions dissolved in the ocean was an efficient loss mechanism. Oxygen began, however, to accumulate in the atmosphere approximately 2 billion years ago, allowing the development of organisms that were able to utilize $O_2$. Cyanbacterial (blue-green algae) photosynthesis and, later, green plant photosynthesis, became the major sources of atmospheric oxygen. Although the precise timing and the reasons for the rise in the atmospheric abundance of oxygen remain unclear, it is believed (see Fig. 16.3) that the level of oxygen reached approximately 1%, 10%, and 100% of its present value 2000, 700, and 350 million years ago, respectively. At the same time, ozone was formed, and provided necessary protection against ultraviolet radiation for the developing land-based life (see Chapter 14).

During the Earth's history, the abundance of many atmospheric trace gases has fluctuated significantly in response to natural climate variations (probably initiated by periodic changes in orbital parameters of the planet and leading to a succession of glacial and interglacial periods). As noted in Chapter 15, the chemical analysis of air bubbles trapped in ice accumulated in the polar regions provides fascinating

**Table 16.1**
*Comparison between Venus, Mars, and the Earth*

| Characteristic | Venus | Earth | Mars |
|---|---|---|---|
| Total mass ($10^{27}$ g) | 5 | 6 | 0.6 |
| Radius (km) | 6049 | 6371 | 3390 |
| Atmospheric mass (ratio) | 100 | 1 | 0.06 |
| Distance from Sun ($10^6$ km) | 108 | 150 | 228 |
| Solar constant (W m$^{-2}$)$^a$ | 2613 | 1367 | 589 |
| Albedo (%) | 75 | 30 | 15 |
| Cloud cover (%) | 100 | 50 | Variable |
| Effective radiative (°C) temperature | −39 | −18 | −56 |
| Surface temperature (°C) | 427 | 15 | −53 |
| Greenhouse warming (°C) | 466 | 33 | 3 |
| $N_2$ (%) | <2 | 78 | <2.5 |
| $O_2$ (%) | <1 ppmv | 21 | <0.25 |
| $CO_2$ (%) | 98 | 0.035 | >96 |
| $H_2O$ (range %) | $1 \times 10^{-4} - 0.3$ | $3 \times 10^{-4} - 4$ | <0.001 |
| $SO_2$ (fraction) | 150 ppmv | <1 ppbv | Nil |
| Cloud composition | $H_2SO_4$ | $H_2O$ | Dust, $H_2O$, $CO_2$ |

$^a$ The intensity of the solar radiation over a square meter of surface at a distance equal to that from the Sun to the planet's orbit.
From Graedel and Crutzen, 1995.

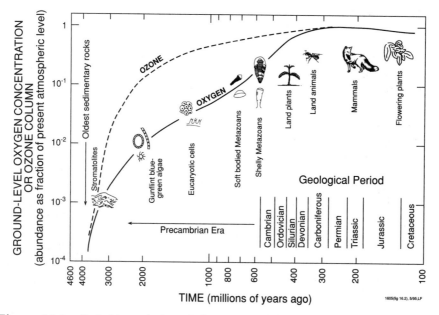

**Figure 16.3.** *Probable evolution of the oxygen and ozone abundance in the atmosphere (fraction of present levels) during the different geological periods of the Earth's history (Wayne, 1991; reprinted by permission of Oxford University Press).*

information about the evolution of the atmospheric chemical composition during the past 200,000 years or so.

As already discussed in Chapter 5, there is growing evidence that the composition of the Earth's atmosphere is, to a high degree, under the control of the terrestrial and marine biosphere. The current atmospheric oxygen content of 20 percent is, for example, the result of photosynthetic activity. Other chemical cycles such as the nitrogen and carbon cycles are also biologically mediated. The Gaia hypothesis, developed by Lovelock (1979, 1988, 1991), suggests that microorganisms, plants, and animals act in such a way that the Earth's environment is adjusted to states optimum for the maintenance of life. Lovelock's theory suggests that over the past 3.5 billion years, the Earth's climate and the chemical composition of the atmosphere have been regulated by life. For example, through changes inferred in the surface albedo of the Earth, the biosphere may have provided a mechanism for regulating the surface temperature along the Earth's history (the so-called "thermostasis effect" that may have taken place in spite of the fact that the radiative energy emitted by the Sun has been increasing with time). It is perhaps the biosphere that has maintained the chemical composition of the atmosphere far away from the thermodynamic equilibrium conditions encountered on Mars and Venus. A key question is whether human activities that have perturbed the chemical composition of the atmosphere in recent decades have the potential to drive the Earth system beyond any hypothetical Gaia repair capability.

## 16.3 Human Influences on the Atmosphere

Although large-scale environmental disruptions of human origin had already occurred more than 2000 years ago, in relation to massive deforestation as well as slash-and-burn agriculture in several regions of Asia, North Africa, and Europe, it is primarily the industrial revolution in the past centuries that produced the most dramatic changes in the chemical composition of the atmosphere at the regional and global scales. High levels of pollution are reported in the industrialized regions of North America, Europe, and Asia; in the past decades, developing countries have also contributed significantly to chemical perturbations in the atmosphere. The changing chemical composition of the troposphere and stratosphere is one of the most dramatic signals of global change.

Several factors have played important roles in the recent atmospheric evolution. Over the past 100 years, the world's industrial production has increased by two orders of magnitude, while the energy consumption is 80 times larger today than in year 1850. Over the past 200 years, the global population has increased by a factor of eight; its absolute growth has been as large between 1950 and 1990 as during the entire previous historical period.

Currently, nearly 80 percent of the world's primary energy is supplied by fossil fuel (coal, petroleum, and natural gas). Until 1975, coal was providing the largest contribution to anthropogenic $CO_2$ sources. Currently, the dominant role is played by petroleum.

Because of their potential environmental consequences, including their effects on the climate system and their impact on living organisms, perturbations in the chemical composition of the stratosphere and the troposphere have drawn the attention of the scientific community, of decision makers, and of the public all around the world.

Climate forcing (see Fig. 16.4) caused by increasing emissions of radiatively active gases such as $CO_2$, $CH_4$, $N_2O$, chlorofluorocarbons, stratospheric and tropospheric ozone (see Table 16.2 and Chapter 15), and by changes in the atmospheric abundance of aerosols is significant, although not entirely quantified (due to large uncertainties in the effects of the different types of aerosols — see IPCC, 1996, and Chapter 15). The impact of human activities on atmospheric ozone is also substantial. For example, it is now well established that the observed depletion in stratospheric ozone is caused primarily by the release into the atmosphere of manmade chlorofluorocarbons

**Table 16.2**
*Mixing Ratios, Growth Rates, and Atmospheric Lifetimes of Several Key Greenhouse Gases*

|  | $CO_2$ (ppmv) | $CH_4$ (ppmv) | CFC-11 (pptv) | CFC-12 (pptv) | $N_2O$ (ppbv) |
| --- | --- | --- | --- | --- | --- |
| Preindustrial (1750-1800) | 280 | 0.79 | 0 | 0 | 288 |
| 1900 | 296 | 0.97 | 0 | 0 | 292 |
| 1960 | 316 | 1.27 | 18 | 30 | 296 |
| 1970 | 325 | 1.42 | 70 | 121 | 299 |
| 1980 | 337 | 1.57 | 158 | 273 | 303 |
| 1990 | 354 | 1.72 | 280 | 484 | 310 |
| Rate of increase in % during 1989-1990 | 0.5 | 0.8 | 4 | 4 | 0.25 |
| Atmospheric lifetime (years) | 50-200 | 10 | 50 | 102 | 121 |

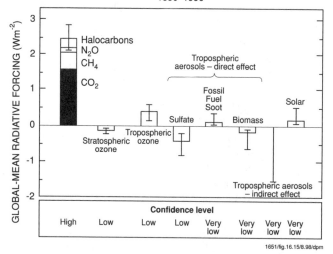

**Figure 16.4.** Radiative forcing resulting from changes since the preindustrial era in the atmospheric concentration of radiatively important gases and of different types of aerosols, as estimated by IPCC (1996). The forcing associated with the estimated change in solar energy intercepted by the Earth is also indicated. Error bars and level of confidence attached to the calculated radiative forcing are also indicated (IPCC, 1996).

(see Chapters 8 and 14). As shown by Figure 16.5, the globally integrated quantity of stratospheric ozone has decreased by approximately five percent over the past 15 years. A dramatic decline is also shown at mid- and high latitudes in both hemispheres. The observations show that much of the trend in the ozone column abundance results from chemical destruction between 10 and 25 km altitude. At midlatitudes, the ozone decline in the 1980s deduced from ozonesonde data was close to seven percent per decade at 20 km. An additional ozone reduction in the upper stratosphere (35-45 km) of 5-10 percent per decade was also reported. Finally, there is strong scientific evidence that the large ozone depletion observed in polar regions during springtime (*e.g.*, the ozone hole in Antarctica) is a consequence of human activities (see Chapter 14). Figure 16.6 shows an estimate of the change in the ozone column abundance during the 1980s, as deduced from the observations by the spaceborne Total Ozone Mapping Spectrometer (TOMS).

*Figure 16.5.* Deviations from 1964-1980 mean ozone level for the global average and for middle and polar latitudes (35°-90°S plus 35°-90°N). Both data sets show a significant decline (about 5%) over the past 15 years. The solid line represents ground-based data (smoothed by 1-2-1 weight function); the dotted line represents TOMS data (Bojkov and Fioletov, 1995).

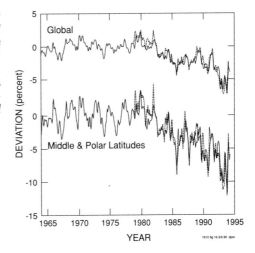

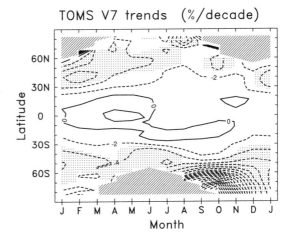

*Figure 16.6.* Relative change (percent per decade) in the ozone column abundance between the late 1970s and the early 1990s deduced from the observations of TOMS (version 7). Values are represented as a function of latitude and month of the year. Unbroken lines represent positive changes and dashed lines represent negative changes. Contour interval is two percent. Light shading represents the areas where the trends are statistically significant (courtesy of W. Randel, 1997).

The negative trends in stratospheric ozone have affected the penetration of solar radiation in the atmosphere, and modified the climate forcing somewhat. Large increases in ultraviolet (UV) radiation have, for example, been observed in Antarctica in association with the ozone hole. There is also theoretical and experimental evidence that, all other things (*e.g.*, cloudiness, aerosol load) remaining equal, the level of UV-B (280-315 nm) radiation has increased at midlatitudes in response to the observed depletion in the ozone column (see Chapter 14). Specifically, anomalously high UV-B levels were observed at midlatitudes in association with the exceptionally low ozone abundance recorded following the volcanic eruption of Mt. Pinatubo. Enhanced exposure of humans to UV-B light is believed to generate a higher number of skin cancers. Calculations performed with general circulation models also suggest that the observed decrease in stratospheric ozone has reduced the calculated warming produced by well-mixed greenhouse gases such as carbon dioxide, methane, nitrous oxide, and the chlorofluorocarbons by approximately 20 percent.

The increase in the abundance of tropospheric ozone reported over the past decades at several locations in industrialized regions of the Northern Hemisphere (see Chapter 13 and Fig. 16.7) seems also to result from human activities. Ozone episodes observed in the boundary layer of Europe and North America in the summertime are associated with stagnant air masses rich in pollutants, including nitrogen oxides, hydrocarbons, and carbon monoxide. Pollution associated with fossil fuel and biomass

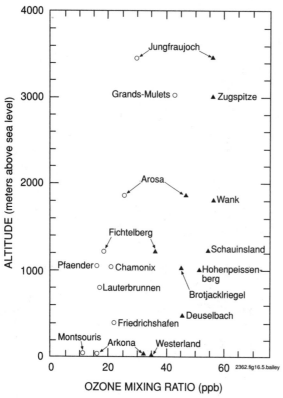

*Figure 16.7. Measurements of surface ozone mixing ratios (ppbv) at different locations in Europe during the months of August and September. The circles represent values observed before the end of the 1950s, while the triangles are representative of recent years (Staehelin et al., 1994).*

combustion also affects the chemical composition of the free troposphere. Furthermore, the release of nitrogen oxides, and other chemical compounds at high altitudes from aircraft engines, has the potential to affect ozone near the tropopause, as well as the radiative forcing of the climate system. An important question is to determine to what extent the "oxidizing capacity" of the atmosphere (*i.e.*, its self-cleansing ability) has changed during this past century as a result of human activities, and is expected to change in the future. Another question is to assess the climatic impact of global chemical changes in the troposphere. Figure 16.8 shows a model estimate of the change in the concentration of surface ozone between 1850 and 1985.

## 16.4 Future Trends

Future changes in the chemical composition of the atmosphere are difficult to predict, because they will be affected by demographic, economic, social, and cultural factors the future evolution of which cannot easily be established. The assessment of these changes is usually based on complex models and on a variety of scenarios that specify plausible evolutions of these factors. Future trends in *stratospheric* ozone, for example, will be determined by the chlorine and bromine loading in the atmosphere, and hence by the future production and emission rates of industrially manufactured halocarbons. In 1987, in recognition of the potential for these compounds to destroy the ozone layer, an international protocol signed in Montreal called for a reduction in the global production of several ozone-depleting substances. More stringent measures were adopted subsequently in London (1990), Copenhagen (1992), and Vienna (1995) as amendments to the original Montreal Protocol. The expected evolution of the equivalent chlorine load under three scenarios is shown in Figure 8.15. Under the current agreements, which reduce and eventually eliminate the production of most ozone-depleting gases, the stratospheric abundance of reactive chlorine and bromine should maximize around the year 2000 before slowly declining. However, the level of chlorine corresponding to the threshold (approximately 2 ppbv) above which the ozone hole was formed should not be reached before the middle of the twenty-first century. The evolution of the ozone column abundance at 50°N (March) predicted by different two-dimensional models (see Chapter 12) for the period 1980-2050 for a scenario following the guidelines of the recent international agreements is shown in Figure 16.9. In these calculations plausible

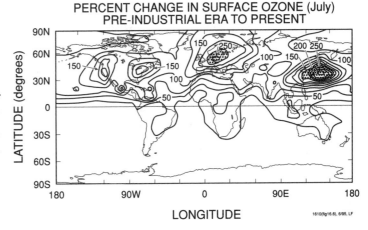

**Figure 16.8.** Change (in percent) of the surface ozone concentration from the preindustrial era to present day for the month of July calculated by the IMAGES 3D chemical-transport model (based on data from Müller and Brasseur, 1995).

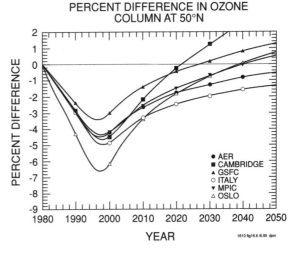

**Figure 16.9.** Calculated change (percent) in the ozone column abundance at 50°N (March) calculated for the period 1980-2050 by several two-dimensional chemical-transport models of the stratosphere, on the basis of the recent international agreements (Montreal Protocol and subsequent amendments) limiting the production of ozone-depleting substances (WMO, 1995).

future increases in the abundance of atmospheric methane and nitrous oxide are specified. Note that the discrepancies between models are significant and associated with different formulations and parameterizations of physical and chemical processes.

Chemical-transport models have also been used to assess the potential impact of a projected fleet of high-altitude (supersonic) aircraft on ozone in the stratosphere. The release of large quantities of nitrogen oxides, water, and sulfur near 15-20 km altitude could affect ozone in various ways. Since a major catalytic destruction mechanism for ozone is driven by the presence of $NO_x$ (see Chapter 14), the emission of nitrogen oxides will tend to reduce the ozone concentration in the stratosphere, especially during summertime. However, the release of $NO_x$ will also affect the partitioning in the $HO_x$ and $ClO_x$ chemical families (which also influence the ozone destruction in the stratosphere). For example, the balance between ClO and $ClONO_2$ is shifted toward $ClONO_2$, so that the efficiency of the ozone-destruction cycle associated with the presence of chlorine is reduced. The release of water vapor by aircraft engines is expected to provide an enhanced source of OH radicals; however, the dominating factor is a reduction in the concentration of $HO_x$ compounds resulting from the enhanced abundance of $HNO_3$ and $HNO_4$ (which destroy OH) associated with the

$NO_x$ emissions. With the OH abundance reduced, the conversion of HCl to Cl is also reduced. Thus the emission of nitrogen oxides by high-altitude transports should result in a more efficient ozone destruction by the $NO_x$ cycle, and a reduced ozone destruction associated with the $ClO_x$ and $HO_x$ cycles. Heterogeneous chemical reactions on sulfate aerosols and polar stratospheric clouds have a significant impact on the calculated net effect: The hydrolysis of $N_2O_5$ to $HNO_3$ on aerosol particles tends to reduce the concentration of $NO_x$ in the lower stratosphere and hence to minimize the impact of aircraft. However, the release of water and nitrogen oxides (rapidly converted into nitric acid) tends to facilitate the formation of polar stratospheric clouds, and hence the destruction of ozone at high latitudes. The emission of sulfur by aircraft engines is expected to cause a slight increase in the surface area density of aerosols near flight corridors and hence the rate at which heterogeneous reactions take place. The potential impact of a fleet of high-altitude aircraft on stratospheric ozone, calculated by a chemical transport model, is shown in Figure 16.10.

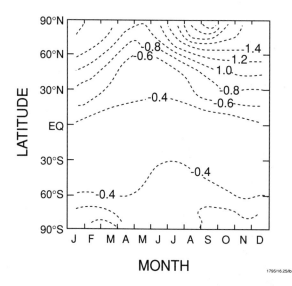

**Figure 16.10.** *Calculated change (percent) in the ozone column abundance (as a function of latitude and season) produced by a potential fleet of 500 high-altitude (supersonic) aircraft flying at Mach 2.4 and releasing approximately $1.2 \times 10^6$ tons yr$^{-1}$ of $NO_x$ (as $NO_2$) near 20 km altitude (assumed total fuel consumption of $82 \times 10^6$ tons yr$^{-1}$ and $NO_2$ emission index of 15 g $NO_2$/kg fuel). The impact of heterogeneous reactions on sulfate aerosols is taken into account. The indirect effect on ozone associated with potential changes in polar stratospheric clouds due to water emissions by aircraft engines is ignored. The results are from the AER 2D model as quoted in NASA (1995) Scientific Assessments of the Effects of Stratospheric Aircraft. As new engine technology develops, the $NO_x$ emission index could drop to perhaps 5 g $NO_2$/kg fuel in several years.*

The future abundance of chemical compounds in the *troposphere* will reflect the evolution of agricultural and industrial activities. Again, predictions are difficult to formulate. As population increases most rapidly in the developing world, and measures to control emissions in the industrial and transportation sectors of developed nations are being implemented, it is likely that the geographical distribution of trends in the level of atmospheric oxidants and of acidic deposition will change in the next decades. Figure 16.11 shows an estimate of plausible changes in surface ozone from present-day conditions to the year 2050, as predicted by a model for a given scenario specifying population growth and energy comsumption in different parts of the world. The model suggests that the most dramatic changes in the oxidant level should occur in the tropics, and more specifically in Southeast Asia. Under the same assumptions, the largest increase in acidic deposition should take place in this region of the world. Finally, Figure 16.12 provides a model calculation of the increase in

Atmospheric Chemistry and Global Change 549

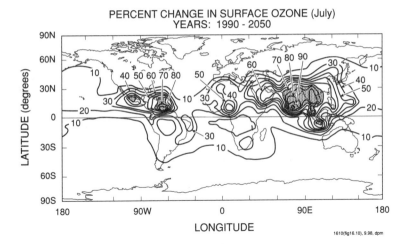

**Figure 16.11.** *Calculated change (percent) in the surface ozone concentration predicted for the period 1990-2050 by the IMAGES model (Müller and Brasseur, 1995), based on plausible assumptions for growth in population, energy consumption, and gross national products (Scenario IS92a of IPCC, 1992). Note the large ozone increase in Southeastern Asia.*

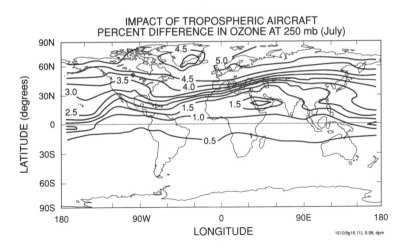

**Figure 16.12.** *Calculated change (percent) in the ozone concentration in the upper troposphere (250 mb) produced by the current fleet of subsonic aircraft, assuming a total $NO_x$ emission of 0.44 Tg N/yr. Because of uncertainties in natural sources of atmospheric $NO_x$ (e.g., $NO_x$ production by lightning), large uncertainties are associated with these model predictions (Brasseur et al., 1996).*

the ozone concentration near 10 km altitude due to the current fleet of commercial (subsonic) aircraft. Based on projected growth in commercial aviation (and ignoring the development of an additional fleet of supersonic transport), the ozone perturbation in midlatitude upper troposphere could double by the year 2020 and could be 5 times larger in 2050.

## 16.5 Global Perspective

The Earth can be regarded as a complex, interactive system in which planetary sources (fossil fuels and nonrenewable resources) are transformed for the benefit of the economic system (population and capital), and eventually converted into heat and waste, which contaminate air, water and soils. Over the long history of the planet, changes produced by human activities have generally remained limited, and the evolution of the chemical composition of the atmosphere has been driven primarily by biological and volcanic activities, as well as by fluctuations in the incoming solar energy. Today, the scale at which resources and energy are being consumed as a result of human activities has become significant compared to natural fluctuations. Food production and industrial activities are growing nearly exponentially, as is the world population.

Today, the health of the planet is worrisome. Humankind is conducting a gigantic experiment using the entire planet as a laboratory. Atmospheric chemistry is a key element—and perhaps a driving force—of global change. The purpose of this book is to provide some insight on some of the complex issues that will determine our common future. As large gaps remain in our understanding of these scientific questions, we are not yet able to predict the future evolution of our planet accurately. But, to quote Antoine de Saint-Exupéry, *As for the future, your task is not to foresee, but to enable it.*

## Further Reading

Firor, J. (1990) *The Changing Atmosphere*, Yale University Press, New Haven, Connecticut.
Graedel, T. E. and P. J. Crutzen (1993) *Atmospheric Change: An Earth System Perspective*, Freeman and Co., New York.
Kasting, J. F. (1993) Earth's early atmosphere, *Science* 259, 920.
Lewis, J. S. and R. G. Prinn (1983) *Planets and Their Atmospheres: Origin and Evolution*, Academic Press, International Geophysics Series, Vol. 33.
Meadows, D. H., D. L. Meadows, and J. Randers (1992) *Beyond the Limits*, Chelsea Green Publishing Company, Post Mills, Vermont.
Mackenzie, F. T. and J. A. Mackenzie (1995) *Our Changing Planet: An Introduction to Earth System Science and Global Environmental Change*, Prentice Hall, Upper Saddle River, New Jersey.
Walker, J. C. G. (1977) *Evolution of the Atmosphere*, Macmillan, New York.
World Commission on Environment and Development (1987) *Our Common Future*, Oxford University Press, Oxford.

Daniel L. Albritton

## The Atmosphere and Humankind: Our Related Futures

*"For the first time in my life, I saw the horizon as a curved line. It was accentuated by a thin seam of dark blue light — our atmosphere. Obviously, this was not the ocean of air I had been told it was so many times in my life. I was terrified by its fragile appearance."*

Ulf Merbold, German Astronaut

Few humans have been privileged with the viewing-port perspective of the Earth's atmosphere that is described above. Clearly, when seen from space, the relative scales of the globe, the atmosphere, and humankind are gained in an instant. Indeed, historians may well conclude that the view gained of our little blue planet from space, first by camera and later by eye, has produced one of the most profound social impacts of the Space Age.

The chapters of this textbook also provide a perspective — one that is much more quantitative and one that is longer in its development — of the atmosphere and our relation to it. Laboratory investigations have quantified the rates of key photochemical reactions; field studies have defined the interplay of chemistry and dynamics; and theory and modeling have assembled the understanding gained into a predictive picture of how the atmosphere "works." That predictive tool — always imperfect, but continually improving — allows us to examine our relation to atmosphere change and consequences quantitatively. Such scientifically based "What if's?" are key input to the decisions faced by our governmental and industrial leaders regarding global change vis-à-vis human well-being.

Actually, it does not take the vista from a spacecraft or a printout from a sophisticated model calculation in order to have an intuitive feel for the role of the atmosphere in global change. A few simple numbers suffice to illustrate the relationships:

| | |
|---|---|
| The mass of the solid Earth is | $\sim 10^{25}$ kg |
| The mass of the oceans is | $\sim 10^{21}$ kg |
| The mass of the atmosphere is | $\sim 10^{18}$ kg |
| The mass of the human population is | $\sim 10^{11}$ kg |

The fourteen orders-of-magnitude difference from the top to the bottom of this list underscores why, although we have scratched the surface of it and drilled into the ground, the human population has had virtually no impact on the solid Earth as a whole. Similarly, while we have polluted rivers, estuaries, and even bays, the oceans have been changed little by our presence on the planet.

However, our relatively recent ability to unlock vast reserves of energy from natural storehouses, to cleverly invent powerful machines that can augment our tiny physical human capabilities, and to innovatively manufacture previously nonexistent chemicals for special uses, all have meant that the human population has bridged the seven orders-of-magnitude gap between ourselves and the "thin seam of dark blue light — our atmosphere." Therefore, it should be no real surprise that, as illustrated in the final chapter and elsewhere in this textbook, it is in the atmosphere that the global footprints of us hominids show up first and most clearly: poor regional-scale air quality, a continental-sized stratospheric ozone "hole," and global perturbations of the radiation budget by greenhouse gases.

It is on such issues that we scientists are most frequently asked to describe what we know. As this textbook illustrates, we know a lot, thanks to the diligent efforts of a large number of very insightful atmospheric scientists. For example, the discoveries regarding the roles of catalytic chemical reactions and heterogeneous processes in the chemistry of the stratosphere have been nothing short of revolutionary.

But perhaps the most fruitful, mind-stretching awareness that has been gained by researchers during the recent decade is the simple point:

*It is one atmosphere.*

Namely, as implied above, we have tended in the past to compartmentalize our atmospheric phenomena/environmental issues, viz., smog, acid rain, ozone-layer depletion, and greenhouse warming. Different "camps" of researchers tackled each phenomenon, and, indeed, different decision makers dealt with each issue under different international agreements, all largely taking independent and sometimes different approaches.

Fortunately, we have begun now to see the couplings more clearly, as these two examples indicate:

1. *Ozone-layer depletion and climate change.* The chlorine- and bromine-induced loss of lower-stratospheric ozone causes a local cooling of that atmospheric layer, which means that less heat is radiated downward to the troposphere/surface system. As a consequence, the climatic cooling tendency due to ozone depletion, once viewed separately by scientists and decision makers, must be factored into the calculations of human-caused surface temperature changes and dealt with in the detection and attribution of those changes.
2. *Surface pollution and climate change.* The hemispheric-scale increases in aerosol concentrations that have occurred because of industrial sulfur emissions of industrial activity also cause a cooling of surface temperatures. As a result, another term in the radiative-forcing, climate-response equation must be incorporated into what must be a comprehensive picture. Therefore, once again, two phenomena/issues, once dealt with very separately, are indeed linked.

As this book also illustrates, the fact that "It is one atmosphere" places a greater challenge to understanding than the ostensible luxury of our past compartmentalized research approaches. There are many things that the atmosphere has not yet revealed to us, and many of these are "interface" phenomena, like convective chemical transport. We also see how a better understanding of these interactions is important to building an improved predictive tool to aid comprehensive, hence more effective, human choices and priority settings.

Can we atmospheric researchers meet this challenge in an era of scarce resources and in a time of competing human issues crying out to be addressed? The answer is very straightforward: We must figure out how. The rationale for this answer is simple and compelling:

- There is only one planet.
- There will be more and more people on it.
- They all will rightly seek a reasonable life style.
- That goal will be pursued via technology, chemistry, power production, transportation, etc.
- Therefore, the information needed to keep the attainment of the goal from being at the expense of the environment will become of higher and higher value.

Indeed, if there is one take-away message of this textbook for us and for the future researchers who will be aided by this book, then it is this:

*Understanding our global environment and our role in it is the first step towards living in harmony with it.*

*Daniel L. Albritton is Director of the NOAA Aeronomy Laboratory. He has carried out laboratory and field studies of the chemical processes occurring throughout the atmosphere — from the ionosphere to the boundary layer. He has also helped lead the international state-of-understanding assessments of stratospheric ozone depletion, which are the scientific input to the U.N. Montréal Protocol.*

# Appendixes

# A  Physical Constants and Other Data

*General and Universal Constants*

| | |
|---|---|
| Base of natural logarithms | 2.7182818285 |
| $\pi$ | 3.14159265 |
| Molar gas constant | 8.3143 J K$^{-1}$ mol$^{-1}$ |
| Boltzmann's constant | $1.38066 \times 10^{-23}$ J K$^{-1}$ |
| Stefan-Boltzmann constant | $5.67032 \times 10^{-8}$ W m$^{-2}$ K$^{-4}$ |
| Planck's constant | $6.626176 \times 10^{-34}$ J s |
| Speed of light (*in vacuo*) | $2.9979246 \times 10^{8}$ m s$^{-1}$ |
| Gravitational constant | $6.67259 \times 10^{-11}$ m$^3$ s$^{-2}$ kg$^{-1}$ |
| Permittivity of free space | $8.85 \times 10^{-12}$ F m$^{-1}$ |
| Mass of electrons | $9.1096 \times 10^{-31}$ kg |
| Charge of electron | $1.6022 \times 10^{-19}$ C |
| Atomic mass unit (amu) | $1.6605402 \times 10^{-27}$ kg |
| Avogadro's number | $6.022137 \times 10^{23}$ mol$^{-1}$ |
| Triple-point temperature of water | 273.16 K |

*Sun*

| | |
|---|---|
| Luminosity | $3.92 \times 10^{26}$ W |
| Mass | $1.99 \times 10^{30}$ kg |
| Equatorial radius | $6.9598 \times 10^{8}$ m |
| Mean angle subtended by photosphere at Earth | 31.988 arc min |
| Emission temperature | 5783 K |

*Earth*

| | |
|---|---|
| Average radius | $6.37 \times 10^{6}$ m |
| Equatorial radius | $6.378388 \times 10^{6}$ m |
| Polar radius | $6.357 \times 10^{6}$ m |
| Average height of land | 840 m |
| Average depth of oceans | 3730 m |
| Area of Earth's surface | $5.10 \times 10^{14}$ m$^2$ |
| Area of Earth's continents | $1.49 \times 10^{14}$ m$^2$ |
| Northern Hemisphere land surface area | $1.03 \times 10^{14}$ m$^2$ |
| Southern Hemisphere land surface area | $0.46 \times 10^{14}$ m$^2$ |
| Area of Earth's oceans | $3.61 \times 10^{14}$ m$^2$ |

## Earth (Cont.)

| | |
|---|---|
| Standard surface gravity | 9.80665 m s$^{-2}$ |
| Mass of Earth | 5.983 × 10$^{24}$ kg |
| Mass of ocean | 1.4 × 10$^{21}$ kg |
| Mass of atmosphere | 5.3 × 10$^{18}$ kg |
| Mass of the biosphere | 1.15 × 10$^{16}$ kg |
| Eccentricity of orbit | 0.016750 |
| Inclination of rotation axis | 23.45 deg or 0.409 rad |
| Mean angular rotation rate | 7.292 × 10$^{-5}$ rad s$^{-1}$ |
| Earth orbital period | 365.25463 days |
| Length of sidereal day | 23.94 hours |
| Lunar orbital period | 27.32 days |
| Solar constant | 1367 ± 2 W m$^{-2}$ |
| Mean distance from sun (1AU) | 1.496 × 10$^{11}$ m |
| Furthest Sun-Earth distance (4 July) | 1.5196 × 10$^{11}$ m |
| Nearest Sun-Earth distance (3 January) | 1.4696 × 10$^{11}$ m |
| Human population (1987) | 5.0 × 10$^9$ |
| Nondairy cattle population | 1.1 × 10$^9$ |
| Mass of plants | |
|     Land | 5.6 × 10$^{14}$ kg C |
|     Oceans | 3.0 × 10$^{12}$ kg C |
| Net primary production | |
|     Land | 6.0 × 10$^{13}$ kg C yr$^{-1}$ |
|     Oceans | 5.0 × 10$^{13}$ kg C yr$^{-1}$ |

## Dry Air

| | |
|---|---|
| Average molecular mass | 28.97 g mol$^{-1}$ |
| Specific gas constant | 287 J K$^{-1}$ kg$^{-1}$ |
| Standard surface pressure | 1.01325×10$^5$ Pa |
| Mass density at 0°C and 101,325 Pa | 1.293 kg m$^{-3}$ |
| Molecular density at 0°C and 101,325 Pa | 2.69 × 10$^{25}$ m$^{-3}$ |
| Molar volume at 0°C and 101,325 Pa | 22.414 × 10$^{-3}$ m$^3$ mol$^{-1}$ |
| Specific heat at constant pressure ($c_p$) | 1004 J K$^{-1}$ kg$^{-1}$ |
| Specific heat at constant volume ($c_v$) | 717 J K$^{-1}$ kg$^{-1}$ |
| Index of refraction for air | 1.000277 |
| Dry adiabatic lapse rate | −9.75 K km$^{-1}$ |
| Speed of sound in standard, calm air | 343.15 m s$^{-1}$ |

## Water

| | |
|---|---|
| Molecular weight | 18.016 g mol$^{-1}$ |
| Gas constant for vapor | 461.5 J K$^{-1}$ kg$^{-1}$ |
| Density of pure water at 0°C | 1000 kg m$^{-3}$ |
| Density of ice at 0°C | 917 kg m$^{-3}$ |
| Density of water vapor at STP | 0.803 kg m$^{-3}$ |
| Specific heat of vapor at constant pressure | 1952 J K$^{-1}$ kg$^{-1}$ |
| Specific heat of vapor at constant volume | 1463 J K$^{-1}$ kg$^{-1}$ |
| Specific heat of liquid water at 0°C | 4218 J K$^{-1}$ kg$^{-1}$ |
| Specific heat of ice at 0°C | 2106 J K$^{-1}$ kg$^{-1}$ |

| | |
|---|---|
| Latent heat of vaporization at 0°C | $2.501 \times 10^6$ J kg$^{-1}$ |
| Latent heat of vaporization at 100°C | $2.25 \times 10^6$ J kg$^{-1}$ |
| Latent heat of fusion at 0°C | $3.34 \times 10^5$ J kg$^{-1}$ |
| Index of refraction for liquid water | 1.336 |
| Index of refraction for ice | 1.312 |

# B  Units, Conversion Factors, and Multiplying Prefixes

## International System of Units

| Quantity | Name of Unit | Symbol | Definition |
|---|---|---|---|
| Length | meter | m | |
| Mass | kilogram | kg | |
| Time | second | s | |
| Electrical current | ampere | A | |
| Temperature | kelvin | K | |
| *Derived Units* | | | |
| Force | newton | N | kg m s$^{-2}$ |
| Pressure | pascal | Pa | N m$^{-2}$ |
| Energy | joule | J | kg m$^2$ s$^{-2}$ |
| Power | watt | W | J s$^{-1}$ |
| Electric potential difference | volt | V | W A$^{-1}$ |
| Electrical charge | coulomb | C | A s |
| Electrical resistance | ohm | $\Omega$ | V A$^{-1}$ |
| Electrical capacitance | farad | F | A s V$^{-1}$ |
| Frequency | hertz | Hz | s$^{-1}$ |

## Conversion Factors

| | |
|---|---|
| Area | 1 ha = $10^4$ m$^2$ |
| Volume | 1 liter = $10^{-3}$ m$^3$ |
| Force | 1 N = $10^5$ dyn |
| Pressure | 1 bar = $10^5$ Pa |
| | 1 atm = $1.01325 \times 10^5$ Pa = 760.0 Torr |
| Energy | 1 cal = 4.1855 J |
| | 1 eV = $1.6021 \times 10^{-19}$ J |
| Power | 1 W = 14.3353 cal min$^{-1}$ |
| Temperature | T(°C) = T(K) - 273.15 |
| | T(°F) = 1.8 T(°C) + 32 |
| Mixing ratios | 1 ppb = $10^{-3}$ ppm |
| | 1 ppt = $10^{-3}$ ppb = $10^{-6}$ ppm |
| Logarithms | $\ln x = 2.3026 \log_{10} x$ |

## Multiplying Prefixes

| Multiple | Prefix | Symbol | Multiple | Prefix | Symbol |
|---|---|---|---|---|---|
| $10^{-1}$ | deci | d | $10$ | deca | da |
| $10^{-2}$ | centi | c | $10^2$ | hecto | h |
| $10^{-3}$ | milli | m | $10^3$ | kilo | k |
| $10^{-6}$ | micro | $\mu$ | $10^6$ | mega | M |
| $10^{-9}$ | nano | n | $10^9$ | giga | G |
| $10^{-12}$ | pico | p | $10^{12}$ | tera | T |
| $10^{-15}$ | femto | f | $10^{15}$ | peta | P |
| $10^{-18}$ | atto | a | $10^{18}$ | exa | E |

The International System of units is used except where usage suggests otherwise. The unit of pressure is the pascal (Pa), but meteorologists commonly use the millibar (mb) or equivalently the hectopascal (hPa). Number densities are often expressed in molecules per cubic centimeter ($cm^{-3}$). Volume mixing ratios are given in percent, parts per million (ppmv), parts per billion (ppbv), or parts per trillion (pptv). Mass mixing ratios are given in kilograms per kilograms or grams per kilograms. Wavelengths are often expressed in micrometers ($\mu$m) or nanometers (nm). Wavenumbers are expressed in inverse centimeters ($cm^{-1}$).

# C  Altitude Profiles and Mixing Ratios of Chemical Consituents

## Altitude Profiles of Pressure and Temperature

| Altitude (km) | Pressure (mb) | Temperature (K) | Air density (cm$^{-3}$) |
|---|---|---|---|
| 0.0 | 1013.000 | 290.0 | 2.53E+19 |
| 2.0 | 761.246 | 278.0 | 1.98E+19 |
| 4.0 | 572.059 | 264.4 | 1.56E+19 |
| 6.0 | 429.890 | 250.0 | 1.24E+19 |
| 8.0 | 323.052 | 235.8 | 9.92E+18 |
| 10.0 | 242.766 | 223.7 | 7.86E+18 |
| 12.0 | 182.434 | 214.2 | 6.17E+18 |
| 14.0 | 137.095 | 208.8 | 4.75E+18 |
| 16.0 | 103.023 | 206.5 | 3.61E+18 |
| 18.0 | 77.420 | 205.6 | 2.72E+18 |
| 20.0 | 58.179 | 205.6 | 2.05E+18 |
| 22.0 | 43.720 | 206.5 | 1.53E+18 |
| 24.0 | 32.855 | 208.1 | 1.14E+18 |
| 26.0 | 24.690 | 210.5 | 8.49E+17 |
| 28.0 | 18.554 | 213.9 | 6.28E+17 |
| 30.0 | 13.943 | 217.6 | 4.64E+17 |
| 32.0 | 10.478 | 221.2 | 3.43E+17 |
| 34.0 | 7.874 | 225.0 | 2.53E+17 |
| 36.0 | 5.917 | 229.2 | 1.87E+17 |
| 38.0 | 4.446 | 234.0 | 1.37E+17 |
| 40.0 | 3.341 | 239.2 | 1.01E+17 |
| 42.0 | 2.511 | 244.6 | 7.43E+16 |
| 44.0 | 1.887 | 249.5 | 5.48E+16 |
| 46.0 | 1.418 | 253.3 | 4.05E+16 |
| 48.0 | 1.066 | 255.5 | 3.02E+16 |
| 50.0 | 0.801 | 255.4 | 2.27E+16 |
| 52.0 | 0.602 | 253.2 | 1.72E+16 |
| 54.0 | 0.452 | 249.3 | 1.31E+16 |
| 56.0 | 0.340 | 244.3 | 1.00E+16 |
| 58.0 | 0.255 | 238.9 | 7.74E+15 |
| 60.0 | 0.192 | 233.6 | 5.95E+15 |
| 80.0 | 1.1 E-02 | 198.6 | 3.83E+14 |
| 100.0 | 3.2 E-04 | 195.1 | 1.19E+13 |
| 140.0 | 7.2 E-06 | 560.0 | 9.32E+10 |
| 200.0 | 8.5 E-07 | 855.0 | 7.18E+09 |
| 500.0 | 3.0 E-09 | 999.2 | 2.19E+07 |

# Mixing Ratio at 30°N in March

| Altitude (km) | $N_2O$ | $CH_4$ | CO | $H_2O$ |
|---|---|---|---|---|
| 0.0  | 3.08E–07 | 1.68E–06 | 1.47E–07 |          |
| 2.0  | 3.07E–07 | 1.67E–06 | 1.31E–07 |          |
| 4.0  | 3.07E–07 | 1.67E–06 | 1.18E–07 |          |
| 6.0  | 3.07E–07 | 1.67E–06 | 1.08E–07 |          |
| 8.0  | 3.06E–07 | 1.66E–06 | 1.00E–07 |          |
| 10.0 | 3.05E–07 | 1.66E–06 | 9.32E–08 | 3.20E–06 |
| 12.0 | 3.03E–07 | 1.65E–06 | 8.36E–08 | 3.21E–06 |
| 14.0 | 2.99E–07 | 1.63E–06 | 6.93E–08 | 3.20E–06 |
| 16.0 | 2.93E–07 | 1.61E–06 | 5.20E–08 | 3.21E–06 |
| 18.0 | 2.85E–07 | 1.58E–06 | 3.89E–08 | 3.24E–06 |
| 20.0 | 2.73E–07 | 1.54E–06 | 2.89E–08 | 3.30E–06 |
| 22.0 | 2.57E–07 | 1.48E–06 | 2.18E–08 | 3.37E–06 |
| 24.0 | 2.36E–07 | 1.42E–06 | 1.75E–08 | 3.47E–06 |
| 26.0 | 2.10E–07 | 1.33E–06 | 1.55E–08 | 3.59E–06 |
| 28.0 | 1.77E–07 | 1.23E–06 | 1.52E–08 | 3.74E–06 |
| 30.0 | 1.41E–07 | 1.12E–06 | 1.58E–08 | 3.91E–06 |
| 32.0 | 1.05E–07 | 1.00E–06 | 1.65E–08 | 4.10E–06 |
| 34.0 | 7.34E–08 | 8.80E–07 | 1.73E–08 | 4.31E–06 |
| 36.0 | 4.80E–08 | 7.54E–07 | 1.79E–08 | 4.52E–06 |
| 38.0 | 2.99E–08 | 6.32E–07 | 1.85E–08 | 4.74E–06 |
| 40.0 | 1.83E–08 | 5.19E–07 | 1.90E–08 | 4.94E–06 |
| 42.0 | 1.14E–08 | 4.25E–07 | 1.94E–08 | 5.11E–06 |
| 44.0 | 7.41E–09 | 3.54E–07 | 2.03E–08 | 5.24E–06 |
| 46.0 | 5.05E–09 | 3.06E–07 | 2.17E–08 | 5.33E–06 |
| 48.0 | 3.66E–09 | 2.77E–07 | 2.31E–08 | 5.39E–06 |
| 50.0 | 2.81E–09 | 2.62E–07 | 2.47E–08 | 5.42E–06 |
| 52.0 | 2.23E–09 | 2.54E–07 | 2.70E–08 | 5.43E–06 |
| 54.0 | 1.82E–09 | 2.51E–07 | 2.97E–08 | 5.44E–06 |
| 56.0 | 1.52E–09 | 2.50E–07 | 3.32E–08 | 5.44E–06 |
| 58.0 | 1.31E–09 | 2.51E–07 | 3.86E–08 | 5.43E–06 |
| 60.0 | 1.16E–09 | 2.52E–07 | 4.63E–08 | 5.42E–06 |

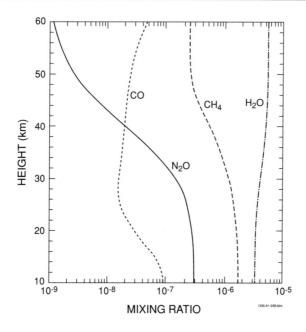

# $O_x$ Mixing Ratio at 30°N in March

| Altitude (km) | $O(^1D)$ | $O(^3P)$ | $O_3$ |
|---|---|---|---|
| 0.0 | 1.97E−22 | 1.03E−16 | 3.82E−08 |
| 2.0 | 2.46E−22 | 1.39E−16 | 3.42E−08 |
| 4.0 | 4.18E−22 | 2.58E−16 | 4.37E−08 |
| 6.0 | 7.80E−22 | 5.27E−16 | 6.30E−08 |
| 8.0 | 1.43E−21 | 1.06E−15 | 9.05E−08 |
| 10.0 | 2.67E−21 | 2.20E−15 | 1.30E−07 |
| 12.0 | 5.33E−21 | 4.94E−15 | 1.97E−07 |
| 14.0 | 1.21E−20 | 1.31E−14 | 3.30E−07 |
| 16.0 | 2.99E−20 | 3.83E−14 | 5.72E−07 |
| 18.0 | 7.24E−20 | 1.09E−13 | 9.39E−07 |
| 20.0 | 1.78E−19 | 3.13E−13 | 1.50E−06 |
| 22.0 | 4.54E−19 | 8.91E−13 | 2.32E−06 |
| 24.0 | 1.17E−18 | 2.46E−12 | 3.44E−06 |
| 26.0 | 3.11E−18 | 6.66E−12 | 4.84E−06 |
| 28.0 | 8.57E−18 | 1.79E−11 | 6.60E−06 |
| 30.0 | 2.38E−17 | 4.64E−11 | 8.34E−06 |
| 32.0 | 5.78E−17 | 1.03E−10 | 8.79E−06 |
| 34.0 | 1.30E−16 | 2.24E−10 | 8.56E−06 |
| 36.0 | 2.69E−16 | 4.71E−10 | 7.61E−06 |
| 38.0 | 5.20E−16 | 9.95E−10 | 6.33E−06 |
| 40.0 | 9.61E−16 | 2.15E−09 | 5.03E−06 |
| 42.0 | 1.68E−15 | 4.71E−09 | 4.01E−06 |
| 44.0 | 2.82E−15 | 1.03E−08 | 3.35E−06 |
| 46.0 | 4.50E−15 | 2.20E−08 | 2.91E−06 |
| 48.0 | 6.76E−15 | 4.38E−08 | 2.57E−06 |
| 50.0 | 9.46E−15 | 8.03E−08 | 2.27E−06 |
| 52.0 | 1.24E−14 | 1.36E−07 | 2.01E−06 |
| 54.0 | 1.53E−14 | 2.11E−07 | 1.75E−06 |
| 56.0 | 1.81E−14 | 3.13E−07 | 1.51E−06 |
| 58.0 | 2.06E−14 | 4.43E−07 | 1.28E−06 |
| 60.0 | 2.25E−14 | 6.04E−07 | 1.06E−06 |

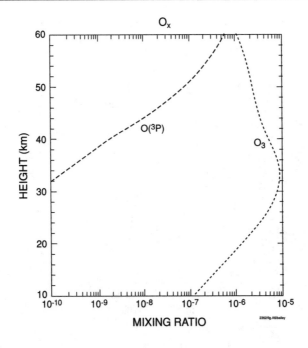

# $HO_x$ Mixing Ratio at 30°N in March

| Altitude (km) | $H_2O_2$ | OH | $HO_2$ | H |
|---|---|---|---|---|
| 10.0 | 6.32E–12 | 2.02E–14 | 1.21E–13 | 2.43E–21 |
| 12.0 | 6.11E–13 | 1.60E–14 | 4.83E–14 | 1.99E–21 |
| 14.0 | 3.89E–14 | 2.18E–14 | 4.63E–14 | 2.71E–21 |
| 16.0 | 4.98E–14 | 4.75E–14 | 1.26E–13 | 5.60E–21 |
| 18.0 | 1.50E–13 | 8.40E–14 | 2.73E–13 | 9.66E–21 |
| 20.0 | 4.37E–13 | 1.41E–13 | 5.55E–13 | 1.63E–20 |
| 22.0 | 1.30E–12 | 2.44E–13 | 1.14E–12 | 3.00E–20 |
| 24.0 | 3.72E–12 | 4.36E–13 | 2.32E–12 | 6.45E–20 |
| 26.0 | 1.02E–11 | 8.25E–13 | 4.71E–12 | 1.69E–19 |
| 28.0 | 2.76E–11 | 1.63E–12 | 9.67E–12 | 5.43E–19 |
| 30.0 | 6.45E–11 | 3.68E–12 | 1.95E–11 | 2.21E–18 |
| 32.0 | 9.82E–11 | 7.78E–12 | 3.20E–11 | 9.11E–18 |
| 34.0 | 1.14E–10 | 1.60E–11 | 4.69E–11 | 4.10E–17 |
| 36.0 | 1.03E–10 | 3.08E–11 | 6.07E–11 | 1.89E–16 |
| 38.0 | 7.94E–11 | 5.65E–11 | 7.27E–11 | 9.14E–16 |
| 40.0 | 5.79E–11 | 9.83E–11 | 8.42E–11 | 4.49E–15 |
| 42.0 | 4.57E–11 | 1.54E–10 | 9.88E–11 | 2.08E–14 |
| 44.0 | 4.16E–11 | 2.20E–10 | 1.20E–10 | 8.80E–14 |
| 46.0 | 4.10E–11 | 2.94E–10 | 1.49E–10 | 3.35E–13 |
| 48.0 | 4.08E–11 | 3.71E–10 | 1.80E–10 | 1.11E–12 |
| 50.0 | 4.00E–11 | 4.46E–10 | 2.10E–10 | 3.19E–12 |
| 52.0 | 3.92E–11 | 5.21E–10 | 2.39E–10 | 8.11E–12 |
| 54.0 | 3.76E–11 | 5.87E–10 | 2.64E–10 | 1.81E–11 |
| 56.0 | 3.54E–11 | 6.43E–10 | 2.86E–10 | 3.71E–11 |
| 58.0 | 3.35E–11 | 7.00E–10 | 3.07E–10 | 7.24E–11 |
| 60.0 | 3.21E–11 | 7.61E–10 | 3.32E–10 | 1.36E–10 |

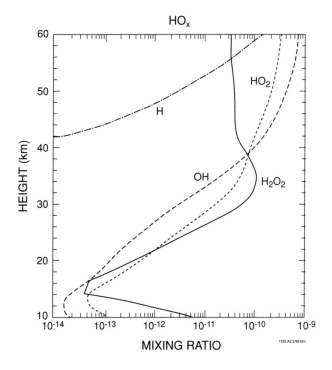

# $NO_y$ Mixing Ratio at 30°N in March

| Altitude (km) | $NO_y$ | $HNO_3$ | $N_2O_5$ | $NO_2$ | NO | $HNO_4$ | N |
|---|---|---|---|---|---|---|---|
| 10.0 | 1.48E−09 | 8.95E−10 | 1.55E−11 | 2.12E−10 | 2.84E−10 | 5.77E−11 | 2.48E−25 |
| 12.0 | 2.13E−09 | 1.18E−09 | 5.62E−11 | 3.35E−10 | 5.09E−10 | 2.99E−11 | 4.57E−24 |
| 14.0 | 3.01E−09 | 1.74E−09 | 9.41E−11 | 4.65E−10 | 6.65E−10 | 2.69E−11 | 4.66E−23 |
| 16.0 | 3.70E−09 | 2.63E−09 | 8.30E−11 | 4.44E−10 | 5.23E−10 | 4.04E−11 | 2.47E−22 |
| 18.0 | 4.37E−09 | 3.34E−09 | 9.44E−11 | 4.88E−10 | 4.83E−10 | 5.45E−11 | 1.34E−21 |
| 20.0 | 5.35E−09 | 4.13E−09 | 1.27E−10 | 6.11E−10 | 5.04E−10 | 7.67E−11 | 7.32E−21 |
| 22.0 | 6.65E−09 | 4.96E−09 | 1.90E−10 | 8.29E−10 | 5.70E−10 | 1.12E−10 | 3.90E−20 |
| 24.0 | 8.30E−09 | 5.76E−09 | 2.95E−10 | 1.19E−09 | 6.98E−10 | 1.61E−10 | 2.06E−19 |
| 26.0 | 1.03E−08 | 6.37E−09 | 4.57E−10 | 1.78E−09 | 9.14E−10 | 2.21E−10 | 1.05E−18 |
| 28.0 | 1.25E−08 | 6.50E−09 | 6.85E−10 | 2.74E−09 | 1.24E−09 | 2.90E−10 | 4.95E−18 |
| 30.0 | 1.49E−08 | 5.70E−09 | 9.22E−10 | 4.43E−09 | 1.93E−09 | 3.34E−10 | 2.36E−17 |
| 32.0 | 1.71E−08 | 5.28E−09 | 8.67E−10 | 6.00E−09 | 2.97E−09 | 2.68E−10 | 1.01E−16 |
| 34.0 | 1.89E−08 | 4.29E−09 | 6.90E−10 | 7.60E−09 | 4.55E−09 | 1.75E−10 | 3.68E−16 |
| 36.0 | 2.01E−08 | 3.29E−09 | 4.65E−10 | 8.45E−09 | 6.57E−09 | 9.24E−11 | 1.10E−15 |
| 38.0 | 2.07E−08 | 2.21E−09 | 2.87E−10 | 8.35E−09 | 9.03E−09 | 4.08E−11 | 2.76E−15 |
| 40.0 | 2.08E−08 | 1.29E−09 | 1.68E−10 | 7.09E−09 | 1.18E−08 | 1.56E−11 | 5.95E−15 |
| 42.0 | 2.04E−08 | 6.29E−10 | 9.76E−11 | 5.11E−09 | 1.43E−08 | 5.51E−12 | 1.12E−14 |
| 44.0 | 1.98E−08 | 2.47E−10 | 5.90E−11 | 3.17E−09 | 1.62E−08 | 1.85E−12 | 1.93E−14 |
| 46.0 | 1.92E−08 | 8.48E−11 | 3.66E−11 | 1.73E−09 | 1.72E−08 | 5.89E−13 | 3.13E−14 |
| 48.0 | 1.85E−08 | 2.70E−11 | 2.27E−11 | 8.70E−10 | 1.75E−08 | 1.79E−13 | 5.02E−14 |
| 50.0 | 1.78E−08 | 8.46E−12 | 1.36E−11 | 4.25E−10 | 1.72E−08 | 5.48E−14 | 8.28E−14 |
| 52.0 | 1.70E−08 | 2.73E−12 | 7.78E−12 | 2.08E−10 | 1.65E−08 | 1.74E−14 | 1.42E−13 |
| 54.0 | 1.59E−08 | 9.04E−13 | 4.07E−12 | 1.03E−10 | 1.55E−08 | 5.75E−15 | 2.50E−13 |
| 56.0 | 1.47E−08 | 3.06E−13 | 2.00E−12 | 5.25E−11 | 1.43E−08 | 1.95E−15 | 4.45E−13 |
| 58.0 | 1.32E−08 | 1.07E−13 | 9.34E−13 | 2.72E−11 | 1.30E−08 | 6.91E−16 | 7.83E−13 |
| 60.0 | 1.17E−08 | 3.91E−14 | 4.17E−13 | 1.45E−11 | 1.16E−08 | 2.54E−16 | 1.33E−12 |

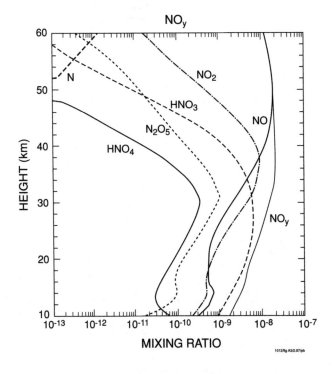

# Chlorocarbon Mixing Ratio at 30°N in March

| Altitude (km) | $CCl_4$ | $CFCl_3$ | $CF_2Cl_2$ | $CH_3CCl_3$ | $CH_3Cl$ | $Cl_y$ |
|---|---|---|---|---|---|---|
| 0.0  | 1.03E–10 | 2.53E–10 | 4.34E–10 | 1.45E–10 | 6.00E–10 | 5.00E–11 |
| 2.0  | 1.02E–10 | 2.52E–10 | 4.33E–10 | 1.44E–10 | 5.85E–10 | 6.88E–11 |
| 4.0  | 1.02E–10 | 2.52E–10 | 4.33E–10 | 1.43E–10 | 5.74E–10 | 8.49E–11 |
| 6.0  | 1.02E–10 | 2.51E–10 | 4.32E–10 | 1.42E–10 | 5.65E–10 | 1.00E–10 |
| 8.0  | 1.01E–10 | 2.50E–10 | 4.31E–10 | 1.41E–10 | 5.57E–10 | 1.17E–10 |
| 10.0 | 1.00E–10 | 2.48E–10 | 4.30E–10 | 1.40E–10 | 5.48E–10 | 1.41E–10 |
| 12.0 | 9.92E–11 | 2.45E–10 | 4.27E–10 | 1.37E–10 | 5.37E–10 | 1.79E–10 |
| 14.0 | 9.61E–11 | 2.39E–10 | 4.22E–10 | 1.33E–10 | 5.17E–10 | 2.51E–10 |
| 16.0 | 9.06E–11 | 2.29E–10 | 4.13E–10 | 1.25E–10 | 4.91E–10 | 3.66E–10 |
| 18.0 | 8.26E–11 | 2.14E–10 | 4.00E–10 | 1.15E–10 | 4.62E–10 | 5.17E–10 |
| 20.0 | 7.11E–11 | 1.93E–10 | 3.83E–10 | 1.01E–10 | 4.30E–10 | 7.22E–10 |
| 22.0 | 5.57E–11 | 1.64E–10 | 3.60E–10 | 8.20E–11 | 3.92E–10 | 9.96E–10 |
| 24.0 | 3.76E–11 | 1.28E–10 | 3.30E–10 | 5.90E–11 | 3.47E–10 | 1.33E–09 |
| 26.0 | 2.02E–11 | 8.75E–11 | 2.92E–10 | 3.52E–11 | 2.96E–10 | 1.72E–09 |
| 28.0 | 7.42E–12 | 4.57E–11 | 2.44E–10 | 1.49E–11 | 2.41E–10 | 2.14E–09 |
| 30.0 | 1.46E–12 | 1.54E–11 | 1.88E–10 | 3.53E–12 | 1.84E–10 | 2.51E–09 |
| 32.0 | 1.54E–13 | 3.60E–12 | 1.35E–10 | 5.16E–13 | 1.31E–10 | 2.77E–09 |
| 34.0 | 7.59E–15 | 5.07E–13 | 8.82E–11 | 4.02E–14 | 8.69E–11 | 2.96E–09 |
| 36.0 | 1.79E–16 | 4.46E–14 | 5.24E–11 | 1.75E–15 | 5.27E–11 | 3.08E–09 |
| 38.0 |          | 2.64E–15 | 2.80E–11 | 4.84E–17 | 2.90E–11 | 3.16E–09 |
| 40.0 |          | 1.16E–16 | 1.37E–11 |          | 1.47E–11 | 3.21E–09 |
| 42.0 |          |          | 6.62E–12 |          | 7.25E–12 | 3.24E–09 |
| 44.0 |          |          | 3.11E–12 |          | 3.69E–12 | 3.25E–09 |
| 46.0 |          |          | 1.49E–12 |          | 2.06E–12 | 3.26E–09 |
| 48.0 |          |          | 7.68E–13 |          | 1.32E–12 | 3.26E–09 |
| 50.0 |          |          | 4.24E–13 |          | 9.66E–13 | 3.26E–09 |

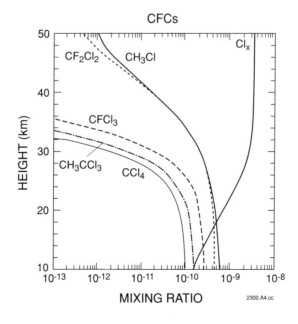

# $Cl_y$ Mixing Ratio at 30°N in March

| Altitude (km) | Cl | ClO | HCl | $ClONO_2$ | HOCl |
|---|---|---|---|---|---|
| 10.0 | 4.89E−17 | 8.70E−15 | 1.38E−10 | 1.33E−12 | 5.98E−16 |
| 12.0 | 5.74E−17 | 7.76E−15 | 1.77E−10 | 1.31E−12 | 1.81E−16 |
| 14.0 | 1.16E−16 | 1.87E−14 | 2.48E−10 | 2.92E−12 | 3.41E−16 |
| 16.0 | 3.83E−16 | 1.32E−13 | 3.57E−10 | 1.21E−11 | 6.43E−15 |
| 18.0 | 9.48E−16 | 5.75E−13 | 4.88E−10 | 3.42E−11 | 3.98E−14 |
| 20.0 | 2.17E−15 | 2.02E−12 | 6.45E−10 | 8.64E−11 | 1.94E−13 |
| 22.0 | 4.79E−15 | 6.17E−12 | 8.07E−10 | 1.98E−10 | 8.55E−13 |
| 24.0 | 1.01E−14 | 1.61E−11 | 9.40E−10 | 4.01E−10 | 3.19E−12 |
| 26.0 | 2.07E−14 | 3.62E−11 | 1.01E−09 | 7.01E−10 | 1.00E−11 |
| 28.0 | 4.08E−14 | 7.31E−11 | 1.00E−09 | 1.07E−09 | 2.80E−11 |
| 30.0 | 8.60E−14 | 1.29E−10 | 9.23E−10 | 1.41E−09 | 6.69E−11 |
| 32.0 | 2.06E−13 | 2.17E−10 | 1.03E−09 | 1.42E−09 | 1.21E−10 |
| 34.0 | 5.12E−13 | 3.48E−10 | 1.23E−09 | 1.20E−09 | 1.90E−10 |
| 36.0 | 1.25E−12 | 5.31E−10 | 1.52E−09 | 7.89E−10 | 2.43E−10 |
| 38.0 | 2.83E−12 | 7.13E−10 | 1.80E−09 | 3.90E−10 | 2.53E−10 |
| 40.0 | 5.88E−12 | 8.52E−10 | 1.99E−09 | 1.42E−10 | 2.19E−10 |
| 42.0 | 1.07E−11 | 8.83E−10 | 2.13E−09 | 3.89E−11 | 1.67E−10 |
| 44.0 | 1.73E−11 | 8.01E−10 | 2.30E−09 | 8.41E−12 | 1.19E−10 |
| 46.0 | 2.55E−11 | 6.36E−10 | 2.51E−09 | 1.48E−12 | 7.80E−11 |
| 48.0 | 3.46E−11 | 4.49E−10 | 2.73E−09 | 2.29E−13 | 4.60E−11 |
| 50.0 | 4.37E−11 | 2.97E−10 | 2.89E−09 | 3.49E−14 | 2.54E−11 |
| 52.0 | 5.30E−11 | 1.94E−10 | 3.00E−09 | 5.73E−15 | 1.40E−11 |
| 54.0 | 6.13E−11 | 1.26E−10 | 3.06E−09 | 1.03E−15 | 7.69E−12 |
| 56.0 | 6.91E−11 | 8.18E−11 | 3.10E−09 | 1.99E−16 | 4.22E−12 |
| 58.0 | 7.60E−11 | 5.28E−11 | 3.12E−09 | 4.14E−17 | 2.35E−12 |
| 60.0 | 8.15E−11 | 3.34E−11 | 3.14E−09 | 9.00E−18 | 1.31E−12 |

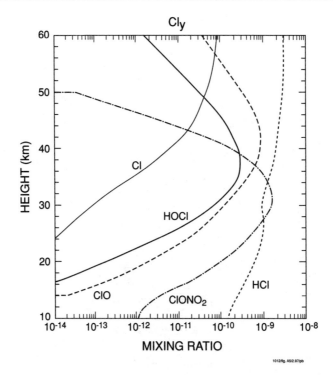

# D  Chemical Species in the Atmosphere

| Formula | Name | Molecular weight (g) |
|---|---|---|
| $O_2$ | molecular oxygen | 32 |
| $O(^3P)$ | atomic oxygen (ground state) | 16 |
| $O(^1D)$ | atomic oxygen (excited state) | 16 |
| $O_3$ | ozone | 48 |
| H | hydrogen | 1 |
| $H_2$ | molecular hydrogen | 2 |
| OH | hydroxyl radical | 17 |
| $HO_2$ | hydroperoxyl radical | 18 |
| $H_2O_2$ | hydrogen peroxide | 34 |
| N | atomic nitrogen | 14 |
| $N_2$ | molecular nitrogen | 28 |
| $N_2O$ | nitrous oxide | 44 |
| NO | nitrogen monoxide (nitric oxide) | 30 |
| $NO_2$ | nitrogen dioxide | 46 |
| $NO_3$ | nitrogen trioxide (nitrate radical) | 62 |
| $HNO_2$ (HONO) | nitrous acid | 47 |
| $HNO_3$ ($HONO_2$) | nitric acid | 63 |
| $HNO_4$ ($HO_2NO_2$) | pernitric acid | 79 |
| $N_2O_5$ | dinitrogen pentoxide | 108 |
| $NH_3$ | ammonia | 17 |
| $CH_3CO_3NO_2$ | peroxyacetylnitrate (PAN) | 121 |
| $CH_2{=}C(CH_3)CO_3NO_2$ | peroxymethacrylic nitrate (MPAN) | 147 |
| $CH_4$ | methane | 16 |
| $C_2H_6$ | ethane | 30 |
| $C_2H_4$ | ethylene (ethene) | 28 |
| $C_3H_6$ | propylene (propene) | 42 |
| $C_5H_8$ | isoprene | 68 |
| $C_{10}H_{16}$ | terpenes (*e.g.*, $\alpha$-pinene) | 136 |
| CO | carbon monoxide | 28 |
| $CO_2$ | carbon dioxide | 44 |
| $CH_3$ | methyl radical | 15 |

| Formula | Name | Molecular weight (g) |
|---|---|---|
| CHO | formyl radical | 29 |
| $CH_3O$ | methoxy radical | 31 |
| $CH_2O$ | formaldehyde | 30 |
| $CH_3O_2$ | methylperoxy radical | 47 |
| $CH_3OOH$ | methyl hydroperoxide | 48 |
| $C_2H_5O_2$ | ethylperoxy radical | 61 |
| $C_2H_5OOH$ | ethyl hydroperoxide | 62 |
| $CH_3C(O)OOH$ | peracetic acid | 76 |
| $CH_3C(O)CHO$ | methylglyoxal | 72 |
| $CH_2{=}CHC(O)CH_3$ | methylvinylketone | 70 |
| $CH_2{=}C(CH_3)CHO$ | methacrolein | 70 |
| $CH_3CO_3$ | acetylperoxy radical | 75 |
| $CH_2{=}C(CH_3)CO_3$ | peroxymethacryloyl radical | 101 |
| $CH_3C(O)CH_3$ | acetone | 58 |
| $CH_3CHO$ | acetaldehyde | 44 |
| $CH_2OHCHO$ | glycolaldehyde | 60 |
| $HC(O)CHO$ | glyoxal | 58 |
| Cl | atomic chlorine | 35.5 |
| $Cl_2$ | molecular chlorine | 71 |
| ClO | chlorine monoxide | 51.5 |
| HCl | hydrogen chloride | 36.5 |
| HOCl | hypochlorous acid | 52.5 |
| ClNO | nitrosyl chloride | 65.5 |
| $ClNO_2$ | nitryl chloride | 81.5 |
| $ClONO_2$ | chlorine nitrate | 97.5 |
| ClOO | chlorine peroxyl radical | 67.5 |
| OClO | chlorine dioxide | 67.5 |
| $Cl_2O_2$ | dichlorine peroxide (ClO dimer) | 103 |
| $CCl_2O$ | carbonyl dichloride (phosgene) | 99 |
| $CH_3Cl$ | methyl chloride | 50.5 |
| $CCl_4$ | carbon tetrachloride | 154 |
| $CH_3CCl_3$ | methylchloroform | 133.5 |
| $CFCl_3$ | trichlorofluoromethane (CFC-11) | 137.5 |
| $CF_2Cl_2$ | dichlorodifluoromethane (CFC-12) | 121 |
| $CCl_2FCClF_2$ | trichlorotrifluoroethane (CFC-113) | 187.5 |
| $CClF_2CClF_2$ | dichlorotetrafluoroethane (CFC-114) | 171 |
| $CClF_2CF_3$ | chloropentafluoroethane (CFC-115) | 154.5 |
| $CHClF_2$ | chlorodifluoromethane (HCFC-22) | 86.5 |
| F | atomic fluorine | 19 |
| FO | fluorine monoxide | 35 |
| HF | hydrogen fluoride | 20 |
| $CF_2O$ | carbonyl difluoride | 66 |
| CClFO | carbonyl chlorofluoride | 82.5 |
| Br | atomic bromine | 80 |
| BrO | bromine monoxide | 96 |
| HBr | hydrobromic acid | 81 |
| HOBr | hypobromous acid | 97 |

| Formula | Name | Molecular weight (g) |
| --- | --- | --- |
| $BrONO_2$ | bromine nitrate | 142 |
| $CHBr_3$ | bromoform | 253 |
| $CH_3Br$ | methylbromide | 95 |
| $CH_2Br_2$ | dibromomethane (methylene bromide) | 174 |
| $CBrClF_2$ | Halon-1211 (bromochlorodifluoromethane) | 165.5 |
| $CF_3Br$ | Halon-1301 (bromotrifluoromethane) | 149 |
| $C_2F_4Br_2$ | Halon-2402 | 260 |
| $CF_2Br_2$ | Halon-1202 (difluorodibromomethane) | 210 |
| S | atomic sulfur | 32 |
| SO | sulfur monoxide | 48 |
| $SO_2$ | sulfur dioxide | 64 |
| $SO_3$ | sulfur trioxide | 80 |
| OCS | carbonyl sulfide | 60 |
| $CH_3SCH_3$ | dimethylsulfide (DMS) | 62 |
| $CH_3SOCH_3$ | dimethylsulfoxide (DMSO) | 78 |
| $CH_3S_2CH_3$ | dimethyldisulfide | 94 |
| $H_2S$ | hydrogen sulfide | 34 |
| $CS_2$ | carbon disulfide | 76 |
| $H_2SO_4$ | sulfuric acid | 98 |
| $SF_6$ | sulfur hexafluoride | 146 |

# E. Rate Coefficients for Second-Order Gas-Phase Reactions[a,b]

| Reaction | A Factor | $E_a/R$ | $k_{298}$ |
|---|---|---|---|
| $O + O_3 \rightarrow O_2 + O_2$ | $8.0 \times 10^{-12}$ | 2060 | $8.0 \times 10^{-15}$ |
| $O(^1D) + O_2 \rightarrow O + O_2$ | $3.2 \times 10^{-11}$ | −70 | $4.0 \times 10^{-11}$ |
| $O(^1D) + O_3 \rightarrow O_2 + O_2$ | $1.2 \times 10^{-10}$ | 0 | $1.2 \times 10^{-10}$ |
| $\rightarrow O_2 + O + O$ | $1.2 \times 10^{-10}$ | 0 | $1.2 \times 10^{-10}$ |
| $O(^1D) + H_2 \rightarrow OH + H$ | $1.1 \times 10^{-10}$ | 0 | $1.1 \times 10^{-10}$ |
| $O(^1D) + H_2O \rightarrow OH + OH$ | $2.2 \times 10^{-10}$ | 0 | $2.2 \times 10^{-10}$ |
| $O(^1D) + N_2 \rightarrow O + N_2$ | $1.8 \times 10^{-11}$ | −110 | $2.6 \times 10^{-11}$ |
| $O(^1D) + N_2O \rightarrow N_2 + O_2$ | $4.9 \times 10^{-11}$ | 0 | $4.9 \times 10^{-11}$ |
| $\rightarrow NO + NO$ | $6.7 \times 10^{-11}$ | 0 | $6.7 \times 10^{-11}$ |
| $O(^1D) + CH_4 \rightarrow$ products | $1.5 \times 10^{-10}$ | 0 | $1.5 \times 10^{-10}$ |
| $O(^1D) + CCl_4 \rightarrow$ products | $3.3 \times 10^{-10}$ | 0 | $3.3 \times 10^{-10}$ |
| $O(^1D) + CH_3Br \rightarrow$ products | $1.8 \times 10^{-10}$ | 0 | $1.8 \times 10^{-10}$ |
| $O(^1D) + CH_2Br_2 \rightarrow$ products | $2.7 \times 10^{-10}$ | 0 | $2.7 \times 10^{-10}$ |
| $O(^1D) + CHBr_3 \rightarrow$ products | $6.6 \times 10^{-10}$ | 0 | $6.6 \times 10^{-10}$ |
| $O(^1D) + CHClF_2 \rightarrow$ products | $1.0 \times 10^{-10}$ | 0 | $1.0 \times 10^{-10}$ |
| $O(^1D) + CCl_3F \rightarrow$ products | $2.3 \times 10^{-10}$ | 0 | $2.3 \times 10^{-10}$ |
| $O(^1D) + CCl_2F_2 \rightarrow$ products | $1.4 \times 10^{-10}$ | 0 | $1.4 \times 10^{-10}$ |
| $O + OH \rightarrow O_2 + H$ | $2.2 \times 10^{-11}$ | −120 | $3.3 \times 10^{-11}$ |
| $O + HO_2 \rightarrow OH + O_2$ | $3.0 \times 10^{-11}$ | −200 | $5.9 \times 10^{-11}$ |
| $H + O_3 \rightarrow OH + O_2$ | $1.4 \times 10^{-10}$ | 470 | $2.9 \times 10^{-11}$ |
| $H + HO_2 \rightarrow$ products | $8.1 \times 10^{-11}$ | 0 | $8.1 \times 10^{-11}$ |
| $OH + O_3 \rightarrow HO_2 + O_2$ | $1.6 \times 10^{-12}$ | 940 | $6.8 \times 10^{-14}$ |
| $OH + H_2 \rightarrow H_2O + H$ | $5.5 \times 10^{-12}$ | 2000 | $6.7 \times 10^{-15}$ |
| $OH + OH \rightarrow H_2O + O$ | $4.2 \times 10^{-12}$ | 240 | $1.9 \times 10^{-12}$ |
| $OH + HO_2 \rightarrow H_2O + O_2$ | $4.8 \times 10^{-11}$ | −250 | $1.1 \times 10^{-10}$ |
| $OH + H_2O_2 \rightarrow H_2O + HO_2$ | $2.9 \times 10^{-12}$ | 160 | $1.7 \times 10^{-12}$ |
| $HO_2 + O_3 \rightarrow OH + 2O_2$ | $1.1 \times 10^{-14}$ | 500 | $2.0 \times 10^{-15}$ |
| $HO_2 + HO_2 \rightarrow H_2O_2 + O_2$ | $2.3 \times 10^{-13}$ | −600 | $1.7 \times 10^{-12}$ |
| $\xrightarrow{M} H_2O_2 + O_2$ | $1.7 \times 10^{-33} [M]$ | −1000 | $4.9 \times 10^{-32} [M]$ |
| $O + NO_2 \rightarrow NO + O_2$ | $6.5 \times 10^{-12}$ | −120 | $9.7 \times 10^{-12}$ |
| $OH + HONO \rightarrow H_2O + NO_2$ | $1.8 \times 10^{-11}$ | 390 | $4.5 \times 10^{-12}$ |
| $OH + HNO_3 \rightarrow H_2O + NO_3$ | (see Note 1) | | |
| $OH + HO_2NO_2 \rightarrow$ products | $1.3 \times 10^{-12}$ | −380 | $4.6 \times 10^{-12}$ |

[a]The rate constant ($cm^3$ molecule$^{-1}$ s$^{-1}$) is expressed by $k = A \exp(-E_a/RT)$, where $A$, the pre-exponential factor, is given in $cm^3$ molecule$^{-1}$ s$^{-1}$, and $E_a/R$ (activation energy of the reaction divided by the gas constant) is given in degrees kelvin. The value of the rate constant $k_{298}$ at 298 K is given in $cm^3$ molecule$^{-1}$ s$^{-1}$.

[b] DeMore, W. B., S. P. Sander, D. M. Golden, R. F. Hampson, M. J. Kurylo, C. J. Howard, A. R. Ravishankara, C. E. Kolb, and M. J. Molina (1997) *Chemical Kinetics and Photochemical Data for Use in Stratospheric Modeling*, Evaluation No. 12, JPL Publication No. 97-4.

| Reaction | A Factor | $E_a/R$ | $k_{298}$ |
|---|---|---|---|
| $OH + NH_3 \to H_2O + NH_2$ | $1.7 \times 10^{-12}$ | 710 | $1.6 \times 10^{-13}$ |
| $NH_2 + O_2 \to$ products | | | $< 6 \times 10^{-21}$ |
| $NH_2 + O_3 \to$ products | $4.3 \times 10^{-12}$ | 930 | $1.9 \times 10^{-13}$ |
| $NH_2 + NO \to$ products | $4.0 \times 10^{-12}$ | $-450$ | $1.8 \times 10^{-11}$ |
| $NH_2 + NO_2 \to$ products | $2.1 \times 10^{-12}$ | $-650$ | $1.9 \times 10^{-11}$ |
| $HO_2 + NO \to NO_2 + OH$ | $3.5 \times 10^{-12}$ | $-250$ | $8.1 \times 10^{-12}$ |
| $HO_2 + NO_3 \to$ products | | | $3.5 \times 10^{-12}$ |
| $N + O_2 \to NO + O$ | $1.5 \times 10^{-11}$ | 3600 | $8.5 \times 10^{-17}$ |
| $N + NO \to N_2 + O$ | $2.1 \times 10^{-11}$ | $-100$ | $3.0 \times 10^{-11}$ |
| $N + NO_2 \to N_2O + O$ | $5.8 \times 10^{-12}$ | $-220$ | $1.2 \times 10^{-11}$ |
| $NO + O_3 \to NO_2 + O_2$ | $2.0 \times 10^{-12}$ | 1400 | $1.8 \times 10^{-14}$ |
| $NO + NO_3 \to 2NO_2$ | $1.5 \times 10^{-11}$ | $-170$ | $2.6 \times 10^{-11}$ |
| $NO_2 + O_3 \to NO_3 + O_2$ | $1.2 \times 10^{-13}$ | 2450 | $3.2 \times 10^{-17}$ |
| $O_3 + C_2H_2 \to$ products | $1.0 \times 10^{-14}$ | 4100 | $1.0 \times 10^{-20}$ |
| $O_3 + C_2H_4 \to$ products | $1.2 \times 10^{-14}$ | 2630 | $1.7 \times 10^{-18}$ |
| $O_3 + C_3H_6 \to$ products | $6.5 \times 10^{-15}$ | 1900 | $1.1 \times 10^{-17}$ |
| $OH + CO \to$ products | $1.5 \times 10^{-13} \times (1 + 0.6 P_{atm})$ | 0 | $1.5 \times 10^{-13} \times (1 + 0.6 P_{atm})$ |
| $OH + CH_4 \to CH_3 + H_2O$ | $2.45 \times 10^{-12}$ | 1775 | $6.3 \times 10^{-15}$ |
| $OH + CH_2O \to H_2O + HCO$ | $1.0 \times 10^{-11}$ | 0 | $1.0 \times 10^{-11}$ |
| $OH + CH_3OH \to$ products | $6.7 \times 10^{-12}$ | 600 | $8.9 \times 10^{-13}$ |
| $OH + CH_3OOH \to$ products | $3.8 \times 10^{-12}$ | $-200$ | $7.4 \times 10^{-12}$ |
| $OH + HCN \to$ products | $1.2 \times 10^{-13}$ | 400 | $3.1 \times 10^{-14}$ |
| $OH + C_2H_6 \to H_2O + C_2H_5$ | $8.7 \times 10^{-12}$ | 1070 | $2.4 \times 10^{-13}$ |
| $OH + C_3H_8 \to H_2O + C_3H_7$ | $1.0 \times 10^{-11}$ | 660 | $1.1 \times 10^{-12}$ |
| $OH + CH_3CHO \to CH_3CO + H_2O$ | $5.6 \times 10^{-12}$ | $-270$ | $1.4 \times 10^{-11}$ |
| $OH + CH_3CN \to$ products | $7.8 \times 10^{-13}$ | 1050 | $2.3 \times 10^{-14}$ |
| $HO_2 + CH_3O_2 \to CH_3OOH + O_2$ | $3.8 \times 10^{-13}$ | $-800$ | $5.6 \times 10^{-12}$ |
| $HO_2 + C_2H_5O_2 \to C_2H_5OOH + O_2$ | $7.5 \times 10^{-13}$ | $-700$ | $8.0 \times 10^{-12}$ |
| $NO_3 + CH_2O \to HCO + HNO_3$ | | | $6.8 \times 10^{-16}$ |
| $NO_3 + CH_3CHO \to CH_3CO + HNO_3$ | $1.4 \times 10^{-12}$ | 1900 | $2.4 \times 10^{-15}$ |
| $HCO + O_2 \to CO + HO_2$ | $3.5 \times 10^{-12}$ | $-140$ | $5.5 \times 10^{-12}$ |
| $CH_2OH + O_2 \to CH_2O + HO_2$ | | | $9.1 \times 10^{-12}$ |
| $CH_3O + O_2 \to CH_2O + HO_2$ | $3.9 \times 10^{-14}$ | 900 | $1.9 \times 10^{-15}$ |
| $CH_3O_2 + CH_3O_2 \to$ products | $2.5 \times 10^{-13}$ | $-190$ | $4.7 \times 10^{-13}$ |
| $CH_3O_2 + NO \to CH_3O + NO_2$ | $4.2 \times 10^{-12}$ | $-180$ | $7.7 \times 10^{-12}$ |
| $C_2H_5O + O_2 \to CH_3CHO + HO_2$ | $6.3 \times 10^{-14}$ | 550 | $1.0 \times 10^{-14}$ |
| $C_2H_5O_2 + C_2H_5O_2 \to$ products | $6.8 \times 10^{-14}$ | 0 | $6.8 \times 10^{-14}$ |
| $C_2H_5O_2 + NO \to$ products | $2.6 \times 10^{-12}$ | $-365$ | $8.7 \times 10^{-12}$ |
| $CH_2C(O)O_2 + NO \to CH_3C(O)O + NO_2$ | $5.3 \times 10^{-12}$ | $-360$ | $1.8 \times 10^{-11}$ |
| $CH_3C(O)O_2 + CH_3C(O)O_2 \to CH_3C(O)O + CH_3C(O)O + O_2$ | $2.9 \times 10^{-12}$ | $-500$ | $1.5 \times 10^{-11}$ |
| $CH_3C(O)O_2 + HO_2 \to$ products | $4.5 \times 10^{-13}$ | $-1000$ | $1.3 \times 10^{-11}$ |
| $CH_3C(O)O_2 + CH_3O_2 \to$ products | $1.3 \times 10^{-11}$ | $-640$ | $1.1 \times 10^{-11}$ |
| $OH +$ isoprene $(C_5H_8) \to$ products | $2.5 \times 10^{-11}$ | $-410$ | $1.0 \times 10^{-10}$ |
| $OH + \alpha$-pinene $(C_{10}H_{16}) \to$ products | $1.2 \times 10^{-11}$ | $-444$ | $5.4 \times 10^{-11}$ |
| $OH + \beta$-pinene $(C_{10}H_{16}) \to$ products | $2.4 \times 10^{-11}$ | $-357$ | $7.9 \times 10^{-11}$ |

| Reaction | A Factor | $E_a/R$ | $k_{298}$ |
|---|---|---|---|
| OH + methylvinylketone → products | $4.1 \times 10^{-12}$ | −452 | $1.9 \times 10^{-11}$ |
| OH + methacrolein → products | $1.9 \times 10^{-11}$ | −175 | $3.4 \times 10^{-11}$ |
| OH + hydroxyacetone → products | | | $3.0 \times 10^{-12}$ |
| OH + methylglyoxal ($CH_3COCHO$) → $CH_3COCO + H_2O$ | $8.4 \times 10^{-13}$ | −830 | $1.3 \times 10^{-11}$ |
| OH + glycolaldehyde ($HOCH_2CHO$) → $HOCH_2CO + H_2O$ | | | $1.0 \times 10^{-11}$ |
| OH + $CH_3C(O)CH_3$ → $CH_3C(O)CH_2 + H_2O$ | $2.2 \times 10^{-12}$ | 685 | $2.3 \times 10^{-13}$ |
| $O_3$ + isoprene → products | $1.2 \times 10^{-14}$ | 2013 | $1.4 \times 10^{-17}$ |
| $O_3$ + α-pinene → products | $1.0 \times 10^{-15}$ | 731 | $8.5 \times 10^{-17}$ |
| O + FO → F + $O_2$ | $2.7 \times 10^{-11}$ | 0 | $2.7 \times 10^{-11}$ |
| F + $O_3$ → FO + $O_2$ | $2.2 \times 10^{-11}$ | 230 | $1.0 \times 10^{-11}$ |
| F + $H_2$ → HF + H | $1.4 \times 10^{-10}$ | 500 | $2.6 \times 10^{-11}$ |
| F + $H_2O$ → HF + OH | $1.4 \times 10^{-11}$ | 0 | $1.4 \times 10^{-11}$ |
| F + $CH_4$ → HF + $CH_3$ | $1.6 \times 10^{-10}$ | 260 | $6.7 \times 10^{-11}$ |
| FO + NO → $NO_2$ + F | $8.2 \times 10^{-12}$ | −300 | $2.2 \times 10^{-11}$ |
| $FO_2$ + NO → FNO + $O_2$ | $7.5 \times 10^{-12}$ | 690 | $7.5 \times 10^{-13}$ |
| $CF_3O + O_3$ → $CF_3O_2 + O_2$ | $2 \times 10^{-12}$ | 1400 | $1.8 \times 10^{-14}$ |
| $CF_3O$ + NO → $CF_2O$ + FNO | $3.7 \times 10^{-11}$ | −110 | $5.4 \times 10^{-11}$ |
| $CF_3O + CH_4$ → $CF_3OH + CH_3$ | $2.6 \times 10^{-12}$ | 1420 | $2.2 \times 10^{-14}$ |
| $CF_3O_2 + O_3$ → $CF_3O + 2O_2$ | | | $< 3 \times 10^{-15}$ |
| $CF_3O_2$ + NO → $CF_3O + NO_2$ | $5.4 \times 10^{-12}$ | −320 | $1.6 \times 10^{-11}$ |
| OH + $CF_3CH_2F$ (HFC-134a) → $H_2O + CF_3CHF$ | $1.5 \times 10^{-12}$ | 1750 | $4.2 \times 10^{-15}$ |
| O + ClO → Cl + $O_2$ | $3.0 \times 10^{-11}$ | −70 | $3.8 \times 10^{-11}$ |
| O + OClO → ClO + $O_2$ | $2.4 \times 10^{-12}$ | 960 | $1.0 \times 10^{-13}$ |
| O + HCl → OH + Cl | $1.0 \times 10^{-11}$ | 3300 | $1.5 \times 10^{-16}$ |
| O + HOCl → OH + ClO | $1.7 \times 10^{-13}$ | 0 | $1.7 \times 10^{-13}$ |
| O + $ClONO_2$ → products | $2.9 \times 10^{-12}$ | 800 | $2.0 \times 10^{-13}$ |
| OH + ClO → products | $1.1 \times 10^{-11}$ | −120 | $1.7 \times 10^{-11}$ |
| OH + OClO → HOCl + $O_2$ | $4.5 \times 10^{-13}$ | −800 | $6.8 \times 10^{-12}$ |
| OH + HCl → $H_2O$ + Cl | $2.6 \times 10^{-12}$ | 350 | $8.0 \times 10^{-13}$ |
| OH + HOCl → $H_2O$ + ClO | $3.0 \times 10^{-12}$ | 500 | $5.0 \times 10^{-13}$ |
| OH + $ClONO_2$ → products | $1.2 \times 10^{-12}$ | 330 | $3.9 \times 10^{-13}$ |
| OH + $CH_3Cl$ → $CH_2Cl + H_2O$ | $4.0 \times 10^{-12}$ | 1400 | $3.6 \times 10^{-14}$ |
| $HO_2$ + Cl → HCl + $O_2$ | $1.8 \times 10^{-11}$ | −170 | $3.2 \times 10^{-11}$ |
| → OH + ClO | $4.1 \times 10^{-11}$ | 450 | $9.1 \times 10^{-12}$ |
| $HO_2$ + ClO → HOCl + $O_2$ | $4.8 \times 10^{-13}$ | −700 | $5.0 \times 10^{-12}$ |
| Cl + $O_3$ → ClO + $O_2$ | $2.9 \times 10^{-11}$ | 260 | $1.2 \times 10^{-11}$ |
| Cl + $H_2$ → HCl + H | $3.7 \times 10^{-11}$ | 2300 | $1.6 \times 10^{-14}$ |
| Cl + $H_2O_2$ → HCl + $HO_2$ | $1.1 \times 10^{-11}$ | 980 | $4.1 \times 10^{-13}$ |
| Cl + $CH_4$ → HCl + $CH_3$ | $1.1 \times 10^{-11}$ | 1400 | $1.0 \times 10^{-13}$ |
| Cl + $CH_2O$ → HCl + HCO | $8.1 \times 10^{-11}$ | 30 | $7.3 \times 10^{-11}$ |
| Cl + OClO → ClO + ClO | $3.4 \times 10^{-11}$ | −160 | $5.8 \times 10^{-11}$ |
| ClO + NO → $NO_2$ + Cl | $6.4 \times 10^{-12}$ | −290 | $1.7 \times 10^{-11}$ |
| ClO + ClO → $Cl_2 + O_2$ | $1.0 \times 10^{-12}$ | 1590 | $4.8 \times 10^{-15}$ |
| → ClOO + Cl | $3.0 \times 10^{-11}$ | 2450 | $8.0 \times 10^{-15}$ |
| → OClO + Cl | $3.5 \times 10^{-13}$ | 1370 | $3.5 \times 10^{-15}$ |

| Reaction | A Factor | $E_a/R$ | $k_{298}$ |
|---|---|---|---|
| OH + CFCl$_3$ (CFC-11) → products | | | $< 5 \times 10^{-18}$ |
| OH + CF$_2$Cl$_2$ (CFC-12) → products | | | $< 6 \times 10^{-18}$ |
| OH + CHF$_2$Cl (HCFC-22) → CF$_2$Cl + H$_2$O | $1.0 \times 10^{-12}$ | 1600 | $4.7 \times 10^{-15}$ |
| OH + CH$_3$CFCl$_2$ (HCFC-141b) → CH$_2$CFCl$_2$ + H$_2$O | $1.7 \times 10^{-12}$ | 1700 | $5.7 \times 10^{-15}$ |
| OH + CH$_3$CF$_2$Cl (HCFC-142b) → CH$_2$CF$_2$Cl + H$_2$O | $1.3 \times 10^{-12}$ | 1800 | $3.1 \times 10^{-15}$ |
| O + BrO → Br + O$_2$ | $1.9 \times 10^{-11}$ | −230 | $4.1 \times 10^{-11}$ |
| OH + HBr → H$_2$O + Br | $1.1 \times 10^{-11}$ | 0 | $1.1 \times 10^{-11}$ |
| OH + CH$_3$Br → CH$_2$Br + H$_2$O | $4.0 \times 10^{-12}$ | 1470 | $2.9 \times 10^{-14}$ |
| HO$_2$ + Br → HBr + O$_2$ | $1.5 \times 10^{-11}$ | 600 | $2.0 \times 10^{-12}$ |
| HO$_2$ + BrO → products | $3.4 \times 10^{-12}$ | −540 | $2.1 \times 10^{-11}$ |
| Br + O$_3$ → BrO + O$_2$ | $1.7 \times 10^{-11}$ | 800 | $1.2 \times 10^{-12}$ |
| Br + H$_2$O$_2$ → HBr + HO$_2$ | $1.0 \times 10^{-11}$ | >3000 | $< 5 \times 10^{-16}$ |
| Br + CH$_2$O → HBr + HCO | $1.7 \times 10^{-11}$ | 800 | $1.1 \times 10^{-12}$ |
| BrO + O$_3$ → Br + 2O$_2$ | $\sim 1.0 \times 10^{-12}$ | >3200 | $< 2 \times 10^{-17}$ |
| BrO + NO → NO$_2$ + Br | $8.8 \times 10^{-12}$ | −260 | $2.1 \times 10^{-11}$ |
| BrO + ClO → Br + OClO | $1.6 \times 10^{-12}$ | −430 | $6.8 \times 10^{-12}$ |
| → Br + ClOO | $2.9 \times 10^{-12}$ | −220 | $6.1 \times 10^{-12}$ |
| → BrCl + O$_2$ | $5.8 \times 10^{-13}$ | −170 | $1.0 \times 10^{-12}$ |
| BrO + BrO → 2Br + O$_2$ | $4.0 \times 10^{-12}$ | 190 | $2.1 \times 10^{-12}$ |
| → Br$_2$ + O$_2$ | $4.2 \times 10^{-14}$ | −660 | $3.8 \times 10^{-13}$ |
| OH + H$_2$S → SH + H$_2$O | $6.0 \times 10^{-12}$ | 75 | $4.7 \times 10^{-12}$ |
| OH + OCS → products | $1.1 \times 10^{-13}$ | 1200 | $1.9 \times 10^{-15}$ |
| OH + CS$_2$ → products | (see Note 2) | | |
| OH + CH$_3$SH → H$_2$O + CH$_3$S | $9.9 \times 10^{-12}$ | −360 | $3.3 \times 10^{-11}$ |
| OH + CH$_3$SCH$_3$ → H$_2$O + CH$_3$SCH$_2$ | $1.2 \times 10^{-11}$ | 260 | $5.0 \times 10^{-12}$ |
| OH + CH$_3$SSCH$_3$ → products | $6.0 \times 10^{-11}$ | −400 | $2.3 \times 10^{-10}$ |
| S + O$_2$ → SO + O | $2.3 \times 10^{-12}$ | 0 | $2.3 \times 10^{-12}$ |
| SO + O$_2$ → SO$_2$ + O | $2.6 \times 10^{-13}$ | 2400 | $8.4 \times 10^{-17}$ |
| SO + O$_3$ → SO$_2$ + O$_2$ | $3.6 \times 10^{-12}$ | 1100 | $9.0 \times 10^{-14}$ |

*Note 1:* The rate coefficient for this reaction is a complex function of temperature and pressure and can be obtained as follows:

$$k(M, T) = k_0 + \frac{k_2 k_3 [M]}{k_2 + k_3 [M]}$$

where $k_0 = 7.2 \times 10^{-15} \exp(785/T)$
$k_2 = 4.1 \times 10^{-16} \exp(1440/T)$
$k_3 = 1.9 \times 10^{-33} \exp(725/T)$

*Note 2:* This reaction proceeds via the reversible formation of an adduct, which further reacts with O$_2$. The effective rate coefficient in 1 atm of air is given by:

$$k = \left[1.25 \times 10^{-16} \exp(4550/T)\right] \Big/ \left[T + 1.81 \times 10^{-3} \exp(3400/T)\right]$$

# F  Rate Coefficients for Gas-Phase Association Reactions[a,b]

| Reaction | Low-Pressure Limit $k_0^{300}$ | $n$ | High-Pressure Limit $k_\infty^{300}$ | $m$ |
| --- | --- | --- | --- | --- |
| $O + O_2 \xrightarrow{M} O_3$ | $6.0 \times 10^{-34}$ | 2.3 | | |
| $H + O_2 \xrightarrow{M} HO_2$ | $5.7 \times 10^{-32}$ | 1.6 | $7.5 \times 10^{-11}$ | 0 |
| $OH + NO \xrightarrow{M} HONO$ | $7.0 \times 10^{-31}$ | 2.6 | $3.6 \times 10^{-11}$ | 0.1 |
| $OH + NO_2 \xrightarrow{M} HNO_3$ | $2.5 \times 10^{-30}$ | 4.4 | $1.6 \times 10^{-11}$ | 1.7 |
| $HO_2 + NO_2 \xrightarrow{M} HO_2NO_2$ | $1.8 \times 10^{-31}$ | 3.2 | $4.7 \times 10^{-12}$ | 1.4 |
| $NO_2 + NO_3 \xrightarrow{M} N_2O_5$ | $2.2 \times 10^{-30}$ | 3.9 | $1.5 \times 10^{-12}$ | 0.7 |
| $CH_3 + O_2 \xrightarrow{M} CH_3O_2$ | $4.5 \times 10^{-31}$ | 3.0 | $1.8 \times 10^{-12}$ | 1.7 |
| $OH + C_2H_4 \xrightarrow{M} HOCH_2CH_2$ | $1.0 \times 10^{-28}$ | 0.8 | $8.8 \times 10^{-12}$ | 0 |
| $CH_3O_2 + NO_2 \xrightarrow{M} CH_3O_2NO_2$ | $1.5 \times 10^{-30}$ | 4.0 | $6.5 \times 10^{-12}$ | 2.0 |
| $CH_3C(O)O_2 + NO_2 \xrightarrow{M} CH_3C(O)O_2NO_2$ | $9.7 \times 10^{-29}$ | 5.6 | $9.3 \times 10^{-12}$ | 1.5 |
| $FO + NO_2 \xrightarrow{M} FONO_2$ | $2.6 \times 10^{-31}$ | 1.3 | $2.0 \times 10^{-11}$ | 1.5 |
| $ClO + NO_2 \xrightarrow{M} ClONO_2$ | $1.8 \times 10^{-31}$ | 3.4 | $1.5 \times 10^{-11}$ | 1.9 |
| $ClO + ClO \xrightarrow{M} Cl_2O_2$ | $2.2 \times 10^{-32}$ | 3.1 | $3.5 \times 10^{-12}$ | 1.0 |
| $BrO + NO_2 \xrightarrow{M} BrONO_2$ | $5.2 \times 10^{-31}$ | 3.2 | $6.9 \times 10^{-12}$ | 2.9 |
| $OH + SO_2 \xrightarrow{M} HOSO_2$ | $3.0 \times 10^{-31}$ | 3.3 | $1.5 \times 10^{-12}$ | 0 |

[a] The rate constant (expressed in $cm^3$ molecule$^{-1}$ s$^{-1}$) is given by

$$k(M,T) = \left( \frac{k_0(T)[M]}{1 + k_0(T)[M]/k_\infty(T)} \right) 0.6^{\{1+[\log_{10}(k_0(T)[M]/k_\infty(T))]^2\}^{-1}}$$

where $[M]$ is the air density (molecules cm$^{-3}$), and $T$ is the temperature (kelvin). $k_0$ is expressed in cm$^6$ molecule$^{-2}$ s$^{-1}$, and $k_\infty$ in cm$^3$ molecule$^{-1}$ s$^{-1}$. $k_0(T) = k_0^{300}(T/300)^{-n}$, $k_\infty(T) = k_\infty^{300}(T/300)^{-m}$.

[b] DeMore, W. B., S. P. Sander, D. M. Golden, R. F. Hampson, M. J. Kurylo, C. J. Howard, A. R. Ravishankara, C. E. Kolb, and M. J. Molina (1997) *Chemical Kinetics and Photochemical Data for Use in Stratospheric Modeling*, Evaluation No. 12, JPL Publication No. 97-4.

# G  Mass Accommodation Coefficients

Selected mass accommodation coefficients ($\alpha$) for the reversible uptake of various gases onto surfaces of specified composition.[a]

| Gas-Phase Species | Surface Type and Composition | Temp. (K) | $\alpha$ | Uncertainty Factors |
|---|---|---|---|---|
| $O_3$ | $H_2O$ (s) | 195-262 | >0.04 | |
| | $H_2O$ (l) | 292 | >0.002 | |
| | $HNO_3 \cdot 3H_2O$ (s) | 195 | $2.5 \times 10^{-4}$ | 3 |
| OH | $H_2O$ (l) | 275 | >0.004 | |
| | $H_2O$ (s) | 205-253 | >0.1 | |
| | $H_2SO_4 \cdot nH_2O$ (l), 28 wt% $H_2SO_4$ | 275 | >0.07 | |
| | $H_2SO_4 \cdot nH_2O$ (l), 97 wt% $H_2SO_4$ | 298 | $>5 \times 10^{-4}$ | |
| $HO_2$ | $H_2O$ (l) | 275 | >0.02 | |
| | $NH_4HSO_4$ (aq) | 293 | >0.2 | |
| | NaCl (s) or KCl (s) | 295 | 0.02 | 5 |
| $H_2O$ | $H_2O$ (s) | 200 | 0.5 | 2 |
| | $H_2SO_4 \cdot nH_2O$ (l), 96 wt% $H_2SO_4$ | 298 | >0.002 | |
| | $HNO_3 \cdot nH_2O$ (l) | 278 | >0.3 | |
| $H_2O_2$ | $H_2O$ (l) | 273 | 0.18 | 2 |
| | $H_2SO_4 \cdot H_2O$ (l), 96 wt% $H_2SO_4$ | 298 | >0.0008 | |
| $HNO_3$ | $H_2O$ (s) | 200 | 0.3 | 3 |
| | $H_2O$ (l) | 268 | 0.2 | 2 |
| | $HNO_3 \cdot 3H_2O$ (s) | 191-200 | 0.4 | 2 |
| | $HNO_3 \cdot nH_2O$ (l) | 278 | 0.6 | 2 |
| | $H_2SO_4 \cdot nH_2O$ (l) | | | |
| | (57.7 wt% $H_2SO_4$) | 191-200 | >0.3 | |
| | (73 wt% $H_2SO_4$) | 283 | 0.1 | 2 |
| | (75 wt% $H_2SO_4$) | 230 | >0.002 | |
| | (97 wt% $H_2SO_4$) | 295 | >0.002 | |
| | $H_2SO_4 \cdot 4H_2O$ (s) | 192 | >0.02 | |
| $HO_2NO_2$ | $H_2O$ (s) | 200 | 0.1 | 3 |
| $CH_2O$ | $H_2O$ (l) | 260-270 | 0.04 | 3 |
| | $H_2SO_4 \cdot nH_2O$ (l) | 235-300 | 0.04 | 3 |
| HCl | $H_2O$ (s) | 191-211 | 0.3 | 3 |
| | $H_2O$ (l) | 274 | 0.2 | 2 |
| | $HNO_3 \cdot 3H_2O$ (s) | 191-211 | 0.3 | 3 |
| HBr | $H_2O$ (s) | 200 | >0.2 | |
| | $HNO_3 \cdot 3H_2O$ (s) | 200 | >0.3 | |
| $SO_2$ | $H_2O$ (l) | 260-292 | 0.11 | 2 |
| $NH_3$ | $H_2O$ (l) | 295 | 0.06 | 3 |
| $CH_3CHO$ | $H_2O$ (l) | 267 | >0.03 | |

[a] Adapted from DeMore, W. B., S. P. Sander, D. M. Golden, R. F. Hampson, M. J. Kurylo, C. J. Howard, A. R. Ravishankara, C. E. Kolb, and M. J. Molina (1997) *Chemical Kinetics and Photochemical Data for Use in Stratospheric Modeling*, Evaluation No. 12, JPL Publication No. 97-4.

# H  Surface Reaction Probability

Selected surface reaction probabilities ($\gamma$) for irreversible reactive uptake of trace gases on condensed surfaces.[a]

| Reaction | Surface Type | Temp. (K) | $\gamma$ | Uncertainty Factor |
|---|---|---|---|---|
| $O_3$ (g) + carbon soot (s) $\to$ products | C (s) | 300 | 0.003 | 20 |
| $N_2O_5$ (g) + $H_2O$ $\to$ 2$HNO_3$ | $H_2O$ (s) | 195-200 | 0.01 | 3 |
| | $H_2O$ (l) | 260-295 | 0.05 | 2 |
| | $HNO_3 \cdot 3H_2O$ (s) | 200 | $3 \times 10^{-4}$ | 3 |
| | $H_2SO_4 \cdot nH_2O$ (l) | 195-300 | 0.1 | 2 |
| | $H_2SO_4 \cdot 4H_2O$ (s) | 195-207 | 0.006 | 2 |
| $N_2O_5$ (g) + HCl (s) $\to$ $ClNO_2$ + $HNO_3$ | $H_2O$/HCl (s) | 190-220 | 0.03 | |
| | $HNO_3 \cdot 3H_2O \cdot$ HCl (s) | 200 | 0.003 | 2 |
| $N_2O_5$ (g) + NaCl (s) $\to$ $ClNO_2$ + $NaNO_3$ (s) | NaCl (s) | 298 | $5 \times 10^{-4}$ | 20 |
| | NaCl (aq) | | >0.02 | |
| HOCl (g) + HCl (s) $\to$ $Cl_2$ + $H_2O$ | $H_2O \cdot$ HCl (s) | 195-200 | 0.3 | 3 |
| | $HNO_3 \cdot 3H_2O \cdot$ HCl (s) | 195-200 | 0.1 | 3 |
| $ClONO_2$ (g) + $H_2O$ (s) $\to$ HOCl + $HNO_3$ | $H_2O$ (s) | 180-200 | 0.3 | 3 |
| | $HNO_3 \cdot 3H_2O$ (s) | 200 | 0.001 | 10 |
| | $H_2SO_4 \cdot nH_2O$ (l) | 200-265 | see Note 1 | |
| $ClONO_2$ (g) + HCl (s) $\to$ $Cl_2$ + $HNO_3$ | $H_2O$ (s) | 180-202 | 0.3 | 5 |
| | $HNO_3 \cdot 3H_2O \cdot$ HCl (s) | 200 | 0.1 | 3 |
| $ClONO_2$ (g) + NaCl (s) $\to$ $Cl_2$ + $NaNO_3$ | NaCl (s) | 200-300 | 0.05 | 10 |
| $ClONO_2$ (g) + HBr (s) $\to$ BrCl + $HNO_3$ | $H_2O$/HBr (s) | 200 | >0.3 | |
| | $HNO_3 \cdot 3H_2O \cdot$ HBr (s) | 200 | >0.3 | |

[a]Adapted from DeMore, W. B., S. P. Sander, D. M. Golden, R. F. Hampson, M. J. Kurylo, C. J. Howard, A. R. Ravishankara, C. E. Kolb, and M. J. Molina (1997) *Chemical Kinetics and Photochemical Data for Use in Stratospheric Modeling*, Evaluation No. 12, JPL Publication No. 97-4.

*Note 1:* Values of $\gamma$ are strongly dependent on the $H_2SO_4$ concentration of the droplet, increasing with decreasing [$H_2SO_4$].

| Reaction | Surface Type | Temp. (K) | $\gamma$ | Uncertainty Factor |
|---|---|---|---|---|
| HOBr (g) + HCl (s)<br>$\rightarrow$ BrCl + H$_2$O | H$_2$O/HBr (s) | 228 | 0.3 | 3 |
| HOBr (g) + HBr (s)<br>$\rightarrow$ Br$_2$ + H$_2$O | H$_2$O/HBr (s) | 228 | 0.1 | 3 |
| BrONO$_2$ (g) + H$_2$O<br>$\rightarrow$ HOBr + HNO$_3$ | H$_2$O (s) | 200 | >0.3 | |
| | H$_2$SO$_4 \cdot n$H$_2$O | 210-298 | >0.8 | 2 |
| CH$_3$C(O)O$_2$ + H$_2$O<br>$\rightarrow$CH$_3$C(O)OH + HO$_2$ | H$_2$O (l) | 225 | $4 \times 10^{-3}$ | 3 |
| | H$_2$SO$_4 \cdot n$H$_2$O | | | |
| | (84 wt% H$_2$SO$_4$) | 246 | $3 \times 10^{-3}$ | 3 |
| | (51 wt% H$_2$SO$_4$) | 223 | $1 \times 10^{-3}$ | 3 |
| | (71 wt% H$_2$SO$_4$) | 298 | $1 \times 10^{-3}$ | 3 |

# I  Atmospheric Humidity

Atmospheric water vapor is produced by evaporation at the Earth's surface. Its abundance is highly variable in space and time, and is expressed in the *meteorological* literature in a variety of ways:

*Partial pressure:* pressure exerted by water vapor in air (expressed, for example, in mbar)

*Absolute humidity:* mass density of water vapor (expressed, for example, in $kg_{H_2O}$ m$^{-3}$)

*Mass mixing ratio:* ratio of the specific mass of water vapor to the specific mass of *dry* air (expressed, for example, in percent or ppmm)

*Specific humidity:* ratio of the specific mass of water vapor to the total specific mass of *moist* air (usually expressed in g kg$^{-1}$)

*Relative humidity:* ratio of actual vapor pressure to the saturation vapor pressure at a given temperature (expressed in percent)

Atmospheric *chemists* usually characterize the water vapor abundance by (see Box 1.4)

*Volume mixing ratio:* ratio of the water vapor number density to the total (moist) air number density (in percent or ppmv)

*Mass mixing ratio:* ratio of the water vapor mass density to the total air mass density (in percent or ppmm)

The saturation vapor pressure of a system $p_w^\circ$ is the pressure at which the vapor is in thermodynamic equilibrium with the liquid (or the solid) phase of the substance. Its value can be derived from the Clausius-Clapeyron equation (see, *e.g.*, Wallace and Hobbs, 1977), or from an empirical formula (Riegel and Bridger, 1992) such as

$$p_w^\circ(mb) = 6.11 \times 10^{aT/(b+T)}$$

with

$$\begin{array}{lll} a = 7.567 & b = 239.7 & T > 0°C \text{ over water} \\ = 7.744 & = 245.2 & T < 0°C \text{ over water} \\ = 9.716 & = 271.5 & T < 0°C \text{ over ice} \end{array}$$

if $T$ is temperature expressed in degrees Celsius. Numerical values of saturation pressure over water and ice are given in the accompanying table. Thus, for a given temperature, the abundance of atmospheric water vapor is limited by a saturation value at which it condenses and gives rise to the formation of clouds. The rapid upward and poleward decline in the specific humidity (illustrated in the accompanying figure) is directly related to the strong temperature dependence of the saturation vapor pressure. In addition, the cold temperatures at the tropopause limit the penetration of water vapor into the stratosphere (where its mixing ratio is on the order of only a few ppm).

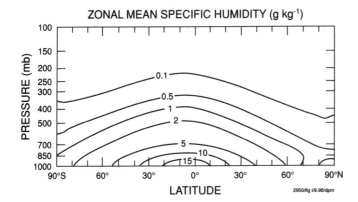

Zonal-mean specific humidity for January, as a function of latitude and pressure (data from European Centre for Medium Range Weather Forecasting analyses: 1986-1989; courtesy K. Trenberth).

## Saturation Vapor Pressure ($p_w^\circ$) over Water or Ice[a]

| Water | | Ice | |
|---|---|---|---|
| Temperature (°C) | $p_w^\circ$ (mbar) | Temperature (°C) | $p_w^\circ$ (mbar) |
| −35 | 3.14 (−1) | −100 | 1.40 (−5) |
| −30 | 5.09 (−1) | −95 | 3.78 (−5) |
| −25 | 8.07 (−1) | −90 | 9.67 (−5) |
| −20 | 1.25 | −85 | 2.35 (−4) |
| −15 | 1.91 | −80 | 5.47 (−4) |
| −10 | 2.86 | −75 | 1.22 (−3) |
| −5 | 4.21 | −70 | 2.62 (−3) |
| 0 | 6.11 | −65 | 5.41 (−3) |
| 5 | 8.72 | −60 | 1.08 (−2) |
| 10 | 1.23 (1) | −55 | 2.09 (−2) |
| 15 | 1.70 (1) | −50 | 3.93 (−2) |
| 20 | 2.34 (1) | −45 | 7.20 (−2) |
| 25 | 3.17 (1) | −40 | 1.28 (−1) |
| 30 | 4.24 (1) | −35 | 2.23 (−1) |
| 35 | 5.62 (1) | −30 | 3.80 (−1) |
| 40 | 7.38 (1) | −25 | 6.32 (−1) |
| 45 | 9.59 (1) | −20 | 1.03 |
| | | −15 | 1.65 |
| | | −10 | 2.6 |
| | | −5 | 4.02 |
| | | 0 | 6.12 |

[a] From List, R. J., ed. (1951) *Smithsonian Meteorological Tables*, Smithsonian Institution Press, Washington, D.C.

# J Henry's Law Coefficients of Gases Dissolving in Liquid Water[a]

| Species | $\mathcal{H}_{298}$(M atm$^{-1}$)[b] | $-\Delta \widehat{H}/R$ (K) |
|---|---|---|
| $SO_2$ | 1.23 | 3120 |
| $H_2O_2$ | $7.45 \times 10^4$ | 6620 |
| $HNO_3$ | $2.1 \times 10^5$ | |
| $HNO_2$ | 49.0 | 4780 |
| $O_3$ | $1.13 \times 10^{-2}$ | 2300 |
| $O_2$ | $1.3 \times 10^{-3}$ | |
| $NO_2$ | $1. \times 10^{-3}$ | 2500 |
| $NO$ | $1.9 \times 10^{-3}$ | 1480 |
| $CH_3O_2$ | 6.0 | 5600 |
| $CH_3OH$ | $2.2 \times 10^2$ | 4900 |
| $CO_2$ | $3.4 \times 10^{-2}$ | 2420 |
| $NH_3$ | 75.0 | 3400 |
| $CH_3C(O)O_2NO_2$ | 2.9 | 5910 |
| HCHO | $6.3 \times 10^3$ | 6460 |
| HCOOH | $3.5 \times 10^3$ | 5740 |
| HCl | $7.27 \times 10^2$ | 2020 |
| $CH_3OOH$ | $2.27 \times 10^2$ | 5610 |
| $CH_3C(O)OOH$ | $4.73 \times 10^2$ | 6170 |
| $NO_3$ | $2.1 \times 10^5$ | 8700 |
| OH | 25.0 | 5280 |
| $HO_2$ | $2.0 \times 10^3$ | 6640 |

[a]Pandis, S. N. and J. H. Seinfeld (1989) Sensitivity analysis of a chemical mechanism for aqueous-phase atmospheric chemistry, *J. Geophys. Res.* 94, 1105.

[b]$\mathcal{H}(T) = \mathcal{H}_{298} \exp \left[ \dfrac{-\Delta \widehat{H}}{R} \left( \dfrac{1}{T} - \dfrac{1}{298 \text{ K}} \right) \right]$.

# K  Aqueous Equilibrium Constants[a]

| Reaction | $K_{298}(M)$[b] | $-\Delta\widehat{H}/R$ K |
|---|---|---|
| $SO_2(aq) \rightleftharpoons HSO_3^- + H^+$ | $1.23 \times 10^{-2}$ | 1960 |
| $HSO_3^- \rightleftharpoons SO_3^{2-} + H^+$ | $6.61 \times 10^{-8}$ | 1500 |
| $H_2SO_4(aq) \rightleftharpoons HSO_4^- + H^+$ | 1000.0 | |
| $HSO_4^- \rightleftharpoons SO_4^{2-} + H^+$ | $1.02 \times 10^{-2}$ | 2720 |
| $H_2O_2(aq) \rightleftharpoons HO_2^- + H^+$ | $2.2 \times 10^{-12}$ | -3730 |
| $HNO_3(aq) \rightleftharpoons NO_3^- + H^+$ | 15.4 | 8700 |
| $HNO_2(aq) \rightleftharpoons NO_2^- + H^+$ | $5.1 \times 10^{-4}$ | -1260 |
| $CO_2(aq) \rightleftharpoons HCO_3^- + H^+$ | $4.46 \times 10^{-7}$ | -1000 |
| $HCO_3^- \rightleftharpoons CO_3^{2-} + H^+$ | $4.68 \times 10^{-11}$ | -1760 |
| $NH_4OH(aq) \rightleftharpoons NH_4^+ + OH^-$ | $1.75 \times 10^{-5}$ | -450 |
| $H_2O \rightleftharpoons OH^- + H^+$ | $1.0 \times 10^{-14}$[c] | -6710 |
| $CH_2O(aq) \rightleftharpoons H_2C(OH)_2(aq)$ | $1.82 \times 10^{3}$[d] | 4020 |
| $HCOOH(aq) \rightleftharpoons HCOO^- + H^+$ | $1.78 \times 10^{-4}$ | -20 |
| $HCl(aq) \rightleftharpoons Cl^- + H^+$ | $1.74 \times 10^{6}$ | 6900 |
| $HO_2(aq) \rightleftharpoons O_2^- + H^+$ | $3.5 \times 10^{-5}$ | |

[a] Pandis, S. N. and J. H. Seinfeld (1989) Sensitivity analysis of a chemical mechanism for aqueous-phase atmospheric chemistry, *J. Geophys. Res.* 94, 1105.

[b] $K(T) = K_{298} \exp\left[\dfrac{-\Delta\widehat{H}}{R}\left(\dfrac{1}{T} - \dfrac{1}{298}\right)\right]$.

[c] Units are $M^2$

[d] unitless

# L    Rate Coefficients for Aqueous-Phase Reactions[a,b]

| Reaction | $k_{298}(M^{-1}\,s^{-1})$ | $E_a/R$ (K) |
|---|---|---|
| $O_3 + h\nu + H_2O \to H_2O_2 + O_2$ | | |
| $H_2O_2 + h\nu \to 2\,OH$ | | |
| $NO_3^- + h\nu + O_2 \to NO_2^- + O_3$ | | |
| $NO_3^- + h\nu + H^+ \to NO_2 + OH$ | | |
| $Fe^{3+} + h\nu \xrightarrow{H_2O} Fe^{2+} + OH + H^+$ | $9.6 \times 10^{-7}\,s^{-1}$ | |
| $Fe(OH)^{2+} + h\nu \to Fe^{2+} + OH$ | $5.9 \times 10^{-4}\,s^{-1}$ | |
| $Fe(OH)_2^+ + h\nu \to Fe^{2+} + OH + OH^-$ | $5.8 \times 10^{-4}\,s^{-1}$ | |
| $H_2O_2 + OH \to HO_2 + H_2O$ | $2.7 \times 10^7$ | 1715 |
| $OH + O_2^- \to OH^- + O_2$ | $1.0 \times 10^{10}$ | 1500 |
| $O_3 + OH \to HO_2 + O_2$ | $2.0 \times 10^9$ | 0 |
| $OH + HO_2 \to H_2O + O_2$ | $7.0 \times 10^9$ | 1500 |
| $HO_2 + HO_2 \to H_2O_2 + O_2$ | $8.6 \times 10^5$ | 2350 |
| $HO_2 + O_2^- \to HO_2^- + O_2$ | $1.0 \times 10^8$ | 1500 |
| $O_3 + O_2^- + H_2O \to OH + 2O_2 + OH^-$ | $1.5 \times 10^9$ | 1500 |
| $O_3 + HO_2^- \to OH + O_2^- + O_2$ | $2.8 \times 10^6$ | 2500 |
| $CH_3OO + O_2^- + H_2O \to CH_3OOH + OH^- + O_2$ | $5.0 \times 10^7$ | 1600 |
| $CH_3OOH + OH \to CH_3OO + H_2O$ | $2.7 \times 10^7$ | 1715 |
| $CH_3OOH + OH \to CH_2(OH)_2 + OH$ | $1.9 \times 10^7$ | 1850 |
| $CH_3OH + OH + O_2 \to CH_2(OH)_2 + HO_2$ | $4.5 \times 10^8$ | 1500 |
| $OH + Cl^- + H^+ \to H_2O + Cl$ | $1.5 \times 10^{10}$ | 1500 |
| $HO_2 + Cl_2^- \to O_2 + 2\,Cl^- + H^+$ | $4.5 \times 10^9$ | 1500 |
| $O_2^- + Cl_2^- \to O_2 + 2\,Cl^-$ | $1.0 \times 10^9$ | 1500 |
| $H_2O_2 + Cl_2^- \to HO_2 + 2\,Cl^- + H^+$ | $1.4 \times 10^5$ | 3350 |
| $CH_2(OH)_2 + OH + O_2 \to HCOOH + HO_2 + H_2O$ | $2.0 \times 10^9$ | 1500 |
| $HCOOH + OH + O_2 \to CO_2 + HO_2 + H_2O$ | $2.0 \times 10^8$ | 1500 |
| $HCOO^- + OH + O_2 \to CO_2 + HO_2 + OH^-$ | $2.5 \times 10^9$ | 1500 |
| $HCOO^- + Cl_2^- + O_2 \to CO_2 + HO_2 + 2\,Cl^-$ | $1.9 \times 10^6$ | 2550 |
| $HCO_3^- + OH \to CO_3^- + H_2O$ | $1.5 \times 10^7$ | 1900 |
| $HCO_3^- + O_2^- \to CO_3^- + HO_2^-$ | $1.5 \times 10^6$ | 0 |
| $CO_3^- + H_2O_2 \to HO_2 + HCO_3^-$ | $8.0 \times 10^5$ | 2800 |
| $CO_3^- + O_2^- \to CO_3^{2-} + O_2$ | $4.0 \times 10^8$ | 1500 |

[a] Jacob, D. J. (1986) Chemistry of OH in remote clouds and its role in the production of formic acid and peroxymonosulfate, *J. Geophys. Res.* 91, 9807.

[b] Jacob, D. J., E. W. Gottlieb, and M. J. Prather (1989) Chemistry of a polluted cloudy boundary layer, *J. Geophys. Res.* 94, 12,975.

| Reaction | $k_{298}(\text{M}^{-1}\text{ s}^{-1})$ | $E_a/R$ (K) |
|---|---|---|
| $CO_3^- + CH_3OH \rightarrow CH_2(OH)_2 + HO_2 + HCO_3^-$ | $2.6 \times 10^3$ | 4500 |
| $CO_3^- + HCOO^- \rightarrow CO_2 + HO_2 + HCO_3^- + OH^-$ | $1.1 \times 10^5$ | 3400 |
| $HNO_2 + OH \rightarrow NO_2 + H_2O$ | $1.0 \times 10^9$ | 1500 |
| $NO_2^- + OH \rightarrow NO_2 + OH^-$ | $1.0 \times 10^{10}$ | 1500 |
| $NO_2^- + Cl_2^- \rightarrow NO_2 + 2\,Cl^-$ | $2.5 \times 10^8$ | 1500 |
| $NO_2 + OH \rightarrow NO_3^- + H^+$ | $1.3 \times 10^9$ | 1500 |
| $NO_3 + H_2O_2 \rightarrow NO_3^- + H^+ + HO_2$ | $1.0 \times 10^6$ | 2750 |
| $NO_3 + Cl^- \rightarrow NO_3^- + Cl$ | $1.0 \times 10^7$ | 4300 |
| $NO_2 + O_2^- \rightarrow NO_2^- + O_2$ | $1 \times 10^8$ | 0 |
| $NO_2 + HO_2 \rightarrow HO_2NO_2$ | $1 \times 10^7$ | 0 |
| $NO_3 + H_2O \rightarrow HNO_3 + OH$ | 6 | 4510 |
| $HSO_3^- + H_2O_2 \rightarrow SO_4^{2-} + H^+ + H_2O$ | $c$ | |
| $HSO_3^- + O_3 \rightarrow SO_4^{2-} + H^+ + O_2$ | $3.2 \times 10^5$ | 4800 |
| $SO_3^{2-} + O_3 \rightarrow SO_4^{2-} + O_2$ | $1.0 \times 10^9$ | 4000 |
| $S(IV) + Fe^{3+} + O_2 \rightarrow SO_4^{2-} + \text{products}$ | $d\,\dfrac{6}{[H^+]}10^{\frac{-2\sqrt{I}}{1+\sqrt{I}}}$ | 0 |
| $HSO_3^- + OH \rightarrow SO_3^- + H_2O$ | $4.5 \times 10^9$ | 1500 |
| $SO_3^- + O_2 \rightarrow SO_5^-$ | $1.5 \times 10^9$ | 1500 |
| $SO_5^- + HSO_3^- \rightarrow SO_4^{2-} + SO_4^- + H^+$ | $7.5 \times 10^4$ | 3500 |
| $SO_5^- + HSO_3^- \rightarrow HSO_5^- + SO_3^-$ | $2.5 \times 10^4$ | 3850 |
| $HSO_5^- + HSO_3^- \rightarrow 2\,SO_4^{2-} + 2\,H^+$ | $7.1 \times 10^6$ | 3100 |
| $SO_4^- + Cl^- \rightarrow SO_4^{2-} + Cl$ | $2.6 \times 10^8$ | 1500 |
| $SO_4^- + HSO_3^- \rightarrow SO_4^{2-} + SO_3^- + H^+$ | $7.5 \times 10^8$ | 1500 |
| $SO_4^- + SO_3^{2-} \rightarrow SO_4^{2-} + SO_3^-$ | $5.3 \times 10^8$ | 1500 |
| $SO_4^- + H_2O_2 \rightarrow SO_4^{2-} + H^+ + HO_2$ | $1.2 \times 10^7$ | 2000 |
| $HSO_3^- + Cl_2^- \rightarrow SO_3^- + H^+ + 2\,Cl^-$ | $3.4 \times 10^8$ | 1500 |
| $Fe(II) + OH \rightarrow Fe(III) + OH^-$ | $3.0 \times 10^8$ | 1500 |
| $Fe(II) + HO_2 \xrightarrow{H_2O} Fe(III) + H_2O_2 + OH^-$ | $1.2 \times 10^6$ | 2700 |
| $Fe(II) + O_2^- \xrightarrow{2\,H_2O} Fe(III) + H_2O_2 + 2\,OH^-$ | $1.0 \times 10^7$ | 2050 |
| $Fe(OH)^+ + H_2O_2 \rightarrow Fe^{3+} + OH + 2\,OH^-$ | $1.9 \times 10^6$ | 6200 |
| $Fe(II) + O_3 \xrightarrow{H_2O} Fe(III) + OH + OH^- + O_2$ | $1.7 \times 10^5$ | 3250 |
| $Fe(II) + Cl_2^- \rightarrow Fe(III) + 2\,Cl^-$ | $1.0 \times 10^7$ | 2700 |
| $\rightarrow Fe(III) + 2\,Cl^-$ | $4.0 \times 10^6$ | 3750 |
| $Fe(III) + HO_2 \rightarrow Fe(II) + H^+ + O_2$ | $2.0 \times 10^4$ | 3900 |
| $Fe(III) + O_2^- \rightarrow Fe(II) + O_2$ | $1.5 \times 10^8$ | 1500 |
| $Mn(II) + OH \rightarrow Mn(III) + OH^-$ | $3.4 \times 10^7$ | 1700 |
| $Mn(II) + HO_2 \xrightarrow{H_2O} Mn(III) + H_2O_2 + OH^-$ | $6.0 \times 10^6$ | 2200 |
| $Mn(II) + O_2^- \xrightarrow{2\,H_2O} Mn(III) + H_2O_2 + 2\,OH^-$ | $1.1 \times 10^8$ | 1500 |
| $Mn(II) + Cl_2^- \rightarrow Mn(III) + 2\,Cl^-$ | $1.4 \times 10^7$ | 4050 |
| $Mn(III) + HO_2 \rightarrow Mn(II) + O_2 + H^+$ | $2.0 \times 10^4$ | 3900 |
| $Mn(III) + O_2^- \rightarrow Mn(II) + O_2$ | $1.5 \times 10^8$ | 1500 |
| $Mn(III) + H_2O_2 \rightarrow Mn(II) + HO_2 + H^+$ | $3.2 \times 10^4$ | 3750 |
| $Mn(III) + Fe(II) \rightarrow Mn(II) + Fe(III)$ | $2.1 \times 10^4$ | 3900 |
| $Mn(III) + Cu(I) \rightarrow Mn(II) + Cu(II)$ | $2.1 \times 10^4$ | 3900 |

$c$: $k_{298} = \dfrac{7.45 \times 10^7 [H^+]}{1+13[H^+]}$  $E_a/R = 4750$.

$d$: $I$ = ionic strength.

|  | $k_{298}(\text{M}^{-1}\text{ s}^{-1})$ | $E_a/R$ (K) |
|---|---|---|
| $\text{Cu(I)} + \text{OH} \rightarrow \text{Cu(II)} + \text{OH}^-$ | $3.0 \times 10^8$ | 1500 |
| $\text{Cu(I)} + \text{HO}_2 \xrightarrow{\text{H}_2\text{O}} \text{Cu(II)} + \text{H}_2\text{O}_2 + \text{OH}^-$ | $1.5 \times 10^9$ | 1500 |
| $\text{Cu(I)} + \text{O}_2^- \xrightarrow{2\text{H}_2\text{O}} \text{Cu(II)} + \text{H}_2\text{O}_2 + 2\text{ OH}^-$ | $1.0 \times 10^{10}$ | 1500 |
| $\text{Cu(I)} + \text{H}_2\text{O}_2 \rightarrow \text{Cu(II)} + \text{OH} + \text{OH}^-$ | $4.0 \times 10^5$ | 3100 |
| $\text{Cu(I)} + \text{Cl}_2^- \rightarrow \text{Cu(II)} + 2\text{ Cl}^-$ | $1.4 \times 10^7$ | 3200 |
| $\text{Cu(II)} + \text{HO}_2 \rightarrow \text{Cu(I)} + \text{O}_2 + \text{H}^+$ | $1.0 \times 10^8$ | 1500 |
| $\text{Cu(II)} + \text{O}_2^- \rightarrow \text{Cu}^+ + \text{O}_2$ | $5.0 \times 10^9$ | 1500 |

# M Spectrum of Solar Extraterrestial Actinic Flux (120-730 nm)

The following table presents the spectral distribution of the solar actinic flux between 120 and 730 nm at the mean Sun-Earth distance (1 AU). The values, expressed as the number of photons and the energy per unit area and unit time, are integrated over wavelength intervals of 5 nm (between 120 and 650 nm) and of 10 nm (between 650 and 730 nm). These fluxes are representative of low solar activity. The values are from the SOLSTICE instrument (Upper Atmosphere Research Satellite) for the 120-400 nm interval, and were kindly provided by G. Rottman (Univ. of Colorado). An estimate of the variability (expressed in percent) between solar minimum and solar maximum conditions over the 11-year solar cycle is also provided. For the 400-730 nm interval, the values of the solar flux are based on the compilations by WMO (1981) and by Brasseur and Simon (1981). The solar variability is low (less than the precision in the measurement), and no estimate is given.

| Wavelength Interval (nm) | (Photons) ($cm^{-2}\ s^{-1}$) | (Energy) ($mW\ m^{-2}$) | Solar Cycle Variability[a] (percent) |
|---|---|---|---|
| 120-125 | 3.59E+11 | 5.87 | 68.19 |
| 125-130 | 7.89E+09 | 0.12 | 37.02 |
| 130-135 | 2.68E+10 | 0.40 | 31.76 |
| 135-140 | 1.43E+10 | 0.21 | 27.47 |
| 140-145 | 1.91E+10 | 0.27 | 19.07 |
| 145-150 | 2.83E+10 | 0.38 | 13.08 |
| 150-155 | 5.13E+10 | 0.67 | 19.37 |
| 155-160 | 7.03E+10 | 0.89 | 14.28 |
| 160-165 | 1.02E+11 | 1.25 | 15.09 |
| 165-170 | 1.93E+11 | 2.29 | 10.77 |
| 170-175 | 3.28E+11 | 3.78 | 8.44 |
| 175-180 | 6.22E+11 | 6.95 | 9.29 |
| 180-185 | 1.01E+12 | 11.03 | 9.28 |
| 185-190 | 1.46E+12 | 15.41 | 6.41 |
| 190-195 | 2.12E+12 | 21.85 | 6.24 |
| 195-200 | 3.14E+12 | 31.53 | 5.89 |
| 200-205 | 4.55E+12 | 44.60 | 5.69 |
| 205-210 | 7.67E+12 | 73.33 | 4.11 |
| 210-215 | 1.89E+13 | 176.91 | 1.53 |
| 215-220 | 2.22E+13 | 202.98 | 1.78 |

[a] Defined as 100 [flux (solar max)-flux (solar min)]/flux (solar min).

| Wavelength Interval (nm) | (Photons) (cm$^{-2}$ s$^{-1}$) | (Energy) (mW m$^{-2}$) | Solar Cycle Variability (percent) |
|---|---|---|---|
| 220-225 | 3.03E+13 | 269.99 | 0.99 |
| 225-230 | 2.74E+13 | 238.87 | 0.72 |
| 230-235 | 2.84E+13 | 243.12 | 0.85 |
| 235-240 | 2.93E+13 | 245.49 | 0.75 |
| 240-245 | 3.64E+13 | 298.23 | 0.04 |
| 245-250 | 3.27E+13 | 262.38 | 1.01 |
| 250-255 | 3.28E+13 | 258.18 | 1.74 |
| 255-260 | 7.26E+13 | 559.67 | 0.77 |
| 260-265 | 9.73E+13 | 734.94 | 0.38 |
| 265-270 | 1.73E+14 | 1282.59 | 1.01 |
| 270-275 | 1.47E+14 | 1074.77 | 0.82 |
| 275-280 | 1.29E+14 | 925.98 | 2.40 |
| 280-285 | 1.73E+14 | 1215.72 | 1.43 |
| 285-290 | 2.42E+14 | 1668.35 | 1.16 |
| 290-295 | 4.21E+14 | 2860.30 | 1.49 |
| 295-300 | 3.95E+14 | 2637.67 | 1.82 |
| 300-305 | 4.03E+14 | 2647.27 | |
| 305-310 | 4.70E+14 | 3036.75 | |
| 310-315 | 5.59E+14 | 3553.80 | |
| 315-320 | 5.74E+14 | 3592.15 | |
| 320-325 | 6.26E+14 | 3859.35 | |
| 325-330 | 8.31E+14 | 5037.69 | |
| 330-335 | 8.47E+14 | 5059.18 | |
| 335-340 | 7.94E+14 | 4673.09 | |
| 340-345 | 8.42E+14 | 4885.65 | |
| 345-350 | 8.30E+14 | 4745.21 | |
| 350-355 | 9.43E+14 | 5311.66 | |
| 355-360 | 8.42E+14 | 4680.46 | |
| 360-365 | 9.19E+14 | 5037.67 | |
| 365-370 | 1.14E+15 | 6148.40 | |
| 370-375 | 1.02E+15 | 5452.38 | |
| 375-380 | 1.16E+15 | 6090.10 | |
| 380-385 | 9.42E+14 | 4895.22 | |
| 385-390 | 1.03E+15 | 5303.67 | |
| 390-395 | 1.07E+15 | 5409.19 | |
| 395-400 | 1.27E+15 | 6319.53 | |
| 400-405 | 1.72E+15 | 8471.69 | |
| 405-410 | 1.77E+15 | 8614.07 | |
| 410-415 | 1.86E+15 | 8979.47 | |
| 415-420 | 1.90E+15 | 9048.08 | |
| 420-425 | 1.81E+15 | 8515.24 | |
| 425-430 | 1.69E+15 | 7880.95 | |
| 430-435 | 1.79E+15 | 8249.42 | |
| 435-440 | 1.98E+15 | 9018.36 | |
| 440-445 | 2.09E+15 | 9388.11 | |
| 445-450 | 2.25E+15 | 9993.89 | |
| 450-455 | 2.30E+15 | 10125.06 | |

| Wavelength Interval (nm) | (Photons) (cm$^{-2}$ s$^{-1}$) | (Energy) (mW m$^{-2}$) | Solar Cycle Variability (percent) |
|---|---|---|---|
| 455-460 | 2.31E+15 | 10057.85 | |
| 460-465 | 2.35E+15 | 10078.05 | |
| 465-470 | 2.34E+15 | 9949.00 | |
| 470-475 | 2.36E+15 | 9927.85 | |
| 475-480 | 2.42E+15 | 10073.66 | |
| 480-485 | 2.35E+15 | 9701.50 | |
| 485-490 | 2.32E+15 | 9459.29 | |
| 490-495 | 2.45E+15 | 9867.74 | |
| 495-500 | 2.45E+15 | 9788.55 | |
| 500-505 | 2.43E+15 | 9612.04 | |
| 505-510 | 2.48E+15 | 9713.17 | |
| 510-515 | 2.43E+15 | 9405.09 | |
| 515-520 | 2.39E+15 | 9160.59 | |
| 520-525 | 2.43E+15 | 9244.11 | |
| 525-530 | 2.52E+15 | 9495.62 | |
| 530-535 | 2.59E+15 | 9686.41 | |
| 535-540 | 2.57E+15 | 9503.86 | |
| 540-545 | 2.58E+15 | 9452.90 | |
| 545-550 | 2.60E+15 | 9439.18 | |
| 550-555 | 2.60E+15 | 9353.76 | |
| 555-560 | 2.59E+15 | 9234.22 | |
| 560-565 | 2.60E+15 | 9205.14 | |
| 565-570 | 2.64E+15 | 9229.11 | |
| 570-575 | 2.66E+15 | 9235.31 | |
| 575-580 | 2.70E+15 | 9275.81 | |
| 580-585 | 2.72E+15 | 9298.56 | |
| 585-590 | 2.69E+15 | 9101.01 | |
| 590-595 | 2.64E+15 | 8873.24 | |
| 595-600 | 2.64E+15 | 8782.36 | |
| 600-605 | 2.66E+15 | 8758.96 | |
| 605-610 | 2.68E+15 | 8785.03 | |
| 610-615 | 2.66E+15 | 8615.96 | |
| 615-620 | 2.67E+15 | 8594.48 | |
| 620-625 | 2.66E+15 | 8493.51 | |
| 625-630 | 2.61E+15 | 8283.29 | |
| 630-635 | 2.63E+15 | 8264.95 | |
| 635-640 | 2.65E+15 | 8262.49 | |
| 640-645 | 2.66E+15 | 8229.13 | |
| 645-650 | 2.66E+15 | 8168.08 | |
| 650-660 | 5.13E+15 | 15567.90 | |
| 660-670 | 5.15E+15 | 15393.26 | |
| 670-680 | 5.25E+15 | 15459.69 | |
| 680-690 | 5.18E+15 | 15030.88 | |
| 690-700 | 5.44E+15 | 15572.50 | |
| 700-710 | 5.43E+15 | 15295.22 | |
| 710-720 | 5.03E+15 | 13969.32 | |
| 720-730 | 4.98E+15 | 13653.26 | |

# N  Photolysis Frequencies

## Some Photolysis Reactions of Importance in the Lower Atmosphere

For species for which more than one photolysis channel exists, approximate branching ratios (applicable in the sunlit lower stratosphere) are given in parentheses in the following table. Note that these values can vary substantially with altitude and solar zenith angle.

In the following figures, the photolysis frequency $j(z)$ (also called the photodissociation coefficient) for selected atmospheric molecules is represented as a function of altitude $z$. Except for molecular oxygen and ozone in the middle atmosphere and for selected molecules in the troposphere, where different solar zenith angles $\chi$ are considered ($\chi = 0°, 60°, 75.5°$, and $87.1°$ or sec $\chi = 1, 2, 4, 20$), the photolysis frequencies are shown for overhead sun conditions ($\chi = 0°$ or sec $\chi = 1$) and represent therefore maximum values at a given height. Atmospheric absorption of solar radiation by ozone and molecular oxygen, as well as multiple scattering by air molecules, are taken into account in the calculation of the $j$ values, but the presence of aerosols and clouds is ignored (clear sky conditions). The adopted Earth's albedo is 0.3 (figures representing $j$ values in the middle atmosphere) or 0.1 (figures representing $j$ values below 20 km).

| | | | |
|---|---|---|---|
| $O_3 + h\nu$ | $\to O(^3P) + O_2(^3\Sigma)(\approx 95\%)$ | $Cl_2O_2 + h\nu$ | $\to Cl + ClOO$ |
| | $\to O(^1D) + O_2(^1\Delta)(\approx 5\%)$ | $HOCl + h\nu$ | $\to OH + Cl$ |
| $H_2O_2 + h\nu$ | $\to OH + OH$ | $ClONO_2 + h\nu$ | $\to ClO + NO_2 \ (\approx 30\%)$ |
| $NO_2 + h\nu$ | $\to NO + O$ | | $\to Cl + NO_3 \ (\approx 70\%)$ |
| $NO_3 + h\nu$ | $\to NO_2 + O \ (\approx 90\%)$ | $CCl_3F(\text{F-11}) + h\nu$ | $\to CCl_2F + Cl$ |
| | $\to NO + O_2 \ (\approx 10\%)$ | $CCl_2F_2(\text{F-12}) + h\nu$ | $\to CClF_2 + Cl$ |
| $N_2O + h\nu$ | $\to N_2 + O(^1D)$ | $HOBr + h\nu$ | $\to OH + Br$ |
| $N_2O_5 + h\nu$ | $\to NO_3 + NO_2$ | $BrO + h\nu$ | $\to Br + O$ |
| $HONO + h\nu$ | $\to OH + NO$ | $BrONO_2 + h\nu$ | $\to Br + NO_3 \ (\approx 100\%)$ |
| $HONO_2 + h\nu$ | $\to OH + NO_2$ | | $\to BrO + NO_2 \ (\text{minor})$ |
| $HO_2NO_2 + h\nu$ | $\to OH + NO_3 \ (\approx 33\%)$ | $CH_3Br + h\nu$ | $\to CH_3 + Br$ |
| | $\to HO_2 + NO_2 \ (\approx 67\%)$ | $CF_3Br + h\nu$ | $\to CF_3 + Br$ |
| $CH_2O + h\nu$ | $\to CO + H_2 \ (\approx 60\%)$ | $CF_3I + h\nu$ | $\to CF_3 + I$ |
| | $\to HCO + H \ (\approx 40\%)$ | $OCS + h\nu$ | $\to CO + S$ |
| $CH_3OOH + h\nu$ | $\to CH_3O + OH$ | | |

# Atmospheric Chemistry and Global Change

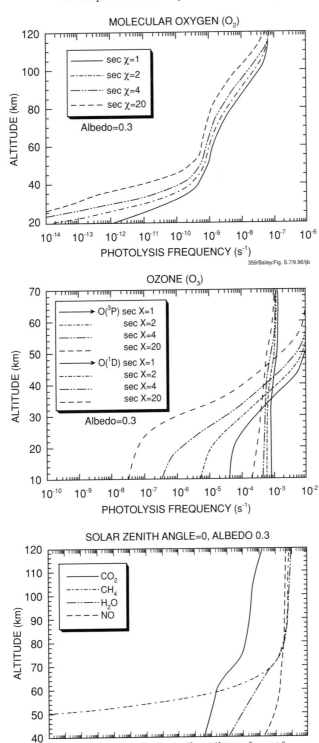

Appendix N

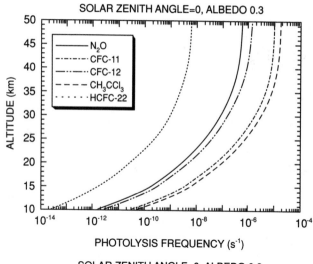

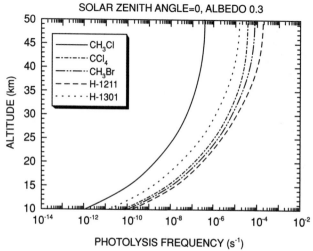

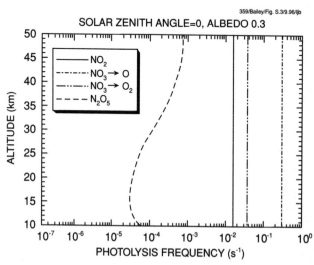

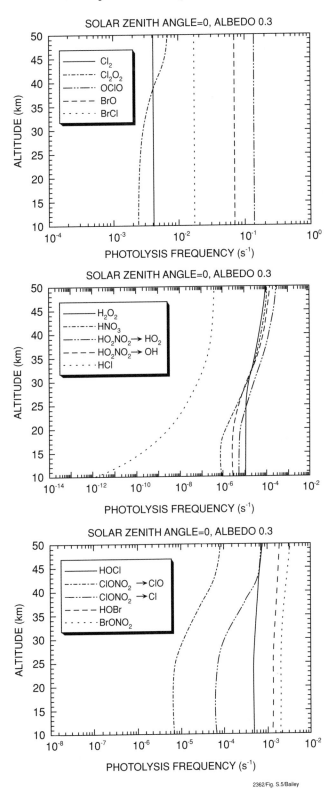

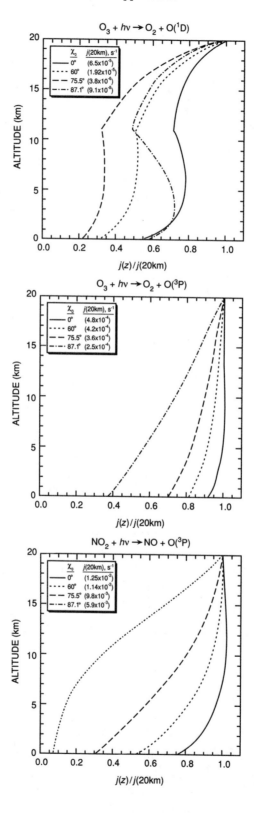

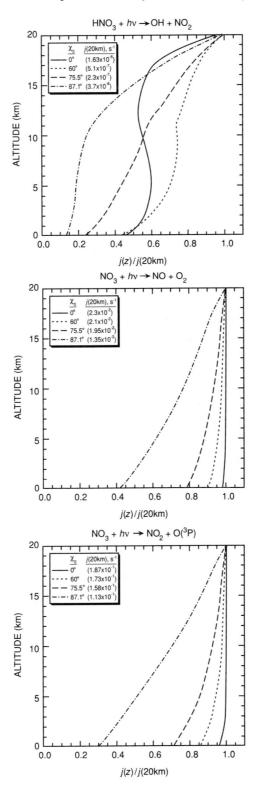

Appendix N

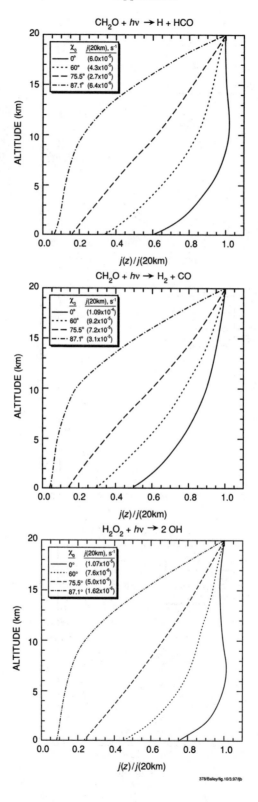

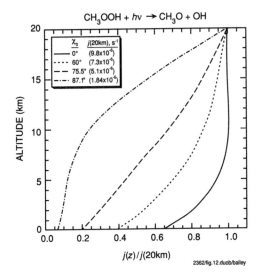

# Sample Problems

# Sample Problems

## 1. Atmospheric Burden and Global Lifetime

The major gases of the terrestrial atmosphere are molecular nitrogen $N_2$ (78%), molecular oxygen $O_2$ (21%), and argon Ar (1%). It is assumed that, to a first approximation, the temperature $T_0$ remains constant as a function of altitude (isothermal atmosphere).

1. Calculate the molar mass $\mathcal{M}$ of the atmospheric gas assumed to be a mixture of $N_2$, $O_2$, and Ar. The molar mass of N is 14 g mol$^{-1}$, that of O is 16 g mol$^{-1}$, and that of Ar is 40 g mol$^{-1}$.

2. Express the atmospheric scale height $H_0$ as a function of $T_0$, $\mathcal{M}$, the Boltzmann constant $k$, the Avogadro number $\mathcal{N}$, and the gravitational acceleration $g$. Derive the "equivalent volume" $V$ of the atmosphere (volume assuming that the whole atmosphere follows STP conditions) as a function of $H_0$ and the Earth's radius $a$. Calculate $H_0$ and $V$, assuming that $a = 6370$ km, $k = 1.38 \times 10^{-23}$ J K$^{-1}$, $\mathcal{N} = 6.02 \times 10^{23}$ molecules per mole, and $T_0 = 273$ K.

3. Calculate the air number density $n_0$ (molecules cm$^{-3}$) at the surface, knowing that, for STP conditions (pressure of 1 atm and temperature of 273 K), the volume of 1 mole is 22,400 cm$^3$. Calculate the number density of molecular oxygen.

4. Express the total mass $M$ of the atmosphere as a function of the equivalent volume $V$, the air density at the surface $n_0$, the molar mass $\mathcal{M}$, and the Avogadro number $\mathcal{N}$. Calculate $M$ as well as the total mass of oxygen $M(O_2)$.

5. Express the surface pressure $p_0$ as a function of the total atmospheric mass $M$, the gravitational acceleration $g$, and the Earth's radius $a$. Compare with the value obtained with the ideal gas law. Note that $g = 9.81$ m$^2$ s$^{-1}$.

6. Explain why the total mass $M_i$ of a gas $i$ can be expressed as a function of its net atmospheric input flux $F_i$ and its overall lifetime $\tau_i$ by

$$\frac{dM_i}{dt} + \frac{M_i}{\tau_i} = F_i \tag{1}$$

   Express the equilibrium burden $M_{\text{eq}}$ as a function of $F_i$ and $\tau_i$.

7. Assuming that the net input flux of oxygen ($O_2$) per unit surface area is $3.8 \times 10^{-6}$ kg m$^{-2}$ s$^{-1}$, calculate the global lifetime of oxygen.

8. Assume that the current volume mixing ratio of methane is 1.75 ppmv and that its concentration increases by $x = 1\%$ yr$^{-1}$. Having calculated above the total mass of the atmosphere, calculate the total mass of methane (knowing that the molar mass of CH$_4$ is 16 g mol$^{-1}$). Derive from Eq. (1) that the global lifetime

of methane can be expressed as

$$\tau(CH_4) = \frac{M(CH_4)}{F(CH_4) - xM(CH_4)}$$

Calculate $\tau(CH_4)$ for a net input flux per unit area of $1.025 \times 10^{-3}$ kg m$^{-2}$ yr$^{-1}$.

## 2. Photodissociation

The variation with height of the ozone number density $n$ above $z_0 = 35$ km can be written

$$n(z) = n(z_0) \exp\left(-\frac{z-z_0}{H}\right)$$

with $H = 5$ km and $n(z_0) = 10^{12}$ molecule cm$^{-3}$.

Assume that ozone absorbs solar radiation with a cross section $\sigma_m = 4 \times 10^{-17}$ cm$^2$ molecule$^{-1}$ at a "mean" wavelength $\lambda_m$. The corresponding solar actinic flux at the top of the atmosphere $F_S = 3.8 \times 10^{14}$ photons cm$^{-2}$ s$^{-1}$. The Sun is assumed to be overhead.

1. Express the solar flux $F(z)$ at altitude $z$ as a function of $F_S$, $\sigma_m$, $n(z_0)$, $H$, and $z$.

2. Deduce the photolysis rate of ozone at altitude $z$ [number of ozone molecules photodissociated per unit volume and time (molecule cm$^{-3}$ s$^{-1}$)]. The quantum yield for photolysis is assumed to be unity.

3. Express, as a function of $\sigma_m$, $n(z_0)$, $H$, and $z_0$, the altitude ($z_m$) at which the photolysis rate reaches a maximum. Calculate $z_m$ and the photolysis rate at this height.

## 3. Ozone Heating

Knowing that the total energy liberated by the reaction leading to ozone formation

$$O + O_2 + M \rightarrow O_3 + M$$

is 24 kcal mol$^{-1}$, calculate the heating rate at 50 km altitude for an average day length of 12 hours.

*Numerical Input:*

- Number of ozone molecules photolyzed at 50 km (per time unit): $3 \times 10^8$ molecules cm$^{-3}$ s$^{-1}$
- Air (mass) density at 50 km: $2 \times 10^{-10}$ kg cm$^{-3}$
- Avogadro number: $6.023 \times 10^{23}$ molecules mol$^{-1}$
- 1 cal = 4.18 J
- Specific heat at constant pressure: 1005 J kg$^{-1}$ K$^{-1}$

## 4. Vertical Eddy Diffusion

In the stratosphere, the vertical distribution of methane results from a balance between vertical transport and photochemical destruction. The vertical flux $\phi$ is expressed by

$$\phi = -K(z)\rho(z)\frac{d\tilde{\mu}(z)}{dz}$$

where $K(z)$ is the so-called eddy diffusion coefficient at altitude $z$, $\rho(z)$ is the air mass density, and $\widetilde{\mu}(z)$ is the mass mixing ratio of methane. Assume that the destruction of $CH_4$ results only from its reaction with the electronically excited $O(^1D)$ atom.

1. Show that, at equilibrium, we have

$$K(z)\rho(z)\frac{d\widetilde{\mu}(z)}{dz} = -\int_z^\infty L(z)\, dz$$

if $L(z)$ is the photochemical loss rate of methane.

2. Observations and model calculations show that
   - Below the tropopause level ($z_0$), the mass mixing ratio of methane is 1 ppmm.
   - Above level ($z_0$), the mass mixing ratio of methane $\widetilde{\mu}(z)$ decreases with a scale height $H_1 = g\, H_0$, if $H_0$ is the atmospheric scale height (8 km).
   - Above level ($z_0$), the mass density of the $O(^1D)$ atom decreases with a scale height $H_2 = 0.9\, H_0$.

   Calculate the eddy diffusion coefficient $K(z_0)$ at the tropopause level ($z_0 = 16$ km) where the methane loss rate is $6.67 \times 10^{-16}$ kg m$^{-3}$ s$^{-1}$ and the air density is $1.75 \times 10^{-1}$ kg m$^{-3}$.

## 5. Atmospheric Reservoirs and Interhemispheric Transfer

The Earth's atmosphere can be described by two boxes representing the Northern and Southern Hemispheres, respectively. The dynamical exchange time between the two hemispheres is noted $\tau_E$. Consider a chemical compound released exclusively in the Northern Hemisphere at a rate $E$ equal to 100 kT yr$^{-1}$. This compound is lost into the ocean, assumed to be located exclusively in the Southern Hemisphere, with a time constant $\tau_D$ (see the accompanying figure).

1. Express the budget equations for the burden (mass) of this compound in the Northern ($M_N$) and Southern ($M_S$) hemispheres, respectively. Express the steady state value of $M_N$ and $M_S$, and the overall residence time as a function of $E$, $\tau_D$, and $\tau_E$.

2. Observations show that the total atmospheric mass of the compound is 1900 kilotons with a north/south ratio of 1.11. The growth in the burden of both hemispheres is 1% yr$^{-1}$. Calculate the hemispheric exchange time $\tau_E$ and the ocean uptake time $\tau_D$.

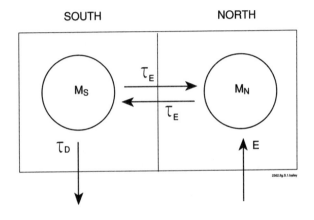

## 6. Carbon Monoxide

Carbon monoxide (CO) is produced in the atmosphere from the oxidation of methane by OH, and is also emitted directly by combustion. The main sink of CO is oxidation by OH. The rate constants for oxidation of methane and CO by OH are $k_1 = 4.0 \times 10^{-15}$ cm$^3$ molecule$^{-1}$ s$^{-1}$ and $k_2 = 2.4 \times 10^{-13}$ cm$^3$ molecule$^{-1}$ s$^{-1}$, respectively. Oxidation of one methane molecule yields one CO molecule. Assume throughout this problem a uniform OH concentration of $7 \times 10^5$ molecules cm$^{-3}$.

1. Calculate the atmospheric lifetimes (in years) of methane and CO. Based on these lifetimes, how well do you expect methane and CO to be mixed in the atmosphere?

2. The atmospheric concentration of methane is 1700 ppb, and the average atmospheric concentration of CO in the Northern Hemisphere is 80 ppb. Assuming steady state for CO, calculate the fraction of CO in the Northern Hemisphere contributed by combustion. Show that this fraction is independent of the assumed OH concentration.

3. Using a two-box model for mass exchange between the Northern and Southern Hemispheres, calculate the steady-state concentration of CO in the Southern Hemisphere assuming that combustion sources in that hemisphere are negligible. The rate constant for mass exchange between the two hemispheres is $k = 1.0$ yr$^{-1}$.

4. There is in fact a large seasonal combustion source of CO in the Southern Hemisphere. What is this source?

5. The use of the $C_2H_2$/CO ratio as a tracer of urban pollution is now examined. Acetylene ($C_2H_2$) is emitted to the atmosphere exclusively by combustion, and is removed from the atmosphere mainly by reaction with OH ($k_3 = 8.3 \times 10^{-13}$ cm$^3$ molecule$^{-1}$ s$^{-1}$). The $C_2H_2$/CO emission ratio in the United States has a remarkably constant value of 0.01 mole/mole. It has been proposed to use the $C_2H_2$/CO concentration ratio measured at the Atlantic island of Bermuda as an indicator of long-range transport of pollution plumes from the United States to the island. Assess the merit of this approach with the following exercises.

    a. Calculate the temporal evolution of the $C_2H_2$/CO concentration ratio in a U.S. pollution plume transported over the Atlantic, assuming that the plume leaves the United States at $t = 0$ and does not dilute with background air during its travel over the Atlantic. A typical travel time from the United States to Bermuda is four days. What is the $C_2H_2$/CO concentration ratio when the U.S. plume arrives at Bermuda at $t = 4$ days?

    b. In fact, dilution of the plume with background air cannot be ignored. Consider a plume initially 100 km wide and containing 150 ppb CO, 1.5 ppb $C_2H_2$. As the plume is transported over the Atlantic, it is diluted with background air containing 80 ppb CO and 0.1 ppb $C_2H_2$. After four days of transport the volume of the plume has increased by a factor of two (50/50 mixture of polluted air and background air). Calculate the $C_2H_2$/CO ratio in the diluted plume.

    c. Conclude: Is the $C_2H_2$/CO concentration ratio measured at Bermuda determined principally by the time it takes for U.S. pollution plumes to reach Bermuda, or by the rate of dilution of U.S. pollution plumes as they travel to Bermuda?

# 7. Formaldehyde and $HO_x$ in the Troposphere

Consider the following reactions (at $p = 1$ atm and $T = 288$ K).

1. $OH + CH_4 \rightarrow$ intermediates $\rightarrow CH_2O$    $k_1 = 7.0 \times 10^{-15}$ cm$^3$ molecule$^{-1}$ s$^{-1}$
2. $OH + CH_2O \rightarrow CHO + H_2O$    $k_2 = 1.0 \times 10^{-11}$ cm$^3$ molecule$^{-1}$ s$^{-1}$
3a. $CH_2O + h\nu \rightarrow 2\, HO_2 + CO$    $k_{3a} = 2.7 \times 10^{-5}$ s$^{-1}$
3b.            $\rightarrow H_2 + CO$    $k_{3b} = 3.5 \times 10^{-5}$ s$^{-1}$
4. $O_3 + h\nu \rightarrow O(^1D) + O_2$    $k_4 = 2.5 \times 10^{-5}$ s$^{-1}$
5. $O(^1D) + M \rightarrow O + M$    $k_5 = 2.9 \times 10^{-11}$ cm$^3$ molecule$^{-1}$ s$^{-1}$
6. $O(^1D) + H_2O \rightarrow 2\, OH$    $k_6 = 2.2 \times 10^{-10}$ cm$^3$ molecule$^{-1}$ s$^{-1}$

Reactions 1 through 3 involve the production and loss of formaldehyde ($CH_2O$). Note that for simplicity, Reaction 1 assumes that each $CH_4$ oxidized produces exactly one $CH_2O$ with the effective rate constant $k_1$. Reactions 4 through 6 involve the production and loss of $O(^1D)$. (*Note:* at $p = 1$ atm and $T = 288$ K, $[M] = 2.55 \times 10^{19}$ molecules cm$^{-3}$.)

1. Under these conditions, derive an expression for the steady-state $[CH_2O]/[CH_4]$ ratio. Evaluate this expression for $[OH] = 1.0 \times 10^6$ molecules cm$^{-3}$. If $[CH_4] = 1700$ ppbv, what is $[CH_2O]$?

2. For the concentration of $CH_2O$ from Part 1, what is the production rate of $HO_x$ radicals via $CH_2O$ photolysis?

3. From Reactions 4 through 6, obtain an expression for $[O(^1D)]$ at steady state. If $[O_3] = 50$ ppb, and $[H_2O] = 15{,}000$ ppm, what is the production rate of $HO_x$ radicals from ozone photolysis?

4. In order for $CH_2O$ photolysis to match the $HO_x$ production calculated in Part 3, what would the concentration of $CH_2O$ need to be?

# 8. Peroxyacetylnitrate (PAN) as a Reservoir for $NO_x$

1. Consider an atmosphere containing 10 ppb $NO_x$ and 100 ppb $O_3$ at $T = 298$ K and $P = 1000$ mb. Calculate the steady-state concentrations of NO and $NO_2$ at noon based on the null cycle:

$$NO + O_3 \rightarrow NO_2 + O_2 \qquad (1)$$

$$NO_2 + h\nu \xrightarrow{O_2} NO + O_3 \qquad (2)$$

with $k_1 = 2.2 \times 10^{-12} \exp(-1430/T)$ cm$^3$ molecule$^{-1}$ s$^{-1}$ and $k_2 = 1 \times 10^{-2}$ s$^{-1}$ (noon). Calculate the $[NO_2]/[NO_x]$ ratio at noon. How would this ratio vary with time of day? How would it be affected by the presence of peroxy radicals?

2. Photolysis of acetone ($CH_3C(O)CH_3$) is an important precursor of PAN in the atmosphere. In a high-$NO_x$ atmosphere, the peroxyacetyl radical $[CH_3C(O)OO]$ produced by photolysis of acetone reacts exclusively with either NO or $NO_2$:

$$CH_3C(O)CH_3 + h\nu \xrightarrow{O_2} CH_3C(O)OO + CH_3 \qquad (3)$$

$$CH_3C(O)OO + NO \rightarrow CH_3 + CO_2 + NO_2 \qquad (4a)$$

$$CH_3C(O)OO + NO_2 \rightarrow PAN \qquad (4b)$$

$$PAN \rightarrow CH_3C(O)OO + NO_2 \qquad (5)$$

Show that in such an atmosphere the steady-state concentration of PAN is independent of $NO_x$ but increases with increasing acetone and $O_3$. Briefly explain why the PAN concentration is independent of $NO_x$.

3. Consider an air parcel ventilated from a city at time $t = 0$ and subsequently transported for 10 days without exchanging air with its surroundings. The air parcel initially contains 100 ppb $NO_x$. To examine the fate of $NO_x$ as the air parcel ages, assume that $[NO] \ll [NO_2]$ in the air parcel at all times (cf. Part 1).

   a. In the absence of PAN formation, the only sink for $NO_x$ is oxidation to $HNO_3$

   $$NO_2 + OH + M \rightarrow HNO_3 + M \qquad (6)$$

   Calculate the rate of Reaction (6) as $d[NO_2]/dt = -k_6[NO_2]$, where $k_6 = 1 \times 10^{-5}$ s$^{-1}$ is a simplified pseudo-first-order rate constant. The main simplification involved in the formulation of $k_6$ is that there is no distinction made between night and day; instead, a 24-hour average OH concentration is built into $k_6$. It is assumed that $HNO_3$ is rapidly removed by deposition and cannot be recycled back to $NO_x$. What is the lifetime of $NO_x$? Calculate the concentration of $NO_x$ in the air parcel as a function of time from $t = 0$ to 10 days. What is the concentration of $NO_x$ remaining after 10 days?

   b. Now examine the effect of PAN formation, assuming a constant concentration $[CH_3C(O)OO] = 1 \times 10^8$ molecules cm$^{-3}$ in the air parcel. Rate constants are $k_{4b} = 1.0 \times 10^{-11}$ cm$^3$ molecule$^{-1}$ s$^{-1}$ and $k_5 = 1.95 \times 10^{16}$ exp($-13{,}543/T$).

   i. What is the lifetime of $NO_x$? What is the lifetime of PAN at 298 K? at 260 K?

   ii. Calculate the concentrations of $NO_x$ and PAN as a function of time for an air parcel transported in the boundary layer ($T = 298$ K) from $t = 0$ to 10 days. (*Hint:* assume that $NO_x$ and PAN are in equilibrium with each other. Why is this assumption reasonable?) What are the concentrations of $NO_x$ and PAN remaining after 10 days?

   iii. Consider now that the air parcel is pumped up to high altitude ($T = 260$ K) at time $t = 0$, and remains at that temperature for $t = 10$ days. Here the assumption of equilibrium between $NO_x$ and PAN is not appropriate (why?). What assumption can you use to calculate $NO_x$ and PAN concentrations in an air parcel? What are the concentrations of $NO_x$ and PAN remaining in the air parcel after 10 days?

   iv. How does formation of PAN at different temperatures affect the long-range transport of $NO_x$?

## 9. Hydrogen Peroxide in the Troposphere

The following reaction mechanism is postulated for the troposphere to explain ozone photolytic decay and hydrogen peroxide ($H_2O_2$) production in the absence of nitrogen oxides:

| Reaction | Rate Constant $k_i$ (25°C) |
|---|---|
| (1) $O_3 + h\nu \rightarrow O(^1D) + O_2$ | $4.5 \times 10^{-5}$ s$^{-1}$ |
| (2) $O(^1D) + M \xrightarrow{O_2} O_3$ | $2.9 \times 10^{-11}$ cm$^3$ molec$^{-1}$ s$^{-1}$ |
| (3) $O(^1D) + H_2O \rightarrow 2$ OH | $2.2 \times 10^{-10}$ cm$^3$ molec$^{-1}$ s$^{-1}$ |
| (4) $OH + O_3 \rightarrow HO_2 + O_2$ | $6.7 \times 10^{-14}$ cm$^3$ molec$^{-1}$ s$^{-1}$ |
| (5) $HO_2 + O_3 \rightarrow OH + 2\ O_2$ | $2.0 \times 10^{-15}$ cm$^3$ molec$^{-1}$ s$^{-1}$ |
| (6) $HO_2 + HO_2 \rightarrow H_2O_2 + O_2$ | $2.7 \times 10^{-12}$ cm$^3$ molec$^{-1}$ s$^{-1}$ |

1. Set up the steady-state expressions for $O(^1D)$, OH, and $HO_2$.
2. Show that $d[H_2O_2]/dt = k_1k_3[O_3][H_2O]/(k_2[M] + k_3[H_2O])$.
3. Calculate the initial rate of $H_2O_2$ production for the condition of 30 ppbv $O_3$ and $1 \times 10^7$ ppbv $H_2O$ at 1 atm pressure and 25°C.

The presence of nitrogen oxides introduces the following additional reactions:

| Reaction | Rate Constant $k_i$ (25°C) |
|---|---|
| (7) $NO_2 + h\nu\ (+O_2) \rightarrow NO + O_3$ | $8.8 \times 10^{-3}$ s$^{-1}$ |
| (8) $NO + O_3 \rightarrow NO_2 + O_2$ | $1.8 \times 10^{-14}$ cm$^3$ molec$^{-1}$ s$^{-1}$ |
| (9) $HO_2 + NO \rightarrow NO_2 + OH$ | $8.6 \times 10^{-12}$ cm$^3$ molec$^{-1}$ s$^{-1}$ |
| (10) $OH + NO_2 \xrightarrow{M} HNO_3$ | $6.4 \times 10^{-12}$ cm$^3$ molec$^{-1}$ s$^{-1}$ |

4. Redevelop the steady-state approximation for $HO_2$ and discuss the implications that the presence of nitrogen oxides have on $H_2O_2$ production.

## 10. Halocarbons

1. Show that the evolution of the mixing ratio $\mu$ of a long-lived (well-mixed) chemical compound can be expressed as a function of time $t$ by

$$\frac{d\mu}{dt} + \frac{\mu}{\tau} = S$$

where $\tau$ is the lifetime of this compound and $S$ its source function (expressed in ppb yr$^{-1}$ if $\mu$ is expressed in ppb and $\tau$ in years).

2. Assume that chlorine atoms present in the stratosphere are released exclusively from the photodecomposition of industrially manufactured CFC-11 (CFCl$_3$). The lifetime of this compound is 70 years and the annual emission is $10^6$ metric tons per year. At time $t = 0$, the volume mixing ratio of chlorine in the stratosphere is equal to 3 ppbv. Calculate the total mass of chlorine contained in the atmosphere at time $t = 0$ and express the source function in ppb yr$^{-1}$ of chlorine equivalents.

3. Show that the evolution with time of the mixing ratio $\mu(t)$ of chlorine equivalents due to CFCl$_3$ can be expressed by

$$\mu(t) = \mu(t=0)\exp\left(-\frac{t}{\tau}\right) + S\tau\left[1 - \exp\left(-\frac{t}{\tau}\right)\right]$$

Assuming that the emissions of CFCl$_3$ remain constant with time, calculate $\mu$ in 50 and 200 years. What is the equilibrium value (for $t = \infty$)?

4. Assuming that, as a result of regulatory measures, the production is instantaneously reduced to zero at $t = 10$ years, calculate the mixing ratio $\mu$ for $t = 50$ and 200 years.

5. It is assumed that at $t = 10$ years, CFC-11 is replaced with an equal source of chlorine equivalents by a substitute whose atmospheric lifetime is 5 years. Calculate the mixing ratio of chlorine equivalents at times $t = 50$ and 200 years. What is the new value at equilibrium ($t = \infty$)?

Note that the atmospheric scale height is 8 km, the air density for STP conditions is $2.5 \times 10^{19}$ molecules cm$^{-3}$, the Earth's radius is 6370 km. The molar masses are 1 for H, 12 for C, 19 for F, and 35.5 for Cl (in g mol$^{-1}$).

## 11. Methyl Bromide

Methyl bromide ($CH_3Br$) is the principal source of bromine in the stratosphere and as such plays a significant role in stratospheric $O_3$ loss. It is emitted to the atmosphere by a number of anthropogenic sources (agricultural fumigants, leaded gasoline, ...) and also has a natural source from biogenic activity in the ocean. There has been much recent interest in quantifying the relative magnitude of the anthropogenic versus natural sources. The current understanding of this issue is surveyed here.

1. Atmospheric lifetime of $CH_3Br$. The main sinks for atmospheric $CH_3Br$ are thought to be oxidation by OH and hydrolysis in the oceans. The corresponding atmospheric lifetime of $CH_3Br$ is estimated here.

    a. The rate constant for oxidation of $CH_3Br$ by OH is $k = 2 \times 10^{-14}$ cm$^3$ molecule$^{-1}$ s$^{-1}$. Based on a global mean OH concentration of $8 \times 10^5$ molecules cm$^{-3}$, show that $CH_3Br$ has an atmospheric lifetime of 2.0 years against oxidation by OH.

    b. The Henry's Law constant for $CH_3Br$ in seawater is

    $$\mathcal{H} = \frac{[CH_3Br(aq)]}{P_{CH_3Br}} = 0.11 \text{ M atm}^{-1}$$

    and the volume of the oceanic mixed layer is $3.6 \times 10^{19}$ liters. Calculate the equilibrium fractionation $n_{ocean}/n_{atm}$ of $CH_3Br$ between the atmosphere and the oceanic mixed layer, where $n_{atm}$ is the total number of moles of $CH_3Br$ in the atmosphere and $n_{ocean}$ is the total number of moles of $CH_3Br$ in the oceanic mixed layer.

    c. In Part (b) it became clear that the oceanic mixed layer contains only a small amount of $CH_3Br$ compared to the atmosphere. However, ocean uptake can still represent an important sink for atmospheric $CH_3Br$ due to rapid hydrolysis of $CH_3Br(aq)$ in the ocean. The accompanying figure shows a two-box model for $CH_3Br$ in the atmosphere-ocean system. The rate constant for hydrolysis of $CH_3Br(aq)$ is $k_0 = 40$ yr$^{-1}$. The transfer rate constants for $CH_3Br$ from the atmosphere to the oceanic mixed layer, and from the oceanic mixed layer to the atmosphere, are $k_1 = 0.5$ yr$^{-1}$ and $k_2 = 22$ yr$^{-1}$. Show that the atmospheric lifetime of $CH_3Br$ against loss by hydrolysis in the oceans is $\tau = (k_0 + k_2)/k_0 k_1 = 3.3$ years.

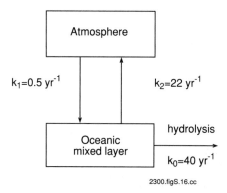

    d. Could significant quantities of $CH_3Br$ be transferred from the oceanic mixed layer to the deep ocean? (*i.e.*, can the deep ocean represent a large reservoir for $CH_3Br$, as it does for $CO_2$?) Briefly explain.

e. By considering the sinks from both atmospheric oxidation and hydrolysis in the oceans, show that the overall atmospheric lifetime of $CH_3Br$ is 1.2 years.

f. Based on the answer to 1.5, and assuming a rate constant $k = 0.14$ yr$^{-1}$ for transfer of air from the troposphere to the stratosphere, estimate the fraction of emitted $CH_3Br$ that enters the stratosphere and is thus active in $O_3$ depletion.

2. A two-box model for atmospheric $CH_3Br$. Data on the atmospheric distribution of $CH_3Br$ are now used to constrain the importance of the anthropogenic source. Observations indicate that the Northern and Southern Hemispheres are individually well mixed with respect to $CH_3Br$, but that there is 30% more $CH_3Br$ in the Northern Hemisphere. This difference is expressed in terms of an interhemispheric ratio $R$

$$R = \frac{m_N}{m_S} = 1.3 \tag{1}$$

where $m_N$ and $m_S$ are the masses of $CH_3Br$ in the Northern and Southern Hemispheres, respectively. Let the ratio be interpreted using a two-box model for the troposphere where the Northern and Southern Hemispheres are individually well mixed, with a transfer rate constant for air between the two hemispheres of $k = 0.9$ yr$^{-1}$ (derived from $^{85}$Kr data). Assume that $CH_3Br$ is at steady state and removed from the atmosphere with a rate constant $k' = 0.8$ yr$^{-1}$ (corresponding to a lifetime of 1.2 years, as derived above).

a. If the source of $CH_3Br$ were exclusively anthropogenic and located in the Northern Hemisphere, show that $R$ would have a value of 1.9, higher than observed.

b. The discrepancy may be explained by the natural biogenic source of $CH_3Br$ from the oceans. Assume that this biogenic source is equally distributed between the two hemispheres, as opposed to the anthropogenic source which is exclusively in the Northern Hemisphere. In order to match the observed value of $R$, what fraction of the global source must be biogenic?

c. Is the hypothesis that the ocean represents a major source of $CH_3Br$ contradictory with the hypothesis that the ocean represents a major sink, as described in Part (c) of the preceding question?

## 12. SO$_2$ AqueousPhase Oxidation

Calculate the pH at which the instantaneous rate of $SO_2$ aqueous-phase oxidation by $H_2O_2$ and $O_3$ are equal ($R_1 = R_2$). The atmospheric conditions are $SO_2 = 5$ ppbv, $H_2O_2 = 1$ ppbv, and $O_3 = 80$ ppbv. The temperature is 25°C. The Henry's Law coefficients are:

for $O_3$: $1.13 \times 10^{-2}$ M atm$^{-1}$
$SO_2$: 1.23 M atm$^{-1}$
$H_2O_2$ $7.45 \times 10^4$ M atm$^{-1}$

The rate of $SO_2$ oxidation in the aqueous form can be expressed by (Hoffmann and Calvert, 1985):

For $H_2O_2$ oxidation:

$$R_1 = \frac{k[H^+][HSO_3^-][H_2O_2(aq)]}{1 + K[H^+]}$$

with $k = 7.45 \times 10^7$ M$^{-2}$ s$^{-1}$ and $K = 13$ M$^{-1}$.

For $O_3$ oxidation:

$$R_2 = \{k_0[SO_2(aq)] + k_1[HSO_3^-] + k_2[SO_3^{2-}]\}[O_3(aq)]$$

with $k_0 = 2.4 \times 10^4$ $M^{-1}$ $s^{-1}$
$k_1 = 3.7 \times 10^5$ $M^{-1}$ $s^{-1}$
$k_2 = 1.5 \times 10^9$ $M^{-1}$ $s^{-1}$

The equilibrium constants for the following reactions are

$$SO_2(aq) \rightleftharpoons H^+ + HSO_3^- \qquad K_1 = 1.29 \times 10^{-2} \text{ M}$$
$$HSO_3^- \rightleftharpoons H^+ + SO_3^{2-} \qquad K_2 = 6.01 \times 10^{-8} \text{ M}$$

Compare your answer with the results in Figure 10.3.

### 13. Ozone in the Free Troposphere

Consider the following 12 reactions with their respective rate constants (or photo-dissociation frequencies):

|  |  |
|---|---|
| $NO + O_3 \rightarrow NO_2 + O_2$ | $(1.8 \times 10^{-14}$ $cm^3$ $molecule^{-1}$ $s^{-1})$ |
| $NO + HO_2 \rightarrow NO_2 + OH$ | $(8.6 \times 10^{-12}$ $cm^3$ $molecule^{-1}$ $s^{-1})$ |
| $NO_2 + h\nu \rightarrow NO + O(^3P)$ | $(8 \times 10^{-3}$ $s^{-1})$ |
| $O(^3P) + O_2 + M \rightarrow O_3 + M$ | $(6.0 \times 10^{-34}$ $cm^6$ $molecule^{-2}$ $s^{-1})$ |
| $O_3 + h\nu \rightarrow O(^3P) + O_2$ | $(5 \times 10^{-4}$ $s^{-1})$ |
| $O_3 + h\nu \rightarrow O(^1D) + O_2$ | $(4 \times 10^{-5}$ $s^{-1})$ |
| $O(^1D) + M \rightarrow O(^3P) + M$ | $(2.9 \times 10^{-11}$ $cm^3$ $molecule^{-1}$ $s^{-1})$ |
| $O(^1D) + H_2O \rightarrow 2OH$ | $(2.2 \times 10^{-10}$ $cm^3$ $molecule^{-1}$ $s^{-1})$ |
| $OH + CO \xrightarrow{O_2} HO_2 + CO_2$ | $(1.5 \times 10^{-13}$ $cm^3$ $molecule^{-1}$ $s^{-1})$ |
| $OH + O_3 \rightarrow HO_2 + O_2$ | $(6.8 \times 10^{-14}$ $cm^3$ $molecule^{-1}$ $s^{-1})$ |
| $HO_2 + O_3 \rightarrow OH + 2O_2$ | $(2.0 \times 10^{-15}$ $cm^3$ $molecule^{-1}$ $s^{-1})$ |
| $HO_2 + HO_2 \rightarrow products$ | $(2.7 \times 10^{-12}$ $cm^3$ $molecule^{-1}$ $s^{-1})$ |

This set of reactions is assumed to represent the key processes that govern the photochemistry of ozone in the free (unpolluted) troposphere. In the real atmosphere, other reactions (*e.g.*, the methane oxidation chain) play an important role, but are ignored in this problem. The $HO_2 + HO_2$ reaction is assumed to be the (irreversible) loss mechanism for $HO_x$ radicals. Formation and destruction mechanisms for nitrogen oxides are ignored.

1. Write the kinetics (rate) equations for $O(^3P)$, $O(^1D)$, $O_3$, NO, $NO_2$, OH, and $HO_2$. Define the following chemical families: $NO_x = NO + NO_2$, $HO_x = OH + HO_2$, and $O_x = O(^3P) + O(^1D) + O_3 + NO_2$ and write the rate equation for each of these families. Express the net tropospheric $O_x$ source as a function of NO, OH, $HO_2$, $H_2O$, $O_3$, and $M$ number densities (where $M$ represents any air molecule). For this purpose, photochemical steady-state conditions are assumed for $O(^1D)$.

2. Derive two expressions for the concentration of $HO_2$: one as a function of $O_3$, NO, and $NO_2$ densities:, $j_{NO_2}$ (and the various rate constants involved), the other one as a function of $O_3$, $H_2O$, and $M$ densities, $j^*_{O_3}$ (and the various rate constants involved). For this purpose, assume photochemical steady-state conditions for

$O(^1D)$, $NO_2$, and $HO_x$. For the following measured concentrations:

$$[O_3] = 40 \text{ ppbv}$$
$$[NO] = 20 \text{ pptv}$$
$$[NO_2] = 30 \text{ pptv}$$
$$[H_2O] = 10^{17} \text{ molecule cm}^{-3}$$
$$[M] = 1.75 \times 10^{19} \text{ molecule cm}^{-3}$$

and the values of the photolysis frequencies given above, calculate the density of $HO_2$ from these two expressions.

3. Express [OH] as a function of [$HO_2$] and calculate the OH density. Derive the local lifetime of carbon monoxide.

4. Calculate the local production and destruction of $O_x$ as well as its net photochemical source. Based on the observed concentrations of the chemical species, can the local environment be regarded as a production or sink region for ozone?

5. Neglecting the $O(^1D)$ + $H_2O$ reaction, express the value of the NO density at which the $O_x$ production rate equals the $O_x$ destruction rate. Calculate the corresponding NO mixing ratio.

## 14. Tropospheric Ozone Budget

The relative importance of stratospheric input versus tropospheric production as sources of tropospheric $O_3$ is examined using a simple model.

1. The total amount of $O_3$ in the atmosphere is $6.9 \times 10^{13}$ moles. Of that total, 10% is in the troposphere. Using $k = 0.8$ yr$^{-1}$ as the rate constant for transport from the stratosphere to the troposphere, derive the magnitude of stratospheric input (moles yr$^{-1}$) of $O_3$ to the troposphere.

2. Now view the troposphere as a well-mixed box containing 30 ppb $O_3$, 30 ppt $NO_x$, and 70 ppb CO (global mean values based on observations). $CH_4$ is neglected; its effect for these purposes is secondary. Consider the following ensemble of reactions to describe $O_3$ photochemistry:

$$O_3 + h\nu \xrightarrow{H_2O} 2\,OH + O_2 \qquad (1)$$
$$CO + OH \xrightarrow{O_2} CO_2 + HO_2 \qquad (2)$$
$$HO_2 + HO_2 \rightarrow H_2O_2 + O_2 \qquad (3)$$
$$HO_2 + NO \rightarrow OH + NO_2 \qquad (4)$$
$$O_3 + NO \rightarrow O_2 + NO_2 \qquad (5)$$
$$NO_2 + h\nu \xrightarrow{O_2} NO + O_3 \qquad (6)$$

with rate constants $k_1 = 1 \times 10^{-6}$ s$^{-1}$, $k_2 = 2 \times 10^{-13}$ cm$^3$ molecule$^{-1}$ s$^{-1}$, $k_3 = 5 \times 10^{-12}$ cm$^3$ molecule$^{-1}$ s$^{-1}$, $k_4 = 8 \times 10^{-12}$ cm$^3$ molecule$^{-1}$ s$^{-1}$, $k_5 = 2 \times 10^{-14}$ cm$^3$ molecule$^{-1}$ s$^{-1}$, $k_6 = 1 \times 10^{-3}$ s$^{-1}$. Assume here that $H_2O_2$ is removed by deposition. Further assume steady state for all radicals: OH, $HO_2$, NO, and $NO_2$.

   a. The rate $P$ (molecules cm$^{-3}$ s$^{-1}$) of tropospheric $O_3$ production is $P = k_4[HO_2][NO]$. Explain why we cannot write $P = k_6[NO_2]$.

   b. Which of reactions (1)-(6) are sources or sinks of odd-hydrogen radicals?

c. Using a chemical steady-state equation for odd hydrogen, show that

$$[\text{HO}_2] = \left(\frac{k_1[\text{O}_3]}{k_3}\right)^{\frac{1}{2}}$$

Calculate $[\text{HO}_2]$ in units of molecules $\text{cm}^{-3}$ (use as conversion factor 1 ppb $= 1.5 \times 10^{10}$ molecules $\text{cm}^{-3}$).

d. Using a chemical steady state equation for NO, show that $[\text{NO}]/[\text{NO}_2] = 0.09$.

e. From knowledge of $[\text{NO}_x] = [\text{NO}] + [\text{NO}_2] = 30$ ppt, calculate [NO] in units of molecules $\text{cm}^{-3}$ (use as conversion factor 1 ppt $= 1.5 \times 10^7$ molecules $\text{cm}^{-3}$).

f. Calculate $P$ in units of moles $\text{yr}^{-1}$ (use $5 \times 10^{24}$ $\text{cm}^3$ as total volume for the troposphere). Compare to the source from stratospheric input calculated in Question 1.

## 15. $\text{NO}_x$ and Hydrocarbon-Limited Regimes for Ozone Production

Imagine a model of the planetary boundary layer (PBL) over the eastern United States as a well-mixed box with a 1000-km dimension in the east-west direction and a vertical dimension of 2 km. The region is ventilated by a constant wind from the west with a speed of 2 m $\text{s}^{-1}$. The mean $\text{NO}_x$ emission flux in the region is $2 \times 10^{11}$ molecules $\text{cm}^{-2}$ $\text{s}^{-1}$, constant throughout the year. Let $P_{\text{HO}_x}$ be the production rate of $\text{HO}_x$ in the region. It can be diagnosed whether $\text{O}_3$ production in the region is hydrocarbon or $\text{NO}_x$ limited by determining which one of the two sinks for $\text{HO}_x$ (1 or 2) is dominant.

$$\text{HO}_2 + \text{HO}_2 \rightarrow \text{H}_2\text{O}_2 + \text{O}_2 \qquad (1)$$

$$\text{NO}_2 + \text{OH} + M \rightarrow \text{HNO}_3 + M \qquad (2)$$

We present here a simple approach for making this diagnostic.

1. The $\text{NO}_x$ emitted in the eastern United States has a lifetime of 12 h against oxidation to $\text{HNO}_3$ by Reaction (2). Calculate the fraction of emitted $\text{NO}_x$ that is oxidized within the region (vs. ventilated out of the region). It will be evident that most of the $\text{NO}_x$ emitted in the eastern United States is oxidized within the region.

2. From a photochemical model calculation, obtain a 24-h average $P_{\text{HO}_x}$ of $4 \times 10^6$ molecules $\text{cm}^{-3}$ $\text{s}^{-1}$ over the eastern United States in July. Compare this source of $\text{HO}_x$ to the source of $\text{NO}_x$. Conclude whether $\text{O}_3$ production over the eastern United States in July is hydrocarbon or $\text{NO}_x$ limited.

3. The same photochemical model calculation indicates a 24-h average $P_{\text{HO}_x}$ of $1.0 \times 10^6$ molecules $\text{cm}^{-3}$ $\text{s}^{-1}$ in October.

   a. Why is $P_{\text{HO}_x}$ lower in October than in July?

   b. Conclude whether ozone production over the eastern United States in October is hydrocarbon or $\text{NO}_x$ limited.

4. As temperatures decrease in the fall, $\text{NO}_x$ may be increasingly removed by the reaction

$$\text{RO}_2 + \text{NO}_2 + M \rightarrow \text{RO}_2\text{NO}_2 + M \qquad (3)$$

where $\text{RO}_2\text{NO}_2$ is an organic peroxynitrate (in summer, the organic peroxynitrates decompose back to $\text{NO}_x$ because of the high temperatures). Consider a situation where Reaction (3) represents the main $\text{HO}_x$ sink.

a. Write an equation for the $O_3$ production rate as a function of $P_{HO_x}$, [NO], and [$NO_2$].

b. Assuming that the [NO]/[$NO_2$] ratio is a constant, show that $O_3$ production is neither hydrocarbon nor $NO_x$ limited.

## 16. Impact of Stratospheric Changes on Tropospheric Chemistry

Assume that the OH density in the troposphere is proportional to the intensity of solar radiation below the tropopause. This radiation is attenuated (Beer-Lambert Law) by ozone as it propagates through the stratosphere. The stratospheric optical depth $\tau$ is assumed to be equal to 2. (Assume an overhead sun and no wavelength dependence.) Calculate the change in the OH density resulting from a 10% decrease in the stratospheric ozone column abundance (*e.g.*, as a result of human-induced perturbations).

## 17. Stratospheric Ozone and the Chapman Mechanism

Consider the four reactions proposed by S. Chapman to describe the photochemical behavior of ozone in the stratosphere:

$$(j_{O_2}); \quad O_2 + h\nu \to O + O$$
$$(k_2); \quad O + O_2 + M \to O_3 + M$$
$$(j_{O_3}); \quad O_3 + h\nu \to O + O_2$$
$$(k_3); \quad O + O_3 \to O_2 + O_2$$

1. Ignore transport processes and write the kinetics equations for atomic oxygen (O), ozone ($O_3$), and odd oxygen ($O_x = O + O_3$). Assume photochemical steady-state conditions for atomic oxygen and derive the [O]/[$O_3$] concentration ratio for daytime. At what altitude are the daytime O and $O_3$ concentrations equal, assuming that $j_{O_3} = 10^{-2}$ s$^{-1}$, $k_2 = 10^{-33}$ cm$^6$ molecule$^{-2}$ s$^{-1}$, and that the altitude $z$ (in km) corresponding to air number density [$M$] (expressed in molecules cm$^{-3}$) is given by $z = 7$ km $\ln\left(2.7 \times 10^{19}/[M]\right)$. Remember that the number density of molecular oxygen is [$O_2$] = 0.2 [$M$]. What is the concentration of atomic oxygen during nighttime, under the assumptions made?

2. Remembering that, in the case of the differential equation

$$\frac{dy}{dt} = \alpha y^2 = P$$

with the initial condition $y(t=0) = 0$, the time $\tau_{50\%}$ needed to reach 50% of the steady-state solution is $\tau_{50\%} = 0.55[\alpha P]^{-1/2}$, calculate $\tau_{50\%}(O_x)$ as a function of the rate coefficients ($k_2, k_3, j_{O_2}, j_{O_3}$) and the air density.

3. Assume now steady-state (ss) conditions for $O_x$ (and $O_3$) and derive the ozone number density [$O_3$]$_{ss}$ as a function of the rate coefficients and the air number density. Express $\tau_{50\%}(O_x)$ as a function of the ozone density [$O_3$]$_{ss}$ and the ozone production rate $P = 2j_{O_2}[O_2]$.

4. Assuming that the ratio of rate constants $k_2/k_3$ varies with temperature $T$ as $\exp(2800/T)$, express the sensitivity of the ozone number density to temperature change $\partial[O_3]/\partial T$ for the following cases: (1) the total air density [$M$] is insensitive to temperature changes, and (2) the air density [$M$] varies with temperature as $1/T$.

5a. Very often, the ozone sensitivity to temperature is expressed by the $\theta$ parameter (in Kelvin) defined as

$$\theta = \frac{\partial \ln[O_3]}{\partial T^{-1}}$$

Calculate $\theta$ when the temperature dependence of the air density is ignored. Express $\theta$ (as a function of temperature) when the temperature dependence of the air density is taken into account.

5b. A similar parameter $\hat{\theta}$ that applies to the ozone mixing ratio $\mu$ is defined as

$$\hat{\theta} = \frac{\partial \ln \mu}{\partial T^{-1}}$$

Derive a simple relationship between $\theta$ and $\hat{\theta}$, taking into account the temperature dependence of the air density.

6. The solar ultraviolet radiation ($\lambda < 300$ nm) varies by a few percent over the 11-year solar cycle, and stratospheric ozone is expected to respond to such solar forcing. Because the period of the solar cycle is substantially longer than the atmospheric (chemical and radiative) lifetimes, equilibrium conditions are assumed. Calculate the relative change in the ozone density for a 2% change in the photolysis frequency of $O_2$, assuming that $j_{O_3}$ is not affected by solar variability and that the temperature remains constant. Evaluate the impact of temperature feedback on the ozone response. For this purpose, assume radiative equilibrium conditions; express the change in the heating rate as $\Delta H = \alpha \Delta [O_3]/[O_3]$ and the change in the cooling rate as $\Delta C = \beta \Delta T$, with $\alpha/\beta = 0.1$ K. Calculate $\Delta[O_3]/[O_3]$ for $\Delta j_{O_2}/j_{O_2} = 2\%$. Assume that $T = 250$ K and $[O_3] = 4 \times 10^{11}$ molecule cm$^{-3}$ (corresponding to the approximate altitude of 40 km).

## 18. Ozone Destruction by Chlorine Compounds at Midlatitudes and in Polar Regions

During the month of September, the following values of the ozone, $NO_2$, and $ClONO_2$ concentrations as well as the total atmospheric density are measured in two specific regions of the atmosphere (45 km at midlatitudes and 16 km in Antarctica).

| molecule cm$^{-3}$ | [$O_3$] | [$NO_2$] | [$ClONO_2$] | [M] |
|---|---|---|---|---|
| Region I:<br>45 km — midlatitudes | $4 \times 10^{11}$ | $5 \times 10^8$ | $2.8 \times 10^6$ | $4 \times 10^{16}$ |
| Region II:<br>16 km — Antarctica | $6 \times 10^{12}$ | $1 \times 10^8$ | $8.5 \times 10^8$ | $2.5 \times 10^{18}$ |

1. First consider Region I.

   a. The steady state between the concentration of chlorine nitrate ($ClONO_2$) and chlorine monoxide (ClO) is determined by the two following reactions:

   $ClO + NO_2 + M \rightarrow ClONO_2 + M$ ($k_1 = 2 \times 10^{-31}$ cm$^6$ molecule$^{-2}$ s$^{-1}$)
   $ClONO_2 + h\nu \rightarrow ClO + NO_2$ ($j_1 = 10^{-6}$ s$^{-1}$)

   Assuming photochemical steady-state conditions, calculate the number density of ClO.

b. Steady-state conditions between atomic oxygen (O) and ozone ($O_3$) is determined by the two following reactions:

$$O + O_2 + M \rightarrow O_3 + M \quad (k_2 = 2 \times 10^{-33} \text{ cm}^6 \text{ molecule}^{-2} \text{ s}^{-1})$$
$$O_3 + h\nu \rightarrow O + O_2 \quad (j_3 = 10^{-4} \text{ s}^{-1})$$

Calculate the number density of atomic oxygen, assuming that the mixing ratio of $O_2$ is 20 percent.

c. We assume that the catalytic destruction of ozone by chlorine is limited in this region by the following reaction:

$$ClO + O \rightarrow Cl + O_2 \quad (k_3 = 2 \times 10^{-11} \text{ cm}^3 \text{ molecule}^{-1} \text{ s}^{-1})$$

Detail the whole destruction cycle. Assuming steady-state conditions between Cl and ClO, calculate the mean destruction rate of ozone in percent per day.

2. Now consider Region II.

a. Acknowledging that the midlatitude concentration of $NO_2$ at 16 km is of the order of $1 \times 10^9$ molecule cm$^{-3}$, explain the low values of the $NO_2$ density measured over Antarctica.

b. Explain qualitatively why, for these conditions, the calculated concentration ratio [ClO]/[ClONO$_2$] is equal to 2. Deduce the number density of ClO.

c. The photolysis ratio of ozone ($j_3$) is now equal to $10^{-3}$ s$^{-1}$. Show, by calculating the destruction rate of ozone (in percent per day), that the "classic cycle" [see Part (1c) above] considered at midlatitudes cannot explain the ozone depletion observed above the Antarctic continent during austral springtime.

d. We assume that the dominant destruction cycle of ozone in Region II is limited by the following reaction (which produces the ClO dimer, $Cl_2O_2$).

$$ClO + ClO + M \rightarrow Cl_2O_2 + M \quad (k_4 = 8 \times 10^{-32} \text{ cm}^6 \text{ molecule}^{-2} \text{ s}^{-1})$$

Detail the whole destruction cycle. Assuming again stationary conditions, calculate the average ozone destruction rate in percent per day.

e. In how many days will the local ozone concentration be depleted by 50%?

## 19. Space Observation of Ozone

We would like to design a space experiment to measure ozone in the terrestrial atmosphere, using different spectrometric methods at different wavelength regions. It is assumed that the ozone column abundance is 300 Dobson Units. 90% of this ozone is located in the stratosphere, 10% in the troposphere. It is also assumed that this layer is horizontally uniform in the domain where the satellite measurement is performed.

Typical values for ozone absorption cross sections in ultraviolet and visible absorption bands are the following:

Hartley band: $10^{-17} - 5 \times 10^{-19}$ cm$^2$

Huggins bands: $5 \times 10^{-19} - 10^{-21}$ cm$^2$

Chappuis bands: $5 \times 10^{-21} - 5 \times 10^{-22}$ cm$^2$

1. Calculate the vertically integrated ozone quantity ($N_0$) in molecules cm$^{-2}$ and express the optical depth along the vertical $\tau_\lambda$ as a function of $N_0$ and of the absorption cross section $\sigma_\lambda$ at wavelength $\lambda$.

2. The ozone quantity is derived from the measurement of the solar intensity $(I_{0,\lambda})$ above the atmosphere and of the solar intensity $(I_\lambda)$ that is attenuated by ozone along an optical path through the atmosphere. If $\tau_\lambda$ is the optical path, $I_\lambda$ and $I_{0,\lambda}$ are related by the Beer-Lambert law

$$I_\lambda = I_{0,\lambda} \exp(-\tau_\lambda)$$

Assume that $S_0$ and $S$, the signals measured by the detector (expressed as a number of light pulses per unit time), are proportional to $I_0$ and $I$, respectively ($S_0 = \alpha I_0$ and $S = \alpha I$). Neglecting the uncertainty of $S_0$, we assume that the precision of signal $S$ is given by a Poisson law

$$\frac{\Delta S}{S} = \frac{1}{\sqrt{S}}$$

If $N$ (molecules cm$^{-2}$) is the integrated ozone quantity corresponding to optical depth $\tau_\lambda$, show that the precision of the quantity $N$ to be measured is given by

$$\frac{\Delta N}{N} = A \frac{\exp(\tau_\lambda/2)}{\tau_\lambda}$$

where $A$ is an experimental constant. Calculate the value of $\tau_\lambda$ for which the measurement of $N$ is optimized.

3. Consider a nadir-viewing experiment by which the solar light scattered by the surface is measured by the spacecraft (see the accompanying figure). (The intensity of scattered light is independent of incident angle.) Calculate as a function of the solar zenith angle $\chi$ the total optical depth along the optical path between the Sun, the Earth's surface, and the spacecraft. (Assume that $\chi < 60°$.) Justify the approximation made. Calculate the maximum and minimum values of the absorption cross sections to optimize the measurement. Establish which spectral region is optimal for this space measurement.

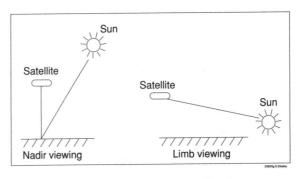

4. Next consider a limb view of the Sun at sunrise or sunset (see the accompanying figure), by which the spacecraft instrument derives the vertical profile of ozone. Assume that the ozone number density at three selected altitudes is the following:

| | |
|---|---|
| 15 km | $1 \times 10^{11}$ molecules cm$^{-3}$ |
| 25 km | $6 \times 10^{12}$ molecules cm$^{-3}$ |
| 40 km | $5 \times 10^{11}$ molecules cm$^{-3}$ |

The length of the optical path corresponding to the ozone absorption at a given altitude is assumed to be 500 km. Calculate the optimal spectral region for the measurements at the different altitudes. Can some of these observations be made in the Chappuis bands? What is the advantage of this spectral domain versus that of the Huggins bands?

# References

# References

Abel P. G., P. J. Ellis, J. T. Houghton, G. Peckham, C. D. Rodgers, S. D. Smith, and E. J. Williamson (1970) Remote sounding of atmospheric temperature from satellites, 2. The selective chopper radiometer for Nimbus D, *Proc. Roy. Soc. Lond.* A320, 35.

Aber, J. D., K. J. Nadelhoffer, P. Steudler, and J. M. Melillo (1989) Nitrogen saturation in northern forest ecosystems, *Bioscience* 39, 378.

Ajtay, G. L., P. Ketner, and P. Duvigneaud (1979) Terrestrial primary production and phytomass, in: *The Global Carbon Cycle: SCOPE 13*, B. Bolin, E. T. Degens, S. Kempe, and P. Ketner, eds., John Wiley and Sons, Chichester.

Allègre, C. J. and S. H. Schneider (1994) The evolution of the Earth, *Scientific American* Oct., 66.

Allen, D., R. Rood, A. Thompson, and R. Hudson (1996) Three-dimensional Rn-222 calculations using assimilated meteorological data and a convective mixing algorithm, *J. Geophys. Res.* 101, 6871.

Amthor, J. S. (1995) Terrestrial higher-plant response to increasing atmospheric [$CO_2$] in relation to the global carbon cycle, *Global Change Biology* 1, 243.

Anderson, J. G. (1978) Measurements of atomic and diatomic radicals in the Earth's stratosphere, *Atmos. Tech.* 9, 55.

Anderson, J. G., D. W. Toohey, and W. H. Brune (1991) Free radicals within the Antarctic vortex: The role of CFCs in Antarctic ozone loss, *Science* 251, 39.

Andreae, M. O. (1985) The emissions of sulfur to the remote atmosphere: Background paper, in: *The Biogeochemical Cycling of Sulfur and Nitrogen in the Remote Atmosphere*, J. N. Galloway, R. J. Charlson, M. O. Andreae, H. Rodhe, eds., D. Reidel Publishing, Boston.

Andreae, M. O., A. Chapuis, B. Cros, J. Fontan, G. Helas, C. Justice, Y. J. Kaufman, A. Minga, and D. Nganga (1992) Ozone and Aitken nuclei over equatorial Africa: Airborne observations during DECAFE 88, *J. Geophys. Res.* 97, 6137.

Andreae, M. O. (1993) Global distribution of fires seen from space, *EOS Trans.* 74, 129.

Andrews, D. G. and M. E. McIntyre (1976) Planetary waves in horizontal and vertical shear: The generalized Eliassen-Palm relation and the mean zonal acceleration, *J. Atmos. Sci.* 33, 2031.

Andrews, D. G. and M. E. McIntyre (1978a) An exact theory of nonlinear waves on a Lagrangian mean flow, *J. Fluid Mech.* 89, 609.

Andrews, D. G. and M. E. McIntyre (1978b) Generalized Eliassen-Palm and Charney-Drazin theorems for waves in axisymmetric mean flows in compressible atmospheres, *J. Atmos. Sci.* 35, 175.

Andrews, D. G., J. R. Holton, and C. B. Leovy (1987) *Middle Atmosphere Dynamics*, Academic Press, London.

Andrews, J. E., P. Brimblecombe, T. D. Jickells, and P. S. Liss (1996) *An Introduction to Environmental Chemistry*, Blackwell Science, Oxford, UK.

Aneja, V. P. (1990) Natural sulfur emissions into the atmosphere, *J. Air & Waste Mgn. Assoc.* 40, 469.

Anlauf, K. G., R. E. Mickle, and N. B. A. Trivett (1994) Measurement of ozone during Polar Sunrise Experiment 1992, *J. Geophys. Res.* 99, 25,345.

Anthes, R. A., E.-Y. Hsie, and Y. H. Kuo (1987) *Description of the Penn State/NCAR Mesoscale Model Version 4 (MM4)*, NCAR Technical Note NCAR/TN-282+STR, National Center for Atmospheric Research, Boulder, Colorado.

Anthony, S. E., R. T. Tisdale, R. S. Disselkamp, M. A. Tolbert, and J. C. Wilson (1995) FTIR studies of low temperature sulfuric acid aerosols, *Geophys. Res. Lett.* 22, 1105.

Antoine, D., J.-M. André, and A. Morel (1996) Oceanic primary production, 2. Estimation at global scale from satellite (coastal zonal color scanner) chlorophyll, *Glob. Biogeochem. Cycles* 10, 57.

Arimoto, R., R. A. Duce, D. L. Savoie, and J. M. Prospero (1992) Trace elements in aerosol particles from Bermuda and Barbados: Concentrations, sources and relationships to aerosol sulfate, *J. Atmos. Chem.* 14, 439.

Arlander, D. W., D. Bruning, U. Schmidt, and D. H. Ehhalt (1995) The tropospheric distribution of formaldehyde during TROPOZ II, *J. Atmos. Chem.* 22, 251.

Asman, W. A. H. and A. J. Janssen (1987) A long-range transport model for ammonia and ammonium for Europe, *Atmos. Environ.* 21, 2099.

Atkins, P. W. (1986) *Physical Chemistry*, 3rd ed., W. H. Freeman and Company, New York.

Atkinson, R. A., S. M. Aschmann, W. P. Carter, A. M. Winer, and J. N. Pitts, Jr. (1982) Alkyl nitrate formation from the $NO_x$-air photooxidation of $C_2$-$C_8$ alkanes, *J. Phys. Chem.* 86, 4563.

Atkinson, R. (1987) A structure-activity relationship for the estimation of rate constants for the gas-phase reactions of OH radicals with organic compounds, *Int. J. Chem. Kinet.* 19, 799.

Atkinson, R. (1990) Gas-phase tropospheric chemistry of organic compounds: A review, *Atmos. Environ.* 24A, 1.

Atkinson, R. (1997a) Atmospheric reactions of alkoxy and $\beta$-hydroxyalkoxy radicals, *Int. J. Chem. Kinet.* 29, 99.

Atkinson, R. (1997b) Gas-phase tropospheric chemistry of volatile organic compounds: 1. Alkanes and alkenes, *J. Phys. Chem. Ref. Data*, 26, 217.

Atkinson, R., D. L. Baulch, R. A. Cox, R. F. Hampson, J. A. Kerr, M. J. Rossi, and J. Troe (1997) Evaluated kinetic, photochemical, and heterogeneous data for atmospheric chemistry, Supplement V. IUPAC subcommittee on gas kinetic data evaluation for atmospheric chemistry, *J. Phys. Chem. Ref. Data* 26, 521.

Atlas, E. L., B. A. Ridley, G. Hübler, J. G. Walega, M. A. Carroll, D. D. Montzka, B. J. Huebert, R. B. Norton, F. E. Grahek, and S. Schauffler (1992) Partitioning and budget of $NO_y$ species during the Mauna Loa Observatory Photochemistry Experiment, *J. Geophys. Res.* 97, 10,449.

Avallone, L. M., D. W. Toohey, W. H. Brune, R. J. Salawitch, A. E. Dessler, and J. G. Anderson (1993) Balloon-borne in situ measurements of ClO and ozone: Implications for heterogeneous chemistry and mid-latitude ozone loss, *Geophys. Res. Lett.* 20, 1795.

Avallone, L. M., D. W. Toohey, S. M. Schauffler, W. H. Pollock, L. E. Heidt, E. L. Atlas, and K. R. Chan (1995) In situ measurements of BrO during AASE II, *Geophys. Res. Lett.* 22, 831.

Balkanski, Y., D. Jacob, R. Arimoto, and M. Kritz (1992) Distribution of $^{222}$Rn over the north Pacific: Implications for continental influences, *J. Atmos. Chem.* 14, 353.

Bandy, A. R., D. L. Scott, B. W. Blomquist, S. M. Chen, and D. C. Thornton (1992a) Low yields of $SO_2$ from dimethyl sulfide oxidation in the marine boundary layer, *Geophys. Res. Lett.* 19, 1125.

Bandy, A. R., D. C. Thornton, D. L. Scott, M. Lalevic, E. E. Lewin, and A. R. Driedger III (1992b) A time series for carbonyl sulfide in the northern hemisphere, *J. Atmos. Chem.* 14, 527.

Bandy, A. R., D. C. Thornton, and A. Driedger III (1993a) Airborne measurements of sulfur dioxide, dimethyl sulfide, carbon disulfide and carbonyl sulfide by isotope dilution gas chromatography/mass spectrometry, *J. Geophys. Res.* 98, 23,423.

Bandy, A. R., D. C. Thornton, and J. E. Johnson (1993b) Carbon disulfide measurements in the atmosphere of the western north Atlantic and the northwestern south Atlantic oceans, *J. Geophys. Res.* 98, 23,449.

Barath, F. T., M. C. Chavez, R. E. Cofield, D. A. Flower, M. A. Frerking, M. B. Gram, W. M. Harris, J. R. Holden, R. F. Jarnot, W. G. Kloezeman, G. J. Klose, G. K. Lau, M. S. Loo, B. J. Maddison, R. J. Mattauch, R. P. McKinney, G. E. Peckham, H. M. Pickett, G. Siebes, F. S. Soltis, R. A. Suttie, J. A. Tarsala, J. W. Waters, and W. J. Wilson (1993) The Upper Atmosphere Research Satellite Microwave Limb Sounder Instrument, *J. Geophys. Res.* 98, 10,751.

Barnes, I. (1993) Overview and atmospheric significance of the results from laboratory kinetic studies performed within the CEC project "OCEANO-NOX," in: *Dimethylsulphide: Oceans, Atmosphere and Climate*, G. Restelli and G. Angeletti, eds., Proc. International Symposium, Belgirate, Italy, Kluwer Academic Publishers, Dordrecht, The Netherlands.

Barnes, I., K. H. Becker, and T. Zhu (1993) Near UV absorption spectra and photolysis products of difunctional organic nitrates: Possible importance as $NO_x$ reservoirs, *J. Atmos. Chem.* 17, 353.

Barnes, I., K. H. Becker, and N. Mihalopoulos (1994) An FTIR product study of the photooxidation of dimethyl disulfide, *J. Atmos. Chem.* 18, 267.

Barnola, J.-M., M. Anklin, J. Porcheron, D. Raynaud, J. Schwander, and B. Stauffer (1994) $CO_2$ evolution during the last millenium as recorded by Antarctic and Greenland ice, *Tellus* 47B, 264.

Barrie, L. A., J. W. Bottenheim, R. C. Schnell, P. J. Crutzen, and R. A. Rasmussen (1988) Ozone destruction and photochemical reactions at polar sunrise in the lower Arctic atmosphere, *Nature* 334, 138.

Barrie, L. A., J. W. Bottenheim, and W. R. Hart (1994) Polar Sunrise Experiment 1992 (PSE 1992): Preface, *J. Geophys. Res.* 99, 25,313.

Bates, D. R. and M. Nicolet (1950) Atmospheric hydrogen, *Publ. Astron. Soc. Pacific* 62, 106.

Bates, T. S., B. K. Lamb, A. Guenther, J. Dignon, and R. E. Stoiber (1992) Sulfur emissions to the atmosphere from natural sources, *J. Atmos. Chem.* 14, 315.

Bates, T. S., K. C. Kelly, J. E. Johnson, and R. H. Gammon (1995) Regional and seasonal variations in the flux of oceanic carbon monoxide to the atmosphere, *J. Geophys. Res.* 100, 23,093.

Baumgardner, D., J. E. Dye, B. G. Gandrud, D. Rogers, K. Weaver, R. G. Knollenberg, R. Newton, and R. Gallant (1994) The multiangle aerosol spectrometer probe: A new instrument for airborne particle research, *Proc. 9th Symposium on Meteorological Observations and Instrumentation*, Charlotte, NC, Amer. Met. Soc., 434.

Beig, G., S. Walters, and G. Brasseur (1993) A two-dimensional model of ion composition in the stratosphere, 2. Negative ions, *J. Geophys. Res.* 98, 12,775.

Bekki, S. and J.A. Pyle (1992) Two-dimensional assessment of the impact of aircraft sulphur emissions on the stratospheric sulphate aerosol layer, *J. Geophys. Res.* 97, 15,839.

Bekki, S. and J. A. Pyle (1993) Potential impact of combined $NO_x$ and $SO_x$ emissions from future high speed civil transport aircraft on stratospheric aerosols and ozone, *Geophys. Res. Lett.* 20, 723.

Bekki, S., R. Toumi, and J. A. Pyle (1993) Role of sulphur photochemistry in tropical ozone changes after the eruption of Mount Pinatubo, *Nature* 362, 331.

Benson, S. W. (1960) *The Foundations of Chemical Kinetics*, McGraw-Hill, New York.

Benson, S. W. (1976) *Thermochemical Kinetics*, Wiley, New York.

Berresheim, H. and W. Jaeschke (1983) The contribution of volcanoes to the global atmospheric sulfur budget, *J. Geophys. Res.* 88, 3732.

Berresheim, H. and V. D. Vulcan (1992) Vertical distributions of COS, $CS_2$, DMS, and other sulfur compounds in a loblolly pine forest, *Atmos. Environ.* 26A, 2031.

Berresheim, H., P. H. Wine, and D. D. Davis (1995) Sulfur in the atmosphere, in: *Composition, Chemistry and Climate of the Atmosphere*, H. B. Singh, ed., van Nostrand Reinhold, New York, 251.

Bingemer, H. G., S. Bürgermeister, R. L. Zimmermann, and H.-W. Georgii (1990) Atmospheric OCS: Evidence for a contribution of anthropogenic sources? *J. Geophys. Res.* 95, 20,617.

Binkley, D., C. T. Driscoll, H. L. Allen, P. Schoeneberger, and D. McAvoy, eds. (1989) *Acidic Deposition and Forest Soils: Context and Case Studies of the Southeastern United States*, Springer-Verlag, New York.

Bjerknes, J. and H. Solberg (1922) Life cycle of cyclones and the polar front theory of atmospheric circulation, *Geofys. Publ.* 3, 1-18.

Blake, D. R. and F. S. Rowland (1988) Continuing worldwide increase in tropospheric methane, 1978-1987, *Science* 239, 1129.

Bohren, C. and D. Huffman (1981) *Scattering and Absorption of Light by Small Particles*, Wiley, New York.

Bojkov, R. D. (1986) Surface ozone during the second half of the nineteenth century, *J. Climate Appl. Meteorol.* 25, 343.

Bojkov, R. D. and V. E. Fioletov (1995) Estimating the global ozone characteristics during the last 30 years, *J. Geophys. Res.* 100, 16,537.

Bolin, B. and E. Ericksson (1959) Changes in the carbon content of the atmosphere and the sea due to fossil fuel combustion, in: *The Atmosphere and the Sea in Motion*, B. Bolin, ed., Rossby Memorial Volume, The Rockefeller Institute Press, New York.

Bolin, B. (1981) Steady-state and response characteristics of a simple model of the carbon cycle, in: *Carbon Cycle Modelling*, B. Bolin, ed., SCOPE 16, Wiley & Sons.

Borys, R. D. (1989) Studies of ice nucleation by arctic aerosol on AGASP-II, *J. Atmos. Chem.* 9, 169.

Bott, A. (1989) A positive definite advection scheme obtained by nonlinear renormalization of the advective fluxes, *Mon. Wea. Rev.* 117, 1006.

Bouwman, A. F., D. S. Lee, W. A. H. Asman, F. J. Dentener, K. W. Van Der Hoek, and J. G. J. Olivier (1997) A global high-resolution emission inventory for ammonia, *Glob. Biogeochem. Cycles* 11, 561.

Boville, B. A. (1993) Sensitivity of simulated climate to model resolution, *J. Climate* 4, 469.

Bowman, F. M., J. R. Odum, J. H. Seinfeld and S. N. Pandis (1997) Mathematical model for gas-particle partitioning of secondary organic aerosols, *Atmos. Environ.* 31, 3921.

Boyd, J. P. (1976) The noninteraction of waves with the zonally-averaged flow on a spherical Earth and the interrelationship of eddy fluxes of energy, heat and momentum, *J. Atmos. Sci.* 33, 2285.

Brasseur, G. P. and P. C. Simon (1981) Stratospheric chemical and thermal response to long-term variability in the solar UV irradiance, *J. Geophys. Res.* 86, 7343.

Brasseur, G. P. and S. Solomon (1986) *Aeronomy of the Middle Atmosphere*, 2nd ed., D. Reidel Publishing Co., Dordrecht.

Brasseur, G. P., M. H. Hitchman, S. Walters, M. Dymek, E. Falise, and M. Pirre (1990) An interactive chemical dynamical radiative two-dimensional model of the middle atmosphere, *J. Geophys. Res.* 95, 5639.

Brasseur, G. and C. Granier (1992) Mount Pinatubo aerosols, chlorofluorocarbons, and ozone depletion, *Science* 257, 1239.

Brasseur, G. P., J.-F. Müller, and C. Granier (1996) Atmospheric impact of $NO_x$ emissions by subsonic aircraft: A three-dimensional model study, *J. Geophys. Res.* 101, 1423.

Brasseur, G. P., X. X. Tie, P. J. Rasch, and F. Lefèvre (1997) A three-dimensional simulation of the Antarctic ozone hole: The impact of anthropogenic chlorine on the lower stratosphere and upper troposphere, *J. Geophys. Res.* 102, 8909.

Brasseur, G. P., D. A. Hauglustaine, S. Walters, P. J. Rasch, J.-F. Müller, C. Granier, and X. X. Tie (1998) MOZART: A global chemical-transport model for ozone and related chemical tracers, 1. Model description, *J. Geophys. Res.* 103.

Brewer, A. W. (1949) Evidence for a world circulation provided by the measurements of helium and water vapor distribution in the stratosphere, *Quart. J. Roy. Meteor. Soc.* 75, 351.

Brewer, A. W. and J. R. Milford (1960) The Oxford Kew ozone sonde, *Proc. Roy. Soc. London* A256, 470.

Brimblecombe, P. and S. L. Clegg (1988) The solubility and behaviour of acid gases in the marine aerosol, *J. Atmos. Chem.* 7, 1.

Broecker, W. S. (1995) Chaotic climate, *Scientific American*, 44.

Brost, R. A. and R. B. Chatfield (1989) Transport of radon in a three-dimensional, subhemispheric model, *J. Geophys. Res.* 94, 5095.

Brost, R. A., J. Feichter, and M. Heimann (1991) Three-dimensional simulation of $^7Be$ in a global climate model, *J. Geophys. Res.* 96, 22,423.

Browell, E. V. (1989) Differential absorption lidar sensing of ozone, *Proc. IEEE* 77, 419.

Browell, E. V., M. A. Fenn, C. F. Butler, W. B. Grant, M. B. Clayton, J. Fishman, A. S. Bachmeier, B. E. Anderson, G. L. Gregory, H. E. Fuelberg, J. D. Bradshaw, S. T. Sandholm, D. R. Blake, B. G. Heikes, G. W. Sachse, H. B. Singh, and R. W. Talbot (1996) Ozone and aerosol distributions and air mass characteristics over the South Atlantic Basin during the burning season, *J. Geophys. Res.* 101, 24,043.

Brueckner, G. E., K. L. Edlow, L. E. Floyd IV, J. L. Lean, and M. E. VanHoosier (1993) The Solar Ultraviolet Spectral Irradiance Monitor (SUSIM) experiment on board the Upper Atmosphere Research Satellite (UARS), *J. Geophys. Res.* 98, 10,695.

Brune, W. H., J. G. Anderson, and K. R. Chan (1989) In situ observations of BrO over Antarctica: ER-2 aircraft results from 54°S to 72°S latitude, *J. Geophys. Res.* 94, 16,639.

Businger, J. A., W. F. Dabberdt, A. C. Delany, T. W. Horst, C. L. Martin, S. P. Oncley, and S. R. Semmer (1990) The NCAR Atmosphere-Surface Turbulent Exchange Research (ASTER) facility, *Bull. Am. Met. Soc.* 71, 1006.

Byrne, G. D. and A. C. Hindmarsh (1975) A polyalgorithm for the numerical solution of ordinary differential equations, *ACM Trans. Math. Software* 1, 71.

Calvert, J. G., A. Lazrus, G. L. Kok, B. G. Heikes, J. G. Walega, J. Lind, and C. A. Cantrell (1985) Chemical mechanisms of acid generation in the troposphere, *Nature* 317, 27.

Cantrell, C. A., R. E. Shetter, A. H. McDaniel, J. G. Calvert, J. A. Davidson, D. C. Lowe, S. C. Tyler, R. J. Cicerone, and J. R. Greenberg (1990) Carbon kinetic isotope effect on the oxidation of methane by the hydroxyl radical, *J. Geophys. Res.* 95, 455-462.

Cantrell, C. A., R. E. Shetter, J. G. Calvert, D. D. Parrish, F. C. Fehsenfeld, P. D. Goldan, W. Kuster, E. J. Williams, H. H. Westberg, G. Allwine, and R. Martin (1993a) Peroxy radicals as measured in ROSE and estimated from photostationary state deviations, *J. Geophys. Res.* 98, 18,355.

Cantrell, C. A., R. E. Shetter, J. A. Lind, A. H. McDaniel, and J. G. Calvert (1993b) An improved chemical amplifier technique for peroxyl radical measurements, *J. Geophys. Res.* 98, 2897.

Cantrell, C. A., R. E. Shetter, and J. G. Calvert (1995) Comparison of peroxy radical concentrations at several contrasting sites, *J. Atmos. Sci.* 52, 3408.

Carli, B. and J. H. Park (1988) Simultaneous measurement of minor stratospheric constituents with emission far infrared spectroscopy, *J. Geophys. Res.* 93, 3851.

Carli, B., M. Carlotti, B. M. Dinelli, F. Mencaraglia, and J. H. Park (1989) The mixing ratio of the stratospheric hydroxyl radical from far infrared emission measurements, *J. Geophys. Res.* 94, 11,049.

Carlotti, M., P. A. R. Ade, B. Carli, P. Ciarpallini, U. Cortesi, M. J. Griffin, G. Lepri, F. Mencaraglia, A. G. Murray, I. G. Nolt, J. H. Park, and J. V. Radostitz (1995) Measurement of stratospheric HBr using high resolution far infrared spectroscopy, *Geophys. Res. Lett.* 22, 3207.

Carslaw, N., L. J. Carpenter, J. M. C. Plane, B. J. Allan, R. A. Burgess, K. C. Clemitshaw, H. Coe, and S. A. Penkett (1997) Simultaneous observations of nitrate and peroxy radicals in the marine boundary layer, *J. Geophys. Res.* 102, 18,917.

Carswell, A. I. (1983) Lidar measurements of the atmosphere, *Can. J. Phys.* 61, 378.

Chameides, W. L. and J. C. G. Walker (1973) A photochemical theory of tropospheric ozone, *J. Geophys. Res.* 78, 8751.

Chameides, W. L., R. W. Lindsay, J. Richardson, and C. S. Kiang (1988) The role of biogenic hydrocarbons in urban photochemical smog: Atlanta as a case study, *Science* 241, 1473.

Chameides, W. L. and A. W. Stelson (1992) Aqueous-phase chemical processes in deliquescent sea-salt aerosols: A mechanism that couples the atmospheric cycles of S and sea salt, *J. Geophys. Res.* 97, 20,565.

Chameides, W. L., P. S. Kasibhatla, J. Yienger, and H. Levy III (1994) Growth of continental-scale metro-agro-plexes, regional ozone production, and world food production, *Science* 264, 74.

Chance, K. V. and W. A. Traub (1987) Evidence for stratospheric hydrogen peroxide, *J. Geophys. Res.* 92, 3061.

Chance, K. V., D. G. Johnson, W. A. Traub, and K. W. Jucks (1991) Measurement of the stratospheric hydrogen peroxide concentration profile using far infrared thermal emission spectroscopy, *Geophys. Res. Lett.* 18, 1003.

Chang, J. S., R. A. Brost, I. S. A. Isaksen, S. Madronich, P. Middleton, W. R. Stockwell, and C. J. Walcek (1987) A three-dimensional Eulerian acid deposition model: Physical concepts and formulation, *J. Geophys. Res.* 92, 14,681.

Chapman, S. (1930) On ozone and atomic oxygen in the upper atmosphere, *Phil. Mag.* 10, 369.

Charlson, R. J., J. E. Lovelock, M. O. Andreae, and S. G. Warren (1987) Oceanic phytoplankton, atmospheric sulphur, cloud albedo and climate, *Nature* 326, 655.

Charlson, R. J., G. H. Orians, G. V. Wolfe, and S. S. Butcher (1992a) Human modification of global biogeochemical cycles, in: *Global Biogeochemical Cycles*, S. S. Butcher, R. J. Charlson, G. H. Orians, and G. V. Wolfe, eds., Academic Press, San Diego, 353.

Charlson, R. J., S. E. Schwartz, J. M. Hales, R. D. Cess, J. A. Coakley, Jr., J. E. Hansen, and D. J. Hofmann (1992b) Climate forcing by anthropogenic aerosols, *Science* 255, 423.

Charney, J. G. and M. E. Stern (1962) On the stability of internal baroclinic jets in a rotating atmosphere, *J. Atmos. Sci.* 19, 159.

Chatfield, R. B. and P. J. Crutzen (1984) Sulfur dioxide in remote oceanic air: Cloud transport of reactive precursors, *J. Geophys. Res.* 89, 7111.

Chaumerliac, N., E. Richard, J.-P. Pinty, and E. C. Nickerson (1987) Sulfur scavenging in a mesoscale model with quasi-spectral microphysics: Two-dimensional results for continental and maritime clouds, *J. Geophys. Res.* 92, 3114.

Chipperfield, M. P., D. Cariolle, P. Simon, R. Ramaroson, and D. J. Lary (1993) A three-dimensional modeling study of trace species in the Arctic lower stratosphere during winter 1989-1990, *J. Geophys. Res.* 98, 7199.

Chock, D. P. (1991) A comparison of numerical methods for solving the advection equation-III, *Atmos. Environ.* 25A, 853.

Ciais, P., P. P. Tans, M. Trolier, J. W. C. White, and R. J. Franey (1995) A large northern hemisphere terrestrial $CO_2$ sink indicated by $^{13}C/^{12}C$ of atmospheric $CO_2$, *Sciences* 269, 1098.

Clarke, A. D. (1993) Atmospheric nuclei in the Pacific midtroposphere: Their nature, concentration and evolution, *J. Geophys. Res.* 98, 20,633.

Clausen, H. B. and C. C. Langway (1989) The ionic deposits in polar ice cores, in: *The Environmental Record in Glaciers and Ice Sheets*, H. Oeschger and C. C. Langway, eds., Wiley and Sons, New York.

Climatic Impact Assessment Program (CIAP) (1975) *The Natural Stratosphere of 1974*, Dept. of Transport, Final Report, DOT-TST-75-51, Sept. 1975 (National Technical Information Service, Springfield, Virginia 22151).

Coffey, M. T. and W. G. Mankin (1993) Observations of the loss of stratospheric $NO_2$ following volcanic eruptions, *Geophys. Res. Lett.* 20, 2873.

Coffman, D. J. and D. A. Hegg (1995) A preliminary study of the effect of ammonia on particle nucleation in the marine boundary layer, *J. Geophys. Res.* 100, 7147.

Cohen, R. C., P. O. Wennberg, R. M. Stimpfle, J. Koplow, J. G. Anderson, D. W. Fahey, E. L. Woodbridge, E. R. Keim, R. Gao, M. H. Proffitt, M. Loewenstein, and K. R. Chan (1994) Are models of catalytic removal of $O_3$ by $HO_x$ accurate? Constraints from in situ measurements of the OH to $HO_2$ ratio, *Geophys. Res. Lett.* 21, 2539.

Connell, P. S., D. J. Wuebbles, and J. S. Chang (1985) Stratospheric hydrogen peroxide: The relationship of theory and observation, *J. Geophys. Res.* 90, 10,726.

Cotton, W. R. and R. A. Anthes (1989) *Storm and Cloud Dynamics,* Academic Press, San Diego.

Coulson, K. L. (1975) *Solar and Terrestrial Radiation*, Academic Press, New York.

Covert, D. S., V. N. Kapustin, P. K. Quinn, and T. S. Bates (1992) New particle formation in the marine boundary layer, *J. Geophys. Res.* 97, 20,581.

Craig, H. (1957) The natural distribution of radiocarbon and the exchange time of carbon dioxide between atmosphere and sea, *Tellus* 9, 1.

Crutzen, P. J. (1970) The influence of nitrogen oxide on the atmospheric ozone content, *Quart. J. Roy. Met. Soc.* 96, 320.

Crutzen, P. J. (1973) A discussion of the chemistry of some minor constituents in the stratosphere and troposphere, *Pure Appl. Geophys.* 106, 1385.

Crutzen, P. J. (1976) The possible importance of COS for the sulfate layer of the stratosphere, *Geophys. Res. Lett.* 3, 73.

Curry, J. A., E. E. Ebert, and J. L. Schramm (1993) Impact of clouds on the surface radiation budget of the Arctic Ocean, *Meteorol. Atmos. Phys.* 57, 197.

Dabberdt, W. F., D. H. Lenschow, T. W. Horst, P. R. Zimmerman, S. P. Oncley, and A. C. Delany (1993) Atmosphere-surface exchange measurements, *Science* 260, 1472.

Daley, R. (1991) *Atmospheric Data Analysis*, Cambridge University Press, Cambridge, UK.

d'Almeida, G. A., P. Koepke, and E. P. Shettle (1991) *Atmospheric Aerosols: Global Climatology and Radiative Characteristics,* A. Deepak Publishing, Hampton, Virginia.

Danielsen, E. F. (1985) Ozone transport, in: *Ozone in the Free Atmosphere*, R. C. Whitten and S. S. Prasad, eds., Van Nostrand Reinhold, New York.

Dasgupta, P. K., K. DeCesare, and J. C. Ullrey (1980) Determination of atmospheric sulfur dioxide without tetrachloromercurate(II) and the mechanism of the Schiff reaction, *Anal. Chem.* 52, 1912.

Davidson, E. A. (1991) Fluxes of nitrous oxide and nitric oxide from terrestrial ecosystems, in: *Microbial Production and Consumption of Greenhouse Gases: Methane, Nitrogen Oxides, and Halomethanes,* J. E. Roger, and W. B. Whitman, eds., American Society for Microbiology, Washington, D.C.

Davis, D. D., W. S. Heaps, D. Philen, M. Rodgers, T. McGee, A. Nelson, and A. J. Moriarty (1979) Airborne laser induced fluorescence system for measuring OH and other trace gases in the parts-per-quadrillion to parts-per-trillion range, *Rev. Sci. Instrum.* 50, 1505.

Davis, D. D., J. Crawford, G. Chen, W. Chameides, S. Liu, J. Bradshaw, S. Sandholm, G. Sachse, G. Gregory, B. Anderson, J. Barrick, A. Bachmeier, J. Collins, E. Browell, D. Blake, S. Rowland, Y. Kondo, H. Singh, R. Talbot, B. Heikes, J. Merrill, J. Rodriguez, and R. E. Newell (1996) Assessment of ozone photochemistry in the western North Pacific as inferred from PEM-West A observations during the Fall 1991, *J. Geophys. Res.* 101, 2111.

Defant, A. and F. Defant (1958) *Physikalische Dynamik der Atmospäre*, Akademische Verlagsgesselschaft GmbH, Frankfurt am Main.

Delmas, R. J. (1992) Free tropospheric reservoir of natural sulfate, *J. Atmos. Chem.* 14, 261.

DeMore, W. B., D. M. Golden, R. F. Hampson, C. J. Howard, M. J. Kurylo, M. J. Molina, A. R. Ravishankara, and S. P. Sander (1990) *Chemical Kinetics and Photochemical Data for Use in Stratospheric Modeling*, Jet Propulsion Laboratory, California Institute of Technology, Evaluation No. 9, JPL Publ. 90-1.

DeMore, W. B., S. P. Sander, D. M. Golden, R. F. Hampson, M. J. Kurylo, C. J. Howard, A. R. Ravishankara, C. E. Kolb, and M. J. Molina (1992) *Chemical Kinetics and Photochemical Data for Use in Stratospheric Modeling*, Jet Propulsion Laboratory, California Institute of Technology, Evaluation No. 10, JPL Publ. 92-20.

DeMore, W. B., S. P. Sander, D. M. Golden, R. F. Hampson, M. J. Kurylo, C. J. Howard, A. R. Ravishankara, C. E. Kolb, and M. J. Molina (1997) *Chemical Kinetics and Photochemical Data for Use in Stratospheric Modeling*, Jet Propulsion Laboratory, California Institute of Technology, Evaluation No. 12, JPL Publ. 97-4.

Dentener, F. J. and P. J. Crutzen (1993) Reaction of $N_2O_5$ on tropospheric aerosols: Impact on the global distributions of $NO_x$, $O_3$ and OH, *J. Geophys. Res.* 98, 7149.

de Zafra, R. L. (1989) New observations of a large concentration of ClO in the springtime lower stratosphere over Antarctica and its implications for ozone-depleting chemistry, *J. Geophys. Res.* 94, 11,423.

Dickerson, R. R. and D. H. Stedman (1980) Measurements of solar ultraviolet radiation and atmospheric photolysis rates, *Atmos. Tech.* 12, 56.

Dixon, R. K., S. Brown, R. A. Houghton, A. M. Solomon, M. C. Trexler, and J. Wisniewski (1994) Carbon pools and flux of global forest ecosystems, *Science* 263, 185.

Dlugokencky, E. J., L. P. Steele, P. M. Lang, and K. A. Masarie (1994) The growth rate and distribution of atmospheric methane, *J. Geophys. Res.* 99, 17,021.

Dobson, G. M. B. (1930) Observations of the amount of ozone in the Earth's atmosphere and its relation to other geophysical conditions, *Proc. Roy. Soc. London, Sec. A* 129, 411.

Dobson, G. M. B. (1956) Origin and distribution of the polyatomic molecules in the atmosphere, *Proc. Roy. Soc. London* A236, 187.

Dobson, G. M. B. (1968) Forty years' research on atmospheric ozone at Oxford: A history, *Appl. Optics* 7, 405.

Dominé, F., A. R. Ravishankara, and C. J. Howard (1992) Kinetics and mechanisms of the reactions of $CH_3S$, $CH_3SO$, and $CH_3SS$ with $O_3$ at 300 K and low pressures, *J. Phys. Chem.* 96, 2171.

Dougherty, E. P., J. T. Hwang, and H. Rabitz (1979) Further developments and applications of the Green's function method and sensitivity analysis in chemical kinetics, *J. Comp. Phys.* 71, 1794.

Draajers, G. P. J., W. P. M. F. Ivens, M. M. Ross, and W. Blouten (1989) The contribution of ammonia emissions from agriculture to the deposition of acidifying and eutrophying compounds into forests, *Environmental Pollution* 60, 55.

Drake, B. G. (1992) The impact of rising $CO_2$ on ecosystem production, *Water, Air, and Soil Pollution* 64, 25.

Drummond, J. W., A. Volz, and D. H. Ehhalt (1985) Optimized chemiluminescence detector for tropospheric NO measurements, *J. Atmos. Chem.* 2, 287.

Drummond, J. R. and G. S. Mand (1996) The measurements of pollution in the troposphere (MOPITT) instrument: Overall performance and calibration requirements, *J. Atmos. Oceanic Tech.* 13, 314.

Duce, R. A., P. S. Liss, J. T. Merrill, E. L. Atlas, P. Buat-Menard, P. Hicks, J. M. Miller, J. M. Prospero, R. Arimoto, T. M. Church, W. Ellis, J. N. Galloway, L. Hansen, T. D. Jidkells, A. H. Knap, K. H. Reinhart, B. Schneider, A. Soudin, J. J. Tokos, S. Tsunogai, R. Wollast, and M. Zhou (1991) The atmospheric input of trace species to the world ocean, *Glob. Biogeochem. Cycles* 5, 193.

Ducros, M. (1845) Observation d'une pluie acide, *J. Pharm. Chem.* 3, 273.

Dunkerton, T. J. and M. P. Baldwin (1991) Quasi-biennial modulation of planetary wave fluxes in the Northern Hemisphere winter, *J. Atmos. Sci.* 48, 1043.

Dutton, E. G. and I. R. Christy (1992) Solar radiative forcing at selected locations and evidence for global lower tropospheric cooling following the eruptions of El Chichón and Pinatubo, *Geophys. Res. Lett.* 19, 2313.

Dutton, J. A. (1986) *The Ceaseless Wind*, Dover Publications, Inc., New York.

Dye, J. E. and D. Baumgardner (1984) Evaluation of the forward scattering spectrometer probe, 1. Electronic and optical studies, *J. Atmos. Ocean. Tech.* 1, 329.

Ehhalt, D. H., L. E. Heidt, R. H. Lueb, and E. A. Martell (1975) Concentrations of $CH_4$, $CO_2$, $H_2$, $H_2O$ and $N_2O$ in the upper stratosphere, *J. Atmos. Sci.* 32, 163.

Ehhalt, D. H., U. Schmidt, and L. E. Heidt (1977) Vertical profiles of molecular hydrogen in the troposphere and stratosphere, *J. Geophys. Res.* 82, 5907.

Ehhalt, D. H. and J. W. Drummond (1982) The tropospheric cycle of $NO_x$, in: *Chemistry of the Unpolluted and Polluted Troposphere*, H. W. Georgii and W. Jaeschke, eds., D. Reidel, Hinham, Massachusetts.

Ehhalt, D. H. (1986) On the consequence of tropospheric $CH_4$ increase to the exospheric density, *J. Geophys. Res.* 91, 2843.

Ehhalt, D. H. and F. Rohrer (1994) The impact of commercial aircraft on tropospheric ozone, *Proc. of the 7th BOC Priestley Conference*, Lewisburg, PA.

Eisele, F. L. and D. J. Tanner (1991) Ion-assisted tropospheric OH measurements, *J. Geophys. Res.* 96, 9295.

Eisele, F. L. and H. Berresheim (1992) High-pressure chemical ionization flow reactor for real-time mass spectrometric detection of sulfur gases and unsaturated hydrocarbons in air, *Anal. Chem.* 64, 283.

Eisele, F. L., G. H. Mount, F. C. Fehsenfeld, J. Harder, E. Marovich, D. D. Parrish, J. Roberts, M. Trainer, and D. Tanner (1994) Intercomparison of tropospheric OH and ancillary trace gas measurements at Fritz Peak Observatory, Colorado, *J. Geophys. Res.* 99, 18,605.

Eisele, F. L. (1995) New insight and questions resulting from recent ion-assisted OH measurements, *J. Atmos. Sci.* 52, 3337.

Elkins, J., J. Butler, S. Montzka, R. Myers, T. Thompson, T. Baring, S. Cummings, G. Dutton, A. Hayden, J. Lobert, G. Holcomb, W. Sturges, and T. Gilpin (1993) Nitrous oxide and halocarbons division, Section 5, in: *Climate Monitoring and Diagnostics Laboratory, Summary Report, 1992*, U.S. Dept. of Commerce, National Oceanic and Atmospheric Administration, 59.

Emanuel, K. A. (1991) A scheme representing cumulus convection in large-scale models, *J. Atmos. Sci.* 48, 2313.

Enting, I. (1985) A classification of some inverse problems in geochemical modeling, *Tellus* 37, 216.

Enting, I. and J. Mansbridge (1989) Seasonal sources and sinks of atmospheric $CO_2$ direct inversion of filtered data, *Tellus* 41, 111.

Enting, I. G., C. M. Trudinger, and R. Francey (1995) A synthesis inversion of the concentration and $\delta^{13}C$ of atmospheric $CO_2$, *Tellus* 47B, 35.

Erickson, D. J. III (1993) A stability dependent theory for air-sea gas exchange, *J. Geophys. Res.* 98, 8471.

Ertel, H. (1942) Ein neuer hydrodynamischer Wirbelsatz, *Meteor. Z.* 59, 271.

Evans, W. F. J., C. T. McElroy, and I. E. Galbally (1985) The conversion of $N_2O_5$ to $HNO_3$ at high latitudes in winter, *Geophys. Res. Lett.* 12, 825.

Fabian, P., J. A. Pyle, and R. J. Wells (1982) Diurnal variations of minor constituents in the stratosphere modeled as a function of latitude and season, *J. Geophys. Res.* 87, 4981.

Fahey, D. W., D. M. Murphy, K. K. Kelly, M. K. W. Ko, M. H. Proffitt, C. S. Eubank, G. V. Ferry, M. Loewenstein, and K. R. Chan (1989) Measurements of nitric oxide and total reactive nitrogen in the Antarctic stratosphere: Observations and chemical implications, *J. Geophys. Res.* 94, 16,665.

Fahey, D. W., S. R. Kawa, and K. R. Chan (1990a) Nitric oxide measurements in the Arctic winter stratosphere, *Geophys. Res. Lett.* 17, 489.

Fahey, D. W., K. K. Kelly, S. R. Kawa, A. F. Tuck, M. Loewenstein, K. R. Chan, and L. E. Heidt (1990b) Observations of denitrification and dehydration in the winter polar stratospheres, *Nature* 344, 321.

Fahey, D. W., S. Solomon, S. R. Kawa, M. Loewenstein, J. R. Podolske, S. E. Strahan, and K. R. Chan (1990c) A diagnostic for denitrification in the winter polar stratosphere, *Nature* 345, 698.

Fahey, D. W., S. R. Kawa, E. L. Woodbridge, P. Tin, J. C. Wilson, H. H. Jonsson, J. E. Dye, D. Baumgardner, S. Borrmann, D. W. Toohey, L. M. Avallone, M. H. Proffitt, J. Margitan, M. Loewenstein, J. R. Podolske, R. J. Salawitch, S. C. Wofsy, M. K. W. Ko, D. E. Anderson, M. R. Schoeberl, and K. R. Chan (1993) In situ measurements constraining the role of sulfate aerosols in mid-latitude ozone depletion, *Nature* 363, 509.

Fan, S. M. and D. J. Jacob (1992) Surface ozone depletion in Arctic spring sustained by bromine reactions on aerosols, *Nature* 359, 522.

Farman, J. C., B. G. Gardiner, and J. D. Shanklin (1985) Large losses of total ozone in Antarctic reveal seasonal $ClO_x/NO_x$ interaction, *Nature* 315, 207.

Farmer, C. B. (1974) Infrared measurements of stratospheric composition, *Can. J. Chem.* 52, 1544.

Farmer, C. B. (1987) High resolution infrared spectroscopy of the sun and the earth's atmosphere from space, *Mikrochim. Acta* 3, 189.

Fehsenfeld, F. C., D. D. Parrish, and D. W. Fahey (1988) The measurement of $NO_x$ in the non-urban troposphere, in *Tropospheric Ozone*, I. S. A. Isaksen, ed., D. Reidel, Dordrecht, The Netherlands.

Fehsenfeld, F., J. Calvert, R. Fall, P. Goldan, A. B. Guenther, C. N. Hewitt, B. Lamb, S. Liu, M. Trainer, H. Westberg, and P. Zimmerman (1992) Emissions of volatile organic compounds from vegetation and the implications for atmospheric chemistry, *Glob. Biogeochem. Cycles* 6, 389.

Fehsenfeld, F. C. and S. C. Liu (1993) Tropospheric ozone: Distribution and sources, in *Global Atmospheric Chemical Change*, C. N. Hewitt and W. T. Sturges, eds., Elsevier Applied Sciences, New York.

Feichter, J. and P. J. Crutzen (1990) Parameterization of vertical tracer transport due to deep cumulus convection in a global transport model and its evaluation with $^{222}$radon measurements, *Tellus* 42B, 100.

Feichter, J., R. A. Brost, and M. Heimann (1991) Three-dimensional modeling of the concentration and deposition of $^{210}$Pb aerosols, *J. Geophys. Res.* 96, 22,447.

Felton, C. C., J. C. Sheppard, and M. J. Campbell (1992) Precision of the radiochemical OH measurement method, *Atmos. Environ.* 26, 2105.

Ferm, M. (1986) A $Na_2CO_3$-coated denuder and filter for determination of gaseous $HNO_3$ and particulate $NO_3^-$ in the atmosphere, *Atmos. Environ.* 20, 1193.

Finlayson-Pitts, B. J., F. E. Livingston, and H. N. Berko (1990) Ozone destruction and bromine photochemistry at ground level in the Arctic spring, *Nature* 343, 622.

Fisher, M. and D. J. Lary (1995) Lagrangian four-dimensional variational data assimilation of chemical species, *Quart. J. Roy. Meteor. Soc.* 121, 1681.

Fishman, J., V. Ramanathan, P. J. Crutzen, and S. C. Liu (1979) Tropospheric ozone and climate, *Nature* 282, 818.

Fishman, J., F. M. Vukovich, and E. V. Browell (1985) The photochemistry of synoptic-scale ozone synthesis: Implication for the global tropospheric budget, *J. Atmos. Chem.* 3, 299.

Fishman, J., C. E. Watson, J. C. Larsen, and J. A. Logan (1990) Distribution of tropospheric ozone determined from satellite data, *J. Geophys. Res.* 95, 3599.

Fishman, J., V. G. Brackett, and K. Fakhruzzaman (1992) Distribution of tropospheric ozone in the tropics from satellite and ozonesonde measurements, *J. Atmos. Terr. Phys.* 54, 589.

Fishman, J. and V. G. Brackett (1997) The climatological distribution of tropospheric ozone derived from satellite measurements using version 7 Total Ozone Mapping Spectrometer and Stratospheric Aerosol and Gas Experiment data sets, *J. Geophys. Res.* 102, 19, 275.

Fitzgerald, J. W. (1991) Marine aerosols: A review. *Atmos. Environ.* 25A, 533.

Fitzgerald, J. W., W. A. Hoppel, and F. Gelbard (1998) A one-dimensional sectional model to simulate multicomponent aerosol dynamics in the marine boundary layer, 1: Model description.

Fletcher, C. A. J. (1988) *Computational Techniques for Fluid Dynamics*, vol. 1-2, Springer Verlag, Berlin.

Flocke, F., A. Volz-Thomas, and D. Kley (1994) The use of alkyl nitrate measurements for the characterization of the ozone balance at TOR-Station No. 11, Schauinsland. *Proc. of the 1994 Annual EUROTRAC Symposium*, Garmisch-Partenkirchen, Germany.

Forester, K. C. (1977) Higher order monotonic convective difference schemes, *J. Comp. Phys.* 23, 1.

Forster, R., M. Frost, D. Fulle, H. F. Hamann, H. Hippler, A. Schlepegrell, and J. Troe (1995) High pressure range of the addition of HO to HO, NO, $NO_2$, and CO, 1. Saturated laser induced fluorescence measurements at 298 K, *J. Chem. Phys.* 103, 2949.

Frew, N. M. (1997) The role of organic films in air-sea gas exchange, in: *The Sea Surface and Global Change*, P. S. Liss and R. A. Duce, eds., Cambridge University Press, Cambridge, UK, 121.

Fried, A., J. R. Drummond, B. Henry, and J. Fox (1991) Versatile integrated tunable diode laser system for ambient measurements of OCS, *Appl. Opt.* 30, 1916.

Friedlander, S. K. (1983) Dynamics of aerosol formation by chemical reaction, *Ann. NY Acad. Sci.* 404, 354.

Fritts, D. C. and W. Lu (1993) Spectral estimates of gravity wave energy and momentum fluxes, II. Parameterization of wave forcing and variability, *J. Atmos. Sci.* 50, 3695.

Fuchs, N. A. (1964) *The Mechanics of Aerosols*, Pergamon Press, Oxford.

Fuchs, N. A. and A. G. Sutugin (1970) *Highly Dispersed Aerosols*, Butterworth Publishing, Stoneham, Massachusetts.

Fung, I., J. John, J. Lerner, E. Matthews, M. Prather, L. Steele, and P. Fraser (1991) Three-dimensional model synthesis of the global methane cycle, *J. Geophys. Res.* 96, 13,033.

Fung, I. (1993) Models of oceanic and terrestrial sinks of anthropogenic $CO_2$: A review of the contemporary carbon cycle, in: *Biogeochemistry and Global Change*, R. S. Oremland, ed., Chapman and Hall, New York, 166.

Garcia, R. R. and S. Solomon (1983) A numerical model of the zonally-averaged dynamical and chemical structure of the middle atmosphere, *J. Geophys. Res.* 88, 1379.

Garcia, R. R. and S. Solomon (1985) The effect of breaking gravity waves on the dynamics and chemical composition of the mesosphere and lower thermosphere, *J. Geophys. Res.* 90, 3850.

Garcia, R. R. (1991) Parameterization of planetary wave breaking in the middle atmosphere, *J. Atmos. Sci.* 48, 1405.

Garcia, R. R., F. Stordal, S. Solomon, and J. T. Kiehl (1992) A new numerical model of the middle atmosphere, 1. Dynamics and transport of tropospheric source gases, *J. Geophys. Res.* 97, 12967.

Garcia, R. R. and S. Solomon (1994) A new numerical model of the middle atmosphere: 2. Ozone and related species, *J. Geophys. Res.* 99, 12,937.

Gardner, C. S. (1986) Lidar studies of the nighttime sodium layer over Urbana, Illinois, 1. Seasonal and nocturnal variations, *J. Geophys. Res.* 91, 13,659.

Gast, P. R. (1961) Thermal radiation, in: *Handbook of Geophysics*, United States Air Force Research Division, 2nd ed., Macmillan Publishing Co., New York.

Gear, C. W. (1971) *Numerical Initial Value Problems in Ordinary Differential Equations*, Prentice Hall, New York.

Geiss, H., and A. Volz-Thomas (1992) *Lokale und regionale Ozonproduktion: Chemie und Transport*, Berichte des Forschungszentrum Julich N. 2764. Institut für Chemie und Dynamik der Geosphäre 2, Julich, Germany.

Gelbard, F. and J. H. Seinfeld (1979) The general dynamic equation for aerosols — Theory and application to aerosol formation and growth, *J. Colloid Interface Sci.* 69, 363.

Gelbard, F. and J. H. Seinfeld (1980) Simulation of multicomponent aerosol dynamics, *J. Colloid Interface Sci.* 78, 485.

Gidel, L. T. and M. A. Shapiro (1980) General circulation model estimates of the net vertical flux of ozone in the lower stratosphere and the implications for the tropospheric ozone budget, *J. Geophys. Res.* 85, 4059.

Gill, A. (1980) *Atmosphere-Ocean Dynamics*, Academic Press, New York.

Gille, J. C. and J. M. Russell III (1984) The limb infrared monitor of the stratosphere: Experiment description, performance, and results, *J. Geophys. Res.* 89, 5125.

Gille, J. C. and L. V. Lyjak (1986) Radiative heating and cooling rates in the middle atmosphere, *J. Atmos. Sci.* 43, 2215.

Gilpin, T., E. Apel, A. Fried, B. Wert, J. Calvert, Z. Genfa, P. Dasgupta, J. W. Harder, B. Heikes, B. Hopkins, H. Westberg, T. Kleindienst, Y.-N. Lee, X. Zhou, W. Lonneman, and S. Sewell (1997) Intercomparison of six ambient [$CH_2O$] measurement techniques, *J. Geophys. Res.* 102, 21,161.

Giorgi, F. and W. L. Chameides (1985) The rainout parameterization in a photochemical model, *J. Geophys. Res.* 90, 7872.

Goldan, P. D., M. Trainer, W. C. Kuster, D. D. Parrish, J. Carpenter, J. M. Roberts, J. E. Yee, and F. C. Fehsenfeld (1995) Measurements of hydrocarbons, oxygenated hydrocarbons, carbon monoxide, and nitrogen oxides in an urban basin in Colorado: Implications for emission inventories, *J. Geophys. Res.* 100, 22,771.

Goody, R. M. and Y. L. Yung (1989) *Atmospheric Radiation. Theoretical Basis*, Oxford University Press, Oxford and New York.

Goody, R. (1995) *Principles of Atmospheric Physics and Chemistry*, Oxford University Press, Oxford, UK.

Götz, F. W. P., A. R. Meetham, and G. M. B. Dobson (1934) The vertical distribution of ozone in the atmosphere, *Proc. Roy. Soc.* A145, 416.

Graedel, T. E. and P. J. Crutzen (1995) *Atmosphere, Climate, and Change*, Scientific American Library, New York.
Graedel, T. E. and W. C. Keene (1995) Tropospheric budget of reactive chlorine, *Global Biogeochemical Cycles* 9, 47.
Granier, C. and G. Brasseur (1992) Impact of heterogeneous chemistry on model predictions of ozone changes, *J. Geophys. Res.* 97, 18,015.
Gray, L. J. and J. A. Pyle (1986) Semi-annual oscillation and equatorial tracer distributions, *Quart. J. Roy. Meteor. Soc.* 112, 387.
Gray, L. J. and J. A. Pyle (1987) Two-dimensional model studies of equatorial dynamics and tracer distributions, *Quart. J. Roy. Meteor. Soc.* 113, 635.
Greenberg, J. P. and P. R. Zimmerman (1984) Nonmethane hydrocarbons in remote tropical, continental, and marine atmospheres, *J. Geophys. Res.* 89, 4767.
Greenberg, J. P., P. R. Zimmerman, and R. B. Chatfield (1985) Hydrocarbons and carbon monoxide in African savannah air, *Geophys. Res. Lett.* 12, 113.
Greenberg, J. P., P. R. Zimmerman, and P. Haagenson (1990) Tropospheric hydrocarbon and CO profiles over the U.S. west coast and Alaska, *J. Geophys. Res.* 95, 14,015.
Greenberg, J. P., D. Helmig, and P. R. Zimmerman (1996) Seasonal measurements of nonmethane hydrocarbons and carbon monoxide at the Mauna Loa Observatory during the Mauna Loa Observatory Photochemistry Experiment 2, *J. Geophys. Res.* 101, 14,581.
Griffith, D. W. T. and G. Schuster (1987) Atmospheric trace gas analysis using matrix isolation-Fourier transform infrared spectroscopy, *J. Atmos. Chem.* 5, 59.
Grose, W. L., R. S. Eckman, R. E. Turner, and W. T. Blackshear (1989) Global modeling of ozone and trace gases, in: *Atmospheric Ozone Research and Its Policy Implications*, T. Schneider, S. D. Lee, G. J. R. Wolters, L. D. Grant, eds., Elsevier, New York.
Guenther, A., B. Lamb, and H. Westberg (1989) U.S. national biogenic sulfur emissions inventory, in: *Biogenic Sulfur in the Environment*, E. S. Saltzman and W. J. Cooper, eds., American Chemical Society, Washington, D.C.
Guenther, A., P. Zimmerman, P. Harley, R. Monson, and R. Fall (1993) Isoprene and monoterpene emission rate variability: Model evaluation and sensitivity analysis, *J. Geophys. Res.* 98, 12,609.
Guenther, A., C. N. Hewitt, D. Erickson, R. Fall, C. Geron, T. Graedel, P. Harley, L. Klinger, M. Lerdau, W. McKay, T. Pierce, B. Scholes, R. Steinbrecher, R. Tallamraju, J. Taylor, and P. Zimmerman (1995) A global model of natural volatile organic compound emissions, *J. Geophys. Res.* 100, 8873.
Hack, J. J. (1994) Parameterization of moist convection in the National Center for Atmospheric Research Community Climate Model (CCM), *J. Geophys. Res.* 99, 5551.
Hall, M. C. and D. G. Cacuci (1983) Physical interpretation of adjoint functions for sensitivity analysis of atmospheric models, *J. Atmos. Sci.* 40, 2537.
Hamill, P. and O. B. Toon (1991) Polar stratospheric clouds and the ozone hole, *Physics Today* 44, 34.
Han, Q., W. B. Rossow, and A. A. Lacis (1994) Near-global survey of effective droplet radii in liquid water clouds using ISCCP data, *J. Climate*, 7, 465.
Hanel, R. A., B. J. Conrath, V. G. Kunde, C. Prabhakara, I. Revah, V. V. Salomonson, and G. Wolford (1972) The Nimbus 4 infrared spectroscopy experiment, 1. Calibrated thermal emissions spectra, *J. Geophys. Res.* 77, 2629.
Hanson, D. R. and K. Mauersberger (1988) Vapor pressures of $HNO_3/H_2O$ solutions at low temperatures, *J. Phys. Chem.* 92, 6167.
Hanson, D. R., A. R. Ravishankara, and S. Solomon (1994) Heterogeneous reactions in sulfuric acid aerosols: A framework for model calculations, *J. Geophys. Res.* 99, 3615.
Hanson, D. R. and A. R. Ravishankara (1995) Heterogeneous chemistry of bromine species in sulfuric acid under stratospheric conditions, *Geophys. Res. Lett.* 22, 385.
Hanson, D. R., A. R. Ravishankara, and E. R. Lovejoy (1996) Reaction of $BrONO_2$ with $H_2O$ on submicron sulfuric acid aerosol and the implications for the lower stratosphere, *J. Geophys. Res.* 101, 9063.
Hao, W. M. and M.-H. Liu (1994) Spatial and temporal distribution of tropical biomass burning, *Glob. Biogeochem. Cycles* 8, 495.
Hartmann, D. L. and R. R. Garcia (1979) A mechanistic model of ozone transport by planetary waves in the stratosphere, *J. Atmos. Sci.* 36, 350.
Hartmann, D. L. (1994) *Global Physical Climatology*, Academic Press, San Diego.

Hartley, D. and R. Prinn (1993) Feasibility of determining surface emissions of trace gases using an inverse method in a three-dimensional chemical transport model, *J. Geophys. Res.* 98, 5183.

Harwood, M. H. and R. L. Jones (1994) Temperature dependent ultraviolet-visible absorption cross sections of $NO_2$ and $N_2O_4$: Low-temperature measurements of the equilibrium constant for $2\ NO_2 \rightleftharpoons N_2O_4$, *J. Geophys. Res.* 99, 22,955.

Hasebe, F. (1983) Interannual variation of global total ozone revealed from NIMBUS 4 BUV and ground-based observations, *J. Geophys. Res.* 88, 6819.

Hauglustaine, D. A., G. P. Brasseur, S. Walters, P. J. Rasch, and L. K. Emmons (1998) MOZART: A global chemical-transport model for ozone and related tracers, 2. Model results and evaluation, *J. Geophys. Res.* 103.

Hauglustaine, D. A., S. Madronich, B. A. Ridley, S. J. Flocke, C. A. Cantrell, F. L. Eisele, R. E. Shetter, D. J. Tanner, P. Ginoux, and E. L. Atlas (1998) Photochemistry and budget of species during the Mauna Loa Observatory Photochemistry Experiment (MLOPEX2), *J. Geophys. Res.* 103.

Haynes, P. H. (1989) The effect of barotropic instability on the nonlinear evolution of a Rossby-wave critical layer, *J. Fluid Mech.* 207, 231.

Haynes, P. H., C. J. Marks, M. E. McIntyre, T. G. Shepherd, and K. P. Shine (1991) On the "downward control" of extratropical diabatic circulations and eddy-induced mean zonal forces, *J. Atmos. Sci.* 48, 651.

Hays, P. B., V. J. Abreu, M. E. Dobbs, D. A. Gell, H. J. Grassl, and W. R. Skinner (1993) The high-resolution Doppler imager on the Upper Atmosphere Research Satellite, *J. Geophys. Res.* 98, 10,713.

Heaps, W. S. and T. J. McGee (1985) Progress in stratospheric hydroxyl measurements by balloon-borne lidar, *J. Geophys. Res.* 90, 7913.

Heath, D. F., A. J. Krueger, H. A. Roeder, and B. D. Henderson (1975) The Solar Backscatter Ultraviolet and Total Ozone Mapping Spectrometer (SBUV/TOMS) for NIMBUS G, *Opt. Eng.* 14, 323.

Hegg, D. A. (1985) The importance of liquid-phase oxidation of $SO_2$ in the troposphere, *J. Geophys. Res.* 90, 3773.

Hegg, D. A., S. A. Rutledge, and P. V. Hobbs (1986) A numerical model for sulfur and nitrogen scavenging in narrow cold-frontal rainbands, 2. Discussion of chemical fields, *J. Geophys. Res.* 91, 14,403.

Heikes, B. G., G. L. Kok, J. G. Walega, and A. L. Lazrus (1987) $H_2O_2$, $O_3$, and $SO_2$ measurements in the lower troposphere over the eastern United States during fall, *J. Geophys. Res.* 92, 915.

Heikes, B. G. (1992) Formaldehyde and hydroperoxides at Mauna Loa Observatory, *J. Geophys. Res.* 97, 18,001.

Heintzenberg, J. (1989) Fine particles in the global troposphere, *Tellus* 41B, 149.

Held, I. M. and A. Y. Hou (1980) Nonlinear axially symmetric circulations in a nearly inviscid atmosphere, *J. Atmos. Sci.* 37, 515.

Herzberg, G. (1945) *Infrared and Raman Spectra*, Van Nostrand Reinhold Company, New York.

Hess, P. G. and S. Madronich (1997) On tropospheric oscillations, *J. Geophys. Res.* 102, 15,949.

Hesstvedt, E., O. Høv, and I. S. A. Isaksen (1978) Photochemistry of mixtures of hydrocarbons and nitrogen oxides in air, *Geophys. Norv.* 31, 27.

Hills, A. J. and P. R. Zimmerman (1990) Isoprene measurement by ozone-induced chemiluminescence, *Anal. Chem.* 62, 1055.

Hinds, W. C. (1982) *Aerosol Technology: Properties, Behavior, and Measurement of Airborne Particles,* Wiley, New York.

Hitchman, M. H. and G. P. Brasseur (1988) Rossby wave activity in a two-dimensional model: Closure for wave driving and meridional eddy diffusivity, *J. Geophys. Res.* 93, 9405.

Hodkinson, J. R. (1966) The optical measurement of aerosols, in: *Aerosol Science*, C. N. Davies, ed., Academic Press, New York.

Hoefs, J. (1980) *Stable Isotope Geochemistry*, 3rd ed., Springer-Verlag, New York.

Hoffmann, M. R. and D. J. Jacob (1984) Kinetics and mechanisms of the catalytic oxidation of dissolved sulfur dioxide in aqueous solution: An application to nighttime fog water chemistry, in: *$SO_2$, NO, and $NO_2$ Oxidation Mechanisms,* J. G. Calvert, ed., Acid Precipitation Series, Vol. 3, Butterworth Publications.

Hoffmann, M. R. and J. G. Calvert (1985) *Chemical Transformation Modules for Eulerian Acid Deposition Models, II. The Aqueous-Phase Chemistry,* National Center for Atmospheric Research, Boulder, CO.

Hoffmann, T., J. R. Odum, F. Bowman, D. Collins, D. Klockow, R. C. Flagan, and J. H. Seinfeld (1997) Formation of organic aerosols from the oxidation of biogenic hydrocarbons, *J. Atmos. Chem.* 26, 189.

Hofmann, D. J. (1987) Perturbations to the global atmosphere associated with the El Chichón volcanic eruption of 1982, *Rev. Geophys.* 25, 743.

Hofmann, D. and S. Solomon (1989) Ozone depletion through heterogeneous chemistry following the eruption of the El Chichón volcano, *J. Geophys. Res.* 94, 5029.

Holsen, T. M. and K. E. Noll (1992) Dry deposition of atmospheric particles: Application of current models to ambient data, *Environ. Sci. and Technol.* 26, 1807.

Holton, J. R. (1972) *An Introduction to Dynamic Meteorology*, Academic Press.

Holton, J. R. (1979) *An Introduction to Dynamic Meteorology*, 2nd ed., Academic Press.

Holton, J. R. (1990) On the global exchange of mass between the stratosphere and troposphere, *J. Atmos Sci.* 47, 392.

Holton, J. R. and J. Austin (1991) The influence of the equatorial QBO on sudden stratospheric warmings, *J. Atmos. Sci.* 48, 607.

Holton, J. R. (1992) *An Introduction to Dynamic Meteorology*, 3rd ed., Academic Press.

Holton, J. R., P. H. Haynes, M. E. McIntyre, A. R. Douglass, R. B. Rood, and L. Pfister (1995) Stratosphere-troposphere exchange, *Rev. Geophys.* 33, 403.

Holtslag, A. A. M. and C.-H. Moeng (1991) Eddy diffusivity and counter gradient transport in the convective atmospheric boundary layer, *J. Atmos. Sci.* 48, 1690.

Houghton, J. T. and S. D. Smith (1970) Remote sounding of atmospheric temperature from satellites, 1. Introduction, *Proc. Roy. Soc. Lond.* A320, 23.

Houghton, J. T., G. J. Jenkins, and J. J. Ephraums, eds. (1990) *Climate Change*, The IPCC Scientific Assessment, Working Group I, Cambridge University Press, Cambridge, UK.

Høv, O., Z. Zlatev, R. Berkowicz, A. Eliassen, and L. P. Prahm (1989) Comparison of numerical techniques for use in air pollution models with non-linear chemical reactions, *Atmos. Environ.* 23, 967.

Howard, C. J. (1979) Kinetic measurements using flow tubes, *J. Phys. Chem.* 83, 3.

Hu, J. and D. H. Stedman (1995) Atmospheric $RO_x$ radicals at an urban site: Comparison to a simple theoretical model, *Envir. Sci. Tech.* 29, 1655.

Huebert, B. J., W. T. Luke, A. C. Delany, and R. A. Brost (1988) Measurements of concentrations and dry surface fluxes of atmospheric nitrates in the presence of ammonia, *J. Geophys. Res.* 93, 7127.

Huebert, B. J., S. Howell, P. Laj, J. E. Johnson, T. S. Bates, P. K. Quinn, V. Yegorov, A. D. Clarke, and J. N. Porter (1993) Observations of the atmospheric sulfur cycle on SAGA-3, *J. Geophys. Res.* 98, 16,985.

Hynes, A. J., P. H. Wine, and D. H. Semmes (1986) Kinetics and mechanism of OH reactions with organic sulfides, *J. Phys. Chem.* 90, 4148.

Hynes, A. J., R. B. Stoker, A. J. Pounds, T. McKay, J. D. Bradshaw, J. M. Nicovich, and P. H. Wine (1995) A mechanistic study of the reaction of OH with dimethyl-$d_6$ sulfide, Direct observation of adduct formation and the kinetics of the adduct reaction with $O_2$, *J. Phys. Chem.* 99, 16,967.

IPCC (1990) *Climate Change: The IPCC Scientific Assessment*, J. T. Houghton, C. J. Jenkins, and J. J. Ephraums, eds., Intergovernmental Panel on Climate Change, Cambridge University Press, Cambridge, UK.

IPCC (1992) *Climate Change, 1992*, J. T. Houghton, B. A. Callander, and S. K. Varney, eds., Intergovernmental Panel on Climate Change, Cambridge University Press, Cambridge, UK.

IPCC (1994) *Climate Change 1994: Radiative Forcing of Climate Change and an Evaluation of the IPCC IS 92 Emission Scenarios*, J. T. Houghton, L. G. Meira Filho, J. Bruce, Hoesung Lee, B. A. Callander, E. F. Haites, N. Harris, and K. Maskell, eds., Intergovernmental Panel on Climate Change, Cambridge University Press, Cambridge, UK.

IPCC (1995) *Radiative Forcing of Climate Change*, J. T. Houghton, L. G. Meira Fieho, J. Bruce, Hoesung Lee, B. A. Callander, E. Haites, N. Harris, and K. Maskell, eds., Intergovernmental Panel on Climate Change, Cambridge University Press, Cambridge, UK.

IPCC (1996) *Climate Change 1995: The Science of Climate Change*, J. T. Houghton, L. G. Meira Filho, B. A. Callander, N. Harris, A. Kattenberg, and K. Maskell, eds., Intergovernmental Panel on Climate Change, Cambridge University Press, Cambridge, UK.

Jackman, C. H., E. L. Fleming, S. Chandra, D. B. Considine, and J. E. Rosenfield (1996) Past, present, and future modeled ozone trends with comparisons to observed trends, *J. Geophys. Res.* 101, 28,753.

Jacob, D. J. and M. J. Prather (1990) Radon-222 as a test of convective transport in a general circulation model, *Tellus* 42B, 118.

Jacob, D. J., B. G. Heikes, S.-M. Fan, J. A. Logan, D. L. Mauzerall, J. D. Bradshaw, H. B. Singh, G. L. Gregory, R. W. Talbot, D. R. Blake, and G. W. Sachse (1996) Origin of ozone and $NO_x$ in the tropical troposphere: A photochemical analysis of aircraft observations over the South Atlantic basin, *J. Geophys. Res.* 101, 24,235.

Jacob, D. J., M. J. Prather, P. J. Rasch, R.-L. Shia, Y. J. Balkanski, S. R. Beagley, D. J. Bergmann, W. T. Blackshear, M. Brown, M. Chiba, M. P. Chipperfield, J. de Grandpré, J. E. Dignon, J. Feichter, C. Genthon, W. L. Grose, P. S. Kasibhatla, I. Köhler, M. A. Kritz, K. Law, J. E. Penner, M. Ramonet, C. E. Reeves, D. A. Rotman, D. Z. Stockwell, P. F. J. Van Velthoven, G. Verver, O. Wild, H. Yang, and P. Zimmermann (1997) Evaluation and intercomparison of global atmospheric transport models using $^{222}Rn$ and other short-lived tracers, *J. Geophys. Res.* 102, 5953.

Jacobson, M. Z. and R. P. Turco (1994) SMVGEAR: A sparse-matrix vectorized Gear code for atmospheric models, *Atmos. Environ.* 28A, 273.

Jaenicke, R. (1988) Aerosol physics and chemistry, in: *Zahlenwerte und Funktionen aus Naturwissenschaften und Technik*, G. Fischer, ed., Springer-Verlag, Berlin.

Jensen, T. L., S. M. Kreidenweis, Y. Kim, H. Sievering, and A. Pszenny (1996) Aerosol distributions in the North Atlantic marine boundary layer during Atlantic Stratocumulus Transition Experiment/Marine Aerosol and Gas Exchange, *J. Geophys. Res.* 101, 4455.

Jiang, Y. and Y. L. Yung (1996) Concentrations of tropospheric ozone from 1979 to 1992 over tropical Pacific South America from TOMS data, *Science* 272, 714.

Jobson, B. T., H. Niki, Y. Yokouchi, J. Bottenheim, F. Hopper, and R. Leaitch (1994) Measurements of $C_2$-$C_6$ hydrocarbons during the Polar Sunrise 1992 Experiment: Evidence for Cl atom and Br atom chemistry, *J. Geophys. Res.* 99, 25,355.

Johnson, D. G., K. W. Jucks, W. A. Traub, and K. V. Chance (1995a) Smithsonian stratospheric far-infrared spectrometer and data reduction system, *J. Geophys. Res.* 100, 3091.

Johnson, D. G., W. A. Traub, K. V. Chance, and K. W. Jucks (1995b) Detection of HBr and upper limit for HOBr: Bromine partitioning in the stratosphere, *Geophys. Res. Lett.* 22, 1373.

Johnston, H. S. (1971) Reduction of stratospheric ozone by nitrogen ozone catalysts from supersonic transport exhaust, *Science* 173, 517.

Johnson, J. E., A. R. Bandy, D. C. Thornton, and T. S. Bates (1993) Measurements of atmospheric carbonyl sulfide during the NASA CITE-3 project: Implications for the global COS budget, *J. Geophys. Res.* 98, 22,443.

Jones, R. L. and J. A. Pyle (1984) Observations of $CH_4$ and $N_2O$ by the Nimbus 7 SAMS: A comparison with in situ data and two-dimensional numerical model calculations, *J. Geophys. Res.* 89, 5263.

Joseph, J. H., W. J. Wiscombe, and J. A. Weinman (1976) The delta-Eddington approximation for radiative flux transfer, *J. Atmos. Sci.* 33, 2452.

Jouzel, J., N. I. Barkov, J. M. Barnola, M. Bender, J. Chappellaz, C. Genthon, V. M. Kotlyakov, V. Lipenkov, C. Lorius, J. R. Petit, D. Raynaud, G. Raisbeck, C. Ritz, T. Sowers, M. Stievenard, F. Yiou, and P. Yiou (1993) Extending the Vostok ice-core record of paleoclimate to the penultimate glacial period, *Nature* 364, 407.

Junge, C. E. and P. E. Gustafson (1957) On the distribution of seasalt over the United States and its removal by precipitation, *Tellus* 9, 164.

Junge, C. E., C. W. Chagnon, and J. E. Manson (1961) Stratospheric aerosols, *J. Meteor.* 18, 81.

Kasting, J. F. and T. P. Ackerman (1985) High atmospheric $NO_x$ levels and multiple photochemical steady-states, *J. Comp. Phys.* 3, 321.

Kasting, J. F. and H. B. Singh (1986) Nonmethane hydrocarbons in the troposphere: Impact of the odd hydrogen and odd nitrogen chemistry, *J. Geophys. Res.* 91, 13,239.

Kasting, J. F. (1990) Bolide impacts and the oxidation state of carbon in the earth's early atmosphere, *Orig. Life* 20, 199.

Kauppi, P. E., K. Mielikäinen, and K. Kuusela (1992) Biomass and carbon budget of European forests, 1971 to 1990, *Science* 256, 70.

Kawa, S. R., D. W. Fahey, L. E. Heidt, W. H. Pollock, S. Solomon, D. E. Anderson, M. Loewenstein, M. H. Proffitt, J. J. Margitan, and K. R. Chan (1992) Photochemical partitioning of the reactive nitrogen and chlorine reservoirs in the high-latitude stratosphere, *J. Geophys. Res.* 97, 7905.

Kawa, S. R., D. W. Fahey, K. K. Kelly, J. E. Dye, D. Baumgardner, B. W. Gandrud, M. Loewenstein, G. V. Ferry, and K. R. Chan (1992) The Arctic polar stratospheric cloud aerosol: Aircraft measurements of reactive nitrogen, total water, and particles, *J. Geophys. Res.* 97, 7925.

Kaye, J. A. and C. H. Jackman (1986) Concentrations and uncertainties of stratospheric trace species inferred from Limb Infrared Monitor of the Stratosphere data, 2. Monthly averaged OH, $HO_2$, $H_2O_2$, and $HO_2NO_2$, *J. Geophys. Res.* 91, 1137.

Kaye, J. A. and R. B. Rood (1989) Chemistry and transport in a three-dimensional stratospheric model: Chlorine species during a simulated stratospheric warming, *J. Geophys. Res.* 94, 1057.

Kaye, J. A., S. A. Penkett, and F. M. Ormund, eds., (1994) *Report on Concentrations, Lifetimes, and Trends of CFCs, Halons, and Related Species*, NASA Reference Publication No. 1339.

Keeling, C. D. (1973) The carbon dioxide cycle, Reservoir models to depict the exchange of atmospheric carbon dioxide with ocean and land plants, in: *Chemistry of the Lower Atmosphere*, S. Rassol, ed., Plenum Press, New York.

Keeling, C. D., S. C. Piper, and M. Heimann (1989) A three-dimensional model of atmospheric $CO_2$ transport based on observed winds: 4. Mean annual gradients and interannual variations, in: *Aspects of Climate Variability in the Pacific and Western Americas*, D. H. Peterson, ed., American Geophysical Union, Washington, D. C.

Khalil, M. A. K. and R. A. Rasmussen (1984) Global sources, lifetimes and balances of carbonyl sulfide (OCS) and carbon disulfide ($CS_2$) in the earth's atmosphere, *Atmos. Environ.* 18, 1805.

Khalil, M. A. K. and R. A. Rasmussen (1990) Carbon monoxide in the Earth's atmosphere: Increasing trend, *Science* 224, 54.

Kiehl, J. T. and S. Solomon (1986) On the radiative balance of the stratosphere, *J. Atmos. Sci.* 43, 1525.

Kiehl, J. T. (1992) Atmospheric general circulation modeling, in: *Climate System Modeling*, K. E. Trenberth, ed., Cambridge University Press, Cambridge, UK.

Kiehl, J. T. and B. P. Briegleb (1993) The relative roles of sulfate aerosols and greenhouse gases in climate forcing, *Science* 260, 311.

Kiehl, J. T. and H. Rodhe (1995) Modeling geographical and seasonal forcing due to aerosols, in: *Aerosol Forcing of Climate*, R. J. Charlson and J. Heintzenberg, eds., Wiley, Chichester, UK, 281.

Kim, J. H. and M. J. Newchurch (1996) Climatology and trends of tropospheric ozone over the eastern Pacific Ocean: The influences of biomass burning and tropospheric dynamics, *Geophys. Res. Lett.* 23, 3723.

Kim, J. H. and M. J. Newchurch (1998) Biomass-burning influence on tropospheric ozone over New Guinea and South America, *J. Geophys. Res.* 103, 1455.

Koch, D., D. J. Jacob, and W. C. Graustein (1996) Vertical transport of tropospheric aerosols as indicated by $^7Be$ and $^{210}Pb$ in a chemical tracer model, *J. Geophys. Res.* 101, 18,651.

Kockarts, G. (1994) Penetration of solar radiation in the Schumann-Runge bands of molecular oxygen: A robust approximation, *Ann. Geophys.* 12, 1207.

Komhyr, W. D. (1969) Electrochemical concentration cells for gas analysis, *Ann. Geophys.* 25, 203.

Kondo, Y., P. Aimedieu, M. Koike, Y. Iwasaka, P. A. Newman, U. Schmidt, W. A. Mathews, M. Hayashi, and W. R. Sheldon (1992) Reactive nitrogen, ozone, and nitrate aerosols observed in the Arctic stratosphere in January 1990, *J. Geophys. Res.* 97, 13,025.

Kwok, E. S. C. and R. Atkinson (1995) Estimation of hydroxyl radical reaction rate constants for gas-phase organic compounds using a structure-reactivity relationship: An update, *Atmos. Environ.* 29, 1685.

Kwok, E. S. C., R. Atkinson, and J. Arey (1995) Observation of hydroxycarbonyls from the OH radical-initiated reaction of isoprene, *Environ. Sci. Tech.* 29, 2467.

Lamb, R. G. (1988) Diagnostic studies of ozone in the northeastern United States based on application of the Regional Oxidant Model (ROM), in: *The Scientific and Technical Issues Facing Post-1987 Ozone Control Strategies*, G. T. Wolff, J. L. Hanisch, and K. Schere, eds., Air Waste Management Association, Pittsburgh, Pennsylvania.

Lambert, G., G. Polian, J. Sanak, B. Ardouin, A. Buisson, A. Jegou, and J. C. Le Roulley (1982) Cycle du radon et de ses descendants: Application à l'étude des échanges troposphère-stratosphère, *Géophys.* 38, 497.

Langner, J. and H. Rodhe (1991) A global three-dimensional model of the tropospheric sulfur cycle, *J. Atmos. Chem.* 13, 225.

Lary, D. J. (1996) Gas phase atmospheric bromine photochemistry, *J. Geophys. Res.* 101, 1505.

Laws, E. A. (1993) *Aquatic Pollution*, 2nd ed., John Wiley and Sons, New York.

Leaitch, W. R., G. A. Isaac, J. W. Strapp, C. M. Banic, and H. A. Wiebe (1992) The relationship between cloud droplet number concentrations and anthropogenic pollution: Observations and climatic implications, *J. Geophys. Res.* 97, 2463.

Lee, Y.-N. and X. Zhou (1993) Method for the determination of some soluble atmospheric carbonyl compounds, *Environ. Sci. Technol.* 27, 749.

Lee, Y. N. and X. Zhou (1994) Aqueous reaction kinetics of ozone and dimethylsulfide and its atmospheric implications, *J. Geophys. Res.* 99, 3597.

Leemans, R. and W. Cramer (1992) IIASA database for mean monthly values of temperature, precipitation, and cloudiness on a global terrestrial grid. Digital raster data on a 30 minute geographic (lat/long) 320 × 720 grid, in: *Global Ecosystems Database Version 1.0: Disc A*, NOAA National Geophysical Data Center, Boulder, Colorado.

Lefèvre, F., G. P. Brasseur, I. Folkins, A. K. Smith, and P. Simon (1994) The chemistry of the 1991-92 stratospheric winter: Three-dimensional model simulation, *J. Geophys. Res.* 99, 8183.

Lelieveld, J. and P. J. Crutzen (1990) Influences of cloud photochemical processes on tropospheric ozone, *Nature* 343, 227.

Lelieveld, J. and P. J. Crutzen (1991) The role of clouds in tropospheric photochemistry, *J. Atmos. Chem.* 12, 229.

Lenoble, J. (1993) *Atmospheric Radiative Transfer*, A. Deepak Publishing, Hampton, Virginia.

Lenschow, D. H. (1995) Micrometeorological techniques for measuring biosphere-atmosphere trace gas exchanges, in: *Biogenic Trace Gases: Emissions from Soil and Water*, P. A. Matson and R. C. Harriss, eds., Blackwell Science, Ltd.

Leuenberger, M., U. Siegenthaler, and C. C. Langway (1992) Carbon isotope composition of atmospheric $CO_2$ during the last ice age from an Antarctic ice core, *Nature* 357, 488.

Le Texier, H., S. Solomon, and R. R. Garcia (1988) The role of molecular hydrogen and methane oxidation in the water vapour budget of the stratosphere, *Quart. J. Roy. Meteor. Soc.* 114, 281.

Levelt, P. F., M. A. F. Allaart, and H. M. Kelder (1996) On the assimilation of total-ozone satellite data, *Ann. Geophys.* 14, 1111.

Levy II, H. (1971) Normal atmosphere: Large radical and formaldehyde concentrations predicted, *Science* 173, 141.

Levy II, H., J. D. Mahlman, and W. J. Moxim (1982) Tropospheric $N_2O$ variability, *J. Geophys. Res.* 87, 3061.

Levy II, H., J. D. Mahlman, W. J. Moxim, and S. C. Liu (1985) Tropospheric ozone: The role of transport, *J. Geophys. Res.*, 90, 3752.

Levy II, H., W. J. Moxim, and P. S. Kasibhatla (1996) A global three-dimensional time-dependent lightning source of tropospheric $NO_x$, *J. Geophys. Res.* 101, 22,911.

Liang, J. and D. J. Jacob (1997) Effect of aqueous phase cloud chemistry on tropospheric ozone, *J. Geophys. Res.* 102, 5993.

Liaw, Y. P., D. L. Sisterson, and N. L. Miller (1990) Comparison of field, laboratory, and theoretical estimates of global nitrogen fixation by lightning, *J. Geophys. Res.* 95, 489.

Lieth, H. (1975) Modeling of primary productivity of the world, in: *Primary Productivity of the Biosphere*, H. Lieth and R. H. Whittaker, eds., Springer-Verlag, New York, 237.

Lightfoot, P. D., R. A. Cox, J. N. Crowley, M. Destriau, G. D. Hayman, M. E. Jenkin, G. K. Moortgat, and F. Zabel (1992) Organic peroxy radicals: Kinetics, spectroscopy and tropospheric chemistry, *Atmos. Environ.* 26A, 1805.

Lin, X., M. Trainer, and S. C. Liu (1988) On the nonlinearity of the tropospheric ozone production, *J. Geophys. Res.* 93, 15,879.

Lindzen, R. S. (1981) Turbulence and stress owing to gravity wave and tidal breakdown, *J. Geophys. Res.* 86, 9707.

Lindzen, R. S. (1990) *Dynamics in Atmospheric Physics*, Cambridge University Press, New York.

Linthurst, R. A., ed. (1984) *Direct and Indirect Effects of Acidic Deposition on Vegetation*, Butterworth Publications.

Liou, K. N. (1980) *An Introduction to Atmospheric Radiation*, International Geophysical Series, Academic Press.

Liss, P. S. and L. Merlivat (1986) Air-sea gas exchange rates: Introduction and synthesis, in: *The Role of Air-Sea Exchange in Geochemical Cycling*, P. Buat-Ménard, ed., 113.

Liu, S. C., D. Kley, M. McFarland, J. D. Mahlman, and H. Levy II (1980) On the origin of tropospheric ozone, *J. Geophys. Res.* 85, 7546.

Liu, S. C., J. R. McAfee, and R. J. Cicerone (1984) $^{222}$Radon and tropospheric vertical transport, *J. Geophys. Res.* 89, 7291.

Liu, S. C., M. Trainer, F. C. Fehsenfeld, D. D. Parrish, E. J. Williams, D. W. Fahey, G. Hübler, and P. C. Murphy (1987) Ozone production in the rural troposphere and the implications for regional and global ozone distributions, *J. Geophys. Res.* 92, 4191.

Liu, S. C., M. Trainer, M. A. Carroll, G. Hübler, D. D. Montzka, R. B. Norton, B. A. Ridley, J. G. Walega, E. L. Atlas, B. G. Heikes, B. J. Huebert, and W. Warren (1992) A study of the photochemistry and ozone budget during the Mauna Loa Observatory Photochemistry Experiment, *J. Geophys. Res.* 97, 10,463.

Logan, J. A., M. J. Prather, S. C. Wofsy, and M. B. McElroy (1981) Tropospheric chemistry: A global perspective, *J. Geophys. Res.* 86, 7210.

Logan, J. A. (1983) Nitrogen oxides in the troposphere: Global and regional budgets, *J. Geophys. Res.* 88, 10,985.

Logan, J. A. (1994) Trends in the vertical distribution of ozone: An analysis of ozonesonde data, *J. Geophys. Res.* 99, 25,553.

London, J. and S. C. Liu (1992) Long-term tropospheric and lower stratospheric ozone variations from ozonesonde observations, *J. Atmos. Terr. Phys.* 54, 599.

Lorenc, A. (1986) Analysis methods for numerical weather prediction, *Quart. J. Roy. Met. Soc.* 112, 1177.

Lowe, D. C., C. A. M. Brenninkmeijer, M. R. Manning, R. Sparks, and G. Wallace (1988) Radiocarbon determination of atmospheric methane at Baring Head, New Zealand, *Nature* 332, 522.

Lovelock, J. E. (1979) *Gaia: A New Look at Life on Earth*, Oxford University Press, Oxford.

Lovelock, J. E. (1988) *The Ages of Gaia*, W. W. Norton and Company, New York.

Lovelock, J. E. (1991) *Healing Gaia*, Harmony Books, New York.

Lovejoy, E. R., T. P. Murrells, A. R. Ravishankara, and C. J. Howard (1990) Oxidation of $CS_2$ by reaction with OH, 2: Yields of $HO_2$ and $SO_2$ in oxygen, *J. Phys. Chem.* 94, 2386.

Lovejoy, E. R., D. R. Hanson, and L. G. Huey (1996) Kinetics and products of the gas-phase reaction of $SO_3$ with water, *J. Phys. Chem.* 100, 19,911.

Lueb, R. A., D. H. Ehhalt, and L. E. Heidt (1975) Balloon-borne low temperature air sampler, *Rev. Sci. Insts.* 46, 702.

Luo, M., R. J. Cicerone, and J. M. Russell III (1995) Analysis of HALogen Occultation Experiment HF versus $CH_4$ correlation plots: Chemistry and transport implications, *J. Geophys. Res.* 100, 13,927.

Lurmann, F. W., A. S. Wexler, S. N. Pandis, S. Musarra, N. Kumar, and J. H. Seinfeld (1997) Modelling urban and regional aerosols, II: Application to California's south coast air basin, *Atmos. Environ.* 31, 2695.

Lyster, P. M., S. E. Cohn, R. Menard, L.-P. Chang, S.-J. Lin, and R. G. Olsen (1997) Parallel implementation of a Kalman filter for constituent data assimilation, *Mon. Wea. Rev.* 125, 1674.

Machta, L. (1971) The role of the oceans and biosphere in the carbon cycle, in: *The Changing Chemistry of the Oceans*, F. Dryssen and D. Jagner, eds., Nobel Symposium 20, Almqvist and Wikrell, Stockholm.

Mackenzie, F. T. and J. A. Mackenzie (1995) *Our Changing Planet, An Introduction to Earth System Science and Global Environmental Change*, Prentice Hall, Upper Saddle River, NJ.

Mackenzie, F. T. (1997) *Our Changing Planet, An Introduction to Earth System Science and Global Environmental Change*, 2nd ed., Prentice Hall, Upper Saddle River, New Jersey.

Madronich, S. (1987) Photodissociation in the atmosphere, I. Actinic flux and the effects of ground reflections and clouds, *J. Geophys. Res.* 92, 9740.

Madronich, S. (1993) The atmosphere and UV-B radiation at ground level, in: *Environmental UV Photobiology*, A. Young et al. eds., Plenum Press, New York.

Madronich, S., R. L. McKenzie, L. Björn, and M. Caldwell (1995) Changes in ultraviolet radiation reaching the Earth's surface, *Ambio* 24, 143.

Mahlman, J. D. and W. J. Moxim (1978) Tracer simulations using a global general circulation model: Results from a mid-latitude instantaneous source experiment, *J. Atmos. Sci.* 35, 1340.

Mahlman, J. D., H. Levy, and W. J. Moxim (1980) Three-dimensional tracer structure and behavior as simulated in two ozone precursor experiments, *J. Atmos. Sci.* 37, 655.

Mahowald, N. M., P. J. Rasch, and R. G. Prinn (1995) Cumulus parameterizations in chemical transport models, *J. Geophys. Res.* 100, 26,173.

Maier-Reimer, E. and K. Hasselmann (1987) Transport and storage in the ocean—An inorganic ocean-circulation carbon cycle model, *Climate Dynamics* 2, 63.

Manabe, S. and R. T. Wetherald (1967) Thermal equilibrium of the atmosphere with a given distribution of relative humidity, *J. Atmos. Sci.* 24, 241.

Mankin, W. G. (1978) Airborne Fourier transform spectroscopy of the upper atmosphere, *Opt. Engr.* 17, 39.

Mankin, W. G. and M. T. Coffey (1984) Increased stratospheric hydrogen chloride in the El Chichón cloud, *Science* 226, 170.

Mankin, W. G., M. T. Coffey, A. Goldman, M. R. Schoeberl, L. R. Lait, and P. A. Newman (1990) Airborne measurements of stratospheric constituents over the Arctic in the winter of 1989, *Geophys. Res. Lett.* 17, 473.

Mankin, W. G., M. T. Coffey, and A. Goldman (1992) Airborne observations of $SO_2$, HCl, and $O_3$ in the stratospheric plume of the Pinatubo volcano in July 1991, *Geophys. Res. Lett.* 19, 179.

Martin, R. L. (1984) Kinetic studies of sulfite oxidation in aqueous solution, in: $SO_2$, NO, and $NO_2$ Oxidation Mechanisms, J. G. Calvert, ed., Acid Precipitation Series, Vol. 3, Butterworth Publications.

Matson, P. A. and P. M. Vitousek (1990) Ecosystem approach to a global nitrous oxide budget, *BioScience* 40, 667.

Matsuno, T. (1966) Quasi-geostrophic motions in the equatorial area, *J. Meteor. Soc. Japan* 44, 25.

Matsuno, T. (1980) Lagrangian motion of air parcels in the stratosphere in the presence of planetary waves, *Pure Appl. Geophys.* 118, 189.

Matthews, E. (1994) Nitrogenous fertilizers: Global distribution of consumption and associated emissions of nitrous oxide and ammonia, *J. Geophys. Res.* 98, 411.

McConnell, J. C., G. S. Henderson, L. Barrie, J. Bottenheim, H. Niki, C. H. Langford, and E. M. Templeton (1992) Photochemical bromine production implicated in Arctic boundary-layer ozone depletion, *Nature* 355, 150.

McCormick, M. P., H. M. Steele, P. Hamill, W. P. Chu, and T. J. Swissler (1982) Polar stratospheric cloud sightings by SAM II, *J. Atmos. Sci.* 39, 1387.

McCormick, M. P. and R. E. Veiga (1992) SAGE II measurements of early Pinatubo aerosols, *Geophys. Res. Lett.* 19, 155.

McDaniel, A. H., C. A. Cantrell, J. A. Davidson, R. E. Shetter, and J. G. Calvert (1991) The temperature dependent, infrared absorption cross-sections for the chlorofluorocarbons CFC-11, CFC-12, CFC-13, CFC-14, CFC-22, CFC-113, CFC-114, and CFC-115, *J. Atmos. Chem.* 12, 211.

McEwan, M. J. and L. F. Phillips (1976) *Chemistry of the Atmosphere*, Wiley, New York.

McGraw, R. (1997) Description of aerosol dynamics by the quadrature method of moments, *Aerosol Sci. Technol.* 27, 255.

McIntyre, M. E. (1980) Towards a Lagrangian-mean description of stratospheric circulations and chemical transports, *Phil. Trans. Roy. Soc. London* A296, 129.

McIntyre, M. E. and T. N. Palmer (1983) Breaking planetary waves in the stratosphere, *Nature* 305, 593.

McIntyre, M. E. and T. N. Palmer (1984) The 'surf-zone' in the stratosphere, *J. Atmos. Terr. Phys.* 46, 825.

McIntyre, M. E. and T. N. Palmer (1985) A note on the general concept of wave breaking for planetary and gravity waves, *Pure App. Geophys.* 123, 964.

McIntyre, M. E. (1992) Atmospheric dynamics: Some fundamentals, with observational implications, in: *Proc. Internat. School Phys.* "Enrico Fermi," CXV Course, J. C. Gille and G. Visconti, eds., North-Holland, Amsterdam, 313.

McKeen, S. A., E. Y. Hsie, M. Trainer, R. Tallamraju, and S. C. Liu (1991a) A regional model study of the ozone budget in the eastern United States, *J. Geophys. Res.* 96, 10,809.

McKeen, S. A., E. Y. Hsie, and S. C. Liu (1991b) A study of the dependence of rural ozone on ozone precursors in the eastern United States, *J. Geophys. Res.* 96, 15,377.

McKeen, S. A., G. Mount, F. Eisele, E. Williams, J. Harder, P. Goldan, W. Kuster, S. C. Liu, K. Baumann, D. Tanner, A. Fried, S. Sewell, C. Cantrell, and R. Shetter (1997) Photochemical modeling of hydroxyl and its relationship to other species during the tropospheric OH photochemistry experiment, *J. Geophys. Res.* 102, 6467.

Meador, W. E. and W. R. Weaver (1980) Two-stream approximations to radiative transfer in planetary atmospheres: A unified description of existing methods and a new improvement, *J. Atmos. Sci.* 37, 630.

Meehl, G. (1987) The tropics and their roles in the global climate system, *Geographical Journal* 153, 21.

Mégie, G. and J. E. Blamont (1977) Laser sounding of atmospheric sodium: Interpretation in terms of global atmospheric parameters, *Plan. Space Sci.* 25, 1093.

Mehrbach, C., C. H. Culberson, S. E. Hawley, and R. M. Pytkowiez (1973) Measurement of the apparent dissociation constants of carbonic acid in sea water at atmospheric pressure, *Limnol. Oceanogr.* 18, 897.

Meng, Z., D. Dabdub, and J. H. Seinfeld (1998) Size-resolved and chemically resolved model of atmospheric aerosol dynamics, *J. Geophys. Res.* 103, 3419.

Menke, W. (1984) *Geophysical Data Analysis: Discrete Inverse Theory*, Academic Press, London.

Mie, G. (1908) Beiträge zur Optiktrübermedien, Speziell Kolloidaler Metallösungen, *Ann. Physik* 25, 377.

Mihalopoulos, N., J.-P. Putaud, and B. C. Nguyen (1993) Seasonal variation of methanesulfonic acid in precipitation at Amsterdam Island in the southern Indian Ocean, *Atmos. Environ.* 27A, 2069.

Mihelcic, D., P. Müsgen, and D. H. Ehhalt (1985) Improved method of measuring tropospheric $NO_2$ and $RO_2$ by matrix isolation and electron spin resonance, *J. Atmos. Chem.* 3, 341.

Mihelcic, D., A. Volz-Thomas, H. W. Pätz, D. Kley, and M. Mihelcic (1990) Numerical analysis of ESR spectra from atmospheric samples, *J. Atmos. Chem.* 11, 271.

Milford, J. B., A. G. Russell, and G. J. McRae (1989) A new approach to photochemical pollution control: Implications of spatial patterns in pollutant responses to reductions in nitrogen oxides and reactive organic gas emissions, *Environ. Sci. Technol.* 23, 1290.

Miller, R. L., A. G. Suits, P. L. Houston, R. Toumi, J. A. Mack, and A. M. Wodtke (1994) The "ozone deficit" problem: $O_2$ $(X, v \geq 26)$ + O $(^3P)$ from 226-nm ozone photodissociation, *Science* 265, 1831.

Miller, S. L. (1953) A production of amino acids under possible primitive earth conditions, *Science* 117, 528.

Miller, S. L. and H. C. Urey (1959) Organic compound synthesis on the primitive earth, *Science* 130, 245.

Minschwaner, K., R. J. Salawitch, and M. B. McElroy (1993) Absorption of solar radiation by $O_2$: Implications for $O_3$ and lifetimes of $N_2O$, $CFCl_3$ and $CF_2Cl_2$, *J. Geophys. Res.* 98, 10,543.

Miyoshi, A., S. Hatakeyama, and N. Washida (1994) OH radical-initiated photooxidation of isoprene: An estimate of global CO production, *J. Geophys. Res.* 99, 18,779.

Mlynczak, M. G., S. Solomon, and D. S. Zaras (1993) An updated model for $O_2(a^1 \Delta g)$ concentrations in the mesosphere and lower thermosphere and implications for remote sensing of ozone at 1.27 $\mu$m, *J. Geophys. Res.* 98, 18,639.

Moeng, C.-H. (1984) A large-eddy-simulation model for the study of planetary boundary-layer turbulence, *J. Atmos. Sci.* 41, 2052.

Molina, M. J. and F. S. Rowland (1974) Stratospheric sink for chlorofluoromethanes: Chlorine atomc-atalysed destruction of ozone, *Nature* 249, 810.

Molina, L. T. and M. J. Molina (1986) Absolute absorption cross sections of ozone in the 185- to 350-nm wavelength range, *J. Geophys. Res.* 91, 14,501.

Molina, L. T. and M. J. Molina (1987) Production of $Cl_2O_2$ from the self-reaction of the ClO radical, *J. Phys. Chem.* 91, 433.

Möller, D. (1984) Estimation of the global man-made sulphur emission, *Atmos. Environ.* 18, 19.

Monfray, P. (1987) Echanges océan-atmosphére du gaz carbonique: Variabilité avec l'état de la mer, Thése d'Etat (Ph.D. thesis), Université de Picardie, France.

Montzka, S. A., R. C. Myers, J. H. Butler, J. W. Elkins, and S. O. Cummings (1993) Global tropospheric distribution and calibration scale of HCFC-22, *Geophys. Res. Lett.* 20, 703.

Montzka, S. A., J. H. Butler, R. C. Myers, T. M. Thompson, T. H. Swanson, A. D. Clarke, L. T. Lock, and J. W. Elkins (1996a) A decline in the tropospheric abundance of ozone-depleting halogen, *Science* 272, 1318.

Montzka, S. A., R. C. Myers, J. H. Butler, J. W. Elkins, L. T. Lock, A. D. Clarke, and A. H. Goldstein (1996b) Observations of HFC-134a in the remote troposphere, *Geophys. Res. Lett.* 23, 169.

Morgan, M. G. and D. W. Keith (1995) Subjective judgements by climate experts, *Environ. Sci. Technol.* 29, 468A.

Morris, R. A., T. M. Miller, A. A. Viggiano, J. F. Paulson, S. Solomon, and G. Reid (1995) Effects of electron and ion reactions on atmospheric lifetimes of fully fluorinated compounds, *J. Geophys. Res.* 100, 1287.

Mote, P. W., J. R. Holton, J. M. Russell III, and B. A. Boville (1993) A comparison of observed (HALOE) and modeled (CCM2) methane and stratospheric water vapor, *Geophys. Res. Lett.* 20, 1419.

Mote, P. W., K. H. Rosenlof, J. R. Holton, R. S. Harwood, and J. W. Waters (1995) Seasonal variations of water vapor in the tropical lower stratosphere, *Geophys. Res. Lett.* 22, 1093.

Mote, P. W., K. H. Rosenlof, M. E. McIntyre, E. S. Carr, J. C. Gille, J. R. Holton, J. S. Kinnersley, H. C. Pumphrey, J. M. Russell, and J. W. Waters (1996) An atmospheric tape recorder: The imprint of tropical tropopause temperatures on stratospheric water vapor, *J. Geophys. Res.* 101, 3989.

Mount, G. H. (1992) The measurement of tropospheric OH by long path absorption, 1. Instrumentation, *J. Geophys. Res.* 97, 2427.

Mount, G. H. and F. L. Eisele (1992) An intercomparison of tropospheric OH measurements at Fritz Peak Observatory, Colorado, *Science* 256, 1187.

Mount, G. H. and J. W. Harder (1995) The measurement of tropospheric trace gases at Fritz Peak Observatory, Colorado, by long-path absorption: OH and ancillary gases, *J. Atmos. Sci.* 52, 3342.

Mozurkewich, M., P. H. McMurry, A. Gupta, and J. G. Calvert (1987) Mass accommodation coefficient for $HO_2$ radicals on aqueous particles, *J. Geophys. Res.* 92, 4163.

Mozurkewich, M. and J. G. Calvert (1988) Reaction probability of $N_2O_5$ on aqueous aerosols, *J. Geophys. Res.* 93, 15,889.

Mozurkewich, M. (1993) The dissociation constant of ammonium nitrate and its dependence on temperature, relative humidity and particle size, *Atmos. Environ.* 27A, 261.

Müller, J.-F. (1992a) *Modélisation tri-dimensionnelle globale de la chimie et du transport des gaz en trace dans la troposphère*, Thesis, Free University of Brussels, Sciences Department.

Müller, J.-F. (1992b) Geographical distribution and seasonal variation of surface emissions and deposition velocities of atmospheric trace gases, *J. Geophys. Res.* 97, 3787.

Müller, J.-F. and G. Brasseur (1995) IMAGES: A three-dimensional chemical transport model of the global troposphere, *J. Geophys. Res.* 100, 16,445.

Murphy, D. M., D. W. Fahey, M. H. Proffitt, S. C. Liu, K. R. Chan, C. S. Eubank, S. R. Kawa, and K. K. Kelly (1993) Reactive nitrogen and its correlation with ozone in the lower stratosphere and upper troposphere, *J. Geophys. Res.* 98, 8751.

Nairn, A. E. W. M. and N. Thorley (1961) The application of geophysics to palaeoclimatology, in: *Descriptive Palaeoclimatology*, A. E. M. Nairn, ed., Interscience Publishers, New York.

Najjar, R. (1992) Marine biogeochemistry, in: *Climate System Modeling*, K. Trenberth, ed., Cambridge University Press, New York.

NASA (1986) *A Program for Global Change*, National Aeronautics and Space Administration, Washington, D.C.

NASA (1995) *1995 Scientific Assessment of Atmospheric Effects of Stratospheric Aircraft*, NASA Reference Publication No. 1381.

National Research Council (1991) *Four-Dimensional Model Assimilation of Data: A Strategy for the Earth System Sciences*, National Research Council, Washington, D.C.

Naudet, J. P., D. Huguenin, P. Rigaud, and D. Cariolle (1981) Stratospheric observations of $NO_3$ and its experimental and theoretical distribution between 20 and 40 km, *Planet. Space Sci.*, 29, 707.

Nevison, C. (1994) A model analysis of the spatial distribution and temporal trends of nitrous oxide sources and sinks, NCAR Cooperative Thesis NCAR/CT-147, Stanford University and the National Center for Atmospheric Research, Boulder, Colorado.

Newman, P., J. E. Gleason, R. D. McPeters, and R. S. Stolarski (1997) Anomalously low ozone over the Arctic, *Geophys. Res. Lett.* 24, 2689.

Nightingale, R. W., A. E. Roche, J. B. Kumer, J. L. Mergenthaler, J. C. Gille, S. T. Massie, P. L. Bailey, D. P. Edwards, M. R. Gunson, G. C. Toon, B. Sen, J.-F. Blavier, and P. S. Connell (1996) Global $CF_2Cl$ measurements by UARS cryogenic limb array etalon spectrometer: Validation by correlative data and a model, *J. Geophys. Res.* 101, 9711.

Nihlgard, B. (1985) The ammonium hypothesis: An additional explanation of the forest dieback in Europe, *Ambio.* 14, 2.

Norton, R. B. (1992) Measurements of gas phase formic and acetic acids at the Mauna Loa, Observatory, Hawaii during the Mauna Loa Observatory Photochemistry Experiment 1988, *J. Geophys. Res.* 97, 10,389.

Norton, R. B., M. A. Carroll, D. D. Montzka, G. Hübler, B. J. Huebert, G. Lee, W. W. Warren, B. A. Ridley, and J. G. Walega (1992) Measurements of nitric acid and aerosol nitrate at the Mauna Loa Observatory during the Mauna Loa Observatory Photochemistry Experiment 1988, *J. Geophys. Res.* 97, 10,415.

Novakov, T. and J. E. Penner (1993) Large contribution of organic aerosols to cloud condensation nuclei concentrations, *Nature* 365, 823.

Novelli, P. C., K. A. Masarie, P. Tans, and P. M. Lang (1994) Recent changes in atmospheric carbon monoxide, *Science* 263, 1587.

Noxon, J. F. (1975) Nitrogen dioxide in the stratosphere and troposphere measured by ground-based absorption spectroscopy, *Science* 189, 547.

Noxon, J. F., R. B. Norton, and W. R. Henderson (1978) Observation of atmospheric $NO_3$, *Geophys. Res. Lett.* 5, 675.

Noxon, J. F. (1979) Stratospheric $NO_2$, 2. Global behavior, *J. Geophys. Res.* 84, 5067. Correction (1980) *J. Geophys. Res.* 85, 4560.

Noxon, J. F., R. B. Norton, and E. Marovich (1980) $NO_3$ in the troposphere, *Geophys. Res. Lett.* 7, 125.

Noxon, J. F. (1983) $NO_3$ and $NO_2$ in the mid-Pacific troposphere, *J. Geophys. Res.* 88, 11,017.

O'Brien, J. M., P. B. Shepson, K. Muthuramu, C. Hao, H. Niki, D. R. Hastie, R. Taylor, and P. B. Roussel (1995) Measurements of alkyl and multifunctional organic nitrates at a rural site in Ontario, *J. Geophys. Res.* 100, 22,795.

Odum, J. R., T. P. W. Jungkamp, R. J. Griffin, R. C. Flagan, and J. H. Seinfeld (1997) The atmospheric aerosol-forming potential of whole gasoline vapor, *Science* 276, 96.

Oeschger, H., U. Siegenthaler, U. Schotterer, and A. Guegelmann (1975) A box diffusion model to study the carbon dioxide exchange in nature, *Tellus* 27, 168.

Ogren, J. A. and R. J. Charlson (1992) Implications for models and measurements of chemical inhomogeneities among cloud droplets, *Tellus* 44B, 208.

Okamura, J. P. and D. T. Sawyer (1978) Gas chromatography, in: *Physical Methods in Modern Chemical Analysis*, T. Kuwana, ed., Academic Press, New York.

Oram, D. E. and S. A. Penkett (1994) Observations in eastern England of elevated methyl iodide concentrations in air of Atlantic origin, *Atmos. Env.* 28, 1159.

Oran, E. S. and J. P. Boris (1987) *Numerical Simulation of Reactive Flows*, Elsevier, Oxford.

Orlando, J. J., J. B. Burkholder, S. A. McKeen, and A. R. Ravishankara (1991) Atmospheric fate of several hydrofluoroethanes and hydrochloroethanes: 2. UV absorption cross sections and atmospheric lifetimes, *J. Geophys. Res.* 96, 5013.

Osborn, M. T., R. J. DeCoursey, C. R. Trepte, D. M. Winker, and D. C. Woods (1995) Evolution of the Pinatubo volcanic cloud over Hampton, Virginia, *Geophys. Res. Lett.* 22, 1101.

Osborn, M. T., G. S. Kent, and C. R. Trepte (1998) Stratospheric aerosol measurements by the lidar in space technology experiment, *J. Geophys. Res.* 103, 11,447.

O'Sullivan, D. and R. E. Young (1992) Modeling the quasi-biennial oscillation effect on the winter stratospheric circulation, *J. Atmos. Sci.* 49, 2437.

O'Sullivan, D. and T. J. Dunkerton (1994) Seasonal development of the extratropical QBO in a numerical model of the middle atmosphere, *J. Atmos. Sci.* 51, 3706.

Palmén, E. and C. W. Newton (1969) *Atmospheric Circulation Systems*, Academic Press, London.

Pandis, S. N. and J. H. Seinfeld (1990) On the interaction between equilibration processes and wet or dry deposition, *Atmos. Environ.* 24A, 2313.

Park, J. H. and B. Carli (1991) Spectroscopic measurement of $HO_2$, $H_2O_2$, and OH in the stratosphere, *J. Geophys. Res.* 96, 22,535.

Parrish, A., R. L. de Zafra, P. M. Solomon, and J. W. Barrett (1988) A ground-based technique for millimeter wave spectroscopic observations of stratospheric trace constituents, *Radio Sci.* 23, 106.

Parrish, A., B. J. Connor, J. J. Tsou, I. S. McDermid, and W. P. Chu (1992) Ground-based microwave monitoring of stratospheric ozone, *J. Geophys. Res.* 97, 2541.

Pasquill, F. and F. B. Smith (1983) *Atmospheric Diffusion*, 3d ed., Wiley & Sons, New York.

Paulson, S. E., R. C. Flagan, and J. H. Seinfeld (1992) Atmospheric photooxidation of isoprene, Part I: The hydroxyl radical and ground state atomic oxygen reactions, *Int. J. Chem. Kinet.* 24, 79.

Paulson, S. E. and J. H. Seinfeld (1992) Development and evaluation of a photooxidation mechanism for isoprene, *J. Geophys. Res.* 97, 20,703.

Paulson, S. E. and J. J. Orlando (1996) The reactions of ozone with alkenes: An important source of $HO_x$ in the boundary layer, *Geophys. Res. Lett.* 23, 3727.

Peaceman, D. W. and H. H. Rachford, Jr. (1955) The numerical solution of parabolic and elliptic differential equations, *J. Soc. Indust. Appl. Math* 3, 28.

Pedlosky, J. (1979) *Geophysical Fluid Dynamics*, Springer Verlag.

Penkett, S. A. and K. A. Brice (1986) The spring maximum in photo-oxidants in the northern hemisphere troposphere, *Nature* 319, 655.

Penner, J. E., R. J. Charlson, J. M. Hales, N. Laulainen, R. Leifer, T. Novakov, J. Ogren, L. F. Radke, S. E. Schwartz, and L. Travis (1993) *Quantifying and Minimizing Uncertainty of Climate Forcing by Anthropogenic Aerosols*, U.S. Dept. of Energy.

Pham, M., G. Mégie, J. F. Müller, G. Brasseur and C. Granier (1995) A three-dimensional study of the tropospheric sulfur cycle, *J. Geophys. Res.* 100, 26,061.

Phillips, N. A. (1973) Principles of large-scale numerical weather prediction, in: *Dynamic Meteorology*, P. Morel, ed., D. Reidel Publishing Co., Dordrecht, The Netherlands.

Pickering, K. E., A. M. Thompson, R. R. Dickerson, W. T. Luke, D. P. McNamara, J. P. Greenberg, and P. R. Zimmerman (1990) Model calculations of tropospheric ozone production potential following observed convective events, *J. Geophys. Res.* 95, 14,049.

Pilling, M. J. and P. W. Seakins (1995) *Reaction Kinetics*, Oxford University Press, Oxford, UK.

Platt, U., D. Perner, and H. W. Pätz (1979) Simultaneous measurement of atmospheric $CH_2O$, $O_3$, and $NO_2$ by differential optical absorption, *J. Geophys. Res.* 84, 6329.

Platt, U. F., A. M. Winer, H. W. Biermann, R. Atkinson, and J. N. Pitts, Jr. (1984) Measurement of nitrate radical concentrations in continental air, *Environ. Sci. Technol.* 18, 365.

Plumb, R. A. and M. K. W. Ko (1992) Interrelationships between mixing ratios of long-lived stratospheric constituents, *J. Geophys. Res.* 97, 10,145.

Poppe, D., J. Zimmermann, R. Bauer, T. Brauers, D. Brüning, J. Callies, H.-P. Dorn, A. Hofzumahaus, F.-J. Johnen, A. Khedin, H. Koch, R. Koppmann, H. London, K.-P. Müller, R. Neuroth, C. Plass-Dülmer, U. Platt, F. Rohrer, E.-P. Roth, J. Rudolph, U. Schmidt, M. Wallasch, and D. H. Ehhalt (1994) Comparison of measured OH concentrations with model calculations, *J. Geophys. Res.* 99, 16,633.

Possiel, N. C., J. A. Tikvart, J. H. Novak, K. L. Schere, and E. L. Meyer (1989) Evaluation of ozone control strategies in the northeastern region of the United States, in: *Atmospheric Ozone Research and its Policy Implications*, T. Schneider, S. D. Lee, G. J. R. Wolters, and E. D. Grant, eds., Proceedings of the 3d U.S.-Dutch International Symposium, Nijmegen, The Netherlands, Elsevier, Amsterdam.

Powell, K. A., C. R. Trepte, and G. S. Kent (1997) Observations of Saharan dust by LITE, in: *Advances in Atmospheric Remote Sensing with Lidar*, A. Ansmann, R. Neuber, P. Rairoux, and U. Wandinger, eds., Springer-Verlag, Berlin, 149.

Prather, M. J. (1986) Numerical advection by conservation of second-order moments, *J. Geophys. Res.* 91, 6671.

Prather, M., M. McElroy, S. Wofsy, G. Russell and D. Rind (1987) Chemistry of the global troposphere: Fluorocarbons as tracers of air motion, *J. Geophys. Res.* 92, 6579.

Prather, M. J. (1994) Lifetimes and eigenstates in atmospheric chemistry, *Geophys. Res. Lett.* 21, 801.

Prinn, R. G. (1994) Global atmospheric-biospheric chemistry: An overview, in: *Global Atmospheric-Biospheric Chemistry*, R. G. Prinn, ed., Environmental Science Research Vol. 48, Plenum Press.

Prinn, R. G., P. G. Simmonds, R. A. Rasmussen, R. D. Rosen, F. N. Alyea, C. A. Cardelino, A. J. Crawford, D. M. Cunnold, P. J. Fraser, and J. E. Lovelock (1983) The atmospheric lifetime experiment, 1. Introduction, instrumentation, and overview, *J. Geophys. Res.* 88, 8353.

Prinn, R. G., R. F. Weiss, B. R. Miller, J. Huang, F. N. Alyea, D. M. Cunnold, P. J. Fraser, D. E. Hartley, and P. G. Simmonds (1995) Atmospheric trends and lifetime of $CH_3CCl_3$ and global OH concentrations, *Science* 269, 187.

Proffitt, M. H. and R. J. McLaughlin (1983) Fast-response dual-beam UV absorption ozone photometer suitable for use on stratospheric balloons, *Rev. Sci. Instrum.* 54, 1719.

Prospero, J. M., R. J. Charlson, V. Mohnen, R. Jaenicke, A. C. Delany, J. Moyers, W. Zoller, and K. Rahn (1983) The atmospheric aerosol system: An overview, *Rev. Geophys. Space Phys.* 21, 1607.

Prospero, J. M., R. T. Nees, and M. Uematsu (1987) Deposition rate of particulate and dissolved aluminum derived from Saharan dust in precipitation at Miami, Florida, *J. Geophys. Res.* 92, 14,723.

Pruppacher, H. R. and J. D. Klett (1980) *Microphysics of Clouds and Precipitation*, D. Reidel, Dordrecht, the Netherlands.

Prusa, J. M., P. K. Smolarkiewicz, and R. R. Garcia (1996) On the propagation and breaking at high altitudes of gravity waves excited by tropospheric forcing, *J. Atmos. Sci.* 53, 2186.

Putaud, J. P., N. Mihalopoulos, B. C. Nguyen, J. M. Campin, and S. Belviso (1992) Seasonal variations of atmospheric sulfur dioxide and dimethylsulfide concentrations at Amsterdam Island in the southern Indian Ocean, *J. Atmos. Chem.* 15, 117.

Qian, J. and C. S. Gardner (1995) Simultaneous lidar measurements of mesospheric Ca, Na, and temperature profiles at Urbana, Illinois, *J. Geophys. Res.* 100, 7453.

Quinn, P. K., R. J. Charlson, and W. H. Zoller (1987) Ammonia: The dominant base in the remote marine atmosphere: A review, *Tellus,* 39, 413.

Quinn, P. K., T. S. Bates, J. E. Johnson, D. S. Covert, and R. J. Charlson (1990) Interaction between the sulfur and reduced nitrogen cycles over the central Pacific ocean, *J. Geophys. Res.* 95, 16,405.

Quinn, P. K., W. E. Asher, and R. J. Charlson (1992) Equilibria of the marine multiphase ammonia system, *J. Atmos. Chem.* 14, 11.

Rabitz, H., M. Kramer, and D. Dacol (1983) Sensitivity analysis in chemical kinetics, *Ann. Rev. Phys. Chem.* 34, 419.

Radke, L. F., J. A. Coakley, and M. D. King (1989) Direct and remote sensing observations of the effects of ships on clouds, *Science* 246, 1146.

Raes, F. and R. Van Dingenen (1995) Comment on "The relationship between DMS flux and CCN concentration in remote marine regions," *J. Geophys. Res.* 100, 14,355.

Ramaroson, R., M. Pirre, and D. Cariolle (1992) A box model for on-line computations of diurnal variations in a 1-D model: Potential for application in multidimensional cases, *Ann. Geophys.* 10, 416.

Randel, W. J., F. Wu, J. M. Russell III, J. W. Waters and L. Froidevaux (1995) Ozone and temperature changes in the stratosphere following the eruption of Mt. Pinatubo, *J. Geophys. Res.* 100, 16,753.

Rasch, P. J., B. A. Boville, and G. P. Brasseur (1995) Three-dimensional general circulation model with coupled chemistry for the middle atmosphere, *J. Geophys. Res.* 100, 9041.

Ravishankara, A. R., P. H. Wine, and J. M. Nicovich (1983) Pulsed laser photolysis study of the reaction between $O(^3P)$ and $HO_2$, *J. Chem. Phys.* 78, 6629.

Ravishankara, A. R., S. Solomon, A. A. Turnipseed, and R. F. Warren (1993) Atmospheric lifetimes of long-lived halogenated species, *Science* 259, 194.

Ravishankara, A. R., A. A. Turnipseed, N. R. Jensen, S. Barone, M. Mills, C. J. Howard, and S. Solomon (1994) Do hydrofluorocarbons destroy stratospheric ozone? *Science* 263, 71.

Ravishankara, A. R., G. Hancock, M. Kawasaki, and Y. Matsumi (1998) Photochemistry of ozone: Surprises and recent lessons, *Science* 280, 60.

Ray, E. A., J. R. Holton, E. F. Fishbein, L. Froidevaux, and J. W. Waters (1994) The tropical semiannual oscillation in temperature and ozone as observed by the MLS, *J. Atmos. Sci.* 51, 3045.

Reber, C. A., C. E. Trevathan, R. J. McNeal, and M. R. Luther (1993) The Upper Atmosphere Research Satellite (UARS) mission, *J. Geophys. Res.* 98, 10,643.

Regener, E. (1949) Ozonschicht und Atmosphärische Turbulenz. *Ber. deut. Wetterdienstes US-Zone* 11, 45.

Reichle, H. G., Jr., V. S. Connors, J. A. Holland, W. D. Hypes, H. A. Wallio, J. C. Casas, B. B. Gormsen, M. S. Saylor, and W. D. Hesketh (1986) Middle and upper tropospheric carbon monoxide mixing ratios as measured by a satellite-borne remote sensor during November 1981, *J. Geophys. Res.* 91, 10,865.

Reid, J., B. K. Garside, J. Shewchun, M. El-Sherbiny, and E. A. Ballik (1978) High sensitivity point monitoring of atmospheric gases employing tunable diode lasers, *Appl. Opt.* 17, 1806.

Reiner, T. and F. Arnold (1997) Stratospheric $SO_3$: Upper limits inferred from ion composition measurements — Implications for $H_2SO_4$ and aerosol formation, *Geophys. Res. Lett.* 24, 1751.

Rennenberg, H., B. Huber, P. Schröder, K. Stahl, W. Haunold, H-W. Georgii, S. Slovik, and H. Pfanz (1990) Emission of volatile sulfur compounds from spruce trees, *Plant Physiol.,* 92, 560.

Revelle, R. and H. E. Suess (1957) Carbon dioxide exchange between atmosphere and ocean and the question of an increase of atmospheric $CO_2$ during the past decades, *Tellus* 9, 18.

Ridley, B. A. and L. C. Howlett (1974) An instrument for nitric oxide measurements in the stratosphere. *Rev. Sci. Insts.* 45, 742.

Ridley, B. A., M. McFarland, J. T. Bruin, H. I. Schiff, and J. C. McConnell (1977) Sunrise measurements of stratospheric nitric oxide, *Can. J. Phys.* 55, 212.

Ridley, B. A., M. McFarland, A. L. Schmeltekopf, M. H. Proffitt, D. L. Albritton, R. H. Winkler, and T. L. Thompson (1987) Seasonal differences in the vertical distributions of NO, $NO_2$, and $O_3$ in the stratosphere near 50°N, *J. Geophys. Res.* 92, 11,919.

Ridley, B. A. (1991) Recent measurements of oxidized nitrogen compounds in the troposphere, *Atmos. Environ.* 25A, 1905.

Ridley, B. A., S. Madronich, R. B. Chatfield, J. G. Walega, R. E. Shetter, M. A. Carroll, and D. D. Montzka (1992) Measurements and model simulations of the photostationary state during the Mauna Loa Observatory Photochemistry Experiment: Implications for radical concentrations and ozone production and loss rates, *J. Geophys. Res.* 97, 10,375.

Ridley, B. A. and E. Robinson (1992) The Mauna Loa Observatory Photochemistry Experiment, *J. Geophys. Res.* 97, 10,285.

Riishøjgaard, L.P. (1996) On four-dimensional variational assimilation ozone data in weather prediction models, *Quart. J. Roy. Meteor. Soc.* 122, 1545.

Rinsland, C. P., R. Zander, E. Mahieu, P. Demoulin, A. Goldman, D. Ehhalt, and J. Rudolph (1992) Ground-based infrared measurements of carbonyl sulfide total column abundances: Long-term trends and variability, *J. Geophys. Res.* 97, 5995.

Rinsland, C. P., M. R. Gunson, M. K. W. Ko, D. W. Weisenstein, R. Zander, M. C. Abrams, A. Goldman, N. D. Sze, and G. K. Yue (1995) $H_2SO_4$ photolysis: A source of sulfur dioxide in the upper stratosphere, *Geophys. Res. Lett.* 22, 1109.

Robert, A. (1981) A stable numerical integration scheme for the primitive meteorological equations, *Atmos.-Ocean* 19, 35.

Roberts, J. M. (1990) The atmospheric chemistry of organic nitrates, *Atmos. Environ.* 24A, 243.

Roche, A. E., J. B. Kumer, J. L. Mergenthaler, G. A. Ely, W. G. Uplinger, J. F. Potter, T. C. James, and L. W. Sterritt (1993) The Cryogenic Limb Array Etalon Spectromeneter (CLAES) on UARS: Experiment description and performance, *J. Geophys. Res.* 98, 10,763.

Roche, A. E., J. B. Kumer, J. L. Mergenthaler, R. W. Nightingale, W. G. Uplinger, G. A. Ely, J. F. Potter, D. J. Wuebbles, P. S. Connell, and D. E. Kinnison (1994) Observations of lower stratospheric $ClONO_2$, $HNO_3$, and aerosol by the UARS CLAES experiment between January 1992 and April 1993, *J. Atmos. Sci.* 51, 2877.

Rodgers, C. D. (1977) Statistical principles of inversion theory, in: *Inversion Methods in Atmospheric Remote Sounding*, A. Deepak, ed., Academic Press, 117.

Rodhe, H. (1992) Modeling biogeochemical cycles, in: *Global Biogeochemical Cycles*, S. S. Butcher, R. J. Charlson, G. H. Orians, G. V. Wolfe, eds., Academic Press, San Diego.

Roelofs, G.-J. and J. Lelieveld (1995) Distribution and budget of $O_3$ in the troposphere calculated with a chemistry-general circulation model, *J. Geophys. Res.* 100, 20,983.

Rood, R. B. (1987) Numerical advection algorithms and their role in atmospheric transport and chemistry models, *Rev. Geophys.* 25, 71.

Root, T. L. and S. H. Schneider (1995) Ecology and climate: Research strategies and implications, *Science* 269, 331.

Rosenlof, K. H. (1995) Seasonal cycle of the residual mean meridional circulation in the stratosphere, *J. Geophys. Res.* 100, 5173.

Rose, K. and G. Brasseur (1989) A three-dimensional model of chemically active trace species in the middle atmosphere during disturbed winter conditions, *J. Geophys. Res.* 94, 16,387.

Ross, H. B. and K. J. Noone (1991) A numerical investigation of the destruction of peroxy radical by Cu ion catalysed reaction on atmospheric particles, *J. Atmos. Chem.* 12, 121.

Rossby, C. G. (1936) Dynamics of steady ocean currents in the light of experimental fluid mechanics, Massachusetts Institute of Technology and Woods Hole Oceanographic Institution, *Papers in Physical Oceanography and Meteorology* 5 (1), 1.

Rossby, C. G. (1940) Planetary flow patterns in the atmosphere, *Quart. J. Roy. Meteor. Soc.* 66, 68.

Rottman, G. J., T. N. Woods, and T. P. Sparn (1993) Solar-stellar irradiance comparison experiment, 1. Instrument design and operation, *J. Geophys. Res.* 98, 10,667.

Rudolph, J. and F. J. Johnen (1990) Measurements of light atmospheric hydrocarbons over the Atlantic in regions of low biological activity, *J. Geophys. Res.* 95, 20,583.

Russell, J. M. III, C. B. Farmer, C. P. Rinsland, R. Zander, L. Froidevaux, G. C. Toon, B. Gao, J. Shaw, and M. Gunson (1988) Measurements of odd nitrogen compounds in the stratosphere by the ATMOS experiment on Spacelab 3, *J. Geophys. Res.* 93, 1718.

Russell, J. M. III, L. L. Gordley, J. H. Park, S. R. Drayson, W. D. Hesketh, R. J. Cicerone, A. F. Tuck, J. E. Fredrick, J. E. Harries, and P. J. Crutzen (1993) The halogen occultation experiment, *J. Geophys. Res.* 98, 10,777.

Russell III, J. M., M. Luo, R. J. Cicerone, and L. E. Deaver (1996) Satellite confirmation of the dominance of chlorofluorocarbons in the global stratospheric chlorine budget, *Nature* 379, 526.

Salby, M. L. (1987) Irregular and diurnal variability in asynoptic measurements of stratospheric trace species, *J. Geophys. Res.* 92, 14,781.

Saltzman, E. S., D. L. Savoie, R. G. Zika, and J. M. Prospero (1983) Methane sulfonic acid in the marine atmosphere, *J. Geophys. Res.* 88, 10,897.

Saltzman, E. S. and D. J. Cooper (1988) Shipboard measurements of atmospheric dimethysulfide and hydrogen sulfide in the Caribbean Gulf of Mexico, *J. Atmos. Chem.* 7, 191.

Sander, S. P. (1986) Temperature dependence of the $NO_3$ absorption spectrum, *J. Phys. Chem.* 90, 4135.

Santee, M. L., W. G. Read, J. W. Waters, L. Froidevaux, G. L. Manney, D. A. Flower, R. F. Janot, R. S. Harwood, and G. E. Peckham (1995) Interhemispheric differences in polar stratospheric $HNO_3$, $H_2O$, $ClO$, and $O_3$ from UARS MLS, *Science* 267, 849.

Saxena, P., L. M. Hildemann, P. H. McMurry, and J. H. Seinfeld (1995) Organics alter hygroscopic behavior of atmospheric particles, *J. Geophys. Res.* 100, 18,755.

Schauffler, S. M., L. E. Heidt, W. H. Pollock, T. M. Gilpin, J. F. Vedder, S. Solomon, R. A. Lueb, and E. L. Atlas (1993) Measurements of halogenated organic compounds near the tropical tropopause, *Geophys. Res. Lett.* 20, 2567.

Schere, K. L. and R. A. Wayland (1989) *EPA Regional Oxidant Model (ROM2.0): Evaluation on 1980 NEROS Data Bases*, Report EPA-600/3-89/057, Research Triangle Park, North Carolina.

Schimel, D. S. (1993) *Theory and Application of Tracers*, Academic Press, New York.

Schimel, D. S. (1995) Terrestrial ecosystems and the carbon cycle, *Global Change Biology* 1, 77.

Schimel, D. S. and C. S. Potter (1995) Process modeling and spatial extrapolation, in: *Biogenic Trace Gases: Measuring Emissions from Soil and Water*, P. A. Matson and R. C. Harriss, eds., Blackwell Science, Ltd., Oxford.

Schindler, D. W. (1988) Effects of acid rain on freshwater ecosystems, *Science* 239, 149.

Schindler, D. W. and S. E. Bayley (1993) The biosphere as an increasing sink for atmospheric carbon: Estimates from increased nitrogen deposition, *Glob. Biogeochem. Cycles* 7, 717.

Schlesinger, W. H. and A. E. Hartley (1992) A global budget for atmospheric $NH_3$, *Biogeochemistry*, 15, 191.

Schmidt, U., G. Kulessa, and E. P. Röth (1980) The atmospheric $H_2$ cycle, in *Proc. NATO Advanced Study Institute on Atmospheric Ozone: Its Variation and Human Influences*, FAA-EE-80-20, 307. National Technical Information Service, Springfield, Virginia

Schneider, S. H. (1992) Introduction to climate modeling, in: *Climate System Modeling*, K. E. Trenberth, ed., Cambridge University Press, Cambridge, UK.

Schneider, S. H. (1994) Detecting climatic change signals: Are there any "fingerprints"? *Science* 263, 341.

Scholz, T. G., D. H. Ehhalt, L. E. Heidt, and E. A. Martell (1970) Water vapor, molecular hydrogen, methane, and tritium concentrations near the stratosphere, *J. Geophys. Res.* 75, 3049.

Schwartz, S. E. and J. E. Freiberg (1981) Mass-transport limitation to the rate of reaction of gases in liquid droplets: Application to oxidation of $SO_2$ in aqueous solutions, *Atmos. Environ.* 15, 1129.

Schwartz, S. E. (1984a) Gas- and aqueous-phase chemistry of $HO_2$ in liquid water clouds, *J. Geophys. Res.* 89, 11589.

Schwartz, S. E. (1984b) Gas-aqueous reactions of sulfur and nitrogen oxides in liquid water clouds, in $SO_2$, $NO$ and $NO_2$ *Oxidation Mechanisms: Atmospheric Considerations*, J. G. Calvert, ed., Butterworth Publications, Boston.

Schwartz, S. E. (1986) Mass-transport considerations pertinent to aqueous phase reactions of gases in liquid-water clouds, in: *Chemistry of Multiphase Atmospheric Systems*, W. Jaeschke, ed., Springer-Verlag, Berlin.

Schwartz, S. E. (1988) Mass-transport limitation to the rate of in-cloud oxidation of $SO_2$: Re-examination in the light of new data, *Atmos. Environ.* 22, 2491.

Sebacher, D. I. (1978) Airborne nondispersive infrared monitor for atmospheric trace gases, *Rev. Sci. Insts.* 49, 1520.

Sedjo, R. A. (1992) Temperate forest ecosystems in the global carbon cycle, *Ambio* 21, 274.

Seeley, J. V., J. T. Jayne, and M. J. Molina (1996) Kinetic studies of chlorine atom reactions using the turbulent flow tube technique, *J. Phys. Chem.* 100, 4019.

Seigneur, C., A. B. Hudischewskyj, J. H. Seinfeld, K. T. Whitby, E. R. Whitby, J. R. Brock, and H. M. Barnes (1986) Simulation of aerosol dynamics: Comparative review of mathematical models, *Aerosol Sci. Technol.* 5, 205.

Seila, R. L., W. Lonneman, and S. Meeks (1989) *Determination of $C_2$ to $C_{12}$ Ambient Hydrocarbon in 39 U.S. Cities From 1986 Through 1989*, U.S. Environ. Prot. Agency Off. Res. Dev. Report EPA/600/53-89/058.

Seinfeld, J. H. and S. N. Pandis (1998) *Atmospheric Chemistry and Physics*, Wiley, New York.

Shapiro, R. (1971) The use of linear filtering as a parameterization of atmospheric diffusion, *J. Atmos. Sci.* 28, 523.

Shepherd, G. G., G. Thuillier, W. A. Gault, B. H. Solheim, C. Hersom, J. M. Alunni, J.-F. Brun, S. Brune, P. Charlot, L. L. Cogger, D.-L. Desaulniers, W. F. J. Evans, R. L. Gattinger, F. Girod, D. Harvie, R. H. Hum, D. J. W. Kendall, E. J. Llewellyn, R. P. Lowe, J. Ohrt, F. Pasternak, O. Peillet, I. Powell, Y. Rochon, W. E. Ward, R. H. Wiens, and J. Wimperis (1993) WINDII, the Wind Imaging Interferometer on the Upper Atmosphere Research Satellite, *J. Geophys. Res.* 98, 10,725.

Shepson, P. B., E. O. Edney, T. E. Kleindienst, J. H. Pittman, G. R. Namie, and L. T. Cupitt (1985) The production of organic nitrates from hydroxyl and nitrate radical reaction with propylene, *Environ. Sci. Technol.* 19, 849.

Shetter, R. E., A. H. McDaniel, C. A. Cantrell, S. Madronich, and J. G. Calvert (1992) Actinometer and Eppley radiometer measurements of the $NO_2$ photolysis rate coefficient during the Mauna Loa Observatory Photochemistry Experiment, *J. Geophys. Res.* 97, 10,349.

Shetter, R. E., C. A. Cantrell, K. O. Lantz, S. J. Flocke, J. J. Orlando, G. S. Tyndall, T. M. Gilpin, C. A. Fisher, S. Madronich, J. G. Calvert, and W. Junkermann (1996) Actinometric and radiometric measurement and modeling of the photolysis rate coefficient of ozone to $O(^1D)$ during Mauna Loa Observatory Photochemistry Experiment 2, *J. Geophys. Res.* 101, 14,631.

Shimazaki, T. (1985) *Minor Constituents in the Middle Atmosphere*, Terra Scientific Publishing, Tokyo, Japan.

Shorter, J. M., C. E. Kolb, P. M. Crill, R. A. Kerwin, R. W. Talbot, M. E. Mines, and R. C. Harriss (1995) Rapid degradation of atmospheric methyl bromide in soils, *Nature* 377, 717.

Siegenthaler, U. (1983) Uptake of excess $CO_2$ by an outcrop-diffusion model of the ocean, *J. Geophys. Res.* 88, 3599.

Siegenthaler, V. and F. Joos (1992) Use of a simple model for studying oceanic tracer distributions and the global carbon cycle, *Tellus* 44B, 186.

Sievering, H., G. Ennis, E. Gorman, and C. Nagamoto (1990) Size distributions and statistical analysis of nitrate, excess sulfate, and chloride deficit in the marine boundary layer during GCE/CASE/WATOX, *Global Biogeochem. Cycles* 4, 395.

Sievering, H., J. Boatman, E. Gorman, Y. Kim, L. Anderson, G. Ennis, M. Luria, and S. Pandis (1992) Removal of sulphur from the marine boundary layer by ozone oxidation in sea-salt aerosols, *Nature* 360, 571.

Simpson, D., A. Guenther, C. N. Hewitt, and R. Steinbrecher (1995) Biogenic emissions in Europe, 1. Estimates and uncertainties, *J. Geophys. Res.* 100, 22,875.

Singh, H. B., E. Condon, J. Vedder, D. O'Hara, B. A. Ridley, B. W. Gandrud, J. D. Shetter, L. J. Salas, B. Huebert, G. Hübler, M. A. Carroll, D. L. Albritton, D. D. Davis, J. D. Bradshaw, S. T. Sandholm, M. O. Rodgers, S. M. Beck, G. L. Gregory, and P. J. LeBel (1990) PAN measurements during CITE 2: Atmospheric distribution and precursor relationships, *J. Geophys. Res.* 95, 10,163.

Singh, H. B. and P. R. Zimmerman (1992) Atmospheric distribution and sources of nonmethane hydrocarbons, in: *Gaseous Pollutants: Characterization and Cycling*, J. O. Nriagu, ed., John Wiley and Sons, New York.

Singh, H. B., D. O'Hara, D. Herlth, W. Sachse, D. R. Blake, J. D. Bradshaw, M. Kanakidou, and P. J. Crutzen (1994) Acetone in the atmosphere: Distribution, sources, and sinks, *J. Geophys. Res.* 99, 1805.

Singh, H. B., M. Kanakidou, P. J. Crutzen, and D. J. Jacob (1995) High concentrations and photochemical fate of oxygenated hydrocarbons in the global troposphere, *Nature* 378, 50.

Singh, H. B., G. L. Gregory, B. Anderson, E. Browell, G. W. Sachse, D. D. Davis, J. Crawford, J. D. Bradshaw, R. Talbot, D. R. Blake, D. Thornton, R. Newell, and J. Merrill (1996) Low ozone in the marine boundary layer of the tropical Pacific Ocean: Photochemical loss, chlorine atoms, and entrainment, *J. Geophys. Res.* 101, 1907.

Smith, A. (1872) *Air and Rain: The Beginnings of a Chemical Climatology*, London.
Smith III, F. L. and C. Smith (1972) Numerical evaluation of Chapman's grazing incidence integral ch $(X, \chi)$, *J. Geophys. Res.* 77, 3592.
Smith, I. W. M. (1980) *Kinetics and Dynamics of Elementary Gas Reactions*, Butterworth Publishing, Boston.
Smolarkiewicz, P. K. (1984) A fully multidimensional positive definite advection transport algorithm with small implicit diffusion, *J. Comput. Phys.* 54, 325.
Sokolik, I. N., A. V. Andronova, and T. C. Johnson (1993) Complex refractive index of atmospheric dust aerosols, *Atmos. Environ.* 27A, 2495.
Solomon, S., R. R. Garcia, F. S. Rowland, and D. J. Wuebbles (1986a) On the depletion of Antarctic ozone, *Nature* 321, 755.
Solomon, S., J. T. Kiehl, R. R. Garcia, and W. Grose (1986b) Tracer transport by the diabatic circulation deduced from satellite observations, *J. Atmos. Sci.* 43, 1603.
Solomon, S., H. L. Miller, J. P. Smith, R. W. Sanders, G. H. Mount, A. L. Schmeltekopf, and J. F. Noxon (1989a) Atmospheric $NO_3$, 1. Measurement technique and the annual cycle at 40°N, *J. Geophys. Res.* 94, 11,041.
Solomon, S., R. W. Sanders, M. A. Carroll, and A. L. Schmeltekopf (1989b) Visible and near-ultraviolet spectroscopy at McMurdo Station, Antarctica: 5. Observations of the diurnal varations of BrO and OClO, *J. Geophys. Res.* 94, 11,393.
Solomon, S. (1990) Progress towards a quantitative understanding of Antarctic ozone depletion, *Nature* 347, 347.
Solomon, S., J. B. Burkholder, A. R. Ravishankara, and R. R. Garcia (1994a) On the ozone depletion and global warming potentials of $CH_3I$, *J. Geophys. Res.* 99, 20,929.
Solomon, S., R. R. Garcia, and A. R. Ravishankara (1994b) On the role of iodine in ozone depletion, *J. Geophys. Res.* 99, 20,491.
Staehelin, J., J. Thudium, R. Bühler, A. Volz-Thomas, and W. Graber (1994) Trends in surface ozone concentrations at Arosa (Switzerland), *Atmos. Environ.* 28, 75.
Staffelbach, T. A., G. L. Kok, B. G. Heikes, B. McCully, G. I Mackay, D. R. Karecki, and H. I. Schiff (1996) Comparison of hydroperoxide measurements made during the Mauna Loa Observatory Photochemistry Experiment 2, *J. Geophys. Res.* 101, 14,729.
Staniforth, A. and J. Côté (1991) Semi-Lagrangian integration schemes for atmospheric models—A review, *Mon. Wea. Rev.* 119, 2206.
Staubes, R. and H.-W. Georgii (1993) Measurements of atmospheric and seawater DMS concentrations in the Atlantic, the Arctic and Antarctic region, in: *Dimethylsulphide: Oceans, Atmosphere, and Climate*, G. Restelli and G. Angeletti, eds., Kluwer Academic Publishers, the Netherlands.
Stecher, H. A. III, G. W. Luther III, D. L. MacTaggart, S. O. Farwell, D. R. Crosley, W. D. Dorko, P. D. Goldan, N. Beltz, U. Krischke, W. T. Luke, D. C. Thornton, R. W. Talbot, B. L. Lefer, E. M. Scheuer, R. L. Benner, J. Wu, E. S. Saltzman, M. S. Gallagher, and R. J. Ferek (1997) Results of the gas-phase sulfur intercomparison experiment (GASIE): Overview of experimental setup, results and general conclusions, *J. Geophys. Res.* 102, 16,219.
Steele, L. P., E. J. Dlugokencky, P. M. Lang, P. P. Tans, R. C. Martin, and K. A. Masarie (1992) Slowing down of the global accumulation of atmospheric methane during the 1980's, *Nature* 358, 313.
Steele, P., P. H. Fraser, R. A. Rasmussen, M. A. K. Khalil, T. J. Conway, A. J. Crawford, R. H. Gammon, K. A. Masarie, and K. W. Shoning (1987) The global distribution of methane in the troposphere, *J. Atmos. Chem* 5, 125.
Stewart, R. W. (1993) Multiple steady states in atmospheric chemistry, *J. Geophys. Res.* 98, 20,601.
Stewart, R. W. (1995) Dynamics of the low to high $NO_x$ transition in a simplified tropospheric photochemical model, *J. Geophys. Res.* 100, 8929.
Stickel, R. E., M. Chin, E. P. Daykin, A. J. Hynes, P. H. Wine, and T. J. Wallington (1993) Mechanistic studies of the OH-initiated oxidation of $CS_2$ in the presence of $O_2$, *J. Phys. Chem.* 97, 13,653.
Stimpfle, R. M., P. O. Wennberg, L. B. Lapson, and J. G. Anderson (1990) Simultaneous, in situ measurements of OH and $HO_2$ in the stratosphere, *Geophys. Res. Lett.* 17, 1906.
Stockwell, W. R. and J. G. Calvert (1983) The mechanism of the $HO-SO_2$ reaction, *Atmos. Environ.* 17, 2231.
Stockwell, W. R. (1986) A homogeneous gas phase mechanism for use in a regional acid deposition model, *Atmos. Environ.* 20, 1615.

Stockwell, W. R., P. Middleton, and J. S. Chang (1990) The second generation regional acid deposition model chemical mechanism for regional air quality modeling, *J. Geophys. Res.* 95, 16,343.

Stolarski, R. S. and R. J. Cicerone (1974) Stratospheric chlorine: A possible sink for ozone, *Can. J. Chem.* 52, 1610.

Strain, B. R. and J. D. Cure, eds. (1985) *Direct Effects of Increasing Carbon Dioxide on Vegetation*, National Technical Information Service, U.S. Dept. of Commerce, Springfield, VA.

Stull, R. B. (1988) *An Introduction to Boundary Layer Meteorology*, Kluwer Academic Publishers, Dordrecht, The Netherlands.

Stull, R. B. (1993) *An Introduction to Boundary Layer Meteorology*, Kluwer Academic Publishers, Dordrecht, The Netherlands.

Suhre, K., M. O. Andreae, and R. Rosset (1995) Biogenic sulfur emissions and aerosols over the tropical south Atlantic. 2: One-dimensional simulation of sulfur chemistry in the marine boundary layer, *J. Geophys. Res.* 100, 11,323.

Summers, M. E., R. R. Conway, D. E. Siskind, M. H. Stevens, D. Offermann, M. Riese, P. Preusse, D. F. Strobel, J. M. Russell III (1996) Implications of satellite OH observations for middle atmospheric $H_2O$ and ozone, *Science* 277, 1967.

Talagrand, O. and P. Courtier (1987) Variational assimilation of meteorological observations with the adjoint vorticity equations, Part 1. Theory, *Quart. J. Roy. Meteor. Soc.* 113, 1311.

Talukdar, R. K., C. A. Longfellow, M. K. Gilles, and A. R. Ravishankara (1998) Quantum yields of $O(^1D)$ in the photolysis of ozone between 289 and 329 nm as a function of temperature, *Geophys. Res. Lett.* 25, 143.

Tang, I. N. (1976) Phase transformation and growth of aerosol particles composed of mixed salts, *J. Aerosol Sci.* 7, 361.

Tang, I. N., H. R. Munkelwitz, and J. G. Davis (1978) Aerosol growth studies—IV. Phase transformation of mixed salt aerosols in a moist atmosphere, *J. Aerosol Sci.* 9, 505.

Tanguay, M., A. Simard, and A. Staniforth (1989) A three-dimensional semi-Lagrangian scheme for the Canadian regional finite-element forecast model, *Mon. Wea. Rev.* 117, 1861.

Tans, P. P., I. Y. Fung, and T. Takahashi (1990) Observational constraints on the global atmospheric carbon dioxide budget, *Science* 247, 1431.

Taylor, F. W., C. D. Rogers, J. G. Whitney, S. T. Werrett, J. J. Barnett, G. D. Peskett, P. Venters, J. Ballard, C. W. P. Palmer, R. J. Knight, P. Morris, T. Nightingale, and A. Dudhia (1993) Remote sensing of atmospheric structure and composition by pressure modulator radiometry from space: The ISAMS experiment on UARS, *J. Geophys. Res.* 98, 10,799.

Taylor, G. (1953) Dispersion of soluble matter in solvent flowing slowly through a tube, *Proc. Roy. Soc. A* 229, 186.

Taylor, J. A., G. P. Brasseur, P. R. Zimmerman, and R. J. Cicerone (1991) A study of the sources and sinks of methane and methyl chloroform using a global three-dimensional Lagrangian tropospheric tracer transport model, *J. Geophys. Res.* 96, 3013.

Thompson, A. M. and R. W. Stewart (1991) Effect of chemical kinetics uncertainties on calculated constituents in a tropospheric photochemical model, *J. Geophys. Res.* 96, 13,089.

Thornton, D. C., A. R. Bandy, and A. R. Driedger III (1987) Sulfur dioxide over the Western Atlantic Ocean, *Glob. Biogeochem. Cycles* 1, 317.

Tie, X.-X. and G. Brasseur (1996) The importance of heterogeneous bromine chemistry in the lower stratosphere, *Geophys. Res. Lett.* 23, 2505.

Tiedtke, M. (1989) A comprehensive mass flux scheme of cumulus parameterization in large-scale models, *Mon. Wea. Rev.* 117, 1779.

Tiwari, S. N. (1978) Models for infrared atmospheric radiation, *Adv. Geophys.* 20, 1.

Tolbert, M. A. (1994) Sulfate aerosols and polar stratospheric cloud formation, *Science* 264, 527.

Toohey, D. W., L. M. Avallone, L. R. Lait, P. A. Newman, M. R. Schoeberl, D. W. Fahey, E. L. Woodbridge, and J. G. Anderson (1993) The seasonal evolution of reactive chlorine in the northern hemisphere stratosphere, *Science* 261, 1134.

Toon, O. B., J. F. Kasting, R. P. Turco, and M. S. Liu (1987) The sulfur cycle in the marine atmosphere, *J. Geophys. Res.* 92, 943.

Toon, O. B., C. P. McKay, T. P. Ackerman, and K. Santhanam (1989) Rapid calculation of radiative heating rates and photodissociation rates in inhomogeneous multiple scattering atmospheres, *J. Geophys. Res.* 94, 16,287.

Trainer, M., E.-Y. Hsie, S. A. McKeen, R. Tallamraju, D. D. Parrish, F. C. Fehsenfeld, and S. C. Liu (1987a) Impact of natural hydrocarbons on hydroxyl and peroxy radicals at a remote site, *J. Geophys. Res.* 92, 11,879.

Trainer, M., E. J. Williams, D. D. Parrish, M. P. Buhr, E. J. Allwine, H. H. Westberg, F. C. Fehsenfeld, and S. C. Liu (1987b) Models and observations of the impact of natural hydrocarbons on rural ozone, *Nature*, 329, 705.

Trainer, M., D. D. Parrish, M. P. Buhr, R. B. Norton, F. C. Fehsenfeld, K. G. Anlauf, J. W. Bottenheim, Y. Z. Tang, H. A. Wiebe, J. M. Roberts, R. L. Tanner, L. Newman, V. C. Bowersox, J. F. Meagher, K. J. Olszyna, M. O. Rodgers, T. Wang, H. Berresheim, K. L. Demerjian, and U. K. Roychowdhury (1993) Correlation of ozone with $NO_y$ in photochemically aged air, *J. Geophys. Res.* 98, 2917.

Traub, W. A., D. G. Johnson, and K. V. Chance (1990) Stratospheric hydroperoxyl measurements, *Science* 247, 446.

Trenberth, K. E. and C. J. Guillemot (1994) The total mass of the atmosphere, *J. Geophys. Res.* 99, 23079.

Troe, J. (1979) Predictive possibilities of unimolecular rate theory, *J. Phys. Chem.* 83, 114.

Tuazon, E. C. and R. Atkinson (1990) A product study of the gas-phase reaction of isoprene with the OH radicals in the presence of $NO_x$, *Int. J. Chem. Kinet.* 22, 1221.

Tuazon, E. C. and R. Atkinson (1993) Tropospheric transformation products of a series of hydrofluorocarbons and hydrochlorofluorocarbons, *J. Atm. Chem.* 17, 179.

Tuck, A. F., J. M. Russell III, and J. E. Harries (1993) Stratospheric dryness: Antiphased desiccation over Micronesia and Antarctica, *Geophys. Res. Lett.* 20, 1227.

Tung, K. K. and H. Yang (1994) Global QBO in circulation and ozone, I. Reexamination of observational evidence, *J. Atmos. Sci.* 51, 2699.

Turekian, K. K., Y. Nozaki, and L. K. Benninger (1977) Geochemistry of atmospheric radon and radon products, *Ann. Rev. Earth Planet. Sci.* 5, 227.

Turnipseed, A. A., S. B. Barone, and A. R. Ravishankara (1993) Reactions of $CH_3S$ and $CH_3SOO$ with $O_3$, $NO_2$, and NO, *J. Phys. Chem.* 97, 5926.

Twomey, S. (1977) *Atmospheric Aerosols*, Elsevier, New York.

Tyler, S. C. (1989) $^{13}C/^{12}C$ ratios in atmospheric methane and some of its sources, in: *Stable Isotopes in Ecological Research*, P. W. Rundel, J. R. Ehleringer, and K. A. Nagy, eds., Springer-Verlag, New York, 395.

Tyndall, G. S. and A. R. Ravishankara (1991) Atmospheric oxidation of reduced sulfur species, *Int. J. Chem. Kinet.* 23, 483.

Tyndall, G. S., T. J. Wallington, and J. C. Ball (1998) FTIR product study of the reactions $CH_3O_2 + CH_3O_2$ and $CH_3O_2 + O_3$, *J. Phys. Chem.* 102, 2547.

Uno, I., S. Wakamatsu, R. A. Wadden, S. Konno, and H. Koshio (1985) Evaluation of hydrocarbon reactivity in urban air, *Atmos. Environ.* 19, 1283.

U.S. Standard Atmosphere (1976) National Oceanic and Atmospheric Administration, National Aeronautics and Space Administration, and United States Air Force, Washington, D.C.

Utter, R. G., J. B. Burkholder, C. J. Howard, and A. R. Ravishankara (1992) Measurement of the mass accommodation coefficient of ozone on aqueous surfaces, *J. Phys. Chem.*, 96, 4973.

van Dingenen, R., N. R. Jensen, J. Hjorth, and F. Raes (1994) Peroxynitrate formation during the night-time oxidation of dimethylsulfide: Its role as a reservoir species for aerosol formation, *J. Atmos. Chem.* 18, 211.

van Doren, J. M., L. R. Watson, P. Davidovits, D. R. Worsnop, M. S. Zahniser, and C. E. Kolb (1990) Temperature dependence of the uptake coefficients of $HNO_3$, HCl, and $N_2O_5$ by water droplets, *J. Phys. Chem.* 94, 3265.

Vitousek, P. M. (1994) Beyond global warming: Ecology and global change, *Ecology* 75, 1861.

Vogt, R., P. J. Crutzen, and R. Sander (1996) A mechanism for halogen release from sea-salt aerosol in the remote marine boundary layer, *Nature* 383, 327.

Volz, A., D. H. Ehhalt, and R. G. Derwent (1981) Seasonal and latitudinal variation of $^{14}CO$ and the tropospheric concentration of OH radicals, *J. Geophys. Res.* 86, 5163.

Volz, A. and D. Kley (1988) Evaluation of the Montsouris series of ozone measurements made in the nineteenth century, *Nature* 332, 260.

Volz-Thomas, A., F. Flocke, H. J. Garthe, H. Geiß, S. Gilge, T. Heil, D. Kley, D. Klemp, F. Kramp, D. Mihelcic, H. W. Pätz, M. Schultz, and Y. Su (1993) Photo-oxidants and precursors at Schauinsland, Black Forest, A contribution to subproject TOR, in: *Photo-Oxidants: Precursors and Products, Proc. EUROTRAC Symposium '92* P. Borrell, P. M. Borrell, and W. Seiler, eds., SPB Academic Publishing, The Hague, The Netherlands, 98.

Wahlen, M., N. Tanaka, R. Henry, B. Deck, J. Zeglen, J. S. Vogel, J. Southon, A. Shemesh, A. Fairbanks, and W. Broecker (1989) Carbon-14 in methane sources and in atmospheric methane: The contribution from fossil carbon, *Science* 245, 286.

Wahner, A., F. Rohrer, D. H. Ehhalt, E. Atlas, and B. A. Ridley (1994) Global measurements of photochemically active compounds, in *Global Atmospheric-Biospheric Chemistry*, R. G. Prinn, ed., *Environmental Science Research* 48, 205, Plenum Press.

Wallace, J. and P. Hobbs (1977) *Atmospheric Science: An Introductory Survey*, Academic Press.

Wallington, T. J., L. M. Skewes, W. O. Siegl, C.-H. Wu, and S. M. Japar (1988) Gas phase reaction of Cl atoms with a series of oxygenated organic species at 295 K, *Intl. J. Chem. Kinetics* 20, 867.

Wallington, T. J., M. D. Hurley, J. C. Ball, T. Ellermann, O. J. Nielsen, and J. Sehested (1994a) Atmospheric chemistry of HFC-152: UV absorption spectrum of $CH_2FCFHO_2$ radicals, kinetics of the reaction $CH_2FCFHO_2 + NO \rightarrow CH_2FCHFO + NO_2$, and fate of the alkoxy radical $CH_2FCFHO$, *J. Phys. Chem.* 98, 5435.

Wallington, T. J., W. F. Schneider, D. R. Worsnop, O. J. Nielsen, J. Sehested, W. J. Debruyn, and J. A. Shorter (1994b) The environmental impact of CFC replacements—HFCs and HCFCs, *Env. Sci. and Technol.* 28, 320A.

Walton, J. J., M. C. MacCracken, and S. J. Ghan (1988) A global-scale Lagrangian trace species model of transport, transformation, and removal processes, *J. Geophys. Res.* 93, 8339.

Wang, C. C. (1981) Improved airborne measurements of OH in the atmosphere using the technique of laser-induced fluorescence, *J. Geophys. Res.* 86, 1181.

Wang, W. C., D. J. Wuebbles, W. M. Washington, R. G. Isaacs, and G. Molnar (1986) Trace gases and other potential perturbations to global climate, *Rev. Geophys.* 24, 110.

Wang, W. C., M. P. Dudek, X. Z. Liang, and J. T. Kiehl (1991) Inadequacy of effective $CO_2$ as a proxy in simulating the greenhouse effect of other radiatively active gases, *Nature* 350, 573.

Wanninkhof, R. (1992) Relationship between wind speed and gas exchange over the ocean, *J. Geophys. Res.* 97, 7373.

Warneck, P. (1988) *Chemistry of the Natural Atmosphere*, Academic Press, San Diego.

Warren, D. R. and J. H. Seinfeld (1985) Simulation of aerosol size distribution evolution in systems with simultaneous nucleation, condensation and coagulation, *Aerosol Sci. Technol.* 4, 31.

Warren, S. G., C. J. Hahn, J. London, R. M. Chervin, and R. Jenne (1986) *Global Distribution of Total Cloud Cover and Cloud Type Amounts Over Land*, NCAR Technical Note TN-273+STR, NCAR, Boulder.

Warren, S. G., C. J. Hahn, J. London, R. M. Chervin, and R. Jenne (1988) *Global Distribution of Total Cloud Cover and Cloud Type Amounts Over the Ocean*, NCAR Technical Note TN-317+STR, NCAR, Boulder.

Watanabe, K., E. C. Y. Inn, and M. Zelikoff (1953) Absorption coefficients of oxygen in the vacuum ultraviolet, *J. Chem. Phys.* 21, 1026.

Waters, J. W., L. Froidevaux, W. G. Read, G. L. Manney, L. S. Elson, D. A. Flower, R. F. Janot, and R. S. Harwood (1993) Stratospheric ClO and ozone from the Microwave Limb Sounder on the Upper Atmosphere Research Satellite, *Nature*, 362, 597.

Watson, R. T., L. G. Meiro Filho, E. Sanhueza, and A. Janetos (1992) Greenhouse gases: Sources and sinks, in: *Climate Change 1992: The Supplementary Report to the IPCC Scientific Assessment*, J. T. Houghton, B. A. Callander, and S. K. Varney, eds., Cambridge University Press, Cambridge.

Wayne, R. P. (1991) *Chemistry of Atmospheres*, 2nd ed., Clarendon Press, Oxford.

Wayne, R. P., I. Barnes, P. Biggs, J. P. Burrows, C. E. Canosa-Mas, C. Hjorth, G. Le Bras, G. K. Moortgat, D. Perner, G. Poulet, G. Restelli, and H. Sidebottom (1991) The nitrate radical: Physics, chemistry, and the atmosphere, *Atmos. Environ.* 25A, 1.

Weaver, A., S. Solomon, R. W. Sanders, K. Arpag, and H. L. Miller, Jr. (1996) Atmospheric $NO_3$, 5. Off-axis measurements at sunrise: Estimates of tropospheric $NO_3$ at 40°N, *J. Geophys. Res.* 101, 18,605.

Weber, R. J., P. H. McMurry, F. L. Eisele, and D. J. Tanner (1995) Measurement of expected nucleation precursor species and 3-500-nm diameter particles at Mauna Loa Observatory, Hawaii, *J. Atmos. Sci.* 52, 2242.

Weber, R. J., J. J. Marti, P. H. McMurry, F. L. Eisele, D. J. Tanner, and A. Jefferson (1997) Measurement of new particle formation and ultrafine particle growth rates at a clean continental site, *J. Geophys. Res.* 102, 4375.

Webster, C. R., R. D. May, D. W. Toohey, L. M. Avallone, J. G. Anderson, P. Newman, L. Lait, M. R. Schoeberl, J. W. Elkins, and K. R. Chan (1993a) Chlorine chemistry on polar stratospheric cloud particles in the Arctic winter, *Science* 261, 1130.

Webster, C. R., R. D. May, D. W. Toohey, L. M. Avallone, J. G. Anderson, and S. Solomon (1993b) In situ measurements of the ClO/HCl ratio: Heterogeneous processing on sulfate aerosols and polar stratospheric clouds, *Geophys. Res. Lett.* 20, 2523.

Weinheimer, A. J., J. G. Walega, B. A. Ridley, B. L. Gary, D. R. Blake, N. J. Blake, F. S. Rowland, G. W. Sachse, B. E. Anderson, and J. E. Collins (1994) Meridional distributions of $NO_x$, $NO_y$, and other species in the lower stratosphere and upper troposphere during AASE II, *Geophys. Res. Lett.* 21, 2583.

Weiss, P. S., J. E. Johnson, R. H. Gammon, and T. S. Bates (1995) Reevaluation of the open ocean source of carbonyl sulfide to the atmosphere, *J. Geophys. Res.* 100, 23,083.

Weiss, R. F. (1994) Changing global concentration of atmospheric nitrous oxide, *Proc. Int. Sym. on Global Cycles of Atmospheric Greenhouse Gases*, Sendai, Japan, 78.

Welander, P. (1955) Studies on the general development of motion in a two-dimensional ideal fluid, *Tellus* 7, 141.

Weller, R. and O. Schrems (1993) $H_2O_2$ in the marine troposphere and seawater of the Atlantic Ocean (48°N-63°S), *Geophys. Res. Lett.* 20, 125.

Wennberg, P. O., R. M. Stimpfle, E. M. Weinstock, A. E. Dessler, S. A. Lloyd, L. B. Lapson, J. J. Schwab, and J. G. Anderson (1990) Simultaneous, in situ measurements of OH, $HO_2$, $O_3$, and $H_2O$: A test of modeled stratospheric $HO_x$ chemistry, *Geophys. Res. Lett.* 17, 1909.

Wennberg, P. O., R. C. Cohen, R. M. Stimpfle, J. P. Koplow, J. G. Anderson, R. J. Salawitch, D. W. Fahey, E. L. Woodbridge, E. R. Keim, R. S. Gao, C. R. Webster, R. D. May, D. W. Toohey, L. M. Avallone, M. H. Proffitt, M. Loewenstein, J. R. Podolske, K. R. Chan, and S. C. Wofsy (1994) Removal of stratospheric $O_3$ by radicals: In situ measurements of OH, $HO_2$, NO, $NO_2$, ClO, and BrO, *Science* 266, 398.

Wesely, M. L. (1989) Parameterization of surface resistances to gaseous dry deposition in regional-scale numerical models, *Atmos. Environ.* 23, 1293.

Wexler, A. S. and J. H. Seinfeld (1992) Analysis of aerosol ammonium nitrate: Departures from equilibrium during SCAQS, *Atmos. Environ.*, 26A, 579.

Wexler, A. S., F. W. Lurmann, and J. H. Seinfeld (1994) Modeling urban and regional aerosols: I. Model development, *Atmos. Environ.* 28, 531.

Whitby, E. R., P. H. McMurry, U. Shankar, and F. S. Binkowski (1991) *Modal Aerosol Dynamics Modeling*, EPA Contact No. 68-01-7365, USEPA, Research Triangle Park, N. Carolina 27711.

Whitby, K. T. and B. Cantrell (1976) Fine particles, *Proc. Internat'l Conf. on Environ. Sensing and Assessment*, Institute of Electrical and Electronic Engineers.

Whitby, K. T. (1978) The physical characterization of sulfur aerosols, *Atmos. Environ.* 12, 135.

White, W. H. and D. Dietz (1984) Does the photochemistry of the troposphere admit more than one steady state? *Nature* 309, 242.

Williams, L. R. and D. M. Golden (1993) Solubility of HCl in sulfuric acid at stratospheric temperatures, *Geophys. Res. Lett.* 20, 2227.

Williamson, D. L. and P. J. Rasch (1989) Two-dimensional semi-Langrangian transport with shape-preserving interpolation, *Mon. Wea. Rev.* 117, 102.

Winchester, J. W. (1988) Aerosol sulfur association with aluminum in eastern North America, Evidence for solubilization of atmospheric trace metals before deposition, in: *Advances in Environmental Science,* D. C. Adriano and S. E. Lindberg, eds., Acid Precipitation, Vol. 3, Springer-Verlag, New York.

Winker, D. M., R. H. Couch, and M. P. McCormick (1996) An overview of LITE: NASA's Lidar In-space Technology Experiment, *Proc. IEEE* 84, 164.

Winningham, J. D., J. R. Sharber, R. A. Frahm, J. L. Burch, N. Eaker, R. K. Black, V. A. Blevins, J. P. Andrews, J. Rudzi, M. J. Sablik, D. L. Chenette, D. W. Datlowe, E. E. Gaines, W. I. Imhof, R. W. Nightingale, J. B. Reagan, R. M. Robinson, T. L. Schumaker, E. G. Shelley, R. R. Vondrak, H. D. Voss, P. F. Bythrow, B. J. Anderson, T. A. Potemra, L. J. Zanetti, D. B. Holland, M. H. Rees, D. Lummerzheim, G. C. Reid, R. G. Roble, C. R. Clauer, and P. M. Banks (1993) The UARS Particle Environment Monitor, *J. Geophys. Res.* 98, 10,649.

Wisniewski, J. and A. E. Lugo, eds. (1992) *Natural Sinks of $CO_2$*, Kluwer Academic Publishers, Dordrecht, The Netherlands.

WMO (1981) *The Stratosphere: 1981, Theory and Measurements*, Report No. 11, Global Ozone Research and Monitoring Project, World Meteorological Organization, Geneva.

WMO (1986) *Atmospheric Ozone: 1985*, Report No. 16, World Meteorological Organization, Geneva.

WMO (1990a) *Report of the International Ozone Trends Panel: 1988*, Report No. 18, World Meteorological Organization, Geneva.

WMO (1990b) *Scientific Assessment of Stratospheric Ozone: 1989*, Global Ozone Research and Monitoring Project, Report No. 20, World Meteorological Organization, Geneva.

WMO (1991) *Scientific Assessment of Ozone Depletion: 1990*, Global Ozone Research and Monitoring Project, Report No. 25, World Meteorological Organization, Geneva.

WMO (1992) *Scientific Assessment of Stratospheric Ozone 1991*, Global Ozone Research and Monitoring Project, Report No. 26, World Meteorological Organization, Geneva.

WMO (1995) *Scientific Assessment of Ozone Depletion, 1994*, Global Ozone Research and Monitoring Project Report No. 37, World Meteorological Organization, Geneva.

Worsnop, D. R., L. E. Fox, M. S. Zahniser, and S. C. Wofsy (1993) Vapor pressures of solid hydrates of nitric acid: Implications for polar stratospheric clouds, *Science* 259, 71.

Wyngaard, J. C. and R. A. Brost (1984) Top-down and bottom-up diffusion of a scalar in the convective boundary layer, *J. Atmos. Sci.* 41, 102.

Yamartino, R. J., J. S. Scire, G. R. Carmichael, and Y. S. Chang (1992) The CALGRID mesoscale photochemical grid model, I. Model Formulation, *Atmos. Environ.* 26A, 1493.

Yulaeva, E., J. R. Holton, and J. M. Wallace (1994) On the cause of the annual cycle in the tropical lower stratospheric temperature, *J. Atmos. Sci.* 51, 169.

Yvon-Lewis, S. A. and J. H. Butler (1997) The potential effect of oceanic biological degradation on the lifetime of atmospheric $CH_3Br$, *Geophys. Res. Lett.* 24, 1227.

Zander, R., M. R. Gunson, and C. B. Farmer (1989) Remote sensing of the earth's atmosphere by infrared absorption spectroscopy — An update of the ATMOS program, in: *Advanced Optical Instrumentation for Remote Sensing of the Earth's Surface from Space, Proc. SPIE* 1129, 10.

Zander R., M. R. Gunson, C. B. Farmer, C. P. Rinsland, F. W. Irion, and E. Mahieu (1992) The 1985 chlorine and fluorine inventories in the stratosphere based on ATMOS observations at 30° north latitude, *J. Atmos. Chem.* 15, 171.

Zander, R., E. Mahieu, M. R. Gunson, M. C. Abrams, A. Y. Chang, M. Abbas, C. Aellig, A. Engel, A. Goldman, F. W. Irion, N. Kämpfer, H. A. Michelsen, M. J. Newchurch, C. P. Rinsland, R. J. Salawitch, G. P. Stiller, and G. C. Toon (1996a) The 1994 northern midlatitude budget of stratospheric chlorine derived from ATMOS/ATLAS-3 observations, *Geophys. Res. Lett.* 23, 2357.

Zander, R., S. Solomon, E. Mahieu, A. Goldman, C. P. Rinsland, M. R. Gunson, M. C. Abrams, A. Y. Chang, R. J. Salawitch, H. A. Michelsen, M. J. Newchurch, and G. P. Stiller (1996b) Increase of stratospheric carbon tetrafluoride ($CF_4$) based on ATMOS observations from space, *Geophys. Res. Lett.* 23, 2353.

Zimmerman, P. R., J. P. Greenberg, and C. E. Westberg (1988) Measurements of atmospheric hydrocarbons and biogenic emission fluxes in the Amazon boundary layer, *J. Geophys. Res.* 93, 1407.

Zimmermann, P. R., J. Feichter, H. K. Rath, P. J. Crutzen, and W. Weiss (1989) A global three-dimensional source-receptor model investigation using $^{85}Kr$, *Atmos. Environ.* 23, 25.

# Index

absorption cross section, 88
absorption of radiation, 86
accommodation coefficient, 104
accumulation mode, 118
acetone, 344
    photolysis, 217
accommodation coefficient, 575
acetonitrile, 370
acid precipitation, 349, 365, 367-368
acid smog, 367
actinic flux, 96, 136, 518, 585-587
activation energy, 99-100
active chlorine, 307
    production of in polar vortex, 503-505
activity, 140
aerodynamic resistance, 163
aerosol
    budget, 121
    composition, 121
    continental, 121
    dry removal, 142
    indirect radiative effects, 533-534
    marine, 121
    measurements of, 405-406
    modeling, 138, 431
    radiative effects, 530-534
    reactions on, 104
    residence time, 142
    size distribution, 118, 130-134, 433
    stratospheric, 123
    wet removal, 142
aerosol dynamic equation, 431
aircraft
    emissions, 193, 271
    atmospheric measurements from, 406-407
Aitken particles, 118
aldehydes, 331
alkane oxidation, 327
alkene oxidation, 328
alkoxy radicals, 330, 336
ammonia budget, 191
ammonium nitrate, 139
ammonium sulfate, 192, 120
Antarctic ozone hole: see ozone hole
aqueous reactions, 151-154
Arctic vortex, ozone loss, 505-506
Arrhenius expression, 98
association reactions, 101-103, 574
ATMOS spectrometer, 306-7
atmospheric circulation, 7
atmospheric composition, 8-11
Atmospheric Lifetime Experiment/
    Atmospheric Gases Experiment
    (ALE/GAGE), 294, 408

atmospheric pressure, 6
atmospheric radiation, measurement of, 403-404
Atmospheric Trace Molecular Spectroscopy Experiment (ATMOS), 409-414
atmospheric waves, 53-64
atmospheric window, 517
balloons, use of for atmospheric measurements, 407
baroclinic instability, 64, 66-68
Beer-Lambert Law, 88, 134
bicarbonate ion, 175
biogenic hydrocarbons
    emissions, 183-186
    oxidation, 335-337
    role in ozone production, 483-484
biogeochemical cycles, 5, 159
biological daily dose of solar radiation, 510-512
biomass burning, 1, 15, 178, 198
    and tropospheric ozone, 469-470
biosphere, 15-17
bisulfite ion, 356-359
boundary layer, 438
Brewer-Mast sonde, 391
bromine
    burden, future prediction, 320-321
    catalytic ozone destruction, 309, 497, 499-500
    chemistry in the stratosphere, 308-310, 497
    source gases, 201, 295-298
bromine nitrate, 307-309
bromine oxide, 308-309
Brunt-Väisälä frequency, 31
carbon cycle, 167-188
carbon dioxide ($CO_2$)
    budget, 170-181
    climate effects, 515-517, 520-523, 525-529
    historic abundance, 541
    mixing ratio, 167
    ocean flux, 14, 174
carbon disulfide ($CS_2$)
    oxidation, 352
    distribution, 362
carbon monoxide (CO)
    budget, 187
    oxidation, 187, 332
    measurements, 340-342
carbon storage, 181
carbonyl sulfide (OCS), 123, 197
    oxidation, 352, 369
    distribution, 361

catalytic cycles, 106-107, 494-501
   chlorine, 304-5
      halogen compounds, 302-313
      polar vortex, 312-313
CFCs: see chlorofluorocarbons
$CH_3S$ radical, 353
Chapman mechanism for ozone production, 106, 487-489, 493-494
Chappuis bands of ozone, 94, 488
chemical actinometers, 403-404
chemical amplification, for detection of peroxy radicals, 383
chemical families, 112
chemical feedbacks, and climate, 526-528
chemical ionization mass spectrometry (CIMS), 387
chemical reactions, 95-106, 570-574
   association, 101-103
   bimolecular, 98-99
   complex, 95
   elementary, 95
   termolecular, 101-103
   unimolecular, 101-103
chemiluminescence, in situ measurements by, 381-382
chlorine
   activation, 248, 252, 503-505
   burden, future prediction, 320-321
   catalytic ozone destruction, 304-306, 487, 489, 496-497, 499-500
   chemistry in the stratosphere, 304-307
   chemistry in the troposphere, 313-314
   source gases, 201, 292-294
chlorine monoxide
   polar vortex, 311-313
   stratospheric chemistry, 304-6
chlorine nitrate, 242, 305-307
   hydrolysis, 151, 248
chlorofluorocarbons (CFCs), 4, 89, 201, 291-292, 496, 506-507
   and climate, 515, 522, 524-526
   photochemical destruction, 298-299
   reaction with $O(^1D)$, 299
chromatography, 385-386
Clean Air Act, 367
climate sensitivity factor, 534-535
climatic response to radiative forcing, 534-535
ClO dimer, ClOOCl, 312-313
cloud condensation nuclei (CCN), 126-127, 533
clouds, 126-129, 331
   acidity, 147
   global energy budget, 522
   occurrence, 126
   reactions, 152
coagulation, 118
coarse mode, 118
column abundance, 12
condensation nuclei, measurement of, 405-406
continental boundary layer, 272
continuity equation, 32, 424
convection, 65, 438
Coriolis force, 29
Courant-Friedricks-Lewy condition, 434

Criegee biradical, 328
critical $NO_x$, 267-268
Cryogenic Limb Array Etalon Spectrometer (CLAES), 410
Cunningham factor, 129
data assimilation, 454
deforestation, 173, 178-180
denitrification (bacteria), 189
denitrification (stratosphere), 239, 250-251
deposition (particles), 118
deposition velocity, 163
deposition velocity (aerosol), 142
desert dust, 122
diabatic circulation, 49-50
differential absorption lidar (DIAL), 397
differential optical absorption spectroscopy (DOAS), 279, 376
diffusion, 103, 104-105, 148-150
diffusion denuder, 384
diffusion equation, 439
dimethyl sulfide
   distribution, 362
   oxidation, 352-355
   sources, 197
dinitrogen pentoxide ($N_2O_5$)
   hydrolysis, 150-151, 154, 248, 279, 281
   in stratosphere, 244, 279
   in troposphere, 278
   photolysis, 244
dipole, 87
direct radiative forcing, 524-526
DNA damage, by UV radiation, 509-510
Dobson spectrometer, 389-390, 488, 491
Dobson unit, 12
downward control principle, 53
dry deposition, 163
dry deposition (aerosol), 142
dust, 200
   radiative forcing, 531
Earth system science, 6
ecosystems, 15-17
eddy-correlation method, 162, 400-401
eddy flux, 445
electrochemical measurements, 391
El Niño, 14
Eliasson-Palm flux, 48, 445
emission experiment, 409
emission of radiation, 86
enclosure method, for flux measurements, 398-400
energy budget, global, 521-522
energy levels
   electronic, 87, 89
   rotational, 87, 89
   vibrational, 87, 89
energy of photon, 86
enthalpy, 99
entropy, 99
equilibrium constant, 99-100, 146, 581
erythema induction, by UV radiation, 509-510
ethyl radical, 328
Euler backward scheme, 427
Euler forward scheme, 427
Eulerian mean, 44-47
Eulerian transport, 434

evolution of the atmosphere, 539-543
excited states, 108-109
extinction, 134-135
faint young sun paradox, 541
family grouping technique, 430
fertilizer, 2, 190
filter radiometer, 393
fine particles, 118
flow methods, 109
fluorescence, 94
fluorine chemistry in the stratosphere, 310
fluorine source gases, 292-294
flux measurements, 399-401
flux of trace gases, 162
formaldehyde, 334, 343
    photolysis, 211, 216
formic acid, 332
fossil fuel combustion, 1, 169, 171
Fourier transform spectrometers, 393, 412-413
free radicals, 100
friction velocity, 162
Gaia hypothesis, 159, 543
gas chromatography, 385
gas correlation radiometers, 379
Gear method, 429
generalized Lagrangian mean, 47
geopotential, 29
geostrophic wind, 35
global change, 1-5
global warming, 88, 89
global warming potential (GWP), 317, 528-530
gradient method for flux measurements, 401
gravity waves, 54-57
greenhouse gases; see radiatively active gases, 2
gross primary production, 177
Hadley cells, 8, 70-71
Halogen Occultation Experiment (HALOE), 410
halogen source gases, 291-292
    abundances, 292-298
    emissions, 292-295
    lifetimes, 292-293
    reaction with OH, 300-301
halons, 291-292
    photolysis, 299-300
    reaction with $O(^1D)$, 300-301
Hartley bands of ozone, 91, 488
Henry's law, 105, 145, 164
Herzberg continuum, 91
heterogeneous reactions, 104-105, 109, 125, 246, 281, 305, 331, 499, 508
    and ozone depletion, 125, 497-499, 502-505
high altitude aircraft, future effects, 547-548
High Resolution Doppler Image (HRDI), 411
High Resolution Dynamics Limb Sounder (HIRDLS), 415
$HO_x$ in stratosphere
    loss, 213
    measurements, 223-225
    production, 212
    partitioning, 213-215, 225
$HO_x$ in troposphere
    loss, 218-220
    measurements, 226-230
    production, 216
    partitioning, 218
    catalytic ozone destruction, 487, 489, 495, 497 499-501
HS radical, 351
Huggins bands of ozone, 488
hydrocarbons
    distributions, 338-344
    $NO_3$ reaction, 327, 329
    $O_3$ reaction, 327-8
    OH reaction, 327-8
    oxidation, 326
    ozone production from, 183, 472-484
    sources, 183-186, 212
hydrochlorofluorocarbons, 291-292, 506-507
    chemistry, 317-320
    global warming potentials, 317
    lifetimes, 316-317
    ozone depletion potentials, 317
    reaction with OH, 317
hydrofluorocarbons, 291-292, 506-507
    chemistry of, 317-320
    global warming potentials, 317
    lifetimes, 316-17
    ODPs, 317
    reaction with OH, 317
hydrogen (molecular)
    distribution, 221
    sources, budget, 210-211
hydrogen bromide (HBr), 307-9
hydrogen chloride (HCl), 201, 305-7
    tropospheric chemistry, 314
hydrogen fluoride (HF), 310
hydrogen peroxide, 219-220
    uptake, 220
    photolysis, 219
hydrogen sulfide ($H_2S$), 199
    oxidation, 351
hydrological cycle, 165-167
hydroperoxy radical (solution), 152
hydrostatic equilibrium, 26-27
hydroxyl radical (OH), 5, 114, 153
    lifetime, 218
    global concentration, 218
    measurements, 223-226, 227-228, 376, 387
hygroscopicity, 139-141
hypochlorous acid, 305-7
ice-forming nuclei (IN), 127, 533
Improved Stratospheric and Mesospheric Sounder (ISAMS), 411
in situ measurements, 375-383
inactive chlorine, 307
industrial revolution, 543
infrared radiation, 87
infrared spectroscopy
    in situ measurements by, 376
    remote sensing, 392-394
inorganic halogen compounds, 292, 306-7
    chemistry in the stratosphere, 301-313
    active forms, 302
    inactive forms, 302
intertropical convergence zone, 70-71
iodine
    source gases, 298
    stratospheric chemistry, 311

ion chromatography, 385
ionic reactions
    in solution, 103, 582-584
    in stratosphere, 370
ionosphere, 498-499
irradiance, net radiant flux, 518
isentropic coordinate system, 31
isoprene, 185, 273
    oxidation mechanism, 335-337
isotopes, 105-106, 167, 180
isotopic composition of gases, 387-388
$j$-value; see also photolysis rates,
    95, 96-97
Jacobian matrix, 428
Junge layer, 123, 197, 248, 349
Kalman-Bucy filter, 457
Kelvin equation, 137, 141, 152
Kelvin waves, 59-60
kinetic isotope effect, 105-106
Lagrangian transport, 436
laser-induced fluorescence, 381
lifetimes, 110, 112-114
light absorption, 134
light detection and ranging (Lidar),
    395-397
lightning, 193
Limb Infrared Monitor of the
    Stratosphere (LIMS), 414-415
limb viewing path, 409
Lindemann mechanism, 101-102
liquid chromatography, 385
liquid water content, 127, 129
liquid water fraction, 147
Lorenz-Mie theory, 134-135
Lyman-$\alpha$ radiation, 94
magnetosphere, 499
Mars, atmospheric composition,
    541-542
mass spectrometry, 385-388
mass transfer rates, 148-150
matrix isolation, 384
mean diameters, 131-133
Measurement of Air Pollution from
    Satellites (MAPS), 417-418
Measurements of Pollution in the
    Troposphere (MOPITT), 418-419
Meinel bands of OH, 109, 498
meridional circulation, 8
mesopause, 7
mesosphere, 7
methacrolein, 336
methane
    and climate, 515, 517, 522, 524-526,
        529
    budget, 182
    isotopes, 168, 182
    lifetime, 181
    measurements, 340-342
    oxidation, 111, 332
    stratospheric, 344
methane sulfonic acid, 353-5, 360
methoxy radicals, 334
methyl bromide, 202
methyl chloride
    abundance, 294-295
    chemistry, 313-314, 496
    flux, 201
methyl chloroform, 114, 218
methyl hydroperoxide, 334
methyl iodide, 202
methyl peroxy radical, 333

methyl vinyl ketone, 336
microwave radiation, 87
Microwave Limb Sounder (MLS),
    410-411
Milankovich forcing, 522
mixing, 63
mixing ratio, 12, 560-566, 578
modeling
    advection, 433
    aerosols, 431
    boundary layer exchanges, 438
    chemistry, 427
    convective transport, 438
    coupled systems, 451
    global budgets, 450
    inverse modeling, 458
    nonlinear systems, 441
    transport, 433
models
    one-dimensional, 442
    three-dimensional, 443
        global, 446
        regional, 448
    two-dimensional, 442
    zero-dimensional, 441
Montreal Protocol, 201, 294, 316, 320
nadir viewing path, 409
NAPAP, 367
net primary production, 177
net primary productivity, 15
Network for Detection of Stratospheric
    Change (NDSC), 408-409
nitrate deposition, 195
nitrate radical ($NO_3$), 277-280
    measurements, 279-280
    photolysis, 278
    stratospheric chemistry, 244, 279
    tropospheric chemistry, 278-279
nitric acid, 264
    lifetime, 281
nitric acid trihydrate, 124
nitrification, 189
nitrogen cycle, 188-195
nitrogen fixation, 188-191
nitrous oxide ($N_2O$), 237
    sources, 190, 192-193
    and climate, 515, 522
$NO_2$ measurement, by visible
    absorption, 390-391
$NO_2$ photolysis measurements,
    403-404
NOAA CMDL, 294, 408
non-sea-salt sulfate, 142, 197, 361
nonacceleration theorem, 45, 48
nonlinear systems, 441
nontransport theorem, 45, 48
$NO_x$, 188
    budget, 193
    catalytic ozone destruction,
        242, 487, 489, 495-499, 501-502
    coupling with other families, 244
    in continental boundary layer,
        272-274
    in remote troposphere, 261-272
    lifetime, 270, 274
    partitioning in stratosphere, 243
    partitioning in troposphere, 261,
        262-264
    role in $CH_4$ oxidation, 261
    role in CO oxidation, 261
    role in $HO_x$ partitioning, 268-270

tropospheric measurements, 255-259
Noxon "cliff," 245-246
$NO_y$
 tropospheric measurements, 255-259
$NO_y$ partitioning
 in stratosphere, 240, 252-254
 in continental boundary layer, 275
$NO_y$-$N_2O$ correlation, 237-240
nucleation, 137
nucleation mode, 118
occultation experiment, 409
ocean circulation, 13-15, 165
ocean upwelling, 14, 166
odd hydrogen: see $HO_x$
odd nitrogen: see $NO_x$
OH; see hydroxyl radical
optical depth, 135
organic aerosols, and climate, 530-531
organic chlorine burden, 292-295
organic nitrates, 282-287
organic peroxides, 219, 330
origin of life, 540
oxidizing capacity, 8, 11
oxygen
 initial (prehistoric) formation, 541
 $O(^1D)$ state, 90, 108, 216, 265, 299, 300, 404, 475, 493
 $O_2(^1\Delta)$ state, 90, 108
ozone depletion potential (ODP), 317, 506-507
ozone hole, 4, 248, 311-313, 487, 489, 493, 501-506
ozone: see also stratospheric ozone and tropospheric ozone
 and climate, 3, 515, 517, 522
 destruction in the stratosphere, 494-500
 measurement of, 376, 391
 photochemical lifetime, 475-476
 photolysis, 90, 592
 photolysis measurements, 404
 relation to tropospheric $HO_x$, 217
ozonesondes, 391
ozonolysis, 217
PAHs, 119-120
PAN, 256, 282, 284-286, 330, 331
Particle Environment Monitor (PEM), 411
particle formation, 361
particles; see aerosols
particulate nitrate, 119, 281-282
past climates, 522-523
perfluorinated organic species, 301
peroxy radicals, 227, 262-263, 273, 329
peroxynitrates, 283, 330
peroxynitric acid, 264, 269-270, 281-282
pH of particles, 118, 146
phase ratio, 147
photochemical processes, 88, 94
photodissociation, 94-97, 136
photolysis rates, 95, 96-97, 403-404, 588-595
photosynthesis, 177
phytoplankton, 197, 201, 363
piston velocity, 164
planetary boundary layer, 6, 65, 72-73, 161

polar stratospheric clouds, 124-126, 150, 248, 311-313
 ozone loss related to, 502-505
polar vortex; see also ozone hole, 76, 502-505
potential temperature, 30
potential vorticity, 38, 40-43
pre-concentration, 384
precipitation, 127
primary aerosol, 118
primitive equations, 28-31, 443
probability density function, 455
proton hydrates, 370, 498-499
pulsed photolysis, 109
quadrupole spectrometers, 387-388
quantum yield, 89, 90-91, 94
quasibiennial oscillation, 76-78, 445
quasi-steady-state approximation, 429
quenching, 94
radiation
 absorption, 86
 emission, 86
 solar, 91-92, 96
radiation amplification factor, 510
radiative forcing, by gases, 543-544
radiative transfer equation, 518-520
radiatively active gases (greenhouse gases), 2, 515-517, 520-522
rate coefficient, 98, 109-110
rate limiting step, 107
Rayleigh scattering, 134-135
reaction kinetics, 109-110
reaction probability, 104, 576-577
regional modeling studies of tropospheric ozone, 479-483
relative humidity, 140, 578
relative radiative forcing, of gases, 526-527
relative rate technique, 110
remote sensing of atmospheric composition, 388-399, 418-419
reservoir species, 213, 246-248, 305
residence time (aerosol), 142
resonance fluorescence, 379-381
Revelle factor, 175, 453
rockets, use of for atmospheric measurements, 407
Rossby radius of deformation, 38
Rossby waves, 57-59
Satellite measurements of atmospheric composition, 409-419
SBUV, 491
scale height, 6
scavenging (particle), 143
Schmidt number, 165
Schönbein, 488
Schumann-Runge bands, 91-92
sea salt, 121, 141, 153, 200
secondary aerosol, 118
sectional models, 432
semi-Lagrangian transport, 437
semiannual oscillation, 76-78, 445
sensitivity analysis, 443
settling velocity, 129
singlet oxygen, 108
size parameter, 134
smog, 254, 272, 466, 472-473, 479-483
soil emissions, 195
solar constant, 516-517
solar cycle, 517

solar radiation, 516-519, 585-587
solar spectrum, 91-92
Solar Ultraviolet Spectra Irradiance Monitor (SUSIM), 411
Solar/Stellar Irradiance Comparison Experiment (SOLSTICE), 411
solid adsorbent cartridges, 384
solubility of gases, 144
soot, 119
specific humidity, 578
spectroradiometers, 404
spectrum
    absorption, 89
    electromagnetic, 86
    electronic, 88
    rotational, 87
    vibrational, 87
spherical radiant flux, 518
static stability, 33
stationary state approximation, 112
stiff systems, 426
stratopause, 6
stratosphere, 6
stratosphere-troposphere exchange, 78-80
stratospheric aerosol, 369
stratospheric heating, by ozone absorption, 490-491
stratospheric ozone
    and climate, 527
    destruction, 494-500
    distribution, 491-493
    future trends, 546-547
    lifetime, 492-494, 501
    loss due to $HO_x$, 214-215
    loss due to $NO_x$, 107, 242
    production, 493-494
sulfate (aqueous production), 152
sulfate aerosol, 118, 123, 197, 248, 251-254
    reactions, 150
    radiative forcing, 530-532
sulfur
    budget, 196
    climate feedback, 349
    emissions, 197-201
    isotopes, 168
    oxidation states, 351
sulfur cycle, 195-201
sulfur dioxide ($SO_2$), 152, 199
    distribution, 364-367
    Henry's Law, 357, 580
    in stratosphere, 369
    oxidation, 356-359
sulfuric acid, 123, 369
supersaturation, 137
surface ozone depletion in the Arctic, 472
surface reactions, 150
surface tension, 137
tape recorder effect, 222
terpenes, 185, 337, 339, 483
terrestrial radiation, 517, 519-522
thermal wind, 36
thermohaline circulation, 14
thermosphere, 7
Total Ozone Mapping Spectrometer (TOMS), 415-418, 491, 510-511
trace gas fluxes, 15
trace gases, 8
transformed Eulerian mean, 47-49, 445

transition state, 99
transport, effects on stratospheric ozone, 491, 501
trifluoroacetic acid, 319-320
Troe equation, 102
tropopause, 6, 221, 579
troposphere, 6
tropospheric ozone, 3
    and climate, 527
    column abundance, 469-470
    control strategies, 479-484
    detrimental effects on human health, 465
    future trends, 548-549
    greenhouse gas, 465
    lifetime, 475-476
    loss mechanisms, 265-267
    mixing ratios, 467-472
    oxidation capacity of the atmosphere, 465
    photochemical production, 259-268, 270-271, 272-276, 465-466
    production and loss, 472-484
    seasonal variation, 467-469
    springtime maximum, 469-470
    surface depletion in the Arctic, 314-316
    transport from the stratosphere, 465
    trends, 471-472, 545-546
tunable diode laser absorption spectroscopy, 376-379
turnover time, 451
ultrafine particles, 123
Umkehr technique, for ozone measurement, 391
unimolecular reactions, 101-103
units, 558-559
Upper Atmosphere Research Satellite, 409-411
upper atmosphere, chemistry, 498-499
ultraviolet radiation (UV), 88
    UV absorption by ozone, 488, 490-491, 509-511
    UV radiation, increases in, 545
    UV-A radiation, 509
    UV-B radiation, 509
vapor pressure, 137, 578-579
Venus, atmospheric composition, 541-542
visibility, 136
volcanic plumes, radiative forcing, 531
volcanoes, 199, 201, 251, 349, 508
von Karman constant, 162
vorticity, 37
water vapor
    and climate, 516-517, 522-523
    photolysis by Lyman-$\alpha$, 498
    stratospheric distribution, 221-223
    vapor pressure, 578
wave breaking, 61-63, 445
wet deposition, 142-144, 163
wetlands, 17
whole air sampling, 384
Wind Imaging Interferometer (WINDII), 411
Younger Dryas, 522
zonal mean, 443

WITHDRAWN FROM
OHIO NORTHERN
UNIVERSITY LIBRARY